ATLANTIS COMPUTATIONAL INTELLIGENCE SYSTEMS

VOLUME 6

SERIES EDITORS: JIE LU, JAVIER MONTERO

Atlantis Computational Intelligence Systems

(ISSN 1875-7650)

Aims and scope of the series

The series 'Atlantis Computational Intelligence Systems' aims at covering state-of-the-art research and development in all fields where computational intelligence is investigated and applied. The series seeks to publish monographs and edited volumes on foundations and new developments in the field of computational intelligence, including fundamental and applied research as well as work describing new, emerging technologies originating from computational intelligence research. Applied CI research may range from CI applications in the industry to research projects in the life sciences, including research in biology, physics, chemistry and the neurosciences.

All books in this series are co-published with Springer.

For more information on this series and our other book series, please visit our website at:

www.atlantis-press.com/publications/books

ATLANTIS PRESS

AMSTERDAM – PARIS – BEIJING

Computational Intelligence Systems in Industrial Engineering

With Recent Theory and Applications

Cengiz Kahraman

Istanbul Technical University
Department of Industrial Engineering
34367 Macka Istanbul, Turkey

ATLANTIS PRESS

AMSTERDAM – PARIS – BEIJING

Atlantis Press

8, square des Bouleaux
75019 Paris, France

For information on all Atlantis Press publications, visit our website at: *www.atlantis-press.com*

Atlantis Computational Intelligence Systems

Volume 1: Linguistic Values Based Intelligent Information Processing: Theory, Methods, and Applications - Da Ruan
Volume 2: Computational Intelligence in Complex Decision Systems - Da Ruan (Editor)
Volume 3: Intelligence for Nonlinear Dynamics and Synchronisation - K. Kyamakya, A. Bouchachia, J.C. Chedjou
Volume 4: Trust Networks for Recommender Systems - P. Victor, C. Cornelis, M. De Cock
Volume 5: Answer Set Programming For Continuous Domains: A Fuzzy Logic Approach - J. Janssen, S. Schockaert, D. Vermeir, M. De Cock

ISBN 978-94-6239-054-6
ISBN 978-94-91216-77-0 (eBook)
ISSN 1875-7650

Softcover reprint of the hardcover 1st edition 2012

I dedicate this book to Industrial Engineering Department of
Istanbul Technical University, which I graduated from.

Preface

Industrial engineering is a branch of engineering dealing with the optimization of complex processes or systems. It is concerned with the development, improvement, implementation and evaluation of integrated systems of people, money, knowledge, information, equipment, energy, materials, analysis and synthesis, as well as the mathematical, physical and social sciences together with the principles and methods of engineering design to specify, predict, and evaluate the results to be obtained from such systems or processes. As the complexity of the production and service systems increase, the classical solution techniques to the problems arising from these systems become insufficient.

In the literature, computational intelligence is defined as a methodology involving computing that exhibits an ability to learn and/or deal with new situations such that the system is perceived to possess one or more attributes of reason, such as generalisation, discovery, association, and abstraction. The output of a computationally intelligent system often includes predictions and/or decisions. Or it is defined as follows: Computational intelligence is the study of the design of intelligent agents. An intelligent agent is a system that acts intelligently. It is flexible to changing environments and changing goals, it learns from experience, and it makes appropriate choices given perceptual limitations and finite computation.

Many computational intelligent systems have been developed and appeared in the literature for about 40 years. These have been accepted excellent tools for the solution of the complex problems of industrial engineering, which the classical techniques are not sufficient to solve them. This book aims at presenting the recent developments in computational intelligent systems with industrial engineering applications. The book is composed of 30 chapters, each is summarized below. First chapter briefly summarizes the computational intelligence systems and their application areas in industrial engineering. The second chapter presents a comparison of some a priori and a posteriori decision making support methods, aimed at aiding the decision maker in the selection of the preferred solutions. The considered methods are compared with respect to their application to a case study concerning the optimization of the test intervals of the components of a safety system of a nuclear power plant. The engine for the multiobjective optimization search is based on genetic algorithms. The third chapter proposes a hybrid fuzzy linguistic recommender system to help the Technology Transfer Office staff in the dissemination of research resources interesting for the users. The system recommends users both specialized and complementary research resources and additionally, it discovers potential collaboration possibilities in order to form multidisciplinary working groups. The fourth chapter presents an extension of its authors' interactive approach to the generation of linguistic summaries through the use of our fuzzy querying interface supporting queries with fuzzy linguistic quantifiers. They concentrate on a specific type of linguistic summaries which parallel specific fuzzy association rules, and show the use of an efficient algorithm for mining such rules. They show an extension to the dynamic case of by using linguistic summaries of times series data. The fifth chapter demonstrates the robustness and uniqueness of fuzzy logic based method with two case studies. The first case study relates to the assessment of the status of ambient air quality in Pune city at the defined locations, while fuzzy description of river water quality in Indrayani River, near Pune for bathing is presented in the second study. The comparison of the results of air/water quality index and the proposed fuzzy logic based method, is an integral part of the paper.

The sixth chapter illustrates the use of TOPSIS method to combine individual safety performance indicators into an overall index of road safety performance for a set of European countries. In this respect, to deal with the subjective kind of uncertainty on data (such as linguistic variables given by experts) which are usually adopted to assess the weights of criteria/indicators, the authors explore an extension of the classical TOPSIS method to fuzzy environments. The seventh chapter proposes and discusses a failure possibility-based reliability algorithm to assess nuclear event reliability data from failure possibilities, which are expressed in qualitative natural languages, mathematically represented by membership functions of fuzzy numbers, and subjectively justified by a group of experts based on their working experience and expertise. The eighth chapter describes the Analytical Network Process, a multi-criteria prioritization method to support decision making in complex and uncertain environments and suggests a fuzzy analytic network process approach for prioritizing decision elements. The proposed fuzzy set theoretic method accommodates fuzziness in the supermatrix computations and thereby provides the opportunity to capture the uncertainty associated with the cumulative influence of each factor on every other factor with which it interacts. The ninth chapter is devoted to swarm intelligence and its engineering applications in nuclear power plants. Particle Swarm Optimization will be used to illustrate the power of such an implementation. The tenth chapter introduces how genetic algorithms are applied to solve the supply chain network design problem. A classification of the recent research in the field provides a valuable insight into current state of literature and outlines directions for future research.

The eleventh chapter proposes and tests the validity and effectiveness of a fuzzy multi-criteria method, called F-Promethee, to help the analyst to compare a set of energy crops and to obtain a ranking of their environmental sustainability. In the twelfth chapter, the previous researches on energy demand forecast are classified and fuzzy techniques are introduced. A fuzzy seasonal time series model that forecasts the energy demand is proposed and illustrated with a real world application. The thirteenth chapter studies a bi-objective re-entrant permutation flowshop scheduling problem in which the jobs have strict due dates. If the manufacturer will not be able to respect any job's due date, that job shall be rejected whether by ignoring it completely or sub-contracting the job. A genetic algorithm coupled with a new dominance concept, different from Pareto, is developed to find a good estimation of non-dominated solutions set. The next chapter provides a general introduction to the structure, algorithm and quality of Self Organizing Maps and presents industrial engineering related applications reported in the literature. The fifteenth chapter presents development and application of a fuzzy mixed integer goal programming (FMIGP) model for rural cooking and heating end-uses. The developed model considers various scenarios such as economical, environmental, social acceptance and local resources to trade off between socio-economical and environmental issues. Due to uncertainty involved in real world energy planning, exact input data is impossible to acquire. Hence, FMIGP model is used to consider four fuzzy objectives. The sixteenth describes chance constrained programming (CCP) to deal with optimization problems in fuzzy environment. CCP models may be solved either by transforming them to their crisp equivalents or by means of fuzzy simulation. This chapter addresses fuzzy simulation which is used for estimating the possibility (credibility) of the constraints, and genetic algorithm (GA) which is applied as an optimization heuristic. The next chapter proposes a computationally effective approach of combining bacterial foraging strategy with particle swarm optimization for solving the minimum makespan problem of job shop scheduling.

The eighteenth chapter uses fuzzy cognitive maps as a modeling tool for traveler satisfaction in public transit systems. The next chapter focuses on optimizing the way of allocating inbound and outbound containers in storage locations, known as the Container Storage Problem (CSP). It proposes a genetic algorithm to solve the CSP for a single and various container types (refrigerated, open side, empty, dry, open top and tank). The main objective of this approach is to find an optimal container arrangement which minimize the re-handle operations of containers at their departure dates (unloading time). The twentieth chapter proposes several fuzzy formulations for fuzzy geometric programming (FGP) by analogy to fuzzy linear programming. Then, the basic GP applications in industrial engineering (IE) are reviewed, and the proposed FGP formulations are adapted to two basic applications of FGP in IE, i.e. to inventory problems and entropy based transportation problems. In the last part of the chapter, a solution approach is proposed for fuzzy inventory problem formulated as a GP model. The twenty first chapter presents a prototype of an automated classification system for thermoplastics that can be integrated on an industrial recycling plant. The main contribution of this new prototype is the use of a novel artificial intelligent classification system based on a multi-sensor strategy. The twenty second chapter provides a mathematical programming basis for distribution network design problem under imprecise data environment. After providing background information about Distribution Network Design (DND) problem as well as fuzzy linear programming, fuzzy linear programming model of DND problem is solved for different a-cut values. The next chapter proposes a new training algorithm to train Convolutional Neural Networks (CNNs) in Optical Character Recognition (OCR) applications. This algorithm is characterized by the use of Gravitational Search Algorithms (GSAs) in combination with the back-propagation algorithm (BP), and a new family of CNNs is suggested. The twenty fourth chapter summarizes the fuzzy techniques, especially the fuzzy clustering and fuzzy programming employed for the cellular manufacturing system design, and their past applications, and offer directions for future research. The twenty fifth chapter tries to provide evidence that joining Self Organizing Maps together with some graphs theory tools (namely: the Minimum Spanning Tree), they can be successfully employed to develop macroeconomic models thus taking both static and dynamic (i.e. over a moving period of time) snapshots of countries financial situations. The twenty sixth chapter aims to present the results of developing a novel swarm optimization method that responds to the need of using multidimensional parameters. This meta-heuristic optimization approach is inspired by the ecological system of animals and their hierarchical relationship. The next chapter presents a hybrid optimization method that aims at maximizing the Net Present Value related to the investment made by Wind Turbines developers in an active distribution network and smart grids. The proposed method combines a Genetic Algorithm with a multi-period optimal power flow. The twenty eighth chapter investigates the multi-faceted re-mining of association mining results, develops and presents a practical methodology, and shows the applicability of the developed methodology through real world data. The next chapter investigates the performance of several Multi-Objective Evolutionary Algorithms (MOEAs) in discovering solutions to the Competitive Facility Location (CFL) problem and a sensitivity analysis of the solutions related to input parameters. The last chapter focuses on the offset of the interdependencies between aspects of Human Resource Management and the phenomena of fatigue; specifically, the implementation of Bayesian Networks (BNs) with the use of evidence(s) leads in a systematic manner to the projection and evaluation of fatigue.

I hope that this book will provide a useful resource of ideas, techniques, and methods for the research on the applications of computational intelligent systems in industrial engineering problems. I am very thankful to my colleagues Dr. Seyda Serdarasan, Dr. Başar Öztayşi, and Dr. Umut Asan for their selfless and careful work in editing the manuscript of this book. I am also grateful to the referees whose valuable and highly appreciated works contributed to select the high quality of chapters published in this book.

Prof. Cengiz Kahraman

Istanbul Technical University
Department of Industrial Engineering
34367 Macka Istanbul Turkey

Tel.: +90 212 296 40 40
Fax: +90 212 240 72 60

E-mail: kahramanc@itu.edu.tr

Contents

Chapter 1

Computational Intelligent Systems in Industrial Engineering

Cengiz Kahraman

Istanbul Technical University, Department of Industrial Engineering, 34367 Maçka Istanbul Turkey

Email: kahramanc@itu.edu.tr

In this chapter, computational intelligence systems are briefly defined and then classified with respect to the interest areas of indutrial engineering. Literature review results for this classification are given for each intelligence system separately. It is seen that each interest area of industrial engineering finds an intelligent system application in the literature. Either integrated or hybrid usages of these systems are observed.

1.1 Introduction

Industrial Engineering (IE) is defined by various ways in the literature. Some of them are as follows: Industrial Engineers integrate people, technology, and information to enhance a globally competing enterprise. IE is concerned with the design, improvement and installation of integrated systems of people, materials, information, equipment and energy. It draws upon specialized knowledge and skill in the mathematical, physical, and social sciences together with the principles and methods of engineering analysis and design, to specify, predict, and evaluate the results to be obtained from such systems. IE is a branch of engineering dealing with the optimization of complex processes or systems. It is concerned with the development, improvement, implementation and evaluation of integrated systems of people, money, knowledge, information, equipment, energy, materials, analysis and synthesis, as well as the mathematical, physical and social sciences together with the principles and methods of engineering design to specify, predict, and evaluate the results to be obtained from such systems or processes.

C. Kahraman (ed.), *Computational Intelligence Systems in Industrial Engineering*,
Atlantis Computational Intelligence Systems 6, DOI: 10.2991/978-94-91216-77-0_1,

Interest areas of IE include operations research & optimization techniques, engineering economics, supply chain management & logistics, systems simulation & stochastic processes, system dynamics & policy planning, system analysis & techniques, manufacturing systems/manufacturing engineering, human factors engineering & ergonomics (safety engineering), production planning and control, management sciences, computer aided manufacturing, facilities design & work space design, quality engineering, reliability engineering & life testing, statistical process control or quality control, time and motion study, operations management, corporate planning, productivity improvement, and materials management.

An intelligent system is a system that has its own main objective, as well as senses and actuators. To reach its objective it chooses an action based on its experiences. It can learn by generalizing the experiences it has stored in its memories. The Intelligent Systems focus on modeling intelligence and on research into and development of highly autonomous automated systems which can perform complex real-world tasks.

Computational intelligence (CI) is a set of Nature-inspired computational methodologies and approaches to address complex problems of the real world applications to which traditional methodologies and approaches are ineffective or infeasible. It primarily includes Fuzzy logic systems, Neural Networks and Evolutionary Computation. In addition, CI also embraces techniques that stem from the above three or gravitate around one or more of them, such as Swarm intelligence and Artificial immune systems which can be seen as a part of Evolutionary Computation; Dempster-Shafer theory, Chaos theory and Multi-valued logic. IEEE Computational Intelligence Society defines its subjects of interest as neural networks, fuzzy systems and evolutionary computation, including swarm intelligence. CI is a branch of computer science studying problems for which there are no effective computational algorithms.

Computational intelligent systems have found many application areas in IE. The main aim of this chapter is to classify these intelligent systems and give application examples from IE. The next section is on the interest areas of IE. Section 1.3 classifies the intelligent systems and the application areas of these intelligent systems in IE by a literature review. The chapter is finalized by a conclusion section.

1.2 Industrial Engineering

Industrial engineering (IE) is the engineering discipline concerned with the planning, organising and operation of industrial facilities and processes for the economic, safe and effective use of physical and human resources. IE is applied design for the integration of material, human and financial resources, and of production sequences and methods, optimum flows and layouts, and of work methods and procedures, labour organisation, and in economic evaluation of facilities, processes or techniques.

Industrial engineers determine the most effective ways to use the basic factors of production (people, machines, materials, information, and energy) to make a product or to provide a service. They are the bridge between management goals and operational performance. They are more concerned with increasing productivity through the management of people, methods of business organization, and technology than are engineers in other specialties, who generally work more with products or processes. Although most industrial engineers work in manufacturing industries, they may also work in consulting services, healthcare, and communications.

To solve organizational, production, and related problems most efficiently, industrial engineers carefully study the product and its requirements, use mathematical methods such as operations research to meet those requirements, and design manufacturing and information systems. They develop management control systems to aid in financial planning and cost analysis and design production planning and control systems to coordinate activities and ensure product quality. They also design or improve systems for the physical distribution of goods and services. Industrial engineers determine which plant location has the best combination of raw materials availability, transportation facilities, and costs. Industrial engineers use computers for simulations and to control various activities and devices, such as assembly lines and robots.

They also develop wage and salary administration systems and job evaluation programs. Many industrial engineers move into management positions because the work is closely related. The work of health and safety engineers is similar to that of industrial engineers in that it deals with the entire production process. Health and safety engineers promote worksite or product safety and health by applying knowledge of industrial processes, as well as mechanical, chemical, and psychological principles. They must be able to anticipate, recognize, and evaluate hazardous conditions as well as develop hazard control methods. They also must be familiar with the application of health and safety regulations. The interest areas of IE can be classified as in Figure 1.1:

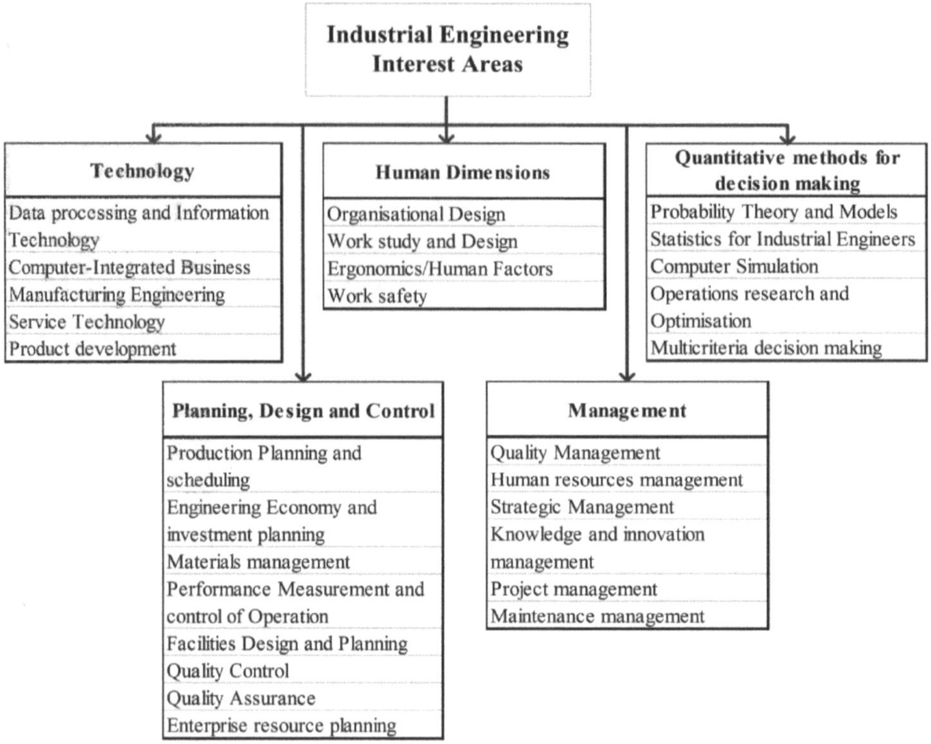

Fig. 1.1 Classification of interest areas of IE

1.3 Computational Intelligent systems in Industrial engineering: A literature Review

In the following, application areas of intelligent systems in IE are presented with respect to the problem areas given in Figure 1.1.

1.3.1 *Fuzzy sets and systems*

Fuzzy sets (FSs) are the basic concept supporting the fuzzy set theory. The main research fields in the fuzzy set theory are fuzzy sets, fuzzy logic, and fuzzy measure. Fuzzy reasoning or approximate reasoning is an application of fuzzy logic to knowledge processing. Fuzzy sets are a generalization of conventional set theory that were introduced by Zadeh in 1965 as a mathematical way to represent vagueness in everyday life [62]. Fuzzy control is an application of fuzzy reasoning to control. One feature of FSs is the ability to realize a complex nonlinear input–output relation as a synthesis of multiple simple input–output relations. Table 1.1 presents some examples of fuzzy sets applications in IE.

Table 1.1. Fuzzy sets and its applications in IE

Technology
Data processing and Information Technology: Cateni and Colla [6] present a preliminary brief review of traditional outlier detection techniques and classification algorithms sutiable for imbalanced dataset.
Computer-Integrated Business: Kahraman *et al.* [25] propose some fuzzy models based on fuzzy present worth to obtain a sensible result in quantifying the manufacturing flexibility in computer integrated manufacturing systems. The fuzzy models based on present worth are basically engineering economics decision models in which the uncertain cash flows and discount rates are specified as triangular fuzzy numbers.
Manufacturing Engineering: Chuu [9] develops a fuzzy multiple attribute decision-making applied in the group decision-making to improving advanced manufacturing technology selection process. Since numerous attributes have been considered in evaluating the manufacturing technology suitability, most information available in this stage is subjective, imprecise and vague, fuzzy sets theory provides a mathematical framework for modeling imprecision and vagueness.
Service Technology: Lu *et al.* [37] aim to offer a thorough introduction and systematic overview of the new field e-service intelligence mainly based on fuzzy set related techniques. It covers the state-of-the-art of the research and development in various aspects including both theorems and applications of e-service intelligence by applying fuzzy set theory. Moreover, it demonstrates how adaptations of existing intelligent technologies benefit from the development of e-service applications in online customer decision, personalised services, web mining, online searching/data retrieval, online pattern recognition/image processing, and web-based e-logistics/planning.
Product development: Wang [59] develops a methodology to schedule product development projects having imprecise temporal information. The research problem is formulated as a fuzzy constraint satisfaction problem and a new method based on the possibility theory is proposed to determine the satisfaction degrees of fuzzy temporal constraints.

Continued on next page

Table 1.1. Fuzzy sets and its applications in IE (continued)

Human Dimensions
Organisational Design: Dimitrov [11] considers the self-organization processes in the human systems under fuzzy management. The principle of Gause-Arrow is presented, according to which in every human system under fuzzy management there is at least one "leader" who possesses the highest ability of generating variety of alternatives in response of fuzzy instructions.
Work study and Design: Hsiao [21] proposes a fuzzy decision-making method for selecting an optimum design from various design alternatives. The evaluation objectives are arranged in a hierarchial structure with several levels. The relative contribution (weighting function) of each objective to the overall value of the solution and the rating or degree of approximation of a solution with respect to a given objective are quantified with the membership functions of a fuzzy set.
Ergonomics/Human Factors: Aluclu *et al.* [1] describe noise–human response and a fuzzy logic model developed by comprehensive field studies on noise measurements (including atmospheric parameters) and control measures. The model has two subsystems constructed on noise reduction quantity in dB.
Work safety: Zheng *et al.* [65] study safety evaluation and early warning rating of the hot and humid environments in order to guarantee workers' health and safety. The fuzzy analytic hierarchy process (AHP) method is proposed to evaluate the work safety in hot and humid environments. Trapezoidal fuzzy numbers are adopted to handle inherent uncertainty and imprecision of the data involved in decision process.
Planning, Design and Control
Production Planning: Torabi *et al.* [51] incorporate the fuzzy set theory into the HPP structure to handle the uncertainty and infeasibility issues. Inspired by a real industrial case, a fuzzy HPP (FHPP) model is proposed which is composed of two decision making levels. At first, an aggregate production plan is determined by solving a fuzzy linear programming model at the product family level and then it is disaggregated through another fuzzy linear programming model at the next level to find a disaggregated production plan in final products level. The FHPP model is implemented for the real industrial case and it is compared with the previously developed crisp model.

Continued on next page

Table 1.1. Fuzzy sets and its applications in IE (continued)

Engineering Economy and investment planning: Tsao [53] presents a series of pragmatic algorithms for calculating the net present values of capital investments in an environment, which are subject to uncertainty from randomness of outcomes and vagueness of estimation. Linguistic terms is adopted to assess the possibilities of future economic scenarios, and fuzzy numbers are then used to represent the linguistic assessments and estimate cash flows and costs of capital.
Quality Control: Pacella *et al.* [44] present a neural network system for quality control which is based on an unsupervised training phase. In particular, the adaptive resonance theory (ART) is investigated in order to implement a model-free quality control system, which can be exploited for recognising changes in the state of a manufacturing process.
Management
Project management: Yakhchali [61] addresses the problem of determining the degree of possible and necessary criticality of activities as well as determining paths in networks that have fuzzy activity durations. In such networks, activities and paths are reported in a fuzzy representation as being critical, with certain degrees of possibility and necessity, instead of being declared critical or not in a binary way.
Supply chain management: Pishvaee and Razmi [45] propose a multi-objective fuzzy mathematical programming model for designing an environmental supply chain under inherent uncertainty of input data.
Quantitative methods for decision making
Statistics for Industrial Engineers: Chang *et al.* [8] propose a new fuzzy Delphi method. The method employs the fuzzy statistics and technique of the conjugate gradient search to fit membership functions. Membership functions besides triangles may be derived for the fuzzy forecasts. A procedure termed as stabilization rather than convergence of the iterative interval-valued surveys is utilized, in which the α-level set of the fuzzy forecasts is used. The testing of stability of the process is conducted on the fuzzy forecasts between two consecutive iterations instead of within each iteration.

Continued on next page

Table 1.1. Fuzzy sets and its applications in IE (continued)

Computer Simulation: Balin [2] addresses parallel machine scheduling problems with fuzzy processing times. A robust genetic algorithm (GA) approach embedded in a simulation model is proposed to minimize the maximum completion time (makespan). The results are compared with those obtained by using the "longest processing time" rule (LPT), which is known as the most appropriate dispatching rule for such problems. This application illustrates the need for efficient and effective heuristics to solve such fuzzy parallel machine scheduling problems (FPMSPs).
Multicriteria decision making: Ho *et al.* [19] develop an integrated approach, combining quality function deployment (QFD), fuzzy set theory, and analytic hierarchy process (AHP) approach, to evaluate and select the optimal third-party logistics service providers (3PLs).

1.3.2 *Artifical Neural Networks*

Neural network (NN) models are inspired by brain processes and structures at almost the lowest level, while symbolic AI models by processes at the highest level. Artificial neural networks (ANN) have been developed as generalizations of mathematical models of biological nervous systems. In other words; ANNs, or simply neural networks, are information processing systems that roughly replicate the behavior of a human brain by emulating the operations and connectivity of biological neurons. Literally, the basic processing elements of neural networks are called artificial neurons, or simply neurons or nodes. In the literature, McCulloch and Pitts [38] firstly introduced simplified neurons. In ANNs, network sees the nodes as artificial neurons. ANNs combine the artificial neurons in order to process information. Simply, an artificial neuron is a computational model inspired in the natural neurons.

ANN models can be used to infer a function from observations. This is particularly useful in applications where the complexity of the data or task makes the design of such a function by hand impractical. ANNs can be trained directly from data. ANNs can be used to extract patterns and detect trends thus it can be applied to data classification and nonlinear functional mapping. Specific application examples include process modeling, control, machine diagnosis, and real-time recognition. Table 1.2 presents some examples of ANN applications in IE.

Table 1.2. ANN and its applications in IE

Technology
Data processing and Information Technology: Blank and Brown [4] use some problems in multivariate calibration and pattern recognition to compare the performance of these neural networks, as trained by back-propagation, with some more well-established chemometric techniques. Simulated data sets are used to probe the capabilities of neural networks in applications where the chemometric methods are known to excel, and to fail.
Product development: Hsu *et al.* [22] formulate and analyze a fairly general product development model based on multiple attribute decision making with emphasis on the treatment of the linguistic and vague aspects by fuzzy logic and up-dating or learning by neural network.
Human Dimensions
Ergonomics/Human Factors: Kolich *et al.* [66] compare two distinct modeling approaches, which are stepwise, linear regression and artificial neural network, and contrast them.
Work safety: Ung *et al.* [55] propose a risk prediction model incorporating the fuzzy set theory and artificial neural network capable of resolving the problem encountered. An algorithm capable of converting the risk-related parameters and the overall risk level from the fuzzy property to the crisp-valued attribute is also developed.
Planning, Design and Control
Engineering Economy and investment planning: Tan *et al.* [49] propose a hybrid intelligent system integrating case-based reasoning and the fuzzy ARTMAP neural network model to support managers in making timely and optimal manufacturing technology investment decisions.
Materials management: Ming and Mak [40] formulate the problems of allocating tolerances to the manufacturing operations and selecting exactly one representative from the alternative sets of manufacturing operations for machining one feature of the part. The purposes to minimize the sum of the costs of the selected sets of manufacturing operations and the dissimilarities in their manufacturing resource requirements. The techniques of the genetic algorithm and the Hopfield neural network are adopted as possible approaches to solve these problems.

Continued on next page

Table 1.2. ANN and its applications in IE (continued)

Management
Quality Management: Köse *et al.* [34] assist users and determine how digital proofing systems used in colour management effect print qualities. They also determine the effects of digital proofing systems used in colour management on print quality by artificial neural network (ANN).
Quantitative methods for decision making
Computer Simulation: Fonseca and Navaresse [14] explore the use of artificial neural networks as a valid alternative to the traditional job-shop simulation approach. Feed forward, multi-layered neural network metamodels are trained through the back-error-propagation (BEP) learning algorithm to provide a versatile job-shop scheduling analysis framework.
Operations research and Optimisation: Kia *et al.* [31] present a novel mixed-integer non-linear programming model for the layout design of a dynamic cellular manufacturing system. nto a linearized.

1.3.3 *Genetic algorithms*

Holland [20] developed genetic algorithms (GAs). GA search methods are rooted in the mechanisms of evolution and natural genetics. GAs are part of the adaptive stochastic optimization algorithms involving search and optimization. GAs provide an alternative to traditional optimization techniques by using directed random searches to locate optimal solutions in complex landscapes.

GAs are a family of computational models inspired by evolution. These algorithms encode a potential solution to a specific problem on a simple chromosome like data structure and apply recombination operators to these structures so as to preserve critical information. GAs are often viewed as function optimizers although the range of problems to which genetic algorithms have been applied is quite broad. The basic idea is to try to mimic a simple picture of natural selection in order to find a good algorithm. The first step is to mutate, or randomly vary, a given collection of sample programs. The second step is a selection step, which is often done through measuring against a fitness function. The process is repeated until a suitable solution is found. Table 1.3 presents some examples of GA applications in IE.

Table 1.3. Genetic Algorithms and its applications in IE

Technology
Manufacturing Engineering: Wu *et al.* [60] develop a hierarchical genetic algorithm to simultaneously form manufacturing cells and determine the group layout of a CMS. The intrinsic features of our proposed algorithm include a hierarchical chromosome structure to encode two important cell design decisions, a new selection scheme to dynamically consider two correlated fitness functions, and a group mutation operator to increase the probability of mutation.
Human Dimensions
Ergonomics/Human Factors: Groissboeck *et al.* [18] deal with an approach allowing to associate visual textures with given human perceptions. Based on a forward model associating human perceptions for given visual textures, the deduction of an reverse process is presented which is able to associate and characterize visual textures for given human perceptions. They propose a constraint-based genetic algorithm approach, which is able to minimize a specific optimization problem containing constraints in form of band-widths for valid individuals (low level features extracted from textures) in a population.
Planning, Design and Control
Production Planning: Qing-dao-er-ji and Wang [43] design some genetic operators to solve the job shop scheduling problem more effectively. In order to increase the diversity of the population, a mixed selection operator based on the fitness value and the concentration value is given. To make full use of the characteristics of the problem itself, new crossover operator based on the machine and mutation operator based on the critical path are specifically designed.
Management
Human resources management: Lin and Gen [67] focus on multi-criteria human resource allocation for solving multistage combinatorial optimization problem. Hence they tackle this problem via a multistage decision-making model and propose a multi-objective hybrid genetic algorithm approach based on the multistage decision-making model for solving combinatorial optimization problems.

Continued on next page

Table 1.3. Genetic Algorithms and its applications in IE (continued)

Quantitative methods for decision making
Computer Simulation: Udhayakumar *et al.* [54] develop a genetic algorithm approach for solving the P-model of chance constrained data envelopment analysis problems, which include the concept of "Satisficing". Problems here include cases in which inputs and outputs are stochastic, as well as cases in which only the outputs are stochastic. The basic solution technique for the above has so far been deriving "deterministic equivalents", which is difficult for all stochastic parameters as there are no compact methods available. In the proposed approach, the stochastic objective function and chance constraints are directly used within the genetic process. The feasibility of chance constraints are checked by stochastic simulation techniques.

1.3.4 *Tabu Search*

The word tabu (or taboo) comes from Tongan, a language of Polynesia, where it was used by theaborigines of Tonga island to indicate things that cannot be touched because they are sacred. According to Webster's Dictionary, the word now also means "a prohibition imposed by social custom as a protective measure" or of something "banned as constituting a risk."

Table 1.4. Tabu search and its applications in IE

Planning, Design and Control
Production Planning: Al-Turki *et al.* [68] develop a tabu search-based solution procedure designed specifically for a certain class of single-machine scheduling problems with a non-regular performance measure. The performance of the developed algorithm is tested for solving the variance minimization problem. Problems from the literature are used to test the performance of the algorithm.
Management
Project management: Tsai and Gemmill [52] propose a higher level heuristic procedure "tabu search" to provide good solutions to resource-constrained, randomized activity duration project scheduling problems. Our adaptation of tabu search uses multiple tabu lists, randomized short-term memory, and multiple starting schedules as a means of search diversification.

Continued on next page

Table 1.4. Tabu search and its applications in IE (continued)

Quantitative methods for decision making
Operations research and Optimisation: Løkketangen and Glover [36] describe a tabu search approach for solving general zero-one mixed integer programming problems that exploits the extreme point property of zero-one solutions. Specialized choice rules and aspiration criteria are identified for the problems, expressed as functions of integer infeasibility measures and objective function values. The first-level TS mechanisms are then extended with advanced level strategies and learning.

Difficulty in optimization problems encountered in practical settings such as telecommunications, logistics, financial planning, transportation and production has motivated in development of optimization techniques. Tabu search (TS) is a higher level heuristic algorithm for solving combinatorial optimization problems. It is an iterative improvement procedure that starts form an initial solution and attempts to determine a better solution. Tabu search applications in the literature are in the areas of Scheduling, Design, Location and Allocation, Logic and Artificial Intelligence, Production, Inventory and Investment, Routing, and General Combinational Optimization. Table 1.4 presents some examples of TS applications in IE.

1.3.5 *Evolutionary Algorithms*

Evolutionary algorithms (EAs) are search methods that take their inspiration from natural selection and survival of the fittest in the biological world. EAs differ from more traditional optimization techniques in that they involve a search from a "population" of solutions, not from a single point. Each iteration of an EA involves a competitive selection that weeds out poor solutions. The solutions with high "fitness" are "recombined" with other solutions by swapping parts of a solution with another. Solutions are also "mutated" by making a small change to a single element of the solution. Recombination and mutation are used to generate new solutions that are biased towards regions of the space for which good solutions have already been seen. Table 1.5 presents some examples of EA applications in IE.

Table 1.5. Evaluationary algorithms and its applications in IE

Planning, Design and Control
Technology
Product development: Velsker *et al.* [58] investigate optimization strategies eligible for solving complex engineering design problems. The developed optimization algorithms allows to predict the performance of the product (structure) for different design and configurations parameters as well as loading conditions.
Management
Project management: Jaskowski and Sobotka [23] attempt to use evolutionary algorithms to solve the problem of minimizing construction project duration in deterministic conditions, with in-time changeable and limited accessibility of renewable resources (workforce, machines, and equipment) Particular construction processes (with various levels of complexity) must be conducted in the established technological order and can be executed with different technological and organizational variants (different contractors, technologies, and ways of using resources).
Quantitative methods for decision making
Computer Simulation: Corno *et al.* [10] describe evolutionary simulation-based validation, a new point in the spectrum of design validation techniques, besides pseudo-random simulation, designer-generated patterns and formal verification. The proposed approach is based on coupling an evolutionary algorithm with a hardware simulator, and it is able to fit painlessly in an existing industrial flow.

1.3.6 *Differential Evolution*

Differential evolution (DE) is introduced by Storn and Price in 1996 [69]. DE is known as population-based optimisation algorithm similar to GAs using similar operators; crossover, mutation and selection. According to Karaboğa and Ökdem [27], the main difference in constructing better solutions is that genetic algorithms rely on crossover while DE relies on mutation operation. This main operation is based on the differences of randomly sampled pairs of solutions in the population. DE algorithm uses mutation operation as a search mechanism and selection operation to direct the search toward the prospective regions in the search space. In addition to this, the DE algorithm uses a non-uniform crossover which can take child vector parameters from one parent more often than it does from others. By using the components of the existing population members to construct trial vectors, the recombination (crossover) operator efficiently shuffles information about successful combinations, enabling the search for a better solution space. An optimization task consisting of D parameters can be represented by a D-dimensional vector. In DE, a population of NP solution vectors is randomly created at the start. This population is successfully improved by applying mutation, crossover and selection operators.

In the area of strategic management, Xu [57] examines various market differential reactions to alliance announcements made by pharmaceutical and biotech firms. This study answers the call for separate examinations of different types of alliance. It also finds evidence on the "size effect" in alliance gains in the harmaceutical/biotech industry, which may provide an explanation for prior empirical cross-industry studies' failure to find such effect.

1.3.7 *Artificial Bee Colony*

Artificial bee colony (ABC) algorithm was proposed by Karaboğa [28], inspired by the foraging and waggles dance behaviors of honeybees. ABC is a relatively new member of swarm intelligence. It has received increasing interest because of its simplicity, wide applicability, and outstanding performance. The basic ABC has been compared with other evolutionary algorithms, such as GA, PSO, and differential evaluation on a limited number of test functions [26, 29].

Honey bees use several mechanisms like waggle dance to optimally locate food sources and to search new ones. This makes them a good candidate for developing new intelligent search algorithms. In the ABC algorithm, the colony of artificial bees contains three groups of bees: employed bees, onlookers and scouts. A bee waiting on the dance area for making decision to choose a food source is called an onlooker and a bee going to the food source visited by itself previously is named an employed bee. A bee carrying out random search is called a scout. In the ABC algorithm, first half of the colony consists of employed artificial bees and the second half constitutes the onlookers. For every food source, there is only one employed bee. In other words, the number of employed bees is equal to the number of food sources around the hive. The employed bee whose food source is exhausted by the employed and onlooker bees becomes a scout [30].

In the area of production planning and scheduling, Tasgetiren *et al.* [50] present a discrete ABC algorithm hybridized with a variant of iterated greedy algorithms to find the permutation that gives the smallest total flowtime. Iterated greedy algorithms are comprised of local search procedures based on insertion and swap neighborhood structures.

In the area of operations research and optimisation, Zhang *et al.* [64] propose a novel approach called "hybrid multi-objective artificial bee colony" to solve a nonlinear constrained multi-objective model. The algorithm is new swarm intelligence based multi-objective optimization technique inspired by the intelligent foraging behavior of honey bees, summation of normalized objective values and diversified selection and nondominated sorting approach.

1.3.8 *Ant Colony Optimization*

Ant colony optimization (ACO) algorithm based on the foraging behavior of ants has been first introduced by Dorigo and Gambardella [13]. The basic idea of ACO is to imitate the cooperative behavior of ant colonies. When searching for food, ants initially explore the area surrounding their nest in a random manner. As soon as an ant finds a food source, it evaluates it and carries some food back to the nest. During the return trip, the ant deposits a pheromone trail on the ground. The pheromone deposited, the amount of which may depend on the quantity and quality of the food, guides other ants to the food source [46]. Quantity of pheromone on the arc is decreased in time due to evaporating. Each ant decides to a path or way according to the quantity of pheromone which has been leaved by other ants. More pheromone trail consists in short path than long path. The first ACO algorithm developed was the ant system (AS) [12], and since then several improvement of the AS have been devised [15, 16, 47].

Berrichi *et al.* [3] present an algorithm based on Ant Colony Optimization paradigm to solve the joint production and maintenance scheduling problem. The model is formulated according to a bi- objective approach to find trade-off solutions between both objectives of production and maintenance. Reliability models are used to take into account the maintenance aspect.

Morin *et al.* [41] study the learning process in an ant colony optimization algorithm designed to solve the problem of ordering cars on an assembly line (car-sequencing problem). A new pheromone trail structure is proposed that is specifically adapted to the type of constraints in the car-sequencing problem.

1.3.9 *Particle Swarm Optimization*

Particle Swarm Optimization (PSO) is a population based stochastic optimization technique inspired by social behavior of bird flocking. PSO applies the concept of social interaction to problem solving. It was developed in 1995 by James Kennedy and Russ Eberhart [70]. In PSO, A group of agents, "particles", are thrown into the search space. The particles communicate either directly or indirectly with one another for search directions (gradients). It is initialized with a population of random solutions and searches for optimal by updating generations. In PSO, the potential solutions (called as particles), move through the problem space by following the current optimum particles. PSO is a simple but powerful search technique. It has been applied successfully to a wide variety of search and optimization problems.

Zhang and Xing [63] present a fuzzy-multi-objective particle swarm optimization to solve the fuzzy time-cost-quality tradeoff problem. The time, cost and quality are described by fuzzy numbers and a fuzzy multiattribute utility methodology incorporated with constrained fuzzy arithmetic operations is adopted to evaluate the selected construction methods. The particle swarm optimization is applied to search for the time-cost-quality tradeoff solutions by incorporating the fuzzy multi-attribute utility methodology.

Lei [35] presents a particle swarm optimization for multi-objective job shop scheduling problem. The objective is to simultaneously minimize makespan and total tardiness of jobs. By constructing the corresponding relation between real vector and the chromosome obtained by using priority rule-based representation method, job shop scheduling is converted into a continuous optimization problem.

1.3.10 *Swarm Intelligence*

Social insects work without supervision. In fact, their teamwork is largely self-organized, and coordination arises from the different interactions among individuals in the colony. Although these interactions might be primitive (one ant merely following the trail left by another, for instance), taken together they result in efficient solutions to difficult problems (such as finding the shortest route to a food source among myriad possible paths). The collective behavior that emerges from a group of social insects has been dubbed swarm intelligence [5]. SI indicates a recent computational and behavioral metaphor for solving distributed problems that originally took its inspiration from the biological examples provided by social insects (ants, termites, bees, wasps) and by swarming, flocking, herding behaviors in vertebrates.

Sundar and Singh [48] describe an application of ABC algorithm, which is a new swarm intelligence approach, for a version of the single machine early/tardy scheduling problem where no unforced machine idle time is allowed. A local search is used inside the ABC algorithm to further improve the schedules obtained through it. A variant of the basic ABC approach is also considered where the best solution obtained through ABC algorithm is improved further via an exhaustive local search.

Kadadevaramath *et al.* [24] specifically deals with the modeling and optimization of a three echelon supply chain network using the particle swarm optimization / intelligence algorithms.

1.3.11 *Simulated Annealing*

Simulated annealing (SA) methods are methods proposed for the problem of finding, numerically, a point of the global minimum of a function defined on a subset of a k-dimensional Euclidean space. The motivation of the methods lies in the physical process of annealing, in which a solid is heated to a liquid state and, when cooled sufficiently slowly, takes up the configuration with minimal inner energy. Metropolis *et al.* [39] described this process mathematically. SA uses this mathematical description for the minimization of other functions than the energy. The first results published by Kirpatrick *et al.* [32], German and German [17], Cerny [7].

SA algorithm is a technique to find a good solution of an optimization problem using a random variation of the current solution. A worse variation is accepted as the new solution with a probability that decreases as the computation proceeds. The slower the cooling schedule, or rate of decrease, the more likely the algorithm is to find an optimal or near-optimal solution [56].

Kouvelis *et al.* [33] address the machine layout problem in the presence of zoning constraints. The Quadratic Assignment Problem formulation is appropriately modified to account for the existence of such constraints. Two distinct implementations of the Simulated Annealing algorithms, the Compulsion and the Penalty procedures, are proposed for the solution of the problem.

Musharavati and Hamouda [42] devise and employ three simulated annealing based algorithms that Exploit Auxiliary Knowledge indifferent ways to handle a manufacturing process planning problem for reconfigurable manufacturing. These algorithms are configured based on a generic combination of the simulated annealing technique with heuristic knowledge, and metaknowledge.

1.4 Conclusion

Computational intelligent systems are very popular since they are excellent devices to handle the complex problems, which are very hard to solve by traditional methodologies and approaches. The number of the proposed intelligent systems is increasing day to day. Integrated use of these techniques is often used in the solution of complex IE problems. Hybrid models also exist in the literature.

IE discipline has many complex problems in real life from manufacturing to human factors. It has to use CI techniques for the solution of these complex problems. Most of the papers using CI techniques come from IE departments of universities. Computer softwares will make their use more frequent in the near future.

Bibliography

[1] Aluclu I., Dalgic A., and Toprak Z.F., (2008). A fuzzy logic-based model for noise control at industrial workplaces, *Applied Ergonomics*, **39**, 368–378.

[2] Balin S., (2011). Parallel machine scheduling with fuzzy processing times using a robust genetic algorithm and simulation, *Information Sciences*, **181**, 3551–3569.

[3] Berrichi A., Yalaoui F., Amodeo L., and Mezghiche M., (2010). Bi-Objective Ant Colony Optimization approach to optimize production and maintenance scheduling, *Computers & Operations Research*, **37**, 1584–1596.

[4] Blank T.B. and Brown S.D., (1993). Data processing using neural networks, *Analytica Chimica Acta*, **277** (2), 273–287.

[5] Bonabeou E. and Meyer C. (Eds.), (2001). *Swarm Intelligence: A Whole New Way to Think About Business.* Harward Business Review.

[6] Cateni S. and Colla V., (2012). Fuzzy Inference System for Data Processing in Industrial Applications, *Fuzzy Inference System - Theory and Applications*, Dr. Mohammad Fazle Azeem (Ed.), ISBN: 978-953-51-0525-1.

[7] Cerny V., (1985). A thermodynamical approach to the traveling salesman problem: An efficient simulation algorithm. *Journal of Optimization Theory and Applications*, **45**, 41–51.

[8] Chang P.-T., Huang L.-C., and Lin H.-J., (2000). The fuzzy Delphi method via fuzzy statistics and membership function fitting and an application to the human resources, *Fuzzy Sets and Systems*, **112**, 511–520.

[9] Chuu S.-J., (2009). Selecting the advanced manufacturing technology using fuzzy multiple attributes group decision making with multiple fuzzy information, *Computers & Industrial Engineering*, **57** (3), 1033–1042.

[10] Corno F., Reorda M.S., and Squillero G., (2004). Evolutionary Simulation-Based Validation, *International Journal on Artificial Intelligence Tools*, **14** (1-2), Dec. 2004, 897–916.

[11] Dimitrov V.D., (1977). Social Choice and Self-Organization under Fuzzy Management, *Kybernetes*, **6** (3) 153–156.

[12] Dorigo M., (1992). *Optimization, learning and natural algorithms. Unpublished doctoral dissertation*, University of Politecnico di Milano, Italy.

[13] Dorigo M. and Gambardella L.M., (1997). Ant colony system: a cooperative learning approach to the traveling salesman problem. *IEEE Transaction on Evolutionary Computation*, **1**, 53–66.

[14] Fonseca D.J. and Navaresse D., (2002). Artificial neural networks for job shop simulation, *Advanced Engineering Informatics*, **16**, 241–246.

[15] Gambardella L.M. and Dorigo M., (1995). Ant-Q: a reinforcement learning approach to the travelling salesman problem. In *Proceedings of the Twelfth International Conference on Machine Learning.* California, USA.

[16] Gambardella L.M. and Dorigo M., (1996). Solving Symmetric and Asymmetric TSPs by Ant Colonies. In *Proceedings of the IEEE Conference on Evolutionary Computation* (pp. 622–627). Nagoya, Japan.

[17] German S. and German D., (1984). Stochastic relaxation, Gibbs distributions, and the Bayesian restoration of images. *IEEE Proceedings Pattern Analysis and Machine Intelligence*, **6** (6), 721–741.

[18] Groissboeck W., Lughofer E., and Thumfart S., (2010). Associating visual textures with human perceptions using genetic algorithms, *Information Sciences*, **180**, 2065–2084.

[19] Ho W., He T., Lee C.K.M., and Emrouznejad A., (2012) Strategic logistics outsourcing: An integrated QFD and fuzzy AHP approach, *Expert Systems with Applications*, **39**, 10841–10850.

[20] Holland J.H. (Ed.), (1975). *Adaptation in Natural and Artificial Systems: An Introductory Analysis with Applications to Biology, Control, and Artificial Intelligence.* Ann Arbor, MI: University of Michigan Press.

[21] Hsia S.-W., (1998). Fuzzy logic based decision model for product design, *International Journal of Industrial Ergonomics*, **21**, 103–116.

[22] Hsu C.-H., Jiang B.C., and Lee E.S., (1999). Fuzzy Neural Network Modeling for Product Development, *Mathematical and Computer Modelling*, **29**, 71–81.

[23] Jaśkowski P. and Sobotka A., Scheduling Construction Projects Using Evolutionary Algorithm, *Journal Of Construction Engineering And Management*, Asce / August 2006 / 861.

[24] Kadadevaramath R.S., Chen J.C.H., Shankar B.L., and Rameshkumar K., (2012). Application of particle swarm intelligence algorithms in supply chain network architecture optimization, *Expert Systems with Applications*, **39**, 10160–10176.

[25] Kahraman C., Beskese A., and Ruan D., (2004). Measuring flexibility of computer integrated manufacturing systems using fuzzy cash flow analysis, *Information Sciences*, **168**, 77–94.

[26] Kang F., Li J., and Ma Z., (2011). Rosenbrock artificial bee colony algorithm for accurate global optimization of numerical functions. *Information Science*, **181** (16), 3508–3531.

[27] Karaboga D. and Ökdem S., (2004). A simple and Global Optimization Algorithm for Engineering Problems: Differential Evolution Algorithm. *Turk J. Elec. Engin.*, **12** (1).

[28] Karaboğa D., (2005). *An idea based on honeybee swarm for numerical optimization.* Technical Report TR06, Erciyes University.

[29] Karaboğa D. and Baştürk B., (2007a). A Powerful and efficient algorithm for numerical optimization: artificial bee colony (ABC) algorithm. *Journal of Global Optimization*, **39** (3), 459–471.

[30] Karaboğa D. and Baştürk B., (2007b). Artificial bee colony (ABC) optimization algorithm for solving constrained optimization problems. *LNCS: Advances in Soft Computing: Foundations of Fuzzy Logic and Soft Computing*, **4529**, 789–798.

[31] Kia R., Baboli A., Javadian N., Tavakkoli-Moghaddam R., Kazemi M., and Khorrami J., (2012). Solving a group layout design model of a dynamic cellular manufacturing system with alternative process routings, lot splitting and flexible reconfiguration by simulated annealing, *Computers & Operations Research*, **39**, 2642–2658.

[32] Kirpatrick S., Gelat Jr. C.D., and Vecchi M.P., (1983). Optimization by simulated annealing. *Science*, **220**, 671–680.

[33] Kouvelis P., Chiang W.-C., and James Fitzsimmons J., (1992). Simulated annealing for machine layout problems in the presence of zoning constraints, *European Journal of Operational Research*, **57** (2) 203–223.

[34] Köse E., Şahinbaşkan T., and Güler İ. (2009). The investigation of effects of digital proofing systems used in colour management on print quality with neural Networks, *Expert Systems with Applications*, **36**, 745–754.

[35] Lei D., (2008) A Pareto archive particle swarm optimization for multi-objective job shop scheduling, *Computers & Industrial Engineering*, **54**, 960–971.

[36] Lokketangen A. and Glover F., (1998). Solving zero-one mixed integer programming problems using tabu search, *European Journal of Operational Research*, **106** (2-3), 624–658.

[37] Lu J., Ruan D., and Zhang G., (2008). Fuzzy Set Techniques in E-Service Applications, *Studies in Fuzziness and Soft Computing*, 2008, **220**, 553–566, DOI: 10.1007/978-3-540-73723-0_28.

[38] McCulloch W.S. and Pitts W.H., (1943). A logical calculus of the ideas immanent in nervous activity. *Bulletin of Mathematical Biophysics*, **5**, 115–133.

[39] Metropolis N., Rosenbluth A., Rosenbluth M., Teller A., and Teller E., (1953). Equation of State Calculations by Fast Computing Machines. *Journal Chemistry Physics*, **21** (6), 1087–1092.

[40] Ming X.G. and Mak K.L., (2001). Intelligent approaches to tolerance allocation and manufacturing operations selection in process planning, *Journal of Materials Processing Technology*, **117**, 75–83.

[41] Morin S., Gagné C., and Gravel M., (2009). Ant colony optimization with a specialized pheromone trail for the car-sequencing problem, *European Journal of Operational Research*, **197**, 1185–1191.

[42] Musharavati F. and Hamouda A.M.S., (2012). Simulated annealing with auxiliary knowledge for process planning optimization inreconfigurable manufacturing, *Robotics and Computer-Integrated Manufacturing*, **28**, 113–131.

[43] Qing-dao-er-ji R. and Wang Y., (2012). A new hybrid genetic algorithm for job shop scheduling problem, *Computers & Operations Research*, **39**, 2291–2299.

[44] Pacellaa M., Semerarob Q., and Anglani A., (2004). Manufacturing quality control by means of a Fuzzy ART network trained on natural process data, *Engineering Applications of Artificial Intelligence*, **17**, 83–96.

[45] Pishvaee M.S. and Razmi J., (2012). Environmental supply chain network design using multi-objective fuzzy mathematical programming, *Applied Mathematical Modelling*, **36**, 3433–3446.

[46] Socha K. and Dorigo M., (2008). Ant colony optimization for continuous domains. *European Journal of Operational Research*, **185**, 1155–1173.

[47] Stützle T. and Hoos H., (2000). MAX-MIN Ant System. *Future Generation Computer Systems*, **16** (8), 889–904.

[48] Sundar S. and Singh A., (2012). A swarm intelligence approach to the early/tardy scheduling problem, *Swarm and Evolutionary Computation*, **4**, 25–32.

[49] Tan K.H., Lim C.P., Platts K., and Koay H.S., (2006). An intelligent decision support system for manufacturing technology investments, *International Journal of Production Economics*, **104**, 179–190.

[50] Tasgetiren M.F., Pan Q.-K., Suganthan P.N., and Chen A.H.-L., (2011). A discrete artificial bee colony algorithm for the total flowtime minimization in permutation flow shops, *Information Sciences*, **181** (16) 3459–3475.

[51] Torabi S.A., Ebadian M., and Tanha R., (2010). Fuzzy hierarchical production planning(with a case study), *Fuzzy Sets and Systems*, **161**, 1511–1529.

[52] Tsai Y.-W. and Gemmill D.D., (1998). Using tabu search to schedule activities of stochastic resource-constrained projects, *European Journal of Operational Research*, **111**, 129–141.

[53] Tsao C.-T., (2012). Fuzzy net present values for capital investments in an uncertain environment, *Computers & Operations Research*, **39**, 1885–1892.

[54] Udhayakumar A., Charles V., and Kumar M., (2011). Stochastic simulation based genetic algorithm for chance constrained data envelopment analysis problems, *Omega*, **39**, 387–397.

[55] Ung S.T., Williams V., Bonsall S., and Wang J., (2006). Test case based risk predictions using artificial neural network, *Journal of Safety Research*, **37**, 245–260.
[56] Xinchao Z., (2011). Simulated annealing algorithm with adaptive neighborhood. *Applied Soft Computing*, **11**, 1827–1836.
[57] Xu B., (2006). Market differential evaluations of strategic alliances in the pharmaceutical/biotech industry, *Journal of High Technology Management Research*, **17**, 43–52.
[58] Velsker T., Eerme M., Majak J., Pohlak M., and Karjust K., (2011). Artificial neural networks and evolutionary algorithms in engineering design, *Journal of Achievements in Materials and Manufacturing Engineering*, **44** (1), 88–95.
[59] Wang J.R., (1999). A Fuzzy Set Approach to Activity Scheduling for Product Development, *The Journal of the Operational Research Society*, **50** (12), 1217–1228.
[60] Wu X., Chu C.-H., Wang Y., and Yan W., (2007). A genetic algorithm for cellular manufacturing design and layout, *European Journal of Operational Research*, **181**, 156–167.
[61] Yakhchali S.H., (2012). A path enumeration approach for the analysis of critical activities in fuzzy networks, *Information Sciences*, **204**, 23–35.
[62] Zadeh L.A., (1965). "Fuzzy Sets," *Information and Control*, **8**, 338–352.
[63] Zhang H. and Xing F., (2010). Fuzzy-multi-objective particle swarm optimization for time-cost-quality tradeoff in construction, *Automation in Construction*, **19**, 1067–1075.
[64] Zhang H., Zhu Y., Zou W., and Yan X., (2012). A hybrid multi-objective artificial bee colony algorithm for burdening optimization of copper strip production, *Applied Mathematical Modelling*, **36**, 2578–2591.
[65] Zheng G., Zhu N., Tian Z., Chen Y., and Sun B., (2012). Application of a trapezoidal fuzzy AHP method for work safety evaluation and early warning rating of hot and humid environments, *Safety Science*, **50**, 228–239.
[66] Kolich M., Seal N., and Taboun S., (2004). Automobile seat comfort prediction: statistical model vs. artificial neural network, *Applied Ergonomics*, **35**, 275–284
[67] Lin C.-M. and Gen M., (2008). Multi-criteria human resource allocation for solving multistage combinatorial optimization problems using multiobjective hybrid genetic algorithm. *Expert Syst. Appl.*, **34** (4), 2480–2490.
[68] Al-Turki U., Fedjki C., and Andijani A., (2001). Tabu search for a class of single-machine scheduling problems, *Computers & Operations Research*, **28** (12), 1223–1230.
[69] Price K. and Storn R., (1996). Minimizing the Real Functions of the ICEC'96 contest by Differential Evolution, *IEEE International Conference on Evolutionary Computation (ICEC'96)*, May 1996, pp. 842–844 .
[70] Kennedy J. and Eberhart R.C., Particle Swarm Optimization. In *Proceedings of IEEE International Conference on In Neural Networks*. 1995, pp. 1942–1948.

Chapter 2

A Comparison of Methods For Selecting Preferred Solutions in Multiobjective Decision Making

Enrico Zio[1,2] and Roberta Bazzo[2]

[1] *Chair "Systems Science and Energetic Challenge" European Foundation for New Energy – EDF, École Centrale Paris – Supelec, Paris, France*
[2] *Dipartimento di Energia, Politecnico di Milano, Milano, Italy*

E-mail: enrico.zio@ecp.fr, enrico.zio@supelec.fr, enrico.zio@polimi.it

In multiobjective optimization problems, the identified Pareto Frontiers and Sets often contain too many solutions, which make it difficult for the decision maker to select a preferred alternative. To facilitate the selection task, decision making support tools can be used in different instances of the multiobjective optimization search to introduce preferences on the objectives or to give a condensed representation of the solutions on the Pareto Frontier, so as to offer to the decision maker a manageable picture of the solution alternatives.

This paper presents a comparison of some a priori and a posteriori decision making support methods, aimed at aiding the decision maker in the selection of the preferred solutions. The considered methods are compared with respect to their application to a case study concerning the optimization of the test intervals of the components of a safety system of a nuclear power plant. The engine for the multiobjective optimization search is based on genetic algorithms.

2.1 Introduction

Multiobjective optimization is central for many reliability and risk analyses in support to the design, operation, maintenance and regulation of complex systems like nuclear power plants. The solutions sought must be optimal with respect to several objectives, generally conflicting: then, one cannot identify a unique, optimal solution satisfying all objectives, but rather a set of possible solutions can be identified where none is best for all objectives. This set of solutions in the space of the decision variables is called the Pareto Set; the corresponding values of the objectives form the Pareto Frontier.

C. Kahraman (ed.), *Computational Intelligence Systems in Industrial Engineering*,
Atlantis Computational Intelligence Systems 6, DOI: 10.2991/978-94-91216-77-0_2,

At the end of a multiobjective optimization, the decision maker (DM) has to select the preferred solutions from the Pareto Frontier and Set; this can be a difficult task for large Pareto Frontiers and Sets. For this reason, decision making support tools are developed to aid the DM in selecting the preferred solutions.

There are different approaches for introducing DM preferences in the optimization process, like the ones presented by ATKOSoft [1], Rachmawati and Srinivasan [14], and Coello Coello [6]; a common classification is based on when the DM is consulted: a priori, a posteriori, or interactively during the search.

In this work, a comparison of some a prioripriori methods and a posteriori methodsposteriori methods is performed, aimed at characterizing the different approaches in terms of their advantages and limitations with respect to the support they provide to the DM in the preferential solution selection process; to this purpose, not just the quality of the results, but also the possible difficulties of the DM in applying the procedures are considered. In order to base the comparison on solid experience, the methods considered have been chosen among some of those most extensively researched by the authors.

The a priori method considered is the Guided Multi-Objective Genetic Algorithm (G-MOGA) by Zio, Baraldi and Pedroni [18], in which the DM preferences are implemented in a genetic algorithm to bias the search of the Pareto optimal solutions.

The first a posteriori method considered has been introduced by the authors [20] and uses subtractive clustering [5] to group the Pareto solutions in homogeneous families; the selection of the most representative solution within each cluster is performed by the analysis of Level Diagrams [2] or by fuzzy preference assignment [19], depending on the decision situation, i.e., depending on the presence or not of defined DM preferences on the objectives. The second procedure, is taken from literature [11] and is a two-step procedure which exploits a Self Organizing Map (SOM) [8] and Data Envelopment Analysis (DEA) [7] to first cluster the Pareto Frontier solutions and then remove the least efficient ones. This procedure is here only synthetically described and critically considered with respect to the feasibility of its application in practice.

Instead, the a priori G-MOGA algorithm and the first a posteriori procedure introduced by the authors in [20], are compared with respect to a case study of literature regarding the optimization of the test intervals of the components of a nuclear power plant safety system; the optimization considers three objectives: system availability to be maximized, cost (from operation & maintenance and safety issues) and workers exposure time to be minimized [9]. The a posteriori procedure of analysis is applied to the Pareto Frontier and Set obtained by a standard Multiobjective Genetic Algorithm [9].

The remainder of the paper is organized as follows: Section 2.2 presents the case study to describe upfront the setting of the typical multiobjective optimization problem of interest; Section 2.3 contains the analysis of the different decision making support methods considered; finally some conclusions are drawn in Section 2.4.

2.2 Optimization of the test intervals of the components of a nuclear power plant safety system

The case study here considered is taken from Giuggioli Busacca, Marseguerra and Zio [9] and regards the optimization of the test intervals (TIs) of the high pressure injection system (HPIS) of a pressurized water reactor (PWR), with respect to three objectives: mean system availability to be maximized, cost and workers time of exposure to radiation to be minimized. For reader's convenience, the description of the system and of the optimization problem is here reported, as taken from the original literature source with only minor modifications.

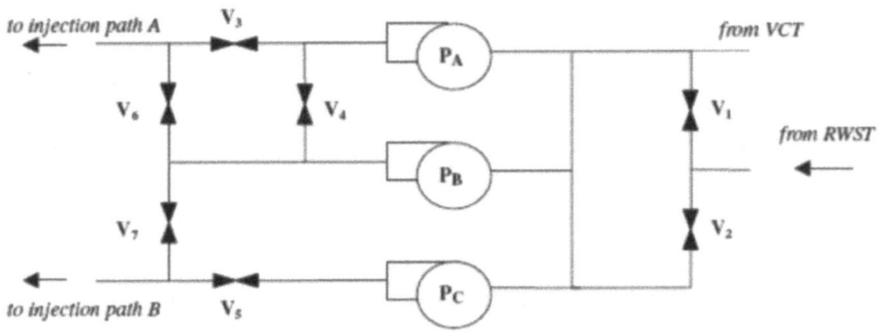

Fig. 2.1 The simplified HPIS system (RWST = radioactive waste storage tank) [9]

Figure 2.1 shows a simplified schematics of a specific HPIS design. The system consists of three pumps and seven valves, for a total of $N_c = 10$ components. During normal reactor operation, one of the three charging pumps draws water from the volume control tank (VCT) in order to maintain the normal level of water in the primary reactor cooling system (RCS) and to provide a small high-pressure flow to the seals of the RCS pumps. Following a small loss of coolant accident (LOCA), the HPIS is required to supply a high pressure flow to the RCS. Moreover, the HPIS can be used to remove heat from the reactor core if the steam generators were completely unavailable. Under normal conditions, the HPIS function is performed by injection through the valves V_3 and V_5 but, for redundancy, crossover valves V_4, V_5 and V_7 provide alternative flow paths if some failure were to occur in one of the nominal paths. This stand-by safety system has to be inspected periodically to test its availability. A TI of 2190 h is specified by the technical specifications (TSs) for both the pumps and the valves. However, there are several restrictions on the maintenance procedures described in the TS, depending on reactor operations.

For this study, the following assumptions are made:

(1) At least one of the flow paths must be open at all times.
(2) If the component is found failed during surveillance and testing, it is returned to an as-good-as-new condition through corrective maintenance or replacement.
(3) If the component is found to be operable during surveillance and testing, it is returned to an as-good-as-new condition through restorative maintenance.
(4) The process of test and testing requires a finite time; while the corrective maintenance (or replacement) requires an additional finite time, the restorative maintenance is supposed to be instantaneous.

The N_c system components are characterized by their failure rate λ_h, $h = 1, \ldots, N_c$, the cost of the yearly test $C_{ht,h}$ and corrective maintenance $C_{hc,h}$, the mean downtime due to corrective maintenance d_h, the mean downtime due to testing t_h and their failure on demand probability ρ_h (Table 2.1). They are also divided in three groups characterized by different test strategies with respect to the TI τ_h between two successive tests, $h = 1, \ldots, N_c$, $N_c = 10$; all the components belonging to a same group undergo testing with the same periodicity T^g, with $g = 1, 2, 3$, i.e., they all have the same test interval ($\tau_h = T^g$, $\forall$ component h in test group g).

Any solution to the optimization problem can be encoded using the following array θ of decision variables:

$$\theta = \begin{bmatrix} T^1 & T^2 & T^3 \end{bmatrix} \tag{2.1}$$

Assuming a mission time (TM) of one year (8760 h), the range of variability of the three TIs is $[1, 8760]$ h.

The search for the optimal test intervals is driven by the following three objective functions $J_i(\theta)$, $i = 1, 2, 3$:

Table 2.1 Characteristics of the system components

Component (j)	Component symbol (Figure 2.1)	λ_h (h^{-1})	$C_{ht,h}$ (\$/h)	$C_{hc,h}$ (\$/h)	d_h(h)	t_h(h)	ρ_h	g
1	V_1	$5.83 \cdot 10^{-6}$	20	15	2.6	0.75	$1.82 \cdot 10^{-4}$	1
2	V_2	$5.83 \cdot 10^{-6}$	20	15	2.6	0.75	$1.82 \cdot 10^{-4}$	1
3	V_3	$5.83 \cdot 10^{-6}$	20	15	2.6	0.75	$1.82 \cdot 10^{-4}$	2
4	V_4	$5.83 \cdot 10^{-6}$	20	15	2.6	0.75	$1.82 \cdot 10^{-4}$	3
5	V_5	$5.83 \cdot 10^{-6}$	20	15	2.6	0.75	$1.82 \cdot 10^{-4}$	2
6	V_6	$5.83 \cdot 10^{-6}$	20	15	2.6	0.75	$1.82 \cdot 10^{-4}$	3
7	V_7	$5.83 \cdot 10^{-6}$	20	15	2.6	0.75	$1.82 \cdot 10^{-4}$	3
8	P_A	$3.89 \cdot 10^{-6}$	20	15	24	4	$5.3 \cdot 10^{-4}$	2
9	P_B	$3.89 \cdot 10^{-6}$	20	15	24	4	$5.3 \cdot 10^{-4}$	2
10	P_c	$3.89 \cdot 10^{-6}$	20	15	24	4	$5.3 \cdot 10^{-4}$	2

Mean Availability, $1-\overline{U}_{HPIS}$:

$$\max_{\theta} J_1(\theta) = \max_{\theta}\left[\left(1-\sum_{v=1}^{N_{MCS}}\prod_{h=1}^{n_v}\overline{u}_h^v(\theta)\right)\right] \tag{2.2}$$

$$\text{Cost, } C\text{:}\quad \min_{\theta} J_2(\theta) = \min_{\theta}\left[C_{\text{accident}}(\theta)+\sum_{h=1}^{N_c} C_{S\&M,h}(\theta)\right] \tag{2.3}$$

$$\text{Exposure Time, } ET\text{:}\quad \min_{\theta} J_3(\theta) = \min_{\theta}\left[\sum_{h=1}^{N_c} ET_h(\theta)\right] \tag{2.4}$$

For every solution alternative θ:

the HPIS mean unavailability $\overline{U}_{HPIS}(\theta)$ is computed from the fault tree for the top event "no flow out of both injection paths A and B" [9]; the boolean reduction of the corresponding structure function allows determining the N_{MCS} system minimal cut sets (MCS); then, the system mean unavailability is expressed as in the argument of the maximization (2.2), where n_v is the number of basic events in the v-th minimal cut set and $\overline{u}_h^v$ is the mean unavailability of the h-th component contained in the v-th MCS, $h = 1,\ldots,n_v$ [12]:

$$\overline{u}_h^v = \rho_h + \frac{1}{2}\lambda_h\tau_h + (\rho_h+\lambda_h\tau_h)\frac{d_h}{\tau_h} + \frac{t_h}{\tau_h} + \gamma_0 \tag{2.5}$$

where γ_0 is the probability of human error. The simple expression in (2.5) is valid for $\rho_h < 0.1$ and $\lambda_h\tau_h < 0.1$, which are reasonable assumptions when considering safety systems.

the cost objective $C(\theta)$ is made up of two major contributions: $C_{S\&M}(\theta)$, the cost associated with the operation of surveillance and maintenance (S&M) and $C_{\text{accident}}(\theta)$, the cost associated with consequences of accidents possibly occurring at the plant.

For a given component h, the S&M cost is computed on the basis of the yearly test and corrective maintenance costs. For a given mission time, TM, the number of tests performed on component h are $\frac{TM}{\tau_h}$; of these, on average, a fraction equal to $(\rho_h + \lambda_h \tau_h)$ demands also a corrective maintenance action of duration d_h; thus, the S&M costs amount to:

$$C_{S\&M,h}(\theta) = C_{ht,h}\frac{TM}{\tau_h}t_h + C_{hc,h}(\rho_h + \lambda_h \tau_h)\frac{TM}{\tau_h}d_h, \quad h = 1,\ldots,N_c \tag{2.6}$$

Concerning the accident cost contribution, it is intended to measure the costs associated to damages of accidents which are not mitigated due to the HPIS failing to intervene. A proper analysis of such costs implies accounting for the probability of the corresponding accident sequences; for simplicity, but with no loss of generality, consideration is here limited only to the accident sequences relative to a small LOCA event tree [17] (Figure 2.2).

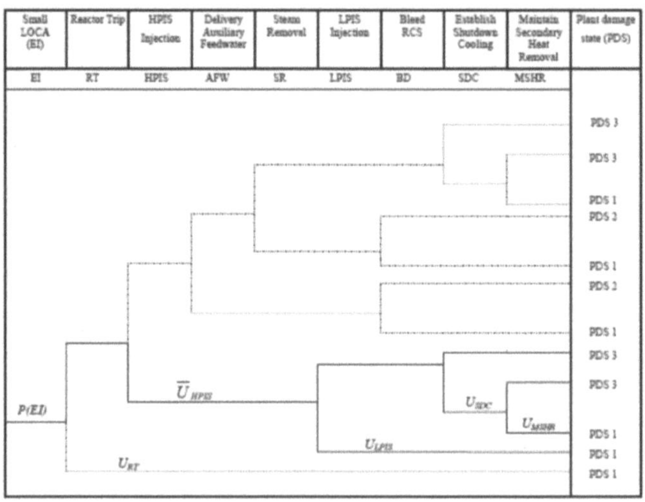

Fig. 2.2 Small LOCA event tree [17]

The accident sequences considered for the quantification of the accident costs are those which involve the failure of the HPIS (thick lines in Figure 2.2), so that the possible Plant Damage States (PDS) are PDS1 and PDS3. Thus:

$$\begin{cases} C_{\text{accident}} = C_1 + C_3 \\ C_1 = P(EI)\cdot(1-U_{RT})\cdot\overline{U}_{HPIS}\cdot\{U_{LPIS}+(1-U_{LPIS})\cdot U_{SDC}\cdot U_{MSHR}\}\cdot C_{PDS1} \\ C_3 = P(EI)\cdot(1-U_{RT})\cdot\overline{U}_{HPIS}\cdot(1-U_{LPIS})\cdot\{(1-U_{MSHR})\cdot U_{SDC}+(1-U_{SDC})\}\cdot C_{PDS3} \end{cases} \tag{2.7}$$

where C_1 and C_3 are the total costs associated with accident sequences leading to damaging states 1 and 3, respectively. These costs depend on the initiating event frequency $P(EI)$

Table 2.2 Accident cost input data [9]

$P(EI)$ (y^{-1})	U_{RT} (y^{-1})	U_{LPIS} (y^{-1})	U_{SDC} (y^{-1})	U_{MSHR} (y^{-1})	C_{PDS1} (\$$\times$event)	C_{PDS2} (\$$\times$event)
$2.43 \cdot 10^{-5}$	$3.6 \cdot 10^{-5}$	$9 \cdot 10^{-3}$	$5 \cdot 10^{-3}$	$5 \cdot 10^{-3}$	$2.1765 \cdot 10^{9}$	$1.375 \cdot 10^{8}$

Table 2.3 MOGA input parameters and rules [9]

Number of chromosomes (N_p)	100
Number of generations (termination criterion)	500
Selection	Standard Roulette
Replacement	Random
Mutation probability	$5 \cdot 10^{-3}$
Crossover probability	1
Number of non-dominated solutions in the archive	100

and on the unavailability values U_i of the safety systems which ought to intervene along the various sequences: these values are taken from the literature [13, 17]. Rates of Initiating Events at United States Nuclear Power Plants: 1987-1995) for all systems except for the SDC and MSHR, which were not available and were arbitrarily assumed of the same order of magnitude of the other safety systems, and for the HPIS for which the unavailability $\overline{U}_{HPIS}$ is calculated from (2.2) and (2.5) and it depends on the TIs of the components. Finally, for the values of C_{PDS1} and C_{PDS3}, the accident costs for PDS1 and PDS3, respectively, are taken as the mean values of the uniform distributions given in Yang, Hwang, Sung and Jin [17]. Table 2.2 summarizes the input data.
the exposure time ET due to the tests and possible maintenance activities on a single component h can be computed as:

$$ET_h(\theta) = \frac{TM}{\tau_h} t_h + (\rho_h + \lambda_h \tau_h) \frac{TM}{\tau_h} d_h, \quad h = 1, \ldots, N_c \tag{2.8}$$

Then,

$$ET(\theta) = \sum_{h=1}^{N_c} ET_h(\theta) \tag{2.9}$$

The multiobjective optimization problem (2.2)–(2.4) has been solved using the MOGA code developed at the Laboratorio di Analisi di Segnale e Analisi di Rischio (LASAR, Laboratory of Signal Analysis and Risk Analysis, `http://lasar.cesnef.polimi.it/`); the input parameters and settings are reported in Table 2.3 [9].

The resulting Pareto Set (Θ) is made of 100 points, and the corresponding Pareto Frontier is showed in Figure 2.3 in the objective functions space.

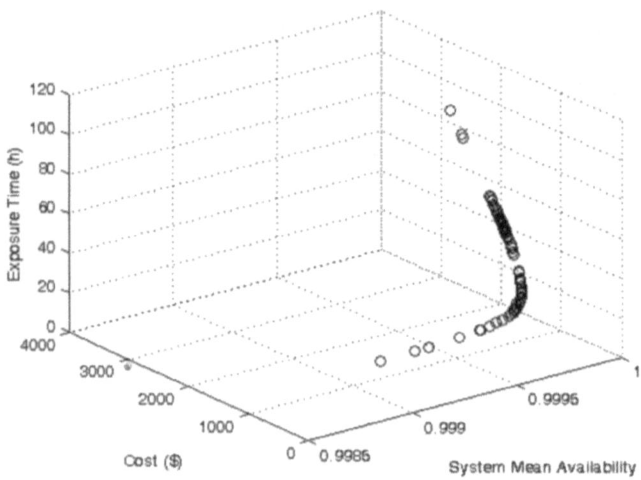

Fig. 2.3 Pareto Frontier, in the objective functions space, obtained by the MOGA code

2.3 Decision support methods

2.3.1 *A posteriori approaches*

When analyzing the Pareto Frontier, the DM either:

- looks for the solution closest to the ideal one, i.e., that which optimizes all the objectives simultaneously;
- applies his or her preferences on the objective functions values to identify the best solution according to these preferences.

The two decision situations, i.e., in presence or not of preferences on the objectives values, may lead to the selection of different solutions and require different procedures of reduction of the solutions in the Pareto Frontier. To this purpose, different a posteriori procedures can be developed to aid the DM in selecting the preferred solution; two of these are synthetically illustrated in the following.

The first a posteriori method presented below, is based on a two-step procedure developed by the authors, for which the availability of the software has rendered possible the comparison on a literature case study.

The second method based on the Self Organizing Maps and Data Envelopment Analysis has been proposed elsewhere in the literature and its application is here critically evaluated.

2.3.1.1 *Subtractive clustering, fuzzy scoring and Level Diagrams for decision making support* [20]

A two-step procedure has been introduced by the authors in Zio and Bazzo [20]. This procedure consists in grouping in "families" by subtractive clustering the non-dominated solutions of the Pareto Set, according to their geometric relative distance in the objective functions space (Pareto Frontier), and then selecting an "head of the family" representative solution within each cluster. Level Diagrams [2] are used to effectively represent and analyze the reduced Pareto Frontiers; they account for the distance of the Pareto Frontier and Set solutions from the ideal (but not feasible) solution, optimal with respect to all the objectives simultaneously.

Considering a multiobjective problem with l objectives to be minimized, m to be maximized (such that $N_{\text{obj}} = l + m$), n solutions in the Pareto Set, and indicating by $\underline{J}(\theta^i) = (J_1(\theta^i) \dots J_s(\theta^i) \dots J_{N_{\text{obj}}}(\theta^i))$ the objective functions values vector corresponding to the solution θ^i in the Pareto Set Θ, $i = 1, \dots, n$, the distance of each Pareto solution from the optimal solution can be measured in terms of the following 1-norm:

$$\text{1-norm:} \quad \left\|\underline{J}(\theta^i)\right\|_1 = \sum_{s=1}^{N_{\text{obj}}} J_{s,\text{norm}}(\theta^i), \text{ with } 0 \leqslant \left\|\underline{J}(\theta^i)\right\|_1 \leqslant s, \quad s = 1, \dots, N_{\text{obj}} \tag{2.10}$$

where each objective value $J_s(\theta^i)$, is normalized with respect to its minimum and maximum values ($J_s^{\min}$ and $J_s^{\max}$) on the Pareto Frontier [2] as follows:

$$J_{s,\text{norm}}(\theta^i) = \frac{J_s(\theta^i) - J_s^{\min}}{J_s^{\max} - J_s^{\min}}, \quad s = 1, \dots, l \tag{2.11}$$

and

$$J_{s,\text{norm}}(\theta^i) = \frac{J_s^{\max} - J_s(\theta^i)}{J_s^{\max} - J_s^{\min}}, \quad s = 1, \dots, m \tag{2.12}$$

Subtractive clustering operates on the normalized objective values $\underline{J}_{\text{norm}}(\theta^i)$, $i = 1, \dots, n$ and groups the non-dominated solutions in "families" according to their geometrical distance; it starts by calculating the following potential $P(\underline{J}_{\text{norm}}(\theta^i))$ [5]:

$$P(\underline{J}_{\text{norm}}(\theta^i)) = \sum_{l=1}^{n} e^{-\alpha \|\underline{J}_{\text{norm}}(\theta^i) - \underline{J}_{\text{norm}}(\theta^l)\|^2}, \quad \alpha = \frac{4}{r_a^2} \tag{2.13}$$

where r_a, the cluster radius, is a parameter which determines the number of clusters that will be identified. The first cluster center $\underline{J}_{\text{norm}}^1$ is selected as the solution with the highest potential value $P(\underline{J}_{\text{norm}}^1)$. All the other $n-1$ solutions potentials $P(\underline{J}_{\text{norm}}(\theta^i))$ are corrected subtracting the potential $P(\underline{J}^1_{\text{norm}})$ multiplied by a factor which considers the distance between the i-th solution and the first cluster center:

$$P(\underline{J}_{\text{norm}}(\theta^i)) = P(\underline{J}_{\text{norm}}(\theta^i)) - P(\underline{J}^1_{\text{norm}})e^{-\beta\|\underline{J}_{\text{norm}}(\theta^i)-\underline{J}^1_{\text{norm}}\|^2},$$

$$\beta = \frac{4}{r_b^2} \text{ and } r_b = qr_a \tag{2.14}$$

where q is an input parameter called squash factor, which indicates the neighborhood with a measurable reduction of potential expressed as a fraction of the cluster radius and is here set equal to 1.25.

Generally, for the the j-th cluster center found $\underline{J}^j_{\text{norm}}$, $j = 1,\ldots,K$, the potentials are reduced as follows:

$$P(\underline{J}_{\text{norm}}(\theta^i)) = P(\underline{J}_{\text{norm}}(\theta^i)) - P(\underline{J}^j_{\text{norm}})e^{-\beta\|\underline{J}_{\text{norm}}(\theta^i)-\underline{J}^j_{\text{norm}}\|^2} \tag{2.15}$$

The process of finding new cluster centers and reducing the potential is repeated until a stopping criterion is reached [5].

The cluster radius r_a is chosen to maximize the quality of the resulting Pareto Frontier partition measured in terms of the silhouette value [15, 16]; for any cluster partition of the Pareto Frontier, a global silhouette index, GS, is computed as follows:

$$GS = \frac{1}{K}\sum_{j=1}^{K} S_j \tag{2.16}$$

where S_j is the cluster silhouette of the j-th cluster F^j, a parameter measuring the heterogeneity and isolation properties of the cluster [15, 16], computed as the average value of the silhouette widths $s(i)$ of its solutions, defined as:

$$s(i) = \frac{b(i)-a(i)}{\max\{a(i),b(i)\}}, \quad i = 1,\ldots,n \tag{2.17}$$

where n is the number of solutions in the Pareto Set, $a(i)$ is the average distance from the i-th solution of all the other solutions in the cluster, and $b(i)$ is the average distance from the i-th solution of all the solutions in the nearest neighbor cluster, containing the solutions of minimum average from the i-th solution, on average.

A head of the family must then be chosen as the best representative solution of each cluster. If no DM preferences are given, the solution with the lowest 1-norm value in each cluster is chosen as the best representative solution; according to the Level Diagrams definition, this means that the selected solution is the closest to the ideal solution, optimal with respect to all objectives. If, on the other hand, the DM preferences on the objective values are available, the best solutions for the DM can be assigned classes of merit with respect to the DM preferences, by setting objective values thresholds. Let us consider the Pareto Set Θ made of n solutions; to the i-th solution θ^i $(i = 1,\dots,n)$ corresponds a vector of objective values

$$\underline{J}(\theta^i) = \left(J_1(\theta^i)\ J_2(\theta^i)\ \dots\ J_{N_{\text{obj}}}(\theta^i)\right) \tag{2.18}$$

where N_{obj} is the number of objective functions of the optimization problem. The objective values thresholds are given in a preference matrix P $(N_{\text{obj}} \times C)$, where C is the number of objective functions thresholds used for the classification, defining $C+1$ preference classes as in Figure 2.4 [2].

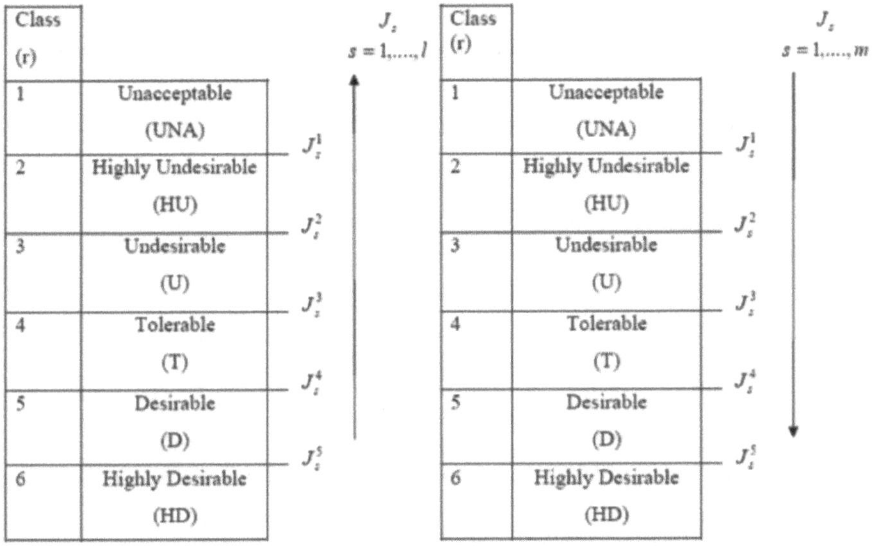

Fig. 2.4 Class Thresholds assignment

where J_s^Z, $Z = 1,\dots,5$, are the thresholds values of the s-th objective, l and m are the number of objectives to be minimized and maximized, respectively.

The fuzzy scoring procedure introduced by the authors in Zio and Bazzo [19] is then applied: each preference class is assigned a score $sv(r)$ [2], $r = 1,\dots,C+1$, such that:

$$sv(C+1) = 0;\ \ sv(r) = N_{\text{obj}} \cdot sv(r+1) + 1, \text{ for } r = C,\dots,1 \tag{2.19}$$

and each objective value $J_s(\theta^i)$, $i = 1,\dots,n$ and $s = 1,\dots,N_{\text{obj}}$, is assigned a membership function $\mu_{A_s^r}(J_s(\theta^i))$ which represents the degree with which $J_s(\theta^i)$ is compatible with the fact of belonging to the r-th preference class, $r = 1,\dots,C+1$.

A vector of $C+1=6$ membership functions is then defined for each objective J_s:

$$\underline{\mu}(J_s(\theta^i)) = \left(\mu_{A_s^1}(J_s(\theta^i))\ \mu_{A_s^2}(J_s(\theta^i))\ \mu_{A_s^3}(J_s(\theta^i))\ \mu_{A_s^4}(J_s(\theta^i))\ \mu_{A_s^5}(J_s(\theta^i))\ \mu_{A_s^6}(J_s(\theta^i))\right)$$

$$i=1,\ldots,n, \quad s=1,\ldots,N_{\text{obj}}. \tag{2.20}$$

The membership-weighted score of each individual objective is then computed; given the scoring vector $\underline{sv} = \left(sv(1)\ sv(2)\ \ldots\ sv(C+1)\right)$, whose components are defined in (2.19), and the membership functions vector $\underline{\mu}\left(J_s(\theta^i)\right)$ in (2.20) for the i-th solution and s-th objective function, the score sv_s^i of the individual objective J_s is obtained by weighting the score $sv(r_s)$ of each class r_s the objective belongs to, by the respective membership function value $\mu_{A_s^{r_s}}\left(J_s(\theta^i)\right)$, $r_s=1,\ldots,6$, and then summing the 6 resulting terms. This can be formulated in terms of the scalar product of the vectors $\underline{\mu}\left(J_s^i\right)$ and $\underline{sv}$ as follows:

$$sv_s^i = \frac{\left\langle \underline{\mu}\left(J_s(\theta^i)\right), \underline{sv}\right\rangle}{\sum\limits_{r_s=1}^{6} \mu_{A_s^{r_s}}\left(J_s(\theta^i)\right)}, \quad i=1,\ldots,n \text{ and } s=1,\ldots,N_{\text{obj}}, \tag{2.21}$$

where the denominator serves as the normalization factor.

Then, the score $S\left(\underline{J}(\theta^i)\right)$ of the i-th solution is the sum of the scores of the individual objectives

$$S\left(\underline{J}(\theta^i)\right) = \sum_{s=1}^{N_{\text{obj}}} sv_s^i, \quad i=1,\ldots,n \tag{2.22}$$

and the lowest score is taken as the most preferred solution.

According to this fuzzy scoring procedure, the head $\underline{H}^j$ of the generic family F^j, $j=1,\ldots,K$, is chosen as the solution in F^j with lowest scores $S\left(\underline{J}(\theta^i)\right)$:

$$S\left(\underline{H}^j\right) = \min S\left(\underline{J}(\theta^i)_k\right), \quad k=1,\ldots,n^j \text{ and } j=1,\ldots,K \tag{2.23}$$

Level Diagrams [2] are finally used to represent and analyse the reduced Pareto Frontier thereby obtained.

With reference to the Pareto Frontier of Figure 2.3 for the test intervals optimization case study, the maximum value of the global silhouette (0.71) is found in correspondence of a cluster radius equal to 0.18 , as showed in Figure 2.5, which results in $K=9$ clusters. For illustration purposes, let us introduce an arbitrary preference matrix P for the test intervals optimization (Table 2.4).

The reduced Pareto Frontier is showed in Figure 2.6: the best solutions (the dark circles) can be easily identified; there are also 4 solutions (the white circles) which have high score values, and thus are unacceptable, i.e., not interesting for the DM.

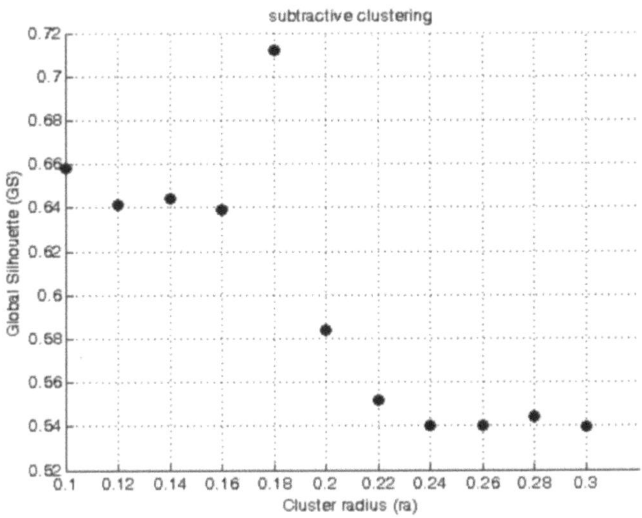

Fig. 2.5 GS for different cluster radius values

Table 2.4 Preference threshold matrix P

	J_s^1	J_s^2	J_s^3	J_s^4	J_s^5
J_1	0.9975	0.998	0.9985	0.999	0.9995
J_2	900	800	700	600	500
J_3	60	50	45	40	30

Note that for the application of the method, the DM only has to select the optimum cluster radius (from Figure 2.5), define the preference matrix and use the Level Diagrams representation to evaluate the solutions according to their distance from the ideal solution, optimal with respect to all objectives.

2.3.1.2 *Self-Organizing Maps solution clustering and Data Envelopment Analysis solution pruning for decision making support* [11]

Another approach to simplifying the decision making in multiobjective optimization problems has been introduced in Li, Liao and Coit [11], based on Self Organizing Maps (SOM) [8] and Data Envelopment Analysis (DEA) [7].

The Pareto optimal solutions are first classified into several clusters by applying the SOM method, an unsupervised classification method based on a particular artificial neural

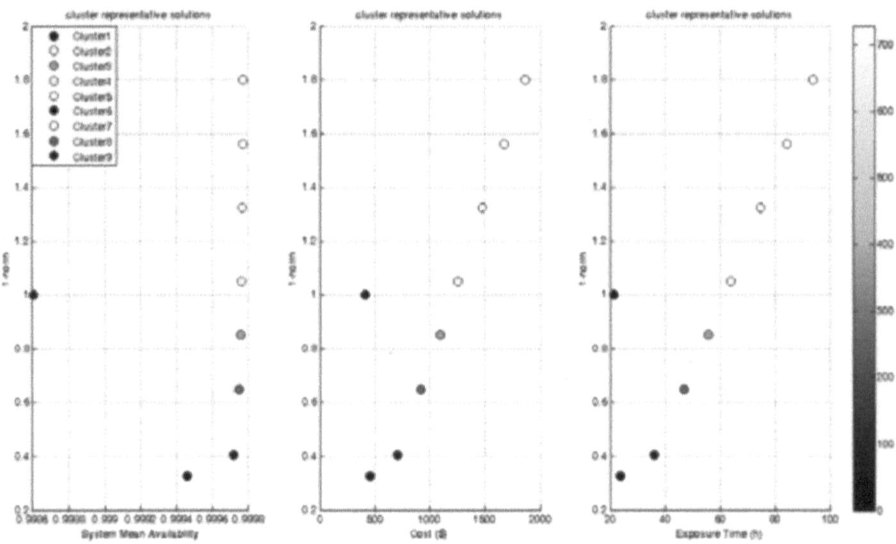

Fig. 2.6 Level Diagrams representation of the family representative solutions with lowest score $S(H^j)$

network with a single layer feedforward structure. Then, non-efficient solutions are eliminated from each cluster and representative efficient solutions are identified, by application of the data envelopment analysis (DEA) method which is a particular multiobjective selection optimization approach. For the efficiency selection, DEA considers an indicator of input/output solution performance based on a predefined relative efficiency criterion: in a multiobjective problem, some objectives can be considered as inputs, e.g., cost, exposure time, which typically have to be minimized, and others can be considered as outputs, e.g., availability, profits, which have to be maximized. Let us consider a problem with l inputs and m outputs; then, for the i-th solution θ^i $(i = 1,\ldots,n)$ in the Pareto Set Θ one can define a relative efficiency as :

$$RE\left(\theta^i\right) = \frac{\text{weighted sum of outputs}}{\text{weighted sum of inputs}} = \frac{\sum_{k=1}^{m} x_{i,k} J_k\left(\theta^i\right)}{\sum_{h=1}^{l} v_{i,h} J_{m+h}\left(\theta^i\right)} \tag{2.24}$$

where $J_k\left(\theta^i\right)$, $k = 1,\ldots,m$, are the outputs, i.e., the objectives to be maximized, $J_{m+h}\left(\theta^i\right)$, $h = 1,\ldots,l$, are the inputs, i.e., the objectives to be minimized, $v_{i,h}$ and $x_{i,k}$ are the weights of the inputs and outputs, respectively. The problem of computing the $RE\left(\theta^i\right)$ values is framed as a particular multiobjective problem for each solution, where the weights are the decision variables and the relative efficiency is the objective function to be maximized:

$$\max_{u_{i,k},v_{i,h}} RE\left(\theta^i\right) = \max_{u_{i,k},v_{i,h}} \frac{\sum_{k=1}^{m} x_{i,k} J_k\left(\theta^i\right)}{\sum_{h=1}^{l} v_{i,h} J_{m+h}\left(\theta^i\right)} \tag{2.25}$$

The Pareto Frontier is then reduced to the solutions with the highest relative efficiency values $RE\left(\theta^i\right)$ and the DM is provided with a small number of most efficient solutions.

This method has been showed to be effective in reducing the number of possible solutions to be presented to the DM in a multiobjective reliability allocation problem [11], but not with the inclusion of the DM preferences. The solution selection is based only on a solution performance criterion (the relative efficiency), but in presence of particular requirements on the objective values, the solutions most preferred by the DM might not be the most efficient ones. Also, the DEA method solves a maximization problem for each solution and this increases the computational time, particularly for large Pareto Frontiers.

2.3.2 *A priori approach*

The a priori approach considered in this work is the Guided Multiobjective Genetic Algorithm (G-MOGA) [18]. The deep knowledge of this method co-developed by one of the authors, makes it a suitable a priori method for detailed comparison on the literature case study.

DM preferences are taken into account by modifying the definition of dominance used for the multiobjective optimization [3, 4]. In general, dominance is determined by pairwise vector comparisons of the multiobjective values corresponding to the pair of solutions under comparison; specifically, solution θ^1 dominates solution θ^2 if

$$\forall i \in \{1,\ldots,s\},\ J_i(\theta^1) \leqslant J_i(\theta^2) \wedge \exists k \in \{1,\ldots,s\} : J_k(\theta^1) < J_k(\theta^2). \tag{2.26}$$

The G-MOGA is based on the idea that the DM is able to provide reasonable trade-offs for each pair of objectives.

For each objective, a weighted utility function of the objective vector $\underline{J}(\theta^i) = (J_1(\theta^i) \ldots J_s(\theta^i) \ldots J_{N_{\text{obj}}}(\theta^i))$ is defined as follows:

$$\Omega_s(\underline{J}(\theta^i)) = J_s(\theta^i) + \sum_{\substack{s=1 \\ p\neq s}}^{N_{\text{obj}}} a_{sp} \cdot J_p(\theta^i), \quad i = 1,\ldots,n \text{ and } s = 1,\ldots,N_{\text{obj}} \tag{2.27}$$

where the coefficients a_{sp} indicate the amount of loss in the s-th objective that the DM is willing to accept for a gain of one unit in the p-th objective, s, $p = 1,\ldots,N_{\text{obj}}$ and $p \neq s$.

Table 2.5 a_{sp} coefficients for the test intervals optimization case study

Preference	G-MOGA trade-offs (a_{sp})	
J_1 much less important than J_2	$a_{12} = 5$,	$a_{21} = 0$
J_1 much less important than J_3	$a_{13} = 100$,	$a_{31} = 0$
J_2 more important than J_3	$a_{23} = 10$,	$a_{32} = 0.1$

Obviously $a_{ss} = 1$. The domination definition is then modified as follows with reference to a minimization problem, for example: θ^1 dominates another solution θ^2 if

$$\forall\, i \in \{1,\ldots,s\},\ \ \Omega_i\big(\underline{J}(\theta^1)\big) \leqslant \Omega_i\big(\underline{J}(\theta^2)\big)\ \wedge\ \exists\, k \in \{1,\ldots,s\}\ :\ \Omega_k\big(\underline{J}(\theta^1)\big) < \Omega_k\big(\underline{J}(\theta^2)\big). \tag{2.28}$$

The guided domination allows the DM to change the shape of the dominance region and to obtain a Pareto Frontier focused on the preferred region, defined by the maximally acceptable trade-offs for each pair of objectives.

The G-MOGA developed at LASAR has been applied to the test interval optimization case study of Section 2.2 and the a_{sp} coefficients are given in Table 2.5.
To obtain results comparable to those of the a posteriori preference assignment, the a priori preferences in the first column of Table 2.5 have been set considering the threshold values assigned in the preference matrix P of Table 2.4. Since the system mean availability unacceptable threshold value (J_1^1) is below the minimum value of the objective in the Pareto Frontier (0.9986), i.e., all the results are at least acceptable, the system mean availability is considered as the least important objective, and thus a_{21} and a_{31}, which indicate the amounts of loss in the cost and exposure time objectives, respectively, that the DM is willing to accept for a gain of one unit in the system mean availability objective, are both set to 0. The cost and the workers' exposure time unacceptable threshold values (900 \$ and 60 h respectively, Table 2.4) are inside the objective values ranges in the Pareto Frontier ([416.23, 2023] and [21.42, 102]). In particular, considering the unacceptable thresholds values normalized by the objective range width

$$\overline{J}_s^1 = \frac{J_s^1}{\max\limits_i \big(J_s(\theta^i)\big) - \min\limits_i \big(J_s(\theta^i)\big)}, \tag{2.29}$$

for these two objectives to be maximized the results are $\overline{J}_2^1 = 0.56$ and $\overline{J}_3^1 = 0.75$, which indicate that the cost objective presents the strongest restrictions on the objective values, because the unacceptable threshold value is closer to the cost minimum value. For this reason, cost is considered a more important objective than the worker's exposure time.

To transform these linguistic preferences into numerical values for the a_{sp} coefficients, $s, p = 1, \ldots, N_{\text{obj}}$ and $p \neq s$, the degradation of the objective J_s ($\Delta^-(J_s)$, in physical units) equivalent to an increment in the objective J_p ($\Delta^+(J_p)$, in physical units) has to be computed; the a_{sp} coefficients can be found as:

$$a_{sp} = \frac{\Delta^-(J_s)}{\Delta^+(J_p)} \tag{2.30}$$

The other G-MOGA settings are the same as those of the standard MOGA applied in Section 2.2.

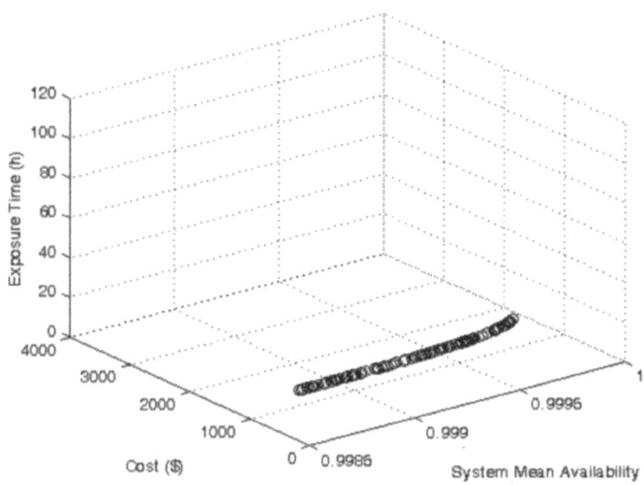

Fig. 2.7 Pareto Frontier obtained with the G-MOGA algorithm

The Pareto Frontier obtained with the G-MOGA (Figure 2.7) is a section of the original Pareto Frontier of Figure 2.3, whose solutions are characterized by low cost and exposure time values. Note that the ranges of these two latter objectives are significantly reduced ([402.98, 497.06] and [20.74, 25.651], respectively), while the range of the system mean availability ([0.9986, 0.996]) is approximately the same; this is due to the lower importance given to the system availability objective.

The Pareto Frontier is dense (still made of 100 solutions) but concentrated in the preferred region of the objective functions space: this means that the algorithm is capable of finding a number of solutions which are preferred according to the DM requirements. This increases the efficiency of the solutions offered to the DM but the decision problem is still difficult because the DM has to choose between very close preferred solutions.

The procedure of solution reduction by clustering illustrated in Section 2.3.1.1 could, then, be applied to the concentrated, preferred Pareto Frontier. In this case, given the narrow objective values ranges, particularly for the cost and the worker's exposure time objectives, it may be difficult to assign preferences on the objectives values. For this reason, the selection of the best representative solution is performed in absence of preferences on the objectives values (Section 2.3.1.1). The optimal cluster radius (r_a) which maximizes the global silhouette value is equal to 0.32, which corresponds to a number of clusters $K = 5$.

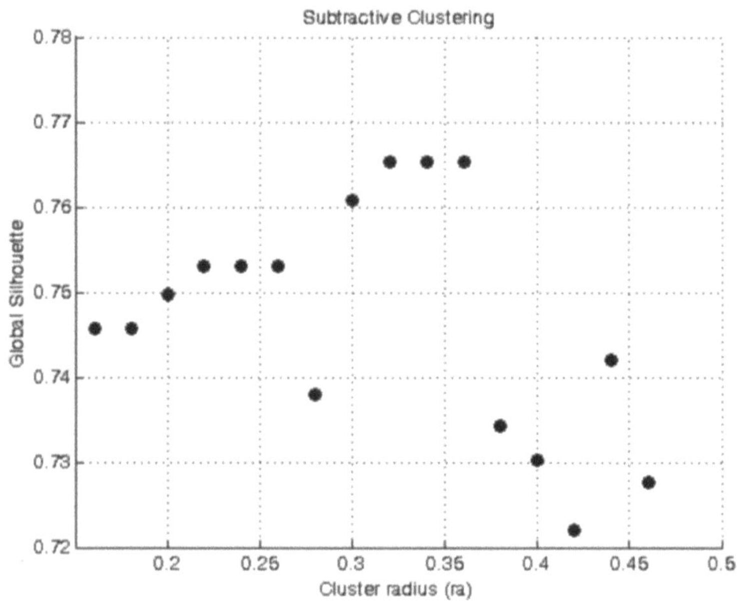

Fig. 2.8 GS for different cluster radius values

The resulting cluster representative solutions, i.e., the solutions in each cluster closest to the optimal point, ideal with respect to all the objectives are showed by Level Diagrams in Figure 2.9.

Given the regular and concentrated Pareto Frontier obtained with the G-MOGA algorithm, the optimal number of clusters, and thus of representative solutions, is smaller than in the previous case; the combined application of the G-MOGA algorithm and clustering procedure is found to provide a small number of preferred solutions, which make it easier for the DM to choose the final solution: the clustering procedure is really effective in re-

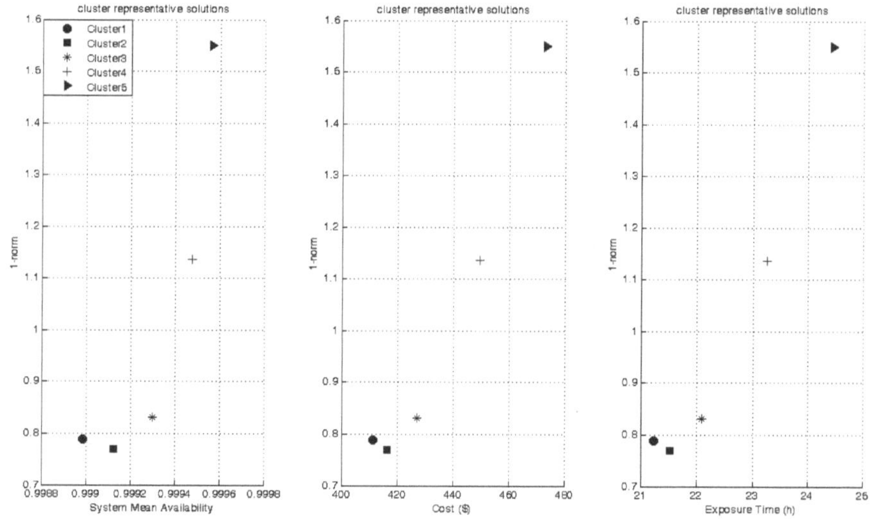

Fig. 2.9 Level Diagrams representation of the family representative solutions closest to the ideal solution optimizing all objectives

ducing the number of solutions to be presented to the DM, overcoming the problem of the crowded Pareto Frontier made of close solutions in the preferred region of the domain.

On the other hand, to compute the a_{sp} coefficients to introduce DM's reasonable trade-offs, one has to know the expressions of the objective functions as implemented in the search algorithm, since, for computational reasons, these expressions might be different from those of the problem statement, e.g., to enhance the procedure of maximization or minimization. Then, if the DM is not satisfied with the resulting Pareto Frontier, he or she has to modify the input parameters of the genetic algorithm. These requests to the DM might be excessive in practical applications because, as showed before, to compute the trade-offs coefficients the DM must, at least, know the orders of magnitude of the objectives. Without any reference value it would be then complicated to define the amount of an objective that the DM accepts to give up for a unitary increase of another objective. Moreover, this task becomes particularly burdensome for problems with more than two objectives, as the required number of trade-offs to be specified increases dramatically with the number of objectives [18].

2.4 Conclusions

The results of algorithms of multiobjective optimization amount to a Pareto Set of non-dominated solutions among which the DM has to select the preferred ones. The selection is difficult because the set of non-dominated solutions is usually large, and the corresponding representative Pareto Frontier in the objective function space crowded.

In the end, the application of DM preferences drives the search of the optimal solution and can be done mainly a priori or a posteriori.

In this work, a comparison of some a priori and a posteriori methods of preference assignment is proposed. The methods have been chosen because the authors have the depth of experience on them necessary for a detailed comparison, here performed on a case study concerning the optimization of the test intervals of the components of a nuclear power plant safety system. The a priori G-MOGA method considered has been showed to lead to a focalized Pareto Frontier, since the DM preferences are embedded in the genetic algorithm to bias the search for non-dominated solutions towards the preferred region; the a posteriori methods considered, on the other hand, have been showed effective in reducing the number of solutions on the Pareto Frontier.

From the results of the comparative analysis, it turns out that the a priori and a posteriori approaches considered are not necessarily in contrast but can be combined to obtain a reduced number of optimal solutions focalized in a preferred region, to be presented to the DM for the decision.

However, the implementation of the a priori method seems more complicated because it requires the assignment of preference trade-offs on the objectives values; this latter task is difficult if the DM has no experience on the specific multiobjective problem, and the complexity increases with the number of the objectives. In these cases, a posteriori procedures can be applied alone, still with satisfactory results. In particular, the two-steps clustering procedure introduced by the authors for identifying a small number of representative solutions to be presented to the DM for the decision, has been showed to be an effective tool which can be applied in different decision situations independently of the Pareto Frontier size and the number of objective functions.

Bibliography

[1] ATKOSoft, *Survey of Visualization Methods and Software Tools*, (1997).

[2] X. Blasco, J.M. Herrero and J. Sanchis, M. Martínez, *Multiobject. Optim., Inf. Sci.*, **178**, 3908–3924 (2008).

[3] J. Branke, T. Kaubler, and H. Schmeck, *Adv. Eng. Software*, **32**, 499 (2001).
[4] J. Branke, T. Kaubler, and H. Schmeck, *Tech. Rep. TR no.399*, Institute AIFB, University of Karlsruhe, Germany (2000).
[5] S. Chiu, *J. of Intell. & Fuzzy Syst.*, **2** (3), 1240 (1994).
[6] C.A. Coello Coello, C, in *2000 Congress on Evolut. Comput.* (IEEE Service Center, Piscataway NJ, 2000), **Vol. 1**, p. 30.
[7] W.W. Cooper, L.M. Seiford, and K. Tone, *Data Envelopment Analysis: a Comprehensive Text with Models, Applications, References, and DEA-Solver Software* (Springer, Berlin, 2006).
[8] L. Fausett, *Fundamentals of Neural Networks: Architectures, Algorithms, and Applications* (Prentice-Hall, Englewood Cliffs, 1994).
[9] P. Giuggioli Busacca, M. Marseguerra, and E. Zio, *Reliab. Eng. Syst. Saf.*, **72**, 59 (2001).
[10] ICRP Publication 60, *Annals of the ICRP*, **21**, 1 (1991).
[11] Z. Li, H. Liao, and D.W. Coit, *Reliab. Eng. Syst. Saf.*, **94**, 1585 (2009).
[12] S. Martorell, S. Carlos, A. Sanchez, and V. Serradell, *Reliab. Eng. Syst. Saf.*, **67**, 215 (2000).
[13] US Nuclear Regulatory Commission, *Rates of Initiating Events at United States Nuclear Power Plants:* 1987-1995, NUREG/CR-5750 (1999).
[14] L. Rachmawati and D. Srinivasan, in *Congress on Evolut. Comput., 2006* (IEEE Conference Pubblications, 2006), p. 962–968.
[15] P.J. Rousseeuw, *J. Comput. Appl. Math.,* **20**, 53 (1987).
[16] P. Rousseeuw, E. Trauwaert, and L. Kaufman, Belgian J. of Oper. Res., Stat. and Comput. Sci., **29** (3), 35 (1989).
[17] J.E. Yang, M.J. Hwang, T.Y. Sung, and Y. Jin, *Reliab. Eng. Syst. Saf.*, **65**, 229 (1999).
[18] E. Zio, P. Baraldi, and N. Pedroni, *Reliab. Eng. Syst. Saf.*, **94**, 432 (2009).
[19] E. Zio and R. Bazzo, *Submitted to Inf. Sci.* (2009).
[20] E. Zio and R. Bazzo, *Eur. J. of Oper. Res.*, **210** (3), 624 (2011).

Chapter 3

Using a Recommender System to Help the Technology Transfer Office Staff to Disseminate Selective Information

C. Porcel[1], A. Tejeda-Lorente[2], M.A. Martínez[2] and E. Herrera-Viedma[2]

[1] *Department of Computer Science, University of Jaen23071 Jaén, Spain*
[2] *Department of Computer Science and Artificial Intelligence, University of Granada 18071 Granada, Spain*

Email: cporcel@ujaen.es, atejeda@decsai.ugr.es, mnglsmartinez@gmail.com, viedma@decsai.ugr.es

Recommender systems evaluate and filter the great amount of information available on the Web, so they could be used to help users in their access processes to relevant information. In the literature we can find a lot of approaches for generating personalized recommendations. Hybrid recommender systems combine in different ways several approaches, so these recommendation strategies represent a promising solution for multiple applications. In this paper we propose a hybrid fuzzy linguistic recommender system to help the Technology Transfer Office staff in the dissemination of research resources interesting for the users. The system recommends users both specialized and complementary research resources and additionally, it discovers potential collaboration possibilities in order to form multidisciplinary working groups.

3.1 Introduction

Theoretical and empirical works in innovation economics suggest that the use of scientific knowledge by setting up and maintaining good industry/science relations positively affects innovation performance [43]. In terms of organizational structure, creating a specialized Technology Transfer Office Technology Transfer Office (TTO) within a university can be instrumental in developing relations with the industry [45]. The TTOs were established to facilitate commercial knowledge transfers from universities to practitioners or university/industry technology transfer [59]. They are responsible for managing and putting into action the activities which generate knowledge and technical and scientific

C. Kahraman (ed.), *Computational Intelligence Systems in Industrial Engineering*,
Atlantis Computational Intelligence Systems 6, DOI: 10.2991/978-94-91216-77-0_3,

collaboration, thus enhancing the interrelation between researchers at the university and the entrepreneurial world and their participation in various support programmes designed to carry out research, development and innovation activities. A service that is particularly important to fulfill this objective is the selective dissemination of information about research resources. But the TTO staff finds difficulties in achieving an effective selective dissemination of information. To solve this problem, automatic techniques are needed in the TTO to facilitate users to selectively access to research resources. In this sense, we consider interesting two different tools to facilitate the access to the information: Information Retrieval Systems [33, 39, 42] and Recommender Systems [4, 17, 40, 54, 60]. The former are focused on information search in a known content repository while the later are focused on information discovery in partially known frameworks.

Recommender systems attempt to discover information items that are likely of interest to a user. They are especially useful when they identify information that a person was previously unaware of. They are becoming popular tools for reducing information overload and to improve the sales in e-commerce web sites [7, 36, 54]. The provision of personalized recommendations requires that the system knows something about every user, such as the ratings provided by the users about the explored items. This knowledge implies that the system must maintain users' profile containing the users' preferences or needs.

From a theoretical point of view, recommender systems have fallen into two main categories: [16, 17, 19, 47, 52, 54, 57, 60]. *Content-based recommender systems* and *Collaborative recommender systems* (see Section 3.2). If we analyze the TTO scope, we find that the collaborative filtering approach is very useful because it allows users to share their experiences, so that popular resources can be easily located or people can receive information items found useful by others with similar profiles. But the collaborative approaches tend to fail when little is known about items, i.e., the system has few ratings. For this reason, we propose to combine the content-based and collaborative approaches to obtain a hybrid recommendation scheme.

The aim of this paper is to present a hybrid fuzzy linguistic recommender system which is applied in the TTO in the University of Granada. In such a way, it allows to help the TTO staff to selectively disseminate research knowledge and the researchers to discover information. The most important novelties of this fuzzy linguistic recommender system are:

- The system implements a hybrid recommendation strategy based in a switching hybrid approach [6], which switches between a content-based recommendation approach and a collaborative one to share the user individual experience and social wisdom.
- The system implements a personalization tool that allows to recommend users three types of items:
- Specialized resources of the own user research area to contribute to his/her specialization.
- Other resources as complementary formation.
- Research collaborators. In this case, it allows researchers to discover new members with complementary profiles, which could provide them real collaboration possibilities to form multidisciplinary working groups and develop common projects.
- The system implements a richer feedback process: when researchers analyze a recommended resource, they provide a satisfaction degree. In such a way, we guarantee that user experiences are taken into account to generate the recommendations done by the system.

The paper is structured as follows. Section 3.2 presents the basic concepts and aspects about the recommender systems. Section 3.3 revises the multi-granular fuzzy linguistic modelling. In Section 3.4 we present the new recommender system to selectively advice research resources in a TTO. Section 3.5 reports the system evaluation and the experimental results. Finally, our concluding remarks are pointed out in Section 3.6.

3.2 Basis of Recommender Systems

The Recommender systems have the effect of guiding the users in a personalized way to relevant or useful objects in a large space of possible options [6]. These applications improve the information access processes for users not having a detailed product domain knowledge. They are becoming popular tools for reducing information overload and to improve the sales in e-commerce web sites [7, 10, 14, 15, 36, 41, 54]. The construction of accurate profiles is a key task and the system's success will depend on a large extent on the ability of the learned profiles to represent the user's preferences. Then, in order to generate personalized recommendations that are tailored to the user's preferences or needs, recommender systems must collect personal preference information, such as user's history of purchase, items which were previously interesting for the user, click-stream data, demographic information, and so on.

Another key aspect to consider when designing the system is the approach used to generate the recommendations. Taking into account the knowledge source, four different approaches can be distinguished: [7, 16, 17, 52, 54, 60]

- *Content-based systems*: They generate the recommendations taking into account the characteristics used to represent the items and the ratings that a user has given to them [5, 11]. These recommender systems tend to fail when little is known about the user information needs. This is called the new user cold-starting problem [38].
- *Collaborative systems*: The system generates recommendations using explicit or implicit preferences from many users, ignoring the items representation. Collaborative systems locate peer users with a rating history similar to the current user and they generate recommendations using this neighborhood. These recommender systems tend to fail when little is known about items, i.e., when new items appear. This is called the new item cold-starting problem [7].
- *Demographic systems*: These systems provide recommendations based on a demographic profile of the user. Recommended products can be produced for different demographic niches, by combining the ratings of users in those niches [51].
- *Knowledge-based systems*: This kind of recommender systems suggest items based on inferences about a users' preferences. This knowledge will sometimes contain explicit knowledge about how the items meet the users' preferences [6].

Each approach has certain advantages and, of course, disadvantages, depending on the scope settings. One solution is to combine different approaches to reduce the disadvantages of each one of them and to exploit their benefits. Using a hybrid strategy, users are provided with more accurate recommendations than those offered by each strategy individually [5, 11, 16]. For this reason, in this paper we propose the use of a hybrid approach.

Moreover, the recommendation activity is followed by a relevance feedback phase. Relevance feedback is a cyclic process whereby the users provide the system with their satisfaction evaluations about the recommended items and the system uses these evaluations to automatically update user profiles in order to generate new recommendations [17, 54].

3.3 Multi-Granular Fuzzy Linguistic Modeling

The Fuzzy linguistic modelling of Fuzzy Sets Theory has given very good results to model qualitative information [62] and it has been proven to be useful in many problems, e.g., decision making [2, 9, 21, 23, 37, 64], quality evaluation [8, 28, 35, 48], information retrieval [26, 27, 29–32], political analysis [3], estimation of student performances [50], etc. It is a tool based on the concept of *linguistic variable* proposed by Zadeh [62].

In any fuzzy linguistic approach, an important parameter to determine is the granularity of uncertainty, i.e., the cardinality of the linguistic term set S. According to the uncertainty degree that an expert qualifying a phenomenon has on it, the linguistic term set chosen to provide his knowledge will have more or less terms. When different experts have different uncertainty degrees on the phenomenon, then several linguistic term sets with a different granularity of uncertainty are necessary [22]. The use of different label sets to assess information is also necessary when an expert has to evaluate different concepts, as it happens in information retrieval problems when users have to evaluate the importance of the query terms and the relevance of the retrieved documents [28]. In such situations, we need tools to manage multi-granular linguistic information [25, 34, 46].

3.3.1 *The 2-Tuple Fuzzy Linguistic Approach*

The 2-tuple fuzzy linguistic modeling [24] is a continuous model of information representation that allows to reduce the loss of information that typically arise when using other fuzzy linguistic approaches (classical and ordinal [20, 62]). To define it both the 2-tuple representation model and the 2-tuple computational model to represent and aggregate the linguistic information have to be established.

Let $S = \{s_0, \ldots, s_g\}$ be a linguistic term set with odd cardinality, where the mid term represents an indifference value and the rest of the terms are symmetric related to it. We assume that the semantics of labels is given by means of fuzzy subsets defined in the $[0,1]$ interval, which are described by their membership functions $\mu_{si} : [0,1] \rightarrow [0,1]$, and we consider all terms distributed on a scale on which a total order is defined, that is, $s_i = s_j \Longleftrightarrow i = j$.

In this fuzzy linguistic context, if a symbolic method [20, 23] aggregating linguistic information obtains a value $\beta \in [0,g]$, and $\beta \notin \{0,\ldots,g\}$, then an approximation function is used to express the result in S.

Definition 3.1 ([24]). Let β be the result of an aggregation of the indexes of a set of labels assessed in a linguistic term set S, i.e., the result of a symbolic aggregation operation, $\beta \in [0,g]$. Let $i = \text{round}(\beta)$ and $\alpha = \beta - i$ be two values, such that, $i \in [0,g]$ and $\alpha \in [-0.5,0.5)$ then s_i represents the linguistic label of the information, and α_i is a numerical value expressing the value of the symbolic translation from the original result β to the closest index label, i, in the linguistic term set ($s_i \in S$).

This model defines a set of transformation functions between numeric values and 2-tuples.

Definition 3.2 ([24]). Let $S = \{s_0, \dots, s_g\}$ be a linguistic term set and $\beta \in [0, g]$ a value representing the result of a symbolic aggregation operation, then the 2-tuple that expresses the equivalent information to β is obtained with the following function:

$$\Delta : [0, g] \longrightarrow S \times [-0.5, 0.5)$$
$$\Delta(\beta) = (s_i, \alpha), \text{ with } \begin{cases} s_i & i = \text{round}(\beta) \\ \alpha = \beta - i, & \alpha \in [-0.5, 0.5) \end{cases} \tag{3.1}$$

where $round(\cdot)$ is the usual round operation, s_i has the closest index label to β and α is the value of the symbolic translation.

For all Δ there exists Δ^{-1}, defined as $\Delta^{-1}(s_i, \alpha) = i + \alpha$. On the other hand, it is obvious that the conversion of a linguistic term into a linguistic 2-tuple consists of adding a symbolic translation value of 0: $s_i \in S \Rightarrow (s_i, 0)$.

The computational model is defined by presenting the Negation operator, Comparison of 2-tuples and Aggregation operators. Using functions Δ and Δ^{-1} any of the existing aggregation operators can be easily extended for dealing with linguistic 2-tuples [61, 63].

Definition 3.3 (Arithmetic mean). Let $x = \{(r_1, \alpha_1), \dots, (r_n, \alpha_n)\}$ be a set of linguistic 2-tuples, the 2-tuple arithmetic mean $\bar{x}^e$ is computed as:

$$\bar{x}^e[(r_1, \alpha_1), \dots, (r_n, \alpha_n)] = \Delta\left(\sum_{i=1}^{n} \frac{1}{n} \Delta^{-1}(r_i, \alpha_i)\right) = \Delta\left(\frac{1}{n} \sum_{i=1}^{n} \beta_i\right) \tag{3.2}$$

Definition 3.4 (Weighted Average Operator). Let $x = \{(r_1, \alpha_1), \dots, (r_n, \alpha_n)\}$ be a set of linguistic 2-tuples, and $W = \{w_1, \dots, w_n\}$ be their associated weights. The 2-tuple weighted average $\bar{x}^w$ is computed as:

$$\bar{x}^w[(r_1, \alpha_1), \dots, (r_n, \alpha_n)] = \Delta\left(\frac{\sum_{i=1}^{n} \Delta^{-1}(r_i, \alpha_i) \cdot w_i}{\sum_{i=1}^{n} w_i}\right) = \Delta\left(\frac{\sum_{i=1}^{n} \beta_i \cdot w_i}{\sum_{i=1}^{n} w_i}\right) \tag{3.3}$$

Definition 3.5 (Linguistic Weighted Average Operator). Let $x = \{(r_1, \alpha_1), \dots, (r_n, \alpha_n)\}$ be a set of linguistic 2-tuples, and $W = \{(w_1, \alpha_1^w), \dots, (r_n, \alpha_n^w)\}$ be their linguistic 2-tuple associated weights. The 2-tuple linguistic weighted average $\bar{x}_l^w$ is computed as:

$$\bar{x}_l^w[((r_1, \alpha_1), (w_1, \alpha_1^w)), \dots, ((r_n, \alpha_n), (w_n, \alpha_n^w))] = \Delta\left(\frac{\sum_{i=1}^{n} \beta_i \cdot \beta w_i}{\sum_{i=1}^{n} \beta w_i}\right) \tag{3.4}$$

with $\beta_i = \Delta^{-1}(r_i, \alpha_i)$ and $\beta w_i = \Delta^{-1}(w_i, \alpha_i^w)$.

3.3.2 *Linguistic Hierarchy to Model Multi-Granular Linguistic Information*

A Linguistic Hierarchy, *LH*, is a set of levels $l(t,n(t))$, i.e., $LH = \bigcup_t l(t,n(t))$, where each level t is a linguistic term set with different granularity $n(t)$ from the remaining of levels of the hierarchy. The levels are ordered according to their granularity, i.e., a level $t+1$ provides a linguistic refinement of the previous level t. We can define a level from its predecessor level as: $l(t,n(t)) \rightarrow l(t+1, 2\cdot n(t)-1)$.

Definition 3.6 ([25]). Let $LH = \bigcup_t l(t,n(t))$ be a linguistic hierarchy whose linguistic term sets are denoted as $S^{n(t)} = \{s_0^{n(t)}, \ldots, s_{n(t)-1}^{n(t)}\}$. The transformation function between a 2-tuple that belongs to level t and another 2-tuple in level $t' \neq t$ is defined as:

$$\begin{aligned} TF_{t'}^{t} &: l(t,n(t)) \longrightarrow l(t',n(t')) \\ TF_{t'}^{t}(s_i^{n(t)}, \alpha^{n(t)}) &= \Delta\left(\frac{\Delta^{-1}(s_i^{n(t)}, \alpha^{n(t)})\cdot(n(t')-1)}{n(t)-1}\right) \end{aligned} \quad (3.5)$$

As it was pointed out in [25] this family of transformation functions is bijective. This result guarantees the transformations between levels of a linguistic hierarchy are carried out without loss of information.

3.4 A Recommender System for the Selective Dissemination of Research Resources in a TTO

In this section, we present a fuzzy linguistic hybrid recommender system to discover both researchers information about research resources and collaboration possibilities. The TTO staff manages and spreads knowledge about research resources such as R&D bulletins, R&D&I, calls, notices, research projects and so on [43, 45]. Nowadays, this amount of information grows continuously and the TTO staff needs automated tools to filter and spread that information to the researchers in a simple and timely manner.

As aforementioned, we present a hybrid recommender system which can be used in a real TTO environment to achieve an effective selective dissemination of research resources. This system works according to a hybrid recommendation strategy based in a switching hybrid approach [6], which switches between a content-based recommendation approach and a collaborative one to share user experiences by generating social recommendations. Basically, the former is applied when a new item is inserted and the latter is applied when a new researcher is registered. Furthermore, we include another feature when suggesting resources to researchers, because the system recommends both specialized and complementary research resources. It improves the recommendation process, allowing researchers to discover real collaboration possibilities in order to form multidisciplinary working groups. In such a way, the system improves the services of a TTO, selectively disseminating research resources, and allowing to share knowledge in an academic context.

We present a multi-granular fuzzy linguistic recommender system that provides high flexibility in the communication processes between users and the system. We use different label sets ($S_1, S_2, \ldots$) to represent the different concepts to be assessed in its filtering activity. These label sets S_i are chosen from those label sets that compose a LH, i.e., $S_i \in LH$. We should point out that the number of different label sets that we can use is limited by the number of levels of LH and therefore, in many cases the label sets S_i and S_j can be associated to a same label set of LH but with different interpretations depending on the concept to be modeled. We consider four concepts that can be assessed in the activity of the recommender system:

- **Importance degree** of a discipline with respect to a resource scope or user interest topic, which is assessed in S_1.
- **Similarity degree** among resources or among users, which is assessed in S_2.
- **Relevance degree** of a resource for a user, which is assessed in S_3.
- **Satisfaction degree** expressed by a user to evaluate a recommended resource, which is assessed in S_4.

We follow a linguistic hierarchy composed by 2 levels, the level 2 (5 labels) to represent importance degrees ($S_1 = S_5$), and the level 3 (9 labels) to represent similarity degrees ($S_2 = S_9$), relevance degrees ($S_3 = S_9$) and satisfaction degrees ($S_4 = S_9$). As the importance degrees are provided by TTO staff, we use a set of five labels to facilitate them the characterization of resource scopes or user interest topics. On the other hand, as the similarity and relevance degrees are computed automatically by the system, we use the set of 9 labels which presents an adequate granularity level to represent the results. Similarly, to provide users with a label set with an adequate granularity level we use the set of 9 labels to express the satisfaction degrees. Using this LH, the linguistic terms in each level are the following:

- $S^5 = \{b_0 = \text{None} = N, b_1 = \text{Low} = L, b_2 = \text{Medium} = M, b_3 = \text{High} = H, b_4 = \text{Total} = T\}$

- $S^9 = \{c_0 = \text{None} = N,\ c_1 = \text{Very\_Low} = VL,\ c_2 = \text{Low} = L,\ c_3 = \text{More\_Less\_Low} = MLL,\ c_4 = \text{Medium} = M,\ c_5 = \text{More\_Less\_High} = MLH,\ c_6 = \text{High} = H,\ c_7 = \text{Very\_High} = VH,\ c_8 = \text{Total} = T\}$.

In Fig. 3.1 we show the basic operating scheme of the recommender system which is based on four main components, which we explain now.

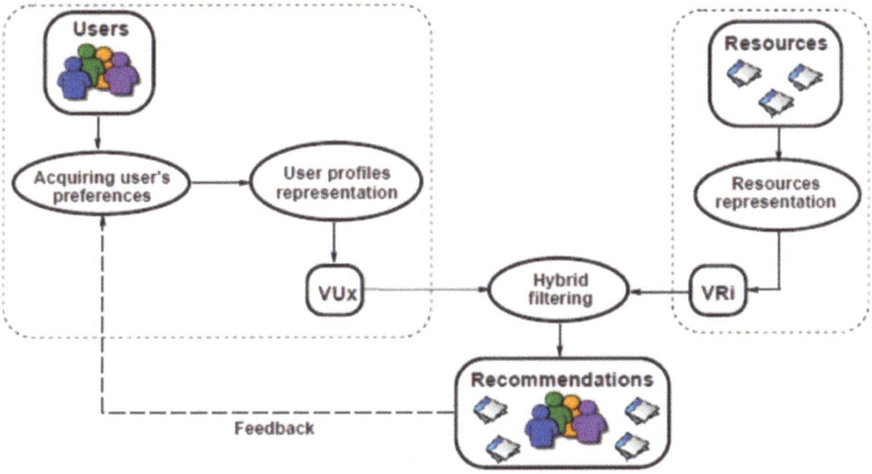

Fig. 3.1 Basic operating scheme

3.4.1 *Resource Representation*

The resources we consider in our system are the research resources such as R&D bulletins, R&D&I, calls, notices or research projects. Once the TTO staff inserts all the available information about a new resource, the system obtains an internal representation mainly based on the resource scope. We use the vector model [39] to represent the resource scope and a classification composed by 25 disciplines (see Fig. 3.2), i.e., a research resource i is represented as $VR_i = (VR_{i1}, VR_{i2}, \ldots, VR_{i25})$ where each component $VR_{ij} \in S_1$ is a linguistic assessment that represents the importance degree of the discipline j with regard to the scope of i. These importance degrees are assigned by the TTO staff when they add new resources.

3.4.2 *User Profiles Representation*

We consider that the users of our system are the researchers of the university and the environment companies. To characterize a researcher the system stores the personal information (login, password, name, phone, email, etc.), research group (it is a string composed by 6 digits, 3 characters indicating the research area and 3 numbers identifying the group) and his/her topics of interest. Similarly, we use the vector model [39] to represent the topics of interest. Then, for a researcher e, we have a vector $VU_e = (VU_{e1}, VU_{e2}, \ldots, VU_{e25})$ where each component $VU_{ej} \in S_1$ is a linguistic assessment that represents the importance degree of the discipline j in the topics of interest of researcher e. Similarly these importance degrees are assigned by the TTO staff when they add a new researcher.

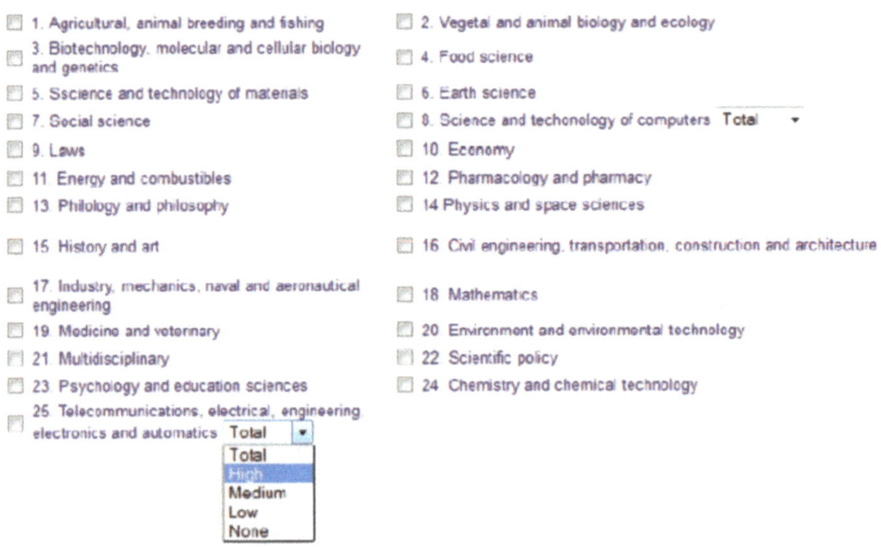

Fig. 3.2 Interface to define the disciplines of the resource scope or user preferences

Furthermore, to avoid the cold-starting problem to handle new items or new users [7, 38], when a new user is inserted, to confirm his/her register it is necessary that he/she assesses some of the resources stored in the system. To do this, the system shows the items randomly and the user assesses what he/she wants.

3.4.3 *Recommendation Strategy*

In this phase the system filters the incoming information to deliver it to the fitting users. This process is based on a matching process developed by similarity measures, such as Euclidean Distance or Cosine Measure [39]. In particular, we use the standard cosine measure but defined in a linguistic framework:

$$\sigma_l(V_1,V_2)=\Delta\left(g\times\frac{\sum_{k=1}^{n}\left(\Delta^{-1}(v_{1k},\alpha_{v1k})\times\Delta^{-1}(v_{2k},\alpha_{v2k})\right)}{\sqrt{\sum_{k=1}^{n}\left(\Delta^{-1}(v_{1k},\alpha_{v1k})\right)^2}\times\sqrt{\sum_{k=1}^{n}\left(\Delta^{-1}(v_{2k},\alpha_{v2k}\right)}}\right)^2 \quad (3.6)$$

with $\sigma_l(V_1,V_2)\in S_2\times[-0.5,0.5)$, and where g is the granularity of the term set used to express the relevance degree, i.e. S_2, n is the number of disciplines and (v_{ik},α_{vik}) is the 2-tuple linguistic value of discipline k in the vector V_i representing the resource scope or user interest topics, depending of the used filtering strategy.

This recommender system works according to a hybrid recommendation strategy. Our proposal is based in a *switching hybrid approach* [6], which uses one technique or another, depending on some criterion. In our system, a content-based recommendation approach is applied when a new item is inserted and a collaborative one is applied when a new researcher is registered. In both cases, the recommender system could send three types of recommendations to a researcher research resources of his/her same area (specialized), research resources of complementary areas, and collaboration possibilities with other researchers.

3.4.3.1 *Content-Based Recommendations*

When a new resource i arrives to the system, the system calculates the content-based recommendations to be sent to a researcher e as follows:

- Compute the linguistic similarity degree between VR_i and VU_e.
- Establish if the resource i could contribute to specialize or complement the researcher's profile. Assuming that $S_2=S^9$, we consider that a resource i is related with the researcher's profile e if $\sigma_l(VR_i,VU_e)>(s_4^9,0)$, i.e., if the linguistic similarity degree is higher than the mid linguistic label. We consider that the resource i could contribute to specialize the researcher's profile e when $\sigma_l(VR_i,VU_e)\geqslant(s_6^9,0)$. On the other hand, we consider that the resource i could contribute to complement the researcher's profile e when $(s_2^9,0)\leqslant\sigma_l(VR_i,VU_e)<(s_6^9,0)$.

- If i is considered a specialization resource for e, then the system recommends this resource i to e with a relevance degree $i(e) \in S_3 \times [-0.5, 0.5)$ which is obtained as follows:
- Look for all specialized research resources stored in the system that were previously assessed by e, i.e., the set of resources $K = \{1, \ldots, k\}$ such that there exists the linguistic satisfaction assessment $e(j)$, $j \in K$ and $\sigma_l(VR_i, VU_e) \geqslant (s_6^9, 0)$.
- Then,

$$i(e) = \bar{x}_l^w(((e(1),0), \sigma_l(VR_i, VR_1)), \ldots, ((e(k),0), \sigma_l(VR_i, VR_k))) \tag{3.7}$$

where $\bar{x}_l^w$ is the linguistic weighted average operator (see Definition 3.5).

- If i is considered a complementary resource for e, then the system recommends this resource i and its authors (community members that could be potential collaborators) to e with a relevance degree $i(e) \in S_3 \times [-0.5, 0.5)$ which is obtained as follows:
- Look for all complementary research resources stored in the system that were previously assessed by e, i.e., the set of resources $K = \{1, \ldots, k\}$ such that there exists the linguistic satisfaction assessment $e(j)$, $j \in K$ and $(s_2^9, 0) \leqslant \sigma_l(VR_i, VU_e) < (s_6^9, 0)$. The latter defines a complementary linguistic interval around mid label that is considered the maximum complementary level.
- Then,

$$i(e) = \bar{x}_l^w(((e(1),0), f(i,1)), \ldots, ((e(k),0), f(i,k))) \tag{3.8}$$

where f is a triangular multidisciplinary matching function that measures the complementary degree between two resources i and j,

$$f(i,j) = \begin{cases} \Delta(2 \times \Delta^{-1}(\sigma_l(VR_i, VR_j))) & \text{if } 0 \leqslant \Delta^{-1}(\sigma_l(VR_i, VR_j)) \leqslant \frac{1}{2} \\ \Delta(2 \times (1 - \Delta^{-1}(\sigma_l(VR_i, VR_j)))) & \text{if } \frac{1}{2} < \Delta^{-1}(\sigma_l(VR_i, VR_j)) \leqslant 1 \end{cases} \tag{3.9}$$

3.4.3.2 *Collaborative Recommendations*

When new users are inserted into the system, they receive recommendations about already inserted resources which may be interesting for them. Usually, new users provide little information about the items that satisfy their topics of interest, so we use the collaborative approach to generate their recommendations. Exactly, we follow a memory-based algorithm or nearest-neighbor algorithm, which generates the recommendations according to the preferences of nearest neighbors. This algorithm has proven good performance [19, 60]. In the following we describe the process in detail.

Given a new researcher e, the recommendations to be sent to e are obtained in the following steps:

- Identify the set of users $\aleph_e$ most similar to that new user e. To do so, we calculate the linguistic similarity degree between the topics of interest vector of the new user (VU_e) against the vectors of all users already inserted into the system (VU_y, $y = 1, \ldots, n$ where n is the number of users), that is, we calculate $\sigma_l(VU_e, VU_y) \in S_2$. As $S_2 = S^9$, we consider that the user y is near neighbor to e if $\sigma_l(VU_e, VU_y) > (s_4^9, 0)$, i.e., if the linguistic similarity degree is higher than the mid linguistic label.
- Look for the resources stored in the system that were previously well assessed by the near neighbors of e, i.e., the set of resources $K = \{1, \ldots, k\}$ such that there exists a linguistic satisfaction assessment $y(j)$, $y \in \aleph_e$, $j \in K$, and $y(j) \geqslant (s_6^9, 0)$.
- Discover if those resources could contribute with specialized or complementary formation. A resource $j \in K$could contribute to specialize the researcher's formation e when $\sigma_l(VR_j, VU_e) \geqslant (s_6^9, 0)$. On the other hand, we consider that the resource j could contribute to complement the researcher's formation e when $(s_2^9, 0) \leqslant \sigma_l(VR_j, VU_e) < (s_6^9, 0)$.
- If j is considered a specialization resource for e, then the system recommends this resource j to e with a relevance degree $j(e) \in S_3 \times [-0.5, 0.5)$ which is obtained as follows:
- To look for all linguistic satisfaction assessments about resources that were well assessed by the nearest neighbors of e. That is, we recovery $y(j)$, $j \in K$ and $y \in \aleph_e$.
- Then,

$$j(e) = \bar{x}_l^w(((y_1(j), 0), \sigma_l(VU_e, VU_{y1})), \ldots, ((y_n(j), 0), \sigma_l(VU_e, VU_{yn}))) \quad (3.10)$$

  where $y_1, \ldots, y_n \in \aleph_e$ and $\bar{x}_l^w$ is the linguistic weighted average operator (see Definition 3.5).
- If j is considered a complementary resource for e, then the system recommends this resource j and its authors (community members that could be potential collaborators) to e with a relevance degree $j(e) \in S_3 \times [-0.5, 0.5)$ which is obtained as follows:
- Look for all complementary research resources stored in the system that previously were well assessed by the nearest neighbors of e, i.e., the set of resources $K = \{1, \ldots, k\}$ such that there exists the linguistic satisfaction assessment $y(j)$, with $j \in K$, $y \in \aleph_e$ and $(s_2^9, 0) \leqslant \sigma_l(VU_y, VR_{je}) < (s_6^9, 0)$. The latter defines a complementary linguistic interval around mid label that is considered the maximum complementary level.

- Then,

$$j(e) = \overline{x}_l^w(((y_1(j),0),h(e,y_1)),\ldots,((y_n(j),0),h(e,y_n))) \tag{3.11}$$

where f is a triangular multidisciplinary matching function that measures the complementary degree between two resources i and j,

$$h(i,j) = \begin{cases} \Delta(2\times\Delta^{-1}(\sigma_l(VU_i,VU_j))) & \text{if } 0 \leqslant \Delta^{-1}(\sigma_l(VU_i,VU_j)) \leqslant \frac{1}{2} \\ \Delta(2\times(1-\Delta^{-1}(\sigma_l(VU_i,VU_j)))) & \text{if } \frac{1}{2} < \Delta^{-1}(\sigma_l(VU_i,VU_j)) \leqslant 1 \end{cases} \tag{3.12}$$

3.4.4 *Feedback Phase*

In this phase the recommender system recalculates and updates the recommendations of the accessed resources. When the system sends recommendations to the users, then they provide a feedback by assessing the relevance of the recommendations, i.e., they supply their opinions about the recommendations received from the system. If they are satisfied with the received recommendation, they shall provide high values and vice versa. This feedback activity is developed in the following steps:

- The system recommends the user U a resource R, and then the system asks him/her his/her opinion or evaluation judgements about recommended resource.
- The user communicates his/her linguistic evaluation judgements to the system, $rc_y \in S_2$.
- This evaluation is registered in the system for future recommendations. The system recalculates the linguistic recommendation of R by aggregating the opinions about R provided by all users. In such a way, the opinion supplied by U is considered. This can be done using the 2-tuple aggregation operator as $\overline{x}^e$ given in Definition 3.3.

3.5 Experiments and Evaluation

In this section we present the evaluation of the proposed recommender system. We propose two kind of experiments, offline and online ones. We begin with an offline setting, where the proposed recommendation approach is compared with other approaches without user interaction, using a standard data set. However, in many applications, accurate predictions are important but insufficient with respect to the user satisfaction. For instance, users may be interested in discovering new items not expected for them, more than getting an exact prediction of their preferences. Consequently, we also propose online experiments, that is, practical studies where a small group of users interact with the system and report us their experiences.

3.5.1 *Evaluation Metrics*

In the scope of recommender systems, precision, recall and F1 are widely used measures to evaluate the quality of the recommendations [10, 12, 56]. To calculate these metrics we need to build a contingency table to categorize the items with respect to the information needs: The items are classified both as relevant or irrelevant and selected (recommended to the user) or not selected.

Definition 3.7. *Precision* is defined as the ratio of the selected relevant items to the selected items, that is, it measures the probability of a selected item to be relevant:

$$P = \frac{N_{rs}}{N_s} \tag{3.13}$$

Definition 3.8. *Recall* is calculated as the ratio of the selected relevant items to the relevant items, that is, it represents the probability of a relevant item to be selected:

$$R = \frac{N_{rs}}{N_r} \tag{3.14}$$

Definition 3.9. *F1* is a combination metric that gives equal weight to both precision and recall, and it is calculated as follows: [10, 56]

$$F1 = \frac{2 \times R \times P}{R + P} \tag{3.15}$$

Besides, in order to test the performance of our model and to compare it with other approaches, we also calculate the system accuracy, that is, its capability to predict users' ratings. We propose to use the Mean Absolute Error.

Definition 3.10. *Mean Absolute Error (MAE)* [19, 58] is a commonly used accuracy metric which considers the average absolute deviation between a predicted rating and the user's true rating:

$$MAE = \frac{\sum_{i=1}^{n} abs(p_i - r_i)}{n} \tag{3.16}$$

where n is the number of cases in the test set, p_i the predicted rating for a item, and r_i the true rating.

3.5.2 *Offline Experiments*

3.5.2.1 *Data Set*

We use MovieLens data sets [18, 49] to develop the offline experiments because the data sets are publicly available and have been usually used to evaluate recommender systems, and in such a way, we could compare our system with other models. MovieLens data sets [49] are related with a cinematographic scope and they were collected by the GroupLens Research Project at the University of Minnesota during the seven-month period from September 19th, 1997 through April 22nd, 1998. Specifically, we use the 100K ratings data set which contains 1682 movies, 943 users and a total of 100 000 ratings on a scale of 1-5 (where 1 = Awful, 2 = Fairly bad, 3 = It's OK, 4 = Will enjoy, 5 = Must see). Each user has rated at least 20 movies. However, to apply this data set to our hybrid recommender system, we need to develop a transformation process in order to adapt the data to the features of our approach. In our system we represent both the resources and the user profiles using vectors. So, we need to transform the MovieLens data sets to this representation avoiding the loss of information. Then, we have to build vectors to represent the users' topics of interest and the movies. The idea is to obtain such vectors from the data stored in the MoviLens data sets.

The 1682 movies are classified into the following 19 genres: unknown, action, adventure, animation, children, comedy, crime, documentary, drama, fantasy, film-noir, horror, musical, mystery, romance, sci-fi, thriller, war and western. In fact, the file *u.item* contains information about the movies, with a tab separated list of the fields movie id, movie title, release date, video release date, IMDb URL, and the last 19 fields are the genres: a value of 1 indicates the movie is of that genre and a value of 0 indicates it is not; movies can be in several genres at once. For each movie we build a vector with 19 positions (one for each genre), following the approach pointed in Subsection 3.4.1 $VR_i = (VR_{i1}, VR_{i2}, \ldots, VR_{i19})$ where each component $VR_{ij} \in S_1$ is a linguistic assessment that represents the importance degree of the genre j with regard to the movie i. Therefore, when the value in the file *u.item* is 1 (the movie is of that genre), we assign the maximum label of S_1 ($(b_4, 0)$ in this case) and when the value is 0 the assigned label is the minimum of S_1 ($(b_0, 0)$).

On the other hand, our system works with the user topics of interest, which are also represented by a vector. So, for each user we need a vector similar to that used to represent the movies. The problem is that MovieLens data sets don't include this information directly, because the file *u.user* only includes demographic information about the users (user id, age, gender, occupation and zip code). However, the information about the topics of interest for each user could be obtained from the available data, aggregating the ratings assigned by the users on each movie with the genre information of the movies. The file *u.data* contains the 100 000 ratings on a scale of 1-5; this is a tab separated list of user id, item id, rating and timestamp. The information about the genres is in the file *u.item*; the movie ids are the ones used in the *u.data* data set. Following the approach pointed in Subsection 3.4.2, for each user e we build a vector with 19 positions $VU_e = (VU_{e1}, VU_{e2}, \ldots, VU_{e19})$ where each component $VU_{ej} \in S_1$ is a linguistic assessment that represents the importance degree of the genre j in the topics of interest related with the user e. These importance degrees are calculated using a weighted average operator:

$$VU_{ej} = \Delta\left(\frac{\sum_{m=1}^{n} r_{em} \cdot g_{mj}}{\sum_{m=1}^{n} r_{em}}\right) \tag{3.17}$$

where r_{em} is the rating assigned by the user e on the movie m and g_{mj} is the value of the genre j for the movie m.

3.5.2.2 *Results of Offline Experiments*

We use the **cross validation** to determine the validity of our model and to analyze the obtained results. Cross validation is typically used to estimate how accurately a predictive model will perform in practice [53]. The data set is divided in complementary subsets, performing the analysis on one subset, called the training set, and validating the analysis on the other subset, called the testing set. To reduce variability, multiple rounds of cross-validation are performed using different partitions, and the validation results are averaged over the different rounds. In k-fold cross validation [53], the original sample is randomly partitioned into k folds. One fold is selected as the testing set, used to estimate the error, and the remaining $k-1$ folds are used as training data set. The cross-validation process is then repeated k times, with each of the k folds used exactly once as the testing set. The k results then can be averaged to produce a single estimation about the deviations between the predictions and the actual ratings.

Values of the folding parameter k commonly assumed are $4, 5, \ldots, 10$. We have chosen a value of $k = 5$. In order to perform 5-fold cross validation, we use the data sets *u1.base* and *u1.test* through *u5.base* and *u5.test* provided by MovieLens which split the collection into 80% for training and 20% for testing, respectively. From the training data sets we build the necessary vectors as we have shown in the previous section. We use them as the input data to predict the unrated ratings. It allows us to measure the system capability in order to predict the users' ratings, calculating the MAE. Besides, to test the effect of the number of neighbors (value of $\aleph_e$ used in the collaborative recommendations) on the accuracy of

Table 3.1 MAE values for our system with MovieLens data sets

$\aleph_e$	u1	u2	u3	u4	u5	Avg MAE
5	0.7405	0.7398	0.7424	0.7432	0.7437	0.7419
10	0.7379	0.7339	0.7353	0.7386	0.7381	0.7368
20	0.7356	0.7351	0.7357	0.7372	0.7370	0.7361
30	0.7434	0.7431	0.7446	0.7452	0.7448	0.7442
50	0.7471	0.7463	0.7468	0.7479	0.7473	0.7471

the system, we have considered the most 5, 10, 20, 30 and 50 similar users. The obtained results are shown in Table 3.1.

As we can see, the performance of the system is quite uniform across the Movielens data set, but considering 10 and 20 similar users we obtain a better average MAE than the rest of configurations. The other three groups (with the most 5, 30 and 50 similar users), present results close to one another. In Table 3.1 we can see that the average MAE increases as the number of neighbors grows or when we consider very few neighbors. When the number of neighbors is between 10 and 20, there is a significant drop in average MAE which indicates a considerable increase in prediction accuracy; in fact, the best average results are obtained considering the most 20 similar users. Therefore, we decide that a number of neighbors between 15 and 20, are the most suitable for our system.

3.5.2.3 *Comparison with other approaches*

In order to compare the results of our system with other, we have implemented several content-based and collaborative models. Firstly, we have implemented a pure content-based approach (CB) [4, 5, 52] in which the similarity between two items is calculated using the cosine measure. We also have implemented the user-based collaborative approach (UBC) [18, 56, 60]. This method uses the ratings of users that are most similar to the target user for predicting the ratings of unrated items; the similarity between users is computed using Pearson's correlation coefficient. Finally, we have implemented the item-based collaborative approach (IBC) [4, 14, 55] in which the similarities of items are used to predict the ratings. The prediction is computed by taking a weighted average of the target user's ratings on similar items. In our experimentation we have used both the cosine and Pearson measure, titled IBC-C and IBC-P, respectively.

To compare the different approaches, we have followed the experimental setting described previously, that is, we perform the 5-fold cross validation, using as training and testing data sets the files *u1.base* and *u1.test* through *u5.base* and *u5.test* provided by

Table 3.2 Average MAE values to compare with other models

	Our system	CB	UBC	IBC-C	IBC-P
Average MAE	0.7412	0.9187	0.7848	0.7705	0.7716
Improvement %		23.94%	5.88%	3.95%	4.10%

MovieLens. With these experiments we calculate the average MAE for all the tests and rounds. To do the comparison, in the case of our system we have used the average MAE for the five values of $\aleph_e$ studied in the previous subsection (see Table 3.1). We prefer to use the average value and not the better MAE, to obtain more significant and realistic results. Table 3.2 presents the MAE results obtained by each approach, where we can see how our system improves the results obtained by the rest of approaches. The row entitled with *Improvement %* presents the improvement percentage obtained with our system over the other approaches.

3.5.3 *Online Experiments*

We have enabled the proposed recommender system for a small group of users, who interact with the system and report us their experience. We test its main features, i.e., its capacities to discover both specialized or complementary resources and collaboration possibilities.

3.5.3.1 *Data Set*

For the online evaluation, we have considered a data set with 200 research resources related with different areas collected by the TTO staff from different information sources. These resources were included into the system following the indications described in Section 3.4.3.1. We assume that the recommender system has to generate recommendations to 15 users and that these users have completed the registration process and evaluated at least 25 resources. From these user assessments, the system is able to build the user profiles.

The resources and the provided user assessments constitute our training data set. Then, we have added 100 new research resources that conform the test data set. The system filtered these 100 resources and it recommended them to the suitable users. To obtain data to compare, these 100 new research resources also were recommended using the advices of the TTO staff.

3.5.3.2 *Results of Online Experiments*

Using the described data set we obtained the contingency table. For example, for user 1, the TTO staff considered 23 relevant resources, of which 16 were specialized and 7 were complementary. Our system selected 26 resources as relevant for user 1, being only 17 really relevant. From these 17 relevant resources, the system classified 15 as specialized and 2 as complementary. Comparing with the recommendations provided by the TTO staff we had 2 resources which are misclassified. So, the success rate for the user 1 was $((17-2)/26)\times 100 = 57.69\%$. Analyzing the contingency table we obtain that the system shows an average precision (success rate) of 61.28%, which is a satisfactory value of face-on system performance.

Similarly, we used the previous scenario to analyze the collaboration possibilities of our recommender system. However, in this case, the items to recommend are not the research resources, but the collaboration opportunities that could appear when the resource is a research project. Thus, we assumed that our system had to recommend research resources to 15 users and a training data set composed by 200 research resources of different areas. Then, we added 100 new resources, of which 30 resources were research projects that constituted the test data set. To compare the collaboration recommendations provided by the system and by the TTO staff we used those 30 projects, and not only the projects considered as relevant by the system or by the TTO staff. So, we can obtain specific measures with regard to collaboration recommendations.

Then, for the 30 projects we compared the collaboration recommendations made by the system with the collaboration recommendations provided by the TTO staff. We classified the collaboration recommendations taking into account the categorization described in Table 3.3. To understand the meaning of this table we provide the following example. Suppose that for project 1, the TTO staff selected user 7 and indicated him/her that he/she could collaborate with users 2, 11 and 12 to develop the project. Our system also selected user 7 for project 1, but in this case it recommended the collaboration with users 2, 3 and 12, and therefore, these recommendations didn't match with the TTO staff recommendations. That is, our system presented 2 hits (for users 2 and 12), 1 failure (user 3) and a non-detected collaboration (user 11). Then, for project 1, $Nchs = 2$, $Nchn = 1$ and $Ncfs = 1$.

Assuming this framework, we obtained the Table 3.4 for the 30 projects, being the average precision of 70.44%, the average recall of 72.50% and an average F1 of 70.40%, which show a satisfactory behavior of our system. The obtained results indicate that the

Table 3.3 Contingency table for the collaboration recommendations

	Selected	Not selected	Total
Considered by TTONot considered by TTO	NchsNcfs	NchnNcfn	NchNcf
Total	Ns	Nn	N

Table 3.4 Contingency table for the collaboration recommendations

Project	1	2	3	4	5	6	7	8	9	10	11	12	13	14	15
Nchs	2	3	4	2	4	2	3	2	3	3	2	2	3	2	2
Nchn	1	1	2	1	1	0	1	1	1	2	2	1	1	0	1
Ncfs	1	1	1	2	2	1	1	1	1	1	1	1	0	2	1
Nch	3	4	6	3	5	2	4	3	4	5	4	3	4	2	3
Ns	3	4	5	4	6	3	4	3	4	4	3	3	3	4	3
Project	16	17	18	19	20	21	22	23	24	25	26	27	28	29	30
Nchs	3	2	2	3	4	3	2	3	3	3	2	3	2	3	1
Nchn	2	1	1	1	1	1	1	2	1	2	0	1	0	1	1
Ncfs	1	1	0	1	1	1	1	1	2	1	1	1	1	2	1
Nch	5	3	3	4	5	4	3	5	4	5	2	4	2	4	2
Ns	4	3	2	4	5	4	3	4	5	4	3	4	3	5	2

collaboration recommendations provided by our system are useful to researchers, and quite similar to those provided by the TTO staff.

3.6 Concluding Remarks

The TTO is responsible for putting into action and managing the activities which generate knowledge and technical and scientific collaboration. A service that is particularly important to fulfill this objective is the selective dissemination of information about research resources. The TTO staff and researchers need tools to assist them in their processes of information discovering because of the large amount of information available on these systems.

We have presented a fuzzy linguistic recommender system to spread selectively research resources in a TTO. Particularly, we propose to use a hybrid approach as recommendation engine, that is, integrating a content-based approach with a collaborative one, in order to take the advantages of both strategies and reduce the disadvantages of each one of them. This recommender system recommends specialized resources, complementary resources and collaboration possibilities that allows the researchers to meet other researchers and to form multidisciplinary groups. Besides, the system improves the feedback process using satisfaction degrees. We have applied our research in a real environment provided by the TTO. The system advices researchers and environment companies about resources that could be interesting for them and collaboration possibilities with other researchers. The experimental results show us that the recommendations provided by our system are useful to researchers.

Analyzing our system, we could conclude that its main limitation is the need for interaction with TTO staff to establish the internal representations for the user profiles and the items. With regard to future research, we believe that a promising direction is to study automatic techniques to establish the representation of user profiles and items. Moreover, we want to explore new improvements of the recommendation approach, exploring new methodologies for the generation of recommendations, as for example, bibliometric tools to enrich the information on the researchers and research resources [1, 13].

Acknowledgments

This book chapter has been developed with the financing of Projects 90/07, TIN2007-61079, PET2007-0460, TIN2010-17876, TIC5299 and TIC-5991.

Bibliography

[1] S. Alonso, F. Cabrerizo, E. Herrera-Viedma, F. Herrera, hg-index: A new index to characterize the scientific output of researchers based on the h- and g- indices, *Scientometrics* **82** (2), 391–400 (2010).

[2] S. Alonso, E. Herrera-Viedma, F. Chiclana, F. Herrera, A web based consensus support system for group decision making problems and incomplete preferences, *Information Sciences* **180** (23), 4477–4495 (2010).

[3] B. Arfi, Fuzzy decision making in politics. a linguistic fuzzy-set approach (lfsa), *Political Analysis* **13** (1), 23–56 (2005).

[4] A. Barragáns-Martínez, E. Costa-Montenegro, J. Burguillo, M. Rey-López, F. Mikic-Fonte, A. Peleteiro, A hybrid content-based and item-based collaborative filtering approach to recommend tv programs enhanced with singular value decomposition, Information Sciences **180** (22), 4290–4311 (2010).

[5] C. Basu, H. Hirsh, W. Cohen, Recommendation as classification: Using social and content-based information in recommendation, in: *Proceedings of the Fifteenth National Conference on Artificial Intelligence*, 1998.

[6] R. Burke, Hybrid recommender systems: Survey and experiments, *User Modeling and User-Adapted Interaction* **12**, 331–370 (2002).

[7] R. Burke, Hybrid web recommender systems, P. Brusilovsky, A. Kobsa, and W. Nejdl (Eds.): *The Adaptive Web*, LNCS **4321**, 377–408 (2007).

[8] F. Cabrerizo, J. López-Gijón, A. Ruíz-Rodríguez, E. Herrera-Viedma, A model based on fuzzy linguistic information to evaluate the quality of digital libraries, *International Journal of Information Technology & Decision Making* **9** (3), 455–472 (2010).

[9] F. Cabrerizo, I. Pérez, E. Herrera-Viedma, Managing the consensus in group decision making in an unbalanced fuzzy linguistic context with incomplete information, *Knowledge-Based Systems* **23** (2), 169–181 (2010).

[10] Y. Cao, Y. Li, An intelligent fuzzy-based recommendation system for consumer electronic products, *Expert Systems with Applications*, 230–240 (2007).

[11] M. Claypool, A. Gokhale, T. Miranda, Combining content-based and collaborative filters in an online newpaper, in: *Proceedings of the ACM SIGIR-99 Workshop on Recommender Systems-Implementation and Evaluation*, 1999.

[12] C. Cleverdon, E. Keen, Factors determining the performance of indexing systems, *ASLIB Cranfield Research Project 2.*

[13] M. Cobo, A. López-Herrera, E. Herrera-Viedma, F. Herrera, An approach for detecting, quantifying, and visualizing the evolution of a research field: A practical application to the fuzzy sets theory field, *Journal of Informetrics* **5** (1), 146–166 (2011).

[14] M. Deshpande, G. Karypis, Item-based top-n recommendation algorithms, ACM Transactions on Information Systems **22**, 143–177 (2004).

[15] L. Duen-Ren, L. Chin-Hui, L.Wang-Jung, A hybrid of sequential rules and collaborative filtering for product recommendation, *Information Sciences* **179**, 3505–3519 (2009).

[16] N. Good, J. Schafer, J. Konstan, A. Borchers, B. Sarwar, J. Herlocker, J. Riedl, Combining collaborative filtering with personal agents for better recommendations, in: *Proceedings of the Sixteenth National Conference on Artificial Intelligence*, 1999.

[17] U. Hanani, B. Shapira, P. Shoval, Information filtering: Overview of issues, research and systems, *User Modeling and User-Adapted Interaction* **11**, 203–259 (2001).

[18] J. Herlocker, J. Konstan, A. Borchers, J. Riedl, An algorithmic framework for performing collaborative filtering, in: *Proceedings of the 22nd Annual International ACM SIGIR Conference on Research and Development in Information Retrieval*, 1999.

[19] J. Herlocker, J. Konstan, L. Terveen, J. Riedl, Evaluating collaborative filtering recommender systems, *ACM Transactions on Information Systems* **22** (1), 5–53 (2004).

[20] F. Herrera, E. Herrera-Viedma, Aggregation operators for linguistic weighted information, *IEEE Transactions on Systems, Man and Cybernetics, Part A: Systems* **27**, 646–656 (1997).

[21] F. Herrera, E. Herrera-Viedma, Choice functions and mechanisms for linguistic preference relations, *European Journal of Operational Research* **120**, 144–161 (2000).

[22] F. Herrera, E. Herrera-Viedma, L. Martínez, A fusion approach for managing multi-granularity linguistic term sets in decision making, *Fuzzy Sets and Systems* **114**, 43–58 (2000).

[23] F. Herrera, E. Herrera-Viedma, J. Verdegay, Direct approach processes in group decision making using linguistic OWA operators, *Fuzzy Sets and Systems* **79**, 175–190 (1996).

[24] F. Herrera, L. Martínez, A 2-tuple fuzzy linguistic representation model for computing with words, *IEEE Transactions on Fuzzy Systems* **8** (6), 746–752 (2000).

[25] F. Herrera, L. Martínez, A model based on linguistic 2-tuples for dealing with multigranularity hierarchical linguistic contexts in multiexpert decision-making, *IEEE Transactions on Systems, Man and Cybernetics. Part B: Cybernetics* **31** (2), 227–234 (2001).

[26] E. Herrera-Viedma, An information retrieval system with ordinal linguistic weighted queries based on two weighting elements, *International Journal of Uncertainty, Fuzziness and Knowledge-Based Systems* **9**, 77–88 (2001).

[27] E. Herrera-Viedma, Modeling the retrieval process of an information retrieval system using an ordinal fuzzy linguistic approach, *Journal of the American Society for Information Science and Technology* **52** (6), 460–475 (2001).

[28] E. Herrera-Viedma, O. Cordón, M. Luque, A. López, A. Muñoz, A model of fuzzy linguistic IRS based on multi-granular linguistic information, *International Journal of Approximate Reasoning* **34** (3), 221–239 (2003).

[29] E. Herrera-Viedma, A. López-Herrera, A model of information retrieval system with unbalanced fuzzy linguistic information, *International Journal of Intelligent Systems* **22** (11), 1197–1214 (2007).

[30] E. Herrera-Viedma, A. López-Herrera, A review on information accessing systems based on fuzzy linguistic modelling, *International Journal of Computational Intelligence Systems* **3** (4), 420–437 (2010).

[31] E. Herrera-Viedma, A. López-Herrera, M. Luque, C. Porcel, A fuzzy linguistic irs model based on a 2-tuple fuzzy linguistic approach, *International Journal of Uncertainty, Fuzziness and Knowledge-based Systems* **15** (2), 225–250 (2007).

[32] E. Herrera-Viedma, A. López-Herrera, C. Porcel, Tuning the matching function for a threshold weighting semantics in a linguistic information retrieval system, *International Journal of Intelligent Systems* **20** (9), 921–937 (2005).

[33] E. Herrera-Viedma, A. G. López-Herrera, S. Alonso, J. Moreno, F. Cabrerizo, C. Porcel, A computer-supported learning system to help teachers to teach fuzzy information retrieval systems, *Information Retrieval* **12** (2), 179–200 (2009).

[34] E. Herrera-Viedma, L. Martínez, F. Mata, F. Chiclana, A consensus support system model for group decision-making problems with multi-granular linguistic preference relations, *IEEE Transactions on Fuzzy Systems* **13** (5), 644–658 (2005).

[35] E. Herrera-Viedma, G. Pasi, A. López-Herrera, C. Porcel, Evaluating the information quality of Web sites: A qualitative methodology based on fuzzy computing with words, *Journal of the American Society for Information Science and Technology* **57** (4), 538–549 (2006).

[36] M. Hsu, A personalized english learning recommender system for esl students, *Expert Systems with Applications* **34**, 377–408 (2008).

[37] Y. Jiang, Z. Fan, J. Ma, A method for group decision making with multi-granularity linguistic assessment information, *Information Sciences* **178** (41), 1098–1109 (2008).

[38] A. Jun, A new similarity measure for collaborative filtering to alleviate the new user cold-starting problem, *Information Sciences* **178** (1), 37–51 (2008).

[39] R. Korfhage, *Information Storage and Retrieval*, New York: Wiley Computer Publishing, 1997.

[40] S. Lee, Y. Cho, S. Kim, Collaborative filtering with ordinal scale-based implicit ratings for mobile music recommendations, *Information Sciences* **180** (11), 2142–2155 (2010).

[41] C. Long-Sheng, H. Fei-Hao, C. Mu-Chen, H. Yuan-Chia, Developing recommender systems with the consideration of product profitability for sellers, *Information Sciences* **178** (4), 1032–1048 (2008).

[42] A. López-Herrera, E. Herrera-Viedma, F. Herrera, Applying multi-objective evolutionary algorithms to the automatic learning of extended boolean queries in fuzzy ordinal linguistic information retrieval systems, *Fuzzy Sets and Systems* **160**, 2192–2205 (2009).

[43] I. Macho-Stadler, D. Perez-Castrillo, and R. Veugelers, Licensing of university inventions: The role of a technology transfer office, *International Journal of Industrial Organization* **25**, 483–510 (2007).

[44] G. Marchionini, Research and development in digital libraries (2000). URL: `http://ils.unc.edu/~march/digital_library_R_and_D.html`

[45] G. Markman, P. Phan, D. Balkin, and P. Gianiodis, Entrepreneurship and university-based technology transfer, *Journal of Business Venturing* **20**, 241–263 (2005).

[46] F. Mata, L. Martínez, E. Herrera-Viedma, An adaptive consensus support model for group decision making problems in a multi-granular fuzzy linguistic context, *IEEE Transactions on Fuzzy Systems* **17** (2), 279–290 (2009).

[47] J. Morales-del Castillo, E. Peis, A. Ruíz-Rodríguez, E. Herrera-Viedma, Recommending biomedical resources: A fuzzy linguistic approach based on semantic web, *International Journal of Intelligent System* **25**, 1143–1157 (2010).

[48] J. Moreno, J. Morales-del Castillo, C. Porcel, E. Herrera-Viedma, A quality evaluation methodology for health-related websites based on a 2-tuple fuzzy linguistic approach, *Soft Computing* **14** (8), 887–897 (2010).

[49] Movielens – movie recommendations. URL: `http://movielens.umn.edu`

[50] E. Nasibov, A. Kinay, An iterative approach for estimation of student performances based on linguistic evaluations, *Information Sciences* **179** (5), 688–698 (2009).

[51] M. Pazzani, A framework for collaborative, content-based and demographic filtering, *AI Review* **13** (56), 393–408 (1999).

[52] A. Popescul, L. Ungar, D. Pennock, S. Lawrence, Probabilistic models for unified collaborative and content-based recommendation in sparce-data environments, in: *Proceedings of the Seventeenth Conference on Uncertainty in Artificial Intelligence* (*UAI*), San Francisco, 2001.

[53] P. Refaeilzadeh, L. Tang, H. Liu, *Cross-validation* (2008). URL: `http://www.public.asu.edu/~ltang9/papers/ency-cross-validation.pdf`

[54] P. Reisnick, H. Varian, Recommender systems, *Special issue of Communications of the ACM* **40** (3), 56–59 (1997).

[55] B. Sarwar, G. Karypis, J. Konstan, R. J., Item-based collaborative filtering recommendation algorithms, in: *Proceedings of the ACM World Wide Web Conference*, 2001.

[56] B. Sarwar, G. Karypis, J. Konstan, J. Riedl, Analysis of recommendation algorithms for ecommerce, *Proceedings of ACM E-Commerce 2000 conference*, 158–167 (2000).

[57] J. Serrano-Guerrero, E. Herrera-Viedma, J. Olivas, A. Cerezo, F. Romero, A google wave-based fuzzy recommender system to disseminate information in university digital libraries 2.0, *Information Sciences* **181**, 1503–1516 (2011).

[58] G. Shani, A. Gunawardana, Recommender Systems Handbook, chap. *Evaluating Recommendation Systems*, Ricci, F., Rokach, L., Shapira, B., and Kantor, P.B. Eds. (Springer), 257–298 (2011).

[59] D. Siegel, D.Waldman, L. Atwater, A. Link, Toward a model of the effective transfer of scientific knowledge from academicians to practitioners: qualitative evidence from the commercialization of university technologies, *Journal of Engineering and Technology Management*, 115–142 (2004).

[60] P. Symeonidis, A. Nanopoulos, A. Papadopoulos, Y. Manolopoulos, Collaborative recommender systems: Combining effectiveness and efficiency, *Expert Systems with Applications* **34** (4), 2995–3013 (2008).

[61] I. Truck, H. Akdag, A tool for aggregation with words, *Information Sciences* **179** (14), 2317–2324 (2009).

[62] L. Zadeh, The concept of a linguistic variable and its applications to approximate reasoning. Part I, *Information Sciences* **8** (1975) 199–249, Part II, *Information Sciences* **8** (1975) 301–357, Part III, *Information Sciences* **9** (1975) 43–80 (1975).

[63] A. Zenebe, A. Nocio, Representation, similarity measures and aggregation methods using fuzzy sets for content-based recommender systems, *Fuzzy Sets and Systems* **160** (1), 76–94 (2009).

[64] G. Zhang, J. Lu, A linguistic intelligent user guide for method selection in multi-objective decision support systems, *Information Sciences* **179** (14), 2299–2308 (2009).

Chapter 4

Power of Linguistic Data Summaries and their Protoforms

Janusz Kacprzyk[1,a] and Sławomir Zadrożny[1,2]

[1] *Systems Research Institute, Polish Academy of Sciences, ul. Newelska 6, 01-447 Warsaw, Poland*

[2] *Warsaw School of Information Technologyul. Newelska 6, 01-447 Warsaw, Poland*

E-mail: kacprzyk@ibspan.waw.pl, zadrozny@wit.edu.pl

> *To the memory of Professor Da Ruan, a younger colleague and a close friend, who has been able to always maintain a proper balance between a formal and analytic elegance, and practical usefulness, and feel of what really matters*

We consider linguistic data(base) exemplified by, for a personnel database, "most employees are young and well paid" (with some degree of truth) and their extensions as a very general tool for a human consistent summarization of large data sets. We advocate the use of the concept of Zadeh's protoforms (prototypical forms) of linguistic summaries, in their more advanced form proposed by Kacprzyk and Zadrożny. Then, we present an extension of our interactive approach to the generation of linguistic summaries through the use of our fuzzy querying interface supporting queries with fuzzy linguistic quantifiers. We concentrate on a specific type of linguistic summaries which parallel specific fuzzy association rules, and show the use of an efficient algorithm for mining such rules. We show an extension to the dynamic case of by using linguistic summaries of times series data. As an example we show the use of linguistic summaries for Web server log analysis, in both the static and dynamic case. The results can be useful for the design and maintenance of computer networks.

4.1 Introduction

The purpose of this paper is to present a unified view of linguistic data summaries as a powerful tool for reaching a higher human consistency of tools and techniques in data

[a] Fellow IEEE

C. Kahraman (ed.), *Computational Intelligence Systems in Industrial Engineering*,
Atlantis Computational Intelligence Systems 6, DOI: 10.2991/978-94-91216-77-0_4,

analysis, data mining, and all kinds of modeling. The basic rationale adopted is that, first, for the human being the only fully natural means of communication and articulation is natural language. Second, the huge amount of data existing around us implies a necessity to develop some tools and techniques which can summarize the volume and content of those data to make them comprehensible, and – in view of the former remark – natural language should be employed as much as possible. Third, an implementation of the above general scheme calls for some formal tools and techniques and Zadeh's computing with words should provide here simple yet effective and efficient means. Fourth, linguistic data summaries based on the ideas of computing with words may be a proper solution. However, their generation may be difficult, and the use of Zadeh's protoforms, some general forms of linguistic summaries, can be of help, notably when their generation is implemented through an efficient fuzzy querying interface and association rule mining. The consecutive steps of the process will be shown to be implementable by using tools and techniques developed by us.

We will now elaborate on the consecutive issues mentioned above. First, data summarization – which is clearly one of basic capabilities of any "intelligent" system – is still to a large extent unsolved a problem. From our perspective it would clearly be desirable to have linguistic data summaries exemplified by, for a data set on employees described as a set of attributes, a statement (linguistic summary) "most young and very well qualified employees are quite well paid".

It is easy to see that this may be viewed as an obvious example of the so-called "computing with words (sometimes 'and with perceptions' is added) paradigm" introduced by Zadeh in the mid-1990s, and presented in Zadeh and Kacprzyk's books [1].

In this paper we will briefly show the use of linguistic database summaries introduced by Yager [2], and then further developed by Kacprzyk and Yager [3], and Kacprzyk, Yager and Zadrożny [4], Zadrożny and Kacprzyk [5], and implemented by Kacprzyk and Zadrożny [6–11]. We derive here linguistic data summaries as linguistically quantified propositions as, e.g., "most of the employees are young and well paid", with a degree of truth (validity), in case of a personnel database.

It may be noticed that the derivation of linguistic summaries can be difficult, mainly because of an inherent difficulty of an automated determination of what the intention and preference of the user is in the sense of which relation, i.e. between which attributes, is of interest to the user. The simplest way is to use Kacprzyk and Zadrożny's [6, 8, 10] interactive approach in which the determination of a class of summaries of interest is done via Kacprzyk and Zadrożny's FQUERY for Access [12, 13], a fuzzy querying add-in to Microsoft Access. In such a way, through a user friendly human computer interface (HCI) we can easily determine the class of summaries of interest.

One can notice here that a standardization of possible classes of linguistic data summaries may be very helpful, sometimes being a prerequisite for the possibility of implementation in non-trivial real applications in which the number of possible summaries may be huge. By extending our previous works, e.g. [10, 11], we show that by relating various forms of linguistic data summaries to various fuzzy queries with linguistic quantifiers, with different known and sought elements, we end up with a hierarchy of prototypical forms, called protoforms, of linguistic data summaries, in Zadeh's terminology [14]. Such a standardization is *per se* a powerful idea but we will elaborate on this later in this paper.

The linguistic data summaries can concern both static and dynamic (time series) data, and as an illustration we will briefly present an application to the summarization of Web logWeb logs, in the static and dynamic (time series) case.

4.2 Linguistic Summaries Using Fuzzy Logic with Linguistic Quantifiers

In the basic Yager's approach [2], notably in its constructive and implementable form by Kacprzyk and Yager [3], and Kacprzyk, Yager and Zadrożny [4], that was implemented in Kacprzyk and Zadrożny [7–11], we have:

- V is a quality (attribute) of interest, e.g. salary in a database of workers,
- $Y = \{y_1, \ldots, y_n\}$ is a set of objects (records) that manifest quality V, e.g. the set of workers; hence $V(y_i)$ are values of quality V for object y_i,
- $D = \{V(y_1), \ldots, V(y_n)\}$ is a set of data (the "database" on question)

A *linguistic summary* of a data set (data base) consists of:

- a summarizer S (e.g. young),
- a quantity in agreement Q (e.g. most),
- truth T – e.g. 0.7,
- a qualifier K (optionally), i.e. another linguistic term (e.g. well-earning), determining a fuzzy subset of Y.

as, e.g., "T(*most* of employees are *young*)$= 0.7$". The truth r may be meant more generally as, e.g., validity.

For a set of data D in question, we can hypothesize some appropriate summarizer S and some appropriate quantity in agreement Q, and T will indicate the truth of the statement that Q data items satisfy S.

We assume that the summarizer S (and qualifier K) is a linguistic expression semantically represented by a fuzzy set as, e.g., "young" would be represented as a fuzzy set in $\{1,2,\ldots,90\}$. Such a simple one-attribute summarizer serves the purpose of introducing the concept of a linguistic summary but it can readily be extended to a confluence of attribute values as, e.g, "*young* and *well paid*". Clearly, the most interesting are non-trivial, *human-consistent* summarizers (concepts) as, e.g.: *productive* workers, involving complicated *combinations of attributes*, e.g.: a hierarchy (not all attributes are of the same importance), the attribute values are ANDed and/or ORed, K out of n, *most*, etc. of them should be accounted for, etc. but they need some specific tools and techniques to be mentioned later.

Basically, two types of a linguistic quantity in agreement can be used: absolute (e.g., "about 5", "more or less 100", "several"), and relative (e.g., "a few", "more or less a half", "most", "almost all"). They are fuzzy linguistic quantifier quantifiers [15, 16] that can be handled by fuzzy logic.

The calculation of truth (validity) boils down to the calculation of the truth value (from $[0,1]$) of a linguistically quantified statement (e.g., "*most* of the employees are *young*") that can be done using Zadeh's [15] calculus of linguistically quantified propositions [1] or Yager's OWA operators [17] (cf. [18]); for a survey, see also Liu and Kerre [19].

In our case, we have a linguistically quantified proposition, written

$$Qy\text{'s are } S \tag{4.1}$$

where Q is a (relative) linguistic quantifier (e.g., most), $Y = \{y\}$ is a set of objects (e.g., employees), and S is some property (e.g., well paid). Another property K may also be added, yielding

$$QKy\text{'s are } S \tag{4.2}$$

e.g., "most (Q) of the highly qualified (K) employees (Y's) are well paid (S)".

We seek the truth values of such statements which are equal to: if S and K are fuzzy sets in Y, and a (proportional, nondecreasing) Q is a fuzzy set in $[0,1]$ exemplified by

$$\mu_Q(x) = \begin{cases} 1 & \text{for } x \geqslant 0.8 \\ 2x - 0.6 & \text{for } 0.3 < x < 0.8 \\ 0 & \text{for } x \leqslant 0.3 \end{cases} \tag{4.3}$$

then, due to Zadeh [15].

$$\text{truth}(Qy\text{'s are } S) = \mu_Q\left[\frac{1}{n}\sum_{i=1}^{n}\mu_S(y_i)\right] \tag{4.4}$$

$$\text{truth}(QKy\text{'s are } S) = \mu_Q\left[\sum_{i=1}^{n}(\mu_K(y_i)\wedge\mu_S(y_i))\Big/\sum_{i=1}^{n}\mu_K(y_i)\right] \tag{4.5}$$

Though the degree of truth (validity) is conceptually most important and intuitively appealing a criterion, it is often insufficient in practice, and some other criteria have been proposed, notably by Kacprzyk and Yager [3], and by Kacprzyk, Yager and Zadrożny [4], like degrees of: imprecision, covering, and appropriateness, and the length of a summary. For more measures, see Kacprzyk, Wilbik and Zadrożny [20, 21].

The problem is therefore how to generate the best summary (or summaries as it is often the case). The brute force solution, i.e. an exhaustive search, is clearly not a scalable choice, and some implicit enumeration type schemes should be used, possibly with some other tools. This will be shown in the next section.

4.3 Fuzzy Queries, Linguistic Summaries, and Their Protoforms

As we have already mentioned, it is difficult to automatically find what the real intention, interest and preference of the user with respect to the linguistic data summaries is, and that is why Kacprzyk and Zadrożny [6] proposed an interactive approach for the definition of elements of an intended linguistic summary via a graphical user interface of a fuzzy querying add-on. The roots of this approach are our previous papers on the use of fuzzy logic in querying databases [22, 23] via imprecise requests which led to our FQUERY for Access package, an add-in to Microsoft Acceess§that makes it possible to use fuzzy linguistic terms in database queries.

FQUERY for Access is an add-in that makes possible to use fuzzy terms in queries, and the following terms, which form a dictionary of the system, are available:

- fuzzy values as by *low* in "profitability is *low*",
- fuzzy relations as by *much greater than* in "income is *much greater than* spending", and
- linguistic quantifiers as by *most* in "*most* conditions have to be met".

The first two terms are elementary building blocks of fuzzy queries in FQUERY for Access and they are meant for numerical fields only. For other type of fields, other elements are employed. To use a field in a query with a fuzzy value, it has to be defined as an attribute through a definition of its: lower (LL) and upper (UL) limit which specify the interval within which the values are to be. A good solution is to universally define fuzzy values as fuzzy sets on $[-10,+10]$. Then, the matching degree $(\cdot,\cdot)$ of a simple condition referring to attribute AT and a fuzzy value FV against a record t is calculated by $md(AT = FV, t) = \mu_{FV}(\tau(t(AT))$, where: $t(AT)$ is the value of attribute AT in record t, μ_{FV} is the membership function of a fuzzy value FV, $\tau : [LLAT, ULAT] \to [-10, 10]$ is a mapping from the interval defining AT onto $[-10, 10]$.

The elicitation (definition) of fuzzy sets corresponding to particular fuzzy values may be done using different methods. Normally, it involves an interface with the user(s) who provide responses to appropriate chosen questions.

Linguistic quantifiers are key elements of both the linguistic data summaries in the sense considered here and in the fuzzy querying interface implemented through FQUERY for Access. They are defined in Zadeh's [15] sense, as fuzzy sets on $[0,1]$. They may be interpreted either using Zadeh's [15] approach or via the OWA operators [17, 18] as mentioned in the previous section.

The matching degree, $md(\cdot,\cdot)$, for the query "Q of N conditions are satisfied" for record T is equal to

$$md(Q, \text{condition}_i, t) = \mu_Q\left[\tau\left(\sum_i md(\text{condition}_i, t)\right)\right] \tag{4.6}$$

We can also assign different importance values to the particular conditions, and the aggregation formula is equivalent to (4.5). Conditions may be given different importance which is represented as a fuzzy set defined on $[0,1]$, and then treated as property K in (4.5) leading to

$$\begin{aligned} &md(QK, \text{condition}_i, t) \\ &\quad = \mu_Q\left[\tau\left(\sum_i (md(\text{condition}_i, t) \wedge \mu_K(\text{condition}_i)) \Big/ \sum_i \mu_K(\text{condition}_i)\right)\right] \end{aligned} \tag{4.7}$$

FQUERY for Access has been designed so that fuzzy queries be syntactically correct queries in Microsoft Access which is done by using the parameters. A parameters naming convention is assumed which provides for their proper interpretation by the FQUERY for Access query execution engine. For example, a parameter named [FfA_FV fuzzy value name] is interpreted as a fuzzy value, while one named [FfA_FQ fuzzy quantifier name] is interpreted as a fuzzy quantifier.

The query set by the user is then automatically transformed and then run as a native query of Microsoft Access. Details can be found in our papers [12, 5, 24, 25].

Obviously, fuzzy queries directly correspond to the linguistic summaries in the form considered here. The derivation of a linguistic summary may therefore proceed interactively in the environment of a fuzzy querying system. In Kacprzyk & Zadrożny's approach [6, 8] interactivity is in the definition of summarizers (indication of attributes and their combinations), via a user interface of a fuzzy querying add-on. The queries (referring to summarizers) allowed are:

- *simple* as, e.g., "salary is *high*"
- *compound* as, e.g., "salary is *low* AND age is *old*"
- *compound with quantifier*, as, e.g., "*most* of {salary is *high*, age is *young*, ..., training is *well above average*}".

Basically, we use a "natural" granulationgranulation of the set of possible values by using some reasonable number ($7 \pm 2!$) of values like: very low, low, medium, high, very high, and also "comprehensible" and intuitively appealing quantifiers as: most, almost all, ..., etc.

The derivation of a linguistic summary may proceed in an interactive way as follows:

- the user formulates a set of linguistic summaries of interest (relevance) using the fuzzy querying add-on described above,
- the system retrieves records from the database and calculates the validity of each summary adopted by matching the records with the summaries, and
- a most appropriate linguistic summary is chosen.

As shown by Kacprzyk and Zadrożny [11, 24], the concept of a protoformprotoform in the sense of Zadeh [14] is highly relevant here. First of all, a protoform is defined as an abstract prototype, that is, given as (4.1) and (4.2). These are the most abstract protoforms. In order to discuss their less abstract forms it is convenient to consider the summarizer (and the qualifier) as an abstract fuzzy logic statement "*X* IS *A*", where *X* is a placeholder for an attribute of objects in *Y* and *A* is a placeholder for a fuzzy set (linguistic term) determining its (fuzzy) value as, for instance, "age IS young", "salary IS low", and also "salary IS *A*". Two former summarizers are fully instantiated, while the latter still contains an abstract form of the attribute value (*A*). Thus, the protoforms may be seen as forming a hierarchy (a tree) a root of which is (4.2), the leaves are actual linguistic summaries with all elements (quantifiers, summarizers and qualifiers) fully instantiated and the intermediate nodes are partly instantiated linguistic summaries with some abstract elements.

Zadeh's protoforms may conveniently be used as a fundamental element of the user interface in that the user selects a protoform of a linguistic summary from that hierarchy and then the system instantiates the selected protoform in all possible ways, replacing abstract symbols denoting its elements with chosen fuzzy values and linguistic quantifiers stored in a dictionary of linguistic terms. A part of such a hierarchy of protoforms is shown in Fig. 4.1. At the top of the hierarchy we have a completely abstract protoform; in a protoform to the right, the qualifier K is instantiated to "age IS young"; in the one to the left summarizer S is first instantiated to "salary IS S^{value}", i.e., the attribute of the summarizer is selected to be "salary" but its value is not determined; then this protoform is further instantiated so that the summarizer is fully specified using "low" as the value of "salary".

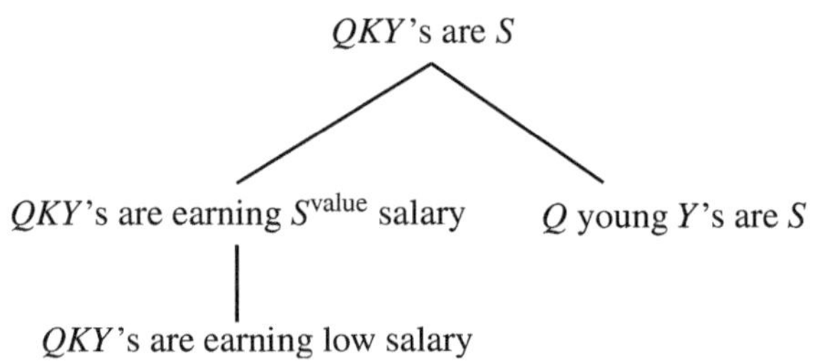

Fig. 4.1 An example of a part of a hierarchy of protoforms

The more abstract protoforms correspond to cases in which we assume less about summaries being sought. There are two extremes when we: (a) assume a totally abstract protoform, or (b) assume that all elements of a protoform are given on the lowest level of abstraction as specific linguistic terms. In case (a) data summarization is extremely time consuming but may produce an interesting, unexpected view on data. In case (b) the user has to guess a good candidate for a summary but the evaluation is fairly simple, equivalent to the answering of a (fuzzy) query. Thus, case (b) refers to ad hoc queries. This classification may be shown as in Table 4.1 in which 5 basic types of linguistic summaries are shown, corresponding to protoforms of a more and more abstract form; $S^{\text{structure}}$ denotes that attributes and the connection of referring to them predicates in a summary are known, while S^{value} denotes the values of the attributes sought.

Table 4.1 A taxonomy of linguistic summaries

Type	Given	Sought	Remarks
1	S	Q	Simple summaries through ad-hoc queries
2	$S\ K$	Q	Conditional summaries through ad-hoc queries
3	$Q\ S^{\text{structure}}$	S^{value}	Simple value oriented summaries
4	$Q\ S^{\text{structure}}\ K$	S^{value}	Conditional value oriented summaries
5	Nothing	$S\ K\ Q$	General fuzzy rules

Type 1 summaries may be easily obtained by a simple extension of fuzzy querying. The user has to construct a query, a candidate summary, and it has to be determined what is the fraction of rows matching this query and what linguistic quantifier best denotes this fraction. A Type 2 summary is a straightforward extension of Type 1. Type 3 summaries require much more effort as their primary goal is to determine typical or exceptional - depending on the quantifier- values of an attribute. A Type 4 summary is meant to find typical (exceptional) values for some, possibly fuzzy, subset of rows (defined by the qualifier K). Computationally, Type 5 summaries represent the most general form considered by us: the fuzzy rules describing dependencies between specific values of particular attributes. The summaries of Type 1 and 3 have been implemented as an extension to Kacprzyk and Zadrożny's FQUERY for Access [12, 5, 13]. Two approaches to Type 5 summaries generation have been proposed. First, a subset of such summaries may be obtained by analogy with association rules concept and employing their efficient algorithms. Second, genetic algorithms may be used to search the space of summaries (cf. [26]). In the next section we discuss in a more detailed way a special case of Type 5 summary, for which algorithms of association rule mining are applicable.

4.4 Derivation of Linguistic Summaries by Association Rule Mining

As we have mentioned above, we will show as an example of an effective and efficient way of derivation of quite general a type of linguistic data summaries, the approach that is based on an intrinsic similarity of Type 5 summaries and *association rules* [27].

An association rule is of the following form:

$$A_1 \wedge A_2 \wedge \cdots \wedge A_n A_{n+1} \tag{4.8}$$

and states that if in a database row all the attributes from $\{A_1, A_2, \ldots, A_n\}$ take on value 1, then alsos attribute A_{n+1} is expected to take on value 1.

A row in a database (table) *supports* a set of attributes $\{A_i\}_{i\in I}$ if all attributes from the set take on in this row value equal to 1. The two main quality measures for the association rule (8) are employed: the *support* which is the fraction of the number of rows supporting the set of attributes $\{A_i\}$, $i \in \{1,\ldots,n+1\}$ in a database (table), and the *confidence* which is the fraction of rows supporting $\{A_i\}$, $i \in \{1,\ldots,n+1\}$ among all rows supporting $\{A_i\}$, $i \in \{1,\ldots,n\}$. While the support determines a statistical significance of a rule, the confidence measures its strength. Usually, we are interested in rules having values of the support above some minimal threshold and a high value of the confidence.

Many algorithms for finding all association rules possessing a required support measure were devised, see, e.g. [27, 28].

One can mention the following extension of the above basic form: (a) the right-hand side, like the left-hand side, may contain a conjunction of the attributes, (b) many-valued scalar values and their hierarchies may be used, (c) numerical, real-valued attributes may be used leading to the *quantitative association rules quantitative association rules*, and (d) some constraints may be imposed on combinations of attributes in rules. In view of this, we can rewrite (4.8) as:

$$A_1 = a_1 \wedge A_2 = a_2 \wedge \cdots \wedge A_n = a_n \longrightarrow A_{n+1} = a_{n+1} \wedge \cdots \wedge A_{n+m} = a_{n+m} \tag{4.9}$$

The association rules may naturally be interpreted as a special case of the linguistic summaries as the antecedent and consequent of (4.8) correspond to the qualifier K and summarizer S of (4.2), respectively. The confidenceconfidence of a rule is related to the combination of the linguistic quantifier and the truth degree of (4.2). A summarizer S is assumed to be a formula, atomic or complex. Therefore, though the structure of the qualifier and the summarizer available in the case of association rules is somehow limited, this simplicity can be beneficial for the derivation of efficient algorithms for rule generation.

As shown in our previous works [7, 9] we implemented the mining of linguistic summaries corresponding to the association rule (4.9) within the framework of our fuzzy querying package FQUERYFQUERY for Access. For that purpose we generalized (4.9) to:

$$A_1 \text{ IS } f_1 \wedge \cdots \wedge A_n \text{ IS } f_n \longrightarrow A_{n+1} \text{ IS } f_{n+1} \wedge \cdots \wedge A_{n+m} \text{ IS } f_{n+m} \tag{4.10}$$

that is, *fuzzy values* f_i instead of crisp values can be used leading to a *fuzzy association rule*.

Then, two extensions have been proposed. First, we enrich the structure of an atomic condition: A_i IS f_i, (an *item* meant in the terminology of the association rules), by allowing it to be:

$$A_i \text{ IS } (f_{j_1} \vee \cdots \vee f_{jk}) \tag{4.11}$$

where f_{jl} are some fuzzy values defined over the domain of A_i so that we have proposed to make it possible to use a range of fuzzy values. Second, we allowed the use of a flexible aggregation operator in the summarizer and/or qualifier formula which implies the following form of the atomic condition:

$$Q \text{ of } (A_1 \text{ IS } f_1, A_2 \text{ IS } f_2, \ldots, A_n \text{ IS } f_n) \tag{4.12}$$

Such an intuitively appealing and highly constructive extension of the fuzzy association rules is to a large extent implied by the capabilities of our FQUERY for Access.

For mining the association rules, use our FQUERY for Access. First, the quantitative association rules usually require discretization of the attributes obtained via a partition of a domain into a number of intervals. Then, each interval is treated as an additional binary attribute and then many known algorithms for the generation of classical association rules may be employed.

Our implementation of association rule mining is based on the Agrawal and Srikant's (1994) Apriori TID Apriori TID algorithm [27, 28] which works in two steps: first it finds *frequent itemsets* and then produces rules from each itemset. The second step is relatively easy, hence we will focus on the first one.

An itemset is a conjunction of the items of the form (4.11) or (4.12). A row in the database (table) is said to *support* an itemset if the corresponding conjunction "is true" (the degree of satisfaction exceed some threshold) for this row. An itemset containing K items is called a K-itemset. The algorithm starts with the evaluation of 1-itemsets. These itemsets which are not supported by sufficient number (*minsup*) of rows are deleted. Previously, we assumed the 1-itemsets only in the form present in (4.10). To implement the items such as (4.11) and (4.12) we have to extend this step. First, only the "regular" 1-itemsets (4.10) are counted, i.e., a full scan of the database (table) is done and the frequency of appearance of all items is calculated. Then, the 1-itemsets of type (4.11) are constructed but only such f_{ij} are taken into account that have the support greater than some value (a parameter of the method, in addition to minsup and minconf) higher than 0 and less than minsup. For example, if a regular 1-itemset "salary IS high" gets a very low support, then we will construct neither "salary IS medium or high" nor "salary IS low or high" 1-itemsets. This helps reduce the time and memory complexity of the algorithm.

Such a reduction is even more important in case of the implementation of the 1-itemsets of type (4.12). Basically, we should take into account all subsets of the regular 1-itemsets and all possible quantifiers Q. This would be computationally intractable and in fact require a kind of a recursive use of AprioriTID in the first step. Thus, in our implementation we limit ourselves to just one, fixed quantifier. Moreover, for obvious reasons, we take into account only such subsets of regular items that: all refer to different attributes, and there is enough number of them to make quantification meaningful. Thus, we will, e.g., neither construct a 1-itemset of the form "*most* (*salary IS high, salary IS low,...*)" nor "*most* (*salary IS high, age IS high*).

Having the 1-itemsets of type (4.11) constructed we calculate their support. We assume that no row supports two different fuzzy values for the same attribute. Then, the support for A_i IS $(f_{j1} \vee \cdots \vee f_{jk})$ is just the sum of supports for A_i IS f_{jl}, $l = 1, \ldots, K$; that we calculated earlier. Now the 1-itemsets of type (4.12) are constructed. They may use both the regular 1-itemsets as well as the 1-itemsets of type (4.11), e.g., "Most (A_1 IS $(f_{11} \vee \cdots \vee f_{1k})$, A_2 IS $f_2, \ldots$)" are allowed. The support for the 1-itemsets of type (4.12) is then calculated. However, due to the use of a priori TID another full scan of the database is not needed. During the first scan we have recorded in some data structures the IDs of rows supporting the particular regular 1-itemsets and now it is enough to operate on these structures. Then, the algorithm proceeds as usual generating and evaluating the K-itemsets for $K = 2, 3, \ldots,$ [27] The only additional effort is needed to guarantee that no itemset produced twice refers to the same attribute, e.g., the 2-itemset "*salary IS high AND salary IS medium*" has to be excluded. Finally, all frequent itemsets found are taken into account when producing association rules of the confidence at least equal to the required value (minconf).

We deal with the real valued attributes so that for each such an attribute and each fuzzy value defined for it we introduce a new items which may be treated as binary, i.e., appearing in a row or not. In this respect, practically only a limited number of fuzzy values per attribute (say 3) leads to computationally tractable mining tasks.

The implementation of the algorithm for the mining of linguistic summaries via extended fuzzy association rules association rules may be presented as follows:

Step 1. Selection of the attributes and fuzzy values

The user chooses the attributes to be used, i.e. builds a query referring to the attributes to be taken into account (this is done via "the navigation" of protoforms hierarchy, as described in Section III). Then, the user initiates the data summarization process, sets the parameters (minsup, minconf, minimam support, ...) and the system automatically performs the rest of the steps.

Step 2. Construction of the items

For each pair – of the selected attributes and fuzzy values - the system creates an item, as described earlier.

Step 3. Forming the data set and starting external application for fuzzy linguistic rules mining

The items constructed are numbered. Then, the data set is produced describing each row with numbers of items supported by it. The calculations proceed by the fuzzy querying module. When the data set is ready, an external application is started with this data set given on input.

Step 4. Calculation of the support for the regular 1-itemsets

A module reads the input data set and immediately calculates support for the regular 1-itemsets. It also records for each 1-itemset numbers (IDs) of rows supporting it.

Step 5. Construction of the 1-itemsets of type (4.11) and calculation of their support

Only the regular 1-itemsets of the support higher than a user-specified threshold are taken into account. The number of 1-itemsets of this type produced for a given attribute depends on the number of fuzzy values defined for it. The support is obtained by summing up the support of the constituent regular 1-itemsets. All new 1-itemsets are numbered.

Step 6. Pruning of the set of 1-itemsets

All itemsets with the support lower than the *support threshold* (minsup) are discarded. Additionally, also itemsets with the support higher than another threshold, an *item omit threshold*, are discarded since the items present in almost all records contribute nothing interesting to the rules produced.

Step 7. Construction of the 1-itemsets of type (4.12), calculation of their support and pruning

Both the regular and 1-itemsets added in Step 5 are considered; we refer to them jointly as simple 1-itemsets. The 1-itemsets constructed are identified with lists of the constituent simple 1-itemsets. The lists are ordered lexicographically which makes the process of generation more efficient. The support is computed for the itemsets generated and those below the minsup threshold are discarded.

All itemsets produced so far and passing the pruning constitute the collection of 1-itemsets.

SET $k = 2$

Step 8. Generate the K-itemsets

They are generated from the frequent $(K-1)$-itemsets as in a priori TID. Pairs of the frequent $(K-1)$ itemsets of the form $A_1 \wedge A_2 \wedge \cdots \wedge A_{k-1}$ and $B_1 \wedge B_2 \wedge \cdots \wedge B_{k-1}$, where

$A_i = B_i$ for $i = 1, \ldots, K-2$, are sought. Then, a new K-itemset of the form $A_1 \wedge A_2 \wedge \cdots \wedge A_{k-1} \wedge B_{k-1}$ is generated. In the original algorithm, the rules generated in such a way are additionally tested and possibly eliminated before Step 7. On the other hand, we add another K-itemset generation limitation, namely the items A_{k-1} and B_{k-1} have to correspond to different original attributes. This is obvious if the items A_{k-1} and B_{k-1} are regular. Otherwise, by identifying an item of type (4.11) or (4.12) with a list (set) of attributes referred to within it, we require the intersection of these sets to be empty.

Step 9. Calculate the support for all the K-itemsets

The calculation is based on the recorded numbers (ID's) of rows supporting the particular $(K-1)$-itemsets. The similar data on the supporting rows is produced for the K-itemsets.

Step 10. Pruning of the set of k-itemsets (as in Step 6)

As a result we obtain the frequent K-itemsets.

IF the set of K-itemsets is void THEN GOTO Step 11.

SET $K = K+1$; GOTO Step 8.

Step 11. Generate rules from the frequent l-itemsets, $l = 1, \ldots, K-1$.

Step 12. Display the results.

The number of the rules produced is usually huge. Some counter-measures have to be undertaken, notably some aggressive pruning schemes.

4.5 An Example

As a simple illustration of Type 5 summaries, an implementation is shown for the summarization of Web server logs, first in the static case and then in a dynamic case.

Each request to a Web server is recorded in one or more of its log files. The recorded information usually comprises the fields as in Table 4.2 (cf. a common log file format `http://www.w3.org/Daemon/User/Config/Logging.html#common-logfile-format`). There is also an extended format which includes more fields but these will be of no interest for us in this paper.

A lot of software is available (e.g., AWStats at `http://awstats.sourceforge.net/`) for reading in a log file and then producing various statistics concerning the usage of, traffic to and from, etc. a given Web server. These statistics usually include the number of requests (or requested Web pages): per month/day/hour of the week, per country or domain of the requesting computer. Moreover, statistics concerning requests from specific sources (mostly search engines), type of browser, type of operating system, etc. are produced, and

Table 4.2 Contents of a web server log file

Field no.	Content
1	the requesting computer name or IP address
2	the username of the user triggering the request (often absent),
3	the user authentication data
4	the date and time of the request
5	the HTTP command related to the request which includes the path to the requested file
6	the status of the request (e.g., to determine whether a resource was correctly transferred, not found, etc.)
7	the number of bytes transferred as a result of the request
8	the software used to issue the request

they may be expressed in terms of the number of requests and/or the number of bytes transferred. Analyses may also concern sessions, i.e. the series of requests send by the same agent. All those analyses may help to model agents' behavior, identify the navigational paths and, e.g., reconstruct the Web site in order to enhance agents' experience.

It is easy to see that though those mostly statistical analyses, are effective, and often powerful and efficient, they are not human consistent enough in the sense of producing numerical results that are often too voluminous and not comprehensible to an average human user for whom the use of natural language, i.e. linguistic summaries of data, may be the right solution.

A Web server log file may be directly interpreted as a table of data with the columns corresponding to the fields listed in Table 4.2 and the rows corresponding to the requests. For instance, for the purposes of linguistic summarization attributes of requests can be as given in Table 4.3.

In this section we will discuss various linguistic summaries that may be derived using this data.

In our context, i.e. while looking for a subclass of linguistic summaries that may be obtained using efficient algorithms for *association rules mining,association rules mining*, the structure of the qualifier and summarizer are clearly limited to the conjunction of simple conditions. The truth value of the summary corresponds to the confidence measure of an association rule

The experiment was run on the log file of one of the Web servers of our institute and we used a two week access request log with about 350 000 requests. We used an imple-

Table 4.3 Attributes of the requests used for their linguistic summarization

Attribute name	Description
Domain	Internet domain extracted from the requesting computer name (if given)
Hour	hour the request arrived; extracted from the date and time of the request
Day of the month	as above
Day of the week	as above
Month	as above
Filename	name of the requested file, including the full path, extracted from the HTTP command
Extension	extension of the requested file extracted as above
Status	Status of the request
Failure	$=1$ if status code is of 4xx or 5xx form, and $=0$ otherwise
Success	$=1$ if status code is of 2xx form, and $=0$ otherwise
Size	number of bytes transferred as a result of the request
Agent	name of the browser used to issue the request (name for major browsers, "other" otherwise)

mentation of the a priori algorithm by Borgelt (cf. `http://www.borgelt.net/apriori.html`). The obtained linguistic summaries are decoded by FQUERY for Access and presented to the user.

In our experiments we have obtained a number of interesting linguistic summaries. Due to space limitations we can only show some of them (T in the parentheses stands for the truth value). First:

All requests with the status code 304 ("not modified") referred to *small* files ($T = 1.0$)

The next summary obtained concerns scalar (non-numerical) attributes:

Most files with the "gif" extension were requested from the domain "pl" ($T = 0.98$)

and it is worth noticing it does not hold that "*Most* files were requested from the domain 'pl' " which is true to degree 0.4 only.

We have obtained more convincing summaries when we add a condition concerning the status code of the request, namely:

Most files with the "gif" extension successfully fetched (with the status code 200) were requested from the domain "pl" ($T = 1$)

Many more interesting linguistic summaries have also been obtained which can give much insight into requests coming into the Web server.

The linguistic data summaries can be extended to the dynamic context, notably to the linguistic summarization of time series as proposed in a series of our papers Kacprzyk, Wilbik and Zadrożny [20, 21] in which the analysis of how trends concerning some numerical attributes evolve over time, how long some types of behavior last, how rapid changes are, etc. is provided.

In the dynamic case the protoforms also play a key role, and the first, basic protoforms used [20, 21] were:

$$\text{Among all segments, } Q \text{ are } S \tag{4.13}$$

e.g.: "Among all segments, most are slowly increasing",

$$\text{Among all } K \text{ segments, } Q \text{ are } S \tag{4.14}$$

e.g.: "Among all short segments, most are slowly increasing".

Then, the above basic protoforms have been extended to so called temporal protoforms [29]:

$$E_T \text{ among all segments, } Q \text{ are } S \tag{4.15}$$

e.g.: "Recently, among all segments, most are slowly increasing",

$$E_T \text{ among all } K \text{ segments, } Q \text{ are } S \tag{4.16}$$

e.g.: "Initially among all short segments, most are slowly increasing".

The calculation of the truth values of linguistic summaries for those protoforms is shown in Kacprzyk, Wilbik and Zadrożny [20, 21], and Kacprzyk and Wilbik [29], and will not be given here for lack of space.

The linguistic summaries of time series provide powerful tools for producing interesting and useful linguistic summaries of Web logs. Just to give some more intuitively appealing examples, these can be:

"In the late evening hours, almost all requests with the status code 304 ('not modified') referred to small files" ($T = 1.0$)

"In almost all days, in evening hours, most files with the *gif* extension were requested from the domain 'pl' " ($T = 1$),

"In recent days, almost all files with the *gif* extension successfully fetched (with the status code equal 200) were requested from the domain 'pl' " ($T = 1$).

4.6 Concluding Remarks

We presented the idea and power of the concept of a linguistic data (base) summary, originated by Yager [2] and further developed first, in a more conventional form, by Kacprzyk and Yager [3], and Kacprzyk, Yager and Zadrożny [4], and then, in a more general context of Zadeh's protoforms [14] by Kacprzyk & Zadrożny [11, 24], and a more implementation oriented context of its relation to fuzzy database querying by Kacprzyk and Zadrożny [7–11]. As an example of application we have shown their use in the summarization of Web server logs, both in the case of static and dynamic mode. The results obtained give much insight into the traffic to/from the Web server and can provide some useful information concerning the design and maintenance of the computer networks.

Acknowledgments

This works was partially supported by the National Science Centre (contract no. UMO-2011/01/B/ST6/06908).

Bibliography

[1] Zadeh L.A., Kacprzyk J. (Eds.), *Computing with Words in Information/Intelligent Systems, 1. Foundations, 2. Applications*. Heidelberg and New York: Physica-Verlag (Springer-Verlag) (1999).

[2] Yager R.R. A new approach to the summarization of data. *Information Sciences*, 28, 69–86 (1982).

[3] Kacprzyk J., Yager R.R., Linguistic summaries of data using fuzzy logic. *International Journal of General Systems*, 30, 133–154 (2001).

[4] Kacprzyk J., Yager R.R., Zadrożny S., A fuzzy logic based approach to linguistic summaries of databases. *International Journal of Applied Mathematics and Computer Science*, 10, 813–834 (2000).

[5] Zadrożny S., Kacprzyk J., On database summarization using a fuzzy querying interface, in *Proc. of IFSA'99 World Congress (Taipei, Taiwan R.O.C.), Vol. 1*, 39–43 (1999).

[6] Kacprzyk J., Zadrożny S., Data mining via linguistic summaries of data: an interactive approach, in T. Yamakawa, G. Matsumoto (Eds.), *Methodologies for the Conception, Design and Application of Soft Computing – Proceedings of IIZUKA'98*, Iizuka, Japan, 668–671 (1998).

[7] Kacprzyk J., Zadrożny S., On linguistic approaches in flexible querying and mining of association rulet, in H.L. Larsen, J. Kacprzyk, S. Zadrożny, T. Andreasen, H. Christiansen (Eds.), *Flexible Query Answering Systems. Recent Advances*, Heidelberg and New York: Springer-Verlag, 475–484 (2001).

[8] Kacprzyk J., Zadrożny S., Data mining via linguistic summaries of databases: an interactive approach, in L. Ding (Ed.), *A New Paradigm of Knowledge Engineering by Soft*, Computing, Singapore: World Scientific, 325–345 (2001).

[9] Kacprzyk J., Zadrożny S., Fuzzy linguistic summaries via association rulet, in A. Kandel, M. Last, H. Bunke (Eds.), *Data Mining and Computational Intelligence*, Heidelberg and New York: Physica-Verlag (Springer-Verlag), 115–139 (2001).

[10] Kacprzyk J., Zadrożny S., Protoforms of linguistic data summaries: towards more general natural-language-based data mining tools, in A. Abraham, J. Ruiz del Solar, M. Koeppen (Eds.), *Soft Computing Systems*, Amsterdam: IOS Press, 417–425 (2002).

[11] Kacprzyk J., Zadrożny S., Linguistic database summaries and their protoforms: towards natural language based knowledge discovery tools. *Information Sciences*, **173** (4),281–304 (2005).

[12] Kacprzyk J. Zadrożny S., FQUERY for Access: fuzzy querying for a Windows-based DBMS, in P. Bosc, J. Kacprzyk (Eds.), *Fuzziness in Database Management Systems*, Heidelberg: Physica-Verlag, 415–433 (1995).

[13] Kacprzyk J., Zadrożny S., Computing with words in intelligent database querying: standalone and Internet-based applications. *Information Sciences*, **34**, 71–109 (2001).

[14] Zadeh L.A. (2002), A prototype-centered approach to adding deduction capabilities to search engines – the concept of a protoform, in *BISC Seminar, 2002, University of California, Berkeley* (2002).

[15] Zadeh L.A., A computational approach to fuzzy quantifiers in natural languages. *Computers and Mathematics with Applications*, **9**, 149–184 (1983).

[16] Zadeh L.A., Syllogistic reasoning in fuzzy logic and its application to usuality and reasoning with dispositions. *IEEE Transaction on Systems, Man and Cybernetics, SMC-15*, 754–763 (1985).

[17] Yager R.R., On ordered weighted avaraging operators in multicriteria decision making. *IEEE Transactions on Systems, Man and Cybernetics, SMC-18*, 183–190 (1988).

[18] Yager R.R., Kacprzyk, J. (Eds.), The Ordered Weighted Averaging Operators: Theory and Applications. Boston: Kluwer (1997).

[19] Liu Y., Kerre E.E., An overview of fuzzy quantifiers. (I) Interpretations. *Fuzzy Sets and Systems*, **95**, 1–21 (1988).

[20] Kacprzyk J., Wilbik A., Zadrożny S., Linguistic summarization of time series using a fuzzy quantifier driven aggregation, *Fuzzy Sets and Systems*, **159**, 1485–1499 (2008).

[21] Kacprzyk J., Wilbik A., Zadrożny S., An approach to the linguistic summarization of time series using a fuzzy quantifier driven aggregation. *International Journal of Intelligent Systems*, **25**, 411–439 (2010).

[22] Kacprzyk J., Ziółkowski A., Database queries with fuzzy linguistic quantifiers. *IEEE Transactions on Systems, Man and Cybernetics, SMC-16*, 474–479 (1986).

[23] Kacprzyk J., Zadrożny S., Ziółkowski, A., FQUERY III+: a 'human consistent' database querying system based on fuzzy logic with linguistic quantifiers. *Information Systems*, **6**, 443–453 (1989).

[24] Kacprzyk J., Zadrożny S., Computing With Words Is an Implementable Paradigm: Fuzzy Queries, Linguistic Data Summaries, and Natural-Language Generation. *IEEE Transactions on Fuzzy Systems*, **18**, 461–472 (2010).

[25] Zadrożny S., Kacprzyk J., Fuzzy querying using the 'query-by-example' option in a Windows-based DBMS, in Proc. of Third European Congress on Intelligent Techniques and Soft Computing – EUFIT'95 (Aachen, Germany), **vol. 2**, 733–736 (1995).

[26] George R. and R. Srikanth R., Data summarization using genetic algorithms and fuzzy logic, in F. Herrera and J.L. Verdegay (eds.): *Genetic Algorithms and Soft Computing*. Physica-Verlag, Heidelberg, 599–611 (1996).

[27] Agrawal R., Srikant R., Fast algorithms for mining association rules, *Proc. of the 20th Int. Conf. on Very Large Databases*, Santiago, Chile, Morgan Kaufmann, San Mateo, CA, USA 1994, 487–499 (1994).

[28] Borgelt Ch., Kruse R., Induction of Association Rules: Apriori Implementation, *15th Conf. on Comp. Statistics,* Berlin, Germany, Physica Verlag, Heidelberg, 395–400 (2002).

[29] Kacprzyk J., Wilbik A., Comparison of time series via classic and temporal protoforms of linguistic summaries: an application to mutual funds and their benchmarks, in Ch. Borgelt *et al.* (eds.): *Combining Soft Computing and Statistical Methods in Data Analysis*. Springer-Verlag, Berlin, 369–377 (2010).

Chapter 5

Can Fuzzy Logic Formalism Bring Complex Environmental Issues into Focus?

Ashok Deshpande

Founding Chair: Berkeley Initiative in Soft Computing (BISC) – Special Interest Group (SIG) – Environment Management Systems (EMS)

E-mail: ashokwd@yahoo.com

Information on the status and changing trends in environmental quality is necessary to formulate sound public policy and efficient implementation of environmental pollution abatement programs. In this quest, air and water quality indices are computed using US-EPA and US-NSF proposed methods for local and regional air / water quality management in many metro cities of the world, respectively. The procedure in vogue in the computation of these indices, however, does not include expert's knowledge. We believe that the development of a method to quantify association between the pollutant and air / water borne diseases is an important step before classifying air / water quality.

There exists aloetary uncertainty in the pollution parametric data and epistemic uncertainty in describing the pollutants by the domain experts in linguistic terms such as *poor, good, very good*, etc. Successes of probability theory have high visibility. But what is not widely not recognized is that these successes mask a fundamental limitation-the inability to operate on what may be called perception - based information. Fuzzy logic based formalism presented in this paper can model the two types of uncertainties, thereby straightway describing air/water quality in linguistic terms with a degree of certainty attached to each term.

5.1 Introduction

You and a friend walk outside on January morning in Pune City. You announce that the weather is *mild*. Your friend declares that it is *cold*. Who is wrong? Or are you both right?

People recognize that language can be imprecise/ fuzzy set theory and that concepts such as *cold, hot, or mild do not* have well-defined boundaries. In 1965, Professor Lotfi Zadeh, introduced fuzzy sets and thereafter fuzzy logic, a means of processing data by extending classical set theory to handle partial membership.

C. Kahraman (ed.), *Computational Intelligence Systems in Industrial Engineering*, Atlantis Computational Intelligence Systems 6, DOI: 10.2991/978-94-91216-77-0_5,

In everyday life and in fields such as environmental health, people deal with concepts that involve factors that defy classification into crisp sets—*safe*, *harmful*, *acceptable*, *unacceptable*, and so on. A classic example is a regulator carefully explaining the result of a detailed quantitative risk assessment to a community group, only to be asked over and over again, *But are we safe?* In this case, *safe* defies crisp classification because it is a multivariate state with gradations that vary among different individuals and groups.

Information on the status and changing trends in environmental quality is necessary to formulate sound public policy and efficient implementation of environmental pollution abatement programs. One of the ways of communicating the information to the policy makers and public at large is with indices. In the computation of air / water quality index (AQI/WQI), first a numerical value is computed and then the air / water quality is described in Linguistic terms. There exists aloetary uncertainty in the pollution parametric data and epistemic uncertainty in describing the pollutants by the domain experts in linguistic terms such as *poor*, *good*, *very good*, etc. Successes of probability theory have high visibility. But what is unrecognized is that these successes mask a fundamental limitation-the inability to operate on what may be called perception-based information. In this paper an attempt has been made to use fuzzy logic based formalism in modelling these two types of uncertainties, thereby straightway describing air/water quality in linguistic terms with a degree of certainty attached to each term.

The rest of the paper is organized as follows: Section 5.2 is a brief account of the theoretical foundation of fuzzy logic based method with brief description of the other mathematical framework used. While Section 5.3 relates to a case study for describing water quality fuzzily and air quality classification is presented. Concluding remarks and future research efforts are covered in Section 5.4.

5.2 Fuzzy Logic Based Formalism

Will I suffer from water borne diseases, if I take a bath in polluted river water? Realizing the complexity in establishing cause–effect relationship between bathing in polluted river and water borne diseases (WBD's), an attempt has been made to present a useful method to address the issue. Some of the important facets of our approach include: interviewing student community (bather/ non bather) with a structured questionnaire, collecting information on belief of the resident medical practitioners about bathing in polluted river and WBDs. Furthermore, modelling of epistemic uncertainty in domain expert's belief in supporting their evidence for various WBDs and alike. Figure 5.1 presents a novel multifaceted formalism for ultimately straightway describing river water quality in linguistic terms with degree of certainty [1]. The technique used in estimating the possible association between bating in polluted river water and water borne disease include: epidemiological study including case control study, river water quality analysis, perception of the resident medical professionals regarding their belief in relation to water borne diseases, Dempster-Shafer (DS) Theory of evidence, bootstrapping along with conventional statistical techniques, and alike. Some of these methods are briefly described in this section.

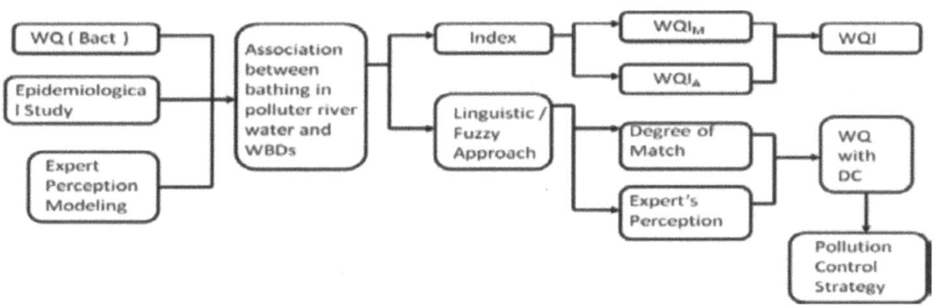

Fig. 5.1 Multifaceted formalism for Water Quality Classification

5.3 Fuzzy measures and Evidence theory

A fuzzy measure describes the vagueness or imprecision in the assignment of an element a to two or more crisp sets. In a fuzzy measure, the concern of attention is to describe the vagueness or imprecision in assigning the point to any of the crisp sets on the power set (Ross, 1997).

Shafer developed Dempster's work and presented an important theory of evidence-called Dempster–Shafer (DS) theory [1] in which DS belief (Bel) and plausibility (Pl) are used to characterize uncertainty. A basic measure in DS theory is a Basic Belief Assignment (BBA). The function (m) is a mapping function to express BBA for a given evidential event A, $m(A)$. BBA is a representation of partial belief not only for a single possible event, but also for a set of possible events with imprecise and incomplete knowledge and data. The main concept of evidence theory is that our knowledge of a given problem can be inherently imprecise. Hence, the bound result, which consists of both belief and plausibility, is presented. BBA expresses the degree of belief in a proposition. BBA is assigned by making use of a mapping function (m) to express our belief with a number in the unit interval $[0,1]$

$$m \,:\, 2^x \longrightarrow [0,1]. \tag{5.1}$$

The number $m(A)$ represents the portion of total belief assigned exactly to proposition A. The measure m, BBA function, must satisfy the following axioms [6]:

$$m(A) \geqslant 0 \;\text{ for any }\; A \in 2^x \tag{5.2}$$

$$m(\varnothing) = 0, \quad \sum_{\in 2^x} mA = 1 \tag{5.3}$$

Though these axioms of evidence theory look similar to those of probability theory, the axioms for the BBA functions are less stringent than those for probability measure.

5.4 Dempster's rule of combining

The information from different sources can be aggregated by Dempster's rule of combining to make a new combined BBA structure as given in the following equation

$$M_{12} = \frac{\sum_{C_i \cap C_j = A} m_1(C_i)m_2(C_j)}{1 - \sum_{C_i \cap C_j = \varnothing} m_1(C_i)m_2(C_j)}, \quad A \neq \varnothing \tag{5.4}$$

where C_i and C_j are propositions from each sources (m_1 and m_2). In equation (5.4), $\sum_{C_i \cap C_j = \varnothing} m_1(C_i)m_2(C_j)$ can be viewed as contradiction or conflict among the information given by the independent knowledge sources [1]. Even when some conflict is found among the information, Dempster's rule disregards every contradiction by normalizing with the complementary degree of contradiction to consider only consistent information. However, this normalization can cause a counter intuitive and numerically unstable combination of information, when the given information from different sources has significant contradiction or conflict [6]. If there is a serious conflict, it is recommended to investigate the given information or to collect more information.

5.5 Belief and plausibility function

Owing to lack of information and various possibilities in constructing BBA structure, it is more reasonable to present a bound of the total degree of belief in a proposition, as opposed to a single value of probability given as a final result in probability theory. The total degree of belief in a proposition 'A' is expressed within bound $[\,\text{bel}(A),\, \text{pl}(A)\,]$, which lies in the unit interval $[0,1]$ as shown in Figure 5.1, where $\mathrm{Bel}(A)$ and $\mathrm{Pl}(A)$ are given as,

$$\mathrm{Bel}(A) = \sum_{C \subset A} m(C) \quad : \quad \text{Belief Function} \tag{5.5}$$

$$\mathrm{Pl}(A) = \sum_{C \cap A \neq \varnothing} m(C) : \quad \text{Plausibility Function} \tag{5.6}$$

$\mathrm{Bel}(A)$ is obtained by the summation of BBAs for proposition, which is included in proposition A fully. $\mathrm{Bel}(A)$ is the total degree of belief. The degree of plausibility $\mathrm{Pl}(A)$ is calculated by adding BBAs of propositions whose intersection with proposition A is not an empty set. That is, every proposition consistent with proposition A at least partially is considered to imply proposition A, because BBA in a proposition is not divided into its subsets. Briefly, $\mathrm{Bel}(A)$ is obtained by adding the BBAs of propositions that totally agree with the proposition A as a measure of belief. Whereas, $\mathrm{Pl}(A)$ plausibility is calculated by adding BBAs of propositions that correspond to the proposition A totally or partially. In a sense, these two measures consist of lower and upper probability bounds.

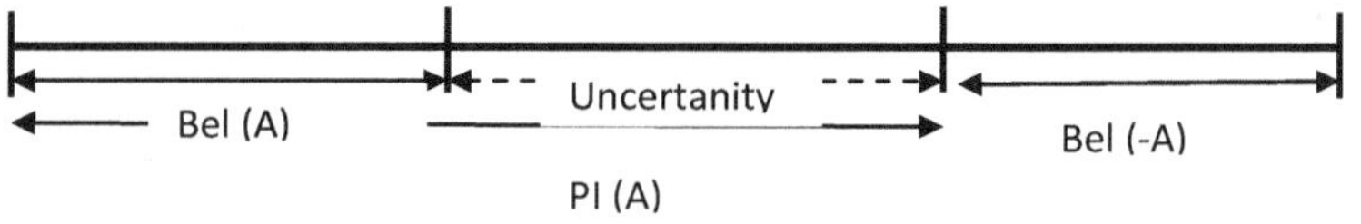

Fig. 5.2 Belief and uncertainty and ignorance

5.6 Bootstrap re-sampling

Large sample size for estimating the concentration of air pollutants from air quality monitoring stations for carrying out probability analysis was not considered necessary as it would have been a drain on the financial resources. Instead, using the well developed bootstrap procedure was considered adequate to arrive at the average parametric concentration.

5.7 What is Bootstrapping?

As an allusion to lifting oneself up by one's own bootstraps, the term *bootstrapping* means using a special process to perform a task that one would be unable to do in general. The term is said to have originated from a tale "The adventures of Baron Munchhausen" who, according to the story, escaped from a swamp by pulling himself up by the straps of his boots (although other versions of the story are available). Bootstrapping is a re-sampling technique used to obtain estimates of summary statistics. Bootstrap procedure is a non parametric statistical techniques which can be used to reduce the bias of point estimate and construct approximate confidence intervals for parameters such as the population mean [1]. Furthermore, bootstrap sampling procedure has been adopted for re-sampling by increasing sample size up to ($n = 1\,000$) with replacement procedure without any loss of generosity of mother sample. The boot estimation was done by using the formula:

$$\overline{x} = \frac{1}{n}\sum_{i=1}^{n} x_i \text{ and for the bootstrap mean } \overline{X}_b = \frac{1}{n}\sum_{i=1}^{n} X(i) \tag{5.7}$$

If the original estimate is biased, then, under certain conditions, part of the bias are removed. Bootstrap procedure, and an estimation of the standard deviation of the bootstrap estimate is given by

$$S = \sqrt{\frac{1}{n+1}\sum_{i=1}^{n}\left(B_i - \overline{B}\right)^2}, \text{ the standard error of } \overline{B},\ \widehat{\sigma} = \frac{S}{\sqrt{n}} \tag{5.8}$$

After re-sampling of the data by Bootstrap procedures, it is believed that data follows normal distribution.

5.8 Fuzzy inference System

Firstly, water quality experts are identified and relevant field data is collected. Additional data generation is a logical step if the available data is inadequate for analysis. Perception of experts about the linguistic description of river water quality for bathing is obtained on interviewing or through a questionnaire. Modeling of uncertainty in the expert's perception by constructing fuzzy sets/ fuzzy number fuzzy numbers and the uncertainty in the field data of water quality parameters using the concept of convex normalized fuzzy number is the next step. The parameters identified for defining bathing water quality by the experts are: Faecal Coliforms (*FC*), Dissolved Oxygen (*DO*), Bio-chemical Oxygen Demand (*BOD*), *pH* and *Turbidity*. The relevant parameters could be considered while over all describing the air quality of a city/region.

Randomness in the water quality data can be transformed into a convex normalized fuzzy number A with membership grade function $\mu_A(x)$ thereby characterizing the dynamic behavior of the water quality parameters can be transformed. If x_i is some point on the parametric domain for which $p(x_j)$ is maximum, then define function $\mu_A(x)$ as [2]

$$\mu_A(x) = \frac{p(x)}{p(x_j)} \tag{5.9}$$

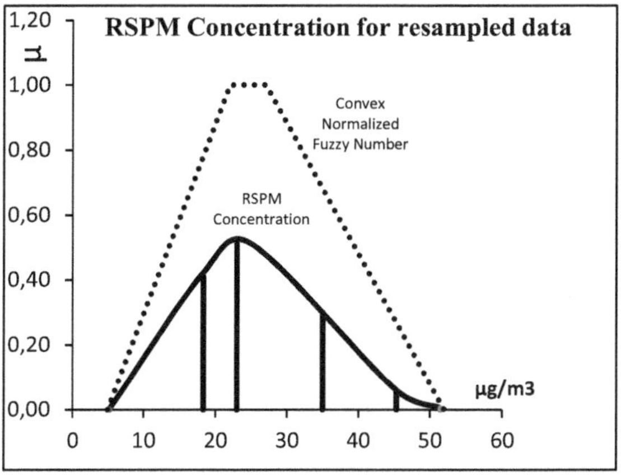

Fig. 5.3 Probability and Possibility Distribution for Respirable Suspended Particulate Matter (RSPM)

Construction of fuzzy number or fuzzy sets for modeling the perception of the experts in classifying each parametric domain in linguistic terms such as *very good, good*, etc. which allows for referencing all possible parametric values to be described. This transforms a random variable into a convex normalized fuzzy number A with membership grade function $\mu_A(x)$, thereby characterizing the dynamic behavior of the water quality parameter. The construction of fuzzy number or fuzzy sets for modeling the perception of the experts in classifying each parametric domain linguistically, involves: Selection of linguistic terms such as *very good, good*, etc. which allows for referencing all possible parametric values to be described; Classification of the parametric domain and assigning linguistic terms to each class linearly by the experts reflecting the imprecision in their perception; the set of values for which all the experts assign the same linguistic term are given $\mu = 1.0$ while none of the expert assigning that term are given $\mu = 0.0$. The break even point membership grade 0.0 and 1.0 are connected by continuous monotonic function which presupposes that the degree of consensus amongst the experts goes on increasing as the parametric values approach the core of fuzzy number for the specified linguistic term.

5.9 Matching between two fuzzy values

The fuzzy number for field data (A) on parameters and the fuzzy number characterizing linguistic terms (A') are matched together to arrive at a measure called Degree of match

(DM). defined by:

$$\mathrm{DM}_{\mathrm{ff}}(A,A') = \frac{f\,\mu_{A\cap A'}(x)\,dx}{f\,\mu_A(x)\,dx}, \quad x \subset X \tag{5.10}$$

in which X denotes the universe, and $\mu_{A\cap A'}(x)$ is membership grade for $A\cap A'$. Furthermore, if A and A' are the discrete possibility distributions the measure is defined as [2]:

$$\mathrm{DM}_{\mathrm{ff}}(A,A') = \frac{\sum \mu_{A\cap A'}(x)}{\sum \mu_A(x)}, \quad x \subset X \tag{5.11}$$

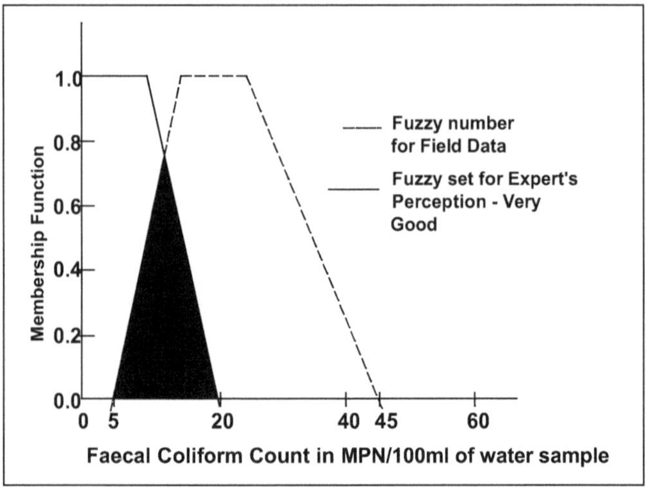

Fig. 5.4 Fuzzy number for very good faecal Coliforms

Fig. 5.4 shows the fuzzy number for *very good faecal Coliforms* reveals that almost all the experts agree that the *faecal coliforms* count between 0-10MPN/100ml of water sample is *very good* for which the indicates the level of presumption $\imath = 1$.

The level of presumption or membership function decreases with the increasing faecal coliforms count. When the count exceeds 20, none of the experts define the parameters as *very good* for bathing purpose. This is indicated by the level of presumption $\imath = 0$.

A set of rules is constructed for classifying air/ water quality as highly acceptable, just acceptable, not acceptable (rejected) in order to aggregate the set of attributes. (Fig. 5.3) Each rule has antecedent propositions connected together using AND operator, resulting in some consequences. The assertions related to its antecedent part are obtained from the users, which are imprecise or fuzzy. Thus a fuzzy rule based system can be developed for the knowledge representation or reasoning process. Here the partial matching is allowed and the analyst can estimate the extent to which the assertion satisfies the antecedent part of the rule contrary to the rule-base system which examines as to whether the antecedent part is satisfied or not [4]. A hierarchical structure for water classification resulting in a set of rules can be constructed (Fig. 5.2). The chemical status of water is judged in the first hierarchical level of knowledge base. The second hierarchical level characterizes bacteriological, chemical and physical status of water to arrive at the ultimate acceptable strategy of water quality for bathing purpose. If needed be, a similar structure can be developed for air quality classification.

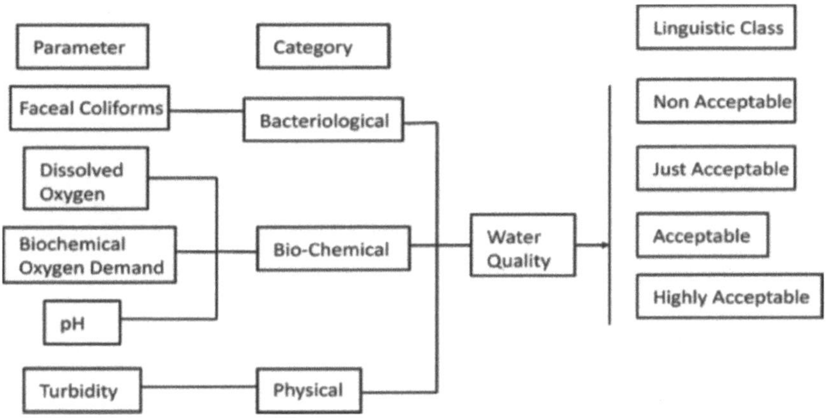

Fig. 5.5 Type 1 Fuzzy Inference System / Fuzzy Expert System

Following are the sample rules stored at two different hierarchical levels of the knowledge base:

Rule 1

If DO is ⟨fair⟩ and

BOD is ⟨good⟩ and

pH is ⟨very good⟩

Then chemical status of water is ⟨good⟩.

The rule at the next level could be

Rule 2

If bacteriological status of water is ⟨fair⟩ and

Chemical status of water is ⟨good⟩ and

Table 5.1 Focal elements and the combined belief measures for water borne diseases

Focal element (A_i)	Expert-1		Expert-2		Combined evidence	
	$m_1(A_i)$	$\text{bel}_1(A_i)$	$m_2(A_i)$	$\text{bel}_2(A_i)$	$m_{12}(A_i)$	$\text{bel}_{12}(A_i)$
Diarrhoea (a)	0.41	0.41	0.39	0.39	0.58	0.58
Cholera (b)	0.04	0.04	0.08	0.08	0.02	0.02
Giardiasis(c)	0.15	0.15	0.16	0.16	0.10	0.10
Diarrhoea and cholera ($a \cup b$)	0.03	0.43	0.01	0.46	0.00	0.60
Diarrhoea and giardiasis ($a \cup c$)	0.02	0.54	0.02	0.54	0.00	0.68
Diarrhoea, cholera and giardiasis ($a \cup b \cup c$)	0.01	0.59	0.04	0.59	0.00	0.71
Others (d)	0.34	0.34	0.03	0.30	0.28	0.28

Physical status of water is $\langle$*fair*$\rangle$

Then Water Quality for bathing is $\langle$just acceptable$\rangle$ (5.12)

5.10 Case Studies

5.10.1 *Case Study 1*

Around 21% of communicable diseases in India are water related, and out of these, diarrhoeal diseases alone killed over 700,000 Indians in the year 1999 [1]. Since time immemorial, sacred bathing on the holy river banks (*ghats*) is practised in India and Nepal. It has been a religious belief of some of the bathers that gulping water at these holy places washes away their sins! The first study, therefore, relates to establishing the cause – effect relationship between bathing in polluted river water and water borne diseases which were carried out at Alandi town near Pune situated on the banks of river Indrayani. We very briefly discuss the final outcome of such an investigation [1]. Table 5.1 present summarises the results wherein of the combined belief of the two domain experts in identifying for the single disease *Diarrheas as* 0.58. The study result infers considerable ambiguity in the belief of the identified two experts. In sum bathing in polluted water can be a cause for the incidence of water borne diseases's in bathers [1].

5.10.2 *Case Study 2*

The present study relates to fuzzy water quality description with the available water quality data from three sampling stations along the Ganga-one of the important rivers in

Table 5.2 Degree of Match at the Sampling Locations

Sampling Location	Parameter	Linguistic Class			
		Very Good	*Good*	*Fair*	*Poor*
Rishikesh	*Faecal Coliforms*	0.21	**0.8**	0.36	0
	Dissolved Oxygen	**0.45**	0.16	0	0
	BOD	0.13	**0.58**	0.4	0.02
	PH	**0.41**	0.3	0.23	0.1
	Turbidity	**0.98**	0.03	0	0
	Faecal Coliforms	0.13	**0.6**	0.4	0
	Dissolved Oxygen	**0.53**	0.08	0.02	0
	Dissolved Oxygen	**0.16**	0	0	0
Varanasi	*BOD*	**0.5**	0.33	0.12	0
	PH	**0.68**	0.47	0.29	0.06
	Turbidity	**0.68**	0.51	0	0

India from religious viewpoint. These locations are: Rishikesh, and Varanasi where the pollution is excessive. The water quality in Ganga has progressively deteriorated due to indiscriminate discharge of municipal sewage and the industrial effluents from the factories along the river. The Government of India had embarked upon the Ganga Action Plan (GAP) with the primary objective of water quality improvement for the various uses. It was, therefore, prudent to evaluate the river water quality at various locations on implementation of the pollution abatement measures. Fifteen water quality experts (including the first author of this paper) were identified and their linguistic description on the respective parametric space of the above parameters was obtained for bathing purpose. Table 5.2 presents the degree of match of field data with the fuzzy terms at the above sampling stations [5].

At Rishikesh, the parameters are described either very good or good, indicating a high DM associated with these terms .It can be inferred that the DM for *faecal coliforms* progressively decreases from Rishikesh to Varanasi indicating phenomenal increase in organic pollution load. The DM computed are the inputs to the fuzzy rule-based system in which the rules are processed resulting in water quality classification of Ganga water for bathing purpose.

According to Fig. 5.5, first the water quality for defining the bio-chemical quality of water are first processed as the first hierarchical structure of the fuzzy rule base using min operator. The outcome of the computation was further processed in conjunction with the rules for faecal coliforms. The computation was continued with the rules for the physical parameter and the final results are obtained (Table 5.3).

Table 5.3 Fuzzy Description of River Water Quality with Degree of Certainty

Water Quality Description	Degree of Certanity	
	Rishikesh	Varanasi
Highly Accepted	0.21	0
Accepted	**0.41**	0
Just Accepted	0.36	0.01
Not Accepted	0.1	**0.98**

It could be stated that the water quality at Rishikesh is *acceptable* with the highest certainty value of 0.41. The next higher certainty value of 0.36 for *just acceptable* indicates that the experts perception is more towards the linguistic term just acceptable than towards highly acceptable. The logical conclusion based on the analysis is that the water quality at *Varanasi for bathing is not acceptable or rejected with certainty value of 0.98.* The study was also carried out at Varanasi in the 25 km stretch of the river Ganga concludes that there is no safe bathing place from the health viewpoint at Varanasi.

5.10.3 *Case Study 3*

The case study relates to fuzzy air quality description with the available air quality data [4] from five monitoring stations in Pune city. These are Pimpri-Chinchwad Municipal Corporation, Karve Raod, Swargate, Bhosari and Nal stop.

In order to bring down progressive deterioration in air quality, the Government of India has enacted Air (Prevention and Control of Pollution) Act in 1981 and further stringent guidelines are promulgated in Environment (Protection) Act, 1986. The need for Ambient air quality monitoring program are needed to determine the existing quality of air, evaluation of the effectiveness of control program and to develop new program was recognised As a follow up action, the Central Pollution Control Board (CPCB) initiated National Air Quality Monitoring (NAQM) Program during 1984-85 at the national level [4].

A well structured air quality monitoring network involves selection of pollutants, selection of locations, frequency, duration of sampling, sampling techniques, infrastructural facilities, man power and operation, maintenance costs, and alike. The network design also depends upon the type of pollutants in the atmosphere through various common sources, called common urban air pollutants, such as Suspended Particulate Matter (SPM), Respirable Suspended Particulate Matter (RSPM), Sulphur Dioxide (SO_2), Oxides of Nitrogen (NO_x), and Carbon Monoxide (CO) etc. The areas to be chosen primarily are such areas which represent high traffic density, industrial growth, human population and its distribution, emission source, public complaints if any and the land use pattern etc. Generally, most of the times the basis of a network design are the pollution source and the pollution present.

Bootstrap method is applied on the original sample which contained approximately 25 to 30 values per month for every parameter (RSPM, NO_x, and SO_2). 1000 bootstrap samples were taken using MATLAB 7.0 software. The mean for the original and bootstrap samples were computed. The month with minimum difference value in the standard error of the original sample and the bootstrap sample was selected for further computation The highlighted row for each parameter is the month selected for the winter season which represents the worst case scenario of the winter season. The worst of the worst months of the winter monthts selected is the one with the minimum standard error amongst the four months of the winter season i.e. October to January [5].

Figure 5.6 is the typical fuzzy numbers for different linguistic hedges (*very good, good, fair, poor, very poor*) of RSPM concentration. According to the expert, RSPM count between 18-57 μg/m^3 is Very Good, 42-75 μg/m^3 is Good, 67-100 μg/m^3 is Fair, 90-130 μg/m^3 is Poor μg/m^3 and above 120 μg/m^3 it is Very Poor The level of membership function decreases with the increasing RSPM level. When it exceeds 57 the expert does not define the parameter as Very Good [4].

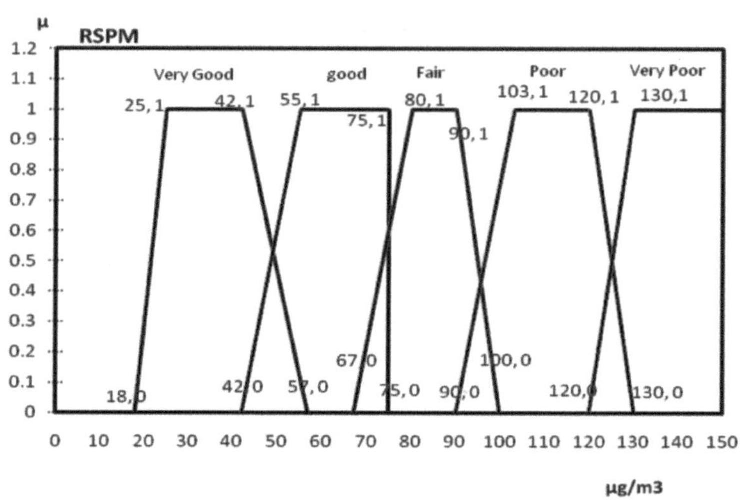

Fig. 5.6 Fuzzy number for RSPM

Table 5.4 Fuzzy Description of Air Quality with Degree of certainty

Air Quality Description	Degree of Certainty (DC)				
	PCMC	Karve Road	Swargate	Nal Stop	Bhosari
Very Good	0.00	0.00	0.00	0.00	0.00
Good	0.00	0.00	0.00	0.00	0.00
Fair	0.00	0.00	0.11	0.00	0.67
Poor	0.91	0.00	0.54	0.10	0.08
Very Poor	0.18	0.45	0.11	0.43	0.00

Table 5.5 Comparison of Conventional AQI, and Fuzzy Description of Air Quality with Degree of Certainty

Monitoring Station	Conventional AQI	AQI with Degree of Certainty
Karve Road	206 Very Poor	0.45 very poor
Bhosari	145.89 Poor	0.43 very poor
Swargate	147.13 Poor	0.54 poor
Nal Stop	120.2 Poor	0.67 fair
PCMC	151.13 Very Poor	0.91 poor

Table 5.4 reveals that the air quality at Pimpri Chinchwad (PCMC) is *Poor* with the highest certainty value of 0.91. The next higher value of 0.18 for *Very Poor* indicates that the expert's perception is more towards the linguistic term Very *Poor* than towards *Fair*. We can also conclude that the air quality could be generally described as very poor or poor at the five monitoring stations which are due high concentration of RSPM pollutant in the air because of the vehicular pollution. It is necessary to contain auto exhaust pollution by reducing the number of vehicular traffic and/ or the use of pollution free vehicles.

Table 5.5 is the comparison between the computed AQI and the proposed fuzzy logic based method. It can be revealed that AQI based on the traditional method does not attach any certainty while describing the air quality. In addition, the method does not consider the aggregated effect and the highest computed AQI is considered as the final decision on the air quality which, in our opinion, is the departure from human thinking. Alternately, using Type 1 Fuzzy Inference system, we can describe the air quality straightway in linguistic terms with some degree of certainty attached to each term.

5.10.4 *Concluding Remarks*

Air Quality Index is a tool to inform the public of the results of air pollutant measurements and is being used extensively. Though agreed in principle, many countries are engaged in developing their own technique- in the present case AQI. However, in view of the uncertainties discussed above, the approach outlined in this paper using fuzzy logic could be a better representation of a dynamic system, and thereby providing a new dimension of monitoring air quality.

5.10.5 *Outlook in environmental policy* [3]

Over the past few decades, soft computing tools such as fuzzy-logic-based methods, neural networks, and genetic algorithms have had significant and growing impacts. But we have seen only limited use of these methods in environmental fields, such as risk assessment, cost-benefit analysis, and life-cycle impact assessment. Because fuzzy methods offer both new opportunities and unforeseen problems relative to current methods, it is difficult to determine how much impact such methods will have on environmental policies in the coming decades. Here, we consider some obvious advantages and limitations.

Quantitative models with explicit and crisp delineations of systems have long been the currency of discourse in engineering and the physical sciences, where basic physical laws form the foundations of analyses. These fields place high value on the causal linkages implicit in model structure and parameterization. But for problems that involve human values, language, control theory, biology, and even environmental systems, researchers have had to rely more on descriptive and empirical approaches. When the goal is to summarize the observations in an efficient and useful manner, fuzzy-logic-based methods should be further investigated as alternative—and perhaps more appropriate—methods for addressing uncertain and complex systems. For the types of complex and imprecise problems that arise in environmental policy, the ability to model complex behaviors as a collection of simple if–then rules makes fuzzy logic an appropriate modeling tool. Because fuzzy arithmetic works well for addressing linguistic variables and poorly characterized parameters, fuzzy methods offer the opportunity to evaluate and communicate assessments on the basis of linguistic terms that could possibly match those of decision makers and the public. Moreover, approximate reasoning methods such as fuzzy arithmetic do not require well-characterized statistical distributions as inputs. Another key advantage of fuzzy logic in risk assessment is the ability to merge multiple objectives with different values and meanings, for example, combining health objectives with aesthetic objectives. It also provides rules for combining qualitative and quantitative objectives (5.5).

But fuzzy logic has at least two limitations for expressing health risks and other environmental impacts. One problem is its strong reliance on subjective inputs. Although this is a problem in any type of assessment, fuzzy methods might provide more opportunity for the misuse of subjective inputs. Although probabilistic assessments based on tools such as Monte Carlo methods are analogous to assessments based on fuzzy logic, these two techniques differ significantly both in approach and in interpretation of results. Fuzzy logic confronts linguistic variables such as "safe", *"hazardous"*, *"acceptable"*, and *"unacceptable"*, whereas Monte Carlo methods are forced to fit linguistic variables for probabilistic assessments. Fuzzy arithmetic combines outcomes from different sets in a way that is analogous to but still different from Monte Carlo methods. Possibility theory can be used as an alternative to probabilistic analysis, but this strategy creates the potential for misuse if membership functions are interpreted as probability distributions.

5.10.6 *No more crisp lines?*

Fuzzy logic represents a significant change in both the approach to and the outcome of environmental evaluations. Currently, risk assessment implicitly assumes that probability theory provides the necessary and sufficient tools for dealing with uncertainty and variability. The key advantage of fuzzy methods is how they reflect the human mind in its remarkable ability to store and process information that is consistently imprecise, uncertain, and resistant to classification. Our case study illustrates the ability of fuzzy logic to integrate statistical measurements with imprecise health goals. But we submit that fuzzy logic and probability theory are complementary and not competitive. In the world of soft computing, fuzzy logic has been widely used and has often been the "smart" behind smart machines. But more effort and further case studies will be required to establish its niche in risk assessment and other types of impact assessment.

Could we adapt to a system that relaxes *"crisp lines"* and sharp demarcations to fuzzy gradations? Would decision makers and the public accept expressions of water- or air-quality goals in linguistic terms with computed degrees of certainty? Resistance is likely. In many regions, such as the United States and EU, both decision makers and members of the public seem more comfortable with the current system—in which government agencies avoid confronting uncertainties by setting guidelines that are crisp and often fail to communicate uncertainty. Perhaps someday a more comprehensive approach that includes exposure surveys, toxicological data, and epidemiological studies coupled with fuzzy modelling will go a long way toward resolving some of the conflict, divisiveness, and controversy in the current regulatory paradigm [3].

Acknowledgement

The author would like to express immense gratitude towards Professor Lotfi Zadeh, the father of fuzzy logic for the motivation. The wholehearted assistance received from D.V. Raje, Kedar Rijal, and Jyoti Yadav for the implementation of the concept developed by the author is gratefully acknowledged. My special thanks to Professor. Thomas McKone who helped the author in many fuzzy ways!

Bibliography

[1] K. Rijal, A. Deshpande, and V. Ghole, Bathing in polluted rivers, Water-Borne Diseases, and fuzzy measures: a case study in India, *Int. J. Environment and Waste Management*, 2009.

[2] Deshpande, A. W, Raje, D.V and Khanna, P; Fuzzy Description of River Water Quality, paper for the International Conference-EUFIT 1996.

[3] E. Mckone Thomas and A.W. Deshpande, Can Fuzzy logic Bring Complex Environmental Problems into Focus? *International Journal of Environmental Science and Technology*, January 15, 2005.

[4] Web site: (`http://mpcb.gov.in`)

[5] J. Yadav, V. Kharat, and A. Deshpande, Fuzzy Description of Air Quality: A Case Study presented in the International Conference on Rough Set and Knowledge Technology 2011 in Banff, Canada.

[6] R.R. Yager, J. Kacprzyk, and M. Fedrizzi, (1994), *Advance in the Dempster-Shafer Theory of Evidence*, John Wiley & Sons, New York.

Chapter 6

TOPSIS and its Extensions: Applications for Road Safety Performance Evaluation

Qiong Bao[1], Da Ruan[1,2], Yongjun Shen[1], Elke Hermans[1], and Davy Janssens[1]

[1] *Transportation Research Institute – Hasselt University, Wetenschapspark 5 bus 6, 3590 Diepenbeek, Belgium*
[2] *Belgian Nuclear Research Centre (SCK·CEN), Boeretang 200, 2400 Mol, Belgium*

E-mail: {qiong.bao, da.ruan, yongjun.shen, elke.hermans, davy.janssens}@uhasselt.be, druan@sckcen.be

Technique for Order Preference by Similarity to Ideal Solution (TOPSIS) is one of the well-known classical multi-criteria decision making (MCDM) techniques. In this chapter we illustrate the use of this method to combine individual safety performance indicators (SPIs) into an overall index of road safety performance for a set of European countries. In this respect, to deal with the subjective kind of uncertainty on data (such as linguistic variables given by experts) which are usually adopted to assess the weights of criteria/indicators, we explore an extension of the classical TOPSIS method to fuzzy environments. Moreover, due to the ever increasing number of SPIs used to reflect each road safety risk factor in a more comprehensive way, we consider a hierarchical structure of the indicators in this study. Accordingly, a hierarchical fuzzy TOPSIS model is realized and applied to combine the multilayer indicators into one overall index. Comparison of the results based on the three models (i.e., the classical TOPSIS, the fuzzy TOPSIS, and the hierarchical fuzzy TOPSIS) demonstrates the effectiveness of applying the hierarchical fuzzy TOPSIS method to handle the problem of linguistic expression instead of crisp values given by experts, and to take the layered hierarchy of the indicators into account which is seldom considered in the current road safety index research.

6.1 Introduction

Performance management systems ranging from economy, education, to transportation, which are seldom assessed nowadays with a single, simple objective or criterion, have resulted in more and more complex performance evaluation. Traditional single criterion decision making or evaluation aimed at maximization of benefits with minimization of

C. Kahraman (ed.), *Computational Intelligence Systems in Industrial Engineering*, Atlantis Computational Intelligence Systems 6, DOI: 10.2991/978-94-91216-77-0_6,

costs therefore no longer satisfies the need of effective management. For instance, with the ever increasing public awareness of the complexity of road safety phenomenon, more detailed aspects of crash and injury causation rather than only crash data (e.g., the number of road fatalities) are extensively investigated in the current road safety management. In this respect, safety performance indicators (SPIs), which are causally related to the number of crashes or to the injury consequences of a crash, are rapidly developed and increasingly used for the road safety performance evaluation at local, regional, and national levels. Moreover, to measure the multi-dimensional concept which cannot be captured by a single indicator, the exploration of a comprehensive composite index is attractive. The index can thus present an overall picture by capturing a multitude of information in one index score, and offers advantages in terms of communication, benchmarking, and decision making.

From a purely mathematical point of view, the aggregation convention used for composite indexes deals with the classical conflictual situation tackled in multi-criteria evaluation. Thus, the use of a multi-criteria framework for composite indexes in different performance management systems is relevant and desirable [1–5]. Amongst others, the multi-criteria decision making (MCDM) is one of the most important techniques, which offers the methodology for decision making analysis when dealing with multiple objectives under the presence of a number of conflicting decision criteria [6]. Over the years, many MCDM approaches have been developed, such as AHP, MAUT, ELECTRE, PROMETHEE, and TOPSIS, focusing on issues such as: analyzing and evaluating incompatible criteria and alternatives; modeling decision makers' preferences; developing MCDM-based decision support systems; and identifying compromising solutions for decision making and performance evaluation problems.

In this chapter, we investigate the TOPSIS (Technique for Order Preference by Similarity to Ideal Solution) method for the construction of composite indexes in road safety study. Moreover, to deal with subjective kinds of uncertainty on data, such as linguistic variables about indicator weights given by experts, we explore an extension of the classical TOPSIS method to fuzzy environments. Furthermore, since the collection of additional data on performance indicators is an ongoing process, more and more indicators are expected to be available and thereby their hierarchical structures should also be taken into account. To this end, we develop a hierarchical fuzzy TOPSIS model to combine multilayer indicators into one overall index. In the applications, we illustrate how the three models (the classical TOPSIS, the fuzzy TOPSIS, and the hierarchical fuzzy TOPSIS) can be used to evaluate the road safety performance of a set of European countries. The results are then compared using the number of road fatalities per million inhabitants as a relevant point of reference.

The remaining of this chapter is structured as follows. In Section 6.2, we elaborate the classical TOPSIS method and its two extensions, i.e., the fuzzy TOPSIS method and the hierarchical fuzzy TOPSIS method. In Section 6.3, we present the applications of these three methods in road safety performance evaluation problems, and the corresponding results are compared subsequently. Finally, we conclude the chapter in Section 6.4.

6.2 The TOPSIS Method and its Extensions

6.2.1 *Classical TOPSIS*

The TOPSIS method, as one of the well-known classical MCDM methods, was first developed by Hwang and Yoon (1981) [7]. It bases upon the concept that the chosen alternative should have the shortest distance from the positive-ideal solution (PIS) and the farthest distance from the negative-ideal solution (NIS), in which the PIS is formed as a composite of the best performance values exhibited in the decision matrix by any alternative for each criterion, and the NIS is the composite of the worst performance values. Proximity to each of these performance poles is measured in the Euclidean sense (e.g., square root of the sum of the squared distances along each axis in the 'criterion space'), with optional weighting of each criterion. The main procedure of the classical TOPSIS method can be described in the following seven steps [6].

Step 1. Identify a decision matrix

To obtain the performance of a set of alternatives on a given set of criteria, a decision table or matrix, **D**, of $m \times n$ dimensions is constructed consisting of: a) alternatives A_i ($i = 1, 2, \ldots, m$), b) criteria C_j ($j = 1, 2, \ldots, n$), and c) measure of performance x_{ij} ($i = 1, 2, \ldots, m$; $j = 1, 2, \ldots, n$) indicating the performance of the alternative A_i with respect to the criterion C_j. Given the decision matrix information and a decision-making method, the task of the decision maker is to find the best alternative and/or to rank the entire set of alternatives.

$$
\mathbf{D} = \begin{array}{c} \\ A_1 \\ A_2 \\ \vdots \\ A_m \end{array}
\begin{array}{c} \begin{array}{cccc} C_1 & C_2 & \cdots & C_n \end{array} \\
\begin{bmatrix} x_{11} & x_{12} & \cdots & x_{1n} \\ x_{21} & x_{22} & \cdots & x_{2n} \\ \vdots & \vdots & \ddots & \vdots \\ x_{m1} & x_{m2} & \cdots & x_{mn} \end{bmatrix} \end{array}
\tag{6.1}
$$

Step 2. Normalize the decision matrix

All the elements in the decision matrix must be normalized to the same units, so that all possible criteria in the decision problem can be considered simultaneously. Here, conversion of the decision matrix to a dimensionless matrix ($\mathbf{D}'$) is realized by using a linear scale transformation as follows:

$$r_{ij} = \begin{cases} \dfrac{x_{ij}}{x_j^*}, & \forall j,\ x_j \text{ is a benefit criterion} \\ \dfrac{x_j^-}{x_{ij}}, & \forall j,\ x_j \text{ is a cost criterion} \end{cases} \tag{6.2}$$

where the r_{ij} are normalized values. x_j^* and x_j^- are the maximum and minimum values of the columns in the decision matrix $\mathbf{D}$, respectively. By applying (6.2), we can rewrite the decision matrix as:

$$\mathbf{D}' = \begin{bmatrix} r_{11} & \cdots & r_{1j} & \cdots & r_{1n} \\ \vdots & & \vdots & & \vdots \\ r_{i1} & \cdots & r_{ij} & \cdots & r_{in} \\ \vdots & & \vdots & & \vdots \\ r_{m1} & \cdots & r_{mj} & \cdots & r_{mn} \end{bmatrix} \tag{6.3}$$

Step 3. Compute the weighted normalized decision matrix

By determining the weights of relative performance of the decision criteria which are denoted as w_j ($j = 1, 2, \ldots, n$), we obtain the weighted normalized decision matrix by multiplying the normalized decision matrix $\mathbf{D}'$ with the weight vector W. The elements of the weighted normalized matrix $\mathbf{V}_{ij}$ are expressed as:

$$v_{ij} = r_{ij} w_j,\ \forall j \tag{6.4}$$

In doing so, determination of the weight vector W, or the relative importance of each of the criteria is critical and several approaches exist. In this study, the geometric mean method is applied to combine different individual weights given by the group decision makers into a single collective weight for each corresponding criterion. For more information about weight determination we refer to Choo and Wedley (1985), Saaty (1995), Nardo *et al.* (2005) [8–10].

Step 4. Identify the positive ideal solution (PIS), $A*$, and the negative ideal solution (NIS), A^-

The PIS and NIS are defined as:

$$A^* = [v_1^*, \ldots, v_n^*] \tag{6.5}$$

$$A^- = [v_1^-, \ldots, v_n^-] \tag{6.6}$$

where $v_j^* = \max_i v_{ij}$ and $v_j^- = \min_i v_{ij}$.

Step 5. Obtain the separation measures S_i^* and S_i^-

By identifying the PIS and the NIS, we obtain the separation measures as:

$$S_i^* = \sum_{j=1}^{n} D_{ij}^*, \quad i = 1, \ldots, n \tag{6.7}$$

$$S_i^- = \sum_{j=1}^{n} D_{ij}^-, \quad i = 1, \ldots, n \tag{6.8}$$

where the difference Euclidean distance D_{ij}^* and D_{ij}^- are given as:

$$D_{ij}^* = \left| v_{ij} - v_j^* \right| \tag{6.9}$$

$$D_{ij}^- = \left| v_{ij} - v_j^- \right| \tag{6.10}$$

Step 6. Compute the relative closeness to ideals

The relative closeness index is used to combine S_i^* and S_i^- indices calculated in Step 5, which is calculated as follows:

$$C_i = \frac{S_i^-}{S_i^* + S_i^-} \tag{6.11}$$

Step 7. Prioritize alternatives

According to the index value C_i, the set of alternatives can be ranked from the most preferred to the least preferred feasible solutions. C_i may also be called the overall or composite performance score of alternative A_i.

In general, the construction process of the classical TOPSIS method is transparent, which makes it easily understood by the general public and can be used to support desired policy. During the past decades, numerous applications of the TOPSIS method exist and have verified the feasibility of using this method to solve various MCDM problems [11–13].

6.2.2 *Fuzzy TOPSIS*

The classical TOPSIS method possesses advantages in that it is easy to compute and easily understood, because the method directly produces a definite index score based on the given indicator values and the specific weights. However, in most real world contexts, MCDM problems at tactical and strategic levels often involve fuzziness in their criteria and decision makers' judgments. Under these conditions, crisp data are inadequate or inappropriate to model real-life situations because of the complexity and constraints of the reality. For example, due to the uncertainty of human cognition and vague judgment, linguistic assessments rather than crisp numerical values are usually given by decision makers

Table 6.1 Linguistic terms and related fuzzy numbers for criteria values

Linguistic **terms**	Fuzzy numbers
Very low (VL)	$\tilde{\beta}_1$
Low (L)	$\tilde{\beta}_2$
Medium low (ML)	$\tilde{\beta}_3$
Medium (M)	$\tilde{\beta}_4$
Medium high (MH)	$\tilde{\beta}_5$
High (H)	$\tilde{\beta}_6$
Very high (VH)	$\tilde{\beta}_7$

Table 6.2 Linguistic terms and related fuzzy numbers for the criteria weights

Linguistic terms	Fuzzy numbers
Absolutely unimportant	$\tilde{\alpha}_1$
Unimportant	$\tilde{\alpha}_2$
Less important	$\tilde{\alpha}_3$
Important	$\tilde{\alpha}_4$
More important	$\tilde{\alpha}_5$
Strongly important	$\tilde{\alpha}_6$
Absolutely important	$\tilde{\alpha}_7$

or experts. Thus, to express the weight of a criterion, terms such as 'important,' 'very important' are commonly used. The concept of linguistic variables is useful in reflecting the uncertainty, inaccuracy and fuzziness of decision makers, and they are regarded as natural representations of preferences/judgments. However, precise mathematical approaches are not enough to tackle such uncertain variables and derive a satisfactory solution. As a result, the application of the classical TOPSIS method may face serious practical problems. During the last two decades, an extension of the classical TOPSIS to the fuzzy environment was widely investigated and a large number of fuzzy TOPSIS methods were developed in the literature [14–16]. In general, they embody the fuzzy nature of the comparison or evaluation process and strengthen the comprehensiveness and rationality of the decision-making process. Compared to the main procedure of the classical TOPSIS method, the main adjustments in applying fuzzy TOPSIS method are presented as follows.

In **Step 1**, since the performance values of criteria and/or their importance weights are now expressed as linguistic variables, they can be expressed in positive triangular fuzzy numbers shown in Table 6.1 and Table 6.2, respectively.

Assume that a decision group has K persons, then the importance of the criteria and the rating of alternatives with respect to each criterion can be calculated based on the geometric mean method expressed as follows:

$$\tilde{x}_{ij} = \sqrt[K]{\tilde{x}_{ij}^{1}(\cdot)\tilde{x}_{ij}^{2}(\cdot)\cdots(\cdot)\tilde{x}_{ij}^{K}} \tag{6.12}$$

$$\tilde{w}_{ij} = \sqrt[K]{\tilde{w}_{ij}^{1}(\cdot)\tilde{w}_{ij}^{2}(\cdot)\cdots(\cdot)\tilde{w}_{ij}^{K}} \tag{6.13}$$

where $\tilde{x}_{ij}^{K}$ and $\tilde{w}_{ij}^{K}$ are the rating and the weight of the Kth decision maker. $(\cdot)$ denotes a fuzzy multiplication operation.

Thus, a fuzzy MCDM problem can be concisely expressed in a matrix format as:

$$\widetilde{\mathbf{D}} = \begin{array}{c} \\ A_1 \\ A_2 \\ \vdots \\ A_m \end{array} \begin{array}{c} \begin{array}{cccc} C_1 & C_2 & \ldots & C_n \end{array} \\ \begin{bmatrix} \tilde{x}_{11} & \tilde{x}_{12} & \ldots & \tilde{x}_{1n} \\ \tilde{x}_{21} & \tilde{x}_{22} & \ldots & \tilde{x}_{2n} \\ \vdots & \vdots & \ddots & \vdots \\ \tilde{x}_{m1} & \tilde{x}_{m2} & \ldots & \tilde{x}_{mn} \end{bmatrix} \end{array} \tag{6.14}$$

$$\widetilde{W} = \begin{bmatrix} \tilde{w}_1 & \tilde{w}_2 & \ldots & \tilde{w}_n \end{bmatrix}$$

where values $\tilde{x}_{ij}$, $\forall i, j$ and weights $\tilde{w}_j$ $(j = 1,2,\ldots,n)$ are linguistic variables that are described by triangular fuzzy numbers, $\tilde{x}_{ij} = (a_{ij}, b_{ij}, c_{ij}) \in (\tilde{\beta}_1, \tilde{\beta}_2, \ldots, \tilde{\beta}_7)$ and $\tilde{w}_j = (w_{j1}, w_{j2}, w_{j3}) \in (\tilde{\alpha}_1, \tilde{\alpha}_2, \ldots, \tilde{\alpha}_7)$.

In **Step 2**, the same linear scale transformation is used to convert the various criteria scales into a comparable scale. However, since $\tilde{x}_{ij}$ is fuzzy, its corresponding normalized value $\tilde{r}_{ij}$ must be fuzzy. (6.2) is then replaced by the following fuzzy operations:

$$\tilde{r}_{ij} = \begin{cases} \tilde{x}_{ij}(:)\tilde{x}_j^* = \left(\dfrac{a_{ij}}{c_j^*}, \dfrac{b_{ij}}{b_j^*}, \dfrac{c_{ij}}{a_j^*}\right), & \forall j,\ \tilde{x}_j \text{ is a benefit criterion} \\ \tilde{x}_j^-(:)\tilde{x}_{ij} = \left(\dfrac{a_j^-}{c_{ij}}, \dfrac{b_j^-}{b_{ij}}, \dfrac{c_j^-}{a_{ij}}\right), & \forall j,\ \tilde{x}_j \text{ is a cost criterion} \end{cases} \tag{6.15}$$

where $(:)$ denotes fuzzy division operation. $\tilde{x}_j^* = (a_j^*, b_j^*, c_j^*)$ and $\tilde{x}_j^- = (a_j^-, b_j^-, c_j^-)$ present the largest and the lowest values for each criterion, respectively.

In **Step 3**, considering the different importance weight of each criterion, we construct the weighted normalized fuzzy decision matrix as:

$$\tilde{v}_{ij} = \tilde{r}_{ij}(\cdot)\tilde{w}_j, \ \forall j \tag{6.16}$$

Since either $\tilde{r}_{ij}$ or $\tilde{w}_j$ or both of them are fuzzy numbers, (6.16) will be calculated by the following fuzzy operations:

$$\tilde{v}_{ij} = \tilde{r}_{ij}(\cdot)\tilde{w}_j = \begin{cases} \left(\dfrac{a_{ij}}{c_j^*}w_{j1}, \dfrac{b_{ij}}{b_j^*}w_{j2}, \dfrac{c_{ij}}{a_j^*}w_{j3}\right), & \forall j,\ \tilde{x}_j \text{ is a benefit criterion} \\ \left(\dfrac{a_j^-}{c_{ij}}w_{j1}, \dfrac{b_j^-}{b_{ij}}w_{j2}, \dfrac{c_j^-}{a_{ij}}w_{j3}\right), & \forall j,\ \tilde{x}_j \text{ is a cost criterion} \end{cases} \tag{6.17}$$

The result of (6.17) can be summarized as:

$$\widetilde{\mathbf{V}} = \begin{bmatrix} \tilde{v}_{11} & \cdots & \tilde{v}_{1j} & \cdots & \tilde{v}_{1n} \\ \vdots & & \vdots & & \vdots \\ \tilde{v}_{i1} & \cdots & \tilde{v}_{ij} & \cdots & \tilde{v}_{in} \\ \vdots & & \vdots & & \vdots \\ \tilde{v}_{m1} & \cdots & \tilde{v}_{mj} & \cdots & \tilde{v}_{mn} \end{bmatrix} \tag{6.18}$$

In **Step 4**, we now define the fuzzy positive ideal solution (FPIS), i.e., $\widetilde{A}^* = \left[\tilde{v}_1^*, \ldots, \tilde{v}_n^*\right]$ and fuzzy negative ideal solution (FNIS), i.e., $\widetilde{A}^- = \left[\tilde{v}_1^-, \ldots, \tilde{v}_n^-\right]$. The $\tilde{v}_j^*$ and $\tilde{v}_j^-$ are the fuzzy numbers with the largest generalized mean and the smallest generalized mean, respectively. The generalized mean for a fuzzy number $\tilde{v}_{ij} = (a_{ij}, b_{ij}, c_{ij})$, $\forall j$, is defined as:

$$M(\tilde{v}_{ij}) = \frac{-a_{ij}^2 + c_{ij}^2 - a_{ij}b_{ij} + b_{ij}c_{ij}}{[3(-a_{ij} + c_{ij})]} \tag{6.19}$$

For each column j, we find the greatest generalized mean as $\tilde{v}_j^*$ and the lowest generalized mean as $\tilde{v}_j^-$. Consequently, the FPIS ($\widetilde{A}^*$) and FNIS ($\widetilde{A}^-$) are obtained.

To acquire the Separation Measures $\widetilde{S}_i^*$ and $\widetilde{S}_i^-$ in **Step 5**, we compute the Euclidean distance $\widetilde{D}_{ij}^*$ and $\widetilde{D}_{ij}^-$ firstly. For fuzzy data, the difference between two fuzzy numbers $\mu_{v_{ij}}(x)$ and $\mu_{v_j^*}(x)$ (according to Zadeh (1965) [17]) is calculated as:

$$\widetilde{D}_{ij}^* = 1 - \left\{ \sup_x \left[\mu_{v_{ij}}(x)^\wedge \mu_{v_j^*}(x)\right] \right\} = 1 - L_{ij}, \quad \forall i, j \tag{6.20}$$

where L_{ij} is the highest degree of similarity of $\tilde{v}_{ij}$ and $\tilde{v}_j^*$. The value of L_{ij} is depicted in Fig. 6.1.

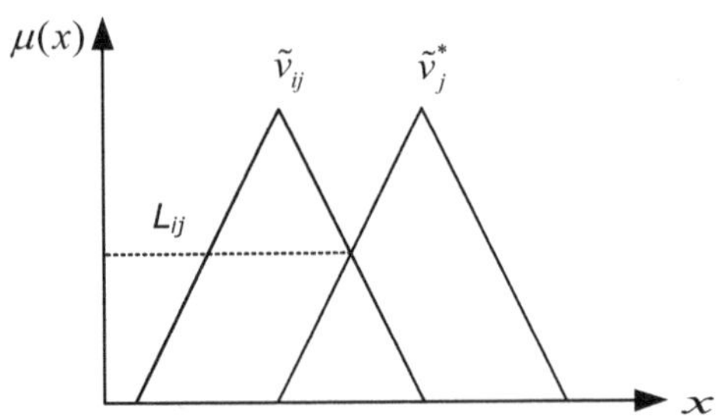

Fig. 6.1 The derivation of L_{ij}

Similarly, the difference between $\mu_{v_{ij}}(x)$ and $\mu_{v_j^-}(x)$ is defined as:

$$\widetilde{D}_{ij}^{-} = 1 - \left\{ \sup_{x} \left[\mu_{v_{ij}}(x)^{\wedge} \mu_{v_j^{-}}(x) \right] \right\} = 1 - L_{ij}, \ \forall i, j \tag{6.21}$$

More specifically, $\widetilde{D}_{ij}^{*}$ and $\widetilde{D}_{ij}^{-}$ are calculated as below, where $\tilde{v}_j^{*} = (a^*, b^*, c^*)$ and $\tilde{v}_j^{-} = (a^-, b^-, c^-)$ are the fuzzy numbers with the largest generalized mean and the smallest generalized mean, respectively.

$$\widetilde{D}_{ij}^{*} = \begin{cases} 1 - \dfrac{c_{ij} - a^*}{b^* + c_{ij} - a^* - b_{ij}} & \text{for } b_{ij} < b^* \\ 1 - \dfrac{c^* - a_{ij}}{b_{ij} + c^* - a_{ij} - b^*} & \text{for } b^* < b_{ij} \end{cases} \quad \forall i, j \tag{6.22}$$

$$\widetilde{D}_{ij}^{-} = \begin{cases} 1 - \dfrac{c^- - a_{ij}}{b_{ij} + c^- - a_{ij} - b^-} & \text{for } b^- < b_{ij} \\ 1 - \dfrac{c_{ij} - a^-}{b^- + c_{ij} - a^- - b_{ij}} & \text{for } b_{ij} < b^- \end{cases} \quad \forall i, j \tag{6.23}$$

Note that both $\widetilde{D}_{ij}^{*}$ and $\widetilde{D}_{ij}^{-}$ are crisp values now. Therefore, the Separation Measures S_i^* and S_i^- can be calculated according to (6.7) and (6.8) in Section 6.2.1, and the relative closeness index C_i obtained as (6.11). Accordingly, the set of alternatives can be ranked from the most preferred to the least preferred feasible solutions.

6.2.3 *Hierarchical fuzzy TOPSIS*

Due to the ever increasing complexity of today's performance evaluation and decision making activities, such as in the road safety context, not only the knowledge from the decision makers or experts, but also the information on the criteria or indicators themselves, e.g., their hierarchical structure, should be taken into account. Therefore, extension of the fuzzy TOPSIS method to a hierarchical one is valuable in completely reflecting the layered hierarchy of all criteria, and it can be treated as a natural generalization of the one-layer fuzzy TOPSIS [18]. The main procedure of the hierarchical fuzzy TOPSIS method is presented in the following steps:

Step 1. Consider a two-layer situation of m alternatives, each including n main criteria (*MC*), r sub-criteria (*SC*). Assume each main criterion has r_j sub-criteria, thereby the total number of sub-criteria r is equal to the sum of r_j $(j = 1, 2, \ldots, n)$. $\tilde{x}_{ijk}$ represents the value of the kth sub-criteria within the jth main criteria of the ith alternative, which can be crisp data or linguistic variables. If the latter, they can be further represented by triangular fuzzy numbers, i.e., $\tilde{x}_{ijk} = (a_{ijk}, b_{ijk}, c_{ijk})$. Thus, a hierarchical fuzzy MCDM problem can be expressed as follows:

$$\widetilde{\mathbf{D}} = \begin{array}{c} \\ \\ A_1 \\ A_2 \\ \vdots \\ A_m \end{array} \begin{array}{c} \qquad MC_1 \qquad\qquad\qquad MC_2 \qquad\qquad\qquad MC_n \\ SC_{11}\ SC_{12}\ \ldots\ SC_{1r_1}\ SC_{21}\ SC_{22}\ \ldots\ SC_{2r_1}\ \ldots\ldots\ SC_{n1}\ SC_{n2}\ \ldots\ SC_{nr_n} \\ \begin{bmatrix} \tilde{x}_{111} & \tilde{x}_{112} & \cdots & \tilde{x}_{11r_1} & \tilde{x}_{121} & \tilde{x}_{122} & \cdots & \tilde{x}_{12r_2} & \cdots\cdots & \tilde{x}_{1n1} & \tilde{x}_{1n2} & \cdots & \tilde{x}_{1nr_n} \\ \tilde{x}_{211} & \tilde{x}_{212} & \cdots & \tilde{x}_{21r_1} & \tilde{x}_{221} & \tilde{x}_{222} & \cdots & \tilde{x}_{22r_2} & \cdots\cdots & \tilde{x}_{2n1} & \tilde{x}_{2n2} & \cdots & \tilde{x}_{2nr_n} \\ \vdots & \vdots & \ddots & \vdots & \vdots & \vdots & \ddots & \vdots & \ddots & \vdots & \vdots & \ddots & \vdots \\ \tilde{x}_{m11} & \tilde{x}_{m12} & \cdots & \tilde{x}_{m1r_1} & \tilde{x}_{m21} & \tilde{x}_{m22} & \cdots & \tilde{x}_{m2r_2} & \cdots\cdots & \tilde{x}_{mn1} & \tilde{x}_{mn2} & \cdots & \tilde{x}_{mnr_n} \end{bmatrix} \end{array} \tag{6.24}$$

Step 2. Convert the decision making matrix ($\widetilde{\mathbf{D}}$) to a dimensionless matrix ($\widetilde{\mathbf{D}}'$). In doing so, the linear scale transformation is used as follows:

$$\tilde{r}_{ijk} = \begin{cases} \tilde{x}_{ijk}(:)\tilde{x}^*_{jk} = \left(\dfrac{a_{ijk}}{c^*_{jk}}, \dfrac{b_{ijk}}{b^*_{jk}}, \dfrac{c_{ijk}}{a^*_{jk}}\right), & \forall\, jk,\ \tilde{x}_{jk} \text{ is a benefit criterion} \\ \tilde{x}^-_{jk}(:)\tilde{x}_{ijk} = \left(\dfrac{a^-_{jk}}{c_{ijk}}, \dfrac{b^-_{jk}}{b_{ijk}}, \dfrac{c^-_{jk}}{a_{ijk}}\right), & \forall\, jk,\ \tilde{x}_{jk} \text{ is a cost criterion} \end{cases} \tag{6.25}$$

where $\tilde{x}^*_{jk} = (a^*_{jk}, b^*_{jk}, c^*_{jk})$ and $\tilde{x}^-_{jk} = (a^-_{jk}, b^-_{jk}, c^-_{jk})$ present the largest and the lowest value of each sub-criterion, respectively.

Step 3. Determine the main criteria weight matrix $\tilde{w}_{MC}$ as follows:

$$\tilde{w}_{MC} = \begin{array}{c} MC_1 \\ MC_2 \\ \vdots \\ MC_j \\ \vdots \\ MC_n \end{array} \begin{bmatrix} \tilde{w}_1 \\ \tilde{w}_2 \\ \vdots \\ \tilde{w}_j \\ \vdots \\ \tilde{w}_n \end{bmatrix} \tag{6.26}$$

where $\tilde{w}_j$ $(j = 1, 2, \ldots, n)$ is the geometric mean of the main criteria weight scores assigned by the K experts, which is calculated the same as in (6.13).

Step 4. Obtain the sub-criteria weight matrix $\tilde{w}_{SC}$ concerning the corresponding main criteria as follows:

$$\tilde{w}_{SC} = \begin{array}{c} \\ \\ SC_{11} \\ SC_{12} \\ \vdots \\ SC_{1r_1} \\ SC_{21} \\ SC_{22} \\ \vdots \\ SC_{2r_2} \\ \vdots \\ SC_{j1} \\ SC_{j2} \\ \vdots \\ SC_{jr_j} \\ \vdots \\ SC_{n1} \\ SC_{n2} \\ \vdots \\ SC_{nr_n} \end{array} \begin{array}{c} \begin{array}{ccccccc} \tilde{w}_1 & \tilde{w}_2 & \dots & \tilde{w}_j & \dots & \tilde{w}_n \\ MC_1 & MC_2 & \dots & MC_j & \dots & MC_n \end{array} \\ \begin{bmatrix} \tilde{w}_{11} & 0 & \dots & 0 & \dots & 0 \\ \tilde{w}_{12} & 0 & \dots & 0 & \dots & 0 \\ \vdots & \vdots & \ddots & \vdots & \ddots & \vdots \\ \tilde{w}_{1r_1} & 0 & \dots & 0 & \dots & 0 \\ 0 & \tilde{w}_{21} & \dots & 0 & \dots & 0 \\ 0 & \tilde{w}_{22} & \dots & 0 & \dots & 0 \\ \vdots & \vdots & \ddots & \vdots & \ddots & \vdots \\ 0 & \tilde{w}_{2r_2} & \dots & 0 & \dots & 0 \\ \vdots & \vdots & \ddots & \vdots & \ddots & \vdots \\ 0 & 0 & \dots & \tilde{w}_{j1} & \dots & 0 \\ 0 & 0 & \dots & \tilde{w}_{j2} & \dots & 0 \\ \vdots & \vdots & \ddots & \vdots & \ddots & \vdots \\ 0 & 0 & \dots & \tilde{w}_{jr_j} & \dots & 0 \\ \vdots & \vdots & \ddots & \vdots & \ddots & \vdots \\ 0 & 0 & \dots & 0 & \dots & \tilde{w}_{n1} \\ 0 & 0 & \dots & 0 & \dots & \tilde{w}_{n2} \\ \vdots & \vdots & \ddots & \vdots & \ddots & \vdots \\ 0 & 0 & \dots & 0 & \dots & \tilde{w}_{nr_n} \end{bmatrix} \end{array} \tag{6.27}$$

where $\tilde{w}_{jr_j}$ $(j = 1, 2, \dots, n)$ is the geometric mean of the sub-criteria weight scores with respect to the corresponding main criteria given by the K experts.

Step 5. Calculate the final weight score for each sub-criterion, which is the product of the main criterion weight score and the sub-criterion weight score with respect to the corresponding main criterion.

$$\tilde{W}_{SCj} = \tilde{w}_{MCj}(\cdot)\tilde{w}_{SCj} = \tilde{w}_j(\cdot) \begin{bmatrix} \tilde{w}_{j1} \\ \tilde{w}_{j2} \\ \vdots \\ \tilde{w}_{jr_j} \end{bmatrix} = \begin{bmatrix} \tilde{w}_{SCj1} \\ \tilde{w}_{SCj2} \\ \vdots \\ \tilde{w}_{SCjr_j} \end{bmatrix}, \quad (j = 1, 2, \dots, n) \tag{6.28}$$

where $\tilde{W}_{SCj}$ represents the final sub-criteria weight matrix concerning the jth main criterion. $\tilde{w}_{MCj}$ and $\tilde{w}_{SCj}$ denote the jth main criterion weight score and the sub-criteria weight scores of this main criterion, respectively.

Step 6. By multiplying the fuzzy criteria values and the fuzzy weight scores of each sub-criterion, the weighted normalized fuzzy decision matrix $\widetilde{\mathbf{V}}$ could be achieved as calculated in (6.17). The results are shown as follows:

$$\widetilde{\mathbf{V}} = \begin{array}{c} \\ \\ A_1 \\ A_2 \\ \vdots \\ A_m \end{array} \begin{array}{c} \begin{array}{cccccccccccc} & MC_1 & & & MC_2 & & \cdots\cdots & & MC_n & \\ SC_{11} & SC_{12} & \cdots & SC_{1r_1} & SC_{21} & SC_{22} & \cdots & SC_{2r_2} & \cdots\cdots & SC_{n1} & SC_{n2} & \cdots & SC_{nr_n} \end{array} \\ \begin{bmatrix} \tilde{v}_{111} & \tilde{v}_{112} & \cdots & \tilde{v}_{11r_1} & \tilde{v}_{121} & \tilde{v}_{122} & \cdots & \tilde{v}_{12r_2} & \cdots\cdots & \tilde{v}_{1n1} & \tilde{v}_{1n2} & \cdots & \tilde{v}_{1nr_n} \\ \tilde{v}_{211} & \tilde{v}_{212} & \cdots & \tilde{v}_{21r_1} & \tilde{v}_{221} & \tilde{v}_{222} & \cdots & \tilde{v}_{22r_2} & \cdots\cdots & \tilde{v}_{2n1} & \tilde{v}_{2n2} & \cdots & \tilde{v}_{2nr_n} \\ \vdots & \vdots & \ddots & \vdots & \vdots & \vdots & \ddots & \vdots & \ddots & \vdots & \vdots & \ddots & \vdots \\ \tilde{v}_{m11} & \tilde{v}_{m12} & \cdots & \tilde{v}_{m1r_1} & \tilde{v}_{m21} & \tilde{v}_{m22} & \cdots & \tilde{v}_{m2r_2} & \cdots\cdots & \tilde{v}_{mn1} & \tilde{v}_{mn2} & \cdots & \tilde{v}_{mnr_n} \end{bmatrix} \end{array} \quad (6.29)$$

Subsequently, the fuzzy addition principle is used to aggregate the values within each main criterion as follows.

$$\tilde{v}'_{ij} = \sum_{k=1}^{r_j} \tilde{v}_{ijk}, \quad j = 1, 2, \ldots, n. \quad (6.30)$$

Thus, the matrix $\widetilde{\mathbf{V}}$ is converted into $\widetilde{\mathbf{V}}'$:

$$\widetilde{\mathbf{V}}' = \begin{array}{c} A_1 \\ A_2 \\ \vdots \\ A_m \end{array} \overset{\begin{array}{cccc} MC_1 & MC_2 & \ldots & MC_n \end{array}}{\begin{bmatrix} \tilde{v}'_{11} & \tilde{v}'_{12} & \cdots & \tilde{v}'_{1n} \\ \tilde{v}'_{21} & \tilde{v}'_{22} & \cdots & \tilde{v}'_{2n} \\ \vdots & \vdots & \ddots & \vdots \\ \tilde{v}'_{m1} & \tilde{v}'_{m2} & \cdots & \tilde{v}'_{mn} \end{bmatrix}} \quad (6.31)$$

By now, the hierarchical algorithm is ready to be included in the computation process of the fuzzy TOPSIS method. Followed by the Step 4 described in Section 6.2.2, the final index score C_i as well as the ranking order of the alternatives can be obtained.

6.3 Applications in Road Safety Performance Evaluation

In this Section, the aforementioned three MCDM methods, i.e., the classical TOPSIS, the fuzzy TOPSIS, and the hierarchical fuzzy TOPSIS, are applied to evaluate the road safety performance of a set of European countries, and the results from different methods are compared afterwards.

6.3.1 *Evaluation using the classical TOPSIS*

To use TOPSIS method in road safety performance evaluation, we consider six safety performance indicators (SPIs) related to the six different road safety risk factors (i.e., Alcohol and drugs, Speed, Protective systems, Vehicle, Roads, and Trauma management) as follows:

- The percentage of surveyed car drivers disrespecting the alcohol limit (A1);
- The percentage of surveyed car drivers exceeding the speed limit in built-up areas (S1);
- The seat belt wearing rate in front seats (P1);
- The share of relatively new passenger cars (i.e., less than 6 years old) (V1);
- The motorway density (R1);
- The expenditure on health care as share of the gross domestic product (T1).

Values related to 2003 are obtained for 21 European countries constituting the decision matrix **D** for the TOPSIS. Subsequently, the dimensionless matrix ($\mathbf{D}'$) can be obtained using (6.2), which is shown in Table 6.3. Thus, all the indicator values are expected to be the higher the better.

Next, to derive the relative importance of each indicator, i.e., the corresponding weight, eight independent experts in road safety field from seven European countries are asked to assign crisp data ranged from 0 to 10 with a higher value representing a more importance of the corresponding SPI. The results are shown in Table 6.4, and their geometric mean values are calculated as shown in the last raw of the table.

Now, by multiplying the normalized decision matrix $\mathbf{D}'$ with the weight vector W, we obtain the weighted normalized matrix **V**. Then the maximum and minimum value in each column of matrix **V** constitutes the PIS and NIS. Based on (6.7)–(6.10) in Section 6.2.1, we calculate the relative Euclidean distance from each country to the ideal solutions and the separation measures S_i^* and S_i^- as well. The relative closeness index C_i can then be computed for each country by combining the S_i^* and S_i^- (see (6.11)), based on which countries can be ranked.

6.3.2 *Evaluation using the fuzzy TOPSIS*

In the application of the fuzzy TOPSIS method, we consider the same SPIs. The difference is that the experts are now asked to indicate the relative importance of these indicators by using linguistic expressions, which are regarded as natural representations of preference and judgment. The results are shown in Table 6.5.

Table 6.3 Normalized decision matrix for TOPSIS

	Alcohol & drugs	Speed	Protective systems	Vehicle	Roads	Trauma management
	A1	S1	P1	V1	R1	T1
Austria	0.038	0.500	0.794	0.655	0.328	0.652
Belgium	0.017	0.250	0.680	0.761	0.934	0.817
Cyprus	0.005	0.250	0.825	0.317	0.475	0.557
Czech Republic	0.050	0.500	0.773	0.362	0.115	0.652
Denmark	0.333	0.750	0.866	0.729	0.393	0.783
Estonia	0.063	0.250	0.773	0.319	0.033	0.461
Finland	0.333	0.500	0.918	0.536	0.033	0.643
France	0.020	0.429	1.000	0.693	0.311	0.878
Germany	0.042	0.429	0.969	0.705	0.557	0.965
Greece	0.013	0.500	0.412	0.685	0.098	0.861
Hungary	0.077	0.250	0.608	0.490	0.098	0.730
Ireland	0.040	1.000	0.876	1.000	0.049	0.635
Italy	0.014	0.250	0.732	0.657	0.361	0.730
the Netherlands	0.053	0.429	0.887	0.725	1.000	0.852
Poland	0.333	0.429	0.732	0.415	0.016	0.565
Portugal	0.024	0.273	0.907	0.535	0.361	0.835
Slovenia	0.037	0.500	0.835	0.933	0.377	0.765
Spain	0.014	0.273	0.887	0.673	0.328	0.670
Sweden	1.000	0.600	0.948	0.735	0.066	0.817
Switzerland	0.024	0.750	0.845	0.738	0.525	1.000
United Kingdom	0.167	0.750	0.959	0.828	0.246	0.696

To handle these linguistic variables, the positive triangular fuzzy numbers are employed, which are listed in Table 6.6.

Thus, the fuzzified weight matrix can be obtained, and the geometric mean method is used to compute the fuzzy weight vector. Then, we calculate the weighted normalized fuzzy decision matrix $\widetilde{\mathbf{V}}$ by multiplying the normalized decision matrix $\mathbf{D}'$ (see Table 6.3) with the fuzzy weight vector based on (6.17). It should be mentioned here that each element in $\widetilde{\mathbf{V}}$ is a fuzzy number $\tilde{v}_{ij} = (a_{ij}, b_{ij}, c_{ij})$, so its generalized mean $M(\tilde{v}_{ij})$ should be figured out according to (6.19). The largest generalized mean and the smallest generalized mean of each indicator could then be picked out, and the countries corresponding to the extremes of $M(\tilde{v}_{ij})$ are the relevant $\tilde{v}_j^*$ and $\tilde{v}_j^-$ for each indicator. Now, the Euclidean distance $\widetilde{D}_{ij}^*$ and $\widetilde{D}_{ij}^-$ could be calculated, and the separation measures $\widetilde{S}_i^*$ and $\widetilde{S}_i^-$, as well as the relative closeness index C_i are subsequently obtained.

Table 6.4 Indicator weights for TOPSIS

Expert	Alcohol & drugs	Speed	Protective Systems	Vehicle	Roads	Trauma management
E1	10	10	8	7	6	6
E2	8	10	10	10	8	10
E3	8	9	8	7	8	6
E4	10	10	8	8	9	10
E5	10	9	7	8	6	5
E6	8	10	8	6	10	8
E7	9	10	9	9	9	9
E8	10	10	10	9	9	9
Geometric Mean (W_j)	9.077	9.740	8.442	7.905	8.001	7.645

Table 6.5 Linguistic expression on indicator weights for fuzzy TOPSIS

Expert	Alcohol & drugs	Speed	Protective systems	Vehicle	Roads	Trauma management
E1	Absolutely Important	Absolutely Important	Strongly Important	More Important	More Important	More Important
E2	Strongly Important	Absolutely Important	Absolutely Important	Absolutely Important	Strongly Important	Absolutely Important
E3	Strongly Important	Strongly Important	Strongly Important	More Important	Strongly Important	More Important
E4	Absolutely Important	Absolutely Important	Strongly Important	Strongly Important	Strongly Important	Absolutely Important
E5	Absolutely Important	Strongly Important	More Important	Strongly Important	More Important	Important
E6	Strongly Important	Absolutely Important	Strongly Important	More Important	Absolutely Important	Strongly Important
E7	Strongly Important	Absolutely Important	Strongly Important	Strongly Important	Strongly Important	Strongly Important
E8	Absolutely Important	Absolutely Important	Absolutely Important	Strongly Important	Strongly Important	Strongly Important

6.3.3 *Evaluation using the hierarchical fuzzy TOPSIS*

Since the collection of additional data on the possible SPIs is an ongoing process, some other available indicators can be formulated to represent some of the six risk factors and served as the complements of the existing six SPIs. In this study, a hierarchical structure of indicators is developed as shown in Fig. 6.2, which includes 11 SPIs with three layers.

Table 6.6 Triangular fuzzy numbers for the linguistic terms of indicator weights

Linguistic terms	Triangular fuzzy numbers
Absolutely unimportant	(0, 0, 1/6)
Unimportant	(0, 1/6, 2/6)
Less important	(1/6, 2/6, 3/6)
Important	(2/6, 3/6, 4/6)
More important	(3/6, 4/6, 5/6)
Strongly important	(4/6, 5/6, 1)
Absolutely important	(5/6, 1, 1)

Correspondingly, a hierarchical fuzzy TOPSIS model should then be applied to combine the multilayer indicators into one overall road safety performance index.

First of all, a new normalized decision matrix $\mathbf{D}'$ is obtained as in Table 6.7. Moreover, the same group of experts is asked to indicate the relative importance of both risk factors and the indicators with respect to the corresponding risk factor by using linguistic expressions. They can then be transferred into positive triangular fuzzy numbers as in Table 6.6.

By calculating the product of each main risk factor weight scores and the (sub-) indicator weight scores with respect to the corresponding risk factor, we obtain the final weight scores for each indicator. Subsequently, the weighted normalized fuzzy decision matrix $\widetilde{\mathbf{V}}$ is achieved by computing the product of the normalized hierarchical decision matrix $\widetilde{\mathbf{D}}'$ and the fuzzy weight scores for each indicator. After aggregating the values belonging to each main risk factor by the fuzzy addition principle as in (6.30), we obtain the final weighted normalized fuzzy decision matrix $\widetilde{\mathbf{V}}'$. Afterwards, the same steps can be followed as in the fuzzy TOPSIS to obtain the final composite road safety performance index scores for the 21 European countries.

6.3.4 *Comparison and discussion*

In this section, we compare the results derived from the classical TOPSIS method, the fuzzy TOPSIS method, and the hierarchical fuzzy TOPSIS method to gain insight into their similarity and difference, and further explore the degree of correlation between each of the above three results and the road safety final outcomes.

Above all, comparing the composite index scores of the 21 European countries based on the classical TOPSIS and the fuzzy TOPSIS in Table 6.8, we find that a relatively high value is derived for each country by using the fuzzy TOPSIS method, which is mainly ow-

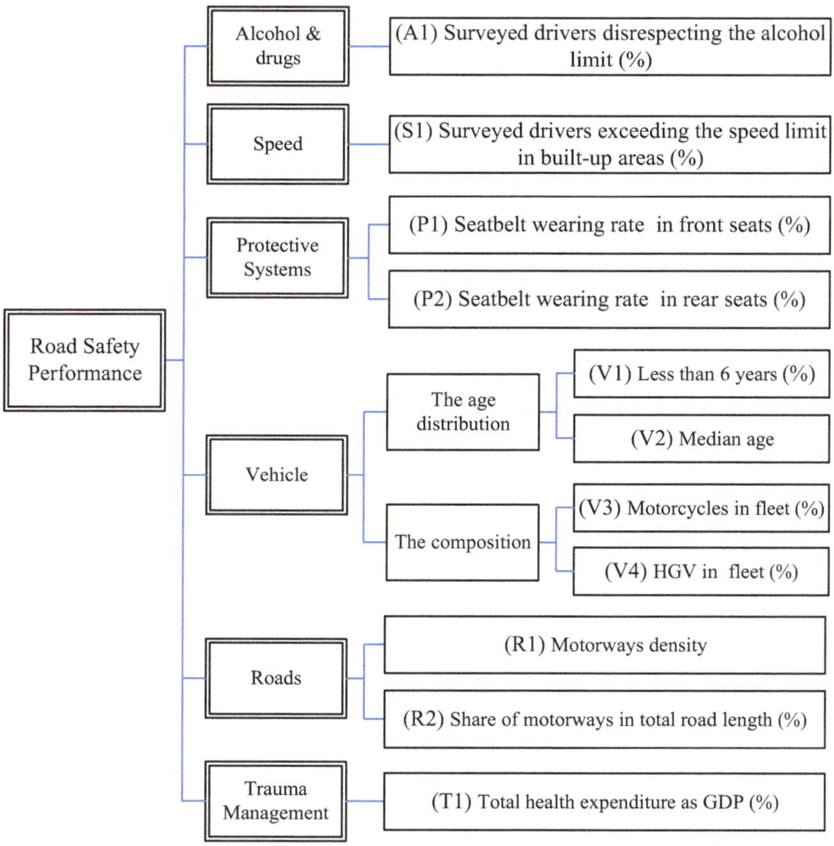

Fig. 6.2 The hierarchical structure of the developed SPIs

ing to the use of fuzzy numbers for representing the importance of each indicator instead of precise numbers. Moreover, by checking the two ranking results, almost all the countries have a difference in rank with a maximum of two positions except for Czech Republic. In other words, the linguistic expressions from the experts can be well handled by integrating the fuzzy logic in the TOPSIS method without losing any important information. We further compare the ranking results based on the fuzzy TOPSIS method and the hierarchical fuzzy TOPSIS. From an overall perspective, they take on an overall high resemblance, since most of the countries have a difference in rank with a maximum of one position. Nevertheless, they disagree most on the first six rankings of countries, which are Sweden, Germany, Denmark, the United Kingdom, Switzerland, and the Netherlands. It implies

Table 6.7 Normalized hierarchical decision matrix

	Alcohol & drugs	Speed	Protective systems		Vehicle				Roads		TM
	A1	S1	P1	P2	V1	V2	V3	V4	R1	R2	T1
AT	0.038	0.500	0.794	0.557	0.655	0.664	0.129	0.729	0.328	0.250	0.652
BE	0.017	0.250	0.680	0.455	0.761	0.751	0.267	0.459	0.934	0.185	0.817
CY	0.005	0.250	0.825	0.341	0.317	0.537	0.211	0.235	0.475	0.365	0.557
CZ	0.050	0.500	0.773	0.466	0.362	0.454	0.106	0.567	0.115	0.065	0.652
DK	0.333	0.750	0.866	0.716	0.729	0.682	0.239	0.273	0.393	0.228	0.783
EE	0.063	0.250	0.773	0.773	0.319	0.392	0.842	0.364	0.033	0.029	0.461
FI	0.333	0.500	0.918	0.909	0.536	0.547	0.152	0.440	0.033	0.101	0.643
FR	0.020	0.429	1.000	0.932	0.693	0.600	0.250	0.372	0.311	0.166	0.878
DE	0.042	0.429	0.969	1.000	0.705	0.832	0.152	1.000	0.557	0.833	0.965
EL	0.013	0.500	0.412	0.284	0.685	0.624	0.093	0.293	0.098	0.048	0.861
HU	0.077	0.250	0.608	0.386	0.490	0.624	0.432	0.408	0.098	0.054	0.730
IE	0.040	1.000	0.876	0.523	1.000	0.972	1.000	0.347	0.049	0.029	0.635
IT	0.014	0.250	0.732	0.341	0.657	0.635	0.078	0.586	0.361	0.127	0.730
NL	0.053	0.429	0.887	0.727	0.725	0.832	0.246	0.451	1.000	0.279	0.852
PL	0.333	0.429	0.732	0.511	0.415	0.423	0.340	0.354	0.016	0.017	0.565
PT	0.024	0.273	0.907	0.511	0.535	0.516	0.178	0.239	0.361	0.160	0.835
SI	0.037	0.500	0.835	0.557	0.933	1.000	0.333	0.810	0.377	0.199	0.765
ES	0.014	0.273	0.887	0.784	0.673	0.657	0.110	0.305	0.328	1.000	0.670
SE	1.000	0.600	0.948	0.909	0.735	0.713	0.167	0.554	0.066	0.182	0.817
CH	0.024	0.750	0.845	0.602	0.738	0.688	0.128	0.797	0.525	0.303	1.000
UK	0.167	0.750	0.959	0.955	0.828	0.998	0.432	0.477	0.246	0.137	0.696

that incorporation of new SPIs will influence the ranking position more on those countries with high rankings, rather than those low ranking ones. Finally, by comparing the ranking results with the ones based on the number of fatalities per million inhabitants, we find that most of the countries' ranking positions are going towards the fatality rankings based on the hierarchical fuzzy TOPSIS method. To make a quantitative comparison and provide a clear insight into the relationship among different methods, the correlation analysis is adopted. In Table 6.9, the Pearson's correlation coefficients among the composite index scores (C) and the number of fatalities per million inhabitants are presented, and Table 6.10 displays the Pearson's correlation coefficients among the corresponding ranking results.

Table 6.8 The results based on the classical TOPSIS, the fuzzy TOPSIS, the hierarchical fuzzy TOPSIS and fatalities

	C TOPSIS	TOPSIS Ranking	*C* Fuzzy TOPSIS	Fuzzy TOPSIS Ranking	*C* Hierarchical Fuzzy TOPSIS	Hierarchical Fuzzy TOPSIS Ranking	No. of fatalities per mln inhab.	Fatality Ranking
AT	0.325	12	0.549	10	0.544	10	115	11
BE	0.412	9	0.523	11	0.501	12	117	12
CY	0.204	19	0.303	21	0.297	21	134	17
CZ	0.220	18	0.471	15	0.463	15	142	18
DK	0.528	2	0.691	1	0.672	3	80	6.5
EE	0.096	21	0.308	20	0.355	20	121	13.5
FI	0.335	11	0.514	12	0.540	11	73	4
FR	0.397	10	0.566	9	0.560	9	101	9
DE	0.466	7	0.617	6	0.679	2	80	6.5
EL	0.236	16	0.442	17	0.396	18	146	19
HU	0.165	20	0.413	18	0.413	17	131	16
IE	0.479	6	0.597	8	0.562	8	84	8
IT	0.266	15	0.452	16	0.420	16	105	10
NL	0.524	3	0.639	5	0.630	6	63	3
PL	0.233	17	0.404	19	0.395	19	149	21
PT	0.308	13	0.497	13	0.469	14	148	20
SI	0.423	8	0.600	7	0.592	7	121	13.5
ES	0.290	14	0.472	14	0.499	13	130	15
SE	0.605	1	0.688	2	0.715	1	59	1
CH	0.522	4	0.656	3	0.653	5	74	5
UK	0.481	5	0.650	4	0.654	4	61	2

As shown in Table 6.9, the composite index scores (C) derived from the classical TOPSIS method, the fuzzy TOPSIS method and the hierarchical fuzzy TOPSIS method present a highly positive-correlated relationship with the highest correlation coefficient of 0.974. Moreover, the index scores and the number of the fatalities per million inhabitants are also highly negative-correlated, and the one generated from the hierarchical fuzzy TOPSIS method produces the highest correlation with the fatalities (-0.810). It means that the development of multilayer SPIs to represent a country's road safety situation is desirable, and the realization of the hierarchical fuzzy TOPSIS method is also valuable in creating an

Table 6.9 The Pearson's correlation coefficients among three Cs and the number of fatalities per million inhabitants

	C (TOPSIS)	C (Fuzzy TOPSIS)	C (Hierarchical Fuzzy TOPSIS)	No. of fatalities per mln inhab.
C (TOPSIS)	1.000	0.952	0.929	−0.802
C (Fuzzy TOPSIS)		1.000	0.974	−0.763
C (Hierarchical Fuzzy TOPSIS)			1.000	−0.810
No. of fatalities per mln inhab.				1.000

Table 6.10 The Pearson's correlation coefficients among the different sets of ranking results

	TOPSIS Ranking	Fuzzy TOPSIS Ranking	Hierarchical Fuzzy TOPSIS Ranking	Fatality Ranking
TOPSIS Ranking	1.000	0.969	0.942	0.804
Fuzzy TOPSIS Ranking		1.000	0.979	0.794
Hierarchical Fuzzy TOPSIS Ranking			1.000	0.814
Fatality Ranking				1.000

overall road safety performance index. The same conclusions can be drawn by considering the corresponding ranking results as in Table 6.10.

6.4 Conclusions

In this chapter, we have investigated the TOPSIS method in combining individual safety performance indicators into an overall index for a set of European countries. To justify the feasibility and effectiveness of this method in road safety area under different requirements, we have also explored a series of extensions and applications in this study. The key research findings are formulated as follows:

(1) The classical TOPSIS method can be effectively used as an alternative way to combine different individual safety performance indicators/criteria into a composite index with the purpose of comparing and ranking given alternatives (countries).

(2) As a nature extension of the classical TOPSIS method, the fuzzy TOPSIS method that integrates fuzzy logic into TOPSIS provides a promising solution to handle the subjective kind of uncertainty on data, which embodies the fuzzy nature of the comparison or evalua-

tion process and strengthens the comprehensiveness and rationality of the decision-making process.

(3) The realization of the hierarchical fuzzy TOPSIS method further extends the use of the TOPSIS method by taking the layered hierarchy of the indicators into account, which is seldom considered in the current studies and difficult to realize in the other traditional weighting methods. The derived composite index scores show a relatively higher correlation with the number of fatalities per million inhabitants than the ones from the single layer TOPSIS methods (both classical and fuzzy), which indicates the usefulness and effectiveness of applying the hierarchical fuzzy TOPSIS method to create a composite index by integrating the indicators with hierarchical structures, and further implied the feasibility of applying this method to a great number of performance evaluation and decision making activities in other wide ranging fields.

Bibliography

[1] B. Ülengin, F. Ülengin, and Ü. Güvenç. A multidimensional approach to urban quality of life: the case of Istanbul. *European Journal of Operational Research.* **130**, pp. 361–374 (2001).

[2] S.O. Funtowicz, J. Martinez-Alier, G. Munda and J. Ravetz. Multicriteria-based environmental policy, in: *H. Abaza and A. Baranzini (eds.), Implementing Sustainable Development.* UNEP/Edward Elgar, Cheltenham, pp. 53–77 (2002).

[3] G. Munda. Multicriteria evaluation as a multidimensional approach to welfare measurement, in: *J. van den Bergh and J. van der Straaten (eds.), Economy and Ecosystems in Change: Analytical and Historical Approaches.* Edward Elgar, Cheltenham, pp. 96–115 (1997).

[4] G. Munda. Social multi-criteria evaluation (SMCE): methodological foundations and operational consequences. *European Journal of Operational Research.* **158** (3), pp. 662–677 (2004).

[5] G. Munda. Measuring sustainability: a multi-criterion framework. *Environment, Development and Sustainability.* **7** (1), pp. 117–134 (2005).

[6] E. Triantaphyllou. *Multi-Criteria Decision Making Methods: A comparative Study*, Kluwer Academic Publishers, Dordrecht (2000).

[7] C.L. Hwang and K. Yoon. *Multiple Attribute Decision Making: Methods and Applications*, Springer-Verlag, Berlin/Heidelberg/New York (1981).

[8] E.U. Choo and W.C. Wedley. Optimal criterion weights in repetitive multicriteria decision-making. *Journal of the Operational Research Society.* **36**, pp. 983–992 (1985).

[9] T.L. Saaty. *The Analytic Hierarchy Process for Decision in a Complex World*, RWS Publication, Pittsburgh, PA, USA (1995).

[10] M. Nardo, M. Saisana, A. Saltelli, S. Tarantola, A. Hoffman, and E. Giovannini. *Handbook on Constructing Composite Indicators: Methodology and User Guide*, OECD Statistics Working Papers, STD/DOC(2005)3 (2005).

[11] Y.J. Lai, T.Y. Liu, and C.L. Hwang. TOPSIS for MODM, *European Journal of Operational Research.* **76** (3), pp. 486–500 (1994).

[12] M.A. Abo-Sinna and A.H. Amer. Extensions of TOPSIS for multi-objective large-scale nonlinear programming problems, *Applied Mathematics and Computation.* **162**, pp. 243–256 (2005).

[13] G.R. Jahanshahloo, L.F. Hosseinzadeh and M. Izadikhah. An algorithmic method to extend TOPSIS for decision-making problems with interval data, *Applied Mathematics and Computation.* **175**, p. 1375–1384 (2006).

[14] S.J. Chen and C.L. Hwang. *Fuzzy Multiple Attribute Decision Making Methods and Applications*, Springer-Verlag, Berlin (1992).

[15] C.T. Chen. Extensions of the TOPSIS for group decision-making under fuzzy environment, *Fuzzy Sets and Systems.* **114**, pp. 1–9 (2000).

[16] Y. Xu and Q. Da. A method for multiple attribute decision making with incomplete weight information under uncertain linguistic environment, *Knowledge-Based Systems.* **21** (8), pp. 837–841 (2008).

[17] L.A. Zadeh. Fuzzy sets. *Information and Control.* **8**, pp. 338–353 (1965).

[18] Q. Bao, D. Ruan, Y. Shen, E. Hermans, and D. Janssens. Improved hierarchical fuzzy TOPSIS for road safety performance evaluation, *Knowledge-Based Systems.* **32**, pp. 84–90 (2012).

Chapter 7

Fuzzy Failure Rate for Nuclear Power Plant Probabilistic Safety Assessment by Fault Tree Analysis

Julwan Hendry Purba, Jie Lu, and Guangquan Zhang

Decision Systems and e-Service Intelligence Laboratory
Centre for Quantum Computation & Intelligent Systems
School of Software, Faculty of Engineering and Information Technology
University of Technology, Sydney (UTS)
P.O. Box 123, Broadway, NSW 2007, Australia

E-mail: {julwan, jielu, zhangg}@it.uts.edu.au

Reliability data is essential for a nuclear power plant probabilistic safety assessment by fault tree analysis to assess the performance of the safety-related systems. The limitation of the conventional reliability data comes from insufficient historical data for probabilistic calculation. This chapter proposes and discusses a failure possibility-based reliability algorithm to assess nuclear event reliability data from failure possibilities, which are expressed in qualitative natural languages, mathematically represented by membership functions of fuzzy numbers, and subjectively justified by a group of experts based on their working experience and expertise. We also discuss an area defuzzification technique, which has been developed, to defuzzify nuclear event failure possibilities into their corresponding fuzzy failure rates, which are similar to the probabilistic failure rates probabilistically calculated from historical failure data. A simplified model of a high pressure core spray system is used to mathematically proof the proposed algorithm and defuzzification technique. The results show that fuzzy failure rates can be used in nuclear power plant probabilistic safety assessment by fault tree analysis as alternatives for probabilistic failure rates when nuclear event historical data are insufficient or unavailable for probabilistic calculation.

7.1 Introduction

System safety is a major requirement for complex systems such as nuclear power plants (NPPs). The functions of the nuclear safety system are to ensure that nuclear facilities can normally operate without an excessive risk exposure to staffs and environment, to prevent accidents, and to mitigate the consequences of accidents if they occur. In 1957, the United

C. Kahraman (ed.), *Computational Intelligence Systems in Industrial Engineering*,
Atlantis Computational Intelligence Systems 6, DOI: 10.2991/978-94-91216-77-0_7,

Nations (UN) set up the International Atomic Energy Agency (IAEA) to act as an auditor of world nuclear safety and work together with nuclear safety inspectorates in member countries, which operates nuclear power plants.

Even though nuclear power plants have been designed to be safe in their operation and safe in the event of any malfunction or accident, incidents and accidents still may happen as in other industries. The most recent nuclear disaster to date is Fukushima Daiichi nuclear power plants, which happen in March 2011. The disaster was affected seriously by a huge tsunami following the Great East Japan Earthquake. Three of six reactors, which were operating at that time, shut down automatically due to this earthquake to prevent potential accidents to the reactors. The control rods had been inserted into the core to stop chain reaction. However, at this point, decay heat at about 7% of the full power heat load under normal operating conditions has to be carried out by the cooling system from the core to avoid fuel meltdown. Unfortunately, the earthquake also destroyed external power supply and hence, backup power system had to work to keep the coolant pumps working. Later, the tsunami induced by the earthquake swamped and flooded the diesel generators to cause pump failure. This situation led to weeks of drama and loss of the reactors. However, Fukushima accidents will have significant implications and provide valuable knowledge to enhance the NPP safety systems.

Over the past few years, probabilistic safety assessment (PSA) has been considered to be an important analysis tool to ensure the safety of a nuclear power plant in relation to potential initiating events that can be caused by random component failures, human errors, internal and external hazards. PSA provides a comprehensive and structured approach to identify and understand key plant vulnerabilities, to develop accident scenarios, to assess the level of the plant safety, and to derive numerical estimates of potential risks. Designers, utility and regulatory personnel use PSA results to verify the NPP design, to assess the possible changes to the plant design or operation, and to assess the potential changes to the plant licensing basis [1–3]. Based on the PSA results, NPPs are subject to change to enhance their safety level. Where the results of the PSA indicate that changes could be made to the design or operation of the plant to reduce risk, the changes should be incorporated where reasonably achievable.

The estimation of failure probabilities of rare events with high consequences is the main focus of the NPP PSA [4]. Fault tree analysis (FTA) is widely used as a deductive tool for PSA to assess the failure probabilities of particular safety functions or safety systems [5–10]. This analysis can be implemented only if the failure rates of all basic events constructing the tree are well known in advance [11]. Failure rates, which are directly taken from the plant being analyzed, are the most appropriate sources [12]. However, plant specific data are not always readily available since nuclear accidents are very rare, the plant may use new components, and the plant environment may change. In case of unavailable plant precise failure data, it is common to use a generic database that can be taken from various sources such as other NPPs, nuclear industries other than NPPs, and non-nuclear industries [4]. Since the used data are not comprehensive into the area under investigation, nuclear safety analysts have to deal with imprecision and uncertainties [13, 14]. Moreover, the results also will not show the real situation of the system function to be used for future recommendations on the safety improvement [15].

In the situation when little quantitative information is available, qualitative justification can be used [16–18]. In addition, qualitative justification is more appropriate for analyzing systems in the presence of uncertainties [19]. Qualitative justification has been widely implemented in nuclear engineering for making engineering decision [20]. The fuzzy reliability theory, which implements failure possibilities and fuzzy sets, has been proposed and developed to overcome the limitation of the conventional reliability theory [11, 21–24]. Failure possibilities can be used to deal with imprecision and uncertainties by implementing fuzzy sets [25, 26]. Safety evaluators can specify a range of values in the context of failure possibility distribution to qualitatively evaluate event failures [25]. The theory of possibility, which is proposed by Zadeh [27] as a further development of fuzzy set theory [28], pointed out that a possibility distribution can be viewed as a fuzzy set, which serves as an elastic constraint on the values that may be assigned to a variable. In technical meaning, a possibility distribution is a fuzzy set and all fuzzy sets are possibility distributions [26]. Possibility distribution numerically equals to the corresponding membership functions, i.e., $\pi_x(x) = \mu_A(x)$, where x is a fuzzy variable and A is the fuzzy set induced by X [11, 29]. Failure possibility is a measure to what extent a value x in the set X to be a member of the subset $A_i \subseteq X$, which can be described by a membership function $\mu_{Ai}(x)$. This membership function is a mathematical representation of a subjective assessment of the failure possibility of an event [30].

This chapter discusses a failure possibility-based reliability algorithm to assess nuclear event fuzzy failure rates as alternative for probabilistic failure rates when quantitative historical failure data is insufficient or unavailable [31–34]. The algorithm utilizes the concept of failure possibilities, which are expressed in qualitative natural languages, to evaluate nuclear events, which do not have quantitative historical failure data. The membership functions of fuzzy numbers are then used to represent those qualitative natural languages in mathematical form. The failure likelihood of an event, which is under investigation, is then subjectively justified by experts. An expert is a very skilful person with considerable training in and knowledge of the nuclear field. The expert is also considered as the person who is familiar with the system and understands its working environment. To minimize bias and incompleteness coming from experts, the subjective justifications are done in a team [35, 36]. The limitation of experts' knowledge and experiences may result in different assessment for the same event [37] and hence, fuzzy aggregation methods are used to aggregate different evaluator opinions [38] to reach a consensus.

An area defuzzification technique to defuzzify the membership functions into nuclear event reliability scores (R_S), which represent the experts' belief of the most value that a nuclear event might happen, is also explained in this chapter. This technique is the most suitable technique to defuzzify nuclear event failure possibilities, which are mathematically represented by the membership functions of fuzzy numbers, into their corresponding reliability data [39, 40]. Finally, the logarithmic function proposed by Onisawa [23], which utilizes the concept of error possibility and likely fault rate, is used to calculate fuzzy failure rates (R), which are similar to probabilistic failure rates. Hence, the results of the proposed failure possibility-based reliability algorithm are suitable for NPP PSA by FTA.

7.2 Fault Tree Analysis

A fault tree is a graphical model representing the combinations of parallel and/or sequential fault events that can lead to the occurrence of the predefined undesired top event. It depicts logical interrelationships amongst basic events to the top event. Boolean gates denote the relationship between inputs and an output. The higher event is the output of the gate and the lower events are the inputs to the gate. In drawing a fault tree, the process starts from the higher faults to the more basic faults. In this analysis, Boolean algebras are used to mathematically represent the tree diagram and calculate the output of every logic gate [10, 11, 41]. The occurrence probability of the undesired top event is a function of the reliability data of primary events, which are also known as basic events [42–44].

In general, an existing NPP PSA by FTA consist of three major analysis types, i.e. qualitative analysis to evaluate minimal cut sets in the tree, quantitative analysis to calculate a cut set probability and a top event probability, and an importance measure to see how far a basic event and a cut set contribute to the top event probability. The typical FTA model is depicted in Fig. 7.1.

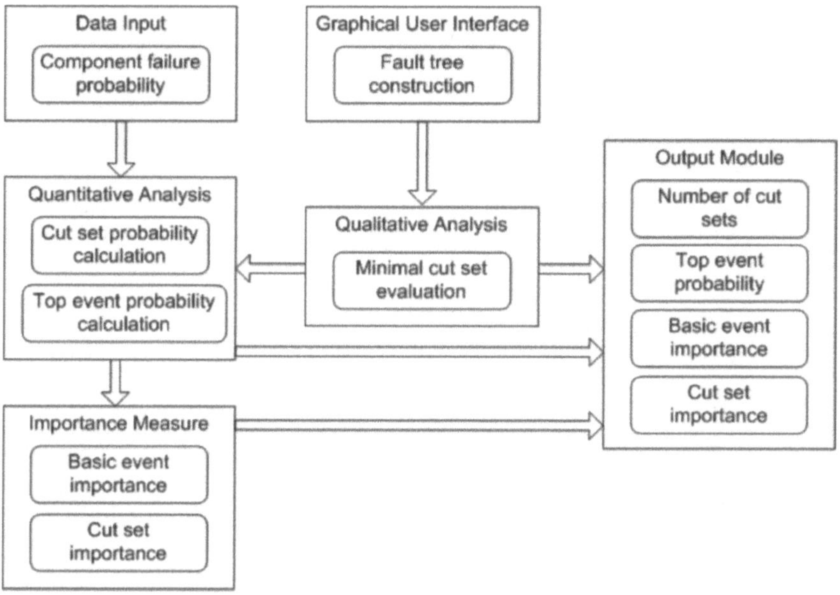

Fig. 7.1 The typical NPP PSA by FTA.

7.2.1 *Fault tree model*

A typical fault tree is composed of a number of symbols to describe events, Boolean gates, and page transfers. Event symbols represent basic events, which may fail in the system to cause the undesired top event to occur. Boolean gate symbols represent relationships between input events and an output event in graphical form. Some Boolean gates will occur if a condition attached to the gates is satisfied. Transfer event symbols are pointers to indicate sub tree branches that are used elsewhere in the tree. All those symbols, together with their name and description, are shown in Tables 7.1, 7.2, and 7.3.

7.2.2 *Repeated fault events, cut sets, and minimal cut sets*

Repeated fault events are events, which appear in the fault tree more than once. To obtain the reliable result in a fault tree analysis, these repeating events must be eliminated prior to calculating the occurrence probability of the undesired top event. Boolean algebra

Table 7.1 Event symbols.

Symbol	Name	Description
	Basic event	A basic event, which do not need further development
	Conditional event	A specific condition is applied to PRIORITY AND and INHIBIT logic gates
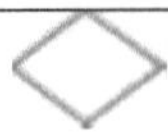	Undeveloped event	A fault event, which cannot be further developed due to lack of information
	Intermediate event	A fault event, which is resulted from an operation of a Boolean gate

Table 7.2 Boolean gate symbols.

Symbol	Name	Description
	AND	The output event occurs when all input events occur
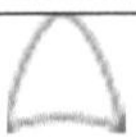	OR	The output event occurs when at least one of input events occur
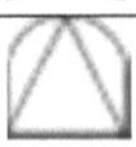	PRIORITY AND	The output event occurs when all input events occur in a specific condition
	EXCLUSIVE OR	The output event occur when exactly only one of input events occurs
	INHIBIT	The output event occur when a single input event occurs within an enabling condition

Table 7.3 Transfer event symbols.

Symbol	Name	Description
	TRANSFER IN	The fault tree is further developed at the corresponding TRANSFER OUT
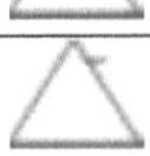	TRANSFER OUT	The fault tree is the attachment of the corresponding TRANSFER IN

properties such as absorption, idempotent, commutative, and distributive law in Table 7.4 can be used to eliminate the repeated fault events.

A cut set is a set of fault events if they occur together can cause the undesired top event to occur. Meanwhile, a minimal cut set is a cut set that has been reduced into the minimum number of fault events to cause the undesired top event to occur. The number of different basic events in a minimal cut set is called the order of the cut set. A cut set of order one

Table 7.4 Boolean algebra properties.

Rules	Engineering symbolism
Idempotent law	$X \cdot X = X$ $X + X = X$
Distributive law	$X \cdot (Y + Z) = X \cdot Y + X \cdot Z$
Commutative law	$X \cdot Y = Y \cdot X$ $X + Y = Y + X$
Absorption law	$X \cdot (X + Y) = X$ $X + (X \cdot Y) = X$

is usually more critical than a cut set of order two or higher. When we have a cut set with only one basic event, the top event will occur as soon as this basic event occurs. When a cut set has two basic events, both of these have to occur at the same time to cause the top event to occur.

7.2.3 *Occurrence probability calculation*

After a fault tree is free from repeated fault events and only consists of minimal cut sets, the occurrence probability of the undesired top event is then calculated. This probability is calculated by first assessing the occurrence probability of the output events of lower and intermediate Boolean gates.

The occurrence probability of an output event from two or more independent input events combined by OR gate as shown in Fig. 7.2 is calculated using Eq. (7.1) and by AND gate as shown in Fig. 7.3 is calculated using Eq. (7.2).

$$P(A_0) = 1 - \prod_{i=1}^{n} \{1 - P(A_i)\} \tag{7.1}$$

$$P(A_0) = \prod_{i=1}^{n} P(A_i) \tag{7.2}$$

where $P(A_i)$ is the failure probability of the basic event A_i and $i = 1, 2, 3, 4, \ldots, n$.

The fault tree representation and the corresponding probability calculation formulas for other three Boolean gates with conditions are shown in Table 7.5 [10].

7.3 An Area Defuzzification Technique

A suitable defuzzification technique is essential in the failure possibility-based reliability algorithm to defuzzify membership functions of fuzzy numbers into nuclear event reliability scores (R_s) to represent the experts' belief of the most value that nuclear events

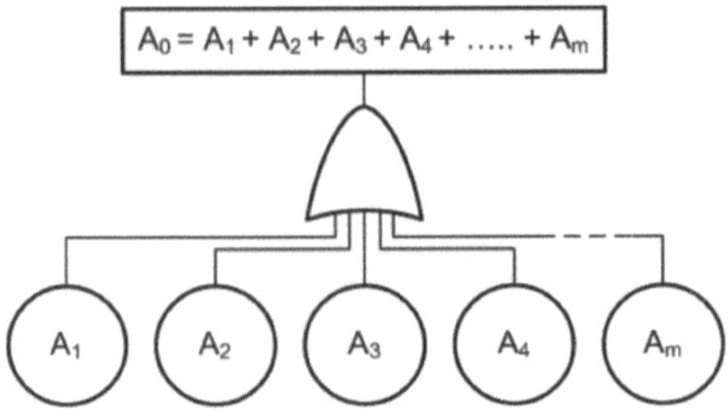

Fig. 7.2 An OR gate of n inputs.

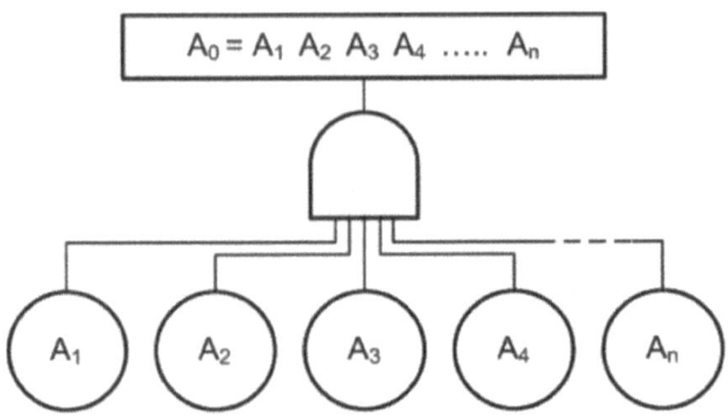

Fig. 7.3 An AND gate of n inputs.

might happen. An area defuzzification technique has been developed and proposed for nuclear power plant probabilistic safety assessment by fault tree analysis to assess nuclear event failure rates without the need for quantitative historical failure data [39, 40]. In this section, we describe the area defuzzification technique, which has been developed to defuzzify a membership function of fuzzy numbers into a nuclear event reliability score.

Definition 7.1 (Real fuzzy number and membership function). The real fuzzy number $\tilde{A} = [a,b,c,d;w]$ is a fuzzy subset of the real R whose membership function $\mu_{\tilde{A}}(x)$ can be continuously map from R into a closed interval $[0,w]$. The membership function $\mu_{\tilde{A}}(x)$ is

Table 7.5 Probability calculation formulas.

Fault tree representation	Probability calculation formulas
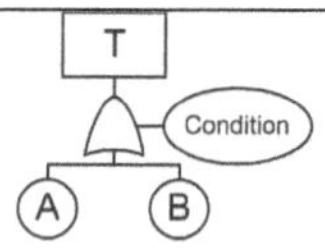	$P(T) = P(A) + P(B) - 2 \times P(A) \times P(B)$
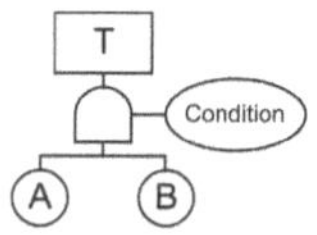	$P(T) = \frac{P(A) \times P(B)}{N!}$ where N is the number of inputs
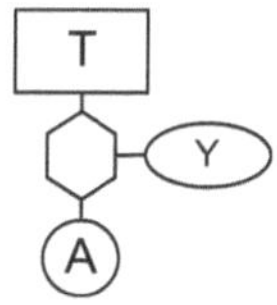	$P(T) = P(A) \times P(B)$

bounded and convex and can be expressed as follows.

$$\mu_{\tilde{A}}(x) = \begin{cases} \mu^L_{\tilde{A}}(x), & a \leqslant x \leqslant b \\ 1, & b \leqslant x \leqslant c \\ \mu^R_{\tilde{A}}(x), & c \leqslant x \leqslant d \\ 0, & \text{otherwise} \end{cases} \tag{7.3}$$

where $\mu^L_{\tilde{A}}(x) : [a,b] \to [0,w]$ and $\mu^R_{\tilde{A}}(x) : [c,d] \to [0,w]$.

Definition 7.2 (Inverse of membership functions). Since both membership functions, $\mu^L_{\tilde{A}}(x)$ and $\mu^R_{\tilde{A}}(x)$, are continuously increasing and decreasing, they also have inverse functions, $\mu^L_{\tilde{A}}(y)$ and $\mu^R_{\tilde{A}}(y)$, which are calculated as follows.

$$\mu^L_{\tilde{A}}(y) = \left(\mu^L_{\tilde{A}}(x)\right)^{-1} \tag{7.4}$$

$$\mu^R_{\tilde{A}}(y) = \left(\mu^R_{\tilde{A}}(x)\right)^{-1} \tag{7.5}$$

where $y \in [0,w]$.

Definition 7.3 (Nuclear Event Reliability Score). A nuclear event reliability score (R_s) in fuzzy reliability approach is a crisp score, which is converted from a membership function of fuzzy numbers using a defuzzification technique, to represent the experts' belief of the most value that a nuclear event might happen.

$$R_s = d\left(\mu_{\tilde{A}}(x)\right) \tag{7.6}$$

Definition 7.4 (Nuclear Event Fuzzy Failure Rate). A nuclear event fuzzy failure rate (R) is a nuclear event reliability data in fuzzy reliability approach, which is similar to a probabilistic failure rate in probabilistic reliability approach. It is calculated from nuclear event reliability score using a logarithmic function proposed by Onisawa [23].

$$R = \begin{cases} \frac{1}{10^m}, & R_s \neq 0 \\ 0, & R_s = 0 \end{cases} \tag{7.7}$$

where $m = \left[\frac{1-R_s}{R_s}\right]^{1/3} \times 2.301$.

Definition 7.5 (Area Defuzzification Technique). An area defuzzification technique (ADT) utilizes the centroid point of the membership functions on the vertical axis and its intersection with the left and the right membership functions as shown by the grayed area in Fig. 7.4.

$$ADT = d\left(\mu_{\tilde{A}}(x)\right) = x_1 y_0 + \int_{x_2}^{d} \mu_{\tilde{A}}^{R}(x)dx \tag{7.8}$$

where y_0 is the centroid point of the real fuzzy number $\tilde{A}$ on the vertical axis, x_1 is the intersection point between the line y_0 and the left membership function $\mu_{\tilde{A}}^{L}(x)$ on the horizontal axis, and x_2 is the intersection point between the line y_0 and the right membership function $\mu_{\tilde{A}}^{R}(x)$ on the horizontal axis. The y_0, x_1 and x_2 are calculated using Eqs. (7.9)–(7.11).

$$y_0 = \frac{\int_0^w y \cdot \mu_{\tilde{A}}^{R}(y)dy - \int_0^w y \cdot \mu_{\tilde{A}}^{L}(y)dy}{\int_0^w \mu_{\tilde{A}}^{R}(y)dy - \int_0^w \mu_{\tilde{A}}^{L}(y)dy} \tag{7.9}$$

$$x_1 = \mu_{\tilde{A}}^{L}(y_0) \tag{7.10}$$

$$x_2 = \mu_{\tilde{A}}^{R}(y_0) \tag{7.11}$$

Theorem 7.1. *If $\tilde{A} = (a,b,c,d;1)$ is a normal trapezoidal fuzzy number, then its centroid point on the vertical axis is shown in Eq.* (7.12)

$$y_0 = \frac{1}{3}\left[\frac{d+2c-a-2b}{d+c-a-b}\right] \tag{7.12}$$

Proof. For a normal trapezoidal fuzzy numbers, the left membership function, the right membership function and their inverse functions are respectively calculated as in Eqs.

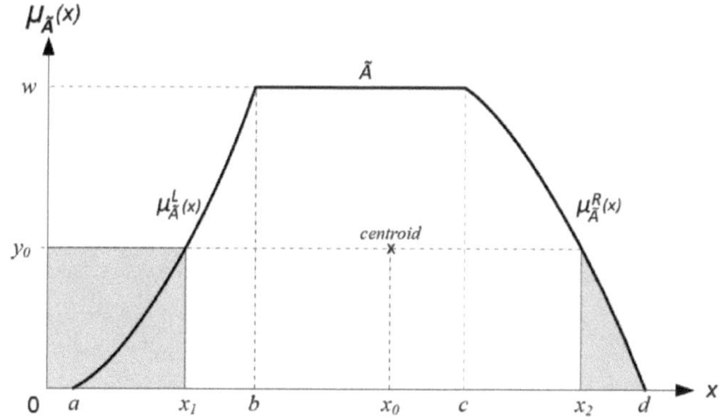

Fig. 7.4 Area defuzzification technique.

(7.13)–(7.16).

$$\mu^L_{\tilde{A}}(x) = \frac{x-a}{b-a} \tag{7.13}$$

$$\mu^R_{\tilde{A}}(x) = \frac{d-x}{d-c} \tag{7.14}$$

$$\mu^L_{\tilde{A}}(y) = a + (b-a)y \tag{7.15}$$

$$\mu^R_{\tilde{A}}(y) = d + (c-d)y \tag{7.16}$$

By substituting Eqs. (7.15)–(7.16) into Eq. (7.9), the centroid point of a normal trapezoidal fuzzy number on the vertical axis is calculated as follows:

$$y_0 = \frac{\int_0^1 [y(d+(c-d)y)]dy - \int_0^1 [y(a+(b-a)y)]dy}{\int_0^1 [d+(c-d)y]dy - \int_0^1 [a+(b-a)y]dy} = \frac{1}{3}\left[\frac{2c+d-a-2b}{c+d-a-b}\right]$$

therefore proof is complete. □

Theorem 7.2. *If $\tilde{A} = (a,b,c,d;1)$ is a normal trapezoidal fuzzy number, y_0 is its centroid point on the vertical axis, then the intersection between the line y_0 and the left membership function $\mu^L_{\tilde{A}}(x)$ on the horizontal axis is shown in Eq. (7.17)*

$$x_1 = \frac{2(a+b)^2 - a(2b+c+2d) - b(2c+d)}{3b-c-d)} \tag{7.17}$$

Proof. By substituting Eq. (7.15) into Eq. (7.10), the intersection point between the line y_0 and the left membership function $\mu^L_{\tilde{A}}(x)$ on the horizontal axis is calculated as follows.

$$x_1 = a + (b-a)\left(\frac{1}{3}\left[\frac{2c+d-a-2b}{c+d-a-b}\right]\right) = \frac{2(a+b)^2 - a(2b+c+2d) - b(2c+d)}{3(a+b-c-d)}$$

therefore proof is complete. □

Theorem 7.3. *If $\tilde{A} = (a,b,c,d;1)$ is a normal trapezoidal fuzzy number, y_0 is its centroid point on the vertical axis, then the intersection between the line y_0 and the right membership function $\mu_{\tilde{A}}^{R}(x)$ on the horizontal axis is shown in Eq.* (7.18).

$$x_2 = \frac{2(c+d)^2 - c(a+2b) - d(2a+b+2c)}{3(c+d-a-b)} \tag{7.18}$$

Proof. By substituting Eq. (7.16) into Eq. (7.11), the intersection point between the line y_0 and the right membership function $\mu_{\tilde{A}}^{R}(x)$ on the horizontal axis is calculated as follows.

$$x_2 = d + (c-d)\frac{1}{3}\left[\frac{d+2c-a-2b}{d+c-a-b}\right] = \frac{2(c+d)^2 - c(a+2b) - d(2a+b+2c)}{3(c+d-a-b)}$$

therefore proof is complete. □

Proposition 7.1. *If $\tilde{A} = (a,b,c,d;1)$ is a normal trapezoidal fuzzy number, then its area defuzzification technique (ADT) is shown in Eq.* (7.19).

$$ADT = \frac{(a+2b-2c-d)\left((2a+2b)^2 + (c+d)(-3a+2c-d) - 2c(3b+d) - 4ab\right)}{18(a+b-c-d)^2} \tag{7.19}$$

7.4 A Failure Possibility-Based Reliability Algorithm

In this section, the failure possibility-based reliability algorithm to calculate component fuzzy failure rates from component failure possibilities, which are expressed in qualitative natural languages, mathematically represented by the membership functions of fuzzy numbers, and subjectively justified by a group of experts based on their expertise and working experiences, is described in detail. The algorithm estimates nuclear event reliability data without the need for historical failure data for probabilistic calculation. Hence, this fuzzy reliability approach is a very good alternative for the probabilistic reliability approach when historical failure data is inadequate or unavailable. The algorithm consists of six steps and each step is explained as a sub section below.

7.4.1 *Define failure possibility distribution*

When historical failure data is inadequate or unavailable for probabilistic failure rate calculation, experts are more comfortable to use qualitative justification such as '*low possibility to be fail*' rather than quantitative justification such as '*1.5E-3*'. The purpose of this step is to define the distribution of the linguistic terms in the context of failure possibilities to quantitatively assess event failure rates. The number of linguistic terms to represent event failure possibilities is called the granularity of the failure possibility distribution.

Let H denote a possibility distribution, h_i represent a failure possibility and m represent the granularity of the failure possibilities to qualitatively justify nuclear events as in Eq. (7.20).

$$H = \{h_i \mid i = 1, 2, \ldots, m\} \tag{7.20}$$

7.4.2 *Build a mathematical model for each failure possibility*

The purpose of this step is to mathematically model every individual event failure possibility (h_i) into a membership function of a fuzzy number. Therefore, each predefined failure possibility in Step 7.4.1 has to have a corresponding mathematical representation. The shapes of the membership functions to mathematically represent linguistic variables in engineering systems should be normal, bounded and convex such as trapezoidal and triangular fuzzy numbers [45]. They are also simple to be implemented in engineering systems and sufficiently accurate [26].

Let μ_i denote the membership function of h_i, the left endpoint, cores and the right endpoint of the membership functions are x_{1i}, x_{2i}, x_{3i}, and x_{4i}, respectively. Then the mathematical form of the h_i can be expressed as in Eq. (7.21).

$$\mu_i(x) = \begin{cases} \dfrac{x - x_{1i}}{x_{2i} - x_{1i}}, & x_{1i} \leqslant x \leqslant x_{2i} \\ 1, & x_{2i} \leqslant x \leqslant x_{3i} \\ \dfrac{x_{4i} - x}{x_{4i} - x_{3i}}, & x_{3i} \leqslant x \leqslant x_{4i} \\ 0, & \text{otherwise} \end{cases} \tag{7.21}$$

where

$\mu = \{\mu_i \mid i = 1, 2, \ldots, m\}$; $\mu_i : X \to [0, 1]$ and $\mu = \{(x, \mu_i(x)) \mid x \in X \text{ and } 0 \leqslant \mu_i(x) \leqslant 1\}$.

7.4.3 *Collect and evaluate component failure possibilities from a group of experts*

The expert-opinion elicitation process is a formal process of obtaining information or answers to specific questions about component failure likelihoods. The purpose of this step is to collect and evaluate the failure likelihood of an event, which is under investigation, from a group of experts. In this step, we consider an expert as a very skilful person, who is familiar with the system, understands system's working environment, and has considerable training in and knowledge of the nuclear field.

Let Y is the event to be evaluated, and then we can ask every expert in the group to subjectively justify the failure likelihood of the event Y using the predefined failure possibility h_i in the first step as follows.

How likely is the event Y to fail? Is it $h_1,\ h_2,\ h_3,\ldots,\ h_m$?

7.4.4 *Calculate component final membership functions*

It is possible that experts who are involved in the elicitation process justify same events using different failure possibilities due to different expertise, different working experiences, and different justification confidential level. Weighted average aggregation is considered as the most appropriate aggregation technique to aggregate all experts' justifications into one final membership function for every event to reach a consensus. It considers experts' knowledge, expertise, and confidential level by giving different justification weights to every individual expert. An expert with the highest justification weight is the expert with the most knowledgeable, expertise, and confident on the system being investigated.

Let $E = \{e_1, e_2, \ldots, e_n\}$ $(n \geqslant 2)$ be a set of experts to evaluate nuclear events using H. Let $W = \{w_{e1}, w_{e2}, w_{e3}, \ldots, w_{en}\}$ $(0 < w_{en} \leqslant 1)$ denote the justification weight of the expert E. Let $\mu_Y\ (x_1,\ x_2,\ x_3,\ x_4)$ denote the final membership function for the basic event Y, which satisfies $\mu_Y : X \to [0,1]$ and $x_1 \leqslant x_2 \leqslant x_3 \leqslant x_4$. The final membership function for the event Y is calculated as in Eq. (7.22).

$$\mu_Y(x) = \frac{w_{e_1} \cdot \mu_{ie_1} \oplus w_{e_2} \cdot \mu_{ie_2} \oplus w_{e_3} \cdot \mu_{ie_3} \oplus \cdots \oplus w_{e_n} \cdot \mu_{ie_n}}{w_{e_1} \oplus w_{e_2} \oplus w_{e_3} \oplus \cdots \oplus w_{e_n}} \tag{7.22}$$

where n is the number of experts justifying an event, μ_{ie_n} is the ith membership function (μ_i) justified by the nth expert (e_n) for the event Y and $i = 0, 1, \ldots, m$, are the granularity of the predefined failure possibilities.

7.4.5 *Calculate nuclear event reliability scores*

In this step, the final membership function in Step 7.4.4 is converted into a nuclear event reliability score (R_S) using the area defuzzification technique described in Section 7.3. This score represents the experts' belief of the most likely score indicating that a nuclear event may occur.

Let $\mu_Y\ (x_1,\ x_2,\ x_3,\ x_4)$ denote the final membership function for the basic event Y, which satisfies $\mu_Y : X \to [0,1]$ and $x_1 \leqslant x_2 \leqslant x_3 \leqslant x_4$ as shown in Step 7.4.4, then the reliability score (R_S) for the event Y is

$$R_S = \frac{(x_1+2x_2-2x_3-x_4)\big((2x_1+2x_2)^2+(x_3+x_4)(-3x_1+2x_3-x_4)-2x_3(3x_2+x_4)-4x_1x_2\big)}{18(x_1+x_2-x_3-x_4)^2} \tag{7.23}$$

7.4.6 *Calculate nuclear event reliability data*

In this step, the event reliability score calculated in Step 7.4.5 is then converted into the corresponding fuzzy failure rate, which is similar to the probabilistic failure rate. The logarithmic function as in Eq. (7.7) utilizing the concept of error possibility and likely fault rate is used. Based on this, the fuzzy failure rate (R) of the event Y is calculated as in Eq. (7.24).

$$R = \begin{cases} \dfrac{1}{10^{\left[\frac{1-R_S}{R_S}\right]^{1/3} \times 2.301}}, & R_s \neq 0 \\ 0, & R_s = 0 \end{cases} \tag{7.24}$$

The obtained fuzzy failure rates (R) will be similar to the probabilistic failure rates, which are calculated from available historical failure data. Hence, they can be applied in NPP PSA by FTA.

7.5 A Case Study

A high pressure core spray system (HPCSS) is an integral part of an emergency core cooling system (ECCS) in boiling water reactors (BWRs). The function of this HPCSS is to depressurize and to supply water to the primary system in case of loss of reactor coolant inventory. If this safety system works well, fuel cladding damage can be avoided. For the sake of simplicity, a simplified model of HPCSS by Paredes *et al.* [46] in Fig. 7.5 is used to mathematically proof the failure possibility-based reliability algorithm described in Section 7.4.

The HPCSS fails if there is no water flowing into the reactor vessel in the event of loss of coolant accident (LOCA). By investigating this system, there are three possibilities causing this system failure. The first possible cause is that the reactor injection valve fails to open. If this valve works well, then the second possible cause is that the pump cannot flow water from a suction system into the reactor injection valve. If the pump also works well, then the third possible cause is that the suction system cannot supply water into the pump. These three possible causes are represented by an OR gate in the fault tree in Fig. 7.6.

The suction system failure still can be further investigated. The suction system fails to flow water from the tank into the pump if both the condensate storage system and the suppression pool system fail together at the same time. This scenario is represented by an AND gate in the fault tree in Fig. 7.6. The condensate storage system fails if the condensate storage suction valve fails to open or the condensate storage tank level is low. This failure

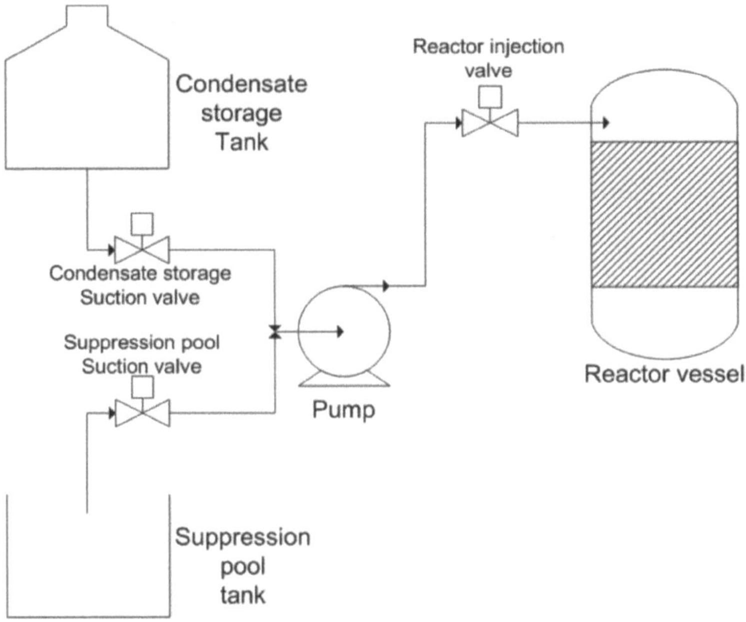

Fig. 7.5 The HPCSS diagram block.

Table 7.6 The meaning of the symbols in the HPCSS.

Basic Event	Legends
A	The pump is failure
B	The reactor injection valve is failure
C	The condensate storage water level is low
D	The condensate storage suction valve is failure
E	The suppression pool water level is low
F	The suppression pool suction valve is failure

scenario is then represented by an OR gate in the fault tree in Fig. 7.6. The same failure scenario occurs to the suppression pool system as well. The complete fault tree, which graphically represents the failure scenario of the simplified HPCSS in Fig. 7.5, is shown in Fig. 7.6 and Table 7.6 lists the meanings of the used symbols in the tree.

The HPCSS fault tree in Fig. 7.6 has been free from repeating events and non-minimal cut sets; hence, it can be directly used to calculate the failure probability of the undesired

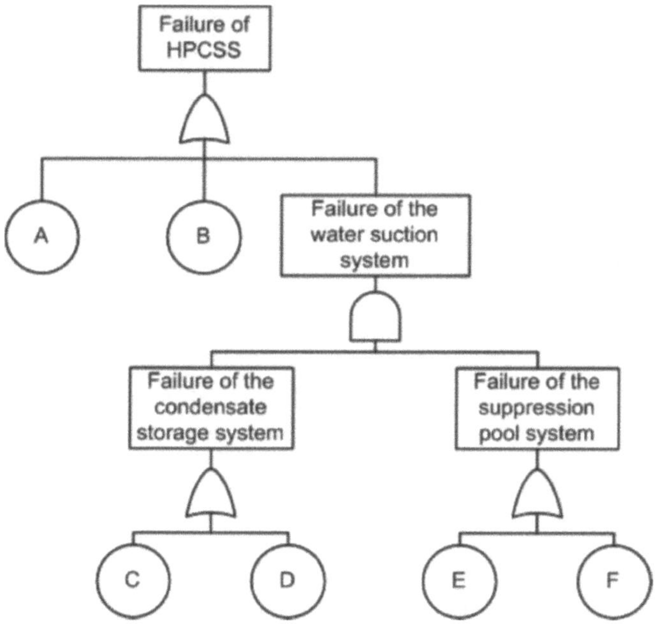

Fig. 7.6 The HPCSS fault tree.

top event in Fig. 7.5, which is the failure of the HPCSS to flow water into the reactor vessel in the event of LOCA.

7.5.1 *Define failure possibility distribution*

Based on the combination between the type of the components and the likely failure occurrences, seven linguistic terms are used to define nuclear event failure possibilities. Events with '*very low*' (*VL*) failure possibilities mean that the events are rigid and very unlikely to become failure, even once. Events with '*very high*' (*VH*) failure possibilities mean that the events have many moving parts and are near certain to become failures several times. Events with '*low*' (*L*), '*reasonably low*' (*RL*), '*moderate*' (*M*), '*reasonably high*' (*RH*), and '*high*' (*H*) failure possibilities are up-graded from '*very low*' (*VL*) to '*very high*' (*VH*) as shown in Eq. (7.25).

$$H = \{h_1, h_2, h_3, h_4, h_5, h_6, h_7\} = \{VL, L, RL, M, RH, H, VH\} \tag{7.25}$$

7.5.2 *Build a mathematical model for each failure possibility*

The mathematical representation of those failure possibilities in Eq. (7.25) are shown in Eqs. (7.26)–(7.32) and their corresponding graphical representations are given in Fig. 7.7.

$$\mu_{VL}(x) = \begin{cases} \dfrac{x}{0.05}, & 0 \leqslant x \leqslant 0.05 \\ \dfrac{0.10 - x}{0.05}, & 0.05 \leqslant x \leqslant 0.10 \\ 0, & \text{otherwise} \end{cases} \tag{7.26}$$

$$\mu_{L}(x) = \begin{cases} \dfrac{x - 0.075}{0.10}, & 0.075 \leqslant x \leqslant 0.175 \\ \dfrac{0.275 - x}{0.10}, & 0.175 \leqslant x \leqslant 0.275 \\ 0, & \text{otherwise} \end{cases} \tag{7.27}$$

$$\mu_{RL}(x) = \begin{cases} \dfrac{x - 0.2375}{0.10}, & 0.2375 \leqslant x \leqslant 0.3375 \\ \dfrac{0.4375 - x}{0.10}, & 0.3375 \leqslant x \leqslant 0.4375 \\ 0, & \text{otherwise} \end{cases} \tag{7.28}$$

$$\mu_{M}(x) = \begin{cases} \dfrac{x - 0.40}{0.10}, & 0.40 \leqslant x \leqslant 0.50 \\ \dfrac{0.60 - x}{0.10}, & 0.50 \leqslant x \leqslant 0.60 \\ 0, & \text{otherwise} \end{cases} \tag{7.29}$$

$$\mu_{RH}(x) = \begin{cases} \dfrac{x - 0.5625}{0.10}, & 0.5625 \leqslant x \leqslant 0.6625 \\ \dfrac{0.7625 - x}{0.10}, & 0.6625 \leqslant x \leqslant 0.7625 \\ 0, & \text{otherwise} \end{cases} \tag{7.30}$$

$$\mu_{H}(x) = \begin{cases} \dfrac{x - 0.725}{0.10}, & 0.725 \leqslant x \leqslant 0.825 \\ \dfrac{0.925 - x}{0.10}, & 0.825 \leqslant x \leqslant 0.925 \\ 0, & \text{otherwise} \end{cases} \tag{7.31}$$

$$\mu_{VH}(x) = \begin{cases} \dfrac{x - 0.90}{0.05}, & 0.90 \leqslant x \leqslant 0.95 \\ \dfrac{1.00 - x}{0.05}, & 0.95 \leqslant x \leqslant 1.00 \\ 0, & \text{otherwise} \end{cases} \tag{7.32}$$

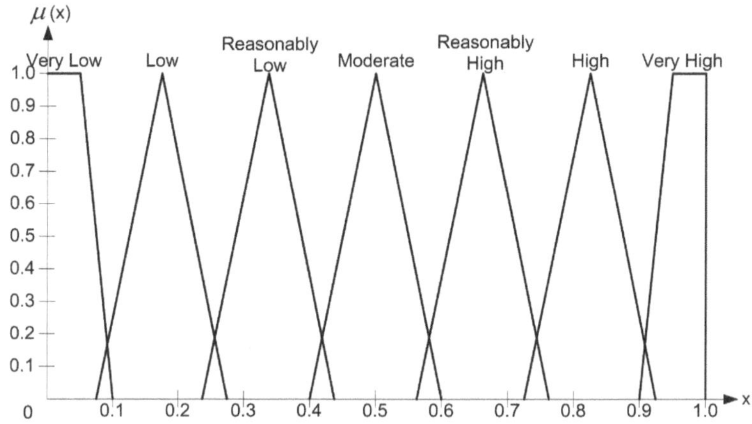

Fig. 7.7 Graphical representation of the failure distribution H.

Table 7.7 List of questionnaires and experts' evaluation results.

Questions	Expert and justification weight				
	1 ($w=1$)	2 ($w=1$)	*3* ($w=1$)	*4* ($w=0.9$)	*5* ($w=0.9$)
How likely is the pump to fail?	*RL*	*RL*	*RL*	*M*	*M*
How likely is the reactor injection valve to fail?	*RL*	*RL*	*RL*	*L*	*L*
How likely is the water level of the condensate storage to be low?	*VL*	*VL*	*VL*	*L*	*L*
How likely is the suction valve of the condensate storage to fail?	*L*	*L*	*L*	*RL*	*RL*
How likely is the water level of the suppression pool to be low?	*VL*	*VL*	*VL*	*L*	*L*
How likely is the suction valve of the suppression pool to fail?	*L*	*L*	*L*	*RL*	*RL*

7.5.3 *Collect and evaluate component failure possibilities from a group of experts*

Let assume, we ask five experts who have different level of expertise, working experience, and justification confidence to evaluate nuclear events in Table 7.6. The questionnaires, the experts' justification weights, and the evaluation results are shown in Table 7.7.

Table 7.8 The final membership function.

Basic Event	Final membership functions
A	(0.2984, 0.3984, 0.3984, 0.4984)
B	(0.1766, 0.2766, 0.2766, 0.3766)
C	(0.0281, 0.0969, 0.0969, 0.1656)
D	(0.1359, 0.2359, 0.2359, 0.3359)
E	(0.0281, 0.0969, 0.0969, 0.1656)
F	(0.1359, 0.2359, 0.2359, 0.3359)

7.5.4 *Calculate component final membership functions*

Using Eq. (7.22), the final membership function for the pump failure possibility (*A*) in the HPCSS is calculated as follows.

$$\mu_A(x_1) = \frac{[1\times(0.2375)]+[1\times(0.2375)]+[1\times(0.2375)]+[0.9\times(0.4)]+[0.9\times(0.4)]}{1+1+1+0.9+0.9} = (0.2984)$$

$$\mu_A(x_2) = \mu_A(x_3) = \frac{[\times(0.3375)]+[1\times(0.3375)]+[1\times(0.3375)]+[0.9\times(0.5)]+[0.9\times(0.5)]}{1+1+1+0.9+0.9} = (0.3984)$$

$$\mu_A(x_4) = \frac{[1\times(0.4375)]+[1\times(0.4375)]+[1\times(0.4375)]+[0.9\times(0.6)]+[0.9\times(0.6)]}{1+1+1+0.9+0.9} = (0.4984)$$

The final membership functions for all other basic event failure possibilities in Table 7.8 are also calculated using the same procedures.

7.5.5 *Calculate nuclear event reliability scores*

Using Eq. (7.23), the reliability score (R_S) for the pump (*A*) is calculated as follows.

$$R_s(A) = \frac{(-0.2)((1.3936)^2+(0.8968)(-0.5968)-(0.7968)(1.69364)-0.4755)}{18(-0.2)^2} = 0.116146$$

The reliability scores for all other basic events in Table 7.9 are also calculated using the same procedure.

7.5.6 *Calculate nuclear event reliability data*

Using Eq. (7.24), the fuzzy failure rate of the pump (*A*) is calculated as follows.

$$R(A) = \frac{1}{10^{\left(\frac{1-0.116146}{0.116146}\right)^{1/3}\times 2.301}} = 2.97892\text{E-05}$$

The fuzzy failure rates for all other basic events in Table 7.9 are also calculated using the same procedure.

Based on the fault tree shown in Fig. 7.6, the failure probability of the HPCSS is then calculated by applying those fuzzy failure rates in Table 7.9 into Eqs. (7.1)–(7.2). The failure probability calculation is started from the bottom of the tree to the top of the tree.

Table 7.9 Basic events' reliability scores and fuzzy failure rates.

Basic event	Event reliability scores (Rs)	Event fuzzy failure rates (R)
A	0.116146	2.97892E-05
B	0.075521	4.97666E-06
C	0.020833	4.96697E-09
D	0.061979	2.03395E-06
E	0.020833	4.96697E-09
F	0.061979	2.03395E-06

Table 7.10 The results of failure probability calculation.

System failure	Failure probability
Condensate storage system failure	2.0389E-06
Suppression pool system failure	2.0389E-06
Water suction system failure	4.1572E-12
High pressure core spray system failure	3.4766E-05

The failure probability of the condensate storage system (P_{CSS}) is calculated by applying fuzzy failure rates of basic events C and D in Table 7.9 into Eq. (7.1) as shown below.

$$P_{CSS} = 1 - [(1 - 4.96697\text{E-}09) \times (1 - 2.03395\text{E-}06)] = 2.0389\text{E-}06$$

The same procedure is applied for the calculation of the failure probability of the suppression pool system (P_{SPS}) in Table 7.10 but using the fuzzy failure rates of the basic events E and F.

The failure probability of the water suction system (P_{WSS}) is calculated by applying the calculated failure probability for condensate storage system (P_{CSS}) and the calculated failure probability for suppression pool system (P_{SPS}) into Eq. (7.2) as shown below.

$$P_{WSS} = 2.0389\text{E-}06 \times 2.0389\text{E-}06 = 4.1572\text{E-}12$$

Finally, the failure probability of the top undesired event, which is the failure of the high pressure core spray system (HPCSS) is calculated by applying the fuzzy failure rates of basic events A and B in Table 7.9 and the calculated P_{WSS} into Eq. (7.1). The results of failure probability calculation for all sub systems and the top undesired event are summarized in Table 7.10.

This case study shows that the failure possibility-based reliability algorithm can be applied to estimate nuclear event failure rates for nuclear power plant probabilistic safety assessment by fault tree analysis. The algorithm is also more intuitive and easy for experts to assess basic event failure rates where historical failure data are insufficient or unavailable for numerical estimation.

7.6 Conclusion and Outlook

This chapter describes a failure possibility-based reliability algorithm to assess nuclear event failure rates without the need for historical failure data as an alternative for probabilistic failure rates to be used for nuclear power plant probabilistic safety assessment by fault tree analysis. Safety analysts define a possibility distribution, which is expressed in qualitative natural languages in the context of failure possibilities and mathematically represented by the membership functions of fuzzy numbers. A group of experts then evaluate nuclear events using those predefined failure possibilities. Since experts may have different expertise, different working background and different confidential level, they may evaluate same events with different failure possibilities. To accommodate this bias, weighted average aggregation is used to aggregate experts' justifications into one final membership function to reach a consensus by giving a justification weight for every expert in the group. An expert with the highest justification weight is the expert with the most knowledgeable and confident on the system being investigated. A case study shows that the failure possibility-based reliability algorithm offers a good alternative of assessing nuclear event failure rates for nuclear power plant probabilistic safety assessment by fault tree analysis when historical failure data is inadequate or unavailable for probabilistic failure rate calculation.

Based on current results, we will conduct more experiments to validate and evaluate this failure possibility-based reliability algorithm. Sensitivity analysis also will be conducted to see how sensitive the results to the variation of the number of experts involved in the justification processes.

Bibliography

[1] Liu, T., J. Tong, and J. Zhao., Probabilistic Risk Assessment Framework Development for Nuclear Power Plant. in *IEEE International Conference on Industrial Engineering and Engineering Management.* 2008.

[2] Delaney, M.J., G.E. Apostolakis, and M.J. Driscoll, Risk-Informed Design Guidance for Future Reactor Systems. *Nucl. Eng. Des.*, 2005. **235** (14): p. 1537–1556.

[3] Kishi, T., H. Kikuchi, S. Miura, M. Fukuda, M. Hirano, and N. Watanabe, Application of Probabilistic Safety Assessment to the Pipe Rupture Incident at Hamaoka Unit-1. *J. Nucl. Sci. Technol.*, 2004. **41** (1): p. 77–85.

[4] Hsu, F. and Z. Musicki, Issues and Insights of PRA Methodology in Nuclear and Space Applications, in *IEEE International Conference on Systems, Man and Cybernetics*. 2005. p. 510–517.

[5] Yuhua, D. and Y. Datao, Estimation of Failure Probability of Oil and Gas Transmission Pipelines by Fuzzy Fault Tree Analysis. *J. Loss Prev. Process Ind.*, 2005. **18**: p. 83–88.

[6] Hadavi, S.M.H., WWER-1000 Shutdown Probabilistic Risk Assessment: An Introductory Insight. *Ann. Nucl. Energy*, 2008. **35** (2): p. 196–208.

[7] Bodansky, D., *Nuclear Reactor Safety, in Nuclear Energy: Principles, Practices, and Prospects*. 2004, Springer-Verlag: New York. p. 371–410.

[8] Dhillon, B.S., *Reliability Evaluation Methods, in Reliability, Quality, and Safety for Engineers*. 2005, CRC Press LLC: Florida. p. 87–105.

[9] Stacey, W.M., *Reactor Safety, in Nuclear Reactor Physics*. 2007, WILEY-VCH Verlag GmbH & Co. KGaA: Weinheim. p. 283–302.

[10] Ericson, C.A., *Fault Tree Analysis, in Hazard Analysis Techniques for System Safety*, Ericson, Editor. 2005, John Wiley & Sons, Inc.: Virginia. p. 183–221.

[11] Huang, H.Z., X. Tonga, and M.J. Zuo, Posbist Fault Tree Analysis of Coherent Systems. *Reliab. Eng. Syst. Saf.*, 2004. **84** (1): p. 141–148.

[12] Wall, I.B., J.J. Haugh, and D.H. Worlege, Recent Applications of PSA for Managing Nuclear Power Plant Safety. *Prog. Nucl. Energy*, 2001. **39** (3-4): p. 367–425.

[13] Shu, L., J. Li, and M. Qiu. Study on Applying Fault Tree Analysis Based on Fuzzy Reasoning in Risk Analysis of Construction Quality. in *International Conference on Risk Management & Engineering Management*. 2008. Beijing.

[14] Song, H., H.Y. Zhang, and C.W. Chan, Fuzzy Fault Tree Analysis Based on T–S Model With Application to INS/GPS Navigation System. Soft Computing – A Fusion of Foundations, *Methodologies and Applications*, 2009. **13** (1): p. 31–40.

[15] NEA, Living PSA and its Use in the Nuclear Safety Decision-making Process, in *Report by a Group of Experts, O.f.E.C.-o.a. Development*, Editor. 2005, Nuclear Energy Agency: Paris.

[16] Lu, J., G. Zhang, D. Ruan, and F. Wu, Multi-Objective Group Decision Making: Methods, Software and Applications with Fuzzy Set Techniques. 1 ed. *Electrical and Computer Engineering*, ed. W.K. Chen. 2007, London: Imperial College Press. 390.

[17] Colettia, G. and R. Scozzafava, Conditional Probability, Fuzzy Sets, and Possibility: a Unifying View. *Fuzzy Sets Syst.*, 2004. **144** (1): p. 227–249.

[18] Suresh, P.V., A.K. Babar, and V. Venkat Raj, Uncertainty in Fault Tree Analysis: A Fuzzy Approach. *Fuzzy Sets Syst.*, 1996. **83** (2): p. 135–141.

[19] Sharma, U. and M. Sudhakar, Use of Recursive Methods in Fuzzy Fault Tree Analysis: an Aid to Quantitative Risk Analysis. *Reliab. Eng. Syst. Saf.*, 1993. **41** (3): p. 231–237.

[20] Moon, J.H. and C.S. Kang, Use of Fuzzy Set Theory in the Aggregation of Expert Judgments. *Ann. Nucl. Energy*, 1999. **26** (6): p. 461–469.

[21] Cai, K.Y., C.Y. Wen, and M.L. Zhang, Fuzzy variables as a basis for a theory of fuzzy reliability in the possibility context. *Fuzzy Sets Syst.*, 1991. **42** (2): p. 145–172.

[22] Huang, H.Z., M.J. Zuo, and Z.Q. Sun, Bayesian reliability analysis for fuzzy lifetime data. *Fuzzy Sets Syst.*, 2006. **157** (12): p. 1674–1686.

[23] Onisawa, T., An Approach to Human Reliability in Man-Machine Systems Using Error Possibility. *Fuzzy Sets Syst.*, 1988. **27** (2): p. 87–103.

[24] Utkin, L.V. and S.V. Gurov, A general formal approach for fuzzy reliability analysis in the possibility context. *Fuzzy Sets Syst.*, 1996. **83** (2): p. 203–213.

[25] Dumitrescu, M., T. Munteanu, I. Voncila, G. Gurguiatu, D. Floricau, and A.P. Ulmeanu, Application of Fuzzy Logic in Safety Computing for a Power Protection System, in *Fuzzy Systems and Knowledge Discovery*, L. Wang, *et al.*, Editors. 2006, Springer Verlag: Berlin Heidelberg. p. 980–989.

[26] Wolkenhauer, O., Fuzzy Mathematics, in *Data Engineering: Fuzzy Mathematics in Systems Theory and Data Analysis*. 2001, John Wiley & Sons, Inc. p. 197–212.

[27] Zadeh, L.A., Fuzzy Sets as a Basis for a Theory of Possibility. *Fuzzy Sets Syst.*, 1978. **1** (1): p. 3–28.

[28] Zadeh, L.A., Fuzzy Sets. *Inform. Control*, 1965. **8**: p. 338–353.

[29] Dubois, D. and H. Prade, Possibility theory and data fusion in poorly informed environments. *Control Eng. Pract.*, 1994. **2** (5): p. 811–823.

[30] Moller, B., M. Beer, W. Graf, and A. Hoffmann, Possibility Theory Based Safety Assessment. *Computer-Aided Civil & Infrastructure Engineering*, 1999. **14** (2): p. 81–91.

[31] Purba, J.H., J. Lu, D. Ruan, and G. Zhang, A Hybrid Approach for Fault Tree Analysis Combining Probabilistic Method with Fuzzy Numbers, in *Artificial Intelligence and Soft Computing*, L. Rutkowski, *et al.*, Editors. 2010, Springer: Berlin / Heidelberg. p. 194–201.

[32] Purba, J.H., J. Lu, D. Ruan, and G. Zhang. Probabilistic safety assessment in nuclear power plants by fuzzy numbers. in *The 9th International FLINS Conference*. 2010. Chengdu: World Scientific.

[33] Purba, J.H., J. Lu, D. Ruan, and G. Zhang, Failure possibilities for nuclear safety assessment by fault tree analysis. *Int. J. Nucl. Knowl. Manag.*, 2011. **5** (2): p. 162–177.

[34] Purba, J.H., J. Lu, D. Ruan, and G. Zhang. A Failure Possibility-Based Reliability Algorithm for Nuclear Safety Assessment by Fault Tree Analysis. in *The 1st International Workshop on Safety & Security Risk Assessment and Organizational Cultures* (SSRAOC2012). 2012. Antwerp: SCKCEN.

[35] Ma, J., J. Lu, and G. Zhang, Team Situation Awareness Measurement Using Group Aggregation and Implication Operators, in *International Conference on Intelligent Systems and Knowledge Engineering*. 2008: Xiamen.

[36] Zhang, G., J. Ma, and J. Lu, Emergency Management Evaluation by a Fuzzy Multi-Criteria Group Decision Support System Stoch. *Env. Res.Risk A.*, 2009. **23** (4): p. 517–527.

[37] Chin, K.S., Y.M. Wang, G.K.K. Poon, and J.B. Yang, Failure Mode and Effects Analysis Using a Group-Based Evidential Reasoning Approach. *Comput. Oper. Res.*, 2009. **36** (6): p. 1768–1779.

[38] Vanicek, J., I. Vrana, and S. Aly, Fuzzy Aggregation and Averaging for Group Decision Making: A Generalization and Survey. *Knowledge-Based Systems*, 2009. **22** (1): p. 79–84.

[39] Purba, J.H., J. Lu, D. Ruan, and G. Zhang, An Area Defuzzification Technique to Assess Nuclear Event Reliability Data from Failure Possibilities. *Int. J. Comput. Intell. Appl.* (accepted), 2012.

[40] Purba, J.H., J. Lu, and G. Zhang. An Area Defuzzification Technique and Essential Fuzzy Rules for Defuzzifying Nuclear Event Failure Possibilities into Reliability Data. in *The 10th International FLINS Conference* (accepted). 2012.

[41] Epstein, S. and A. Rauzy, Can We Trust PRA? Reliab. *Eng. Syst. Saf.*, 2005. **88** (3): p. 195–205.

[42] Yang, G., Potential Failure Mode Avoidance, in *Life Cycle Reliability Engineering*. 2007, John Wiley & Sons, Inc.: Hoboken, New Jersey. p. 194–235.

[43] Verma, A.K., A. Srividya, and D.R. Karanki, System Reliability Modeling, in *Reliability and Safety Engineering*. 2010, Springer-Verlag: London. p. 71–168.

[44] IAEA, IAEA Safety Glossary, in Terminology Used in Nuclear Safety and Radiation Protection. 2007, International Atomic Energy Agency: Vienna, Austria.

[45] Ferdous, R., F. Khan, R. Sadiq, P. Amyotte, and B. Veitch, Fault and Event Tree Analyses for Process Systems Risk Analysis: Uncertainty Handling Formulations. *Risk Anal.*, 2011. **31** (1): p. 86–107.

[46] Paredes, G.E., A.N. Carrera, A.V. Rodriguez, and E.G.E. Martinez, Modeling of the High Pressure Core Spray Systems with Fuzzy Cognitive Maps for Operational Transient Analysis in Nuclear Power Reactors. *Prog. Nucl. Energy*, 2009. **51** (3): p. 434–442.

Chapter 8

A Fuzzy Analytic Network Process Approach

Umut Asan, Ayberk Soyer, and Seyda Serdarasan

Istanbul Technical University, Industrial Engineering Department 34357, Maçka/Istanbul, Turkey

Email: {*asanu, ayberk, serdars*}*@itu.edu.tr*

This chapter describes the Analytical Network Process (ANP), a multi-criteria prioritization method to support decision making in complex and uncertain environments and suggests a fuzzy analytic network process (FANP) approach for prioritizing decision elements. The proposed fuzzy set theoretic method accommodates fuzziness in the supermatrix computations and thereby provides the opportunity to capture the uncertainty associated with the cumulative influence of each factor on every other factor with which it interacts. As its comparison to current methods demonstrates, the method successfully derives meaningful priorities from complex and uncertain decision structures.

8.1 Introduction

As stated in the IIE (`www.iienet2.org`) official definition, "Industrial engineering is concerned with the design, improvement and installation of integrated systems of people, materials, information, equipment and energy". This requires the Industrial engineer (IE) typically to evaluate multiple, even conflicting, criteria when making decisions. The problem situations that the IE is concerned with are not only complex because they involve multiple interacting criteria, but also there are multiple parties who take part in decision making process and would bear the consequences of the decisions made. Multi-criteria decision analysis (MCDA) provides a structured way to deal with these complex IE problems. The Analytic Network Process (ANP) is a MCDA approach that is proved to be effective when dealing with complex decision situations that involve interactions and feedbacks between decision elements. Another source of complexity in IE problems is uncertainty – es-

C. Kahraman (ed.), *Computational Intelligence Systems in Industrial Engineering*, Atlantis Computational Intelligence Systems 6, DOI: 10.2991/978-94-91216-77-0_8,

pecially in form of subjectivity, vagueness and imprecision –, however conventional ANP becomes less effective when dealing with uncertainty. As a way to address uncertainty, use of fuzzy set theoretic approaches has been extensively suggested in the literature. Accordingly, several fuzzy analytic network process (FANP) methods that use the concepts of fuzzy set theory to handle the uncertainty have been introduced in the existing literature. This chapter is aimed at presenting a FANP approach for prioritizing decision elements involved in IE problems. The proposed method accommodates fuzziness in the supermatrix computations and derives meaningful priorities from complex and uncertain decision structures.

The rest of this chapter is organized as follows. Section 8.2 provides an outline of the steps of the conventional ANP. In Section 8.3, the steps of the proposed FANP and its advantages are explained. In Section 8.4, a numerical example is given to demonstrate the ability of the proposed method in dealing with uncertainty and to examine its validity. Finally, conclusions are presented in Section 8.5.

8.2 The Analytic Network Process

The Analytic Network Process (ANP) provides a comprehensive framework to analyze a variety of decisions by deriving priority scales from individual judgments [1] representing the relative dependence among the elements of a system. Unlike the Analytic Hierarchy Process (AHP), ANP makes no assumptions about the independence of higher level elements from lower level elements and about the independence of the elements within a level [2]. In other words, there is no need to specify levels as in a hierarchy. ANP allows both interaction and feedback within and between clusters of elements [3]. The difference between a hierarchy and a network is illustrated in Figure 8.1. As it can be seen in Figure 8.1 (a), a hierarchical model consists of a goal, criteria, sub-criteria and alternatives, and imposes a linear structure, while a network model has cycles connecting its clusters of elements (outer dependence) and loops that connect a cluster to itself (inner dependence), as shown in Figure 8.1 (b). The four major steps of ANP are briefly introduced as follows:

Step 1: Model Construction (Forming the Network Structure)

The process of developing an ANP model begins with a comprehensive analysis of the decision problem and the relevant features. Following the statement of the problem, the elements in terms of criteria, sub-criteria and alternatives are defined and clusters of these elements are respectively formed. A network model is then developed based on the relationships between and within these clusters of elements. The constructed model should allow influence to be transmitted from a cluster to another one and back either directly or indirectly from the second cluster [2].

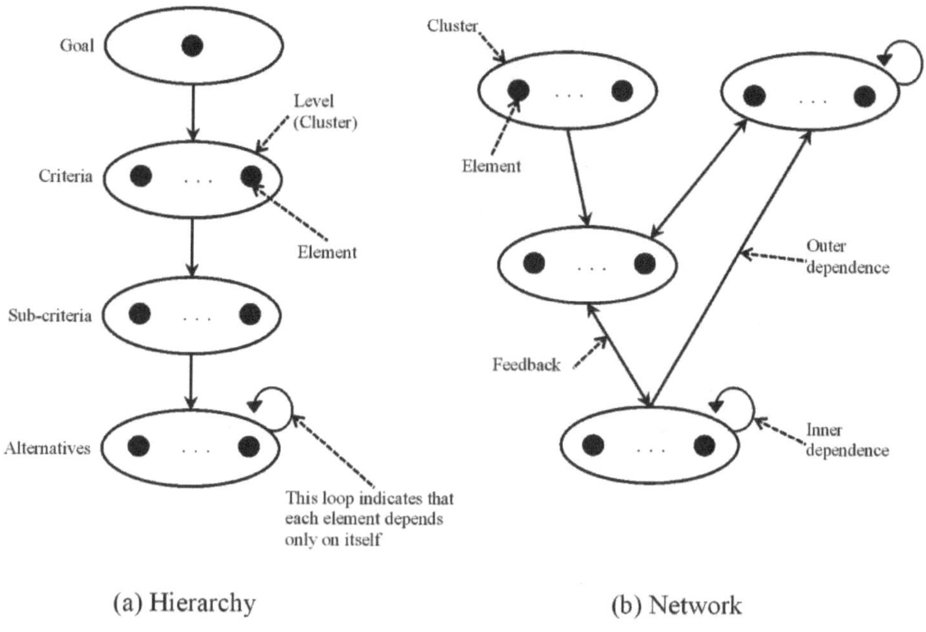

Fig. 8.1 Structural difference between a hierarchy and a network

Step 2: Pairwise Comparison and Local Priority Estimation

In this step, judgments are elicited to estimate the local priorities of elements; i.e., criteria, sub-criteria and alternatives. Decision makers are asked to perform a series of pairwise comparisons where two elements or two clusters at a time are compared in terms of dominance [1]. These comparisons are performed by answering the generic question: Given an element (in the same cluster or in another cluster) or a cluster, how much more does a given element (cluster) of a pair influence that element (cluster) with respect to a criterion? [3]. The responses to the pairwise comparison questions can be scaled on the basis of Saaty's 1-9 scale [4], where a score of 1 indicates indifference between the two elements and a score of 9 indicates overwhelming dominance of the element under consideration over the comparison element [5]. For the inverse comparisons a reciprocal value is automatically assigned, that is, $a_{ij} = 1/a_{ji}$, where a_{ij} denotes the relative dominance of the ith element compared to the jth element.

After the elicitation of subjective judgments from each decision maker in terms of pairwise comparisons, methodologies such as consensus voting or combined individual judgments [6, 7], which typically employs the weighted arithmetic mean method [8], geometric mean method [8–11], weighted geometric mean method [12], or logarithmic goal programming method proposed by Bryson and Joseph [13], can be used to aggregate individual judgments into a group judgment. During or reasonably after the pairwise comparisons, the inconsistency of individual or group judgments is measured by the consistency ratio (CR) proposed by Saaty [14] as a test of reliability and is tried to be reduced to an acceptable level.

Once the pairwise comparisons are completed, a local priority vector is derived for all comparison matrices, as an estimate of relative dominance associated with the elements (or clusters) being compared, by solving the following equation:

$$A \cdot w = \lambda_{\max} \cdot w \tag{8.1}$$

where A is the positive reciprocal matrix of pairwise comparisons, w is the principal eigenvector (priority vector) and $\lambda_{\max}$ is the largest eigenvalue of A. There are several algorithms for approximating w; such as power method, additive normalization method, weighted least-squares method, logarithmic least-squares method, and so on. For a review of alternative algorithms, the reader should refer to Srdjevic [15].

Step 3: Supermatrix Formation and Analysis

As described above, a local priority vector derived from the paired comparisons represents the dominance relations among a given set of elements (e.g., sub-criteria) in a cluster and, thereby, the influence of these elements on another element (e.g., a criterion) in the network [16]. A network, however, involves more complex interactions among the decision elements than offered by local priorities. In other words, the interactions are not restricted to direct impacts; there may be many indirect impacts between the elements. To capture the transmission of influence along all paths defined in the network and to obtain the overall priorities of the elements, Saaty [1] proposes the supermatrix approach. In this approach, each of the local priority vectors are entered as a part of some column of a matrix, known as a supermatrix [1]. This partitioned matrix represents the influence of an element on the left of the matrix on an element at the top of the matrix. The supermatrix, thus, serves as a unifying framework [17]. The general form of the supermatrix along with one of its block entries are shown in Figure 8.2.

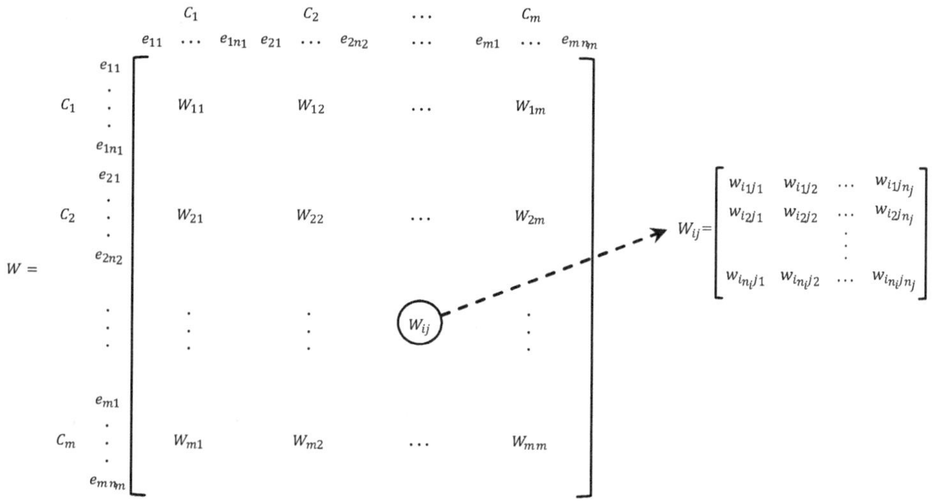

Fig. 8.2 A generalized supermatrix and an example of one of its block matrices

In the figure above, C_m denotes the mth cluster with n_m elements symbolized as $e_{m1}, e_{m2}, \ldots, e_{mn_m}$, where e_{mn} represents the nth element in the mth cluster, and W_{ij} is a block matrix consisting of principal eigenvectors representing the influence of the elements in the ith cluster on the elements in the jth cluster. There is no requirement that every element of a cluster has an influence on an element in another cluster. If this is the case, i.e., the ith cluster has no influence to the jth cluster, then $W_{ij} = 0$. For the sake of clarity, Figure 8.3 presents a sample representation of a supermatrix for a simple hypothetical network (cf. Yu and Tzeng [18]).

Once the supermatrix is formed, it is raised to limiting powers [19] to yield the cumulative influence of each element on every other element with which it interacts. But before taking the limit, it must first be reduced to a column stochastic matrix (i.e. weighted supermatrix), each of whose column sums to unity. This is necessary to obtain meaningful limiting priorities [1]. Two approaches are widely used to derive the stochastic supermatrix: 1) the clusters in the network are compared according to their impact on each other and the resulting weights are used to weight the corresponding blocks in the supermatrix;

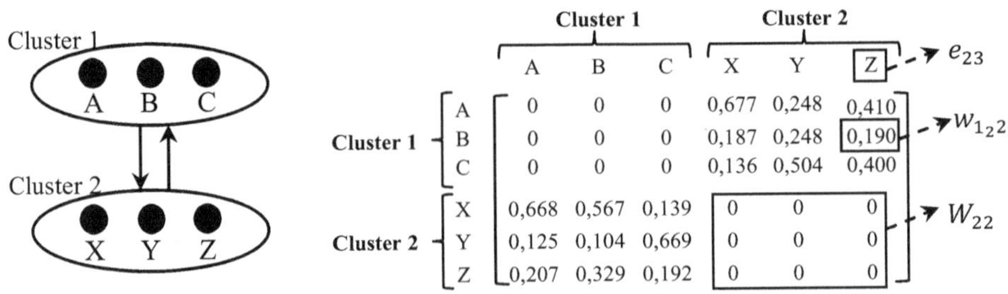

(a) Simple (b) Sample Supermatrix

Fig. 8.3 Supermatrix representation of a simple network

or simply 2) each column of the supermatrix is normalized to sum to one which is based on the assumption that all the clusters are equally dominant. For the synthesis of all the interactions, the weighted supermatrix is raised to large powers:

$$\lim_{k\to\infty} W^k \tag{8.2}$$

where k is an arbitrarily large number. This yields the convergence of the local priorities to steady state global priorities. Note that, the limiting priorities of a stochastic supermatrix depend on its reducibility, primitivity, and cyclicity [1]. If the successive powers do not converge to a single matrix and produce different limit outcomes, W is called cyclic [1]. In this case, the average of the successive matrices of the entire cycle (the Cesaro sum) gives the limiting priorities [1]:

$$\frac{1}{N}\sum_{n=1}^{N} W^{n-1} \lim_{k\to\infty}\left(W^N\right)^k \qquad N \geqslant 2 \tag{8.3}$$

where N indicates the length of the cycle. The values in the limit supermatrix represent the overall relative dominance of one element over another in the network [17]. For a detailed discussion on the calculations of the limit supermatrix the reader should refer to Saaty [2], Saaty [1] and Saaty and Vargas [19].

Step 4: Ranking and Selection of Alternatives

The total ranking of the elements, i.e., the overall priorities of the elements can be found in the relevant rows of the normalized limit supermatrix and those decision elements receiving high overall priorities deserve more attention.

8.3 A Fuzzy Multi-Criteria Prioritization Method For Decision Making In Complex And Uncertain Environments

Decision making in complex and uncertain environments generally requires dealing with problems in which descriptions of the environment and decision elements as well as judgments are highly subjective, vague and/or imprecise by nature. Especially the use of qualitative evaluation criteria, which is typically the case in expert decision making, may cause the decision makers to interpret the same information in different ways. In such problems, it may become unrealistic and impractical to acquire exact judgments, i.e. point estimates [20]. An extensive literature on decision making has evolved over the past 15 years that suggests using fuzzy set theoretic approaches to address these types of uncertainties. As pointed out by Bozdag *et al.* [21], decision-makers usually find it more convenient to express interval judgments than fixed value judgments. Here, fuzzy-based techniques can be viewed as a generalized form of interval analysis to handle uncertain information [22]. In comparison to the crisp methods in the literature, it is also evident that fuzzy-set theoretic approaches provide more flexibility to deal with variations among decision makers' judgments. The involvement of fuzzy theory allows an effective synthesis of group evaluations and deriving meaningful and reliable priorities from heterogeneous groups. In order to address these issues, this chapter presents a fuzzy approach to the conventional analytic network process with the following additional advantages:

- It explicitly models uncertainty in terms of imprecision and vagueness inherent in the judgment process.
- It captures the uncertainty associated with the cumulative influence of each element on every other element with which it interacts along paths of prescribed lengths in the network.
- The method allows an adequate reflection of a group of decision makers' individual preferences (based on individual experiences, knowledge, etc.) with less information loss.

8.3.1 *Outline of the Proposed Fuzzy Analytic Network Process Approach*

There are several fuzzy ANP methods suggested in the literature. These methods are systematic approaches to the alternative selection and justification problem that use the concepts of fuzzy set theory and network structure analysis. The way they cope with uncertain judgments is by expressing the comparison ratios as fuzzy numbers, which incorporate the imprecise and subjective nature of human judgments. To derive priorities from fuzzy comparison matrices they apply different methods such as the eigenvector method [17, 23], logarithmic least-square method [24], non-linear programming [25], and fuzzy additive normalization method [26]. However, all the fuzzy methods mentioned above perform supermatrix computations with nonfuzzy numbers which means they (except the method by Promentilla *et al.* [17]) neglect the uncertainty associated with the cumulative influence along indirect paths. Although Promentilla *et al.* [17] also suggest transforming fuzzy comparative judgments into crisp values before the supermatrix calculations, they estimate limiting priorities for different alpha-cut values, which allows them to represent the uncertainty in the estimates to some extent. The reason for the preference of crisp values in the supermatrix computations is the problem of convergence resulting from standard fuzzy arithmetic operations [27, 28]. In other words, raising a fuzzy supermatrix to limiting powers produces unsatisfactory fuzzy numbers which have wide supports and overlap over a wide range [29, 30].

The first attempt in the literature to overcome this problem was undertaken by Buckley and Eslami [27]. They suggest employing constrained fuzzy arithmetic operations to overcome the convergence problem in fuzzy Markov chains. Although they were able to develop closed form solutions for simple cases (up to 3×3 regular Markov chains), for more complex problems they suggest employing directed search algorithms which are not necessarily practical and often require an infinite number of trials. Another Markov-based approach was proposed by Huang [28], which converts the supermatrix problem into a fully fuzzy linear system problem. To search for approximate solutions of the linear system, Huang [28] also suggests using heuristic algorithms such as genetic algorithm, ant algorithm, or simulated annealing without providing any detail.

This chapter attempts to elaborate the Markov-based approach that explicitly accommodates fuzziness in the supermatrix calculations and suggests a simple yet satisfactory method to solve the resulting fuzzy linear system. Before describing the details of the proposed method, it may be useful to provide an overview of some basic definitions of fuzzy sets, triangular fuzzy numbers and operations related to this study. Fuzzy set theory, first introduced by Zadeh [31], has been developed to meet the objective of solving problems in which descriptions of activities, observations and judgments are by nature subjective, vague and imprecise. In general, the term 'fuzzy' refers to the situation in which no boundary for the set of observations or judgments can be well defined [32]. The theory provides numerous methods to represent the qualitative judgment of the decision maker as quantitative data. In this study, due to their simplicity in both design and implementation based on little information, triangular fuzzy numbers are used to assess the preferences of decision makers.

Definition 8.1. A fuzzy number is a special fuzzy set $\tilde{F} = \{(x, \mu_{\tilde{F}}(x)),\ x \in \mathbb{R}\}$ where x takes its values on the real line $\mathbb{R} : -\infty < x < +\infty$ and $\mu_{\tilde{F}}(x)$ is a continuous mapping from $\mathbb{R}$ to the close interval $[0,1]$. A tilde will be placed above a symbol if the symbol represents a fuzzy set.

Definition 8.2. A triangular fuzzy number (TFN) can be denoted as $\tilde{M} = (l,m,u)$ standing often for the expressions "about m" or "approximately m". Its membership function $\mu_{\tilde{M}}(x)$: $\mathbb{R} \to [0,1]$ is equal to:

$$\mu_{\tilde{M}}(x) = \begin{cases} 0, & x < l \text{ or } x > u \\ (x-l)/(m-l), & l \leqslant x \leqslant m \\ (x-u)/(m-u), & m \leqslant x \leqslant u \end{cases} \tag{8.4}$$

where $l \leqslant m \leqslant u$, l and u stand for the lower and upper value of the support of $\tilde{M}$, respectively, and m is the midvalue of $\tilde{M}$. When $l = m = u$, it is a nonfuzzy number by convention. Alternatively, a TFN can be designated by using its alpha-cut.

Definition 8.3. An alpha-cut ($[\tilde{M}]_\alpha$ or alternatively M_α) of a fuzzy set $\tilde{M}$ is the crisp set of all elements of the universal set X that belong to the fuzzy set $\tilde{M}$ at least to the degree $\alpha \in [0,\ 1]$:

$$[\tilde{M}]_\alpha = M_\alpha = \{x \in X \mid \mu_{\tilde{M}}(x) \geqslant \alpha\}$$

which can be derived as follows

$$[\tilde{M}]_\alpha = [l(\alpha),\ u(\alpha)] = [l + (m-l)\alpha, u - (u-m)\alpha] \tag{8.5}$$

Based on the extension principle, the main operational laws for two triangular fuzzy numbers $\tilde{M}_1$ and $\tilde{M}_2$ are as follows [33]:

$$\tilde{M}_1 + \tilde{M}_2 = (l_1 + l_2,\ m_1 + m_2,\ u_1 + u_2)$$

$$\tilde{M}_1 \otimes \tilde{M}_2 = (l_1 l_2,\ m_1 m_2,\ u_1 u_2) \tag{8.6}$$

$$\lambda \otimes \tilde{M}_1 = (\lambda l_1,\ \lambda m_1,\ \lambda u_1), \quad \lambda > 0, \quad \lambda \in \mathbb{R}$$

$$\tilde{M}_1^{-1} \approx (1/u_1,\ 1/m_1,\ 1/l_1)$$

Note that, for TFNs, addition and subtraction are closed operations i.e., the sum and difference of two TFNs are still TFNs. But multiplication and division of TFNs only produce approximate TFNs, which, as mentioned above, may complicate even simple matrix operations such as raising a matrix to powers. For a more complete treatment the reader should refer to Kaufmann and Gupta [33]. The steps of the proposed FANP method for calculating the global priority vector are as follows.

Step 1. Pairwise Comparisons

In the first step, judgments are elicited to establish the fuzzy pairwise comparison matrices of the decision elements. Decision makers are asked to perform a series of pairwise comparisons where two elements (e_{n_i}, e_{n_k}) at a time are compared in terms of dominance with respect to a common aspect or property (E_{m_j}). This judgment can be expressed verbally or by a corresponding fuzzy number $(\tilde{a}_{i_{n_i n_k}}^{j_{m_j}})$ to derive the ratio of comparison. For example, if e_{n_i} is perceived to dominate e_{n_k} with respect to a property E_{m_j}, then a fuzzy scale ranging from $\tilde{2}$ (about 2) to $\tilde{9}$ (about 9) is being used to represent the intensity of dominance (e.g., moderate, strong, very strong, absolute) that reflects the decision maker's judgment. In this study, the responses to the pairwise comparison questions are scaled on the basis of the flexible fuzzy dominance scale (see Table 8.1) suggested by Promentilla *et al.* [17]. The scale is characterized by the degree of fuzziness (δ) which can be adjusted according to the uncertainty inherent in the decision making process. It simply defines the absolute distance from the lower or upper bound to the modal value of the fuzzy number.

For the inverse comparisons a reciprocal value is automatically assigned, that is, $\tilde{a}_{i_{n_i n_k}}^{j_{m_j}} = 1/\tilde{a}_{i_{n_k n_i}}^{j_{m_j}}$, where $\tilde{a}_{i_{n_i n_k}}^{j_{m_j}}$ $(\tilde{a}_{i_{n_k n_i}}^{j_{m_j}})$ denotes the relative dominance of the n_ith (n_kth) element compared to the n_kth (n_ith) element in the ith cluster with respect to a common aspect m_j in the jth cluster. The resultant matrix including all fuzzy pairwise comparisons is referred to as the fuzzy pairwise comparison matrix $\tilde{A} = \left[\tilde{a}_{i_{n_i n_k}}^{j_{m_j}}\right]$.

Step 2. Aggregation of Individual Judgments

Each decision maker involved in the comparative judgment process provides valuable information that, in fact, represents his/her own subjective opinions and preferences, and may include some uncertainty. However, when individuals are willing to, or must make compromises with respect to their own preferences for the sake of the organization, they

Table 8.1 The Fuzzy Dominance Scale for Pairwise Comparative Judgment [17]

Numerical scale	**Linguistic scale**[a]	**Fuzzy scale**$(\mathbf{l}, \mathbf{m}, \mathbf{u})$[b]
1	Just equal	(1, 1, 1)
2	Equal to moderate	$(\max(L_B, 2-\delta), 2, \min(U_B, 2+\delta))$
3	Moderate dominance	$(\max(L_B, 3-\delta), 3, \min(U_B, 3+\delta))$
4	Moderate to strong	$(\max(L_B, 4-\delta), 4, \min(U_B, 4+\delta))$
5	Strong dominance	$(\max(L_B, 5-\delta), 5, \min(U_B, 5+\delta))$
6	Strong to very strong	$(\max(L_B, 6-\delta), 6, \min(U_B, 6+\delta))$
7	Very strong dominance	$(\max(L_B, 7-\delta), 7, \min(U_B, 7+\delta))$
8	Very strong to absolute	$(\max(L_B, 8-\delta), 8, \min(U_B, 8+\delta))$
9	Absolute dominance	$(\max(L_B, 9-\delta), 9, \min(U_B, 9+\delta))$

[a] For pairwise verbal comparisons, dominance of element n_i over element n_k.
[b] L_B and U_B refers to the lower bound and upper bound of the scale, respectively.
δ indicates the degree of fuzziness.

pool their judgments in such a way that the group becomes a new 'individual' and behaves like one [10]. In this sense, the so-called aggregated judgments may not reflect any one decision maker's thoughts perfectly. Therefore, in any aggregation process such an information loss is inevitable.

Several authors (see Wu *et al.* [23]; Chang *et al.* [34]; Chen and Wang [35]; among others) have demonstrated that fuzzy numbers work best to aggregate fragmented expert opinions. We suggest using the algorithms below to aggregate fuzzy pairwise comparison judgments of different decision makers [23, 36]:

$$\tilde{c}_{i_{n_i}n_k}^{j_{m_j}} = \left(l_{i_{n_i}n_k}^{j_{m_j}},\ m_{i_{n_i}n_k}^{j_{m_j}},\ u_{i_{n_i}n_k}^{j_{m_j}}\right)$$

$$l_{i_{n_i}n_k}^{j_{m_j}} = \min\left(l_{i_{n_i}n_k p}^{j_{m_j}}\right),\ m_{i_{n_i}n_k}^{j_{m_j}} = \left(\prod_{p=1}^{P} m_{i_{n_i}n_k p}^{j_{m_j}}\right)^{1/P},\ u_{i_{n_i}n_k}^{j_{m_j}} = \max\left(u_{i_{n_i}n_k p}^{j_{m_j}}\right) \tag{8.7}$$

where $\tilde{c}_{i_{n_i}n_k}^{j_{m_j}}$ denotes an aggregated fuzzy pairwise comparison of elements n_i and n_k with respect to a common aspect m_j, across multiple $(p = 1, 2, \ldots, P)$ decision makers. The minimum and maximum operations are especially suitable if the judgments of a (large) group of decision makers are homogeneous, at least to a certain degree, and the primary motivation for the use of multiple decision makers is simply to obtain as much information as possible. An information loss could, perhaps, be mitigated by the appropriate use of the min-max operations.

When one or few decision makers in a (small) group deliver extreme judgments which result in a huge span of fuzzy numbers, the min-max approach may cause the aggregated fuzzy judgments to exceed a tolerable range during the required multiplication and addition operations. In such cases, it is of advantage to use the geometric mean instead of the min-max approach to obtain more satisfactory results. The geometric mean is a synergistic aggregation of individual judgments, which can deal with inhomogeneous evaluations and outlier values [36]:

$$l_{i_n i_n k}^{j m_j} = \left(\prod_{p=1}^{P} l_{i_n i_n k p}^{j m_j}\right)^{1/P}, \quad m_{i_n i_n k}^{j m_j} = \left(\prod_{p=1}^{P} m_{i_n i_n k p}^{j m_j}\right)^{1/P}, \quad u_{i_n i_n k}^{j m_j} = \left(\prod_{p=1}^{P} u_{i_n i_n k p}^{j m_j}\right)^{1/P} \tag{8.8}$$

In this approach individual identities may be lost. However, such an information loss should be tolerable, especially when the main contribution of the approach is taken into account [36].

Step 3. Local Priority Estimation

Once the aggregate fuzzy pairwise comparison matrices representing the dominance relations given in the decision problem are established, the local priority vectors are computed. In this study, a fuzzy version of the additive normalization method, also known as mean of normalized values method, is used to approximate the fuzzy local priority. It applies the following simple three-step procedure [4, 37, 38]: 1) the sum of the values in each column of the fuzzy pairwise comparison matrix is calculated; 2) then, each column element is divided by the sum of its respective column, the resultant matrix is referred to as the normalized fuzzy pairwise comparison matrix; 3) finally, arithmetic means of each row of the normalized comparison matrix is calculated. These final fuzzy numbers provide an estimate of the local priorities for the decision elements being compared.

Then, the value of an element's fuzzy local priority, i.e., relative dominance, with respect to a certain aspect m_j, defined as $\tilde{s}_{i_{n_i}}^{j m_j} = \left(l_{i_{n_i}}^{j m_j}, m_{i_{n_i}}^{j m_j}, u_{i_{n_i}}^{j m_j}\right)$, can be obtained using the following fuzzy arithmetic operations:

$$\tilde{s}_{i_{n_i}}^{j m_j} = \sum_{n_k=1}^{n} \tilde{c}_{i_{n_i} n_k}^{j m_j} \otimes \left[\sum_{n_l=1}^{n} \sum_{n_k=1}^{n} \tilde{c}_{i_{n_l} n_k}^{j m_j}\right]^{-1} \quad n_i = 1, \ldots, n \tag{8.9}$$

which can be expressed in the following form as suggested by Wang, Luo [39]:

$$\tilde{s}_{i_{n_i}}^{j m_j} = \left(\frac{\sum\limits_{n_k=1}^{n} l_{i_{n_i} n_k}^{j m_j}}{\sum\limits_{n_k=1}^{n} l_{i_{n_i} n_k}^{j m_j} + \sum\limits_{\substack{n_l=1 \\ n_l \neq n_k}}^{n} \sum\limits_{n_k=1}^{n} u_{i_{n_l} n_k}^{j m_j}}, \frac{\sum\limits_{n_k=1}^{n} m_{i_{n_i} n_k}^{j m_j}}{\sum\limits_{n_l}^{n} \sum\limits_{n_k=1}^{n} m_{i_{n_l} n_k}^{j m_j}}, \frac{\sum\limits_{n_k=1}^{n} u_{i_{n_i} n_k}^{j m_j}}{\sum\limits_{n_k=1}^{n} u_{i_{n_i} n_k}^{j m_j} + \sum\limits_{\substack{n_l=1 \\ n_l \neq n_k}}^{n} \sum\limits_{n_k=1}^{n} l_{i_{n_l} n_k}^{j m_j}} \right)$$

$$n_i = 1, \ldots, n \tag{8.10}$$

Using Eq. (8.10), fuzzy priority vectors are derived for all fuzzy comparison matrices.

Consistency

The quality of the estimation of local priorities highly depends on the consistency of judgments that the decision makers performed throughout the pairwise comparisons. Thus, to avoid misleading solutions and thereby wrong decisions, the consistency of decision makers' judgments needs to be examined and, if necessary, improved. Several methods have been developed in the literature to determine whether a fuzzy comparison matrix is consistent. Here, the method proposed by Leung and Cao [40] is adopted and will be presented in detail.

In practice a certain level of inconsistency ought to be allowed [41] which is realistic in most decision problems and may lead to useful new information. Leung and Cao [40] describe this as a tolerable deviation and define fuzzy consistency as the existence of relative weights (priorities) within a feasible region assuming a certain tolerance deviation. In other words, a fuzzy comparison matrix $[\tilde{A}]_\alpha$ is defined to be consistent within tolerance deviation δ, if the α-level cut feasible region R'_α is not empty [40]:

$$R'_\alpha = \left\{ w:\ (1-\delta) l^{jm_j}_{i_n i_{n_k}}(\alpha) \leqslant \frac{w^{jm_j}_{i_s}}{w^{jm_j}_{i_t}} \leqslant (1+\delta) u^{jm_j}_{i_n i_{n_k}}(\alpha), \right.$$

$$\left. s \neq t = 1,\ldots,n_i,\ w^{jm_j}_{i_t} \geqslant 0,\ \sum_{t=1,\ldots,n_i} w^{jm_j}_{i_t} = 1 \right\} \tag{8.11}$$

where $w^{jm_j}_{i_s}$ and $w^{jm_j}_{i_t}$ are the relative weights of the sth and t^{th} elements, respectively and δ represents deviations from the upper bound $u^{jm_j}_{i_{n_i}}(\alpha)$ and the lower bound $l^{jm_j}_{i_{n_i}}(\alpha)$. Here, $\alpha = 1$ represents the most stringent case which implies that if a solution exists in R'_1, then a solution will exist in R'_α, for all lower levels of α. Therefore it is sufficient to examine whether the feasible region at $\alpha = 1$, R'_1, is non-empty. A practical way to measure the consistency within a tolerable deviation δ is to solve the following linear program [40]:

$$\min \beta = \beta_1 + \beta_2,$$

subject to

$$\ln(1-\delta)\, l^{jm_j}_{i_n i_{n_k}}(1) \leqslant \ln w^{jm_j}_{i_s} - \ln w^{jm_j}_{i_t} + \beta_{1_{st}} - \beta_{2_{st}} \leqslant \ln(1+\delta) u^{jm_j}_{i_n i_{n_k}}(1)$$

$$s \neq t = 1,\ldots,n_i,\quad \beta_1 \geqslant \beta_{1_{st}},\quad \beta_2 \geqslant \beta_{2_{st}},\quad \beta_{1_{st}}, \beta_{2_{st}} \geqslant 0 \tag{8.12}$$

where $\ln w^{jm_j}_{i_s}$, $\beta_{1_{st}}$, $\beta_{2_{st}}$, β_1, β_2 are decision variables. If $\beta = 0$, the fuzzy comparison matrix will be regarded as consistent within the predefined tolerance deviation δ. If $\beta > 0$, the feasible region is empty ($R'_1 = \emptyset$), and the fuzzy comparison matrix will be considered as inconsistent. Such a result would require the decision maker to revise his/her judgments to improve the consistency.

Step 4. Supermatrix Formation and Analysis

As described above, to analyze the transmission of influence along all paths defined in the network and to obtain the overall fuzzy priorities of the elements, the proposed Markov based approach is used. This approach accommodates fuzziness in the supermatrix calculations and provides the opportunity to capture the uncertainty associated with the cumulative influence in form of fuzzy numbers. The uncertainty inherent in the input data is thereby reflected in the output of the decision model. This is certainly not true for most of the current FANP methods in the literature where the arguments involve fuzziness but not the results.

In this step, first, each of the fuzzy local priority vectors computed in Step 3 is entered as a part of a relevant column of the fuzzy supermatrix as shown in Figure 8.4. The initial fuzzy supermatrix ($\tilde{S}$) is defined as follows

$$\tilde{S} = \begin{bmatrix} \tilde{S}_1^1 & \cdots & \tilde{S}_1^q \\ \vdots & \ddots & \vdots \\ \tilde{S}_q^1 & \cdots & \tilde{S}_q^q \end{bmatrix} \qquad \tilde{S}_i^j = \begin{bmatrix} \tilde{s}_{i_1}^{j_1} & \cdots & \tilde{s}_{i_1}^{j_{m_j}} \\ \vdots & \ddots & \vdots \\ \tilde{s}_{i_{n_i}}^{j_1} & \cdots & \tilde{s}_{i_{n_i}}^{j_{m_j}} \end{bmatrix} \tag{8.13}$$

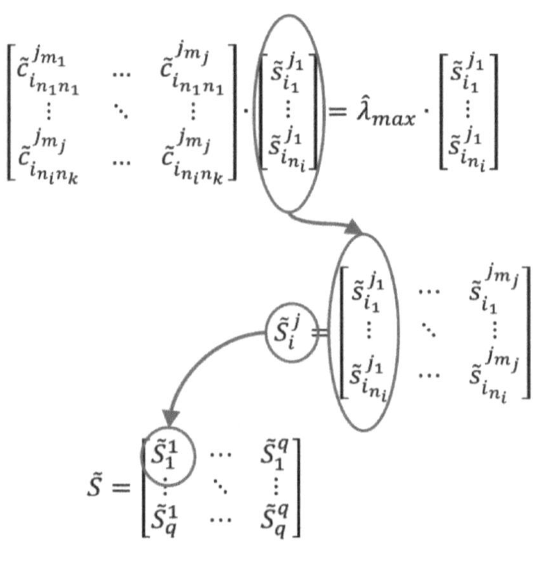

Fig. 8.4 Formation of the Initial Fuzzy Supermatrix (Note that both the fuzzy local priority vector and the maximum lambda values are approximated)

Before the cumulative influence of each element on every other element with which it interacts is calculated, the supermatrix must first be reduced to a fuzzy column stochastic matrix (i.e., weighted supermatrix), where the sum of midvalues in each column equals to unity, as shown below:

$$\tilde{s}_{in_i}^{*jm_j} = \tilde{s}_{in_i}^{jm_j} \otimes \left[\sum_{i=1}^{q} \sum_{n_i=1}^{n} \tilde{s}_{in_i}^{jm_j} \right]^{-1}, \quad m_j = 1,\ldots,m, \quad n_i = 1,\ldots,n \tag{8.14}$$

According to the normalization formula suggested by Wang *et al.* [39] Eq. (8.14) can be expressed as:

$$\tilde{s}_{in_i}^{*jm_j} = \left(\frac{l_{in_i}^{jm_j}}{l_{in_i}^{jm_j} + \sum\limits_{i=1}^{q} \sum\limits_{\substack{n_k=1 \\ in_i \neq in_k}}^{n} u_{in_k}^{jm_j}}, \frac{m_{in_i}^{jm_j}}{\sum\limits_{i=1}^{q} \sum\limits_{n_k=1}^{n} m_{in_k}^{jm_j}}, \frac{u_{in_i}^{jm_j}}{u_{in_i}^{jm_j} + \sum\limits_{i=1}^{q} \sum\limits_{\substack{n_k=1 \\ in_i \neq in_k}}^{n} l_{in_k}^{jm_j}} \right) \tag{8.15}$$

$$m_j = 1,\ldots,m; \quad n_i = 1,\ldots,n; \quad i = 1,\ldots,q; \quad j = 1,\ldots,q$$

Thus, the maximum eigenvalue of the midvalues of a nonnegative fuzzy column stochastic supermatrix is one, which is a necessary condition to derive the desired limiting priorities [1]. The existence and uniqueness of the limit matrix is guaranteed if the supermatrix is irreducible and acyclic. For further detail on the necessary properties of the limit supermatrix the reader should refer to Saaty [1] and Saaty [41]. If the transition matrix $\tilde{s}^*$ is irreducible and acyclic, then this matrix, raised to an arbitrary large power, should tend to an ergodic matrix $\tilde{E}$ which has similar columns, i.e.,

$$\lim_{k \to \infty} \tilde{S}^{*k} = \tilde{E} = \tilde{s}_{in_i} \cdot \widetilde{\mathbf{1}}' \tag{8.16}$$

where $\widetilde{\mathbf{1}}'$ is a q-dimensional fuzzy row vector containing ones, and $\tilde{s}_{in_i}$ is a fuzzy column vector whose entries are positive and midvalues sum to 1. Here, $\tilde{s}_{in_i}$ will be regarded as the fuzzy global priority vector [28] satisfying the following equation:

$$\tilde{S}^* \cdot \tilde{s}_{in_i} = \tilde{s}_{in_i} \text{ or } \left(\tilde{I} - \tilde{S}^*\right) \cdot \tilde{s}_{in_i} = \tilde{0} \tag{8.17}$$

where $\tilde{0}$ is a column vector containing zeros. Obviously solving Eq. (8.17) will not yield a unique solution. The following additional constraint, which describes that the sum of the steady state priorities equals one, is necessary to guarantee the uniqueness of the solution $\tilde{s}_{in_i}' \cdot \widetilde{\mathbf{1}} = \tilde{1}$. In order to take into account the additional constraint, the last row of the matrix $\tilde{I} - \tilde{S}$ is replaced by a row of ones and the right hand side is replaced by a column vector containing zeros except its last element which is one (cf. Kirytopoulos *et al.* [42]). Then, Eq. (8.17) can be represented as the following fully fuzzy linear system in matrix form:

$$\tilde{G} \cdot \tilde{s}_{in_i} = \tilde{e} \tag{8.18}$$

where $\tilde{G} = \begin{bmatrix} 1-\tilde{s}_{i_1}^{*j_1} & -\tilde{s}_{i_1}^{*j_2} & \dots & -\tilde{s}_{i_1}^{*j_q} \\ -\tilde{s}_{i_2}^{*j_1} & 1-\tilde{s}_{i_2}^{*j_2} & \dots & -\tilde{s}_{i_2}^{*j_q} \\ \vdots & \vdots & \ddots & \vdots \\ 1 & 1 & \dots & 1 \end{bmatrix}$, $\tilde{s}_{in_i} = \begin{bmatrix} \tilde{s}_{1n_1} \\ \tilde{s}_{2n_2} \\ \vdots \\ \tilde{s}_{qn_q} \end{bmatrix}$, and $\tilde{e} = \begin{bmatrix} (0,\ 0,\ 0) \\ (0,\ 0,\ 0) \\ \vdots \\ (1,\ 1,\ 1) \end{bmatrix}$

It should be noted that the resulting fully fuzzy linear system (FFLS) can be solved by exploiting different computational methods, supposing that $\tilde{G}$ is a nonsingular matrix.

A Heuristic Based Method

There are several heuristic based methods in the literature for finding approximated solutions of a FFLS (see Dehghan *et al.* [43]). However, many of these methods do not deliver satisfactory non-negative solutions. The method proposed in this study is based on splitting the fuzzy matrix $[\tilde{G}]_\alpha$ into two crisp matrices denoted by $L = \underline{G}(\alpha)$ and $U = \overline{G}(\alpha)$. Assuming that L and U are nonsingular crisp matrices, we can write the following linear systems:

$$\underline{G}(\alpha) \cdot l_{in_i}(\alpha) = \underline{e}(\alpha) \quad \Longrightarrow \quad l_{in_i}(\alpha) = \underline{G}(\alpha)^{-1} \cdot \underline{e}(\alpha) \tag{8.19}$$

$$\overline{G}(\alpha) \cdot u_{in_i}(\alpha) = \overline{e}(\alpha) \quad \Longrightarrow \quad u_{in_i}(\alpha) = \overline{G}(\alpha)^{-1} \cdot \overline{e}(\alpha) \tag{8.20}$$

The approximation of the α-cuts of the fuzzy global priority vector, $[\underline{s}_{in_i}(\alpha),\ \overline{s}_{in_i}(\alpha)]$, is easily obtained by the following equation:

$$\tilde{s}_{in_i}(\alpha) = \left[\min\left\{l_{in_i}(\alpha),\ u_{in_i}(\alpha)\right\},\ \max\left\{l_{in_i}(\alpha),\ u_{in_i}(\alpha)\right\}\right] \tag{8.21}$$

for $\alpha \in [0,\ 1]$. This method suggests solutions to an arbitrary FFLS and provides practical results. Later, by means of an illustrative example, it is shown that the suggested fuzzy supermatrix approach does yield satisfactory results in comparison to the supermatrix computations with crisp values.

Step 5. Defuzzification and Ranking

To allow the comparison of decision elements in terms of their priorities $(\tilde{s}_{in_i})$ the total integral value method developed by Liou and Wang [44] is employed. This method, which is independent of the type of membership functions used and the normality of the functions, can rank more than two fuzzy numbers simultaneously. It is relatively simple in computation, especially in ranking triangular fuzzy numbers [44]. According to this method, the fuzzy numbers are transformed into crisp global priority values $\left(I^{\lambda}\left(\tilde{s}_{in_i}\right)\right)$ by the following equation:

$$I^{\lambda}\left(\tilde{s}_{in_i}\right)=\frac{1}{2}\left[\lambda\left(m_{in_i}+u_{in_i}\right)+(1-\lambda)\left(l_{in_i}+m_{in_i}\right)\right] \tag{8.22}$$

where $\lambda \in [0,1]$ is an index of optimism which represents the decision maker's risk tolerance. A larger λ specifies a higher degree of optimism. Finally, the normalized relative importance vector $w_i=\left(w_{i_1},\ldots,w_{in_i}\right)^T$ for a particular cluster i (composed of decision elements) is derived by the following equation:

$$w_{in_i}=\frac{I^{\lambda}\left(\tilde{s}_{in_i}\right)}{\sum\limits_{n_i=1}^{n} I^{\lambda}\left(\tilde{s}_{in_i}\right)}, \quad i=1,\ldots,q \tag{8.23}$$

The elements with the highest normalized importance values, and thereby having the highest rankings, are selected to establish the set of the most critical (dominant) elements in the decision problem.

8.4 An Illustrative Example

The aim of this section is to demonstrate the proposed method's ability to deal with uncertainty and examine its validity by a numerical example where the results of the method are compared to the results of fuzzy state-of-the-art ANP methods; see Figure 8.5. To do this, the example provided by Huang [28] is used as a reference. According to the example the market share of a food company is believed to be influenced by three main criteria (clusters): Advertising (A), Quality (Q) and Attraction (T). Each of these criteria is represented by three sub-criteria including Creativity (C), Promotion (P) and Frequency (F) for Advertising; Nutrition (N), Taste (Ts) and Cleanliness (Cl) for Quality; and Price (Pr), Location (L) and Reputation (R) for Attraction. A schematic representation of the network structure, which involves only outer dependence, is presented in Figure 8.6. Using this model, the decision maker wants to determine the priorities of the criteria so that he/she can allocate the appropriate budgets to maximize the market share.

To focus on the supermatrix computations and the proposed heuristic, the previous steps including elicitation of judgments, analysis of consistency and estimation of fuzzy local priorities are supposed to be carried out and will therefore not be explained here. For a more complete treatment the reader should refer to Huang [28]. The fuzzy supermatrix

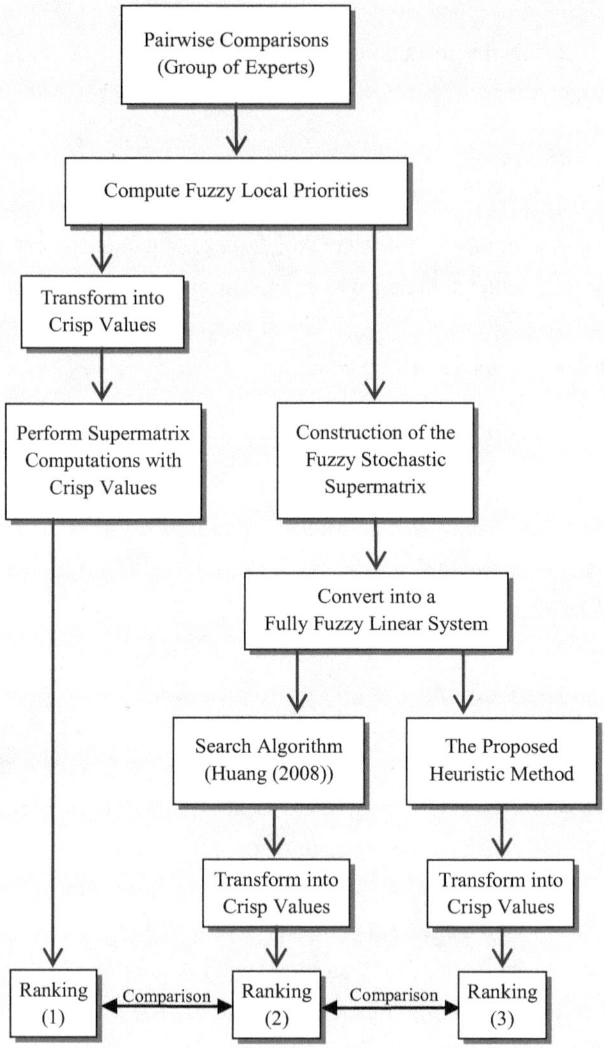

Fig. 8.5 Outline of the Comparative Analysis

containing the fuzzy local priorities derived from pairwise comparisons throughout the network can be formulated as follows:

$$\tilde{S} = \begin{matrix} A \\ Q \\ T \end{matrix} \begin{bmatrix} \mathbf{0} & \mathbf{0} & \tilde{S}_A^T \\ \tilde{S}_Q^A & \mathbf{0} & \mathbf{0} \\ \mathbf{0} & \tilde{S}_T^Q & \mathbf{0} \end{bmatrix}, \quad \text{where } \mathbf{0} = \begin{bmatrix} 0\,0\,0 \\ 0\,0\,0 \\ 0\,0\,0 \end{bmatrix}, \text{ and}$$

$$\tilde{S}_A^T = \begin{bmatrix} (0.2000, 0.2800, 0.3750) & (0.0960, 0.1220, 0.1724) & (0.6154, 0.6800, 0.7273) \\ (0.5000, 0.6000, 0.6857) & (0.4138, 0.5853, 0.6779) & (0.1818, 0.2400, 0.3077) \\ (0.0952, 0.1200, 0.1538) & (0.2105, 0.2927, 0.4494) & (0.0556, 0.0800, 0.1250) \end{bmatrix}$$

$$\tilde{S}_Q^A = \begin{bmatrix} (0.2632, 0.3514, 0.4737) & (0.6250, 0.6842, 0.7333) & (0.1429, 0.2000, 0.3333) \\ (0.4211, 0.5405, 0.6316) & (0.1333, 0.2105, 0.2500) & (0.2222, 0.4000, 0.4286) \\ (0.0800, 0.1081, 0.1395) & (0.0909, 0.1053, 0.1818) & (0.3333, 0.4000, 0.5714) \end{bmatrix}$$

$$\tilde{S}_T^Q = \begin{bmatrix} (0.1358, 0.2131, 0.3023) & (0.2000, 0.2800, 0.3750) & (0.2800, 0.3750, 0.5000) \\ (0.1053, 0.1312, 0.1695) & (0.0952, 0.1200, 0.1538) & (0.0909, 0.1250, 0.1667) \\ (0.5581, 0.6557, 0.7407) & (0.5000, 0.6000, 0.6857) & (0.3750, 0.5000, 0.6000) \end{bmatrix}$$

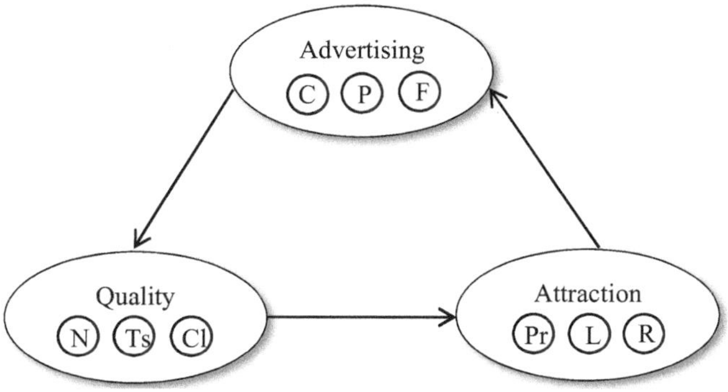

Fig. 8.6 The Network Model (adapted from Huang [28])

Note that the unweighted fuzzy supermatrix $\tilde{S}$ is already stochastic since the sum of the midvalues in each of its columns equals to one. For the given supermatrix the condition that $\left(I + [\tilde{S}]_1\right)^{n-1} > 0$, where n is the total number of elements in the supermatrix and I denotes an $n \times n$ identity matrix, holds which is a necessary and sufficient condition for the supermatrix to be irreducible. This indicates that the given network is strongly connected. When the supermatrix $[\tilde{S}]_1$ is raised to large powers it can be easily observed that it has no single limit and passes through three different cyclic forms ($c = 3$). In the crisp ANP, in order to obtain the limiting outcome the average of the three limits (i.e. Cesaro sum) is calculated which however is not necessary in the proposed method.

Next, the fuzzy stochastic supermatrix is converted into a fully fuzzy linear system ($\tilde{G}\cdot\tilde{s}_{in_i}=\tilde{e}$) by using the Eqs. (8.16), (8.17) and (8.18):

$$\tilde{G}=\begin{matrix}A\\Q\\T\end{matrix}\begin{bmatrix}I & \mathbf{0} & -\tilde{S}_A^T\\ -\tilde{S}_Q^A & I & \mathbf{0}\\ \mathbf{0}^* & -\tilde{S}_T^{*Q} & I^*\end{bmatrix},\ \text{where } \mathbf{0}^*=\begin{bmatrix}0\,0\,0\\0\,0\,0\\1\,1\,1\end{bmatrix},\ I^*=\begin{bmatrix}1\,0\,0\\0\,1\,0\\1\,1\,1\end{bmatrix},$$

$$-\tilde{S}_A^T=\begin{bmatrix}(-0.3750,-0.2800,-0.2000) & (-0.1724,-0.1220,-0.0960) & (-0.7273,-0.6800,-0.6154)\\(-0.6857,-0.6000,-0.5000) & (-0.6779,-0.5853,-0.4138) & (-0.3077,-0.2400,-0.1818)\\(-0.1538,-0.1200,-0.0952) & (-0.4494,-0.2927,-0.2105) & (-0.1250,-0.0800,-0.0556)\end{bmatrix}$$

$$-\tilde{S}_Q^A=\begin{bmatrix}(-0.4737,-0.3514,-0.2632) & (-0.7333,-0.6842,-0.6250) & (-0.3333,-0.2000,-0.1429)\\(-0.6316,-0.5405,-0.4211) & (-0.2500,-0.2105,-0.1333) & (-0.4286,-0.4000,-0.2222)\\(-0.1395,-0.1081,-0.0800) & (-0.1818,-0.1053,-0.0909) & (-0.5714,-0.4000,-0.3333)\end{bmatrix}$$

$$-\tilde{S}_T^{*Q}=\begin{bmatrix}(-0.3023,-0.2131,-0.1358) & (-0.3750,-0.2800,-0.2000) & (-0.5000,-0.3750,-0.2800)\\(-0.1695,-0.1312,-0.1053) & (-0.1538,-0.1200,-0.0952) & (-0.1667,-0.1250,-0.0909)\\(1,1,1) & (1,1,1) & (1,1,1)\end{bmatrix}$$

$$\tilde{s}_{in_i}=\begin{bmatrix}\tilde{s}_{AC}\\\tilde{s}_{AP}\\\tilde{s}_{AF}\\\tilde{s}_{QN}\\\tilde{s}_{QTs}\\\tilde{s}_{QCl}\\\tilde{s}_{TPr}\\\tilde{s}_{TL}\\\tilde{s}_{QR}\end{bmatrix},\ \text{and}\ \tilde{e}=\begin{bmatrix}(0,\,0,\,0)\\(0,\,0,\,0)\\(0,\,0,\,0)\\(0,\,0,\,0)\\(0,\,0,\,0)\\(0,\,0,\,0)\\(0,\,0,\,0)\\(0,\,0,\,0)\\(1,\,1,\,1)\end{bmatrix}$$

This fully fuzzy linear system is then solved by using the proposed heuristic based method. Initially, the α-cut operation (see Eq. (8.15)) is applied to the fuzzy numbers in the FFLS to obtain linear systems of interval equations. In this example, for the ease of understanding, the α-cut is set to 0 and 1. The linear system is then solved for the lower ($L=\underline{G}(0)$) and upper ($U=\overline{G}(0)$) end points of the intervals, separately. The nonsingular crisp matrices L and U, for α-cut$=0$, are given below:

$$L=\begin{bmatrix}1 & 0 & 0 & 0 & 0 & 0 & -0.3750 & -0.1724 & -0.7273\\0 & 1 & 0 & 0 & 0 & 0 & -0.6857 & -0.6779 & -0.3077\\0 & 0 & 1 & 0 & 0 & 0 & -0.1538 & -0.4494 & -0.1250\\-0.4737 & -0.7333 & -0.3333 & 1 & 0 & 0 & 0 & 0 & 0\\-0.6316 & -0.2500 & -0.4286 & 0 & 1 & 0 & 0 & 0 & 0\\-0.1395 & -0.1818 & -0.5714 & 0 & 0 & 1 & 0 & 0 & 0\\0 & 0 & 0 & -0.3023 & -0.3750 & -0.5000 & 1 & 0 & 0\\0 & 0 & 0 & -0.1695 & -0.1538 & -0.1667 & 0 & 1 & 0\\1 & 1 & 1 & 1 & 1 & 1 & 1 & 1 & 1\end{bmatrix}$$

$$U = \begin{bmatrix} 1 & 0 & 0 & 0 & 0 & 0 & -0.2000 & -0.0960 & -0.6154 \\ 0 & 1 & 0 & 0 & 0 & 0 & -0.5000 & -0.4138 & -0.1818 \\ 0 & 0 & 1 & 0 & 0 & 0 & -0.0952 & -0.2105 & -0.0556 \\ -0.2632 & -0.6250 & -0.1429 & 1 & 0 & 0 & 0 & 0 & 0 \\ -0.4211 & -0.1333 & -0.2222 & 0 & 1 & 0 & 0 & 0 & 0 \\ -0.0800 & -0.0909 & -0.3333 & 0 & 0 & 1 & 0 & 0 & 0 \\ 0 & 0 & 0 & -0.1358 & -0.2000 & -0.2800 & 1 & 0 & 0 \\ 0 & 0 & 0 & -0.1053 & -0.0952 & -0.0909 & 0 & 1 & 0 \\ 1 & 1 & 1 & 1 & 1 & 1 & 1 & 1 & 1 \end{bmatrix}$$

The approximation of the fuzzy global priority vector for α-cut= 0 is easily obtained by the following calculations (using Eq. (8.19)–(8.21)):

$$l_{in_i}(0) = \begin{bmatrix} 0.4638 & -0.5066 & -0.5927 & -0.3278 & -0.3448 & -0.3996 & -0.3720 & -0.6372 & 0.1074 \\ -0.1265 & 0.8676 & -0.0988 & 0.0101 & 0.0313 & 0.0822 & 0.3697 & 0.3593 & 0.1626 \\ -0.0573 & -0.0562 & 0.9454 & 0.0026 & -0.0005 & 0.0069 & 0.0262 & 0.3177 & 0.0592 \\ 0.1078 & 0.3775 & -0.0382 & 0.8530 & -0.1406 & -0.1267 & 0.1036 & 0.0676 & 0.1898 \\ 0.2367 & -0.1271 & 0.0061 & -0.2034 & 0.7898 & -0.2289 & -0.1313 & -0.1765 & 0.1338 \\ 0.0089 & 0.0550 & 0.4395 & -0.0424 & -0.0427 & 0.9631 & 0.0303 & 0.1580 & 0.0784 \\ 0.1258 & 0.0939 & 0.2105 & 0.1604 & 0.2323 & 0.3574 & 0.9972 & 0.0332 & 0.1467 \\ 0.0562 & 0.0536 & 0.0677 & 0.1062 & 0.0905 & 0.1039 & 0.0024 & 1.0106 & 0.0658 \\ -0.8154 & -0.7577 & -0.9396 & -0.5586 & -0.6154 & -0.7583 & -1.0262 & -1.1328 & 0.0564 \end{bmatrix} \cdot \begin{bmatrix} 0 \\ 0 \\ 0 \\ 0 \\ 0 \\ 0 \\ 0 \\ 0 \\ 1 \end{bmatrix}$$

$$u_{in_i}(0) = \begin{bmatrix} 0.5325 & -0.4889 & -0.4553 & -0.3156 & -0.3363 & -0.3659 & -0.3948 & -0.4605 & 0.2135 \\ -0.1019 & 0.8913 & -0.0930 & -0.0246 & -0.0064 & 0.0183 & 0.3222 & 0.2453 & 0.0942 \\ -0.0320 & -0.0328 & 0.9691 & -0.0059 & -0.0048 & -0.0022 & 0.0412 & 0.1591 & 0.0282 \\ 0.0719 & 0.4237 & -0.0395 & 0.9007 & -0.0932 & -0.0852 & 0.1034 & 0.0548 & 0.1191 \\ 0.2036 & -0.0943 & 0.0112 & -0.1375 & 0.8565 & -0.1521 & -0.1141 & -0.1259 & 0.1087 \\ 0.0227 & 0.0310 & 0.2781 & -0.0294 & -0.0291 & 0.9717 & 0.0115 & 0.0385 & 0.0351 \\ 0.0568 & 0.0473 & 0.0748 & 0.0866 & 0.1505 & 0.2301 & 0.9944 & -0.0070 & 0.0477 \\ 0.0290 & 0.0385 & 0.0222 & 0.0791 & 0.0691 & 0.0649 & 0.0011 & 0.9973 & 0.0261 \\ -0.7826 & -0.8158 & -0.7676 & -0.5533 & -0.6062 & -0.6795 & -0.9649 & -0.9017 & 0.3274 \end{bmatrix} \cdot \begin{bmatrix} 0 \\ 0 \\ 0 \\ 0 \\ 0 \\ 0 \\ 0 \\ 0 \\ 1 \end{bmatrix}$$

where

$$\underline{G}(0)^{-1} = \begin{bmatrix} 0.4638 & -0.5066 & -0.5927 & -0.3278 & -0.3448 & -0.3996 & -0.3720 & -0.6372 & 0.1074 \\ -0.1265 & 0.8676 & -0.0988 & 0.0101 & 0.0313 & 0.0822 & 0.3697 & 0.3593 & 0.1626 \\ -0.0573 & -0.0562 & 0.9454 & 0.0026 & -0.0005 & 0.0069 & 0.0262 & 0.3177 & 0.0592 \\ 0.1078 & 0.3775 & -0.0382 & 0.8530 & -0.1406 & -0.1267 & 0.1036 & 0.0676 & 0.1898 \\ 0.2367 & -0.1271 & 0.0061 & -0.2034 & 0.7898 & -0.2289 & -0.1313 & -0.1765 & 0.1338 \\ 0.0089 & 0.0550 & 0.4395 & -0.0424 & -0.0427 & 0.9631 & 0.0303 & 0.1580 & 0.0784 \\ 0.1258 & 0.0939 & 0.2105 & 0.1604 & 0.2323 & 0.3574 & 0.9972 & 0.0332 & 0.1467 \\ 0.0562 & 0.0536 & 0.0677 & 0.1062 & 0.0905 & 0.1039 & 0.0024 & 1.0106 & 0.0658 \\ -0.8154 & -0.7577 & -0.9396 & -0.5586 & -0.6154 & -0.7583 & -1.0262 & -1.1328 & 0.0564 \end{bmatrix}$$

$$\overline{G}(0)^{-1} = \begin{bmatrix} 0.5325 & -0.4889 & -0.4553 & -0.3156 & -0.3363 & -0.3659 & -0.3948 & -0.4605 & 0.2135 \\ -0.1019 & 0.8913 & -0.0930 & -0.0246 & -0.0064 & 0.0183 & 0.3222 & 0.2453 & 0.0942 \\ -0.0320 & -0.0328 & 0.9691 & -0.0059 & -0.0048 & -0.0022 & 0.0412 & 0.1591 & 0.0282 \\ 0.0719 & 0.4237 & -0.0395 & 0.9007 & -0.0932 & -0.0852 & 0.1034 & 0.0548 & 0.1191 \\ 0.2036 & -0.0943 & 0.0112 & -0.1375 & 0.8565 & -0.1521 & -0.1141 & -0.1259 & 0.1087 \\ 0.0227 & 0.0310 & 0.2781 & -0.0294 & -0.0291 & 0.9717 & 0.0115 & 0.0385 & 0.0351 \\ 0.0568 & 0.0473 & 0.0748 & 0.0866 & 0.1505 & 0.2301 & 0.9944 & -0.0070 & 0.0477 \\ 0.0290 & 0.0385 & 0.0222 & 0.0791 & 0.0691 & 0.0649 & 0.0011 & 0.9973 & 0.0261 \\ -0.7826 & -0.8158 & -0.7676 & -0.5533 & -0.6062 & -0.6795 & -0.9649 & -0.9017 & 0.3274 \end{bmatrix}$$

Note that for both equations the solutions, $l_{in_i}(0)$ and $u_{in_i}(0)$, are equal to the last column vector of the inverse matrices $\underline{G}(0)^{-1}$ and $\overline{G}(0)^{-1}$, respectively. Finally, the global priority values for α-cut$= 0$ is determined as follows:

$$\tilde{s}_{in_i}(0) = \begin{bmatrix} \min\{0.1074,\ 0.2135\},\ \max\{0.1074,\ 0.2135\} \\ \min\{0.1626,\ 0.0942\},\ \max\{0.1626,\ 0.0942\} \\ \min\{0.0592,\ 0.0282\},\ \max\{0.0592,\ 0.0282\} \\ \min\{0.1898,\ 0.1191\},\ \max\{0.1898,\ 0.1191\} \\ \min\{0.1338,\ 0.1087\},\ \max\{0.1338,\ 0.1087\} \\ \min\{0.0784,\ 0.0351\},\ \max\{0.0784,\ 0.0351\} \\ \min\{0.1467,\ 0.0477\},\ \max\{0.1467,\ 0.0477\} \\ \min\{0.0658,\ 0.0261\},\ \max\{0.0658,\ 0.0261\} \\ \min\{0.0564,\ 0.3274\},\ \max\{0.0564,\ 0.3274\} \end{bmatrix} = \begin{bmatrix} 0.1074,\ 0.2135 \\ 0.0942,\ 0.1626 \\ 0.0282,\ 0.0592 \\ 0.1191,\ 0.1898 \\ 0.1087,\ 0.1338 \\ 0.0351,\ 0.0784 \\ 0.0477,\ 0.1467 \\ 0.0261,\ 0.0658 \\ 0.0564,\ 0.3274 \end{bmatrix}$$

Similar calculations are done for α-cut$= 1$. The results of the proposed method are finally compared to the results of the conventional fuzzy ANP approach with crisp supermatrix computations and Huang's method. The global priority values obtained by all three methods are summarized in Table 8.2. After defuzzifying the global priorities according to Eq. (8.22) with an optimism index of $\lambda = 0.5$, all three methods end up with the same ranking. However, slight differences are observed by setting the index of optimism to its upper bounds which can be explained by the uncertainty inherent in limiting priorities. Note that the size of the interval of the last decision element in the linear system is commonly overestimated or underestimated. This is due to the need to satisfy the additional sum-to-unity constraint incorporated into the linear system. In our example, the resulting interval of the sub-criteria Reputation also indicates an overestimation. This issue may be addressed by replacing any equation in the linear system with the sum-to-unity constraint and taking the average of the results for all possible replacements. Another important issue which needs to be mentioned is that if the network is not strongly connected the proposed solution to an FFLS often fails to exist.

Nevertheless, the results indicate that the suggested fuzzy supermatrix approach does yield satisfactory results. In comparison to the current methods in the literature with crisp supermatrix calculations, it is evident that the suggested method provides additional information and more flexibility to the decision maker, and is more appropriate to represent uncertainties.

Table 8.2 Comparison of three fuzzy ANP methods in terms of global priority values

		Fuzzy ANP with	Huang's Method			Proposed Method		
Criteria	Sub-citeria	Crisp Supermatrix	α-cut = 0	α-cut = 1	λ*=0.5	α-cut = 0	α-cut = 1	λ*=0.5
Advertising	Creativity	0.1682	[0.1257, 0.2047]	[0.1682, 0.1682]	0.1667	[0.1074, 0.2135]	[0.1682, 0.1682]	0.1643
	Promotion	0.1260	[0.0860, 0.1700]	[0.1260, 0.1260]	0.1270	[0.0942, 0.1626]	[0.1260, 0.1260]	0.1272
	Frequency	0.0391	[0.0250, 0.0651]	[0.0391, 0.0391]	0.0421	[0.0282, 0.0592]	[0.0391, 0.0391]	0.0414
Quality	Nutrition	0.1532	[0.1106, 0.2005]	[0.1532, 0.1532]	0.1544	[0.1191, 0.1898]	[0.1532, 0.1532]	0.1538
	Taste	0.1331	[0.0840, 0.1704]	[0.1331, 0.1331]	0.1301	[0.1087, 0.1338]	[0.1331, 0.1331]	0.1272
	Cleanliness	0.0471	[0.0323, 0.0829]	[0.0471, 0.0471]	0.0523	[0.0351, 0.0784]	[0.0471, 0.0471]	0.0519
Attraction	Price	0.0876	[0.0547, 0.1310]	[0.0876, 0.0876]	0.0902	[0.0477, 0.1467]	[0.0876, 0.0876]	0.0924
	Location	0.0420	[0.0289, 0.0613]	[0.0420, 0.0420]	0.0435	[0.0261, 0.0658]	[0.0420, 0.0420]	0.0439
	Reputation	0.2038	[0.1549, 0.2493]	[0.2038, 0.2038]	0.2030	[0.0564, 0.3274]	[0.2038, 0.2038]	0.1978

* Values are defuzzified using the total integral value method (Liou and Wang, 1992)

8.5 Conclusion

Many precision-based methods for the evaluation of decision elements have been investigated, and most of them have been developed on the basis of accurate measurements and crisp evaluation. However, probable deviations caused by the characteristics of information and decision makers are entirely ignored in many of these methods. In this study we propose a fuzzy multi-criteria decision method that is capable of handling not only the complexity of a problem, but also the inherent uncertainty associated with the subjectivity of human judgment. Thus, the suggested method extends the ANP method by performing network based computations with fuzzy numbers. Considering fuzziness in the supermatrix calculations allows capturing the uncertainty associated with the cumulative influence of each element on every other element with which it interacts in the network. The proposed method is demonstrated by an illustrative example.

Bibliography

[1] Saaty T.L., *Decision making with dependence and feedback: The Analytic network process*, 2001. RWS Pub.: Pittsburg.

[2] Saaty T.L., Fundamentals of the analytic network process - Dependence and feedback in decision-making with a single network. *Journal of Systems Science and Systems Engineering*, 2004; **13** (2), 129–157.

[3] Saaty T.L., Fundamentals of the Analytic Network Process. In *International Society of the Analytic Hierarchy Process* (*ISAHP*), 1999: Kobe, Japan.

[4] Saaty T.L., *Multicriteria decision making: The analytic hierarchy process*, 1988. RWS Pub.: Pittsburgh.

[5] Meade L.M., Sarkis J., Analyzing organizational project alternatives for agile manufacturing processes: an analytical network approach. *International Journal of Production Research*, 1999; **37** (2), 241–261.

[6] Saaty T.L., *The Analytic hierarchy process: applications and studies*, (eds. Golden BL, Wasil EA, Harker PT, Alexander JM), 1989; pp. 59–67. Springer-Verlag: Berlin.

[7] Harker P.T., Vargas L.G., Reply to "remarks on the Analytic Hierarchy Process" by *J.S. Dyer. Management Science*, 1990; **36** (3), 269–273.
[8] Bolloju N., Aggregation of analytic hierarchy process models based on similarities in decision makers' preferences. *European Journal of Operational Research*, 2001; **128** (3), 499–508.
[9] Aczel J., Saaty T.L., Procedures for synthesizing ratio judgements. *Journal of Mathematical Psychology*, 1983; **27** (1), 93–102.
[10] Forman E., Peniwati K., Aggregating individual judgments and priorities with the analytic hierarchy process. *European Journal of Operational Research*, 1998; **108** (1), 165–169.
[11] Saaty T.L., Peniwati K., *Group Decision Making: Drawing Out and Reconciling Differences*, 2008. RWS Pub.: Pittsburgh, PA.
[12] Basak I., Saaty T., Group decision making using the analytic hierarchy process. *Mathematical and Computer Modelling*, 1993; **17** (4-5), 101–109.
[13] Bryson N., Joseph A., Generating consensus priority point vectors: a logarithmic goal programming approach. *Computers & Operations Research*, 1999; **26** (6), 637–643.
[14] Saaty T.L., *The Analytic Hierarchy Process*, 1980, McGraw-Hill, Inc.: New York, USA.
[15] Srdjevic B., Combining different prioritization methods in the analytic hierarchy process synthesis. *Computers & Operations Research*, 2005; **32** (7), 1897–1919.
[16] Asan U., Soyer A., Identifying strategic management concepts: An analytic network process approach. *Computers & Industrial Engineering*, 2009; **56** (2), 600–615.
[17] Promentilla MAB, Furuichi T., Ishii K., Tanikawa N., A fuzzy analytic network process for multi-criteria evaluation of contaminated site remedial countermeasures. *Journal of Environmental Management*, 2008; **88** (3), 479–495.
[18] Yu R., Tzeng G.-H., A soft computing method for multi-criteria decision making with dependence and feedback. *Applied Mathematics and Computation*, 2006; **180** (1), 63–75.
[19] Saaty T.L., Vargas L.G., Diagnosis with dependent symptoms: Bayes theorem and the analytic hierarchy process. *Operations Research*, 1998; **46** (4), 491–502.
[20] Wang Y.-M., Yang J.-B., Xu D.-L., Interval weight generation approaches based on consistency test and interval comparison matrices. *Applied Mathematics and Computation*, 2005; **167** (1), 252–273.
[21] Bozdag C.E., Kahraman C., Ruan D., Fuzzy group decision making for selection among computer integrated manufacturing systems. *Computers in Industry*, 2003; **51** (1), 13–29.
[22] Sadiq R., Husain T., Veitch B., Bose N., Risk-based decision-making for drilling waste discharges using a fuzzy synthetic evaluation technique. *Ocean Engineering*, 2004; **31** (16), 1929–1953.
[23] Wu C.R., Chang C.W., Lin H.L., A fuzzy ANP-based approach to evaluate medical organizational performance. *Information and Management Sciences*, 2008; **10** (1), 53–74.
[24] Tuzkaya U.R., Önüt S., A fuzzy analytic network process based approach to transportation-mode selection between Turkey and Germany: A case study. *Information Sciences*, 2008; **178** (15), 3133–3146.
[25] Razmi J., Rafiei H., Hashemi M., Designing a decision support system to evaluate and select suppliers using fuzzy analytic network process. *Computers & Industrial Engineering*, 2009; **57** (4), 1282-1290.
[26] Dağdeviren M., Yüksel Ý., Kurt M., A fuzzy analytic network process (ANP) model to identify faulty behavior risk (FBR) in work system. *Safety Science*, 2008; **46** (5), 771–783.
[27] Buckley J.J., Eslami E., Fuzzy Markov chains: uncertain probabilities. *Mathware and Soft Computing*, 2002; **9** (1), 33–41.
[28] Huang J.-J. A matrix method for the fuzzy analytic network process. International Journal of Uncertainty, *Fuzziness and Knowledge-Based Systems*, 2008; **16** (6), 863–878.
[29] Asan U., Serdar-Asan S., Soyer A., Alternative Methods For Calculating The Fuzzy Limit Matrix In The Analytic Network Process. In *41 International Computers and Industrial Engi-*

neering Conference, 2011; pp. 68–73: Los Angeles, USA.

[30] Mikhailov L., Deriving priorities from fuzzy pairwise comparison judgements. *Fuzzy Sets and Systems*, 2003; **134** (3), 365–385.

[31] Zadeh L.A., Fuzzy sets. *Information and Control*, 1965; **8** (3), 338–353.

[32] Chan L.K., Kao H.P., Wu M.L., Rating the importance of customer needs in quality function deployment by fuzzy and entropy methods. *International Journal of Production Research*, 1999; **37** (11), 2499–2518.

[33] Kaufmann A., Gupta M.M., *Introduction to fuzzy arithmetic: Theory and applications*, 1991. Van Nostrand Reinhold: New York.

[34] Chang C.-W., Wu C.-R., Lin H.-L., Applying fuzzy hierarchy multiple attributes to construct an expert decision making process. *Expert Systems with Applications*, 2009; **36** (4), 7363–7368.

[35] Chen M.-K., Wang S.-C., A hybrid Delphi-Bayesian method to establish business data integrity policy: A benchmark data center case study. *Kybernetes*, 2010; **39** (5), 800–824.

[36] Meixner O., Fuzzy AHP group decision analysis and its application for the evaluation of energy sources. In *10th International Symposium on the Analytic Hierarchy/Network Process*, 2009: Pittsburgh, USA.

[37] Meade L.M., Presley A., R& D project selection using ANP... the analytic network process. *Potentials, IEEE*, 2002; **21** (2), 22-28.

[38] Chung S-H, Lee AHI, Pearn WL. Analytic network process (ANP) approach for product mix planning in semiconductor fabricator. *International Journal of Production Economics*, 2005; **96** (1), 15–36.

[39] Wang Y.-M., Luo Y., Hua Z., On the extent analysis method for fuzzy AHP and its applications. *European Journal of Operational Research*, 2008; **186** (2), 735–747.

[40] Leung L.C., Cao D., On consistency and ranking of alternatives in fuzzy AHP. *European Journal of Operational Research*, 2000; **124** (1), 102–113.

[41] Saaty T.L., Decision making – The Analytic Hierarchy and Network Processes (AHP/ANP). *Journal of Systems Science and Systems Engineering*, 2004; **13** (1), 1–35.

[42] Kirytopoulos K., Voulgaridou D., Platis A, Leopoulos V., An effective Markov based approach for calculating the Limit Matrix in the analytic network process. *European Journal of Operational Research*, 2011; **214** (1), 85–90.

[43] Dehghan M., Hashemi B., Ghatee M., Computational methods for solving fully fuzzy linear systems. *Applied Mathematics and Computation*, 2006; **179** (1), 328–343.

[44] Liou T.-S., Wang M.-J.J., Ranking fuzzy numbers with integral value. *Fuzzy Sets and Systems*, 1992; **50** (3), 247–255.

Chapter 9

Swarm Intelligence Optimization: Applications of Particle Swarms in Industrial Engineering and Nuclear Power Plants

Miltiadis Alamaniotis[1], Andreas Ikonomopoulos[2], Lefteri H. Tsoukalas[1]

[1] *School of Nuclear Engineering, Purdue University400 Central Dr., 47907, West Lafayette, IN, USA*

[2] *Institute of Nuclear & Radiological Sciences & Technology, Energy & Safety, National Center for Scientific Research DEMOKRITOS Aghia Paraskevi, 15310, Athens, Greece*

E-mail: {malamani,tsoukala}@ecn.purdue.edu, anikon@ipta.demokritos.gr

Swarm-based intelligence is a recently developed area of computational intelligence that offers a powerful framework for solving complex optimization problems. It has found applications in various scientific fields where optimization of one or more functions is required. There are several methods developed under the umbrella of swarm intelligence, and *Particle Swarm Optimization* (PSO) is one of them. This chapter presents swarm intelligence and its applications in industrial engineering as well as nuclear power plants and PSO is used to illustrate the potential of such an implementation. The roadmap of the chapter is as follows: Sec. 9.1 provides an introduction to swarm intelligence and the following section presents particle swarm optimization. A discussion on the implementation of PSO in industrial engineering problems is given in Sec. 9.3 while PSO applications in nuclear power plants (NPP) are presented in Sec. 9.4. Section 9.5 concludes the chapter.

9.1 Introduction to Swarm Intelligence

Computational Intelligence (CI) is a broad field comprised of a number of methods that include neural networks, fuzzy logic and evolutionary computation. It encompasses methods derived from topics such as chaos theory, genetic algorithms, swarm intelligence and artificial immune systems [1]. All these areas have their origin in biology or nature and encompass techniques that attempt to replicate natural behavior. Swarm intelligence is presented in this chapter with a focus on Particle Swarm Optimization.

Social behavior among people or animals in a group has been studied extensively and inspired the development of a set of algorithms known as swarm intelligence [2]. Accord-

C. Kahraman (ed.), *Computational Intelligence Systems in Industrial Engineering*, Atlantis Computational Intelligence Systems 6, DOI: 10.2991/978-94-91216-77-0_9,

ingly, a reference to a group can be done with the more abstract term *swarm*. In artificial intelligence that includes computational intelligence, a swarm is comprised of a set of software agents interacting with each other with the intention to reach a list of predetermined objectives. In this sense, computational swarm intelligence refers to the modeling of natural societal behavioral patterns of individuals observed through group interaction for development of problem solving algorithms [2].

The goal of information processing systems is to make inferences regarding a problem, given model parameters and all available data [3]. Choosing the right parameters to evaluate a task is challenging due to the imprecision involved in modeling and the data obtaining process. In addition, constraints are applied to most engineering systems such as solution optimization that is based on one, or more, design objectives. In cases where the solution exists in a multidimensional space the optimal model localization becomes difficult and computationally expensive. As the dimension of the search space increases the complexity of the search exacerbates in a situation known as the *curse of dimensionality* [4]. Thus, algorithms utilized to find the solution of complex problems must be computationally efficient.

Swarm intelligence allows the utilization of distributive methodologies for problem solving. It exploits the capability of each swarm agent to collect information locally and share it with others through interaction. The exchange process of locally collected information in a swarm brings into light a global problem-solving approach. The distributed collection of information in order to emerge a global behavior justifies the term *collective intelligence* as an alternative way to refer to swarm intelligence [5].

The global behavior of the swarm emerges from the interactions among swarm members and it evolves over time. The overall swarm behavior is not an inherent property of any single individual but it is determined by the interactions among them. Thus, it is not easily predictable and cannot be inferred by observing exclusively the behavior of any swarm member. This way the recognition of new patterns in complex systems out of a multiplicity of simpler interactions is expressed by the term *emergence* [6].

Simple interactions form the basic mean of cooperation among swarm members and can take place in a direct or indirect way. Direct interaction denotes the presence of a direct link between two individuals while indirect interaction implies the existence of mediator links between two individuals. Thus, a term used in swarm intelligence with respect to cooperation is *stigmergy* that expresses the indirect interaction among swarm agents [6].

Swarm systems that exist in nature and have inspired the development of several computational models include ants, termites, bees, birds, predator animals, bacteria and others. The objective of computational swarm models is to mimic the simple interactions of swarm members either with their environment or with other members. A sample of swarm intelligence paradigms are:

Particle Swarm Optimization: Particle swarm optimization is a global optimization algorithm searching for the best solution in a multidimensional space. A set of particles is plotted in the space having an initial velocity and communication channel [7]. Particles move through the solution space according to a fitness function. As time passes, particles move towards particles with better fitness values. The main PSO advantage is that a large population does not drive the technique to local minima. A detailed PSO description appears in the next section.

Ant Colony Optimization: Ant colony optimization (ACO) is a group of optimization algorithms based on the behavior of an ant colony. They are particularly useful when there is a need to find paths to goals. Artificial ants search for an optimal solution by moving in a parameter space where all possible solutions are represented. In nature, ants lay down pheromones leading each other to resources while exploring their environment. In computational swarm intelligence, artificial ants keep track of their positions and the quality of their solutions so that in subsequent time steps other ants might be aided to locate better solutions [8].

Artificial Immune Systems: Artificial immune systems (AIS) is a technique based on mimicking the function of the immune system in order to build smart computational systems [9]. Several researchers investigate the application of artificial immune systems for solving complex computational problems in various scientific and engineering areas. AIS belong to the wider area of Artificial Intelligence but at the same time are identified as members of other fields such as biologically inspired computing and natural computation though from a broader point of view belong to machine learning.

Artificial Bee Colony: The Artificial Bee Colony (ABC) algorithm simulates the foraging behavior of honey bees [10]. It is comprised of three stages: i) employed bee, ii) onlooker bee, and iii) scout bee. In the first two stages colony members perform local searches in the neighborhood of a group of preselected solutions deterministically selected in the employed bee stage and probabilistically selected in the onlooker bee stage. During the third stage that mimics the "abandoning exhausted food sources" behavior in the foraging process, solutions that are not useful and do not contribute in the search progress are abandoned and replaced by new solutions so as to promote search at new regions. The ABC swarm method goes through the above steps until the stopping criterion is satisfied.

Multi-Swarm Optimization: Multi-swarm optimization (MSO) is a more general PSO formulation. As its name indicates it is based on the use of multiple swarms (i.e., sub-swarms) instead of a single swarm. The underlying idea in multi-swarm optimization is the use of each sub-swarm at a different region of the search space while an independent algorithm decides the place and time frame for initializing the sub-swarms. MSO is beneficial for optimizing multi-modal problems where more than one optimum exists.

Stochastic Diffusion Search: Stochastic diffusion search (SDS) is an optimization technique for problems where the objective function can be decomposed into multiple, independent, partial-functions. It utilizes an agent-based probabilistic global search where each agent maintains a hypothesis that is tested using an iterative method. At each iteration step a randomly selected partial-function is evaluated. In the basic SDS implementation the partial function evaluation follows a binary form allowing each agent to receive an active, or inactive, mode. The basis of the SDS algorithm is the diffusion of an agent's hypothesis across the remaining population through inter-agent interaction and communication. In SDS systems *stigmergy* does not apply since agent communication is utilized via a direct one-to-one interaction. In addition, a positive feedback mechanism forces the agent population to approach the globally best solution as time passes. SDS has been shown to be not only an efficient but also a robust search and optimization algorithm [11].

Gravitational Search Algorithm: The development of the gravitational search algorithm (GSA) was inspired by both the gravity law and the notion of mass interactions. The underlying idea in the GSA algorithm is to adopt the theory of Newtonian physics to perform solution search. The searcher agents are assumed to be the collection of masses while the system of masses is treated as being isolated from the remaining environment. Using the gravitational force every mass in the system can transfer information to other masses [12]. In general, agents are considered as objects with masses proportional to their performance. The attraction forces that take place between objects are modeled as gravity forces and drive objects to move towards the heavier mass objects. It should be mentioned that the heavier masses represent a set of good solutions to the problem. Each agent finds itself in a position in the problem space that denotes a possible solution while the mass is evaluated by the fitness function. As time passes the lightest masses are attracted by the heaviest ones. In the most optimistic case, the heaviest among all masses will attract all remaining agents and become the optimum solution in the search space. The whole set of agents is considered as an isolated system of masses where only Neutonian gravitational forces are exerted in the system. Recently, Nobahari and Nikusokhan [13] developed a multi-objective version of GSA known as the Non-dominated Sorting Gravitational Search Algorithm (NSGSA).

Charged System Search: Charged System Search (CSS) is a swarm optimization algorithm based on the governing laws of Coulomb in physics and the laws of Gauss in mechanics [14]. CSS is a multi-agent approach like GSA where the agents are assumed to take the form of charged particles. The interactions among charged particles depend on the particles' fitness values and their separation distance. The intensity of the interaction force can be evaluated from the electrostatics laws while the movement intensity is estimated using Newtonian mechanics laws. Although CSS is a powerful algorithm that may be applied in many optimization problems it is highly effective at non-smooth or non-convex domains. The algorithm is considered as a good optimizer globally and locally [14].

Self-Propelled Particles: The self-propelled particles (SPP) algorithm was presented as a special case of the boids model [15] introduced in 1986 by Reynolds [16]. Swarm members in SPP are modeled as a group of particles that move with a constant speed. When the motion of a particle is perturbed, the particle adopts the average direction of motion of the other particles in its neighborhood [17]. SPP systems induce emergent behaviors at various points in space and a few of them turn out to be universal. Thus, the goal is to find the minimal sufficient statistic that captures these behaviors [18].

River Formation Dynamics: The river formation dynamics (RFD) algorithm is a heuristic method that can be seen as the gradient extension of the ant colony optimization approach. RFD was inspired from the river formation process: it first erodes the ground and then deposits the sediments. The river environment is transformed because the water flow first forces the altitudes of different places to be dynamically modified and then constructs decreasing gradients. Consequently, the gradients give rise to new gradients while approaching the best ones and the problem optimal solutions are given in the form of decreasing altitudes [19].

Cuckoo Search: This algorithm utilizes a search in the problem space by taking into consideration the brooding behavior of some cuckoo species. These cuckoo species use host birds to hatch their eggs and raise their chicks. One of the properties of the cuckoo search algorithm is that it can be extended by adding Levy flights with jump steps taken from the Levy distribution as explained in [20].

Firefly Algorithm: Another swarm-based algorithm that was recently developed is the Firefly Algorithm (FA) that was inspired by the flashing ability of fireflies. Light intensity is correlated with the firefly attraction and according to the light intensity fireflies are divided into subgroups with each group placed around its mode. FA has been found suitable for a variety of multimodal problems such as continuous optimization and clustering [21].

The objective of all swarm intelligence paradigms presented is to model the simple behavior of the swarm members in order to emerge more complex behaviors. It is apparent that all algorithms use member dispersion in the search space and converge to a global point through iterative member interaction. Hopefully, that global point will be the optimal solution.

9.2 Particle Swarm Optimization

9.2.1 *Introduction*

Observation of biological swarm behavior was the source of inspiration for developing the Particle Swarm Optimization by Kennedy and Eberhart [22]. The underlying PSO idea was to emulate the social behavior of birds in a flock. In addition, a point of interest was the ability of birds to flight in a synchronous way and regroup fast in case of a sudden direction change without affecting the overall flock movement.

In particle swarm optimization the swarm members are called particles. The particles are considered to "fly" (or move) in an n-dimensional space which represents the parameter space of the problem. Their goal is to reach a point which represents an optimal, or a-near-optimal, solution of the problem. Particles perform two types of movements: i) toward the best neighboring particle and ii) back to the previously identified best solution. From a social point of view, the particle movements are influenced by the experience of its neighbors as well as its own experience. The result is the emergence of a global behavior that moves the swarm towards the best solution region.

The position of a particle denotes a candidate solution. The search space is determined by the fitness (or objective) function and its dimension equals the number of fitness parameters. Particle transition from one position to another depends on a velocity formula. Particle velocity is an attribute which determines the direction and position change.

The two factors that affect the particle movement in the search space are inserted into PSO through the velocity formula and are expressed as a *social component* that expresses the neighbor experience and a *cognitive component* that expresses a particle's own experience. In PSO context the term experience refers to the best position (solution) that a particle has discovered.

Assuming the position of particle k at time t is denoted as $\mathbf{x}_k(t)$ and its velocity as $\mathbf{v}_k(t)$ the equation that governs the particle movement in the search space is:

$$\mathbf{x}_k(t+1) = \mathbf{x}_k(t) + \mathbf{v}_k(t+1) \tag{9.1}$$

whose value for $t = 0$ is taken from:

$$\mathbf{x}_k(0) \sim U(\mathbf{x}_{\text{min}}, \mathbf{x}_{\text{max}}) \tag{9.2}$$

with $\mathbf{x}_{\text{min}}$ and $\mathbf{x}_{\text{max}}$ expressing the minimum and the maximum vector of the search space and $U(\ )$ being the uniform distribution [23]. The initial position of each particle is a random vector whose elements are obtained from a uniform distribution which ensures that the particles are placed within the search space.

The velocity vector in Eq. (9.1) is expressed as the sum of three terms: the previous velocity at time t, the social velocity and the cognitive velocity as shown below:

$$\mathbf{v}_k(t+1) = \mathbf{v}_k(t) + \mathbf{v}_k^{\text{cogn}} + \mathbf{v}_k^{\text{soc}} \tag{9.3}$$

The velocity at time t serves as a memory component and stores the particle's previous movement. Its role is to avoid a drastic change in the particle direction and it is also known as the momentum or the inertial term. The cognitive term works as the performance memory for particle k that records and stores the best position it has visited so far. t is also called the "nostalgia term" [22, 24].

The social velocity term depends on the particle interaction with the neighborhood. The neighborhood size has given rise to the development of two main PSO algorithms; the local best (lbest) and the global best (gbest) [6] explained in the following sections.

9.2.2 *Local Best Particle Swarm Optimization*

The local best PSO adopts the use of smaller (i.e., not comprised of all particles) neighborhoods for each particle [25]. An informal definition of the local best PSO could be: "*At least two particles do not share the same neighborhood*". On one hand the above informal definition ensures that there is no single neighborhood that encompasses all swarm members. On the other hand it allows two, or more, particles to belong to partially overlapped neighborhoods or even share the same one. An illustrative example of particle neighborhoods in lbest PSO is shown in Fig. 9.1.

The social component of the velocity is the algorithm part that is influenced by the information exchange in the particle neighborhood. The best solution discovered in the

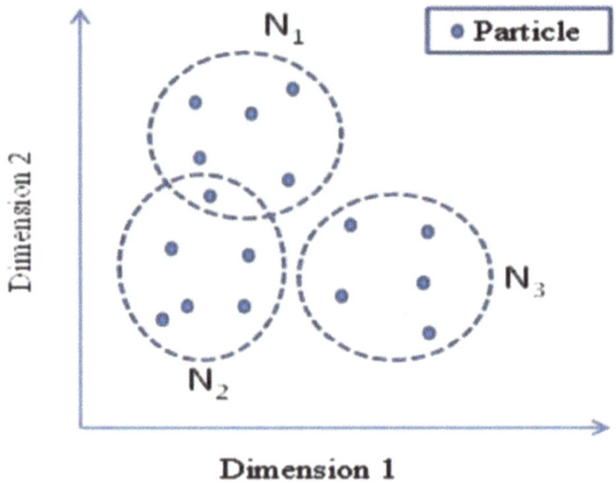

Fig. 9.1 Illustration of the neighborhood notion for the local best PSO. Three neighborhoods (N_1, N_2, and N_3) are observed while N_1 and N_2 partially overlap.

neighborhood is distributed to all particles within that neighborhood. Hence, the velocity value for the lbest PSO at time $t+1$ is given by:

$$v_{kj}(t+1) = v_{kj}(t) + c_1 r_{1j}(t)[b_{kj}(t) - x_{kj}(t)] + c_2 r_{2j}(t)[h_{kj}(t) - x_{kj}(t)] \tag{9.4}$$

where k denotes the k-particle of the neighborhood, j expresses the dimension of the vector in which the calculation takes place, h_{kj} is the best position found in the neighborhood (neighborhood best) and b_{kj} is the best position found by particle j during the search (personal best). The constants c_1 and c_2 are always positive and known as the "acceleration constants" [6] while r_{1j} and r_{2j} are random numbers evaluated from the uniform distribution $U(0,1)$. If we denote the neighborhood of particle k as N_k with size n_k and the problem fitness function as $f(\,)$, then the best position of the neighborhood discovered up to $t+1$ time step is defined as:

$$h_k(t+1) \in \{N_k \mid f(h_k(t+1)) = \min[f(x)],\ \forall \in N_k\} \tag{9.5}$$

It should be mentioned that particles belonging to more than one neighborhood share and distribute information among neighborhoods. As a result, the swarm is able to "fly" towards the global best position through information flow among neighborhoods.

The personal best position is calculated with regard to the fitness function where the fitness value at $t+1$ is compared with the personal best position found up to step t [6]. For minimization problems it is computed as:

$$\mathbf{y}_k(t+1) = \begin{cases} \mathbf{y}_k(t) & \text{if } f(\mathbf{x}_k(t+1)) \geqslant f(\mathbf{x}_k(t)) \\ \mathbf{x}_k(t+1) & \text{if } f(\mathbf{x}_k(t+1)) < f(\mathbf{x}_k(t)) \end{cases} \tag{9.6}$$

and for maximization problems:

$$\mathbf{y}_k(t+1) = \begin{cases} \mathbf{y}_k(t) & \text{if } f(\mathbf{x}_k(t+1)) \leqslant f(\mathbf{x}_k(t)) \\ \mathbf{x}_k(t+1) & \text{if } f(\mathbf{x}_k(t+1)) > f(\mathbf{x}_k(t)) \end{cases} \tag{9.7}$$

The movement of a particle in the search space depends on Eqs. (9.1) and (9.4). Initially, the lbest algorithm uses Eq. (9.4) to update the velocity vector. Subsequently, the updated velocity is plugged into Eq. (9.1) and a new position for the particle is obtained. This two-step procedure takes place for each particle in the swarm and leads after a number of iterations to the best solution.

9.2.3 *Global Best Particle Swarm Optimization*

This second implementation of the particle swarm optimization algorithm encompasses all particles in the velocity update procedure. The gbest PSO can be considered as a general form of lbest since there is a single neighborhood that contains all swarm particles (Fig. 9.2).

The absence of smaller neighborhoods, as it is the case in lbest, dictates the update of the velocity equation using the global best position instead of the neighborhood best. The term global best position denotes the best solution that has been discovered by the whole swarm [26]. An illustrative example of velocity update for a single particle is presented in Fig. 9.2 where the three terms are added to give the "new" vector. The velocity update analytic formula is equal to:

$$v_{kj}(t+1) = v_{kj}(t) + c_1 r_{1j}(t)[b_{kj}(t) - x_{kj}(t)] + c_2 r_{2j}(t)[g_j(t) - x_{kj}(t)] \tag{9.8}$$

where all terms and variables share the same definition as in Eq. (9.4). The only exception is the social component where the global best (g_j) has replaced the neighborhood best position variable [27]. In a similar way, the personal best position is estimated with Eqs. (9.6) and (9.7) for minimization and maximization problems, respectively.

Overall, a particle movement in the gbest algorithm is similar to that in lbest. Initially, the velocity vector is updated using Eq. (9.8) at each time step. Next, the updated velocity is plugged into Eq. (9.1) and a new position for the particle is computed. The process induces all particles to come closer to the global solution once an update takes place. Velocity updates are performed for all particles at each time step while particles independently update their own position in a parallel manner.

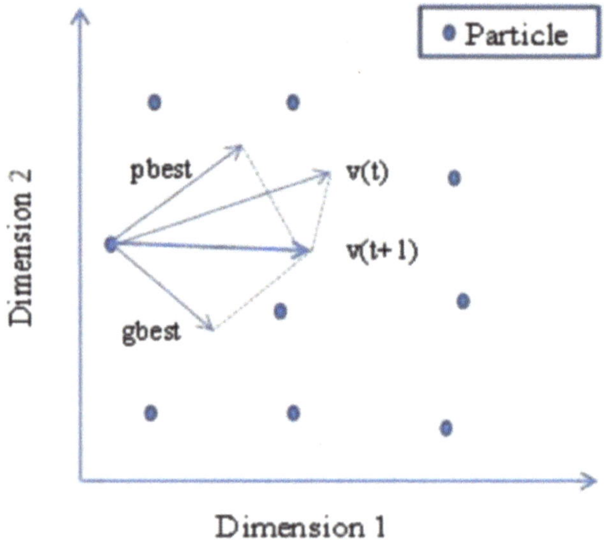

Fig. 9.2 Illustration of the gbest velocity update where three component vectors contribute to the velocity update and there are no neighborhoods.

PSO algorithms can be classified as iterative processes and are listed in Fig. 9.3 for comparative purposes.

Particle Swarm Optimization	
Global Best PSO	Local Best PSO
1. Initialize the *n*-Particle Swarm 2. Set the Personal Best Position for each particle 3. Set the Global Best Position 4. Update the velocity for each particle 5. Update the position for each particle 6. Stopping criterion satisfied?	1. Initialize the *n*-Particle Swarm 2. Set the Personal Best Position for each particle 3. Set the Neighborhood Best Position 4. Update the velocity for each particle 5. Update the position for each particle 6. Stopping criterion satisfied?

Fig. 9.3 Algorithmic steps of both global and local best particle swarm optimization algorithms.

9.2.4 *PSO Stopping Criteria*

The iterative form of the PSO algorithms mandates the utilization of a stopping criterion. If the criterion is satisfied the PSO is terminated and the current position is promoted as the problem solution.

A stopping criterion should be carefully selected to prevent from converging to a suboptimal solution and should maintain computational complexity at low levels. A few of the simple criteria proposed and applied to PSO cases are [6]:

Maximum Number of Iterations: If the number of iterations exceeds a pre-specified threshold then PSO terminates. The use of a low threshold might force the algorithm in premature termination and prevent it from reaching a good solution.

Maximum Number of Fitness Evaluation: If the number of fitness evaluations in iteration exceeds a pre-specified threshold then PSO is forced to terminate. A low number of fitness evaluations might prevent the algorithm from reaching a good solution.

Acceptable Solution is found: If the algorithm reaches a solution that satisfies the condition

$$f(\mathbf{x}_k) \leqslant |f(\mathbf{x}_{\text{optimum}}) - \varepsilon| \tag{9.9}$$

implying that the solution is within the limits of an acceptable error ε then PSO terminates. Selection of ε should be careful so as not to be very high or very low [28]. It should be mentioned that this criterion assumes the prior knowledge of the optimum point (e.g., zero error in pattern fitting) [29].

No improvement over a number of iterations: This criterion measures the PSO improvement over a number of iterations. A measure that defines the performance and a threshold that sets the acceptable performance improvement are required. In addition, the maximum number of iterations over which performance is monitored should be determined in advance. It should be mentioned that convergence in the above criteria does not necessarily imply that the indicated solution is optimal.

9.2.5 *PSO Parameters*

Particle swarm optimization algorithms are controlled by a set of parameters. Control parameters are crucial for PSO performance in terms of convergence speed and quality of solutions. Two types of PSO parameters have been identified: i) basic and ii) modifying-PSO parameters [6, 30]. Beyond that, modified velocity models have also been proposed for controlling the search process.

Basic Parameters The basic parameters are met in all PSO algorithms and are essential for controlling the search process. Their initial values significantly affect the diversity and speed of the search. Those are:

Swarm Size: The number of particles in a swarm defines the swarm size. Large number of particles allows exploration of a large portion of the search space and identification of better solutions, yet increases computational complexity. The optimal swarm size for finding the best solution depends on the specific application and the smoothness of the search space. It has been empirically demonstrated [31, 32] that a number between ten and thirty is sufficient for finding the optimal solution.

Number of Iterations: A large number of algorithm iterations might add unnecessary complexity while a small number might not allow the algorithm to explore large part of the search space and miss optimal solutions.

Neighborhood Size: This parameter is used only by the lbest PSO. Small size neighborhoods allow a small number of interactions among particles resulting in slower convergence. Big-sized neighborhoods allow faster convergence but limit the part of the search space explored [33].

Acceleration Coefficients: The acceleration coefficients c_1 and c_2 found in Eqs. (9.4) and (9.8) play an important role in controlling the cognitive and social velocity components. Their behavior has been studied [34] and several methods have been proposed for evaluating acceleration coefficients [35, 36].

Modifying-Model Parameters In order to overcome the difficulties encountered in various applications several PSO modifications have been introduced. A group of modifications include the use of parameters to control the velocity updates [5]. A few of the proposed parameters that modify PSO are given below.

Inertia Weight: The inertia weight is a single parameter which multiplies the momentum term in Eq. (9.3) (and respectively in Eqs. (9.4) and (9.8)). Its objective is to control the amount of memory from the previous particle flight that will be used to update the velocity in the current time step. Evaluation of the inertia weight is crucial for the overall swarm convergence to a solution and random adjustment [37], linear decreasing weight [38], nonlinear decreasing weight [39], and fuzzy adaptive inertia [40] have been proposed to calculate the inertia value.

Velocity Clamping: Consecutive updates often result in very high velocity values lengthening the position updates and leading to position values that exceed the boundaries of the search space. Such a situation indicates the failure of the optimization process. The goal of velocity clamping is to control the sequential velocity updates and subsequently keep the particles within the search space. Velocity clamping takes the form of an upper velocity limit. All values exceeding clamping limit are made equal to that upper limit [41]. Thus, if the velocity upper limit for the i dimension is denoted as $V_{\max}$ the notion of velocity clumping is analytically expressed as:

$$v_{kj}(t+1) = \begin{cases} v^*_{kj}(t+1) & \text{if } v^*_{kj}(t+1) < V_{\max,j} \\ V_{\max,j} & \text{if } v^*_{kj}(t+1) \geqslant V_{\max,j} \end{cases} \tag{9.10}$$

where $v_{kj}(t+1)$ is the velocity update calculated from the respective equations of the gbest and lbest algorithms. Selection of the clumping threshold should be careful since large values might lead to overpassing good solutions and small values might lead to long-term searches.

Constriction Coefficient: Similar to inertia is the constriction coefficient that is used to control the velocity update equation as well. If we denote the constriction coefficient with c the velocity update equation becomes:

$$\mathbf{v}_k(t+1) = c\left[\mathbf{v}_k(t) + \mathbf{v}_k^{\text{cogn}} + \mathbf{v}_k^{\text{soc}}\right] \tag{9.11}$$

with

$$c = \frac{2k}{|2 - (c_1r_1 + c_2r_2)\sqrt{(c_1r_1 + c_2r_2)((c_1r_1 + c_2r_2) - 4)}|} \tag{9.12}$$

where c_1, c_2, r_1, r_2 are taken from the velocity update formulas, i.e. Eq. (9.4) or Eq. (9.8). Details on the derivation of the above formulas may be found at [42].

Velocity Models Beyond the parameters described above there is a group of modified PSO models proposed to control the search process. These models consist of modifications of the PSO velocity update equations [6]:

$$\mathbf{v}_k(t+1) = \mathbf{v}_k(t) + \mathbf{v}_k^{\text{soc}} \tag{9.13}$$

Clearly the velocity update is not affected by the particle's best position but only from the neighborhood/global best position [44]. In other words, the model does not return to a previous best position since it has no memory of its previous search steps.

Selfless Model: The selfless model is expressed by Eq. (9.13), as well. The difference with the social-only model lies in the restriction for a particle to become the neighborhood/global best in its own update equation [45].

Cognition-Only Model: The cognition model allows for the update of the velocity taking into account only the momentum and the cognitive component of the particle:

$$\mathbf{v}_k(t+1) = \mathbf{v}_k(t) + \mathbf{v}_k^{\text{cogn}} \tag{9.14}$$

It should be mentioned that the cognition-only model restricts the search around the initial particle positions [43] and it admittedly performs worse than the two velocity models presented above [45].

9.3 PSO Examples in Industrial Engineering Applications

Computational intelligence tools have been extensively used in industrial engineering applications. This section presents some of the PSO paradigms applied in various industrial engineering areas.

9.3.1 *PSO Job Shop Scheduling*

The task of assigning a set of jobs on a number of available machines is known as the job shop scheduling problem (JSP). From a mathematical point of view, JSP is a combinatorial optimization problem. The optimization process is subject to a set of constraints regarding the specified processing order of each job.

Several approaches that encompass particle swarm optimization have been proposed for solving the job-shop problem. Liu *et al.* [46] proposed a PSO-based approach to the job schedule problem while they researched for optimal methods on problem coding and PSO operation. Results show that PSO is suitable for job-shop scheduling optimization problems under an optimal problem coding. In addition, a hybrid approach of particle swarm optimization with a local search algorithm was proposed [47]. The method is applied in solving multiobjective problems where PSO explores the solution space, finds a solution and adopts a local search algorithm to reassign the machines to operations. Application results demonstrate the methodology potential in flexible job-shop scheduling and show the significant increase in the PSO convergence speed.

PSO is the main tool in an approach [48] where Lei converted the job-shop scheduling problem into a continuous problem using the following trick: he represented the search space with the help of a set of rules. He used a modified PSO algorithm in which the global best position search is combined with a Pareto archive maintenance strategy. The obtained results showed that a set of Pareto optimal schedules can be estimated with the proposed method. In addition, Lei [49] applied the above approach in the special case of fuzzy processing time and fuzzy due-date scheduling. The results obtained from eight different cases demonstrated the methodology potential in fuzzy-based multiobjective problems.

Other approaches to job shop scheduling problems include the use of PSO in combination with artificial immune systems [50], simulated annealing [51], or multiple swarms [52].

9.3.2 *PSO in Facility Layout Problems*

The goal of a Facility Layout Problem (FLP) is to find the optimal department locations for a production system in order to minimize an objective function. PSO has been proposed in several approaches as an efficient method for optimizing facility layouts.

Ohmori *et al.* [53] used the basic PSO in combination with a constriction model for finding the optimal facility layout. The proposed method was applied in five different test cases each having a different number of production departments. In the majority of the tests PSO outperformed the simulated annealing method which was adopted for benchmarking purposes. Onut *et al.* [54] utilized particle swarm optimization for the optimal design of a multi-level warehouse with the objective to minimize the annual-product-carrying costs. PSO managed to find an optimal solution faster than the branch-and-bound (B&B) algorithm in the vast majority of the test cases. Furthermore, for the same type of problems a combination of PSO and Tabu Search (TS) has been proposed [55]. The synergism of PSO with TS was proven efficient since an optimal solution was found in all test cases while the combined PSO-TS was shown to be faster than the individual PSO and TS.

Other paradigms include the use of PSO in the single row facility layout problem [56] and the dynamic facility layout [57].

9.3.3 *PSO in Assembly Sequence Planning*

Significant role in a product's lifecycle has the assembly sequence planning (ASP) which is a NP-hard to complete problem [58]. The optimal determination of an assembly sequence plan aims in reducing the assembly unit cost and the assembly time required in a manufacturing processes.

An application of the basic PSO in an assembly sequence planning problem was presented by Yu *et al.* [58]. The problem was to find the optimal assembly sequence of a valve comprised of twenty nine components. The objective function was formulated with respect to geometrical constraints, assembly stability and changing times of assembly directions [58]. The PSO approach avoided premature convergence to a non-optimal solution and succeeded in finding an optimal point in a short period of time. An approach employing a discrete particle swarm optimization algorithm coupled with appropriate for the problem coding techniques has been applied in ASP problems [59].

The combination of PSO with other tools intends to improve the algorithm performance for assembly sequence planning in situations where the number of components is large. Wang and Liu [60] have utilized the combination of PSO with chaos methods in order to optimize the assembly of four generators consisting of twenty, thirty, forty and fifty parts. The results obtained clearly promoted the use of the chaos based PSO over the basic PSO approach. Other hybrid methods encompass the PSO synergism with soft computing tools (i.e. artificial neural networks, fuzzy logic, genetic algorithms, etc.) [61], simulated annealing [62] and ant colony algorithms [63].

It is of great interest to determine a product assembly sequence which takes place in multiple plants. In that case, optimal assembly planning is challenging due to the large number of constraints that ought to be satisfied. Utilization of PSO for multi-plant assembly sequence planning has been presented in Cheng *et al.* [64] and shown to discover an optimal plan.

9.3.4 *Multiobjective Design*

Designing a system, or a product, often depends on the use of one, or more, cost (i.e., objective) functions. The majority of industrial applications require multiple objectives to be simultaneously optimized and belong to an area known as multiobjective design.

Particle swarm optimization has been extensively used in a variety of industrial engineering problems requiring multiple objectives to be optimized. For example, it has been applied in the optimal design of multi-sectional hydrogenerators [65], three phase induction machines [66], sustainable construction projects [67], alloy steels [68], laminate composites [69], and collaborative design of products [70]. In all cases, the results reported promote PSO as a method of high potential in finding vector solutions (i.e., a solution for multiple objectives).

9.3.5 *Other Applications*

The list of PSO industrial engineering applications is not limited in the aforementioned cases. There are other areas such as production [71] and flow-shop scheduling [72] where PSO has been employed with satisfactory results.

PSO applicability has been extended from the industrial engineering realm to nuclear power plant design, operation and maintenance tasks. This set of implementations is discussed in detail in the following section.

9.4 Application of PSO in Nuclear Power Plants

Nuclear power plants are large installations of high complexity as they are composed of a large number of interacting systems and subsystems. Thus, optimization of processes related to nuclear plant design, operation and maintenance is required in order to improve its safety and efficiency.

9.4.1 *Reactor Core Design*

The optimal design of a reactor core is of great significance for the overall NPP performance. The design goal is to maximize the energy production while keeping the operation cost low and utilizing as much of the nuclear fuel in the core as possible [73, 74].

Domingos and Schirru [73] proposed the use of particle swarm optimization in light water reactor core designing. They set up an optimization problem in which the objective was to minimize the average power peak factor subject to a set of eleven constraints. PSO was tested on ten different experiments and compared to three established genetic algorithms. It provided the lowest peak factor in all ten experiments and outperformed the three genetic algorithms.

In a similar application Lima *et al.* [74] applied the basic PSO in designing the core of a two-loop pressurized water reactor. The swarm algorithm was used to search for a solution in an eight-dimensional space and its outcome was compared with basic genetic algorithms. Both methods proposed optimal solutions but PSO proved to be much faster and consistent in converging to the optimal solution while it employed less computational resources. Overall, PSO outperforms genetic algorithms in core design since it provides better design parameters.

9.4.2 *Reactor Core Loading Pattern Design*

The nuclear core reloading operation is a challenging and difficult procedure. The ultimate goal in a reactor is to maximize its performance by optimizing the position of the fuel assemblies. The arrangement of the core fuel assemblies in a loading pattern should satisfy a set of engineering constraints (i.e., the number of fresh fuel elements, the number of burnt fuel elements, neutron absorbers etc.). In other words, it should be decided where the fresh and burnt core components ought to be located in order to improve its performance [75, 76]. The major obstacle in determining the fuel assembly position is imposed by the large number of possible combinations of the fuel loading patterns in the core.

PSO has found application in the load refueling problem. Khoshahval *et al.* [75] developed a method based on PSO equipped with an objective function which expressed the power flattening in the reactor core. PSO performance was compared with that of a genetic algorithm where both algorithms were proven capable of providing an optimal solution, though PSO was significantly faster. A continuous PSO version was tested on the same reactor type [76] providing the same conclusions.

Similar conclusions were reached by Meneses *et al.* [77] who applied PSO for fuel reloading at a pressurized water reactor located in Brazil. Furthermore, Babazadeh *et al.* [78] have adopted PSO for fuel reloading in VVER reactors. The results reported do not differ from those provided in the other reports and PSO has come to be recognized as an efficient method for estimating optimum nuclear fuel reloading patterns.

9.4.3 *Panel Layout*

The control room is the center of operations in nuclear power plants. Thus, it is necessary for the panel layout in the control room to satisfy the optimum layout principles in order to improve the man-machine interoperability and enhance the system safety and operation. It is generally true that panel designers create the panel layout based upon their own preferences and experience failing to meet the operator's needs.

This is the reason that several automated panel layout optimization methods have been proposed. Among others there is a short list of PSO based methods specialized in NPP panel layouts. Xu and Yang [79] used PSO setting as a single optimization objective the sum of the distances an operator's hands move on the panel. The method was tested on a set of twelve basic components in NPP panels with satisfactory results.

In the work reported by Xu *et al.* [80, 81] PSO was applied in designing an NPP panel layout having as an optimization objective the operational efficiency defined as a set of pre-specified evaluation indices (e.g., performance was evaluated with more than one objectives). The authors have shown that the total distance an operator's hand moved was shorter compared to a traditional panel layout.

9.4.4 *Maintenance Scheduling*

The optimal maintenance scheduling is crucial for the safe and efficient NPP operation. Maintenance should be done on time to minimize, if not eliminate, the risk of accidents. Furthermore, maintenance should be done in short-term intervals since a long pause of power generation will have serious economic impact to the plant operator. Thus, several strategies for determining the optimal time for plant maintenance have been developed.

A maintenance scheduling strategy based on PSO has been proposed by Carlos *et al.* [82]. The authors state that maintenance planning should be based not only on prescheduled maintenance intervals but also on resource availability. Their planning strategy is modeled as a multiobjective problem and PSO is adopted to find an optimal solution. The test output indicated that PSO was successful in finding optimal solutions and the overall strategy efficiency was increasing when the human resources were incorporated in the objective function of the optimization problem.

An approach developed for preventive maintenance appears in [83]. The authors have utilized a non-periodic preventive maintenance scheduling approach using PSO as the optimization tool. Schedule optimization was based on a set of six factors (i.e., objectives) that include repair probabilities and the respective costs. The results obtained showed that the PSO succeeded in reaching optimal solutions despite the lack of prior information.

9.5 Conclusions

Swarm intelligence is a part of computational intelligence based on modeling the social behavior of humans or animals in nature as it has been discussed in Sec. 9.1. Particle swarm intelligence, which was described in Sec. 9.2, is an optimization technique that mimics the movement of birds in a flock. It uses a group of individual particles which interact with each other and through particle interaction a swarm behavior emerges. PSO is a popular method and has been extensively used in many optimization problems in various fields. This chapter serves as starting point to the interested reader. Its goal is to present a survey of PSO approaches in industrial engineering implementations and nuclear power plants.

In industrial engineering PSO has been widely applied to solve various problems. In Sec. 9.3 PSO implementations in a few industrial engineering areas were presented to demonstrate its broad applicability and highlight the variety of PSO-based methods.

PSO has found limited applications in nuclear power plant design, operation and maintenance when compared to the wealth of implementations proposed in industrial engineering. The most significant applications deal with core design, core refueling and maintenance scheduling problems. As the requirements for safe and efficient plant operation become more strict it is anticipated that PSO will find wider applicability in the nuclear industry.

Bibliography

[1] L.H. Tsoukalas and R.E. Uhrig, *Fuzzy and Neural Approaches in Engineering* (Wiley and Sons, New York, 1997).
[2] J.F. Kennedy, *Swarm Intelligence* (Morgan Kaufmann, Waltham, MA, 2001).
[3] S.J. Russell, *Artificial Intelligence: A modern Approach* (Prentice Hall, New York, 2009).
[4] C. Bishop, *Pattern Recognition and Machine Learning* (Springer, New York, 2006).
[5] A.P. Engelbrecht, *Fundamentals of Computational Swarm Intelligence* (Wiley, New York, 2006).
[6] A.P. Engelbrecht, *Computational Intelligence: An Introduction* (Wiley, New York, 2007).
[7] K.E. Parsopoulos and M.N. Vrahatis, *Nat. Computing* **1** (2-3), 235 (2002).
[8] M. Dorigo, *Ant colony optimization* (Bradford Book, Boston, MA, 2004).
[9] J. Timmis, M. Neal and J. Hunt, *Biosys.*, **55** (1-3), 143 (2000).
[10] D. Karaboga and B. Akay, *Applied Math. Comput.* **214**(1), 108 (2009).
[11] S. Nasuto and M. Bishop, *Parallel Alg. Applic.* **14** (2), 89 (2007).
[12] E. Rashedi, H. Nezamabadi-pour and S. Saryazdi, *Information Sci.* **179** (13), 2232 (2009).
[13] H. Nobahari and M. Nikusokhan, in *Proceedings of the International Conference on Swarm Intelligence* (Chongqing, China, 2011), pp. 1.
[14] A. Kaveh and S. Talatahari, *Acta Mechanica,* **213** (3-4), 267 (2010).
[15] P. Degond and S. Motsch, *Math. Models Methods Appl. Sci.*, **18**, 1193 (2008).
[16] C.W. Reynolds, *Computer Graphics*, **21**(4), 25 (1987).
[17] A. Czirók and T. Vicsek, *Physica A*, **281**, 17 (2006).
[18] Y.X. Li, R. Lukeman and L. Edelstein-Keshet, *Physica D: Nonlinear Phenomena*, **237**(5), 699 (2007).
[19] P. Rabanal, I. Rodríguez and F. Rubio, *Unconventional Computation, UC'07,LNCS 4618.* Springer, 163 (2007).
[20] X.S. Yang and S. Deb, in *Proceedings of the World Congress on Nature & Biologically Inspired Computing, NaBIC* (IEEE, Coimbatore, India, 2009), p. 210.
[21] X.S. Yang, *Nature-Inspired Metaheuristic Algorithms* (Luniver Press, UK, 2008).
[22] J.F. Kennedy and R.C. Eberhart, in *Proceedings of the IEEE International Conference on Neural Networks* (IEEE, New York, 1995), p. 1942.
[23] A. Papoulis and U.S. Pillai, *Probability, Random Variables and Stochastic Processes* (McGraw-Hill, New York, 2002).
[24] G. Ciuprina, D. Ioan and I. Munteanu, *IEEE Trans. Magnetics*, **38** (2), 1037 (2002).
[25] F. Van den Bergh and A.P. Engelbrecht, *IEEE Trans. Evol. Comp.*, **8**(3), 225 (2004).
[26] M. Clerk, *Particle Swarm Optimization* (John Wiley and Sons, New York, 2010).
[27] A. Banks, J. Vincent and C. Anyakoha, *Natural Comp.*, **6**(4), 467 (2007).
[28] T. Hendtlass and M. Randall, in *Proceedings of the Inaugural Workshop Artificial Life*, (Adelaide, Australia, 2001), p. 15.
[29] J.F. Schutte and A.A. Groenwold, *J. Global Optim.*, **31**(1), 93 (2001).
[30] W. Zhang and Y. Liu, in *Proceedings of the Power Engineering Society General Meeting* (IEEE, Denver, 2004), p. 239.

[31] F. Van den Bergh and A.P. Engelbrecht, in *Proceedings of the Genetic Evolutionary Computation Conference* (San Francisco, 2001), p. 892.
[32] R. Brits, A.P. Engelbrecht and F. Van de Bergh, in *Proceedings of the 4th Asia-Pacific Conference on Simulated Evolution and Learning* (Singapore, 2002), p. 692.
[33] W.F. Leong and G.G. Yen, in *Proceedings of the IEEE Congress on Evolutionary Computation* (IEEE, Vancouver, 2006), p. 1718.
[34] F. Van den Bergh and A.P. Engelbrecht, *Inform. Sci.*, **176**(8), 937 (2006).
[35] A. Ratneweera, S. Halgamuge and H. Watson, H. (2003), in *Proceedings of the 1st International Conference on Fuzzy Systems and Knowledge Discovery* (IEEE, Singapore, 2003), p. 264.
[36] P.N. Suganthan, in *Proceedings of the Congress on Evolutionary. Computation* (IEEE, Washington D.C., 1999), p. 1958.
[37] J. Peng, Y. Chen and R.C. Eberhart, in *Proceedings of the Annual Battery Conference on Applications and Advances* (IEEE, Long Beach, CA, 2000), p. 173.
[38] S. Naka, T. Genji, T. Yura and Y. Fukuyama, in *Proceedings of the IEEE Power Engineering Society Winter Meeting* (IEEE, Columbus, OH, 2001), p. 815.
[39] T. Peram, K. Veeramachaneni C.K. and Mohan, in *Proceedings of the IEEE Swarm Intelligence. Symposium* (IEEE, Indianapolis, IN, 2003), p. 174.
[40] Y. Shi, Y. and R.C. Eberhart, in *Proceedings of the IEEE Congress Evolutionary Computation* (IEEE, Seoul, Korea, 2001), p. 101.
[41] R.C. Eberhart, P.K. Simpson and R.W. Dobbins, *Computational Intelligence PC Tools* (Academic Press Professional, New York, 1996).
[42] M. Clerc and J. Kennedy, *IEEE Trans. Evolut. Comput.*, **6**(1), 58 (2002).
[43] J.F. Kennedy, in *Proceedings of the International Conference on Evolutionary Computation* (IEEE, Indianapolis, IN, 1997), p. 303.
[44] E. Ozcan and C. K. Mohan, in *Proceedings of the IEEE Congress Evolutionary Computation* (IEEE, Washington D.C., 1999), p. 1939.
[45] A. Carlisle and G. Dozier, in *Proceedings of the International Conference on Artificial Intelligence* (IEEE, Las Vegas, 2000), p. 429.
[46] Y. Liu, X. Liu and J. Zhao, *Int. J. Adv. Manuf. Technol.*, **38**, 386 (2008).
[47] G. Moslehi and M. Mahnam, *Int. J. Production Economics*, **129**, 14 (2011).
[48] D. Lei, *Comp. Ind. Eng.*, **54**, 960 (2008).
[49] D. Lei, *Int. J. Adv. Manuf. Technol.*, **37**, 157 (2008).
[50] H.W. Ge, L. Sun, Y.C. Liang and F. Qian, *IEEE Trans. Syst. Man Cyber.*, **38**(2), 358 (2008).
[51] W. Xia and Z. Wu, *Comp. Ind. Eng.*, **48**, 409 (2005).
[52] G.G. Yen and B. Ivers, *Intern. J. Intell. Comput. Cybern.*, **2** (1), 5 (2009).
[53] S. Ohmori, K. Yoshimoto and K. Ogawa, in *Proceedings of the IEEE Third International Joint Conference on Computational Science and Optimization* (IEEE, Huanghan, China, 2010), p. 409.
[54] S. Onut, R.U. Tuzkaya and B. Dogac, *Comput. Instr. Eng.*, **54**, 783 (2008).
[55] W.C. Chiang, G. Mudunuri, C. Gangshu, W. Zhu and X. Xu, in *Proceedings of the IEEE Congress on Evolutionary Computation* (IEEE, New Orleans, 2011), p.1679.
[56] H. Samadghandi, P. Taabayan and F.F. Jahantigh, *Comp. Indust. Eng.*, **58**, 529 (2010).
[57] H. Rezazadeh, M. Ghazanfari, M. Saidi-Mehrabad and S. Jafar Sadjadi, *J. Zhejiang Univ. Sci. A.*, **10** (4), 520 (2009).
[58] H. Yu, J. Yu, and W. Zhang, *Trans. Tech. Publications*, **16-19**, 1228 (2009).
[59] H.G. Lv and C. Lu, *Int. J. Adv. Manuf. Technol.* **50**, 761 (2010).
[60] Y. Wang and J.H. Liu, *Robot. Comp. Integr. Manuf.*, **26**, 212 (2010).
[61] M.F.F. Rashid, H. Windo and A. Tiwari, *Int. J. Adv. Manuf. Technol.*, **59**, 349 (2012).
[62] H.G. Lv, C. Lu, and J. Zha, *IEEE International Conference Mechatronics and Automation* (IEEE, Xi'an, China, 2010), p. 1998.

[63] B. Shuang, J. Chen and Z. Li, *Int. J. Manuf. Technol.*, **38**, 1227 (2008).
[64] Y.J. Cheng, J.Y. Chen and F.Y. Huang, *Int. J. Prod. Res.*, **10**, 2765 (2010).
[65] S.L. Ho, S. Yang, G. Ni, E.W.C. Lo and H.C. Wong, *IEEE Trans. Magn.*, **41**, 1756 (2005).
[66] Y. Duan and R. Harley, *IEEE Trans. Indust. Appl.*, **47** (4), 1707 (2011).
[67] J.S. Chou and T.S. Le, in the *Proceedings of the Conference on Industrial Engineering and Engineering Management* (IEEE, Hong Kong, 2011), p. 625.
[68] Q. Zhang and M. Mahfouf, *IEEE Congress Evolutionary Computation* (IEEE, Trontheim, Norway, 2009), p. 3241.
[69] A.R.M. Rao and K. Lakshmi, *J. Reinforced Plast. Comp.*, **30** (20), 1703 (2011).
[70] J. Liu, M. Chen, Y. Yao and Q. Kong , *Adv. Mater. Res.*, **97-101**, 3593 (2010).
[71] F. Zhao, A. Zhu, D. Yu, and Y. Yang, in *Proceedings of the 6th World Congress on Intelligent Control and Automation* (IEEE, Dalian, China, 2006), p. 6772.
[72] Q.K. Pan, M.F. Tasgetiren and Y.C. Liang, *Comp. Oper. Res.*, **35**, 2807 (2008).
[73] R.P. Domingos and R. Schirru, *Nucl. Sci. Eng.*, **152**, 197 (2006).
[74] C.A.S. Lima, C.M.F. Lapa, C.M.N.A. Pereira, J.J. Da Cunha, and A.C.M. Alvim, *Annals Nucl. Energy*, **38**, 1339 (2011).
[75] F. Khoshahval, H. Minuchehr, and A. Zolfaghari, *Nucl. Eng. Des.*, **241**, 799 (2011).
[76] F. Khoshahval, A. Zolfaghari, H. Minuchehr, M. Sadighi, and A. Norouzi, *Annals Nucl. Energy*, **37**, 1263 (2010).
[77] A.A.M. Meneses, M.D. Machado, and R. Schirru, *Prog. Nucl. Energy*, **51**, 319 (2009).
[78] D. Babazadeh, M. Boroushaki and C. Lucas, *Annals Nucl. Energy*, **36**, 923 (2009).
[79] Y. Xu and Z. Yang, *Key Eng. Materials*, **486**, 41 (2011).
[80] Y. Xu, Z. Yang and Q. Meng, Q, in *Proceedings of the 2nd International Conference on Advanced Computer Control* (IEEE, Shenyang, China, 2010), p. 541.
[81] Y. Xu, Z. Yang and Q. Meng, *Key Eng. Materials*, **450**, 308 (2011).
[82] S. Carlos, A. Sanchez and S. Martorell, *Math. Comp. Model.*, **54**, 1808 (2011).
[83] C.M.N. Perreira, C.M.N., Lapa, C.M.A. Mol and A.F. Luz, *Progr. Nucl. Energy*, **52**, 710 (2010).

Chapter 10

A Review of Genetic Algorithm Applications in Supply Chain Network Design

Cagatay Iris and Seyda Serdar Asan

Industrial Engineering Department, Istanbul Technical University, Macka Campus, 34367, Istanbul, Turkey

E-mail: ciris@itu.edu.tr, serdars@itu.edu.tr

Strategic decisions related to the design and planning of the supply chain revolve around design of the network. This chapter introduces how genetic algorithms are applied to solve the supply chain network design problem. A classification of the recent research in the field provides a valuable insight into current state of literature and outlines directions for future research.

10.1 Introduction

Supply chain management (SCM) has become one of the major Industrial Engineering application and research areas over the last two decades. A supply chain, which is the subject of interest in SCM, is a complex network of business entities (e.g., suppliers, manufacturers, distributors, service providers, warehouses, customers) involved in the upstream and downstream flows of material, information and finances [1–3]. Efficient and effective management of these flows is a key to the success of supply chain. Therefore, the management of supply chain flows need to be supported by a coherent supply chain strategy and an appropriate design and planning of the supply chain for execution of the flows.

Traditionally, strategic decisions related to the design and planning of the supply chain revolve around design of the network. A supply chain network consists of a series of linked levels, such as suppliers, production facilities (plants), warehouses, distribution centers and customers, that are involved in processes associated with the flow of products/services. Figure 10.1 illustrates a generic supply chain network with forward and reverse flows. Supply

C. Kahraman (ed.), *Computational Intelligence Systems in Industrial Engineering*, Atlantis Computational Intelligence Systems 6, DOI: 10.2991/978-94-91216-77-0_10,

chain network design (SCND) involves decisions about the configuration of the network, i.e. locating facilities, deciding their capacity, optimizing the transportation of material among them. The network design problem is often solved using large optimization models. SCND problems are combinatorial, complex in nature and thus not well suited to be handled by conventional optimization techniques. Metaheuristic optimization algorithms, such as genetic algorithms, often outperform conventional optimization methods when applied to these problems.

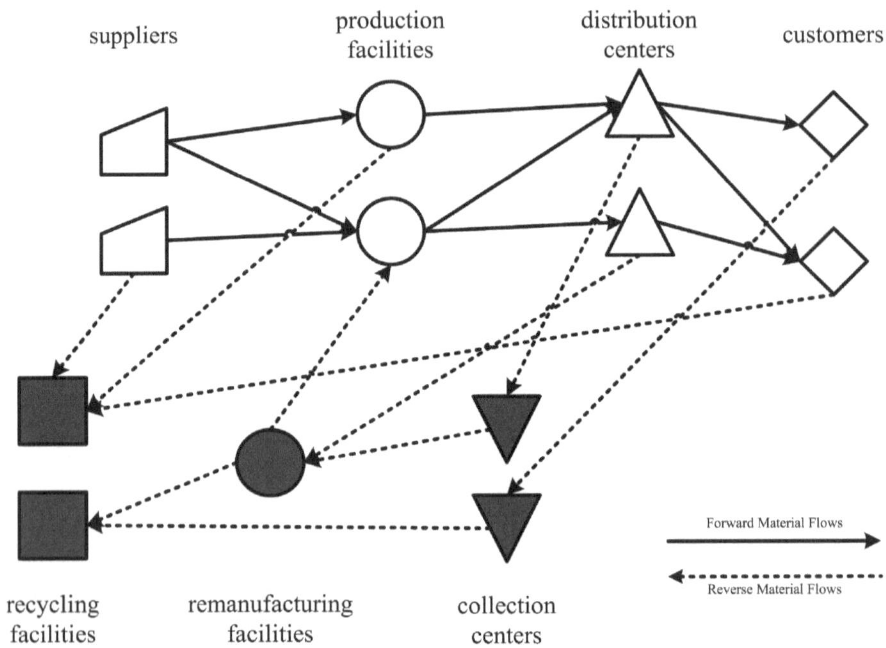

Fig. 10.1 A generic supply chain network

This chapter explains the use of genetic algorithms (GAs) as an optimization tool to solve SCND problems. GA is an evolutionary search algorithm that uses mechanics of natural selection and genetics (e.g. crossover, mutation) to find near-optimal solutions to problems.

The remainder of the chapter is organized into four sections. The next section describes the basic characteristics of the SCND problem. The third section examines how GA is applied to solve SCND problems. The fourth section classifies the studies in the recent literature and outlines research opportunities. Finally the last section provides a summary of the key points of this chapter and a future outlook.

10.2 Supply Chain Network Design Problem

SCND problem has also been studied under the terms logistics network design or distribution network design and has been referred to as *facility location models* in the literature. The facility location problem in the context of SCND is primarily concerned with minimizing the total costs of opening and operating a facility and shipping materials/goods while satisfying demands of the customers. Here, the term facility is used in its broadest sense and includes entities operating at different levels of the supply chain such as plants, warehouses, distribution centers, retail stores, customers and the like. Ideally when modeling a network design, information about the location of supply sources and customers, location of potential facility locations, demand forecast by customer/market, facility, labor and material costs by facility, inventory costs by facility, transportation costs between each pair of facilities, sales price of the product, taxes, tariffs, desired customer response time, and other required service factors is available [4].

The simplest form of the facility location models is the uncapacitated facility location problem (UFLP) that deals with minimizing the network cost for a single commodity with unlimited facility and flow capacities and linear costs. The corresponding Integer Programming model is formulated as (refer to Balinski [5]):

$$\begin{aligned}
&\text{Min} \sum_{j\in D}\sum_{i\in F} c_{ij}x_{ij} + \sum_{i\in F} f_i y_i \\
&\text{subject to} \\
&\sum_{i\in F} x_{ij} = 1, \quad j\in D \\
&x_{ij} \leqslant y_j, \qquad i\in F,\ j\in D \\
&x_{ij} \geqslant 0, \qquad i\in F,\ j\in D \\
&y_i \in \{0,1\}, \quad i\in F
\end{aligned} \tag{10.1}$$

The formulation can be described as follows. F is the set of potential facility locations and building/operating a facility at location $i \in F$ has a fixed cost of f_i. Any open facility can provide an unlimited amount of the certain commodity. D is the set of customers (demand points) and customer $j \in D$ has a demand for the commodity. The decision variable x_{ij} is the fraction of the demand of jth customer supplied from i^{th} facility. Due to the single assignment property of UFLP, we can define x_{ij} as $x_{ij} \in \{0,1\}$, $i \in F$, $j \in D$; since a customer is always entirely supplied from its nearest facility. Hence, $x_{ij} = 1$ means that j^{th} customer is entirely supplied from i^{th} facility. y_i is a binary location variable: $y_i = 1$ indicates a facility is located at $i \in F$, 0 otherwise. The total cost of service or distribution c_{ij} is incurred for supplying j^{th} customer from i^{th} facility; c_{ij} is a variable cost that includes the cost of moving the product between different levels in the network. The goal is to open a subset of facilities and assign customers to these facilities at minimal total cost, that is the total cost of service plus the fixed cost of opening the facilities.

Models of the SCND problem defined through various attributes such as: objective function (single/ multiple), capacity limitation (capacitated/ uncapacitated), number of stages/echelons (single/ multiple) number of commodities/products (single/ multiple), number of sourcing facilities (single/ multiple), planning horizon (single-period/ multi-period), capacity flexibility (fixed/expandable), and type of input data (deterministic/ stochastic). The basic model is extended to accommodate the varying attributes, by introducing additional parameters, variables and constraints. There is a vast literature concerning the development and testing of algorithms that search for a solution to variants of the facility location problem, for a review of contributions and classifications the reader may refer to studies of ReVelle *et al.* [6], Klose and Drexl [7], Melo *et al.* [8], Aikens [9], among others.

The solution approaches for the SCND problem fall into the generic categories of exact and heuristic solutions. Since SCND problems are combinatorial and complex in nature, heuristic solution approaches, such as genetic algorithms, that produce good but not necessarily optimal results are rapidly becoming the preferred way to solve these problems.

10.3 Genetic Algorithms in Supply Chain Network Design

Evolutionary algorithms (EA) are stochastic optimization techniques that utilize principles of natural evolution in finding new search directions. Although, each of the evolutionary techniques, e.g. Genetic algorithms (GA), Memetic algorithms (MA), Cultural algorithms (CA), Differential search strategies (DS), Genetic programming (GP) etc., have their own strengths and weaknesses, for the SCND problem the most common algorithms studied in the literature are variants of GAs [10].

Directed random search techniques such as Genetic Algorithms (GA) can be employed to find a near-optimal solution for many problems in complex multi-dimensional search spaces [11]. The fundamental driver underlying GAs is natural selection which is inspired by the natural evolution process. The process exactly works for biological organism. The key term which is transferred from one generation to another is the changing chromosome combinations to survive in the living world according to the Darwinian principle *survival-of-the-fittest*. After coding the solution in an appropriate way, genetic algorithm iteratively works and evolves to get the global optimum without getting stuck at a local optimum. In the search for global optimum, individuals (chromosomes) in a population are manipulated by genetic operators to improve their fitness values [11]. Unlike many other heuristics or metaheuristics, GAs are much more independent of the problem and more flexible in respect to formulation of objectives and constraints.

The basic steps of GAs may be listed as follows [11]. Note that applying some advanced hybrid methods may change the flow accordingly.

1. Choice of a representation scheme for a possible solution (coding or chromosome representation.)
2. Decision on how to create the initial population.
3. Definition of the fitness function.
4. Definition of the genetic operators to be used (reproduction, mutation, crossover, elitism)
5. Choice of the parameters such as population size, genetic operators probabilities etc.
6. Definition of the termination rule.

A set of initial population must be formed to start GA. Related solutions are illustrated as chromosomes in the search space at this stage. These solutions are evaluated by fitness functions and selected with pre-determined techniques. After this, there are two operators, crossover and mutation, to direct the population to the global optimum. Crossover allows exchanging information between different chromosomes and mutation increases the variety in the population. Next, the new population is formed. This process is continued until a termination criterion is reached.

10.3.1 *Evolutionary Algorithms in SCND problem*

Evolutionary algorithms (EAs) have become quite popular for optimization, machine learning, systems/operations design, financial planning, etc. This rapid increase in the number of studies also affected the use of EA as a SCND optimizer positively. Solution space associated with network design alternatives may be globally searched by using selection strategies which are imposed by chromosome genetic. Fundamental property of EA is maintaining a good population of solution which is coded as genes from one generation to another. In order to understand applicability of EAs in SCND, main characteristics of the problem should be examined.

SCND problem may be described as special extensions of minimal spanning tree, transportation problem, and location-allocation problem. Here, arc values represent amount of items to be carried along the network, and location of DCs and facilities is formulated as open or not binary variable. Evolutionary algorithms help problem structure to obtain 0-1 values, i.e. warehouse/plant enclosures (locations), customer-DC assignment. Additionally, they are capable of representing the interdependency between establishing a new facility and the amounts to be shipped between facilities. Hence, they have the potential to be more efficient than conventional algorithms and could better handle different complexity levels of the SCND problem. However, especially for the problems that can't be illustrated as a network, application of the traditional EA approach becomes quite challenging [12].

As mentioned above, GAs are the most popular evolutionary algorithms used to solve the SCND problem. However, Memetic algorithms, which are similar to GAs, have been proved to have practical success, in particular for an approximate solution of NP-hard optimization problems [13]. One of the main reasons for considering a Memetic algorithm is the complexity of the mathematical model. Extensions to the basic SCND problem, such as multiple capacity levels for each facility, cost savings associated with combined distribution centers and collection/inspection centers require advanced local search strategies in evolutionary aspects [14]. Yeh [13] has mentioned that the lack of local search in GA, would result in more use of MA in the future. An interesting future quest may be the hybridization of MA with different types of metaheuristics [13]. Another evolutionary algorithm that is used to solve the SCND problem is co-evolutionary techniques. *Co-evolutionary strategies* that refer to evolution of two or more interdependent sets of data simultaneously are quite appropriate for SCND. Co-evolutionary algorithm takes the fitness value as the evaluation criterion and seeks to satisfy this criterion through evolution toward a satisfactory/better generation based on the interactive evaluation of chromosomes and criteria [15]. The reason for using co-evolution in a SCND problem is to be able to reconcile the conflict between variable cost of transportation and fixed cost of opening facilities and to provide a faster algorithm Another evolutionary algorithm proposed to solve the SCND problem in the literature is differential evolution algorithm, which is originally developed for global optimization over continuous spaces [16]. When major supplements such as mapping, transformation are required to model given SCND problem, differential evolution is used to solve mathematical models with these major supplements. In the rest of this chapter, genetic algorithms will be examined in detail.

10.3.2 *Applying Genetic Algorithms to SCND problem*

The evolution of GA usually starts with an initiation policy and continues in generations. In each generation, the fitness value of every individual (which are combined of genes) in the population is evaluated, multiple individuals are randomly selected from the current population (based on their fitness value) and modified (recombined/crossover and possibly randomly mutated) to form a new population. The new population is then used in the next iteration of the algorithm [17]. When the pre-defined termination conditions are met, algorithm stops and proposes related solution as global optimum result.

GA is known as a problem-independent approach in terms of execution; however, the encoding depends on the problem under study. In other words, the encoding reflects the objective function proposed by the authors. In the literature both single and multi-objective genetic algorithms are widely studied. Single objective studies mostly focus on the total cost aspect. In these studies, chromosome fitness may be directly linked (parallel or inversely) to the objective function and the search space may be explored with justification of encoding by using a linear vector. Algorithms depend on exploiting special structures are present in the objective function. Hence, when a multi-objective mathematical model is introduced, almost always a multi-objective GA is proposed.

Multiple directional and global search by maintaining a population of potential solutions from generation to generation (population-to-population approach) are the main reasons why GAs may be an appropriate approach for multi-objective optimization [12]. There are primarily two goals to be achieved in multi-objective GAs: (i) convergence to a Pareto-optimal (non-dominated solution) set, and (ii) maintenance of population diversity in a Pareto-optimal set. This can be done by proposing two separate algorithms [18] or by evolving the genetic search toward the ideal Pareto solutions in which all individual objectives are simultaneously achieved at the highest possible level [19]. It is noted in most of the reviewed papers that multiple directional search characteristic of GA is vital to obtain Pareto optimal solution within a convex hull [14, 20]. For multi-objective cases, fitness value should be updated to consider decision maker's attitude towards alternatives, i.e. Random weighted, Adaptive weighted, Interactive adaptive-weighted, Pareto sorting etc. [12]. This would require a change in the chromosome representation as well. Different objective values may require different illustrations or encoding of genes (in different parts of the solution).

In next sections, each step of genetic algorithms will be examined in detail and its merits that have captured the attention of researchers related to its applications to SCND problem will be discussed.

10.3.2.1 *Genetic Representation and Encoding*

A genetic algorithm is a combination of different parameters and an illustration (representation) set. An illustration is a way of representing a solution of a given problem in the chromosome. Encoding, which concentrates on direct gene and individual illustration, is also directly related to representation. It is crucial to illustrate the chromosome structure of the problem under study accurately. It is not sufficient that only the encoding is reflecting the main chromosome structure, but the decoding of the encoded chromosome is critical as well, since it directly affects the results that are obtained from the algorithmic structure. The parameters other than the encoding/decoding are mostly related to the algorithmic structure. The representation of a problem using GAs is both dependent on encoding technique and characteristics of the problem.

In this sense, illustration (representation), encoding/decoding techniques which are associated to SCND problem will be discussed, and applications in the literature will be explained in detail with advantages obtained and requirements that are previously listed due to the problem structure.

Solution encoding in GAs is mostly structured as a finite-length string over some finite alphabet. It should be noted that the choice of chromosome representation can itself affect the performance of a GA-based function optimizer [12].

Chromosome representation and encoding of related individual is crucial to fully characterize the algorithm and the problem type which is formulated. Each chromosome in a SCND problem may be a variant of binary vector or matrix structure. Due to the features such as assignment of DCs to plants, distribution alternatives, customer clusters, capacity, transportation amount alternatives etc. the chromosome illustration generally consists of more than one part. When different aspects are incorporated into the problem structure, such as planning horizon, different commodities (items), alternative routing, etc., illustration may be clustered as matrices composed of vectors [21–23]. For SCND problem, it is also possible to illustrate set of solutions as a tree. Tree based representation has found limited application in the literature. Xu *et al.* [20] have used tree-based illustration to represent the transportation/distribution pattern for stages. Another tree based representation is proposed due to the fact that SCND problem is considered as an extension of the spanning tree problem [24].

Another tree-based illustration may be obtained by using binary vectors to represent routes between SCN facilities. The resulting illustration should contain basic properties of spanning trees with pre-defined characteristics. In this sense, GA may obtain binary vectors that represent a network that is cyclic, disconnected or both. In these cases, a repair mechanism that evaluates the relative importance of the links is required, in order to remove the infeasibilities and obtaining a tree, at the cost of additional computational effort [25].

Encoding of chromosomes may be done with binary encoding, Prüfer number based encoding, random key based encoding, priority based encoding, and permutation encoding. Different encoding approaches are used in the literature for SCND problem. Chang [26] has emphasized that if decision variables of SCND problem are constrained as a positive integer, then the chromosome should be encoded in integer characteristic. However, different aspects of the main problem structure may require use of multiple encoding approaches along with real encoding techniques.

Extensions to the SCND problem may include the stages of a supply chain. Each of the segments on a chromosome may be used to obtain the transportation tree of a stage on the supply chain, i.e. the nth segment of a chromosome matches the nth stage of the SCN. Altiparmak *et al.* [27] have proposed two different representation types in a chromosome. Due to the single assignment property, i.e. each customer should be assigned to one DC, an integer-encoding is used at final stage. For the rest of the stages, a priority-based encoding is used. In priority based encoding, position of genes is the set of facilities, while priorities are used to determine which facility will be involved in the next generation solution. In the final stage's integer-based encoding, the position of a gene is the order of customers, and its value represents the DC that corresponding customer will be assigned to. At the same time, gene values show which DCs will be opened. For this solution type, priority based encoding is used, plants with higher priority are opened consecutively until their number reaches to the maximum number of plants to be opened or their total capacity is greater than or equal to total demand [27]. Figure 10.2 illustrates a typical SCND with the network configuration that corresponds to solution encoding given in Figure 10.3. In the first two stages, priorities of each facility are calculated. Distribution center with highest priority will get in solution in the next generation. Additionally, dummy facilities have a priority of one. For the last stage, customers are served by 1st and 4th DCs. Hence, integer-encoding represents the pairs of DC and customer. There also studies that hybridize the priority-based encoding (for traditional multistage SCND) with random number-based encoding (direct shipment and delivery) by dividing a chromosome into two segments [15].

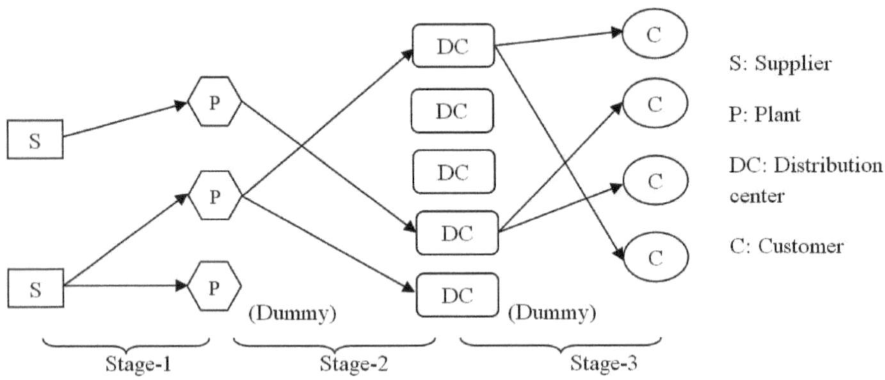

Fig. 10.2 A sample SCND configuration

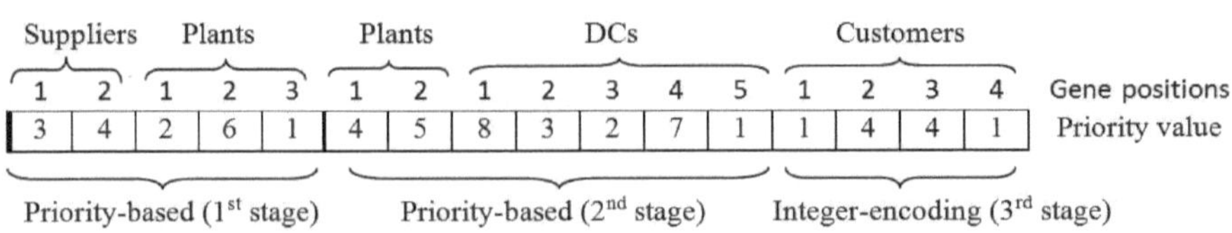

Fig. 10.3 Hybrid priority based encoding

Another binary illustration and encoding was conducted as genes with length $3 \times L$, where L is the total number of customers [28]. Each trio (a set of three genes) represents a delivery route to a customer from a plant via DCs and retailers. The values on the first position of a trio is the ID of a plant, i.e. starting node of the delivery path, the second is the ID of DCs, and the last one indicates the ID of retailers on the path. Figure 10.4 can be used to represent the technique in Lin *et al.*'s study [28]. It also illustrates the related flows: for example, Customer-2 is supplied via a channel that starts with Supplier-1, then Plant-1 and finally DC4.

Binary-integer based representations may also require another type of encoding structure. Ding [16] has proposed a closed loop network model of facilities with expansion

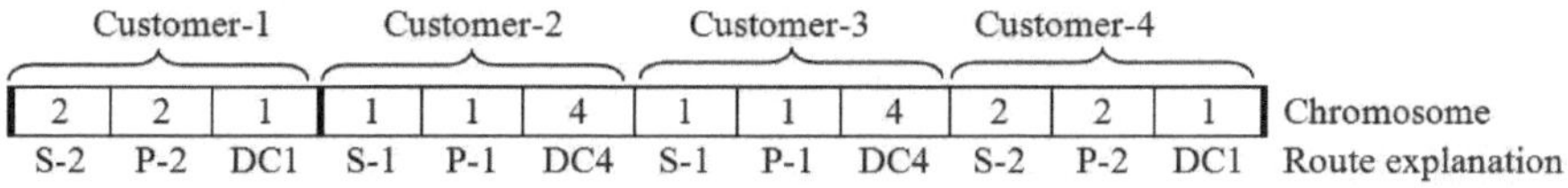

Fig. 10.4 Delivery routes based encoding

potential, where the facilities (i.e. collection points, stores and plants) are 0-1 variables and the expansion multiples of these are integers in the range of 0-5. If the collection points, stores, plants are set to 0, then the expansion multiple is 0.

If the number of items (products) is increased in multi-stage systems, the length of the segment on a chromosome should be enlarged with the number of facilities in that stage. Altiparmak *et al.* [22] have extended priority-based representation to multi-item cases. In this study, each item is represented by different segments. Figure 10.5 shows chromosome representation of network in Figure 10.2 with respect to two different raw materials. For example, the position of a gene on the first segment represents the SCND of raw material-1, its value represents the priority of suppliers or plants that the corresponding raw material will be assigned to. At the same time, gene values show which DCs will be opened [22].

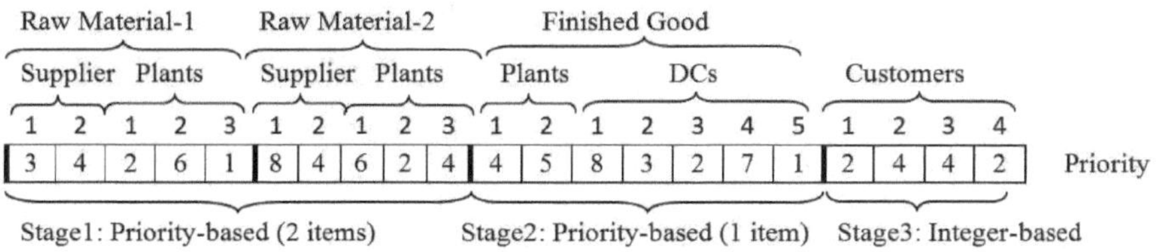

Fig. 10.5 Hybrid priority based encoding for multiple-items

When the investigated problem considers not only the establishment of facilities but also the routings, then the chromosome can be encoded as a binary string with two parts [29]. The first part of the string determines whether the facility (i.e. DC) is opened or not, and the second part represents whether a high demand customer is included in the first level routing or not [29]. Figure 10.6 illustrates binary encoding of supply chain network in Figure 10.2. The set of facilities corresponds to first part of string, while set of customers gives information about customer involvement in solution. The representation may be formulated for each type of customer separately.

Set of facilities										Set of customers			
1	1	1	1	1	1	0	0	1	0	1	1	1	1
S-1	S-2	P-1	P-2	P-3	DC1	DC2	DC3	DC4	DC5	C-1	C-2	C-3	C-4

Fig. 10.6 Binary encoding for multiple facilities

Another encoding strategy is use of Prüfer number (PN) [20, 24, 30]. In Figure 10.7 Prüfer number based encoding of Figure 10.2 is presented. The first and second substrings are binary digits representing opened/closed facilities, respectively. The last four substrings are Prüfer numbers consisting of 6, 4, 5 and 6 digits to represent the distribution pattern for each stage. The representation shows that two suppliers, three plants, and three DCs are opened [20]. It should be noted that use of Prüfer number may result in infeasible sub-trees in spanning tree problem. Since transportation tree in SCND is a special type of spanning tree, Prüfer number encoding may lead to an infeasible solution. Hence, use of Prüfer number to illustrate a logistics network requires some repair mechanisms [31].

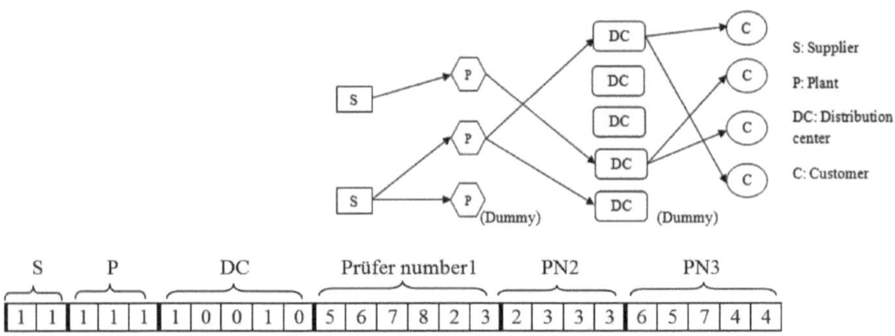

Fig. 10.7 Prüfer number based encoding

Permutation encoding is another way that is used in binary vector representations. In permutation encoding, every chromosome can be described by a feasible binary variable vector with 1 when transportation takes place between DCs and plants [13]. Another encoding representation consists of a permutation chromosome $v(t)$ including only $|\mathrm{P}| + |\mathrm{DC}|$ genes, where $|\mathrm{P}|$ and $|\mathrm{DC}|$ are the number of plants and DCs, respectively. Illustration used as $v(t)$ is a priority-based encoding which is a special type of permutation encoding [32]. Overall chromosome represents plant to DC encoding, while each of two segments is supplier to plants, and DCs to customers, respectively. Figure 10.8 gives a permutation encoding based representation of the network illustrated in Figure 10.2. For the first segment of chromosome, priorities give information about which facility will be opened in next generation and coupled with a supplier in its supply chain.

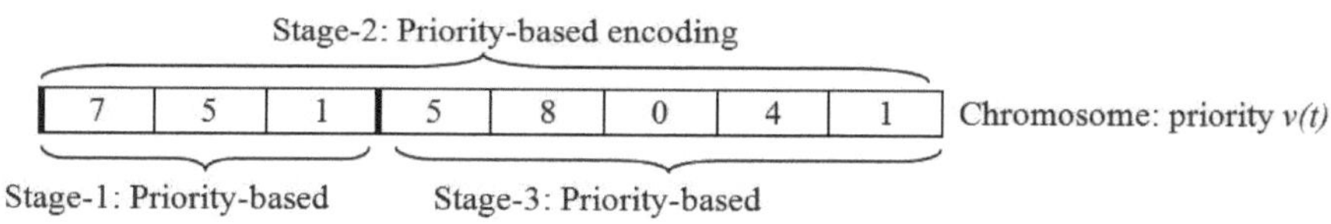

Fig. 10.8 Hybrid permutation encoding

10.3.2.2 *Initialization*

Encoding and decoding aspects are directly linked to problem initiation. Initiation is used to generate a starting pool which is a collection of parent chromosomes. In the literature, there are basically two different strategies to initialize SCND problem individuals successfully. The initial solution may be adopted by a randomized procedure or a well-written heuristic algorithm. Starting with a good solution helps obtain an accelerated convergence. However, there is a tradeoff between speed of convergence and quality of the resulting solution [13]. Most of the studies in the literature start with a randomly chosen solution between lower and upper bounds. Some authors have proposed greedy algorithms to have a better initial start [13, 21, 22, 31]. There is an obvious requirement to test whether the initial solution is in the feasible solution set. Additionally, it is observed that most of the encoding structures are embedded in the initialization procedure. Therefore, obtaining a non-feasible solution becomes impossible because encoding structures are direct variants of chromosome representations.

10.3.2.3 *Fitness Function*

At the end of each generation, all solutions should be evaluated with respect to the overall aim of the SCND problem. Thus, a performance measure known as fitness function should be defined to test the convergence of the solution to the desired optimum solution. The determination of the fitness function is vital, because it must be able to accurately measure the desirability of the features described by the chromosome. Furthermore, it must be computationally efficient, since it is used repeatedly to evaluate all solutions.

In general, the objective function itself provides a mechanism for evaluating each solution (individual) [17, 22, 25, 30, 32]. However, its range of values may become very high in different SCND alternatives. To maintain uniformity over various problem domains, a fitness function may be used to normalize the value of the objective function to a range of 0 to 1. The normalized value of the objective function is the fitness of the individual [12]. Additionally, there are some multiobjective studies where fitness values were not formed directly related to one of the objectives [18, 33]. Six main strategies have been observed in the literature. Four of them are linked to objectives (inverse of obj. function, weighted obj., gap between objective functions, inverse of weighted obj. etc.), rest are not (i.e., total network flow, penalty associated with inverse conversion to non-feasible area etc.). Fitness function is used to reproduce relatively good solutions and to eliminate the bad solutions so that an offspring composed of the good solutions are reproduced.

Due to the fact that, minimizing the total cost of the network is the primary aim of SCND problems, the reciprocal of the objective function is popularly used as fitness function [21, 24, 26]. The fitness value of a chromosome may be the gap between objective function of the worst solution in the current generation and objective value of analyzed chromosome. This means inverse characteristics of objective function is reflected in fitness function [29]. Lin *et al.* [15] have emphasized that higher number of open facilities results in greater fixed cost value, hence fitness of chromosome in their study is described as the number of DCs and facilities that are kept open. A parameter of regret value has been also proposed as fitness function by Gen *et al.* [34]. Regret of move in this study represents the gap between best biggest and lowest regret values of generation. Here, regret of move should be the inverse of fitness. Because, when regret of move is increased, fitness is decreased [34].

Multiobjective genetic algorithms require a fitness function with a different character. For this type of GA, equilibrium between costs, customer satisfaction and transportation times (resulting in earliness/tardiness) is reflected as gaps between objectives to fitness function. A widely used technique is weighted sums of objectives [14, 20, 27]. Besides, distance from Pareto optimal solutions may be used as fitness values [19, 33]. For such types, each non-dominated solution represents best approach in respect to individual objective functions. Chen *et al.* [33] have formulated profit and social welfare maximization as objectives and have found out in analysis that in first iterations there are huge distances between Pareto optimal solutions and generated chromosomes. When the algorithm fitness values are increased, convergence to optimality has been accelerated. Whenever adapted genetic algorithms are used, weights of objectives should be updated [34].

10.3.2.4 *Selection*

Fitness evaluation seeks to associate each individual with a fitness value so that it can reflect the goodness of fit for an individual. This evaluation process intended to compare one individual with other individuals in the population [31]. After this phase, accumulated information should be exploited by the selection mechanism, while new regions of the search space are explored by means of genetic operators, such as crossover and mutation. Selection aims to increase the quality of solutions by selecting better individuals to get copied in next generations. Selection may be linked to basic operators of algorithm. An elaborate selection mechanism requires application of a wider exploration. The selection process directs the genetic search toward promising regions in the search space [12].

The most popular selection techniques found in the literature are tournament selection, random selection, elitist selection, proportional selection (most commonly roulette wheel), $\mu + \lambda$ selection, and crowded comparison selection. Each technique has its own distinctive characteristics. When a decision maker aims to generate the next generation via two best chromosomes of a randomly selected network representation, tournament selection is applicable. Tournament selection is based on selecting chromosomes having better fitness values by assigning higher probabilities to them. This technique is very helpful in SCND problem, because encoding/decoding characteristics of network design directly fit to selection of chromosomes [17, 22, 33]. It is observed that value of the current best SCND cost solution for one instance from each different generation provides a good representation of the algorithm behavior over the entire set of solutions.

It is also emphasized by many authors that the best network design of each generation should be kept in order to obtain individuals with a lower cost. Because, when the solution with a best aspect (cost in most cases) is kept, the resulting offsprings may be in the global neighborhood of the best solution. Thus, elitist selection strategies are implemented. The elitist approach is adopted in order to preserve the best chromosome and overcome probabilistic error of sampling [15]. If best chromosome in the pool has not been reproduced, it is forced to appear in the next generation [15, 20]. A combination of Elitism and $\mu + \lambda$ has been used as a hybrid technique that can enforce best chromosomes into the next generation and avoid the premature convergence of the evolutionary process [12]. This way of selecting the best chromosomes from parents and offspring with elitism produces good computational results [24]. In another study, elitist selection is utilized as a population control technique to overcome the problem of getting stuck with a network design alternative that is local optimum [32].

In most of the SCND studies, fitness value is derived from the objective function. Thus, linking survival of chromosomes to fitness values may be an effective selection approach. Gen *et al.* [21] have proposed a model of roulette wheel displaying proportional probabilities. In this technique, stochastic nature of wheel spinning is used. The number of trials is equal to population size so that each time a new chromosome is selected. The fitness of the chromosome implies the number of times each chromosome will be represented in the mating pool. Proportional selection and tournament selection may be combined in order to get both fittest chromosome and best within the chromosome set [19].

10.3.2.5 *Crossover and Mutation*

Selection of the chromosomes that will be represented in genetic operations constitutes the input stage for crossover and mutation. These two operators (i.e. crossover and mutation) help to enhance the ability of producing new solutions known as offspring. The goal of these operators is to enlarge the search space and increase the speed of search. Crossover helps to obtain global optimum, while mutation searches for local fluctuations in design configuration alternatives [12].

Crossover is a recombination mechanism to explore the new solution space. It is a means of exchanging genetic materials between two parent chromosomes which result in offsprings that contain characteristics of its parents [19]. Basically there are four types of crossover techniques: single point, multiple point, cut-splice, and uniform. Many derivatives of these techniques also have been used in the literature [15, 21, 32, 34].

As mentioned before, it is quite popular to represent SCND problem as an integer string. Use of integer strings to represent chromosomes has encouraged authors to adopt single point crossover strategies [29]. Figure 10.9 illustrates a simple crossover strategy used in SCND problem. The crossover in Figure 10.9 is performed for a two-stage supply chain model with a relatively easy representation. Determining offspring pool and selection among them is crucial for further operations.

Extensions of single-point crossover strategy are considered to be applicable for SCND problem, because matrix illustration and convenient encoding structure helps to divide a chromosome into meaningful patterns in SCND problem. A commonly used extension of

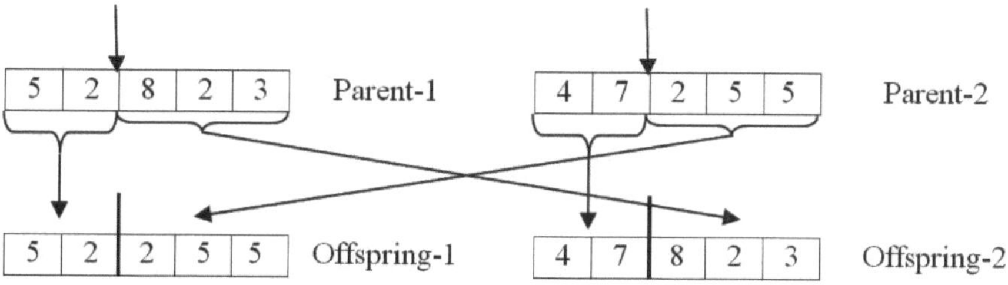

Fig. 10.9 Conventional single point crossover

single-point crossover is the weight mapping technique [31]. This technique is based on remapping the right segment based on the weight of each parent of right segment.

When both selection and encoding illustrations are in form of permutation strings, resulting crossover strategy is most commonly a position-based crossover, which is a variant of multiple point crossover [32]. It is discussed in the literature that the usage of Prüfer number in encoding requires a special crossover treatment. Since use of Prüfer number may result in a non-feasible solution, some special strategies such as feasibility criteria, tight constraints should be embedded in the crossover procedure [20, 30].

As stated by Sourirajan *et al.* [25] empirical studies in network design problems show that the uniform crossover strategy outperforms the single point crossover. In the uniform crossover, a crossover mask consisting of 0's and 1's with the same length as the chromosome structure, determines the positions between the parent chromosomes to exchange genetic materials [14, 22, 33]. Figure 10.10 represents crossover procedure for chromosome representation illustrated in Figure 10.3. In this technique, length of chromosome is equal to the number of the facilities in the network. In the mask, a value of "0" means that the first parent will transfer its genetic materials to the offspring, "1" means that the offspring will take genetic materials from the second parent for the corresponding segment. A special version of uniform crossover is randomly selecting each segment of the offspring with equal chance among the corresponding segments of parents [27].

Similar to crossover, mutation is used to prevent the premature convergence and explore new solution space. However, unlike crossover, mutation is usually done by modifying the gene within a chromosome [15]. Mutation is a background operator which produces spontaneous random changes in various chromosomes. Hence, it helps to obtain local optimum in search space. Most of the studies in the literature use single-point swap operator as illustrated in Figure 10.11. There also some other extensions of single point crossover that

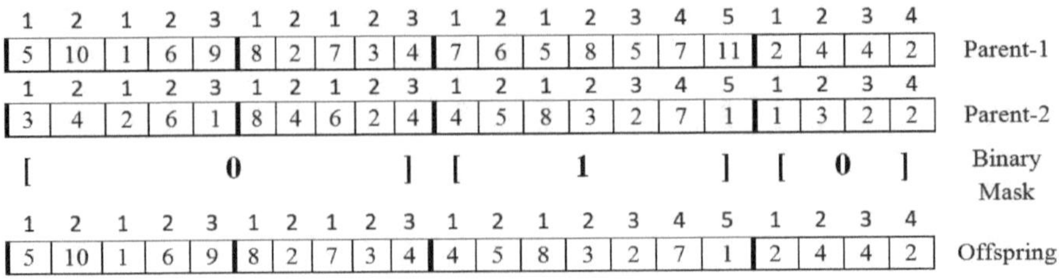

Fig. 10.10 Uniform crossover

are integrated with various techniques of mutation [25, 27, 30, 34]. It is worth noting that in most of the papers the mutation operators share the same mechanism of the adopted crossover techniques. This way it becomes easier to understand effects of each separate operator. Additionally, it allows different operator couples ("single point mutation and crossover", "uniform crossover and multiple-point mutation") to be compared in respect to performance measures.

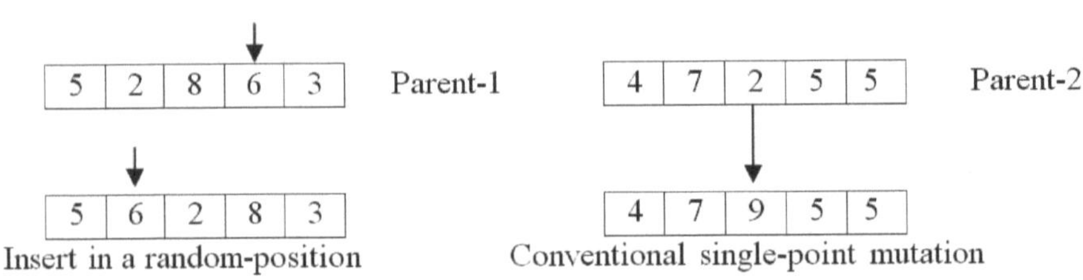

Fig. 10.11 Single point mutation

When more than one type of illustration is used within a chromosome, different mutation operators may be employed to each of these individuals [15, 22]. Altiparmak *et al.* [22] have used a segment-based mutation which divides chromosome into its segments. After randomly determining the segment to which mutation will be applied, authors have launched a pre-determined operator. They used a swap operator to two segments where priority-based encoding was used. A conventional mutation operator (see Figure 10.11) is used for the last segment of the chromosome. In this operation, the value of randomly selected gene is replaced with a new one which is selected between 1 and upper bound of gene value [22].

From our literature survey, it may be deduced that when the representation of the solution is unique to the problem type, the problem specific genetic operators are used to reflect the peculiarities of the problem. This statement emphasizes that for an effective GA design, taking into account the genetic operators together with the chromosome representation is a better idea [35].

Inversion and insertion operators are other types of mutation. These two techniques randomly select two positions within a chromosome and then inverts/inserts the sub-string between these two positions. Variants of insertion may work as selecting a digit randomly and inserting it in a randomly selected new position [20, 31]. A comparison is made between different alternatives of crossover, mutation and encoding techniques in SCND problem by Gen *et al.* [21]. Authors state that operator performance of priority based GA outperforms the standard GA (i.e. single point crossover, and single point swap mutation). However, they have also emphasized that standard GA has a better overall performance than other GAs with respect to computational time for SCND.

10.3.2.6 *Parameter Settings*

The performance of the GAs depends on the rates of the parameters such as the population size, number of generations, crossover rate, mutation rate and termination criteria. However, the optimization of the parameter set has not attracted the attention of the researchers. There are a limited number of studies which conducted an experimental design model for parameter settings [19, 28]. However, lack of parameter optimization for SCND problem demonstrates a potential research area.

Population size of genetic algorithm is related to search efficiency of the algorithm. If the population size is too small, the genetic algorithm may not explore enough of the solution space consistently to find good solutions. On the other hand, a large population size may cause the algorithm to get lost in the search space. For this reason, each problem type should be treated separately. Most of the studies in literature use a population size varying between 20 and 100 [17, 18, 25, 28, 33]. There are also studies that focus on the effect of selecting bigger population size [20, 27, 36]. Xu *et al.* [20] have used population sizes up to 400, and found out that average computation time increases with increasing population size. However, required unit memory space to represent the chromosome doesn't change. Number of generations which slightly depends on population size usually varies between 750 and 5000 for each generation. When population size of GA increases, number of generations in each run increases as well [21].

Genetic operators are main source of variability when solving a SCND problem. Establishing the correct sequence of these operators is also very important for an efficient search in the solution space. For this reason, crossover and mutation probabilities are assigned for each generation. Most commonly, these values differ in between the range of %50–%95 for crossover, %5-%70 for mutation. The rate of crossover is generally greater than the rate of mutation, because mutation which is based on local search is less efficient to find a global optimum for SCND problem. However, findings from studies that investigated the relationship between operator probabilities and solution efficiency show that there is no obvious relationship between better performing algorithm and operator probabilities [15, 28]. The crossover and mutation probabilities increase when the population's performance range tends to get stuck at a local optimum [11].

Each algorithm has its own condition of termination [12]. Search may continue until a proper solution is found. However, there is a limitation in computational time; hence, termination criterion should be set properly. Some authors [15, 25] have determined an iteration threshold for termination criterion. It is reasonable to assign this value to a maximum generation parameter which is previously set. Chen *et al.* [33] have formulated a convergence formula to check whether termination criterion is reached. Another strategy is to control the number of generations at which the solution remains the same. If the best solution remains constant for t generations, the algorithm stops. In the literature, this t value is distributed between 50 and 200 [25]. The termination condition may also work when fitness values of chromosomes in a population are the same [26]. Another technique is to use CPU times. In this technique [22], problem is solved by using a preliminary heuristic (i.e. Lagrangean relaxation) and the CPU time of heuristic is set as a termination value for GA.

A new trend in parameter setting in GAs is adaptive genetic operators in which the rates of genetic operators change during the search. Gen *et al.* [34] have analyzed different type of SCND problems with adapted GAs. Adaptation is made by modifying the chromosome representation and applying adapted genetic operators.

10.3.2.7 *Computational Tests for GA*

In terms of computational complexity theory, the SCND problem and its variations derived from the fundamental structure are typically categorized as NP-hard. Thus, designing efficient algorithms suitable to solve the NP-hard optimization problem using the GA technique becomes a current research focus in SCND studies.

The GA that is proposed to solve the studied problem should be tested by using different performance measures, because due to its stochastic nature the algorithm produces different performance outcomes in each run. For problems of small size, deviation (gap) from optimality is utilized as a test parameter. When the size of test problems is increased, some benchmarks have been presented [22, 24, 26]. In order to evaluate the overall performance, use of algorithmic efficiency is also widely discussed by authors. For large problems, computational time of algorithm which is linked to convergence capability is another parameter that has been utilized [15, 22, 28, 32]. It is also observed in the literature that some authors [24, 28] have used real logistics data to optimize SCND problem. When the SNCD problem is formulated as a multi-objective problem, performance evaluation of the algorithm becomes much more difficult. Then, distances from Pareto optimal solutions are proposed as the most effective performance criteria [27, 33].

Determination of the parameters and performance measures requires validation of proposed algorithm [37]. In the process of algorithm design, each step is checked to reflect characteristics of the problem correctly [21]. This will help to have a validated flow of algorithm. Validation is also based on showing applicability of the solution that is found by GA. Using real data and comparison with data libraries help authors to validate their models. Another technique [36] that is used to validate proposed algorithm is comparison with other techniques in the literature. Such an approach helps authors to assess their model with respect to other techniques.

10.4 Classification of Studies Applying Genetic Algorithms to SCND Problem

Genetic algorithms are increasingly used to model problems related to supply chain management. One of the most popular research topics is the SCND problem. In order to understand the current research issues in solving SCND problems using GAs, this section outlines the principle characteristics of the problems studied and discusses the features of the proposed GAs to deal with these.

The studies reviewed in this chapter are classified to observe the nature of current literature. For this aim, a coding scheme was developed as seen in Figure 10.12. Each distinct feature of GAs is coded with different letters and these letters have been used in Table 10.1 to examine the studies.

Table 10.1 provides detailed information on the studies about the SCND problem. The represented studies correspond to a valuable part of GA related SCND literature and cover most of the GA features. The classification reveals a number of possible areas of research that need to be addressed in the future.

A: Objectives
A1: Single
A2: Multiple

B: Representation
B1: Binary vector
B2: Tree structure
B3: Matrix structure
B4: Real structure

C: Encoding
C1: Binary and integer encoding
C2: Prüfer number (vertex)
C3: Random key based
C4: Priority based
C5: Binary and real coding
C6: Permutation encoding

D: Initiation
D1: Random
D2: Heuristic start
D3: Convergence

E: Selection
E1: Tournament selection
E2: Random selection
E3: Crowded-comparison
E4: $\mu+\lambda$ selection
E5: Elitism
E6: Proportional selection

F: Fitness
F1: Objective function
F2: Gap of objectives
F3: Weighted sum
F4: Network character
F5: Reciprocal of obj. function
F6: Distance from Pareto Opt.

G: Crossover
G1: Extensions of G2,G4
G2: Multiple point
G3: Cut and splice
G4: Single point
G5: Uniform

H: Mutation
H1: Swap (Single point)
H2: Bit string
H3: Insert (Single point)
H4: Multiple points
H5: Inversion mutation

Fig. 10.12 Coding scheme for classification of the studies

When we examine the studies regarding the number of objective function, we observe that unlike the single objective GA where researchers have formulated a common objective of cost minimization consisting of transportation, opening facilities etc., multiple objective GA consider interdependency among objectives in function. Most of the studies in the literature focus on modeling single objective SCND problems. When an additional objective is embedded in GA structure, it is proved that computational performance of the algorithm decreases [20].

Table 10.1 Classification of the studies according to the coding scheme

Authors	A		B				C						D			E						F						G					H				
	A1	A2	B1	B2	B3	B4	C1	C2	C3	C4	C5	C6	D1	D2	D3	E1	E2	E3	E4	E5	E6	F1	F2	F3	F4	F5	F6	G1	G2	G3	G4	G5	H1	H2	H3	H4	H5
Gen *et al.* [34]	*		*									*	*						*						*		*	*			*		*				*
Syarif *et al.* [30]	*		*			*		*						*		*										*					*		*				*
Zhou *et al.* [24]	*			*		*		*					*							*					*							*	*				
Altıparmak *et al.* [27]		*		*			*			*			*					*							*				*			*	*			*	
Chen *et al.* [33]		*				*					*		*								*	*										*	*				
Gen *et al.* [21]	*				*					*				*						*						*	*	*							*		
Yeh [13]	*		*									*		*		*							*			*					*		*				
Zhou *et al.* [38]	*																																				
Ko and Evans [36]	*				*		*						*			*						*							*				*				
Lin *et al.* [28]	*		*								*		*							*							*		*	*			*		*		
Pongcharoen *et al.* [39]	*				*							*		*							*	*									*		*				
Xu *et al.* [20]		*			*	*		*						*				*								*					*		*				*
Altıparmak *et al.* [22]	*				*		*			*					*	*						*										*	*			*	
Gen *et al.* [23]	*				*					*			*							*						*			*				*				
Lee *et al.* [31]	*				*					*				*						*						*	*	*									*
Lin and Lei [29]	*		*				*						*							*							*				*		*				
Lin *et al.* [15]	*					*			*	*			*						*	*						*		*	*						*		*
Sourirajan *et al.* [25]	*		*				*		*				*			*						*				*					*	*	*		*		
Xu *et al.* [17]	*				*						*		*			*						*									*		*				
Ataka *et al.* [35]	*					*		*	*				*							*							*	*			*		*				
Chen *et al.* [19]		*				*					*			*							*						*					*	*		*		
Costa *et al.* [32]	*				*				*			*	*			*										*		*	*				*				
Ding [16]	*		*				*				*		*																		*				*		
Pisvahee *et al.* [14]		*			*					*			*					*									*		*				*				
Chang [26]	*		*				*						*							*							*		*				*				
Liao *et al.* [18]		*	*				*						*			*	*							*		*					*		*				

With respect to representation and encoding, studies are clustered into basic subgroups. Some of them use Prüfer number which requires an illustration with real number representations. This illustration and encoding approaches are valuable for segment based representations, because initial genes may be encoded for binary enclosure variable, and rest may be encoded by using Prüfer numbers. Another popular subgroup is using priority based encoding to assign a weight for opening and closure of each supply chain stage. This type of encoding is not used to represent relation of facilities with customers, because there is no proper correspondence of assignments of facilities to customers.

In most of the studies the method random initialization of starting population is of concern, whereas only little attention has been given to formulate different algorithmic methods to select initiation population. Most of the techniques other than randomization are based on determining sequence of parameters on chromosome and assigning value of each gene iteratively. An important concern in initialization is to produce an infeasible solution. There are three basic strategies to cope with infeasibility; (1) discard all of the infeasibilities by assigning an infinite cost to these solutions, (2) use some finite penalty costs, and reject infeasibility in final generations, (3) introduce some repair mechanisms to deal with infeasibility. Random generators mostly use a control variable to check whether resulting chromosome is feasible, if not some repair mechanisms are introduced.

Fitness evaluation and selection of chromosomes are made by using various techniques in the literature. Each technique has its own strengths and weaknesses. For the ones that utilize proportional selection mechanism, there is an obvious problem with convergence, hence ranking and scaling methods are used. In this phase, a survival strategy, which removes the individual having low fitness values, is initiated. It is deduced from the literature that Elitism, which preserve best chromosome from generation to generation, is a very popular survival strategy. Roulette wheel is another widely used technique. However, there are some disadvantages of using it such as premature convergence, computation of global statistics which reduces potential for parallel processing etc. Authors have gone beyond the risks, and alternatively formulated some precautions for the disadvantages. Each selection technique corresponds to a combination with fitness evaluation, genetic representation, and reproduction method.

Genetic operators which contain crossover, mutation, migration etc. have many different applications in literature. Due to the fact that chromosome representation of SCND configurations may contain different encodings for each gene cluster, different crossover and mutation strategies may be adopted into algorithm structure. Single and multiple point crossover techniques are widely used. However, in respect to mutation most of the studies conduct a swap operator based single-point mutation technique. Optimum set of genetic operators are tested in a number of studies [26, 39]. It may be deduced that there is no unique technique (crossover and mutation characteristics) which dominates other genetic operator sets for all types of SCND problem. However, depending on problem size of SCND configuration, probabilities of crossover and mutation may be changed. Mostly, when problem size is increased, probability and rate of crossover and mutation is also increased [15, 21, 39].

It should be emphasized that it is important to have a set of robust GA parameters which work well for different data sets under different problem characteristics. Because, while one parameter set might give good results for one data set, it might fail for others. Moreover, the population initialization has not taken much attention since most of the studies preferred randomization in constructing the initial population [11]. Use of randomized data may be required to formulate parameters with probability distribution functions. Stochastic nature of input data of GAs is ignored by researchers of the field. Parameters of GA may also be represented as fuzzy decision variables with proper membership functions.

To get more realistic results regarding the performance of the GAs, the focus might be on the real-world SCND problems. Furthermore, to show the effectiveness of GAs in solving supply chain network problems, more complex SCND problems including features such as multiple items, multiple transportation modes, variable cost of opening/transportation, GIS based optimization values can be taken into consideration. An important issue in solving SCND problem using GAs is to deal with probabilistic parameters. It has been noted that combining specially designed GAs with simulation has quite potential to find sufficiently good solutions, because simulation methodology encompasses decision variables that are represented by different probability distributions [11].

Another important gap in the literature is a problem-dependent reformulation. Most of the studies that apply GA in SCND ignore the relationship between SCND problem and relevant areas. These areas may be listed as; inventory planning, vehicle routing, fleet scheduling, assignment of customers to zones (clustering) etc. When SCND problem is analyzed together with these areas, problem formulation becomes larger.

In recent years, with the increasing trend of hybridization of algorithms, researchers exploit the advantage of different techniques and obtain better search algorithms that unite the advantages of the individual pure strategies. Due to the fact that, SCND may be formulated as a mixed integer-based model, MIP-based heuristics may be embedded to GA structures to obtain an upgraded version of global search within the alternatives space.

With the growth of the published literature in recent years, we believe that the use of GAs for solving SCND problems will continue to attract the attention of the researchers in this area.

10.5 Conclusion

In this chapter, we presented how genetic algorithms are applied to solve the SCND problem. We introduced different GA approaches that are used at different steps of the algorithm, such as encoding/decoding techniques, initialization strategies, fitness function determination, selection techniques, crossover and mutation strategies. Merits and limitations of each technique are discussed in the light of recent related research.

The chapter also provides a classification scheme to study the related research. This examination reveals gaps to be addressed by future work. The most common issue is that no one technique is inferior or superior to the others when solving a SCND problem, thus by employing hybrids we can exploit the respective strengths and eliminate weaknesses of the individual approaches. Along with hybridization of different techniques and algorithms, dealing with more complex SCND problems with additional decision variables and constraints, or larger problems involving other SCND decisions such as inventory planning [40], vehicle routing, fleet scheduling, Just-in-time delivery systems [41, 42] will be a future direction of research in this area.

Bibliography

[1] Beamon B.M., Supply chain design and analysis: Models and methods. *International Journal of Production Economics*, 1998; **55** (3), 281–294.

[2] Lambert D.M., Cooper M.C., Pagh J.D., Supply Chain Management: Implementation Issues and Research Opportunities. *International Journal of Logistics Management*, 1998; **9** (2), 1–19.

[3] Mentzer J.T., DeWitt W., Keebler J.S., *et al.*, Defining Supply Chain Management. *Journal of Business Logistics*, 2001; **22** (2), 1–25.

[4] Chopra S., Meindl P., *Supply Chain Management: Strategy, Planning and Operations*, 2010, Prentice Hall: New Jersey.

[5] Balinski M.L., Integer Programming: Methods, Uses, Computations. *Management Science*. 1965 **12** (3), 253–313.

[6] ReVelle C.S., Eiselt H.A., Daskin M.S., A bibliography for some fundamental problem categories in discrete location science. *European Journal of Operational Research*. 2008; **184** (3), 817–848.

[7] Klose A., Drexl A., Facility location models for distribution system design. *European Journal of Operational Research*. 2005; **162** (1), 4–29.

[8] Melo M.T., Nickel S., Saldanha-da-Gama F., Facility location and supply chain management – A review. *European Journal of Operational Research*, 2009; 196 (2), 401–412.

[9] Aikens C.H., Facility Location Models for Distribution Planning. *European Journal of Operational Research*, 1985; **22** (3), 263–279.

[10] Whitley D. An overview of evolutionary algorithms: practical issues and common pitfalls. *Information and Software Technology*, 2001; **43** (13).

[11] Guner Goren H., Tunali S., Jans R., A review of applications of genetic algorithms in lot sizing. *J. Intell. Manuf.*, 2010; **21** (4), 575–590.

[12] Gen M., Cheng R., Lin L., *Network Models and Optimization – Multiobjective Genetic Algorithm Approach*, 2008 Springer-Verlag.

[13] Yeh W.C., An efficient memetic algorithm for the multi-stage supply chain network problem. *International Journal of Advanced Manufacturing Technology*, 2006; **29** (7-8), 803–813.

[14] Pishvaee M.S., Farahani R.Z., Dullaert W., A memetic algorithm for bi-objective integrated forward/reverse logistics network design. *Computers & Operations Research*, 2010; **37** (6), 1100–1112.

[15] Lin L., Gen M.S., Wang X.G., Integrated multistage logistics network design by using hybrid evolutionary algorithm. *Computers & Industrial Engineering*, 2009; **56** (3), 854–873.

[16] Ding S.B., Logistics Network Design Optimization Based on Differential Evolution Algorithm. *Proceedings of 2010 International Conference on Logistics Systems and Intelligent Management*, **Vols. 1–3**. 2010, 1064–1068.

[17] Xu T., Wei H., Wang Z.-D., Study on continuous network design problem using simulated annealing and genetic algorithm. *Expert Systems with Applications*, 2009; **36** (2), 2735–2741.

[18] Liao S.H., Hsieh C.L., Lai P.J., An evolutionary approach for multi-objective optimization of the integrated location-inventory distribution network problem in vendor-managed inventory. *Expert Systems with Applications*, 2011; **38** (6), 6768–6776.

[19] Chen A., Kim J., Lee S., Kim Y., Stochastic multi-objective models for network design problem. *Expert Systems with Applications*, 2010; **37** (2), 1608–1619.

[20] Xu J.P., Liu Q., Wang R., A class of multi-objective supply chain networks optimal model under random fuzzy environment and its application to the industry of Chinese liquor. *Information Sciences*, 2008; **178** (8), 2022–2043.

[21] Gen M., Altiparmak F., Lin L., A genetic algorithm for two-stage transportation problem using priority-based encoding. *OR Spectrum*, 2006; **28** (3), 337–354.

[22] Altiparmak F., Gen M., Lin L., Karaoglan I., A steady-state genetic algorithm for multi-product supply chain network design. *Computers & Industrial Engineering*, 2009; **56** (2), 521–537.

[23] Gen M., Lin L., Jo J.-B., Hybrid Genetic Algorithm for Designing Logistics Network, VRP and AGV Problems – Intelligent and Evolutionary Systems, (eds. Gen M., Green D., Katai O., *et al.*), 2009; pp. 123–139. Springer Berlin / Heidelberg.

[24] Zhou G.G., Min H., Gen M., The balanced allocation of customers to multiple distribution centers in the supply chain network: a genetic algorithm approach. *Computers & Industrial Engineering*, 2002; **43** (1-2), 251–261.

[25] Sourirajan K., Ozsen L., Uzsoy R., A genetic algorithm for a single product network design model with lead time and safety stock considerations. *European Journal of Operational Research*, 2009; **197** (2), 599–608.

[26] Chang Y.H., Adopting co-evolution and constraint-satisfaction concept on genetic algorithms to solve supply chain network design problems. *Expert Systems with Applications*, 2010; **37** (10), 6919–6930.

[27] Altiparmak F., Gen M., Lin L., Paksoy T., A genetic algorithm approach for multi-objective optimization of supply chain networks. *Computers & Industrial Engineering*, 2006; **51** (1),

196–215.

[28] Lin L., Gen M., Wang X., A Hybrid Genetic Algorithm for Logistics Network Design with Flexible Multistage Model. *International Journal of Information Systems for Logistics and Management*, 2007; **3** (1), 1–12.

[29] Lin J.-R., Lei H.-C., Distribution systems design with two-level routing considerations. *Annals of Operations Research*, 2009; 172 (1), 329–347.

[30] Syarif A., Yun Y., Gen M., Study on multi-stage logistic chain network: a spanning tree-based genetic algorithm approach. *Computers & Industrial Engineering*, 2002; **43** (1-2), 299–314.

[31] Lee J.-E., Gen M., Rhee K.-G., Network model and optimization of reverse logistics by hybrid genetic algorithm. *Computers & Industrial Engineering*, 2009; **56** (3), 951–964.

[32] Costa A., Celano G., Fichera S., Trovato E., A new efficient encoding/decoding procedure for the design of a supply chain network with genetic algorithms. *Computers & Industrial Engineering*, 2010; **59** (4), 986–999.

[33] Chen A., Subprasom K., Ji Z., A simulation-based multi-objective genetic algorithm (SMOGA) procedure for BOT network design problem. *Optimization and Engineering*, 2006; **7** (3), 225–247.

[34] Gen M., Cheng R., Oren S.S., Network design techniques using adapted genetic algorithms. *Advances in Engineering Software*, 2001; **32** (9), 731–744.

[35] Ataka S., Kim B., Gen M., Optimal Design of Two-stage Logistics Network Considered Inventory by Boltzmann Random Key-based GA. *IEEJ Transactions on Electrical and Electronic Engineering*, 2010; **5** (2), 195–202.

[36] Ko H.J., Evans G.W., A genetic algorithm-based heuristic for the dynamic integrated forward/reverse logistics network for 3PLs. *Computers & Operations Research*, 2007; **34** (2), 346–366.

[37] Langerman J.J., Ehlers E.M., The validation of evolutionary algorithms. In *Proceedings of 21st International Conference on Computers and Industrial Engineering*, 1997; pp. 204–206: Egypt.

[38] Zhou G., Cao Z., Qi F., Cao J., A genetic algorithm approach on a logistics distribution system with uncertain demand and product return. *World Journal of Modelling and Simulation*, 2006; **2** (2), 99–108.

[39] Pongcharoen P., Khadwilard A., Klakankhai A., Multi-matrix Real-coded Genetic Algorithm for Minimising Total Costs in Logistics Chain Network. *Proceedings of World Academy of Science, Engineering and Technology*, 2007; **26**, 458–463.

[40] Jawahar N., Balaji A.N., A genetic algorithm for the two-stage supply chain distribution problem associated with a fixed charge. *European Journal of Operational Research*, 2009; **194** (2), 496–537.

[41] Farahani R.Z., Elahipanah M., A genetic algorithm to optimize the total cost and service level for just-in-time distribution in a supply chain. *International Journal of Production Economics*, 2008; **111** (2), 229–243.

[42] Nachiappan S.P., Jawahar N., A genetic algorithm for optimal operating parameters of VMI system in a two-echelon supply chain. *European Journal of Operational Research*, 2007; **182** (3), 1433–1452.

Chapter 11

Fuzzy PROMETHEE for the Environmental Quality Assessment of Energy Dedicated Crops

Fausto Cavallaro[1] and Luigi Ciraolo[2]

[1] *Department of Economics, Management and Social Sciences (SEGeS)-University of Molise Via De Sanctis, 86100 Campobasso, Italy*
[2] *Department SEA- University of Messina, Piazza Pugliatti - 98121 Messina, Italy*

E-mail: cavallaro@unimol.it, luigi.ciraolo@unime.it

Energy crops are positioned as the most promising renewable energy sources. They are specifically targeted at the production of biofuels (solid, liquid and gaseous) and for the development of vegetal products with specific attributes that make them suitable for industrial processing and transformation into energy. The prospect of using "alternative crops" renders attractive the opportunity of introducing certain plant species where land, including mountains, hills and plains, has gradually been abandoned or fallen into disuse.

One of the aspects of the use of biomass for energy purposes that is still rather controversial regards their full environmental sustainability. Unfortunately in many cases, when an environmental evaluation is developed, using various methodologies, we immediately encounter a certain degree of uncertainty in input data, making it difficult to obtain a meaningful result. One possible solution for correct management of data uncertainty, as we know, is that of using fuzzy-sets. The main objective of this study is to propose and to test the validity and effectiveness of a fuzzy multi-criteria method, called F-Promethee, to help the analyst to compare a set of energy crops and to obtain a ranking of their environmental sustainability.

11.1 Introduction

As we already know, controlling European energy consumption and greater reliance on energy from renewable sources, together with energy saving and increased energy efficiency, are all important parts of the series of necessary measures in order to reduce greenhouse gas emissions and to meet the requirements of the Kyoto agreement on climate change. Over recent years the use of biomass has been growing significantly, especially in

C. Kahraman (ed.), *Computational Intelligence Systems in Industrial Engineering*, Atlantis Computational Intelligence Systems 6, DOI: 10.2991/978-94-91216-77-0_11,

countries that have made a strong commitment to renewable sources in their energy policies. This growing interest in bioenergies also originates from recent EU policies, such as Directive 2009/28/EC, which promotes the use of renewable energy sources. Among the measures contained in the Directive, there is the obligation for each Member State to use a minimum quota of 10% of renewable energy in the transport sector by 2020. It is easy to imagine, therefore, that biofuels will play a leading role in the future.

Energy crops are specifically targeted at the production of biofuels (solid, liquid and gaseous) and the development of vegetal products with specific attributes that make them suitable for industrial processing and transformation into energy. The EU has been involved in the promotion of non-food crops for years, in part via the set-aside scheme introduced by the European Community in 1988 as part of Common Agricultural Policy (CAP) reform aimed at reducing excess production. According to EC lawmakers the set-aside mechanism should have favoured the creation of new non traditional food-based markets. The cultivation of crops in set-aside areas was therefore authorized provided that the product was destined for industrial or energy purposes [1]. Their level of use however is extremely modest and this is basically due to the difficulties linked to production costs.

Energy crops, as we have already said, can be considered among the most promising sources of renewable energy; however, we have to add that their environmental sustainability is still a rather controversial topic [2, 3]. Considering the crucial importance of this problem, numerous authors have carried out evaluations of the environmental impact of the various types of biomass, mainly by means of a life cycle analysis (LCA) approach. LCA is a powerful tool for compiling and assessing incoming and outgoing flows, as well as potential environmental performances throughout the life cycle of a product (in this case the product being the biomass). The literature in this field is substantial; indeed, there are numerous examples of papers applying this methodology to the various kinds of biomass. In particular see: Blengini *et al.* [4] Buratti and Fantozzi [5], Rafaschieri *et al.* [6], Fazio and Monti [7], Kim and Dale [8], Djomo and Blumberga [9], Cherubini and Stromman [3], Cherubini and Ulgiati [10]. Although LCA is an excellent environmental evaluation tool, it is not able to manage uncertain input data.

This paper sets out an approach for dealing with uncertainty of environmental data based on Fuzzy logic. The main objective of this study is to propose and to test the validity and effectiveness of a fuzzy multi-criteria method, called F-Promethee, to help analysts to compare a set of energy crops and to obtain a ranking of the environmental sustainability of particular energy crops. A variety of interesting approaches to renewable energy issues have been developed to illustrate the potential applications of the multi-criteria approach as follows: for renewable energy planning, policy and evaluation [11–27]. More specifically, regarding biomass Sultana and Kumar [28] developed a multi-criteria assessment of different biomass feedstock-based pellets employing environmental, economic and technical criteria; Kumar *et al.* [29], applied a multi criteria approach to rank alternatives for biomass collection and transportation systems; Buchholz *et al.* [30], analysed the suitability of four MCA tools to assess sustainability of bioenergy systems; Damiano *et al.* [31], proposed a Decision Support System (DSS) based on a multi-criteria approach to determine the most suitable biomass energy basins for smart grid; Čuček *et al.* [32], developed an analysis of biomass and bioenergy supply chains using both life cycle assessment and mixed integer non-linear programming model (MINLP); Perimenis *et al.* [33], offer a decision support tool for the assessment of biofuels; finally, Dorini *et al.* [34], proposed managing sustainability uncertainty by using a probabilistic approach and Monte Carlo simulation.

This paper is organised as follows: Section 11.2 introduces energy crops, Section 11.3 contains the main principles of original Promethee and fuzzy Promethee. Finally, Section 11.4 is dedicated to an empirical case study of energy crop assessment.

11.2 Energy Crops

When we speak of biomass we mean a substance of organic matrice, that is to say vegetal matter of non-fossil biological origin, both wild and grown by man, produced as a result of the process of chlorophyll photosynthesis with the contribution of solar radiation energy. In general, the classification of biomass includes all organic residues and carbon and hydrogen by-products, residues from zootechnic production (sludge and animal faeces), from agriculture, forestry and ancillary industries.

Energy crops are specifically targeted at the production of some form of energy and the development of vegetal products with specific attributes that make them suitable for industrial processing and transformation into energy. Energy may be generated through direct combustion or gasification of the crops in order to create electricity and heat, or by converting them into liquid fuels such as ethanol for use in vehicles.

"Dedicated" crops can be grouped into three main types [35]:

1. Lignocellulosic biomass crops These are crops producing high levels of dry substances that can be subjected to various processes: combustion, pyrolysis, gasification, liquefaction. These group of crops can, in turn, be divided into 3 distinct categories:

Short Rotation Forestry (*SRF*), that is to say fast growing arboreous plants which, using appropriate growing techniques, are cut and harvested at frequent intervals for the production of ligneous material. These crops include the false acacia (*Robinia pseudoacacia*), broom (*Spartium junceum*), eucalyptus (*Eucalyptus*), willow (*Salix*) and poplar (*Populus*);

Perennial herbaceous crops, including interesting species such as miscanthus (*Miscanthus sinensis*), giant reed (*Arundo donax*), switchgrass (*Panicum virgatum*) cardoon (*Cynara cardunculus*) (especially in areas with limited availability of water);

Annual herbaceous crops, particularly sorghum (*Sorghum bicolor*), kenaf (*Hibiscus cannabinus*) and hemp.

2. Oil-seed crops The vegetal oil of certain species can be used, following an esterification process, to produce biodiesel, with similar properties and performance to those of mineral diesel but with the advantage of containing no sulphur or aromatic compounds. This type of crops include rapeseed (*Brassica napus)*, sunflowers (*Helianthus annuus L.*), and soybean (*Glycine max L.*) .

3. Carbohydrate crops Sugary plants are used, through the fermentation of carbohydrates, to produce bioethanol, which is added to petrol following transformation into Ethyl Tertiary Butyl Ether. Of the species that can be used to this end, the most widespread tried and tested ones are sugar cane (*Saccharum officinarum)*, durum wheat (*Triticum durum*), sweet sorghum (*Sorghum bicolour)*, maize (*Zea mays*) and sugar beet (*Beta vulgaris*).

Fig. 11.1 illustrates a schematic picture of the possible uses and enhancement systems of agricultural and forestry biomass. Within the production and management systems, perennial species, though less flexible than annual ones, involve lower management costs overall. Indeed, while in the first year, during the planting phase, the cost is higher (though spread over several years of production), in following years management costs are lower, due to the fact that perennial species are wild and thus require less intensive processing and more limited use of fertilizers, herbicides and pesticides. The fact that they need less processing also results in a more limited environmental impact; moreover, the use of these plants also reduces soil erosion. Certain perennial species are of particular interest today in the production of ligno-cellulosic biomass.

In recent years certain plants previously viewed as weeds have been re-considered. Such plants include ones that are able to grow in extreme conditions on arid and unproductive soil, e.g. cardoon, broom, robinia, and others typical of the subtropical belt, such as sweet sorghum. The prospect of using "alternative crops" renders attractive the opportunity of introducing certain plant species onto farms in southern Italy where land, including mountains, hills and plains, has gradually been abandoned or fallen into disuse since the post-war years due to the migration of the population towards large industrial cities. This land is generally unfertile and not easy to tend and is therefore unsuited to traditional farming [36]. It is likely that on the best farmlands (wet plains) the higher and more secure earnings power of "traditional" farming leaves no room for the development of non-food crops with the exception of oleaginous and sugar crops in which a more profitable and competitive market is forecast compared with the products derived from traditional farming. Land situated high on mountains and on inaccessible hills does not offer the conditions that are technically and economically sustainable for starting up energy crops. Energy crops are still little used and cost more than biomass from waste.

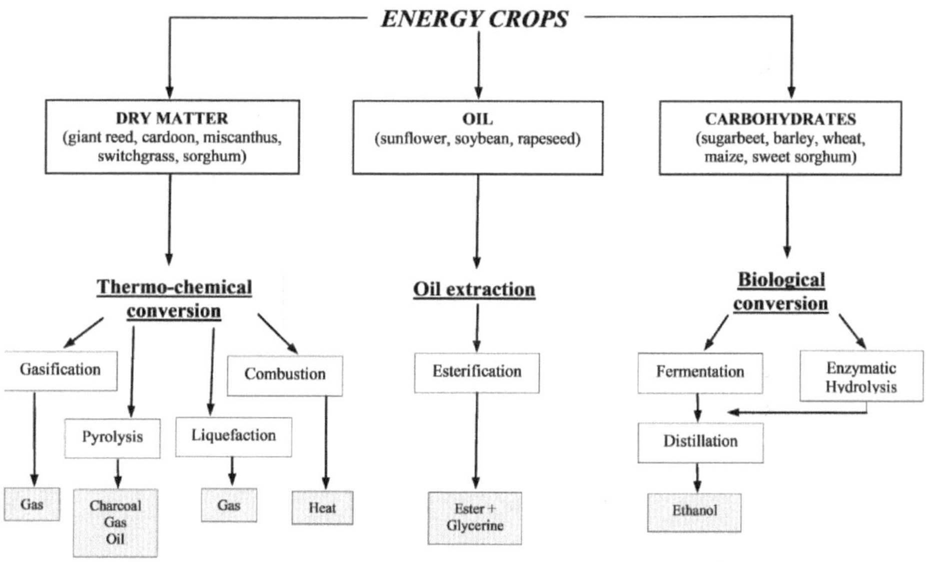

Fig. 11.1 Energy crops and conversion processes (Venturi and Venturi, readapted)

At present farming biomass for energy purposes is only being done experimentally or being used on a small scale. What is indispensable for the launch of a widespread energy production system based on the use of energy crops is the development of systems that can farm on a large enough scale to be sustainable with contractual forms offering profitability.

11.3 Methodology

11.3.1 *The original PROMETHEE methodology*

The introduction highlights why there is a need to develop a planning and management tool that can assist the decision maker. Multicriteria methods can rightly be considered a powerful tool for the assessment of energy sustainable options and to support the decision making procedure. First of all, the original promethee method will be described.

The PROMETHEE (Preference Ranking Organization Method of Enrichment Evaluation) method was devised by Brans J.P. *et al.* [37–40]. This technique, besides possessing all the advantages of B. Roy's outranking methods, is also easy to use and its level of complexity is low. It is based on ranking and is well-suited to problems in which there are a finite number of actions to be assessed on the basis of a range of conflicting criteria. The following procedure is recommended to implement the method:

Identification of alternatives. Under MCDA the decision procedure is normally carried out by choosing between different elements that the decision maker has to examine and then to assess them using a set of criteria. These elements are called actions and they make up part of a global set labelled actions or alternatives

$$A = \{a_1, \ldots, a_2, \ldots, a_m\} \tag{11.1}$$

Defining a set of criteria. The criteria represent the tools which enable alternatives to be compared from a specific point of view. It must be remembered that the selection of criteria is of prime importance in the resolution of a given problem, meaning that it is vital to identify a coherent family of criteria

$$C = \{c_1, \ldots, c_j, \ldots, c_k\} \tag{11.2}$$

The alternatives are compared pairwise under each criterion and the decision maker, faced with the two actions a_i and a_m, can express: an outright preference $(a_i \mathrm{P} a_m)$; a weak preference, if it is less marked, $(a_i \mathrm{Q} a_m)$; indifference $(a_i \mathrm{I} a_m)$; or incomparability $(a_i \mathrm{R} a_m)$ if none of the former apply. For each criterion the decision maker can choose from a set of six different types of preference functions to model the decision maker's preferences [38–40].

Evaluation matrix. Once the set of criteria and the alternatives have been selected then the payoff matrix is built. This matrix tabulates, for each criterion-alternative pair, the quantitative and qualitative measures of the effect produced by that alternative with respect to that criterion. The matrix may contain data measured on a cardinal or an ordinal scale.

Each alternative

$$A_i = \{a_{i,1}, \ldots, a_{i,j}, \ldots, a_{i,m}\} \tag{11.3}$$

is composed of a group of evaluations a_{ij} representing the evaluation given to the alternative i with respect to the criteria j.

Determining the multi-criteria preference index. The degree of preference of an alternative a_i in comparison to a_m is expressed by a number between 0 and 1 (from 0 indicating no preference or indifference up to 1 for an outright preference). When the pairs of alternatives a_i and a_m are compared the outcome of the comparison must be expressed in terms of preference in the following way [38]:

$\mathrm{P}_k(d) = 0$ means there is *indifference* between a_i and a_m or *no preference*;

$\mathrm{P}_k(d) \cong 0$ expresses a *weak preference* for a_i over a_m;

$\mathrm{P}_k(d) \cong 1$ strong preference for a_i over a_m;

$\mathrm{P}_k(d) = 1$ outright preference for a_i over a_m.

In practice this *preference function* $\mathrm{P}_k(d)$ represents the difference between the evaluation of the two alternatives, thus it can be expressed as follows [40]:

$$\begin{aligned} P_k(a_j, a_m) &= P_k[d(a_j, a_m)] \\ P_k(c_k(a_j) - c_k(a_m)) &= P_k(d) \in [0, 1] \end{aligned} \tag{11.4}$$

Once the decision maker has described the preference function P_k ($k = 1, 2, 3, \ldots, n$ represent the criteria) then a vector containing the weights of each criterion must be defined

$$W^T = [w_1, \ldots, w_k] \tag{11.5}$$

The weights w represent the relative importance of the criteria used for the assessment; if all criteria are equally important then the value assigned to each of them will be identical. A variety of techniques exist to determine weights, the simplest but also the most arbitrary is direct assignment where weights are set by the decision maker. Other techniques require that the decision maker and an analyst work together to obtain a vector of weights that conforms as closely as possible to the decision maker's preferences. In addition to weighting, the method involves setting thresholds that delineate the decision maker's preferences for each criterion and the critical thresholds are thus: the indifference threshold q_i and the preference threshold p_i (a more exhaustive description of the procedure can be found in the literature).

The index of preference Π is calculated for each pair of actions a_i and a_m as the weighted average of preferences calculated for each criterion. The index Π is therefore defined as follows [38]:

$$\prod(a_i, a_m) = \frac{\sum_{k=1}^{K} w_k \cdot P_k(c_k(a_i) - c_k(a_m))}{\sum_{k=1}^{K} W_k} \tag{11.6}$$

$\Pi(a_i,a_m)$ represents the strength of the decision maker's preference for action a_i over action a_m considering all criteria simultaneously and $\Pi(a_m,a_i)$ how much a_m is preferred above a_i. Its value falls between 0 and 1 whereby:

$\Pi(a_i,a_m) \cong 0$ indicates a weak preference for a_i over a_m for all criteria;

$\Pi(a_i,a_m) \cong 1$ indicates a strong preference for a_i over a_m for all criteria.

Ranking the alternatives. The traditionally non-compensatory and methodologically important models include ones in which preferences are aggregated by means of outranking relations. Outranking is a binary relation S defined in A such that $a_i \mathrm{S}\, a_m$ if, given the information relating to the decision maker's preferences there are enough arguments to decide that "a_i is at least as good as a_m" while there is no reason to refute this statement, i.e. $a_i \mathrm{Sj}\, a_m$ implies $a_m \mathrm{Sj}\, a_i$. Let us consider how each alternative $a_i \in A$ is evaluated against $(n-1)$ another in A and thereby define the two following outranking flows [38, 39]:

$$\Phi^+(a_i) = \frac{1}{n-1} \cdot \sum_{x \in A} \Pi(a_i,a_m) \tag{11.7}$$

This indicates a preference for action a_i above all others and shows how 'good' action a_i is (positive outranking flow);

$$\Phi^-(a_i) = \frac{1}{n-1} \cdot \sum_{x \in A} \Pi(a_m,a_i) \tag{11.8}$$

This indicates a preference for all the other actions compared with a_i and shows how weak action a_i is (negative outranking flow).

According to PROMETHEE I, a_i is superior to a_m if the leaving flow of a_i is greater than the leaving flow of a_m and the entering flow of a_i is smaller than the entering flow of a_m. In the PROMETHEE I alternative a_i outranks alternative a_m
$(a_i P^I a_m)$ if at least one of the elements of the following equation is satisfied [38]:

$$\Phi^+(a_i) > \Phi^+(a_m) \tag{11.9}$$

and

$$\Phi^-(a_i) < \Phi^-(a_m) \tag{11.10}$$

or

$$\Phi^+(a_i) = \Phi^+(a_m) \tag{11.11}$$

and

$$\Phi^-(a_i) < \Phi^-(a_m) \tag{11.12}$$

or

$$\Phi^+(a_i) > \Phi^+(a_m) \tag{11.13}$$

and

$$\Phi^-(a_i) = \Phi^-(a_m) \tag{11.14}$$

Alternative a_i is indifferent to a_m if they have the same leaving and entering flows:

$$a_i I^I a_m : \Phi^+(a_i) = \Phi^+(a_m) \quad \text{and} \quad \Phi^-(a_i) = \Phi^-(a_m) \tag{11.15}$$

Alternative a_i and a_m are not comparable (R^I) in any other case. Finally, a_i outranks a_m if:

$$\Phi^+(a_i) \geqslant \Phi^+(a_m) \tag{11.16}$$

and

$$\Phi^-(a_i) \leqslant \Phi^-(a_m) \tag{11.17}$$

Equality in Φ^+ and Φ^- indicates indifference between the two compared alternatives. Under the Promethee I method some actions remain incomparable, in this case a complete preorder is required that eliminates any incomparable items, then Promethee II can give a complete ranking as follows [39]:

$$\Phi^{\text{net}}(a_i) = \Phi^+(a_i) - \Phi^-(a_i) \tag{11.18}$$

The net flow is the difference between the out-flow and the in-flow.

11.3.2 *Fuzzy PROMETHEE*

The first studies in the literature which developed an integration between Promethee and fuzzy numbers were proposed by Le Teno and Mareschal [41], Geldermann *et al.* [42], Goumas and Lygerou [43]. Other interesting applications have been developed in recent years by Bilsel *et al.* [44], Tuzkaya *et al.* [45], Chou Lin *et al.* [46], Li and Li [47], Giannopoulos and Founti [48], Oberschmidt *et al.* [49].

In this paper we adopt the approach illustrated by Goumas and Lygerou [43] described in the following part of this section.

According to Dubois and Prade [50] the representation of a TFN can be presented in the following form $x = (m, a, b)LR$. If the variable x is equivalent to the value m, its membership function is $f(x) = 1$. Where its value is smaller than $(m - a)$ and larger than $(m + b)$ then it does not belong to the set and $f(x) = 0$. If its value falls within the interval between $m - a < x < m + b$, its degree of membership is a number between 0 and 1. The letters L and R will be used to refer to the left and right spreads of m. Based on the above, the triangular L-R fuzzy number $x = (1, 0.1, 0.1)$ appears as presented in Fig. 11.2.

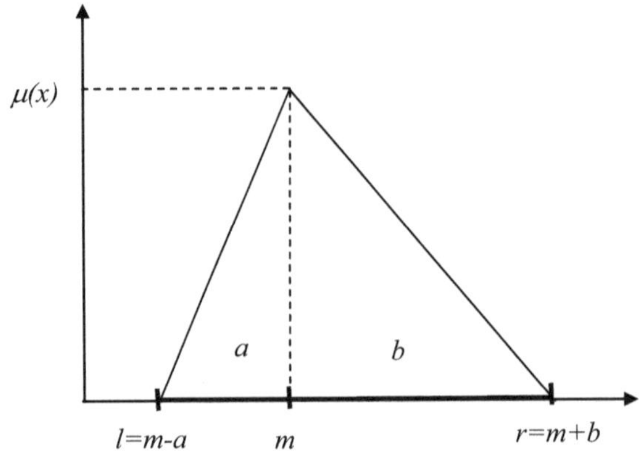

Fig. 11.2 Graphical representation of the L-R fuzzy number $x = (m,a,b)$

The following F-Promethee equations are based on the representation of a triangular fuzzy number such as (m,a,b). A linear preference function, with preference p and indifference q thresholds, was selected (type V), on introducing the fuzzy numbers the evaluation function becomes as follows [43]:

$$P_k(a_i,a_m) = \left\{ \begin{array}{cl} 0 & \text{if } (n-c) \leqslant q \\ \dfrac{(n,c,d)-q}{p-q} & \text{if } q \leqslant n-c \text{ and } n+d \leqslant p \\ 1 & \text{if } n+d \geqslant p \end{array} \right\} \tag{11.19}$$

The decision parameters p and q, are considered as crisp numbers, due to the inherent risk of excessive fuzziness or approximation and the limited benefit of possible fuzzy modelling [43, 48]. As in the case of criterion parameters, it was decided to introduce the weighting factors also as crisp values.

In literature there are several techniques to compare two fuzzy numbers but the defuzzification process is a commonly used approach to convert fuzzy numbers into appropriate crisp values. As suggested by Giannopoulos and Founti in the paper by Deng *et al.* [51], the most popular defuzzification methods are reported here. Geldermann *et al.* [42], for her analysis applied one of the most popular methods (COA – center of area) while Goumas and Lygerou [43] and Le Téno and Mareschal [41] used the Yager Index [52], which is the equivalent of the COA for triangular numbers [48]. The Yager index is the most transparent and "easy to use" defuzzification method. It calculates the weighted average of the fuzzy number that corresponds to the center of the triangular fuzzy numbers as:

$$F(m,a,b) = (3m-a+b)/3 \tag{11.20}$$

Thus, fuzzy numbers with a higher Yager index will be greater than the ones having a smaller Yager index [44, 45]. After converting the triangular fuzzy number (m,a,b) into a magnitude, the steps of the F-Promethee become the same as those of the original PROMETHEE method. The fuzzy Promethee method is preferable because crisp numbers are not adequate to express accurately the ambiguous data used for the application analysed. In this paper the performance of criteria are considered as fuzzy numbers.

11.4 An Application for the Assessment of Biomass Crops Sustainability

11.4.1 *Energy crops selected*

The empirical case proposes a comparative evaluation between a series of alternative crops, using the above described approach, in order to obtain a ranking of their level of sustainability. The energy crops selected are the following: Giant reed (a), Cardoon (b) Miscanthus (c), and Sorghum (d).

Giant reed (*Arundo donax*) is a fast growing perennial grass, native to East Asia, and widespread throughout the Mediterranean area. It can be considered one of the most promising crops in terms of energy production in southern Europe [53–55]. It is also considered to be an environmentally sustainable crop due to its very low pesticide and fertilizer requirements [53]. Currently, experimental farming of giant reed is showing good results. The planting phase, however, is still too expensive and use is made of transplanting rhizomes for this species. If, however, it is possible to spread planting costs over 10-12 years, this may lead to lower management costs [56]. Giant reed can be considered a relatively low input crop. The productivity and stability of this perennial grass, even without fertilization, depends largely upon the growth and the persistence of rhizomes and their seasonal turnover.

Another crop selected for this analysis is cardoon (*Cynara cardunculus L.*) a member of the *Asteraceae* family [57]. Cardoon is a perennial herbaceous species native to the Mediterranean region and it has been cultivated for horticultural purposes since ancient times. This crop is very well adapted to the Mediterranean climate and thus to the hot summers and low annual rainfall. The land area suitable for its growth is mainly located in Spain, Italy, France and Greece [57]. This crop grows very vigorously; a single plant can reach a height of 3 m and cover an area with a diameter of 1.5 m [58, 59]. Due to its adaptation to southern European environmental conditions, characterised by hot, dry summers, and because of its high yield, cardoon could be considered for a variety of industrial applications, in particular for lignocellulosic biomass (solid biofuel) [60–62].

Another very important crop is miscanthus (*Miscanthus sinensis)* a graminaceous plant to which about twenty species belong. It originates from Far Eastern countries and grows naturally in a tropical climate [63, 64]. Originally it grew across south-eastern Asia, China and Japan, reaching as far as Polynesia. Miscanthus is now found widely across Europe, where it was introduced approximately 65 years ago as an ornamental plant. It can grow at a range of altitudes: from sea level up to as high as 3,000 metres, therefore it adapts extremely well to a cold climate [63]; the ideal conditions for sowing are however fairly sunny areas and cool garden soil. Trials conducted found that miscanthus has low resistance to the rigours of winter in the year in which it is planted and that the crop also requires a fair amount of water and thus periods of drought can affect final product yield [65].

The limited use of miscanthus arises essentially from the high costs of planting. Another characteristic that adds to costs is linked to the needs for watering. The plant originates from a tropical climate which is damp all year round whereas the Mediterranean climate is dry over the summer and this is exactly the time in which the plant requires most watering [65].

Finally, we have sorghum (*Sorghum bicolor L. Moench*), a graminaceous plant traditionally grown for zootechnic uses, which includes grain, fibre and sugar species. Sweet sorghum has been used for nearly 150 years to produce forage and silage for animal feed [66]. Over recent years sorghum has gained considerable interest as a source of either fermentable or lignocellulosic feedstock with the potential to produce energy, food and a variety of other products [67]. It can principally be used for the production of electricity or heat through direct combustion of biomass or indirectly by gas and oil derived from it. Moreover, from the sweet types with a high yield of fermentable carbohydrates, it is possible to produce ethyl alcohol (ethanol) as a fuel or fuel additive [68]. Sorghum usually grows well, even in conditions of limited water supply, because of its capacity to overcome short dry periods by temporarily slowing down growth and development [69–71]. However, plants respond very positively to irrigation.

11.4.2 *Criteria Selected and evaluation matrix*

The criteria are the tools that enable alternatives to be compared from a specific viewpoint. Undoubtedly, selecting criteria is the most delicate part in formulating the problem facing the decision maker, and thus it requires the utmost care and attention. The number of criteria is heavily dependent on the availability of information and data.

The first criterion (C_1) used in this study regards the potential amount of biomass producible from the cultivated land. The unit of measurement for this criterion is ha^{-1} per year^{-1} of dry matter. Biomass production data may vary greatly, depending on the agronomic conditions in which it is grown (soil quality, water supply, climatic conditions etc.). Data obtained from Venturi and Venturi [71] indicates that we may expect production levels of between 15 and 30 t ha^{-1} per year^{-1} for miscanthus, between 15 and 35 t ha^{-1} per year^{-1} for the giant reed, more modest levels of between 10 and 15 t ha^{-1} per year^{-1} for cardoon and, finally, for sorghum we can reasonably anticipate values between 20 and 30 t ha^{-1} per year^{-1}. The second criterion (C_2) expresses the lower heating value (LHV) of the biomass in GJ t^{-1} and is obtained from literature. For the giant reed this value varies between 16.5 and 17.4, for the cardoon between 15.6 and 16.8, for the miscanthus between 17.6 and 17.7 and, finally, for sorghum between 16.7 and 16.9. Naturally, one important criterion is that of energy output (C_3), that is to say biomass production in energy terms expressed in GJ ha^{-1}. Data from literature [71] attributes an energy output value of between 260 and 530 GJ ha^{-1} for miscanthus, between 240 and 600 GJ ha^{-1} for the giant reed, of between 155 and 252 GJ ha^{-1} for the cardoon and, finally, between 334 and 507 GJ ha^{-1}for sorghum. The fourth criterion (C_4), closely linked to C_3, is the relationship between output and input, that is to say, the connection between the entire amount of energy consumed in order to produce the biomass (fuel used in machines, etc.) and the quantity of energy that can potentially be extracted from the biomass obtained. Criterion C_5 supplies information regarding the relationship between the quantity of fertilizers needed to sustain and develop the crops. From data in literature it can be seen that, in fact, there is no ideal or "standard" quantity of fertilizers to be used on crops. Indeed, this quantity may vary greatly according to the type of soil and the climatic conditions encountered. Various authors refer to the results of crop experimentation with very different levels of fertilizer use. For our analysis we used data collected from the results of the experimentation carried out within the Bioenergy Farm project (ARSIA) [67]. The ARSIA researchers, on the basis of the results obtained from experimentations carried out on the *pianura pisana* (Italy), supply the following data for miscanthus [67]: between 190 and 210 kg/ha^{-1} of N, between 80 and 90 kg/ha^{-1} of P_2O_5 and between 80 and 85 kg/ha^{-1} of K_2O. For the giant reed Bonari *et al.* [67], while underlining that some authors have not found significant production responses as a result of increased nitrogenous fertilization, report on the other hand

Table 11.1 Evaluation matrix (a)

	Yield dry biomass	Lower heating value	Energy output
	t ha^{-1} Max	GJ t^{-1} Max	GJ ha^{-1} Max
Giant reed	(25,10,10)	(16.95,0.45,0.45)	(428,181,181)
Cardoon	(12.5,2.5,2.5)	(16.15,0.65,0.65)	(203.5,48.5,48.5)
Miscanthus	(22.5,7.5,7.5)	(17.65,0.05,0.05)	(397.5,133.5,133.5)
Sorghum	(25,5,5)	(16.8,0.10,0.10)	(420.5,86.5,86.5)
w	0.1	0.1	0.1
p	5	0.1	30
q	0	0	0
Function	Linear	Linear	Linear

Table 11.2 Evaluation matrix (b)

	Energy balance output/input	Fertilization ($N+P_2O_5+K_2O$)	CO_2 emissions avoided	Ash content
	GJ ha^{-1} Max	kg ha^{-1} Min	t ha^{-1} Max	% Min
Giant reed	(43.5,32.5,32.5)	(480,30,30)	(37.5,15.5,15.5)	(6.10,1.3,1.3)
Cardoon	(19.5,12.5,12.5)	(300,40,40)	(16.5,5.5,5.5)	(14.25,1.25,1.25)
Miscanthus	(39,27,27)	(370,15,15)	(34.5,12.5,12.5)	(2.8,1.2,1.2)
Sorghum	(26,13,13)	(270,30,30)	(32.5,2.5,2.5)	(6.5,2.5,2.5)
w	0.1	0.1	0.1	0.1
p	3	50	5	2.5
q	0	0	0	0
Function	Linear	Linear	Linear	Linear

that the production of large amounts of biomass (including leaves) will lead to high levels of removal of nutritive substances. We will thus obtain approximately 190-210 kg/ha^{-1} of N, 70-90 kg/ha^{-1} of P_2O_5 and 190-210 kg/ha^{-1} of K_2O. For cardoon the levels of removal of nutritive elements are not high and the data contained in literature refers to values of between 100 and 120 kg/ha^{-1} of N, between 60 and 70 kg/ha^{-1} of P_2O_5 and between 100 and 150 kg/ha^{-1} of K_2O. Finally, for sorghum, in the normal course of events we can expect to see values of around 100-120 kg/ha^{-1} of N, 60-80 kg/ha^{-1} of P_2O_5 and 80-100 kg/ha^{-1} of K_2O. For the purposes of calculation we selected one criterion, C_5, reported in the evaluation matrix, which expresses in synthesis the total quantity of fertilizers ($N+K_2O+P_2O_5$) used on the crops.

For the purposes of the overall evaluation of environmental sustainability, one crucial piece of information is undoubtedly that regarding any CO_2 emissions. To this end, we selected a criterion (C_6) that indicates the tonnes of CO_2 emissions saved for each ha^{-1} of biomass produced and used in place of fossil fuels.

Another important qualitative characteristic of biomass is its ash content (C_7). The presence of ashes, mainly silicates and chlorate compounds, is probably the most significant technological barrier in the thermoelectric transformation process of herbacious biomasses. A high ash content may damage certain mechanical components (e.g. turbines) and compromise the functioning of the conversion plant, pushing up maintenance costs and reducing overall efficiency. The lowest level of ash, between 1.6% and 4%, is to be found in miscanthus, while for the giant reed it is between 4.8% and 7.4%, for sorghum around 4% to 9%, and the highest level from 13% to 15.5%, is to be found in the cardoon. Although ashes do not have a direct effect on environmental sustainability, this is an important criterion to take into account when making an overall qualitative evaluation of a biomass.

Lastly, we build the evaluation matrix (see Tab. 11.1 and Tab. 11.2), which contains the options under consideration and the fuzzy values of the parameters selected. Normally, this matrix also contains the weights assigned to the various criteria, that is the importance of each criterion in the evaluation process. How to attribute weights to the criteria remains of the greatest weaknesses of this methodology. Indeed, an arbitrary and subjective assignment of weights can greatly affect the outcome of the analysis. In our case, after reflecting long and hard on this issue, we decided (in the first instance) not to assign any weights to the parameters chosen.

11.4.3 *Results and sensitivity analysis*

An evaluation matrix has been constructed containing the 4 energy crops selected and the environmental evaluation criteria. Having carried out all the calculations and applied the procedures of the methodology chosen, the final ranking that emerges is: Miscanthus ($\Phi^{net}0.573$) > Giant reed ($\Phi^{net}0.325$) > Sorghum ($\Phi^{net}0.296$) > Cardoon ($\Phi^{net} - 1.450$).

Even when the calculations have been completed, the decision maker must regard the results obtained merely as a starting point for further analysis of the problem. It is, therefore, important to carry out a sensitivity analysis in order to ascertain the effect on the original results that is generated by a change in the parameters of the model. In our case we modified the threshold value of p ($\pm 10\%$ and $\pm 25\%$) of the initial values set and the vector of weights w_i. Thus, by varying the threshold value, although we obviously obtained a different value for Φ^{net}, the ranking was stable and confirmed the initial result, as can be seen from Tab. 11.3 and Fig. 11.3.

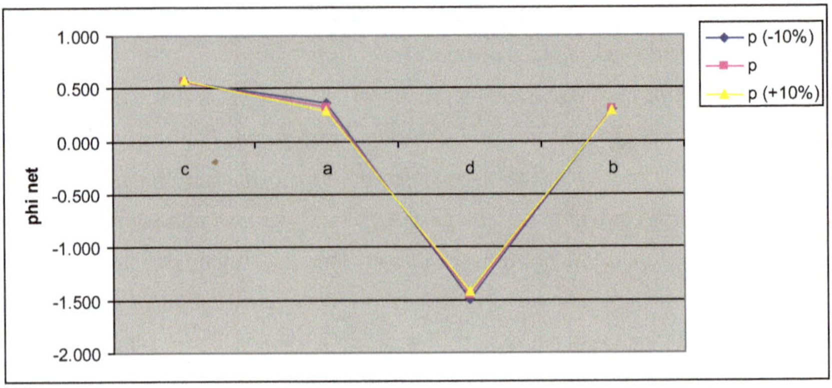

Fig. 11.3 Results of sensitivity analysis (change in threshold $p \pm 10\%$)

Table 11.3 Results of sensitivity analysis (change in threshold $p \pm 25\%$)

	p_α (−25%)		p_β (+25%)	
	Φ^{net}	Rank (stable)	Φ^{net}	Rank (stable)
Giant reed (a)	0.361	2	0.288	2
Cardoon (b)	−1.489	4	−1.418	4
Miscanthus (c)	0.570	1	0.583	1
Sorghum (d)	0.306	3	0.287	3

Table 11.4 Weight vectors

Criteria	w_o	w_σ	w_τ	w_ψ
C_1	1	0.2	0.4	0.2
C_2	1	0.1	0.1	0.2
C_3	1	0.3	0.5	0.3
C_4	1	0.3	0.5	0.5
C_5	1	0.1	0.1	0.3
C_6	1	0.5	0.5	0.5
C_7	1	0.3	0.4	0.3

Subsequently, we attempted to modify the weights vector: in this case, some alterations occur, although they are not substantial. After carrying out the calculations and using the three different vectors (w_σ, w_τ, w_ψ) indicated in Table 11.4, the same ranking is achieved for vectors w_σ and w_ψ as that achieved using w_0.

However, we need to report that for vector w_σ the gap Φ^{net} between miscanthus and the giant reed is absolutely minimal. With vector w_τ, on the other hand, the ranking undergoes the following variation: Giant reed>Miscanthus>Sorghum>Cardoon (see Fig. 11.4).

On the basis of the criteria selected, miscanthus proves to be a better performing crop in terms of environmental sustainability than the other three types of biomass.

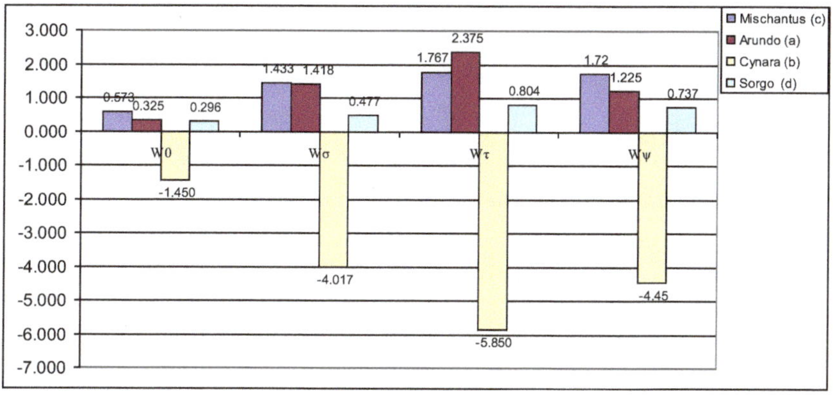

Fig. 11.4 Results of sensitivity analysis (change in weights vector)

A further element that strengthens the hypothesis of the greater ecological sustainability of miscanthus is provided by McCarthy and Mooney [72]. They claim that, on the basis of the experiments carried out in field trials, miscanthus presents a high level of protection against attack by parasites and pathogens, therefore the need for pesticides is low enough to lower the risk of contaminating aquifers and marine organisms and the soil. As far as biodiversity is concerned, these crops have a positive impact on wildlife compared to the traditional crops they could replace.

11.5 Conclusions

The prospects for the development of energy crops, within the wider framework of renewable sources, appear to be extremely significant and promising. However, there are still some areas of doubts regarding their full environmental sustainability and their industrialization on a large scale. A decisive move to enhance these biomasses, in our opinion, should strengthen and highlight their environmental value by means of adequate evaluation procedures.

From the calculations carried out, on the basis of the 7 criteria selected, the highest ranked crop was Miscanthus (Φ^{net} 0.573), followed by the Giant reed (Φ^{net} 0.325), by Sorghum (Φ^{net} 0.296) and with Cardoon ranked last (Φ^{net} -1.450). The sensitivity analysis, performed by modifying the value of the preference threshold p, demonstrated substantial stability of the result. Even when modifying the weights vector, no changes in ranking were observed, except in the case of applying the vector known as w_{τ}, which produced a ranking with the Giant reed at the top.

As this work demonstrates, fuzzy sets are suitable for uncertain approximate reasoning and allow decision making with estimated values where information is incomplete or uncertain. The F-Promethee method produces a more realistic ranking in which imprecision of data is taken into account. This approach is able to provide a useful technical-scientific decision making tool that can be efficiently integrated with fuzzy triangular numbers and can support analyses in real life situations, such as the energy assessment process.

Bibliography

[1] V. Bartolelli, Il contesto agricolo nazionale e la politica europea, in Proceedings of the Conference Valorizzazione energetica e industriale delle biomasse lignocellulosiche, Centro Ricerche Enea-Trisaia 25-27 giugno (1997).
[2] F. Cherubini, *Renewable Energy*, **35** (2010).
[3] F. Cherubini and A.H. Strømman, *Bioresource Technology,* **102** (2011).
[4] G.A. Blengini, E. Brizio, M. Cibrario, and G. Genon, *Resources, Conservation and Recycling*, **57** (2011).
[5] C. Buratti and F. Fantozzi, *Biomass and Bioenergy*, **34** (2010).
[6] A. Rafaschieri, M. Rapaccini, and G. Manfrida, *Energy Conversion & Management*, **40** (1999).
[7] S. Fazio and A. Monti, *Biomass and Bioenergy,* **35** (2011).
[8] S. Kim and B.E. Dale, *Biomass Bioenergy* **29** (2005).
[9] S.N. Djomo and D. Blumberga, *Bioresource Technology,* **102** (2011).
[10] F. Cherubini and S. Ulgiati, *Applied Energy,* **87** (2010).
[11] E. Georgopoulou, D. Lalas, and L. Papagiannakis, *European Journal of Operational Research*, **103** (1997).
[12] K. Kowalski, S. Stagl, R. Madlener, and I. Omann, *European Journal of Operational Research*, **197** (2009).
[13] Y.I. Topcu and F. Ulengin, *Energy*, **29** (2004).
[14] A.R. Borges and C. H. Antunes, *European Journal of Operational Research*, **145** (2003).
[15] C. Kahraman and I. Kaya, *Expert Systems with Applications*, **37** (2010).
[16] S.K. Lee, G. Mogi, J. W. Kim and B. J. Gim, *International Journal of Hydrogen Energy*, **33** (2008).
[17] J. Siskos and P.H. Hubert, *European Journal of Operational Research*, **13** (1983).
[18] J. Oberschmidt, J. Geldermann, J. Ludwig, and M. Schmehl, *International Journal of Energy Sector Management*, **4** (2010).
[19] D.A. Haralambopoulos and H. Polatidis, *Renewable Energy*, **28** (2003).
[20] C. Kahraman, I. Kaya, and S. Cebi, *Energy*, **34** (2009) 1603–1616.

[21] Y.P. Cai, G.H. Huang, Q. Tan, and Z.F. Yang, *Renewable Energy*, **34** (2009).
[22] R. Baños, F. Manzano-Agugliaro, F.G. Montoya, C. Gil, A. Alcayde, and J. Gómez, *Renewable and Sustainable Energy Reviews*, **15** (2011).
[23] E. Heo, J. Kim, and K.J. Boo, *Renewable and Sustainable Energy Reviews*, **14** (2010).
[24] F. Cavallaro, *Energy Policy*, **38** (2010).
[25] F. Cavallaro, *Renewable Energy*, **34** (2009).
[26] T. Kaya and C. Kahraman, *Energy*, **35** (2010).
[27] F. Cavallaro, *Applied Energy*, **87** (2010).
[28] A. Sultana and A. Kumar, Biomass and Bioenergy **39** (2012).
[29] A. Kumar, S. Sokhansanj and P. C. Flynn FLYNN, Applied Biochemistry and Biotechnology, **129** (2006).
[30] T. Buchholz, E. Rametsteiner, T.A. Volk, and V.A. Luzadis, Energy Policy **37** (2009).
[31] A. Damiano, G. Gatto, and I. Marongiu, Decision support system for biomass energy exploitation in smart grid planning, International Symposium on Power Electronics Electrical Drives Automation and Motion (SPEEDAM), 2010, 1183–1188.
[32] L. Čuček, J.J. Klemeš, P.S. Varbanov, and Z. Kravanja, *Chemical Engineering Transaction*, **25** (2011).
[33] A. Perimenis, H. Walimwipi, S. Zinoviev, F. Muller-Langer, and S. Miertus, *Energy Policy*, **39** (2011).
[34] G. Dorini, Z. Kapelan and A. Azapagic, *Clean Techn Environ Policy*, **13** (2011).
[35] P. Venturi and G. Venturi, *Biomass and Bioenergy,* **25** (2003).
[36] F. Pisani and S. Coronati, *Agricoltura e Innovazione*, ENEA, **6** (1989).
[37] J.P. Brans and Ph. Vincke, *Management Science*, **31** (1985).
[38] J.P. Brans, Ph. Vincke and B. Mareschal, *European Journal of Operational Research*, **24** (1986).
[39] J.P. Brans and B. Mareschal, *Decision Support System*, **12** (1994).
[40] J.P. Brans and B. Mareschal, Multicriteria decision aid the promethee-gaia solution. Working paper, (1998) Vrije Universiteit Brussel.
[41] J.F. Le Téno and B. Mareschal, *European Journal of Operational Research*, **109** (1998).
[42] J. Geldermann, T. Spengler and O. Rentz, *Fuzzy Sets and Systems*, **115** (2000).
[43] M. Goumas and V. Lygerou, *European Journal of Operational Research*, **123** (2000).
[44] R.U. Bilsel, G. Büyüközkan and D. Ruan, *International Journal of Intelligent Systems*, **21** (2006).
[45] G. Tuzkaya, B. Gülsün, C. Kahraman, and D. Özgen, *Expert Systems with Applications*, **37** (2010).
[46] T.Y Chou, W.T. Lin, C.Y. Lin, and W.C. Chou, *Ecological Engineering*, **31** (2007).
[47] W. Li and B. Li, Proceedings of 2009 IEEE International Conference on Grey Systems and Intelligent Services, November 10-12, 2009, Nanjing, China.
[48] D. Giannopoulos and M. Founti, *International Journal of Multicriteria Decision Making*, **1** (2010).
[49] J. Oberschmidt, J. Geldermann, J. Ludwig and M. Schmehl, *International Journal of Energy Sector Management*, **4** (2010).
[50] D. Dubois and H. Prade, *International Journal of Systems Science*, **9** (1978).
[51] H. Deng, M. Lau and K. Millar, Defuzzification in fuzzy multicriteria analysis, Paper presented at the 19th International Conference of the North American Fuzzy Information Processing Society (NAFIPS), 13-15 July 2000, Atlanta, USA.
[52] R.R. Yager, *Information Science*, **24** (1981).
[53] I. Lewandowski, J.M.O. Scurlock, E. Lindvall, and M. Christou,. *Biomass Bioenergy*, **25** (2003).
[54] G.L. Angelini, L. Ceccarini, and E. Bonari, *Eur. J. Agron.* **22** (2005).

[55] G.L. Angelini, L. Ceccarini, N. Nassi o Di Nasso, and E. Bonari, Long term evaluation of biomass production of giant reed (Arundo donax L.) to different fertilisation input, plant density and harvest time in a Mediterranean environment. In: Grassi, A., Maniatis, K. (Eds.), Proceeding of the 14th European Biomass Conference and Exhibition. Paris, October 17–21.
[56] G. Candolo, *Agronomica* 4 (2006).
[57] A. Ierna, R.P. Mauro, and G. Mauromicale, *Biomass and Bioenergy,* **36** (2012).
[58] J. Fernández J. Lignocellulosic biomass production from annual energy crops. Report EUR 12631 EN-C. Commission of the European Communities, Luxembourg (1990).
[59] J. Fernández and P. Manzanares. Cynara cardunculus L. a new crop for oil, paper pulp and energy. In: Grassy G., Gosse G., dos Santos G. (Eds). Proceedings of the 5th International Conference on Biomass for Energy and Industry. Barking: Elsevier Science Publishers Ltd., (1990).
[60] J. Fernández, M.D. Curt, and P.L. Aguado, *Industrial Crops and Products*, **24** (2006).
[61] S. Foti, G. Mauromicale, S.A. Raccuia, B. Fallico, F. Fanella and E. Maccarone. *Industrial Crops and Products,* **10** (1999).
[62] I. Piscioneri, N. Sharma, G. Baviello and S. Orlandini, *Energy Conversion & Management,* 41 (1999).
[63] L. Scally, T. Hodkinson, and M.B Jones, Origins and Taxonomy of Miscanthus. Miscanthus for energy and fibre (edited by M.B. Jones and M. Walsh) James&James, London, (2001).
[64] I. Lewandoski, J.C. Clifton-Brown, J.M.O. Scurlock, and W. Huisman, *Biomass and Energy,* **19** (2000).
[65] S. Foti, S.L. Cosentino, C. Patanè, and C. Foti Cuzzola, Effetti del regime idrico e della somministrazione di azoto in Miscanthus x giganteus Greef. Et Deu. in ambiente mediterraneo, Proceedings of XXXIII Congress of Società italiana di Agronomia, Legnaro (Italy), (1999).
[66] R. Janssen, D. Rutz, S. Braconnier, B. Reddy, S. Rao, R. Schaffert, R. Parella, A. Zaccharias, N. Rettenmaier, G. Reinhardt, A. Monti, S. Amaducci, A. Marocco, W. Snijman, H. Terblanche and F. Zavala-Garcia, Sweet sorghum – an alternative energy crop, Proceedings of the 18th European Biomass Conference and Exhibition, 2010.
[67] E. Bonari, G. Picchi, W. Guidi, E. Piccioni, A. Fraga, R. Villani and M. Ginanni, Le colture dedicate ad uso energetico: il progetto Bioenergy Farm. Quaderno ARSIA 6/2004.
[68] A. Monti and G. Venturi, *European Journal of Agronomy*, **19** (2003).
[69] S.L. Cosentino, Crop physiology of sweet sorghum. Proc. of First European Seminar on Sorghum for energy and industry. Tulose, 1–3 Apr. 1996.
[70] S. Foti, S.L. Cosentino, V. Copani, C. Patanè, and P. Guarnaccia, 1996. Growth and yield of C4 species for biomass production in Mediterranean environment. Proc. 9th European Bioenergy Conference Biomass for energy and environment, pp. 616–621
[71] P. Venturi and G. Venturi, *Biomass and Bioenergy*, **25** (2003).
[72] S. McCarthy, M. Mooney, European miscanthus Network. In: Ph. Chartier, A.A.C.M. Beenackers, G. Grassi, (eds.) Biomass for Energy, Environment and Agriculture and Industry - Proceedings of 8th E.C. Conference. 3-5 October, 1994 Vienna Austria, Elsevier Scienze Ltd., Oxford, (1995), pp. 380–8.

Chapter 12

Forecasting Energy Demand Using Fuzzy Seasonal Time Series

İrem Uçal Sarı and Başar Öztayşi

Industrial Engineering Department,Istanbul Technical University, Macka Campus, 34367, Istanbul, Turkey

E-mail: ucal@itu.edu.tr

Demand side energy management has become an important issue for energy management. In order to support energy planning and policy decisions forecasting the future demand is very important. Thus, forecasting the future energy demand has gained attention of both academic and professional world. In this chapter, the previous researches on energy demand forecast are classified and fuzzy techniques are introduced. A fuzzy seasonal time series model that forecasts the energy demand is proposed and illustrated with a real world application.

12.1 Introduction

Energy planning and management has been as serious topic both in national and organizational level. National energy planning problems deal with the energy allocation problems, energy investments and sustainability. On the other hand energy manufacturing organizations tend to develop computer aided, energy management systems to monitor, control and optimize the generation and transmission systems. As the third layer, the end users show high interest on energy efficiency and prone to use automated building energy control systems to monitor and control the energy consumption. These automated systems, can be used by a single building or group of buildings such as office buildings, retail stores networks or factories. The data obtained from the system can be used by the end users to perform self-diagnostics and trend analysis.

Forecasting is a form of predictive analytic methods that deals with predicting the future generally based on data and models [1]. In case of energy management and planning

C. Kahraman (ed.), *Computational Intelligence Systems in Industrial Engineering*, Atlantis Computational Intelligence Systems 6, DOI: 10.2991/978-94-91216-77-0_12,

forecasting is widely used by all related parties. The forecasting studies focus on three main areas namely, energy consumption (or demand), energy price and energy production. The energy consumption forecasts are accomplished and published by governmental agencies both for short and long term. Energy consumption forecast is also an important issue for energy production plants, energy suppliers and the end user especially if the energy market is liberated. In the energy market, since the price can change by time, predicting the future price of energy is very important both for producers and consumers. The producers can increase their income by producing energy at the time that the price is high. The end users can decrease the costs by reducing the energy consumption when the price is high planning the energy production. The final area for forecasting aims to predict the amount of energy production in a specified time. While some sources of energy can be controlled by the energy plant, especially the renewable energy sources cannot be controlled since the main source of energy are natural factors such as wind, wave, rain and water lows. Most of the companies and governments try to build a sustainable energy strategy in order to generate the highest benefit from renewable energy sources so forecasting the production amount is vital for energy planning and investment decisions.

The fuzzy set theory, developed by Zadeh [2], is designed to mathematically represent uncertainty and vagueness. Formalized tools and methods are proposed for dealing with the imprecision in decision making problems. The methods that are based on fuzzy sets can handle the imprecise date with imprecise values instead of precise ones, so the process are more powerful and its results are more credible [3]. Fuzzy set theory is being recognized as an important problem modeling and solution technique because they use approximate information and uncertainty to generate decisions [4]. Fuzzy time series (FTS) combine conventional time series models, which has been used for forecasting problems, with fuzzy set theory. A fuzzy time series essentially consists of steps such as fuzzification, the establishment of fuzzy relationships, and defuzzification [5]. The applications of FTS include; air pollution [6], tourism demand [7], stock index forecasting [8], and temperature [9].

In this study, forecasting issues and techniques related to energy sector is summarized and a real life application of fuzzy seasonal time series technique on the energy consumption is presented. The forecasting application is done for 10 supermarkets in Hatay, one of the cities of Turkey. The originality of this chapter comes from the proposed methodology which is used to forecast the time series data that has a seasonal pattern. The remaining of the chapter is organized as follows; In Section 12.2 the back ground information about energy planning and management is given. Section 12.3 gives a brief literature review about forecasting techniques and previous studies about energy consumption forecasting. Section 12.4 contains the brief information about fuzzy sets, fuzzy time series and the proposed method. A real world numerical application is given in Section 12.5 and finally the conclusion and further steps are discussed in conclusions.

12.2 Energy Demand Management and Demand Forecasting

The energy system consists of both supply-side and demand-side activities. Before 1970s, the focus of energy management was on the supply side, the demand was treated as a given data and the objective was to arrange for adequate supply to satisfy the demand. However, with rising prices in the 1970s, researchers, governments and the utilities started to investigate the problem from demand side.

Energy demand management (or demand side management) is defined as "systematic utility and government activities designed to change the amount and/or timing of customer's use of energy" [10] and involves effective utilization of the energy resources, reliability in supply, efficient management of energy resources, energy conservation, combined heat and power systems, renewable energy systems, integrated energy systems, independent power delivery systems [11].

Demand management consists of planning, implementing, and monitoring activities of energy utilization that are designed to encourage consumers to modify their level and pattern of energy usage. Energy demand management includes various categories of activities such as [10]: load management, energy conservation, fuel substitution, load building. Load management's objective is to reduce or change the size or timing of the demand. Energy conservation aims at reducing the demand, through technical efficiency improvements. Fuel substitution targets to replace one fuel by another and thereby modifies the demand. Load building implies developing load for strategic purposes which could help manage the system better.

Bhattacharyya [12] defines a number of reasons that the demand management is essential. First of all, because of the system losses, any reduction in demand places lower pressure on system expansion. Similarly, since the conversion process has low efficiency, the pressure on resource requirement reduces through demand reduction. Demand management improves the utilization of the available infrastructure by allocating the demand over time and can reduce overcrowding or improve reliability of the supply. Since many countries depend on imports to meet their supply needs, a reduction in demand also reduces the

Table 12.1 Classification of energy forecasting application areas

Application Area	**References**
World level forecasting	Maggio G. and Cacciola G. [14], Valero A. and Valero A. [15]
National level forecasting	Forouzanfar *et al.* [16]; Erdoğdu E. [17]
Regional level forecasting	Jiang *et al.* [18]; Sailor D. and Munoz R. [19]
Individual customers level	Vondracek *et al.* [20]; Brabec *et al.* [21]

import dependence and as a result, the vulnerability to price fluctuations is reduced. And finally integration of the demand-side response in the market operation leads to improved resource utilization and therefore expands market operation.

Energy demand is found to be closely related to energy price, GDP and population. Energy demand management helps providing a sustainable economic development by achieving self-sufficiency and cost effectiveness. Suganthi and Samuel [11] state that energy demand management should help in the following issues;

- planning for the future requirement
- Identification and prioritization of energy resources
- outlining policy decisions
- Identification of strategies for reduced emission

As a result of the mentioned importance of energy demand management, researchers have showed high attention to this area. In order to support planning and policy decisions, the models are developed that tries to forecast the future demand levels.

12.3 Literature Review on Energy Consumption Forecasting

Forecasting energy consumption has been widely investigated in the literature. The studies can be classified in three different perspectives; application area, forecasting horizon and the methods used. Soldo [13] classifies the application area of the forecasting in to groups shown in Table 12.1.

The studies in the literature vary depending on the time horizon used for forecasting. Most of the studies use annual and monthly basis on the other hand is some limited research different combinations of time horizons are also used. Table 12.2 presents the sample references for each time horizon.

Table 12.2 Classification of energy forecasting time intervals

Time Interval	References
Annual	Aydinalp and Ugursal [22], Jiang *et al.* [18]
Monthly	Kizilaslan R. and Karlik B. [23], Yoo *et al.* [24]
Daily	Potocnik *et al.* [25], Vondracek *et al.* [20];
Hourly	Dombayci [26], Thaler *et al.* [27]
Combined	Potocnik *et al.* [28], Sanchez-Ubeda and Berzosa (2007) [29]

Table 12.3 The historical electricity consumption data of supermarkets

Forecasting model	Sample ReferencesFrom the literature
Time Series Models	Gori *et al.* [30], Himanshu *et al.* [31], Al-Shobaki and Mohsen [32], Amarawickrama and Hunt [33]
Regression Models	Bianco *et al.* [34], Bessec and Fouquau [35], Al-Ghandoor *et al.* [36], Lam *et al.* [37], Jónsson [38]
Econometric models	Inglesi [39], Bhattacharyya and Timilsina [40], Zhang *et al.* [41], Al-Ghandoor [42], Meng and Niu [43]
ARIMA models	Ediger and Akar [44], Erdoğdu [45], Erdoğdu [17], Pappas *et al.* [46]
Simulation	Azadeh *et al.* [47], Azadeh and Tarverdian [48]
ANN	Kavaklıoğlu *et al.* [49], Geem and Roper [50], Ekonomou and Greek [51]
Genetic Algorithms	Canyurt and Ozturk H.K. [52], Azadeh and Tarverdian [48], Cankuş and Taşkıran [53], Ceylan and Öztürk [54]
Simulated Annealing	Özçelik and Hepbaşlı [55], Pai and Hong [56]
Particle Swarm Optimization	Kiran *et al.* [57], Ünler [58], AlRashidi and EL-Naggar [59], Hong [60]
Ant Colony Optimization	Toksarı [61], Toksarı [62]
Grey Methods	Wang [63], Akay and Atak [64], Lee and Tong [65], Lee and Shih [66].

There have been various forecasting methods used for predicting energy consumption. The review of major energy demand forecasting models are categorized as shown in Table 12.1 and representative references are also supplied.

Besides the methods listed in Table 12.3, there are also other methods used in the literature. These methods include; Winters' Exponential Smoothing Method and Cycle Analysis [67], Modeling Based on Degree-day Concept [68], Harmony Search Algorithm [69], Markal [70] and Times G5 [71].

There are also other studies that focus on classification of methods used in energy demand forecast. On one of the most comprehensive study, Suganthi and Samuel [11] classify the models in to 11 groups. Another important classification is given by Soldo [13] and Ünler [58].

In the field of energy consumption forecasting, fuzzy set theory is used combined with other methods, such as regression, and time series. Pai [72], propose a hybrid ellipsoidal fuzzy system for time series forecasting (HEFST) in order to forecast regional electricity loads. Song [73] use fuzzy regression analysis to reduce the load forecasting error in the short-term load forecasting problem. Azadeh *et al.* [74] presents an integrated fuzzy regression and time series framework to estimate and predict electricity demand for seasonal and monthly changes in electricity consumption. Che [75], presents an adaptive fuzzy combination model based on the self-organizing map, the support vector regression and the fuzzy inference method for electric load forecasting. Kucukali and Baris [76] forecast Turkey's short-term gross annual electricity demand by applying fuzzy logic methodology take in economic, political and electricity market conditions in to consideration.

Fuzzy neural networks are also used in the research area. Chang [77] presents a weighted evolving fuzzy neural network for monthly electricity demand forecasting in Taiwan. Viet and Mandziuk [78] use neural and fuzzy neural networks to analyze several approaches to predict natural gas consumption in two different regions in Poland

Fuzzy inference systems and adaptive network based fuzzy inference systems are used to make better forecast. Behrouznia *et al.* [79] present an adaptive network based fuzzy inference system and fuzzy data envelopment analysis for gas consumption forecasting and analysis.

Li and Su [80], developed a the hybrid genetic algorithm-hierarchical adaptive network-based fuzzy inference system (GA-HANFIS) model to forecast building energy consumption loads in Taiwan. Azadeh *et al.* [81] propose an adaptive network-based fuzzy inference system (ANFIS) for estimation of demand using daily natural gas consumption. Mamlook *et al.* [82], propose a methodology to decrease the forecasted error and the processing time by using fuzzy logic controller on an hourly base. Ying and Pan [83], presents the adaptive network based fuzzy inference system (ANFIS) model to forecast the regional electricity. Lau *et al.* [84] proposed a fuzzy rule reasoning mechanism that helps the manufacturer forecast the energy consumption change in the plant when certain production input factors are varied.

12.4 Methodology

A time series is a set of regular time-ordered observations of a quantitative characteristic of an individual or collective event. In time series analysis, the trend of data can be the basis of detecting events' occurrence such as increasing, decreasing, seasonal cycles or outliers [85]. The fuzzy time series is a method combining linguistic variables with the analyzing process of applying fuzzy logic into time series to solve the fuzziness of data [85]. The traditional time series approaches require having the linearity assumption and at least 50 observations. In fuzzy time series approaches, there is not only a limitation for the number of observations but also there is no need for the linearity assumption [86]. The fuzzy time series approaches consist of three steps. The first step is the fuzzification of observations. In the second step, fuzzy relationships are established and the defuzzification is done in the third step [87].

There are lots of forecasting methods developed based on fuzzy time series. Fuzzy time series are introduced first by Song and Chissom [88–90]. Chen [91], Huarng [92], Chen [93], Lee and Chou [94], Huarng and Yu [95], Own and Yu [96], Yu [97], Tsaur *et al.* [98] Lee *et al.* [99], Chen and Chung [100], Chen *et al.* [101], Li and Cheng [102], Singh [103] and Cheng *et al.* [104] proposed forecasting methods often form fuzzy logical relationships and group them as heuristic rules to derive the forecasted values [105].

12.4.1 *Fuzzy Time Series and Singh's Method*

Fuzzy set theory was introduced by Zadeh in 1965 to solve problems involving the lack of precisely defined criteria. A fuzzy set is a class of objects with a continuum of grade of membership. Let U be the universe of discourse with $U = \{u_1, u_2, \ldots, u_n\}$, where u_i are possible linguistic values of U, then a fuzzy set of linguistic variables of is defined by Eq. (12.1) where μ_{A_i} is the membership function of the fuzzy set A_i.

$$A_i = \frac{\mu_{A_i}(u_1)}{u_1} + \frac{\mu_{A_i}(u_2)}{u_2} + \cdots + \frac{\mu_{A_i}(u_n)}{u_n} \tag{12.1}$$

Let $Y(t)$ $(t = \ldots, 0, 1, 2, 3, \ldots)$ is a subset of $\mathbb{R}$, be the universe of discourse on which fuzzy sets $f_i(t)$ $(i = 1, 2, 3, \ldots)$ are defined and $F(t)$ is defined as fuzzy time series on $Y(t)$. Suppose $F(t)$ is caused only by $F(t-1)$ and is denoted by $F(t-1) \rightarrow F(t)$; then there is a fuzzy relationship between $F(t)$ and $F(t-1)$ and can be expressed as the relational equation where "$\circ$" is max-min composition operator. The relation $\mathbb{R}$ is called first order model of $F(t)$:

$$F(t) = F(t-1) \circ R(t, t-1) \tag{12.2}$$

If $F(t)$ caused by more fuzzy sets, $F(t-n), F(t-n+1), \ldots, F(t-1)$, the fuzzy relationship is represented by Eq. (12.3) where $F(t-n) = A_{i1}, F(t-n+1) = A_{i2}, \ldots, F(t-1) = A_{in}$:

$$A_{i1}, A_{i2}, \ldots, A_{in} = A_j \tag{12.3}$$

Suppose $F(t)$ is caused by an $F(t-1), F(t-2), \ldots, F(t-m)$ where $m > 0$ simultaneously and the relations are time variant. The $F(t)$ is said to be time variant fuzzy time series and the relation can be expressed as the fuzzy relational equation where $w > 1$ is a time parameter by which the forecast $F(t)$ is being affected [88–90]:

$$F(t) = F(t-1) \circ R^{w}(t, t-1) \tag{12.4}$$

There are two main steps in a fuzzy time series forecasting method. The first one is setting up the fuzzy logical model according to the historical data or experience knowledge. The second step is to interpret the results from the forecasting model. The procedures that are used in forecasting are made up of two steps which are fuzzylising the universe on which the historical data are based and the interpretation of the output results [89]. The fuzzy time series forecasting method which is proposed by Song and Chisson [89] consists of the following steps:

(i) Fuzzylising the universe. Universe is constructed as an interval which is larger than the invertal formed by the minimum and the maximum of the data. To fuzzylise the universe the overall interval is departmentalize into seven even lengthy interval.
(ii) Define the linguistic variables as fuzzy sets on the universe and determine the memberships for each element in the respective fuzzy sets.
(iii) Find out the memberships of each year's data.
(iv) Obtain the historical experience knowledge about the evolution of the data.
(v) Calculate the output of the model.

Singh [103] proposed a new method of fuzzy time series forecasting based on difference parameters. Singh's method consists of the following steps:

(a) Define the Universe of discourse, U based on the range of available historical time series data, by rule $U = [D_{\min} - D_1, D_{\max} - D_2]$ where D_2 and D_1 are two proper positive numbers.
(b) Partition the Universe of discourse into equal length of intervals: $u_1, u_2, \ldots, u_m$. The number of intervals will be in accordance with the number of linguistic variables (fuzzy sets) $A_1, A_2, \ldots, A_m$ to be considered.

(c) Construct the fuzzy sets A_i in accordance with the intervals in Step 2 and apply the triangular membership rule to each interval in each fuzzy set.

(d) Fuzzify the historical data and establish the fuzzy logical relationships by the rule: If A_i is the fuzzy production of period n and A_j is the fuzzify production of period $n+1$, then the fuzzy logical relation is denoted as $A_i \rightarrow A_j$. Here A_i is called current state and A_j is called next state.

(e) Apply the rules for forecasting (for detailed information see [103]).

12.4.2 *A proposed fuzzy time series forecasting method for forecasting the seasonal pattern of data*

Generally, there are two different approaches for forecasting seasonal time series. The first approach is to directly forecast what the seasonal time series will be in the future. The second approach is to remove the seasonal variation from the time series data, and then to apply ordinary forecasting methods [105]. The method most commonly used to compute the typical seasonal pattern is called the ratio-to-moving-average method which eliminates the trend, cyclical, and irregular components from the original data In proposed method, ratio to moving average method [106] is used to remove the seasonal variation from the time series data and then Singh's method is used after some modifications to compute forecasted values. An additional step is defined to take into account expert predictions.

The research steps of the proposed method are described as follows:

Step 1. Deseasonalize the historical data. Apply the ratio-to-moving-average method to remove the seasonality from the time series data. This method consists of six steps to determine the seasonal indexes:

1. The first step is to determine the k period moving average when the seasonal time period is k.
2. If the seasonal time period k is odd go to the next step, otherwise compute the centered moving average of the two consecutive moving averages.
3. Divide the actual value of each time period by its corresponding (centered) moving average.
4. Calculate the median of the ratios for each time period.
5. Compute the seasonal index by adjusting each period's median so that the mean value of the k medians is equal to 1.
6. Divide the actual value by its corresponding seasonal index to obtain the deseasonalized value.

Step 2. Define the Universe of discourse, U based on the range of available historical time series data, by rule $U = [D_{\min} - D_1, D_{\max} - D_2]$ where D_2 and D_1 are two proper positive numbers.

Step 3. Partition the Universe of discourse into equal length of intervals: $u_1, u_2, \ldots, u_m$. The number of intervals will be in accordance with the number of linguistic variables (fuzzy sets) $A_1, A_2, \ldots, A_m$ to be considered.

Step 4. Construct the fuzzy sets A_i in accordance with the intervals in Step 2 and apply the triangular membership rule to each interval in each fuzzy set.

Step 5. Fuzzify the historical data and establish the fuzzy logical relationships by the rule: If A_i is the fuzzy data of period n and A_j is the fuzzy data of period $n+1$, then the fuzzy logical relation is denoted as $A_i \to A_j$. Here A_i is called current state and A_j is called next state.

Step 6. Fuzzify linguistic expectations of experts for the production of the year $n+1$. For more than one expert simply calculate the interval by Eq. (12.5) where A_{ix} $(x = 1, 2, \ldots, z)$ denotes linguistic the demand expectations of z different experts:

$$u_{ix} = \left[\min_{x=1,2,\ldots,z} L[*A_{ix}], \max_{x=1,2,\ldots,z} U[*A_{ix}] \right] \tag{12.5}$$

Step 7. Apply computational algorithm. The notations which are used in the algorithm are as follows: $[*A_j]$ is corresponding interval u_j for which membership in A_j is supremum. $L[*A_j]$ and $U[*A_j]$ are the lower and the upper bounds of interval u_j respectively. $l[*A_j]$ is the length of interval $u_j$whose membership in A_j is supremum and $M[*A_j]$ is the mid value of the interval u_j in having supremum value A_j. A_i is the fuzzified enrollments of period n. A_j is the fuzzified enrollments of period $n+1$. E_i is the actual enrollments of period n. E_{i-1} is the actual enrollments of period $n-1$. E_{i-2} is the actual enrollments of period $n-2$. F_j is the crisp forecasted enrollments of the period $n+1$.

Algorithm: Forecasting enrollments for period $n+1$ and onwards.

For $k = 3$ to … K (end of time series data)

Obtained fuzzy logical relation for period k to $k+1$

$A_i \to A_j$

Compute

$D_i = \big| |(E_i - E_{i-1})| - |(E_{i-1} - E_{i-2})| \big|$

$X_i = E_i - D_i/2$

$XX_i = E_i + D_i/2$

$Y_i = E_i + D_i$

$YY_i = E_i - D_i$

For $I = 1$ to 4;
If $X_i \geqslant L[*A_j]$ and $X_i \leqslant U[*A_j]$
    Then $P_1 = X_i$; $n = 1$
Else $P_1 = 0$; $n = 0$
    Next I
If $XX_i \geqslant L[*A_j]$ and $XX_i \leqslant U[*A_j]$
    Then $P_2 = XX_i$; $m = 1$
Else $P_2 = 0$; $m = 0$
    Next I
If $Y_i \geqslant L[*A_j]$ and $Y_i \leqslant U[*A_j]$
    Then $P_3 = Y_i$; $o = 1$
Else $P_3 = 0$; $o = 0$
    Next I
If $YY_i \geqslant L[*A_j]$ and $YY_i \leqslant U[*A_j]$
    Then $P_4 = YY_i$; $p = 1$
Else $P_4 = 0$; $p = 0$
    $B = P_1 + P_2 + P_3 + P_4$
If $B = 0$
    Then $F_j = M(*A_j)$
Else $F_j = (B + M(*A_j))/(m + n + o + p + 1)$
    Next k

12.5 Numerical Application

As a numerical application, the proposed model is applied to the energy consumption date from 10 supermarkets which operates in Hatay, one of the 81 cities of Turkey in the Mediterranean Region. The data used in the model is shown in Table 12.4 in terms of kilowatts for two years period. An electricity distribution company wants to forecast the electricity demand of each supermarket for the next period to prioritize the supermarkets to construct the best subscription strategy.

Step 1. The historical data is deseasonalized by the ratio-to-moving-average method to remove the seasonality from the time series data. The moving average period is determined as 12 months. Due to the seasonal time period is even the centered moving averages of the two consecutive moving averages are computed. The actual values of each time period

Table 12.4 The historical electricity consumption data of supermarkets

Term	M1	M2	M3	M4	M5	M6	M7	M8	M9	M10
01.2010	4018	5149	6404	4535	4426	3943	6031	5190	5713	4894
02.2010	4712	4271	5613	4136	3848	3301	7580	5483	4992	3756
03.2010	4202	4393	5921	4216	4239	3847	6813	4795	5171	3603
04.2010	4370	4173	6001	4180	5002	3480	5262	4362	5277	3631
05.2010	5921	4592	5277	5693	5629	4267	5609	5339	6489	4811
06.2010	6747	6263	6651	6999	6592	5342	7311	5772	7590	5253
07.2010	8096	6079	7854	7646	7531	5840	8535	6171	8760	6184
08.2010	7731	7024	7841	7012	7534	6238	7651	6325	9313	5894
09.2010	7328	6449	4972	6761	7164	5780	6848	6136	8788	5645
10.2010	4937	5273	7187	4672	6016	4872	6090	4508	7179	3912
11.2010	4208	3929	3851	4474	4270	3457	6072	3853	5553	3390
12.2010	4199	5257	4228	4050	4691	3828	7823	4351	5499	3909
01.2011	3979	4491	5405	4536	5193	4269	5426	4853	6322	4986
02.2011	4662	4616	4741	4122	4519	3580	6800	5131	5262	3775
03.2011	4129	4753	4889	4184	4954	4152	6048	4496	5457	3605
04.2011	4290	4510	4951	4145	5838	3748	4528	4013	5535	3624
05.2011	5323	6831	5900	4437	6162	4410	6646	5024	6062	4766
06.2011	5944	6251	6777	5332	7032	5082	6819	5842	6619	4981
07.2011	8319	7607	8111	7067	8939	6296	6662	6115	8221	6827
08.2011	7880	7266	8602	7731	9788	6167	9079	6868	8693	6496
09.2011	6412	5653	7158	6051	9006	5631	8324	5642	7834	4952
10.2011	4297	4625	9986	4213	7568	4776	7420	4143	6412	3450
11.2011	3713	3438	5574	4084	5359	3352	7392	3547	4938	2983
12.2011	3659	4615	6082	3669	5867	3734	9501	4006	4909	3461

are divided by their corresponding moving averages. Then the medians of the ratios for each time period are calculated. Finally the seasonal indices are computed and the actual values are divided by their corresponding seasonal indices. The deseasonalized data for the electricity consumption of the supermarkets are given in Table 12.5.

Step 2. The universes of discourse are determined for each supermarket data as $U_1 = [4700, 5960]$, $U_2 = [3720, 6625]$, $U_3 = [5440, 8065]$, $U_4 = [4610, 6465]$, $U_5 = [5000, 7261]$, $U_6 = [4275, 4905]$, $U_7 = [5228, 8028]$, $U_8 = [4632, 5577]$, $U_9 = [6048, 7413]$ and $U_{10} = [3995, 5045]$.

Table 12.5 Deaseasonalized data

Term	M1	M2	M3	M4	M5	M6	M7	M8	M9	M10
01.2010	5427	6413	6564	4998	5001	4277	7175	5383	6066	4475
02.2010	5448	5245	6608	5021	5124	4285	7175	5401	6322	4588
03.2010	5453	5217	6903	5043	5293	4297	7386	5392	6252	4605
04.2010	5393	5167	7153	4999	5419	4297	7754	5457	6222	4568
05.2010	5836	3721	5441	6315	5876	4469	5723	5306	6924	4566
06.2010	5933	5526	6038	6440	6072	4851	7329	4920	7387	4752
07.2010	5500	5281	5748	5268	5517	4537	6700	5153	6759	4554
08.2010	5496	5267	5673	5268	5576	4562	6643	5125	6796	4558
09.2010	5491	5297	5596	5266	5633	4587	6580	5098	6819	4559
10.2010	5485	5327	5512	5263	5696	4610	6518	5071	6842	4559
11.2010	5457	5435	5495	5210	5752	4628	6530	5043	6835	4557
12.2010	5399	5529	5525	5091	5792	4623	6553	5033	6776	4544
01.2011	5375	5593	5540	4999	5868	4631	6455	5034	6713	4559
02.2011	5390	5668	5581	5004	6018	4647	6437	5054	6664	4611
03.2011	5358	5645	5700	5005	6186	4638	6557	5056	6598	4607
04.2011	5294	5584	5901	4957	6325	4628	6672	5020	6526	4559
05.2011	5247	5536	6083	4922	6433	4619	6781	4993	6468	4523
06.2011	5227	5515	6153	4906	6477	4615	6836	4980	6442	4506
07.2011	5651	6608	5936	4869	6548	4891	5230	5107	6343	5027
08.2011	5602	5449	6223	5808	7244	4510	7883	5565	6343	5024
09.2011	4805	4643	8056	4713	7081	4468	7998	4687	6079	3999
10.2011	4774	4672	7659	4746	7165	4520	7941	4660	6111	4021
11.2011	4815	4756	7953	4756	7219	4487	7950	4643	6078	4009
12.2011	4704	4854	7948	4612	7244	4509	7958	4634	6049	4023

Step 3. The universes of discourse are partitioned into equal length of 7 intervals: $u_1, u_2, \ldots, u_7$. For an example the intervals for the first supermarket are determined as $u_1 = [4700, 4880]$, $u_2 = [4880, 5060]$, $u_3 = [5060, 5240]$, $u_4 = [5240, 5420]$, $u_5 = [5420, 5600]$, $u_6 = [5600, 5780]$, $u_7 = [5780, 5960]$.

Step 4. The fuzzy sets A_i are determined as given in Table 12.6.

Step 5. The historical data are fuzzified and the fuzzy logical relationships are established.

Step 6. Linguistic expectations of five experts for the consumption of January, 2012 which are average and below average are fuzzified. The interval is determined by Eq. (12.5). For

Table 12.6 Linguistic variables

Symbol	Linguistic variable
A_1	Very low
A_2	Low
A_3	Below average
A_4	Average
A_5	Over average
A_6	High
A_7	Very high

Table 12.7 Forecasted data for each supermarket

Term	M1	M2	M3	M4	M5	M6	M7	M8	M9	M10
01.2012	5240	4876	6565	5405	5969	4516	6428	5037	6633	4445

an example the calculation of the interval for the first supermarket is given below:

$$u_{i5} = \left[\min_{x=1,2,\ldots,5} 5060; 5240, \max_{x=1,2,\ldots,5} 5240; 5420\right] = [5060, 5420]$$

Step 7. The computational algorithm is applied and the crisp forecasted enrollments of January 2012 are found as given in Table 12.7 for each supermarket.

12.6 Conclusion

The energy management consists of both supply-side and demand-side activities. Although in the early times, the demand side management was ignored but as the energy prices rise, demand side management has increased importance. Energy demand forecast is a critical issue for energy demand management which is systematic utility and government activities designed to change the amount of customer's use of energy.

Various techniques such as time series, regression, econometric models, ARIMA, simulation, artificial neural networks, particle swarm optimization and genetic algorithm, have been used to build demand forecasting models. Fuzzy set theory, which is known for its ability to deal with uncertainty, is used in combination with other methods to form models including fuzzy neural networks, fuzzy inference systems, and fuzzy regression models.

In this study, a novel fuzzy seasonal time series model is proposed to forecast the energy demand. A numerical example is illustrated with the data from 10 supermarkets that operates in Turkey.

As further study the same data can be used with other techniques, which are mentioned in the literature review section, to benchmark the predicting performance of the proposed methodology.

Bibliography

[1] Turban E., Sharda R., Delen D., *Decision Support and Business Intelligence Systems*, 2010. Prentice Hall.

[2] Zadeh L.A., Fuzzy sets. *Information and Control.* 1965; **8** (3), 338–353.

[3] Kahraman C., Gülbay M., Kabak Ö., Application of Fuzzy Sets in Industrial Engineering: A Topical Classification. In *Fuzzy Applications in Industrial Engineering.* (ed. Kahraman C), 2006; pp. 1–55. Springer-Verlag: Nederland.

[4] Kahraman C., Kaya İ., A fuzzy multicriteria methodology for selection among energy alternatives. *Expert Systems with Applications.* 2010; **37** (9), 6270–6281.

[5] Huarng K., Hui-Kuang T., The application of neural networks to forecast fuzzy time series. *Physica A: Statistical Mechanics and its Applications.* 2006; **363** (2), 481–491.

[6] Domañska D., Wojtylak M., Application of fuzzy time series models for forecasting pollution concentrations. *Expert Systems with Applications.* 2012; **39** (19), 7673–7679.

[7] Wang C.H., Hsu L.C., Constructing and applying an improved fuzzy time series model:Taking the tourism industry for example. *Expert Systems with Applications.* 2008; **34**, 2732–2738.

[8] Huarng K., Yu H.-K., A Type 2 fuzzy time series model for stock index forecasting. *Physica A: Statistical Mechanics and its Applications.* 2005; **353** (0), 445–462.

[9] Chen S.M., Hwang J.R., Temperature prediction using fuzzy time series. *IEEE Trans Syst Man Cybern B.* 2000; **30** (2), 263–275.

[10] CRA., *Primer on demand-side management, with an emphasis on price-responsive programs.* California: Charles River Associates 2005 Februrary 2005. Report No.: Contract No.: D06090.

[11] Suganthi L., Samuel A., Energy models for demand forecasting-A review. *Renewable and Sustainable Energy Reviews.* 2012; **16**, 1223–1240.

[12] Bhattacharyya S.C., *Energy Economics: Concepts, Issues, Markets and Governance.* 2011. Springer.

[13] Soldo B., Forecasting natural gas consumption. *Applied Energy.* 2012; **92**, 26–37.

[14] Maggio G., Cacciola G., A variant of the Hubbert curve for world oil production forecasts. *Energy Policy.* 2009; **37** (11), 4761–4770.

[15] Valero A., Valero A., Physical geonomics: combining the exergy and Hubbert peak analysis for predicting mineral resources depletion. *Resources Conserv. Recycl.* 2010; **54** (12), 1074–1083.

[16] Forouzanfar M., Doustmohammadi A., Menhaj M.B., Hasanzadeh S., Modeling and estimation of the natural gas consumption for residential and commercial sectors in Iran. *Appl. Energy.* 2010; **87** (1), 268–274.

[17] Erdoğdu E., Natural gas demand in Turkey. *Appl. Energy.* 2010; **87** (1), 211–219.

[18] Jiang B., Wenying C., Yuefeng Y., Lemin Z., Victor D., The future of natural gas consumption in Beijing, Guangdong and Shanghai: An assessment utilizing MARKAL. *Energy Policy.* 2008; **36** (9), 3286–3299.

[19] Sailor D., Munoz R., Sensitivity of electricity and natural gas consumption to climate in the USA-methodology and results for eight states. *Energy.* 1997; **22** (10), 987–998.

[20] Vondracek J., Pelikan E., Konar O., Cermakova J., Eben K., Maly M., A statistical model for the estimation of natural gas consumption. *Appl. Energy.* 2008; **85** (5), 362–370.

[21] Brabec M., Konar O., Maly M., Pelikan E., Vondracek J., A statistical model for natural gas standardized load profiles. *J. Roy. Statist. Soc. Series C: Appl. Statist.* 2009; **58** (1), 123–139.

[22] Aydinalp K.M., Ugursal V.I., Comparison of neural network, conditional demand analysis, and engineering approaches for modeling end-use energy consumption in the residential sector. *Appl. Energy.* 2008; **85** (4), 271–296.

[23] Kizilaslan R., Karlik B., Combination of neural networks forecasters for monthly natural gas consumption prediction. *Neural Network World.* 2009; **19** (2), 191–199.

[24] Yoo S.H., Lim H.J., Kwak S.J., Estimating the residential demand function for natural gas in Seoul with correction for sample selection bias. *Appl. Energy.* 2009; **86** (4), 460–465.

[25] Potocnik P., Thaler M., Govekar E., Grabec I., Poredos A., Forecasting risks of natural gas consumption in Slovenia. *Energy Policy.* 2007; **35**, 4271–4282.

[26] Dombayci O.A., The prediction of heating energy consumption in a model house by using artificial neural networks in Denizli-Turkey. *Adv. Eng. Softw.* 2010; **41** (2), 141–147.

[27] Thaler M., Grabec I., Poredos A., Prediction of energy consumption and risk of excess demand in a distribution system. *Physica A: Statistical Mechanics and its Applications.* 2005; **355**, 46–53.

[28] Potocnik P., Govekar E., Grabec I., Building forecasting applications for natural gas market. In *Natural gas research progress.* (ed. David N., Michel, T.), 2008; pp. 505–530. Nova Science Publishers: New York.

[29] Sanchez-Ubeda E.F., Berzosa A., Modeling and forecasting industrial end-use natural gas consumption. *Energy Economics.* 2007; **29** (4), 710–742.

[30] Gori F., Ludovisi D., Cerritelli P.F., Forecast of oil price and consumption in the short term under three scenarios: parabolic, linear and chaotic behaviour. *Energy.* 2007; **32**, 1291–1296.

[31] Himanshu A.A., Lester C.H., Electricity demand for Sri Lanka: a time series analysis. *Energy.* 2008; **33**, 724–739.

[32] Al-Shobaki S., Mohsen M., Modeling and forecasting of electrical power demands for capacity planning. *Energy Conversion and Management.* 2008; **49** (11), 3367–3375.

[33] Amarawickrama H.A., Hunt L.C., Electricity demand for Sri Lanka: a time series analysis. *Energy.* 2003; **33** (5), 724–739.

[34] Bianco V., Manca O., Nardini S., Electricity consumption forecasting in Italy using linear regression models. *Energy.* 2009; **34** (9), 1413–1421.

[35] Bessec M., Fouquau J., The non-linear link between electricity consumption and temperature in Europe: a threshold panel approach. *Energy Economics.* 2008; **30** (5), 2705–2721.

[36] Al-Ghandoor A., Al-Hinti I., Jaber J.O., Sawalha S.A., Electricity consumption and associated GHG emissions of the Jordanian industrial sector: empirical analysis and future projection. *Energy Policy.* 2008; **36** (1), 258–267.

[37] Lam J.C., Tang H.L., Li D.H.W., Seasonal variations in residential and commercial sector electricity consumption in Hong Kong. *Energy.* 2008; **33** (3), 513–523.

[38] Jónsson T., Pinson P., Madsen H., On the market impact of wind energy forecasts. *Energy Economics.* 2010; **32** (2), 313–320.

[39] Inglesi R., Aggregate electricity demand in South Africa: conditional forecasts to 2030. *Appl. Energy.* 2010; **87** (1), 197–204.

[40] Bhattacharyya S.C., Timilsina G.R., Modelling energy demand of developing countries: are the specific features adequately captured? *Energy Policy.* 2010; **38** (4), 1979–1990.

[41] Zhang M., Mu H., Li G., Ning Y., Forecasting the transport energy demand based on PLSR method in China. *Energy.* 2009; **34** (9), 1396–1400.

[42] Al-Ghandoor A. J.J.O., Al-Hinti I., Mansour I.M., Residential past and future energy consumption: potential savings and environmental impact. *Renewable and Sustainable Energy Reviews.* 2009; **13** (6-7), 1262–1274.

[43] Meng M., Niu D., Annual electricity consumption analysis and forecasting of China based on few observations methods. *Energy Conversion and Management.* 2011; **52**, 953–957.

[44] Ediger V.S., Akar S., ARIMA forecasting of primary energy demand by fuel in Turkey. *Energy Policy.* 2007; **35** (3), 1701–1708.

[45] Erdoğdu E., Electricity demand analysis using co-integration and ARIMA modeling: a case study of Turkey. *Energy Policy.* 2007; **35**, 1129–1146.

[46] Pappas S.S., Ekonomou L., Karamousantas D.C., Chatzarakis G.E., Katsikas S.K., Liatsis P., Electricity demand loads modeling using auto regressive moving average (ARMA) models. *Energy.* 2008; **33**, 1353–1360.

[47] Azadeh A., Ghaderi S.F., Sohrabkhani S., A simulated-based neural network algorithm for forecasting electrical energy consumption in Iran. *Energy Policy.* 2008; **36** (7), 2637–2644.

[48] Azadeh A., Tarverdian S., Integration of genetic algorithm, computer simulation and design of experiments for forecasting electrical energy consumption. *Energy Policy.* 2007; **35** (10), 5229–5241.

[49] Kavaklıoğlu K., Ceylan H., Ozturk H.K., Canyurt O.E., Modeling and prediction of Turkey's electricity consumption using artificial neural networks. *Energy Convers. Manage.* 2009; **50**, 2719–2727.

[50] Geem Z.W., Roper W.E., Energy demand estimation of South Korea using artificial neural network. *Energy Policy.* 2009; **37** (10), 4049–4054.

[51] Ekonomou L., Greek long-term energy consumption prediction using artificial neural networks. *Energy.* 2010; **35**, 512–517.

[52] Canyurt O.E., Ozturk H.K., Application of genetic algorithm (GA) technique on demand estimation of fossil fuels in Turkey. *Energy Policy.* 2008; **36** (7), 2562–2569.

[53] Cankuş M., Taşkıran U., Turkey's Electricity Consumption Forecasting Using Genetic algorithm. *Energy Sources Part B.* 2011; **6**, 406–416.

[54] Ceylan H., Öztürk H.K., Estimating energy demand of Turkey based on economic indicators using genetic algorithm approach. *Energy Convers. Manage.* 2004; **45** (15-16), 2525–2537.

[55] Özçelik Y., Hepbaşlı A., Estimating petroleum energy production and consumption using a simulated annealing approach. *Energy Sources B: Econ. Plan Policy.* 2006; **1** (3), 255–265.

[56] Pai P.F., Hong W.C., Support vector machines with simulated annealing algorithms in electricity load forecasting. *Energy Conversion and Management.* 2005; **46** (17), 2669–2688.

[57] Kiran M.S., Ozceylan E., Gunduz M., Paksoy T., A novel hybrid approach based on Particle Swarm Optimization and Ant Colony Algorithm to forecast energy demand of Turkey. *Energy Conversion and Management Volume.* 2012; **53** (1), 75–83.

[58] Ünler A., Improvement of energy demand forecasts using swarm intelligence: the case of Turkey with projections to 2025. *Energy Policy.* 2008; **36**, 1937–1944.

[59] Al-Rashidi M.R., EL-Naggar K.M., Long term electric load forecasting based on particle swarm optimization. *Applied Energy.* 2010; **87**, 320–326.

[60] Hong W.C., Chaotic particle swarm optimization algorithm in a support vector regression electric load forecasting model. *Energy Conversion and Management.* 2009; **50** (1), 105–117.

[61] Toksarı M.D., Ant colony optimization approach to estimate energy demand of Turkey. *Energy Policy.* 2007; **35**, 3984–3990.

[62] Toksarı M.D., Estimating the net electricity energy generation and demand using the ant colony optimization approach: case of Turkey. *Energy Policy.* 2009; **37**, 1181–1187.

[63] Wang Q., Grey Prediction Model and Multivariate Statistical Techniques Forecasting Electrical Energy Consumption in Wenzhou, China. In *Second International Symposium on Intelligent Information Technology and Security Informatics.* 2009; pp. 167–170.

[64] Akay D., Atak M., Grey prediction with rolling mechanism for electricity demand forecasting of Turkey. *Energy.* 2007; **32** (9), 1670–1675.

[65] Lee S.C., Shih L.H., Forecasting of electricity costs based on an enhanced grey-based learning model: a case study of renewable energy in Taiwan. *Technological Forecasting and Social Change.* 2011; **78** (7), 1242–1253.

[66] Lee Y.S., Tong L.I., deneme. deneme. 1009; deneme.
[67] Ediger V.S., Tatlıdil H., Forecasting the primary energy demand in Turkey and analysis of cyclic patterns. *Energy Convers. Manage.* 2002; **43**, 473–487.
[68] Sarak H., Satman A., The degree-day method to estimate the residential heating natural gas consumption in Turkey: a case study. *Energy.* 2003; **28** (9), 929–939.
[69] Ceylan H., Ceylan H., Haldenbilen S., Başkan Ö., Transport energy modeling with meta-heuristic harmony search algorithm, an application to Turkey. *Energy Policy.* 2008; **36** (7), 2527–2535.
[70] Kanudia A., Labriet M., Loulou R., Vaillancourt K., Waaub J.P., The world-MARKAL model and its application to cost-effectiveness, permit sharing and cost benefit analyses. In *Energy and environment.* (ed. Loulou R. W.J., Zaccour G.), 2005; pp. 111–148. Springer.
[71] Rout U.K., Vo A., Singh A., Fahl U., Blesl M., Gallachóir B.P.O., Energy and emissions forecast of China over a long-time horizon. *Energy.* 2011; **36**, 1–11.
[72] Pai P.F., Hybrid ellipsoidal fuzzy systems in forecasting regional electricity loads. Energy Conversion and Management 2006; **47** (15-16), 2283–2289.
[73] Song K.B., Baek Y.S., Hong D.H., Jang G., Short-term load forecasting for the holidays using fuzzy linear regression method. *IEEE Transactions on Power Systems.* 2005; **20**, 96–101.
[74] Azadeh A., Saberi M., Seraj O., An integrated fuzzy regression algorithm for energy consumption estimation with non-stationary data: A case study of Iran. *Energy.* 2010; **35** (6), 2351–2366.
[75] Che J., Wang J., Wang G., An adaptive fuzzy combination model based on self-organizing map and support vector regression for electric load forecasting. *Energy.* 2012; **37** (1), 657–664.
[76] Kucukali S., Baris K., Turkey's short-term gross annual electricity demand forecast by fuzzy logic approach. *Energy Policy.* 2010; **38**, 2438–2445.
[77] Chang P.C., Fan C.Y., Monthly electricity demand forecasting based on a weighted evolving fuzzy neural network approach. *Int. J. Electr. Power Energy Syst.* 2011; **33** (1), 17–27.
[78] Viet N., Mandziuk J., Neural and fuzzy neural networks in prediction of natural gas consumption. *Neural Parallel Scientific Comput.* 2005; **13** (3-4), 265–286.
[79] Behrouznia A., Saberi M., Azadeh A., Asadzadeh S.M., Pazhoheshfar P., An adaptive network based fuzzy inference system-fuzzy data envelopment analysis for gas consumption forecasting and analysis: the case of South America. In *International conference on intelligent and advanced systems*; ICIAS 2010. 2010.
[80] Li K., Su H., Forecasting building energy consumption with hybrid genetic algorithm-hierarchical adaptive network-based fuzzy inference system. *Energy and Buildings.* 2010; **42** (11), 2070–2076.
[81] Azadeh A., Asadzadeh S.M., Ghanbari A., An adaptive network-based fuzzy inference system for short-term natural gas demand estimation: uncertain and complex environments. *Expert Systems with Applications.* 2010; **38** (3), 2224–2234.
[82] Mamlook R., Badran O., Abdulhadi E. A fuzzy inference model for short-term load forecasting. *Energy Policy.* 2009; **37** (4), 1239–1248.
[83] Ying L.C., Pan M.C., Using adaptive network based fuzzy inference system to forecast regional electricity loads. *Energy Conversion and Management.* 2008; **49** (2), 205–211.
[84] Lau H.C.W., Cheng E.N.M., Lee C.K.M., Ho G.T.S., A fuzzy logic approach to forecast energy consumption change in a manufacturing system. *Expert Systems with Applications.* 2008; **34** (3), 1813–1824.
[85] Nguyen HT, Wu B., *Fundamentals of statistics with fuzzy data*, 2006. Springer.
[86] Kahraman C., Yavuz M., Kaya İ., Fuzzy and Grey Forecasting Techniques and Their Applications in Production Systems, *Production Engineering and Management under Fuzziness.* (eds. Kahraman C., Yavuz M.), 2010; pp. 1–24. Springer Berlin / Heidelberg.
[87] Yolcu U., Egrioglu E., Uslu V.R., Basaran M.A., Aladag C.H., A new approach for determining the length of intervals for fuzzy time series. *Applied Soft Computing.* 2009; **9** (2), 647–651.

[88] Song Q., Chissom B.S., Fuzzy time series and its models. *Fuzzy Sets and Systems.* 1993; **54** (3), 269–277.

[89] Song Q., Chissom B.S., Forecasting enrollments with fuzzy time series — Part I. *Fuzzy Sets and Systems.* 1993; **54** (1), 1–9.

[90] Song Q., Chissom B.S., Forecasting enrollments with fuzzy time series — Part II. *Fuzzy Sets and Systems.* 1994; **62** (1), 1–8.

[91] Chen S.-M., Forecasting enrollments based on fuzzy time series. *Fuzzy Sets and Systems.* 1996; **81** (3), 311-319.

[92] Huarng K., Heuristic models of fuzzy time series for forecasting. *Fuzzy Sets and Systems.* 2001; **123** (3), 369–386.

[93] Chen S.-M., Forecasting Enrollments Based on High-Order Fuzzy Time Series. *Cybernetics and Systems.* 2002; **33** (1), 1–16.

[94] Lee H.-S., Chou M.-T., Fuzzy forecasting based on fuzzy time series. *International Journal of Computer Mathematics.* 2004; **81** (7), 781–789.

[95] Huarng K., Yu H.K., A Type 2 fuzzy time series model for stock index forecasting. *Physica A: Statistical Mechanics and its Applications.* 2005; **353**, 445–462.

[96] Own C.-M., Yu P.-T., Forecasting Fuzzy Time Series on a Heuristic High-Order Model. *Cybernetics and Systems.* 2005; **36** (7), 705–717.

[97] Yu H.-K., A refined fuzzy time-series model for forecasting. *Physica A: Statistical Mechanics and its Applications.* 2005; **346** (3-4), 657–681.

[98] Tsaur R.-C., O Yang J.-C., Wang H.-F., Fuzzy relation analysis in fuzzy time series model. *Computers & Mathematics with Applications.* 2005; **49** (4), 539–548.

[99] Li-Wei L., Li-Hui W., Shyi-Ming C., Yung-Ho L., Handling forecasting problems based on two-factors high-order fuzzy time series. *Fuzzy Systems, IEEE Transactions on.* 2006; **14** (3), 468–477.

[100] Chen S.-M., Chung N.-Y., Forecasting enrollments using high-order fuzzy time series and genetic algorithms. *International Journal of Intelligent Systems.* 2006; **21** (5), 485–501.

[101] Chen T.-L., Cheng C.-H., Jong Teoh H., Fuzzy time-series based on Fibonacci sequence for stock price forecasting. *Physica A: Statistical Mechanics and its Applications.* 2007; **380** (0), 377–390.

[102] Li S.-T., Cheng Y.-C., Deterministic fuzzy time series model for forecasting enrollments. *Computers & Mathematics with Applications.* 2007; **53** (12), 1904–1920.

[103] Singh S.R., A simple method of forecasting based on fuzzy time series. *Applied Mathematics and Computation.* 2007; **186** (1), 330-339.

[104] Cheng C.-H., Chen T.-L., Teoh H.J., Chiang C.-H., Fuzzy time-series based on adaptive expectation model for TAIEX forecasting. *Expert Systems with Applications.* 2008; **34** (2), 1126–1132.

[105] Liu H.-T., Wei M.-L., An improved fuzzy forecasting method for seasonal time series. *Expert Systems with Applications.* 2010; **37** (9), 6310–6318.

[106] Edwin M., *Statistics for Business and Economics: Methods and Applications*, 1994. Norton.

Chapter 13

Multi-objective Scheduling Problems for Re-entrant Shops

Atefeh Moghaddam, Farouk Yalaoui, and Lionel Amodeo

Charles Delaunay Institute (ICD-LOSI), University of Technology of Troyes, STMR, UMR CNRS 6279, 12, rue Marie-Curie, 10010, Troyes, France

Production scheduling problems are multi-objective by nature which means that several criteria, in conflict with each other, have to be considered at the same time. In this chapter, we study a bi-objective re-entrant permutation flowshop scheduling problem in which the jobs have strict due dates. If the manufacturer will not be able to respect any job's due date, that job shall be rejected whether by ignoring it completely or sub-contracting the job. In both cases, the company is charged a fee called "opportunity loss" or "out-sourcing cost". After a brief introduction, we present the existing research works in this area, then we describe the problem in more details by giving a mixed-integer linear programming model. Since the problem is NP-hard, a genetic algorithm coupled with a new dominance concept, different from Pareto, is developed to find a good estimation of non-dominated solutions set. Finally the computational results are discussed and some directions for future research are presented.

13.1 Introduction

Flowshop scheduling problems (FSP) have attracted a great deal of attention since last decades due to many industrial applications [1]. In classical FSP, we have a set of n jobs $(J_1, J_2, \ldots, J_n)$ to be processed on m machines $(M_1, M_2, \ldots, M_m)$. All jobs have the same process route. First they meet M_1, then M_2, and so on up to the last stage in which they are processed by M_m and finally leave the system. For such a problem, there are $(n!)^m$ different alternatives for sequencing the jobs on the machines [2].

Permutation flowshop scheduling problem (PFSP) is a special case of FSP in which the sequence of jobs on all machines is the same. In another word once the job sequence on first machine is decided, all other machines process the jobs in the same order. Therefore,

C. Kahraman (ed.), *Computational Intelligence Systems in Industrial Engineering*, Atlantis Computational Intelligence Systems 6, DOI: 10.2991/978-94-91216-77-0_13,

in PFSP the number of alternatives is reduced to $(n!)$. Permutation schedules are preferred to non permutation in most real systems because of the ease of implementation and material flow management. In many industrial cases, because of technical constraints of material handling systems, non-permutation schedules are not feasible [3].

As it is described before, in FSP/PFSP, each job visits each machine one time, however, in industrial applications, we can find the cases in which at least one machine or work station may be visited by some/all jobs more that once. This type of configuration, we call it re-entrant scheduling problem, causes many conflicts among the jobs visiting the machines in different cycles. There are two main reasons for visiting a work station more than once which are explained by giving two examples as follows: in semiconductor manufacturing industry, for fabrication of wafers or multi-layered printed circuit boards, the machines are usually able to do more than one process on each job or even the same process shall be done more than once so each job visits each work station several times. In this case, usually the number of visiting each work station by each job (which is called cycle, layer or level) is constant. On the other hand, in a production line, for quality assurance, some work stations may be dedicated to detect defective items and then make appropriate treatment to them. Unlike deterministic re-entrant flow where jobs visit each work station in a fixed number of cycles, in this case, re-entrant process is subjected to certain probabilities based on inspection results (especially for re-works).

In this chapter we study a deterministic re-entrant permutation flowshop scheduling problem (RPFSP) which is defined in more details later.

In today's competitive manufacturing environment, one of the key survival factors for an organization is customers' satisfaction. The customers usually expect to get their orders in high quality at the right time with reasonable price. One of the main objectives of the scheduling is the completion of all jobs before their due dates. Failure to keep up this promise has negative effects on the credibility of the company. Regardless of traditional scheduling problems in which all jobs are assumed to be processed and the best sequence is found regarding a scheduling criterion, here the scheduler may be obliged to reject some of the jobs if he cannot meet their strict due dates. The idea is to build a schedule in a way to minimize rejection cost from one side and also total weighted completion time on the other side. Since the rejection of a job is due to its late completion time, and the completion time is directly related to the jobs' sequence, these different but interrelated decisions shall be made simultaneously.

13.2 Literature Review

Re-entrant scheduling problem has been studied for the first time by Graves *et al.* [4]. Since then, this problem attracts many researches due to its large applications in production environment such as wafer fabrication [5], textile manufacturing [6] and weapon production [7]. Semiconductor field is one of the most famous areas of re-entrant scheduling problem which has been investigated by many researches due to its complexity. Extensive surveys of different scheduling models and problems involving re-entrant configuration can be found in Uzsoy *et al.* [5] and Gupta and Sivakumar [8].

Generally, the existing research works in re-entrant scheduling are different one to another regarding problem definition or proposed solution techniques.

In some of the manufacturing systems the process routing for each job is deterministic and the number of cycles is fixed [9–14]. This is the case in which re-entry is happened because of process requirements. Choi and Kim [6] studied a system in which each job visits each machine two times. While Chen *et al.* [12], and Choi and Kim [14] considered a more general case with l cycles. In second group, re-entry happens after quality control station. In all manufacturing systems, quality control is an essential task for tracking products quality. Unlike deterministic re-entrant flow where jobs visit each work station in a fixed number of cycle, in this group, re-entrant process is subjected to certain probabilities based on inspection results and manufacturing policy; so the number of re-visiting of each work station is dynamic [15–17].

Based on shop layout, machines' position and process routing of the jobs, we can categorize most of the studies on re-entrant scheduling problem into two main groups: flowshop and job shop. In flowshop scheduling, there are m stages, all the jobs have the same process routing which starts from stage one and ends with stage m [6, 12]. Some researchers studied RPFSP where apart from the same process routing for all jobs, all machines have also the same job sequence [10, 11, 13, 14]. In hybrid flowshop configuration, at least one stage has several parallel machines. Choi S.-W. *et al.* [18], Choi H.-S. *et al.* [19], Dugardin *et al.* [20] and Yalaoui *et al.* [21] studied the re-entrant scheduling problem in hybrid flowshop configuration with identical parallel machines, while Alfieri [22] investigated the same problem with non-identical machines and Bertel and Billaut [23] with uniform parallel machines.

In job shop scheduling problem, the process routing of each job is different from others. Hwang and Sun [24] studied two-stage job shop scheduling problem. In some researches [25–27] one machine at each stage was assumed while Mason *et al.* [28] and Liu and Wu [29] studied the case of several parallel machines in each stage. Chen J.C. *et al.* [7] studied flexible job-shop scheduling in which jobs had alternative operation sequences and also alternative machines could realize the operations.

Regarding objective function, the literature is heavily biased towards single-objective optimization due to the complexity of the problem. Therefore multi-objective scheduling for re-entrant configuration is a necessary venue of research that has not been explored extensively. To the best of our knowledge, the only previous works on multi-objective re-entrant scheduling problem are those of Dugardin *et al.* [20] and Cho *et al.* [30]. Dugardin *et al.* [20] integrated Lorenz dominance into NSGA-II to find an estimation of Pareto non-dominated solutions for a problem of re-entrant hybrid flowshop with makespan minimization and maximization of the utilization rate of the bottleneck. Cho *et al.* [30] proposed a genetic-based multi-objective algorithm for solving re-entrant hybrid flowshop scheduling problem with minimization of makespan and total tardiness. In single-objective re-entrant scheduling problem, most of the researches addressed makespan as the objective function [10, 13]. Re-entrant flowshop problems with other objective functions are less studied in the literature. Choi and Kim [6] and Choi *et al.* [18] studied re-entrant problems with the objective of minimizing total tardiness. However, very little work has been done in the exploration of re-entrant flowshop scheduling problem with the objective of minimizing total completion time [31].

Finally, based on solution methods proposed for solving re-entrant scheduling problems, the existing works can be categorized into two main groups: exact methods and approximate algorithms. In exact methods, the researches mostly proposed mathematical programming [10, 15] and branch and bound [6, 11] while in second group, heuristics [14, 27] and meta-heuristics [12, 13, 17, 20] were developed. Recently Danping and Lee [32] provided a classification of re-entrant shop scheduling and the corresponding research methodology related to each category of this problem.

The term "rejection" has been studied for the first time by Bartal *et al.* [33] Since then, the complexity of different types of scheduling problems with rejection has been fairly extensively studied [34–41] but same as re-entrant scheduling problems, very few works have been done with multi-objective optimization approach [42–45]. Recent literature survey on scheduling with rejection can be found on Moghaddam *et al.* [44] and Slotnick [46].

13.3 Problem Definition

We study a re-entrant permutation flowshop scheduling problem (RPFSP). Let $J = \{J_1, J_2, \ldots, J_n\}$ be the set of n jobs and $M = \{M_1, M_2, \ldots, M_m\}$ the set of m machines which are available at the beginning of scheduling time horizon. The processing of each job should start from machine 1, then machine 2, and at last finish on machine m for two times (cycles). So the processing route for each job will be $(M_1, M_2, \ldots, M_m, M_1, M_2, \ldots, M_m)$. Since the number of cycles is fixed to 2, we split each job into two sub-jobs with precedence relationships among them so the re-entrant flowshop problem with two cycles can be reduced into a non-re-entrant flowshop problem with $2n$ sub-jobs and precedence relationship.

As mentioned earlier, we only consider permutation schedules for the re-entrant flowshop scheduling problem (RPFS). We define a permutation schedule in the m-machine re-entrant flowshop problem as a schedule in which sequences of sub-jobs are the same through the m machines. In this system, for a problem with n jobs, m machines and two cycles, there are $(2n)!/2^n$ different alternatives for sequencing the jobs on the machines. Figure 13.1 shows the process flow for each job. In this configuration second cycle is shown by dash lines.

The processing time of job i on machine k at cycle l is denoted by p^i_{lk}, which is fixed and nonnegative. We suppose that during high-demand period, manufacturer may not be able to meet some orders' due dates so it is obliged to reject them. Weight value, due date and rejection penalty for job i are indicated as w_i, d_i and e_i, respectively. Consider a schedule S; if for job i in S, $C_i > d_i$, where C_i is completion time of job i, schedule S will not be a feasible solution so this job shall be rejected and removed from current schedule. If a job cannot be accepted, it is excluded from schedule upon arrival. It is assumed that the processing of each job cannot be interrupted. At any time, each job can be processed on at most one machine, and each machine can process at most one job.

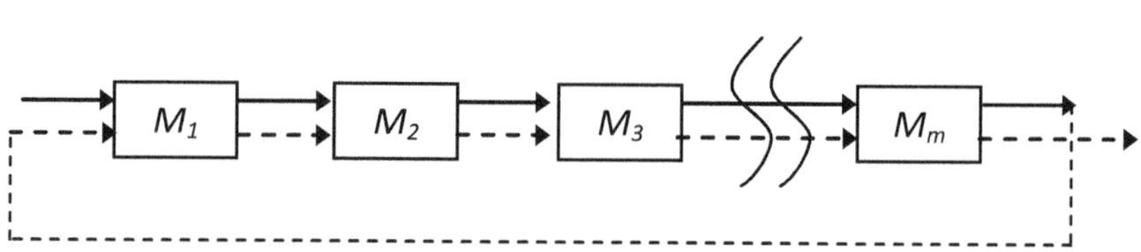

Fig. 13.1 Re-entrant flowshop scheme with two cycles

The aim is to select a set of jobs and schedule them in a way to respect their due dates while minimizing total weighted completion time of accepted jobs and total penalty of rejected jobs simultaneously. The first objective function is production oriented and the second one belongs to customer oriented objective function.

13.3.1 *Problem Formulation*

In the following, we present a mixed-integer linear programming (MILP) model for two-cycle re-entrant permutation flowshop scheduling problem with n jobs and m machines in which no delay is acceptable. We define two dummy jobs: 0 and $n+1$ which are positioned at the beginning and at the end of schedule respectively. For both dummy jobs, we consider just one cycle and their processing times on both machines are set to 0. If a job is rejected, it will be excluded from schedule and automatically its completion time is set to zero. In this case, total rejection cost increases. In contrary, by accepting a job, total weighted completion time will be increased.

Parameters

p^i_{lk} : processing time of job i on machine k at cycle l

w_i : weight value of job i

e_i : rejection cost of job i

d_i : due date for job i

Decision variables

C^i_{lk} : completion time for job i on machine k at cycle l

$$s_i = \begin{cases} 1 & \text{if job } i \text{ is accepted} \\ 0 & \text{otherwise} \end{cases}$$

$$z^{ii'}_{ll'} = \begin{cases} 1 & \text{if job } i \text{ at cycle } l \text{ precedes job } i' \text{ at cycle } l' \\ 0 & \text{otherwise} \end{cases}$$

Model

$$\text{Min}(F_1, F_2) \tag{13.1}$$

$$F_1 = \sum_{i=1}^{n} w_i C^i_{2m} \tag{13.2}$$

$$F_2 = \sum_{i=1}^{n} (1 - s_i)\, e_i \tag{13.3}$$

Subject to:

$$\sum_{l'=1}^{2}\sum_{i'=1}^{n+1} z_{ll'}^{ii'} = s_i \quad \forall i=0,\dots,n, \quad \forall l=1,2, \quad (i,l)\neq(i',l') \tag{13.4}$$

$$\sum_{l'=1}^{2}\sum_{i'=0}^{n} z_{l'l}^{i'i} = s_i \quad \forall i=1,\dots,n+1, \quad \forall l=1,2, \quad (i,l)\neq(i',l') \tag{13.5}$$

$$C_{lk}^{i} + p_{l'k}^{i'} z_{ll'}^{ii'} + d_i(z_{ll'}^{ii'} - 1) \leqslant C_{l'k}^{i'} \quad \forall i=0,\dots,n, \quad \forall i'=1,\dots,n+1, \quad i\neq i', \quad \forall k=1,\dots,m, \quad \forall l,l'=1,2 \tag{13.6}$$

$$\sum_{l'=l}^{2} z_{l'l}^{ii} = 0 \quad \forall i=1,\dots,n, \quad \forall l=1,2 \tag{13.7}$$

$$p_{lk}^{i} s_i \leqslant C_{lk}^{i} \quad \forall i=1,\dots,n, \quad \forall k=1,\dots,m, \quad \forall l=1,2 \tag{13.8}$$

$$C_{lk}^{i} \leqslant d_i s_i \quad \forall i=1,\dots,n, \quad \forall k=1,\dots,m, \quad \forall l=1,2 \tag{13.9}$$

$$C_{l(k+1)}^{i} \geqslant C_{lk}^{i} + p_{l(k+1)}^{i} s_i \quad \forall i=1,\dots,n, \quad \forall k=1,\dots,m-1, \quad \forall l=1,2 \tag{13.10}$$

$$C_{21}^{i} \geqslant C_{1m}^{i} + p_{21}^{i} s_i \quad \forall i=1,\dots,n \tag{13.11}$$

$$\sum_{i=1}^{n} s_i \geqslant 1 \tag{13.12}$$

$$s_0 = s_{n+1} = 1 \tag{13.13}$$

$$s_i \in \{0,1\} \quad \forall i=0,\dots,n \tag{13.14}$$

$$z_{ll'}^{ii'} \in \{0,1\} \quad \forall i=0,\dots,n, \quad \forall i'=1,\dots,n+1, \quad \forall l,l'=1,2 \tag{13.15}$$

$$C_{lk}^{i} \geqslant 0 \quad \forall i=0,\dots,n, \quad \forall k=1,\dots,m, \quad \forall l=1,2 \tag{13.16}$$

Since our problem is bi-objective one, we minimize total weighted completion time by Eq. (13.2) and total rejection cost by Eq. (13.3). Constraints (13.4) and (13.5) ensure that an accepted job precedes just one job and is succeeded by only one job. Constraint (13.6) guarantees that machine k can not start processing a job, before finishing the process of previous job in the schedule. Constraint (13.7) ensures the precedence requirement i.e. for each job, the second cycle will not be realized before the first cycle. Constraints (13.8) and (13.9) together guarantee that if the due date of a job cannot be met, that job will be rejected and its completion time on each machine at each cycle will be set as zero. So by these constraints, the model does not permit late scheduled jobs. Constraint (13.10) shows that each job must respect machines order, so when processing a job on previous machine is not finished, processing on next machine cannot be started. Constraint (13.11) ensures that processing job i on machine one at second cycle cannot be started since its processing on last machine (m) at first cycle has not been finished yet. Constraint (13.12) guarantees that at least one job will be accepted. In our problem a job will be rejected due to its late completion time (comparing to its strict due date) so if we have at least job i with $\sum_{k=1}^{m}\sum_{l=1}^{2} p_{lk}^{i} \leqslant d_i$, we should accept it.

Constraint (13.13) is for initializing the dummy jobs 0 and $n+1$. The binary and non-negativity restrictions on s_i, $z_{ll'}^{ii'}$ and C_{lk}^{i} are specified in (13.14)–(13.16).

Since minimization of total completion time in a two-machine flowshop that is the simpler variant of our problem is NP-hard [47], our problem is also NP-hard so it is impossible to find all exact non-dominated solutions with reasonable computational effort. In next section we propose a genetic-based algorithm to be able to find a good estimation of the set of true non-dominated solutions but before that, we give two properties of the problem as stated below which help the proposed algorithm to accelerate exploring search space:

Prop 13.3.1. In the problem of RPFS with strict due dates and rejection, if $\exists i \in \{1,2,\ldots,n\}$ with $\sum_{k=1}^{m}\sum_{l=1}^{2} p_{lk}^{i} > d_i$, then job i shall be rejected automatically.

Proof. If the total processing time of job i on all machines at both cycles is greater than its due date, then this job will be tardy regardless of its position in the schedule because in the best case, all the machines are available to process it. Even in this case, the summation of processing time is greater than the due date. □

Prop 13.3.2. In the problem of RPFS with strict due dates and rejection, if $\forall i \in \{1,2,\ldots,n\}$, $\sum_{i=1}^{n}\sum_{k=1}^{m}\sum_{l=1}^{2} p_{lk}^{i} < d_i$, then all jobs have to be accepted and the multi-objective optimization problem is reduced to single objective.

Proof. If the due date of each job is greater than total processing time of all jobs on all machines at both cycles, then all jobs shall be accepted. In other word, the jobs have loose due dates so even in the worst case, if job i is scheduled as the last job, it will not be a tardy one. Since the reason for rejecting a job is not to be able to respect its due date, no job will be rejected. So the rejection cost will be zero and the bi-objective optimization problem is reduced to single objective. □

Note. If Property 13.3.2 holds true, in the case of minimization of makespan with $m = 2$, the problem is polynomially solvable based on Johnson's rule.

13.4 Solution Method

Since many years ago, different meta-heuristic algorithms have been developed to find a good estimation of non-dominated solutions for multi-objective optimization problems. Excellent literature surveys on the topic of multi-objective evolutionary algorithms are provided by Coello *et al.* [48] and Deb [49]. During past decades, Genetic Algorithm (GA) has been successfully implemented for solving scheduling problems. As GA is population-based, it is able to find a set of solutions in a single run, so it can be adapted to multi-objective optimization problems easily.

In this part, we develop a GA-based algorithm inspired from NSGA-II proposed by Deb *et al.* [50] but with different dominance concept.

13.4.1 *Multi-objective Optimization*

In multi-objective optimization problems, usually we find a set of non-dominated solutions instead of a single optimum one. Without loss of generality, let us consider a bi-objective minimization problem (*P*) as stated below:

$$\text{Min } F(X) = (f_1(X), f_2(X))^{\mathrm{T}} \tag{13.17}$$

$$X \in D \tag{13.18}$$

where $X = (x_1, x_2, \ldots, x_m)^{\mathrm{T}}$ is the decision variable vector, f_i $(i = 1, 2)$ are the objective functions, and D is the feasible region.

Definition 13.1 (Pareto-dominance). In problem (P), solution X dominates solution Y based on Pareto properties ($X \prec_{\mathrm{P}} Y$) if *both* statements below are true:

1) $f_i(X) \leqslant f_i(Y), \forall i \in \{1,2\}$ and
2) $f_i(X) < f_i(Y), \exists i \in \{1,2\}$.

In another word, the solutions in Pareto optimal set can not be improved in any objective without causing degradation in at least one of the objectives. The image of Pareto optimal set in the objective function space is called the Pareto front.

In Fig. 13.2, the Pareto front for a bi-objective minimization problem is illustrated. In fact, based on the Pareto concept, non-of the solutions on the front has priority one to another. So the aim is to find the set of Pareto non-dominated solutions and give it to the decision maker in order to choose the one he prefers.

However, as it is shown in Fig. 13.2, we can partition the Pareto front into three regions 1, 2 and 3. Regarding the real industrial cases, we believe that more often the decision makers are interested to know the solutions placed in region 2 and they pay less attention to the solutions located in region 1 and 3. So we are motivated to define a new dominance concept to focus our search engine into this region.

Definition 13.2 (Lorenz-dominance). In problem (P), solution X dominates solution Y based on Lorenz properties ($X \prec_L Y$) if Lorenz vector of X dominates Lorenz vector of Y based on Pareto dominance ($L(X) \prec_P L(Y)$). Lorenz vector of solution X is calculated as $L(X) = (f_{(1)}(X), f_{(1)}(X) + f_{(2)}(X))$ in which $f_{(1)}(X) = \max(f_1(X), f_2(X))$.

Giving this definition, Lorenz non-dominated solution set is a sub-set of Pareto non-dominated solution set. For review of Lorenz-dominance concept, readers may refer to Kostreva *et al.* [51].

Dugardin *et al.* [20] integrated Lorenz dominance into NSGA-II and showed that they could find better estimation of non-dominated solutions for a bi-objective hybrid flowshop re-entrant scheduling problem.

Definition 13.3 (MYA-dominance). We define a new dominance concept called MYA which is inspired from r-dominance properties proposed by Ben Said *et al.* [52].

For each solution X, the weighted Euclidean distance from the ideal point $O = (0,0)$ proposed by Deb [53] is calculated as below:

$$\text{Dist}(X,O) = \sqrt{\sum_{i=1}^{2} w_i \left(\frac{f_i(X)}{f_i^{\max} - f_i^{\min}} \right)^2}, \quad w_i \in (0,1), \quad \sum_{i=1}^{2} w_i = 1 \tag{13.19}$$

where $f_i^{\max} = \max_{\text{Pop}}\{f_i\}$, $f_i^{\min} = \min_{\text{Pop}}\{f_i\}$ and Pop is the set of individuals that we want to evaluate their fitness (population).

In problem (P), solution X, MYA-dominates solution Y ($X \prec_{\text{MYA}} Y$) if one of the following statements holds true:

1) X dominates Y in Pareto sense ($X \prec_P Y$) or
2) X and Y are Pareto-equivalent and $D(X,Y,O) < -\eta$ where $\eta \in [0,1]$.

$$D(X,Y,O) = \frac{\text{Dist}(X,O) - \text{Dist}(Y,O)}{\text{Dist}_{\text{Pop}}^{\max} - \text{Dist}_{\text{Pop}}^{\min}} \tag{13.20}$$

$$\text{Dist}_{\text{Pop}}^{\max} = \underset{Z \in \text{Pop}}{Max}\{\text{Dist}(Z,O)\} \tag{13.21}$$

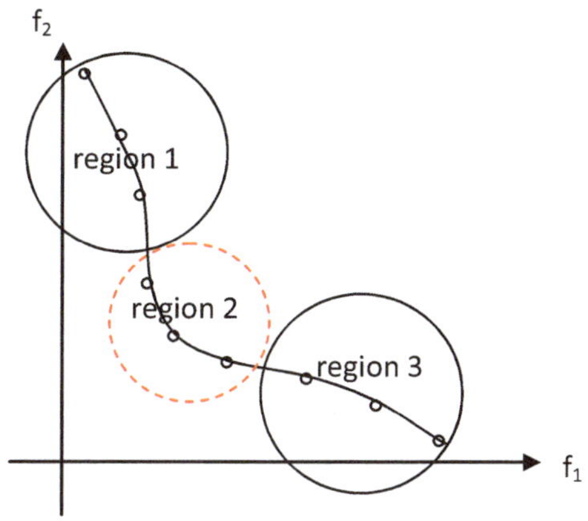

Fig. 13.2 Pareto front partitioning

$$\mathrm{Dist}_{\mathrm{Pop}}^{\min} = \underset{Z \in \mathrm{Pop}}{Min} \{\mathrm{Dist}(Z, O)\} \tag{13.22}$$

η is a threshold calculated as follows:

$$\eta = \frac{\mathrm{Dist}_{\mathrm{ParetoSet}}^{\max} - \mathrm{Dist}_{\mathrm{ParetoSet}-\{Q\}}^{\max}}{\mathrm{Dist}_{\mathrm{Pop}}^{\max} - \mathrm{Dist}_{\mathrm{Pop}}^{\min}} + \varepsilon \tag{13.23}$$

$$\mathrm{Dist}_{\mathrm{ParetoSet}}^{\max} = \underset{R \in \mathrm{ParetoSet}}{Max} \{\mathrm{Dist}(R, O)\} = \mathrm{Dist}(Q, O) \tag{13.24}$$

where ε is a tiny constant.

In contrary to r-dominance, in MYA-dominance the value of η changes dynamically from one generation to another generation and as defined by Eq. (13.23), this value is set in a way that the extrema in each Pareto non-dominated front will be dominated. This dominance concept is irreflexive, asymmetric and transitive [52].

13.4.2 *Proposed Algorithm*

As it is mentioned before, our proposed algorithm is based on NSGA-II coupled with MYA-dominance. Each module of the algorithm is described in details as follows:

13.4.2.1 *Solution Encoding*

We consider each cycle of jobs as sub-jobs. In our problem, since the number of cycles is fixed as 2, so for a problem with n jobs, we will have $2n$ sub-jobs which is considered as chromosome size. The position of genes in the chromosome shows the permutation of sub-jobs on the machine. We remind that the scheduling problem studied here is permutation flowshop with the definition given in part 13.3, so we need to define just the sequence of sub-jobs on first machine.

Figure 13.3 illustrates the sub-jobs' order on all machines for a problem with 3 jobs. As it is shown, each job is appeared twice. Each machine first processes first cycle of job 2, then first cycle of job 3, again waits for job 3 to process it for the second time, after finishing job 3, it processes first cycle of job 1, then second cycle of job 2 and finally second cycle of job 1. As it is shown in this example, regarding the objective functions, sometimes we may prefer to process both cycles of a job successively although machines idle time will be increased.

13.4.2.2 *Initial Population*

In most researches, the initial population is generated randomly to maximize the diversity of the solutions. But as it is the starting point, good initial population can help algorithm to explore the search space more systematically. So we try to generate it in a way to have both diverse and good individuals. For this aim, we generate the initial population partially in random base and partially by applying some heuristics as below:

- **IND 1:** the jobs are sorted in increasing order of their due dates (Earliest Due Dates-EDD) and both cycles of each job are executed successively.
- **IND 2:** for each cycle (sub-job), we calculate the due dates. For job i, $d_i' = d_i - \sum_{k=1}^{m} p_{2k}^i$ is a rough approximation of the due date for first cycle of job i. For second cycle, the due date is d_i which is given as input data. All sub-jobs are sorted in increasing order of their due dates.
- **IND 3:** we calculate total processing time of each sub-job on all machines ($\sum_{k=1}^{m} p_{1k}^i$ for first cycle and $\sum_{k=1}^{m} p_{2k}^i$ for second cycle). Then we sort all first-cycle-sub-jobs in increasing order (Shortest Processing Time-SPT) which is followed by the sorted second-cycle-sub-jobs in the same manner. In this case, we start processing the jobs for second time after finishing the first cycle of all jobs.

2	3	3	1	2	1

Fig. 13.3 Chromosome coding for a problem with 3 jobs

- **IND 4, 5:** total processing time for each job on all machines at both cycles is calculated ($\sum_{l=1}^{2} \sum_{k=1}^{m} p_{lk}^i$). The jobs are sorted based on SPT. In IND 4, both cycles of each job are executed successively while in IND 5, all jobs are processed for the first time and then the second cycles are started.

- **IND 6:** the jobs are sorted in decreasing order of their rejection costs and both cycles of each job are executed successively.
- **IND 7:** we consider each sub-job independently without considering the precedence constraint. For each sub-job, we calculate slop index as suggested by Palmer [54] in Eq. (13.25):

$$S_i = -\sum_{k=1}^{m}[m-(2k-1)]p_{lk}^{i} \quad \forall i = 1,2,\ldots,2n, \quad \forall l = 1,2 \tag{13.25}$$

  The sequence is then determined by ranking the sub-jobs on the basis of the slope index in decreasing order.
- **IND 8:** the same as IND 7, we calculate S_i index for all sub-jobs, then the first-cycle sub-jobs are sorted which are followed by second-cycle sorted ones.
- **IND 9:** here, we consider the number of machines as $2 \times m$, since there are two cycles. For each job, we calculate S_i based on Eq. (13.25) in which we replace m with $2m$. The first-cycle sub-jobs are sorted in decreasing order of S_i. We consider the same order for second-cycle sub-jobs which are placed after first cycle.
- The remaining individuals up to the population size are generated randomly.

After population initialization, we have to check whether each individual is a feasible schedule or not. A schedule is feasible if there is no tardy job among accepted jobs. If there is at least one tardy job, the chromosome needs to be repaired. We propose to reject the job which is scheduled before the first tardy job and has the least rejection cost. We repair the chromosome by rejecting the jobs up to the time we reach into a feasible schedule in which there is no tardy job.

13.4.2.3 *Fitness Function*

Fitness function can be considered as a criterion for classifying different multi-objective meta-heuristics. In some algorithms, the individuals are sorted based on Pareto dominance concept into each front such as NSGA-II [50] and SPEA2 [55]. The first front being completely non-dominant set in the current population and the second front being dominated by the individuals in the first front only and the front goes so on. To each individual in each front, a fitness value is assigned based on front in which it belongs to. Individuals in first front are given a fitness value of 1 and individuals in second are assigned fitness value as 2 and so on.

As it is mentioned, Pareto dominance concept has been considered in NSGA-II to assign a fitness value to each individual. However, in our proposed algorithm, we integrate MYA-dominance properties into GA in order to rank the individuals.

Since all individuals in same front have the same fitness value, a secondary criterion is defined for the individuals belonging to the same front. In NSGA-II, crowding distance has been introduced. The crowding distance is a measure of how close an individual is to its neighbors. In our algorithm, we consider another criterion called angle-based measure in order to emphasize on the knee solutions [56]. In Angle-based measure, the slopes of two lines through an individual and its two neighbours are calculated and then the angle between these slopes is considered as an indication (Fig. 13.4).

13.4.2.4 *Parent Selection*

In this step, we create a mating pool of individuals to which crossover and mutation operators will be applied. We set pool size equal to population size. The individuals belonging to the population set are sorted in increasing order of their fitness value. If two individuals have the same value, they are sorted in decreasing order of angle-based measure. Mating pool is filled with the individuals in that order. The idea is to apply crossover and mutation operators to each pair of fittest individuals.

For crossover, a parameter as crossover rate (p_c) is defined. For each pair of individuals in mating pool, a random number within 0 and 1 is generated. If this random number is less that crossover rate, crossover operator is applied to those individuals to produce the offspring.

13.4.2.5 *Crossover and Mutation Operators*

During two decades, different crossover operators have been proposed. Murata *et al.* [57] showed that the two-point crossover is effective for flowshop problems. Hence we use it in this study. Two random integer numbers are generated regarding the chromosome size. The genes between these two positions in each parent are copied to each offspring in the same positions. The remaining positions are filled regarding the order of gens in other parent.

Mutation is another operator of GA which prevents premature convergence and enlarges the area of search space. A parameter as mutation rate (p_m) is set between 0 and 1. For each individual in the population, a random number within 0 and 1 is generated. If this random number is less than mutation rate, this operator is applied to that individual. In our problem, we use exchange mutation operator (EM) in which two genes in a chromosome are selected randomly and their positions are swapped.

It is evident that after applying crossover and mutation, the offspring may not be feasible solutions so repairing will be needed.

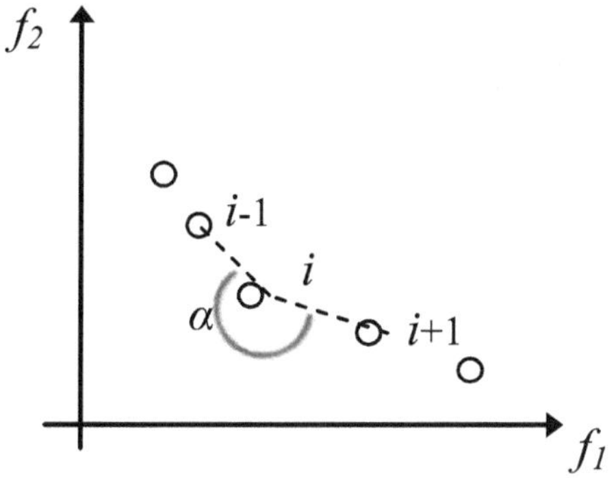

Fig. 13.4 Angle-based measure for solution *i*

13.4.2.6 *New Generation Selection*

In this step we need to choose a set of individuals from parents and offspring for next generation. For this purpose, parents and children create an integrated population set, and the fittest individuals in this set are survived for next generation.

13.4.2.7 *Termination rule*

In GA, more often, fixed number of generation is considered as stopping criterion. Here, we define adaptive generation size proposed by Tan *et al.* [58] as stopping criterion to assess the convergence through non-dominated front.

We call progress ratio $pr^{(n)}$ at generation n as the ratio between the number of non-dominated individuals at generation n (the solutions belongs to the rank 1) dominating any non-dominated individuals at generation $(n-1)$ over the total number of non-dominated individuals at generation n.

$$pr^{(n)} = \frac{\textit{nondom-indiv}^{(n)}\text{dominating } \textit{nondom-indiv}^{(n-1)}}{\textit{nondom-indiv}^{(n)}} \tag{13.26}$$

For tuning progress ratio at any generation n, we calculate the average of last t progress ratio in order to minimize the irregular value that may happen: $pr^{(n)} = (1/t)\sum_{i=n+1-t}^{n} pr^{(i)}$.

At generation n, the evolution is nearly converged if $pr^{(n)} \approx 0$, which could be used as the stopping criterion.

13.5 Computational Experiments

To test the algorithm, a number of problem instances are generated randomly. The processing time of each job on each machine at both cycles (p^i_{lk}) is generated from the discrete uniform distribution within a range of [1, 100]. The job weights (w_i) are uniformly generated in the interval [1, 20]. The job due dates have been randomly generated in the range of $[X(1-T-R/2), X(1-T+R/2)]$ where X denotes a lower bound for the makespan, parameter T is the tardiness factor while R stands for relative range of due dates. This is the standard method used in the scheduling literature to generate due dates for computational experiments.

In this study, we consider each sub-job independently and the value of X is calculated based on Eq. (13.27) proposed by Kim [59] and adapted for our problem:

$$X = 2 \times \max_{1<k<m} \left\{ \sum_{i=1}^{2n} p^i_k + \min_i \sum_{j=1}^{k-1} p^i_j + \min_i \sum_{j=k+1}^{m} p^i_j \right\} \tag{13.27}$$

We also set $T = 0.3$ and $R = 1.4$.

The rejection costs are generated based on the function $\exp(5 + \sqrt{a} \times b)$, where a is a random integer number within $[1, 80]$ and b is a random number within $[0, 1]$.

We conducted experiments on the problems with 20, 30, 40, 80, 100 and 140 sub-jobs and with 5, 10, 15, 20 machines to test the performance of the algorithm. Each experiment is repeated 5 times.

The algorithm is coded in Matlab and the tests are done on a workstation with CPU 2 GB RAM and processor 3 GHz.

13.5.1 *Performance Analysis*

To evaluate the performance of our proposed algorithm which is based on GA coupled with MYA-dominance concept and angle-based measure (MYA-NSGAII), we define 4 different metrics and compare the results with the GA, coupled with Lorenz dominance and crowding distance (L-NSGAII).

- Coverage of two sets (C) [60]: Let A, B be two approximate Pareto-optimal sets. $C_{A,B}$ is the ratio of solutions found by algorithm A which are (Pareto) dominated by at least one solution found by algorithm B. $C_{A,B} = 1$ implies that all solutions in A are dominated by at least one solution in B while the value $C_{A,B} = 0$ means none of the solutions in A is dominated by B. Note that both directions have to be considered because $C_{A,B}$ is not necessarily equal to $1 - C_{B,A}$.

- Hypervolume (H) [61]: Hypervolume calculates the volume of the objective space portion that is weakly dominated by non-dominated solutions found by an algorithm. This indicator needs the specification of a reference point that denotes an upper bound over all the objectives. For each objective function, we set maximum value among all non-dominated solutions found by both algorithms.
- Spacing metric (S) [62]: This metric measures how the solutions are evenly distributed. Let A be the final approximate Pareto optimal set obtained by one of the algorithms with N non-dominated solutions. The function S is defined by:

$$S_A = \frac{\sum_{i=1}^{N-1} |d_{\text{mean}} - d_i|}{(N-1) \times d_{\text{mean}}} \tag{13.28}$$

where d_i is the Euclidean distance between consecutive solutions in the set A, and d_{mean} is the average of these distances. A value of zero for this metric indicates all the non-dominated solutions found are equidistantly spaced in objective space.

- Spread metric (D): This metric measures spread of the solutions. For a final approximate Pareto optimal set A with n non-dominated solutions, D_A is defined as follows:

$$D_A = \max_{i \in A} \left(\max_{j \in A} (\|x_i - x_j\|) \right) \tag{13.29}$$

where $\|x_i - x_j\|$ is the Euclidean distance between two solutions x_i and x_j.

After conducting pilot runs, the tuned parameters resulted in good performance of L-NSGAII with reasonable computational time are as follows: Population size (pop_size): 50, Crossover probability (p_c): 0.9, Mutation probability (p_m): 0.2, No. of progress ration(t): 5, Stopping criterion $(pr^{(n)})$: 0.04.

MYA-NSGAII is applied to the test problems and its performance compared with L-NSGAII. Tables 13.1 and 13.2 list the average values of the comparison metrics and show the superiority of the proposed algorithm in each test problem.

In both tables, the first column shows the number of sub-jobs, and the second column number of machines. In Table 13.1, the third and forth columns indicated the coverage metrics. As it is shown, in all test problems, $C_{\text{L-NSGAII, MYA-NSGAII}} > C_{\text{MYA-NSGAII, L-NSGAII}}$ which implies that the quality of non-dominated solutions found by MYA-NSGAII is superior to the solutions found by L-NSGAII. In fifth column, hypervolume ration is reported.

Table 13.1 Comparing MYA-NSGAII and L-NSGAII regarding coverage and hypervolume metrics

No. of sub-jobs	No. of machines	Coverage		Hypervolume ratio
		C12	C21	H1/H2
20	5	0.00	0.43	1.10
	10	0.06	0.63	1.13
	15	0.04	0.13	1.20
	20	0.03	0.42	1.66
30	5	0.08	0.50	1.10
	10	0.03	0.17	1.04
	15	0.00	0.75	1.12
	20	0.07	0.43	1.20
40	5	0.05	0.67	1.06
	10	0.02	0.33	1.09
	15	0.06	0.59	1.10
	20	0.23	0.48	1.03
80	5	0.07	0.42	1.08
	10	0.15	0.37	1.18
	15	0.10	0.40	1.55
	20	0.01	0.46	1.26
100	5	0.05	0.35	1.47
	10	0.07	0.54	1.44
	15	0.12	0.44	1.45
	20	0.04	0.29	1.34
140	5	0.02	0.32	1.23
	10	0.12	0.18	1.49
	15	0.06	0.33	1.68
	20	0.08	0.30	1.50

* 1: MYA-NSGAII
2: L-NSGAII

Since in all test problems, the value of ratio is greater than one, it means that the objective space portion that is weakly dominated by Pareto solutions found by MYA-NSGAII is larger than the same portion found by L-NSGAII.

Regarding space metric, as reported in Table 13.2, in most of the test problems, the non-dominated solutions found by L-NSGAII are distributed more evenly. The reason may rely on the crowding distance module.

Finally the non-dominated front found by MYA-NSGA-II is stretched more than the front found by L-NSGAII.

In Fig. 13.5, three non-dominated fronts found by NSGA-II with MYA-dominance, NSGA-II with Lorenz dominance and the original one with Pareto dominance concept for a problem instance with 40 sub-jobs and 10 machines are shown. This Figure shows very

Table 13.2 Comparing MYA-NSGAII and L-NSGAII regarding space and spread metrics

No. of sub-jobs	No. of machines	Space		Spread	
		S1	S2	D1	D2
20	5	0.26	0.23	43126.66	24261.88
	10	0.34	0.37	65264.66	28194.97
	15	0.19	0.17	65228.56	37115.69
	20	0.21	0.47	145702.39	50090.38
30	5	0.63	0.56	81474.41	20509.71
	10	0.65	0.62	89056.83	51237.50
	15	0.64	0.65	121735.67	78416.41
	20	0.27	0.42	105310.18	57288.16
40	5	0.52	0.32	135296.98	63507.82
	10	0.81	0.67	151594.96	58018.02
	15	0.42	0.50	190758.24	158143.75
	20	0.40	0.37	175631.01	136808.22
80	5	0.51	0.62	469085.19	214036.12
	10	0.48	0.30	784093.56	198596.48
	15	0.56	0.38	603471.81	90120.80
	20	0.50	0.56	402420.36	151106.33
100	5	0.73	0.59	735985.52	168545.72
	10	0.47	0.43	1097438.47	212724.15
	15	0.54	0.46	1021633.44	156650.70
	20	0.56	0.51	1165857.85	337121.30
140	5	0.62	0.68	1700274.18	398155.80
	10	0.52	0.56	2210142.21	339823.59
	15	0.65	0.36	2226866.53	282687.56
	20	0.57	0.55	2524498.10	415304.32

* 1: MYA-NSGAII
2: L-NSGAII

well how both Lorenz and MYA dominance focus their search in middle part of Pareto front.

13.6 Concluding Remarks

In this chapter, we studied a bi-objective re-entrant permutation flowshop scheduling problem with rejection. Due to the customers' strict duedates, manufacturer was obliged to reject an order if he was not able to respect its due date. The objective functions were minimization of total weighted completion time of accepted jobs and minimization of total penalty of rejected jobs.

We proposed a new dominance concept based on the one proposed by Ben Said *et al.* [52]. In contrary to the most of researches done in multi-objective optimization, in

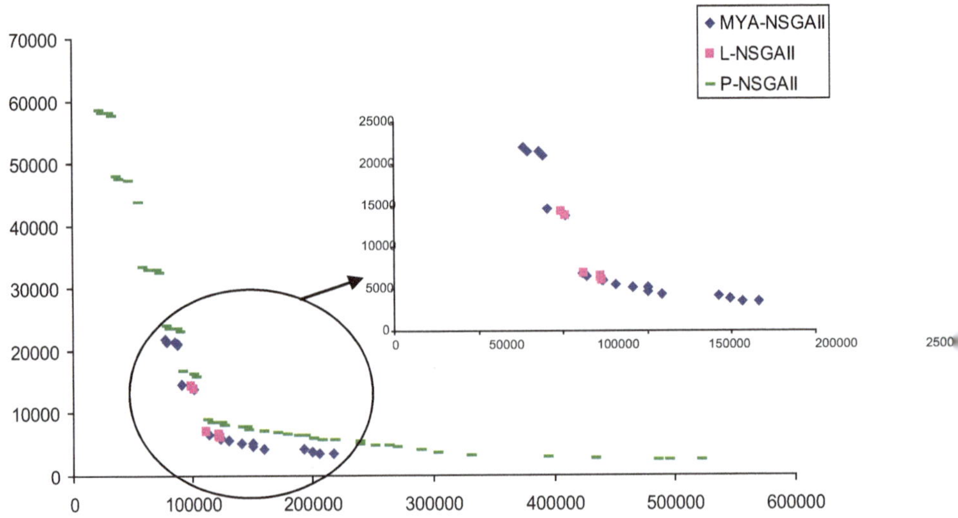

Fig. 13.5 The non-dominated fronts found by three different algorithms for a problem with 40 sub-jobs and 10 machines

which finding all non-dominated solutions is the main objective, in this study we focused our search to the middle part of Pareto front as we believe that this area is of more interest for decision maker.

The computational results confirm that the proposed algorithm works well for our problem.

Studying more general case of the problem such as hybrid flowshop re-entrant configuration with more layers and integrating the proposed ideas into other meta-heuristic approaches rather than genetic would be interesting for further investigations.

Bibliography

[1] Gupta, J.N.D. and Stafford Jr., E.F. (2006). Flowshop scheduling research after five decades, *European Journal of Operational Research*, **169** (3), pp. 699–711.

[2] Baker, K.R. (1974). *Introduction to sequencing and scheduling*, Wiley: New York.

[3] Choi, S.-W., and Kim, Y.-D. (2007). Minimizing makespan on a two-machine re-entrant flowshop, *Journal of the Operational Research Society*, **58** (7), pp. 972–981.

[4] Graves, S.C., Meal, H.C., Stefek, D., and Hamid, A. (1983). Scheduling of re-entrant flow shop, *Journal of Operations Management*, **3** (4), pp. 197–207.

[5] Uzsoy, R., Lee, C.Y. and Martin-Vega, L.A. (1992). A review of production planning and scheduling models in the semiconductor industry. I: System characteristics, performance evaluation and production planning. *IIE Transactions*, **24** (4), pp. 47–60.

[6] Choi, S.-W., and Kim, Y.-D. (2009). Minimizing total tardiness on a two-machine re-entrant flowshop. *European Journal of Operational Research*, **199**, pp. 375–384.
[7] Chen, J.C., Chen, K.H., Wu, J.J., and Chen, C.W. (2008). A study of the flexible job shop scheduling problem with parallel machines and re-entrant process, *International Journal of Advanced Manufacturing Technology*, **39** (3-4), pp. 344–354.
[8] Gupta, A.K. and Sivakumar, A.I. (2006). Job shop scheduling techniques in semiconductor manufacturing, *International Journal of Advanced Manufacturing Technology*, **27** (11-12), pp. 1163–1169.
[9] Narahari, Y. and Khan, L.M. (1996). Performance analysis of scheduling policies in re-entrant manufacturing Systems, *Computers and Operations Research*, **23** (1), pp. 37–51.
[10] Pan, J.C.-H. and Chen, J.-S. (2003). Minimizing makespan in re-entrant permutation flowshops, *Journal of the Operational Research Society*, **54** (6), pp. 642–653.
[11] Chen, J.-S. (2006). A branch and bound procedure for the reentrant permutation flow-shop scheduling problem, *International Journal of Advanced Manufacturing Technology*, **29** (11), pp. 1186–1193.
[12] Chen, J.-S., Pan, J.C.-H., and Lin, C-M. (2008a). A hybrid genetic algorithm for the re-entrant flow-shop scheduling problem. *Expert Systems with Applications*, **34** (1), pp. 570–577.
[13] Chen, J.-S., Pan, J.C.-H. and Wu, C.-K. (2008b). Hybrid tabu search for re-entrant permutation flow-shop scheduling problem. *Expert Systems with Applications*, **34** (3), pp. 1924–1930.
[14] Choi, S.-W. and Kim, Y.-D. (2008). Minimizing makespan on an m-machine re-entrant flowshop. *Computers & Operations Research,* **35**, pp. 1684–1696.
[15] Rau, H., Chu, Y.H., and Cho, K.H. (2005). Layer modeling for the inspection allocation problem in re-entrant production systems. *International Journal of Production Research*, **43** (17), pp. 3633–3655.
[16] Kumar, S. (2008). Modeling of a probabilistic re-rntrant line bounded by limited operation utilization time. *Journal of Industrial and Systems Engineering*, **2** (1), pp. 28–40.
[17] Rau, H. and Cho, K. H. (2009). Genetic algorithm modeling for the inspection allocation in re-entrant production systems, *Expert Systems with Applications*, **36** (8), pp. 11287–11295.
[18] Choi, S.-W., Kim, Y.-D., and Lee, G-C. (2005). Minimizing total tardiness of orders with reentrant lots in a hybrid flowshop, International Journal of Production Research, **43** (11), pp. 2149–2167.
[19] Choi, H.-S., Kim, H.-W., Lee, D.-H., Yoon, J., Yun, C., and Chae, K. (2009). Scheduling algorithms for two-stage reentrant hybrid flow shops: minimizing makespan under the maximum allowable due dates, *International Journal of Advanced Manufacturing Technology*, **42** (9), pp. 963–973.
[20] Dugardin, F., Yalaoui, F., and Amodeo, L. (2010). New multi-objective method to solve reentrant hybrid flow shop scheduling problem, *European Journal of Operational Research*, **203**, pp. 22–31.
[21] Yalaoui, N., Dugardin, F., Yalaoui, F., Amodeo, L., and Mahdi, H. (2010). Fuzzy project scheduling", in "*Production Engineering and Management under Fuzziness*", edited by Kahraman C. and Yavuz M., Springer Publishing, ISBN 978-3-642-12051-0, Production Engineering and Management under Fuzziness Studies in Fuzziness and Soft Computing, **252**, pp. 143–170.
[22] Alfieri, A. (2009). Workload simulation and optimisation in multi-criteria hybrid flowshop scheduling: a case study, International Journal of Production Research, **47** (18), pp. 5129–5145.
[23] Bertel, S. and Billaut, J.-C. (2004). A genetic algorithm for an industrial multiprocessor flow shop scheduling problem with recirculation, *European Journal of Operational Research*, **159** (3), pp. 651–662.
[24] Hwang, H. and Sun, J.U. (1998). Production sequencing problem with re-entrant work flows and sequence dependent setup times, *International Journal of Production Research*, **36** (9), pp. 2435–2450.

[25] Low, C., Wu, T.-H. and Hsu, C.-M. (2005). Mathematical modelling of multi-objective job shop scheduling with dependent setups and re-entrant operations, *International Journal of Advanced Manufacturing Technology*, **27** (1), 181–189.

[26] Pan, J.C.-H. and Chen, J.-S. (2005). Mixed binary integer programming formulations for the reentrant job shop scheduling problem, *Computers &. Operations Research*, **32**, pp. 1197–1212.

[27] Sun, J.U. (2009). A genetic algorithm for a re-entrant job-shop scheduling problem with sequence-dependent setup times, *Engineering Optimization*, **41** (6), pp. 505–520.

[28] Mason, S.J., Fowler, J.W., and Carlyle, W.M. (2002). A modified shifting bottleneck heuristic for minimizing total weighted tardiness in complex job shops, *Journal of Scheduling*, **5** (3), pp. 247–262.

[29] Liu, M. and Wu, C. (2004). Genetic algorithm using sequence rule chain for multi-objective optimization in re-entrant micro-electronic production line, *Robotics and Computer-Integrated Manufacturing*, **20** (3), pp. 225–236.

[30] Cho, H.M., Bae, S.J., Kim, J., and Jeong, I.J. (2011). Bi-objective scheduling for re-entrant hybrid flow shop using Pareto genetic algorithm, *Computers & Industrial Engineering*, **61** (3), pp. 529–541.

[31] Jing, C., Huang, W., and Tang, G. (2011). Minimizing total completion time for re-entrant flow shop scheduling problems, *Theoretical Computer Science*, **412** (48), pp. 6712–6719.

[32] Danping, L. and Lee, C.K.M. (2011). A review of the research methodology for the re-entrant scheduling problem. *International Journal of Production Research*, **49** (8), pp. 2221–2242.

[33] Bartal, Y., Leonardi, S., Marchetti-Spaccamela, A., Sgall, J., and Stougie, L. (2000). Multiprocessor scheduling with rejection, *SIAM Journal on Discrete Mathematics*, **13**, pp. 64–78.

[34] He, Y. and Min, X. (2000). On-line uniform machine scheduling with rejection, *Computing*, **65**, 1–12.

[35] Seiden, S. (2001). Preemptive multiprocessor scheduling with rejection, *Theoretical Computer Science*, **262**, pp. 437–458.

[36] Epstein, L., Noga, J., and Woeginger, G.J. (2002). Online scheduling of unit time jobs with rejection: minimization the total completion time, *Operations Research Letter*, **30**, pp. 415–420.

[37] Engels, D.W., Karger, D.R., Kolliopoulos, S.G., Sengupta, S., Uma, R.N., and Wein, J. (2003). Techniques for scheduling with rejection, *Journal of Algorithms*, **49**, pp. 175-191.

[38] Hoogeveen, H., Skutella, M., and Woeginger, G.J. (2003). Preemptive scheduling with rejection, *Mathematical Programming*, **94**, pp. 361–374.

[39] Dòsa, G. and He, Y. (2006). Scheduling with machine cost and rejection, *Journal of Combinatorial Optimization*, **12**, pp. 337–350.

[40] Lu, L., Cheng, T.C.E., Yuan, J., and Zhang, L. (2009). Bounded single-machine parallel-batch scheduling with release dates and rejection, *Computers and Operations Research*, **36**, pp. 2748–2751.

[41] Lu, L., Ng, C.T., and Zhang, L. (2011). Optimal algorithms for single-machine scheduling with rejection to minimize the makespan, *International Journal of Production Economics*, **130**, pp. 153–158.

[42] Cao, Z., Wang, Z., Zhang, Y., and Liu, S. (2006). On several scheduling problems with rejection or discretely compressible processing times, *Lecture Notes in Computer Science*, **3959**, pp. 90–98.

[43] Moghaddam, A., Yalaoui, F., and Amodeo L. (2011). Lorenz versus Pareto dominance in a single machine scheduling problem with rejection, *Lecture Notes in Computer Science*, **6576**, pp. 520–534.

[44] Moghaddam, A., Amodeo L., Yalaoui, F., and Karimi, B. (2012), Single machine scheduling with rejection: minimizing total weighted completion time and rejection cost, *International*

Journal of Applied Evolutionary Computation, **3** (2), pp. 42–61.

[45] Shabtay, D. and Gasper, N. (2012). Two-machine flow-shop scheduling with rejection, *Computers and Operations Research*, **39** (5), pp. 1087–1096.

[46] Slotnick, S.A. (2011). Order acceptance and scheduling: A taxonomy and review, *European Journal of Operational Research*, **212** (1), pp. 1–11.

[47] Garey, M.R., Johnson, D.S., and Sethi, R. (1976). The complexity of flowshop and jobshop scheduling, *Mathematics of Operations Research*, **1** (2), pp. 117–129.

[48] Coello, C.A.C., Lamont, G.B., and Van Veldhuizen, D.A. (2007). Evolutionary algorithms for solving multi-objective problems. Springer, second edition.

[49] Deb, K. (2009). *Multi-objective optimization using Evolutionary Algorithms*. John Wiley & Sons, ISBN 0470743611.

[50] Deb, K., Pratap, A., Agarwal, S., and Meyarivan, T. (2002). A fast and elitist multiobjective genetic algorithm: NSGA-II. *IEEE Transaction on Evolutionary Computation*, **6** (2), pp. 182–97.

[51] Kostreva, M.M., Ogryczak, W., and Wierzbicki, A. (2004). Equitable aggregation and multiple criteria analysis. *European Journal of Operational Research*, **158**, pp. 362–377.

[52] Ben Said, L., Bechikh, S., and Ghédira, K. (2010). The r-Dominance: a new dominance relation for interactive evolutionary multicriteria decision making, *IEEE Transaction in Evolutionary Computation*, **14** (5), pp. 801–818.

[53] Deb, K., Sundar, J., Bhaskara, U., and Chaudhuri, S. (2006). Reference point based multi-objective optimization using evolutionary algorithms. *International Journal of Computational Intelligence Research*, **2** (3), pp. 273–286.

[54] Palmer, D.S. (1965). Sequencing jobs through a multi-stage process in the minimum total time – A quick method of obtaining a near optimum, *Journal of the Operational Research Society*, **16** (1), pp. 101–107.

[55] Zitzler, E., Laumanns, M., and Thiele, L. (2001). *SPEA2: improving the strength pareto evolutionary algorithm*. Swiss Federal Institute Technology: Zurich, Switzerland.

[56] Branke, J., Deb, K., Dierolf, H., and Osswald, M. (2004). Finding Knees in Multi-objective Optimization. *Lecture Notes in Computer Science*, **3242**, pp. 722–731.

[57] Murata, T., Ishibuchi, H., and Tanaka, H. (1996). Multi-objective genetic algorithm and its applications to flowshop scheduling. *Computers & Industrial Engineering*, **30** (4), pp. 957–968.

[58] Tan, K.C., Lee, T.H., and Khor, E.F. (2001). Evolutionary algorithms with dynamic population size and local exploration for multiobjective optimization. *IEEE Transactions on Evolutionary Computation*, **5** (6), pp. 565–588.

[59] Kim, Y.D. (1995). Minimizing total tardiness in permutation flowshops. *European Journal of Operational Research*, **85**, pp. 541–55.

[60] Zitzler, E., Thiele, L., Laumanns, M., Fonseca, C.M., and da Fonseca V., G. (2003). Performance assessment of multi-objective optimizers: an analysis and review. *IEEE Transactions on Evolutionary Computation*, **7**, 117–132.

[61] Zitzler, E. and Thiele, L. (1998). Multiobjective optimization using evolutionary algorithms: A comparative case study. *Lecture Notes in Computer Science*, **1498**, 292–301.

[62] Schott, J.R. (1995). *Fault tolerant design using single and multicriteria genetic algorithm optimization*. Masters thesis, Massachusetts Institute of Technology, Cambridge, MA.

Chapter 14

An Introduction to Self-Organizing Maps

Umut Asan and Secil Ercan

Department of Industrial Engineering, Istanbul Technical University, 34357, Macka, Istanbul, Turkey

E-mail: asanu@itu.edu.tr, ercansec@itu.edu.tr

As a special class of artificial neural networks the Self Organizing Map is used extensively as a clustering and visualization technique in exploratory data analysis. This chapter provides a general introduction to the structure, algorithm and quality of Self Organizing Maps and presents industrial engineering related applications reported in the literature.

14.1 Introduction

One of the apparent trends today in all disciplines of engineering sciences has the goal of developing computationally intelligent systems that "enable or facilitate intelligent behavior in complex and changing environments" [1]. The concepts, algorithms and models suggested by this new paradigm of knowledge-based processing should have the potential to solve real world problems, be able to learn from experience, have the capability of self-organization, and exhibit an ability to adapt to new situations [2]. One of the methods belonging to the field of computational intelligence which has proven to be a very powerful tool for data analysis is the neurobiologically inspired self-organizing map (SOM). The most common model of SOMs, also known as the Kohonen network, is the topology-preserving map proposed by the Finnish researcher Teuvo Kohonen in 1982 [3, 4]. The method is a special class of artificial neural networks and is used extensively as a clustering and visualization tool in exploratory data analysis. Some of the major practical application areas in which SOMs have been effectively applied are industrial instrumentation, pattern

C. Kahraman (ed.), *Computational Intelligence Systems in Industrial Engineering*, Atlantis Computational Intelligence Systems 6, DOI: 10.2991/978-94-91216-77-0_14,

recognition, data mining and processing, process control, robotics, telecommunications, medical applications, optimization, and product management.

The principal objective of SOMs is to transform a complex high-dimensional input space into a simpler low-dimensional (typically two-dimensional) discrete output space by preserving the relationships (i.e. topology) in the data, but not the actual distances [5, 6]. The spatial locations (i.e., coordinates) of the nodes in the output space are indicative of inherent statistical features contained in the input space [7]. In that respect, they are similar to the well-known dimension reduction techniques such as principal component analysis and multidimensional scaling. However, in comparison to these multivariate techniques, the nonparametric SOM procedures have a number of advantages. First, they do not make assumptions regarding the distributions of variables and nor do they require independence among variables [8, 9]. Second, they are easier to implement and are able to solve nonlinear problems of very high complexity [10]. Last, they more effectively cope with noisy and missing data [11], very small dimensionality [8] and samples of unlimited size [12].

Among the architectures and algorithms suggested for artificial neural networks, the SOM is trained using an unsupervised learning scheme. That means, unlike supervised networks, learning in SOM does not rely on predefined target outcomes that would guide the process. Thus, a form of learning by observation, rather than learning by examples takes place to discover the underlying hidden patterns in the data set [13]. In order to learn without a teacher, SOMs apply a competitive learning rule where the output nodes compete among themselves for the opportunity to represent distinct patterns within the input space [7]. During the learning, the feedforward nature of SOMs allows an information flow in only one direction, without looping or cycling, from the input nodes to the output nodes. Note that every node in the input layer is linked (with weights) to every node in the output layer, which makes an SOM a completely connected network [6].

The formation of an SOM involves three characteristic processes, which can be summarized as follows [7]:

(i) *Competition*: The output nodes (neurons) in a self-organizing map compete with each other to best represent the particular input sample. The success of representation is measured using a discriminant function, where an input vector is compared with the weight vector of each output node. The particular node with its connection weights most similar to the input sample is declared winner of the competition. There are a number of different functions to determine the winner, i.e. the best-matching unit (BMU) on the map. The most commonly used one is the Euclidean distance.

(ii) *Cooperation*: Similar to "[human] neurons dealing with closely related pieces of information are close together so that they can interact via short synaptic connections" [7], SOM is a topographic organization in which nearby locations in the output space represent inputs with similar properties. This is possible in the presence of neighborhood information. The winning node determines the spatial location of a neighborhood of cooperating nodes. These nodes, sharing common features, activate each other to learn something from the same input.

(iii) *Adaptation*: The weight vectors of the winner and its neighboring units in the map are adjusted in favor of higher values of their discriminant functions. Through this learning process the relevant nodes become more similar to the input sample. Thus, nodes which have a strong response to a particular piece of input data will have an increased chance of responding to similar input data in the future.

These processes are repeated for all training samples as they are passed sequentially to the SOM. In this way, different locations on the map are trained to produce a strong response to different kinds of data [14]. Later, we will examine in detail an example of how these processes are carried out.

The aim of this chapter is to provide a general introduction to self-organizing maps and give an overview of the technique. It is not the intention of this chapter to give all theoretical details of the SOM algorithm, which can be found in Kohonen [4]. The material presented in the rest of this chapter is organized as follows. After providing a brief overview of the general structure of an SOM, the chapter continues with a detailed description and an illustrative example of the basic version of the SOM algorithm. In Section 3, the validation of SOMs is discussed and important quality measures proposed in the literature are presented. The following section focuses on the applications of the SOM algorithm relevant to industrial engineering. The chapter concludes with some final remarks in Section 14.5.

14.2 Self-Organizing Map

14.2.1 *Structure*

The SOM network typically consists of two layers of nodes, the input layer and the output layer, see Figure 14.1. Different than most of the other neural networks, in SOM the input layer of source nodes is directly connected to the output layer of computation nodes without any hidden layer [6]. The nodes in the input layer denote the attributes (features) or, more generally, the variables contained in the input data. Each piece of input data is represented by an m-dimensional input vector, $x = (x_1 x_2 \ldots x_m)'$, whose elements indicate the attribute values of a particular dataset. If there are large differences among the attribute values in the data set, normalization of the data is required in order to avoid the dominance of a particular attribute or subset of attributes [6]. This will give equal chances to all the attributes [15] and improve the numerical accuracy [16]. Several methods have been proposed in the literature for normalization. One of the most common methods is the z-score transformation which converts each attribute value to standard scores – with a zero mean and a unit standard deviation - by subtracting the mean and dividing by standard deviation for each attribute ($(x_j - \mu_{x_j})/\sigma_{x_j}$). Another widely used method is to divide each attribute value to the maximum of all values [15]. Unit length where all attributes will be scaled to the same length, the min-max transformation which scales the values between [0, 1], and the logarithmic transformation appropriate for exponentially distributed variables are further methods used for normalization [17].

The output layer, also known as "Kohonen layer" or "SOM layer", represents a low-dimensional visualization of the data. Nodes in the output layer are arranged in form of a topological architecture. This predefined structure is usually two-dimensional and organized as a grid that consists of rows and columns [18]. The number of nodes in the output layer denotes the maximum number of clusters and influences the accuracy and generalization capability of the SOM. According to a practical rule, the number of map units corresponds to 10 percent of the number of attribute values in a data set [15], which, however, should not be considered as a strict rule. The topologic structure can also be one-dimensional [19]. In one-dimensional string, the only parameter that must be decided is the number of nodes in the output layer; there is no need for the decision on the number of rows or columns. However, it is generally agreed that a two-dimensional map provides a better representation of the clusters.

Arrangement of the nodes in the output layer sets up the neighborhood relations which are studied in detail in the following subsection. The spaces between the neighbors are arbitrary and not important for the system. Focus is on the adjacency for two clusters that shows the similarity [8].

Typically the network topology is arranged in either a rectangular or hexagonal grid in the traditional SOM [4, 20]. As shown in Figure 14.2 these topologies have different properties. In the rectangular topology each internal node has four neighbors, while in the hexagonal topology six. Among these topologies, the hexagonal structure is usually more preferred, since it displays greater variance in neighborhood size. In both topologies, number of rows and columns are allowed to differentiate. Yet, research on the variety of topologies shows that the square-type topologies with $n \times n$ network size perform better [21]. Even though both shapes are frequently used, they cause problems on the edges. The neurons at the edge of a network are less central than other neurons, and therefore spherical and geodesic topologies are chosen to decrease the edge effect [22].

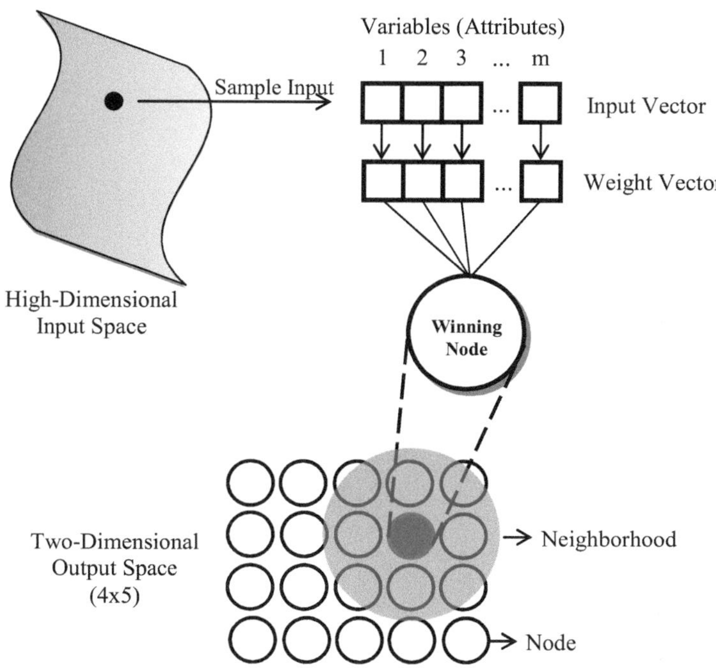

Fig. 14.1 Graphical illustration of a self-organizing map

Instead of using pre-defined structures with fixed number of neurons and grid dimensionality, it is possible to create gradually growing structures to better capture the fine-structure of the input distribution [23]. Several algorithms have been developed for growing SOM structures. For example, the Growing Grid algorithm starts with a 2×2 map and extends a column or a row for the next step. The Incremental Grid Growing (IGG) algorithm also starts with a small number of output nodes but generates any node irregularly when a new node is required. Another algorithm similar to IGG is the Growing SOM which additionally defines and uses a "spread factor" to control the growing [24]. All these growing structures give SOM a dynamic character.

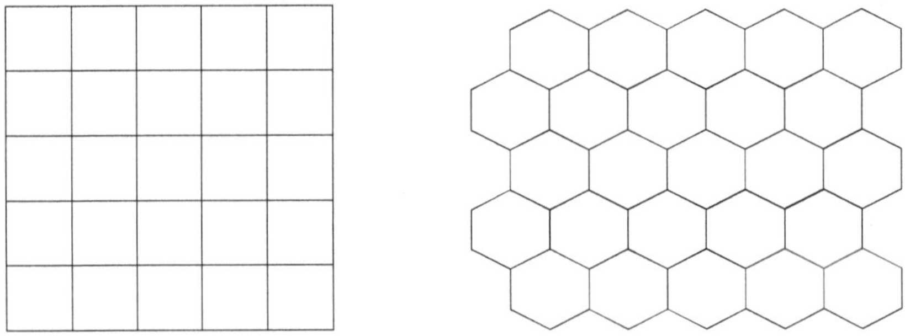

(a) Rectangular topology (b) Hexagonal topology

Fig. 14.2 Typical topologies used in SOM

Table 14.1 Input space

	Personality Traits (Attributes)					
Customer	x_1	x_2	x_3	x_4	x_5	x_6
1	1	0.25	0.75	0	0,50	0.50
2	0.25	0.75	0.75	0.50	0.25	0.50
3	0.75	0.25	0.50	0.50	0.50	0.25
4	0.50	1	1	0.75	0.75	0
...	...	...	...	...	...	...
20	0	0.25	0.75	0.25	0.50	0.75

Example An illustrative SOM example about psychographic segmentation analysis will be presented throughout this section. The aim of using SOM is to divide a heterogeneous market into smaller homogeneous sub-sets consisting of customers who have relatively similar characteristics. The data collected through a survey consists of customer responses ($k = 20$) to items of personality traits ($j = 6$) including being social (x_1), innovative (x_2), confident (x_3), logical (x_4), cheerful (x_5), and practical (x_6). Table 14.1 presents the customer responses as input data. Input values range from 0 to 1 where 1 denotes "strongly agree" and 0 denotes "strongly disagree".

Since there are 20 input samples, at least $20 * (10\%) = 2$ clusters can be used to represent the output layer. However, in order to construct a square-type two dimensional map, a topologic structure with 2×2 network size has been chosen. As shown in Figure 14.3, the SOM in this case consists of 6 input nodes and 4 output nodes arranged in form of a 2×2 map. Because all input nodes are measured on the same scale, input vectors do not need to be normalized.

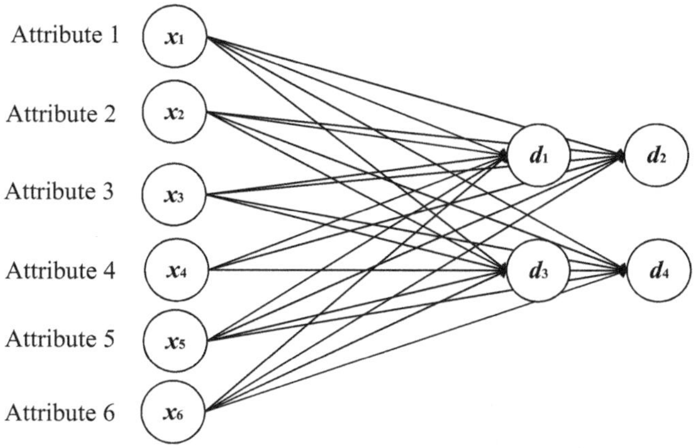

Fig. 14.3 The SOM architecture of the market segmentation example

14.2.2 *Analysis*

14.2.2.1 *Initial Weights*

The formation of the SOM starts first by initializing the weight vectors $w_i = (w_{i1} w_{i2} \dots\ w_{im})'$ where $i = 1, 2, \dots, n$ and denotes the number of output nodes in the network. Weights are links that connect the input nodes to the output nodes and are updated through the learning process. Mazanec [8] states that the final results are somewhat subject to the chosen initial weight distribution which should be examined carefully. The initial weights, $w_i(t)$ for $t = 0$, of the nodes can be determined either arbitrarily or systematically (e.g. linearly). The most common initialization method assigns preferably small random values to the weight vectors of nodes in the network. That means, no prior order is imposed on the network in the initialization operation [7]. The only restriction imposed by this method is to avoid similar weights [16]. Note that, if the initial weights are randomly set around 0, the learning-rate and neighborhood parameters should be chosen reasonably large for a certain number of iterations (at least for 1000 steps) [15]. In practice, however, such random assignments can sometimes cause the algorithm to converge in a longer time. Another way of initializing the weights, which overcomes this problem, is to assign sample vectors drawn randomly from the training set to the weight vectors [7].

Alternatively, a more effective approach is to initialize the weight vectors along the two-dimensional linear subspace that is spanned by the eigenvectors corresponding to the largest eigenvalues computed from the input data [4]. Since the initial weights approximate the probability density function of the input vector (x), the algorithm will converge faster. However, it is clear that this linear initialization method requires more calculations than the random initialization approaches.

14.2.2.2 *Similarity Matching*

As mentioned before, learning in SOM does not rely on predefined target outcomes that would guide the process. The output nodes compete among themselves to become activated. Only the node whose weight vector is most similar to the input vector will be activated and declared as the winner [20]. To find this best matching node, the distances between an input data (x) and all the weight vectors (w_i) of the SOM are computed using different measurement methods, such as Manhattan distance (city-block distance), Chebyshev distance, Euclidean distance, Mahalanobis distance [25] or vector dot product [20]. Of all these methods, as Kohonen [4] points out, Euclidean distance is more favorable in visual representations like those in SOMs, because a more isotropic display is obtained using it. For example, the algorithm in hexagonal lattice does not so much prefer horizontal and vertical directions. That's why the Euclidean distance is a generally accepted measure [4] and will be briefly explained here.

The Euclidean distance between a sample x, chosen randomly from the input dataset, and all the weight vectors at iteration t is calculated by using the following formula:

$$d_i(t) = \|x(t) - w_i(t)\| = \sqrt{\sum_{j=1}^{m} (x_{tj} - w_{tji})^2} \quad i = 1, 2, \ldots, n \tag{14.1}$$

where $\|\cdot\|$ denotes the Euclidean norm. Note that the weight vector w_i of each output node i has the same dimension as the input vector x. At the end of the similarity matching process, the best-matching (winning) node c at iteration t is determined by using the minimum-distance Euclidean criterion:

$$c(t) = \arg\min_i \{\|x(t) - w_i(t)\|\} \tag{14.2}$$

Example (continued) In our market segmentation example, for all four output nodes the initial weight vectors are assigned random values as shown in Table 14.2. The node whose weight vector is most similar to the customer response data, randomly chosen from the input space, will be activated and adjusted together with its neighboring units. For the sake of illustration, the Euclidean distance is applied to calculate the distances between the input data and all the weight vectors. Suppose that the first sample record $x(0) = (0.25\ 0.75\ 0.75\ 0.50\ 0.25\ 0.50)'$ randomly selected is the response of customer number

Table 14.2 Initial weights

j	w_1	w_2	w_3	w_4
1	0.15	0.06	0.11	0.38
2	0.97	0.84	1.00	0.78
3	0.65	0.50	0.99	0.79
4	0.27	0.35	0.64	0.68
5	0.09	0.39	0.19	0.44
6	0.46	0.85	0.47	0.22

two. The Euclidean distances for the four output nodes are calculated using Eq. (14.1):

$$d_1(0) = \sqrt{(0.25-0.15)^2+(0.75-0.97)^2+(0.75-0.65)^2+(0.5-0.27)^2+(0.25-0.09)^2+(0.5-0.46)^2} = 0.385$$

$$d_2(0) = \sqrt{(0.25-0.06)^2+(0.75-0.84)^2+(0.75-0.50)^2+(0.5-0.35)^2+(0.25-0.39)^2+(0.5-0.85)^2} = 0.521$$

$$d_3(0) = \sqrt{(0.25-0.11)^2+(0.75-1.00)^2+(0.75-0.99)^2+(0.5-0.64)^2+(0.25-0.19)^2+(0.5-0.47)^2} = 0.405$$

$$d_4(0) = \sqrt{(0.25-0.38)^2+(0.75-0.78)^2+(0.75-0.79)^2+(0.5-0.68)^2+(0.25-0.44)^2+(0.5-0.22)^2} = 0.408$$

For the first input record, the winning node is the first output node ($c(0) = 1$) that has the minimum Euclidean distance, $d_1(0) = 0.385$. In other words, the first output node is most similar to the given input node.

14.2.2.3 *Weight Updating*

The weight vectors of the winner and its neighboring units in the output space are adjusted to become more representative of the features that characterize the input space. This updating towards the input sample requires the consideration of the following two parameters: learning rate and neighborhood size. The learning rate, $\alpha(t)$, controls the rate of change of the weight vectors and, as in all neural networks, it takes values between 0 and 1. In SOMs, the learning rate gradually decreases as a function of the iteration step index t. That means, while the step index increases, the learning rate might decrease linearly, exponentially or geometrically [6], or it can be inversely proportional to t [4, 17]. Especially in cases where the initial weights are assigned random values, the learning rate should start with a reasonably high value that is close to unity and progressively reduce to small values. This procedure corresponds to larger corrections at the beginning of the training process (i.e. the ordering phase) than at the end where fine-tuning of the map takes place (i.e. the convergence phase) [4, 26]. The ordering phase may take as many as 1000 iterations of the SOM algorithm, and possibly more, while the convergence phase of the adaptive process must be at least 500 times the number of neurons in the network [4, 7].

Using a discrete-time formalism, given the weight vector $w_i(t)$ of the winning neuron i at iteration t, the updated weight vector $w_i(t+1)$ at iteration $t+1$ is defined by [4]:

$$w_i(t+1) = w_i(t) + \alpha(t)[x(t) - w_i(t)] \tag{14.3}$$

As mentioned above, the topological properties of the original data can only be preserved in the output space by considering neighborhood information. Thus, the reference vectors of the nodes in the neighborhood of the BMU also participate in the learning process. The rate of adaptation of the weights decreases away from the winning node, according to a spatio-temporal decay function [11], i.e. a neighborhood function $h_{ci}(t)$ where i denotes the index of the neighboring unit. In the literature, two fundamentally different types of neighborhood functions can be distinguished [16, 18]. The simpler of them, the discrete neighborhood function, defines a neighborhood set N_c of adjacent units around the winning node, whereby $h_{ci}(t) = \beta(t)$ if $i \in N_c$ and $h_{ci}(t) = 0$ if $i \notin N_c$. The value of $\beta(t)$ represents the degree of participation in the weight-updating procedure [18]. Examples of different discrete neighborhoods for rectangular and hexagonal topologies are shown in Figure 14.4. The other type of function widely applied for updating the neighborhood is the continuous neighborhood function which leads to a smoother neighborhood kernel. The most preferred function of this type is the Gaussian function which decreases in both the spatial domain and time domain:

$$h_{ci}(t) = \exp\left(-\frac{d_{ci}^2}{2\sigma^2(t)}\right) \tag{14.4}$$

where d_{ci}^2 denotes the lateral distance between the winning neuron c and the excited neuron i, and $\sigma(t)$ represents the effective width or radius of neighborhood at iteration t. The Gaussian function is symmetric about the maximum point defined by $d_{ci} = 0$, and decrease monotonically to zero with increasing lateral distance (d_{ci}) [7]. Here, $\sigma(t)$ is some monotonically decreasing function of time leading to a decrease in the value of the neighborhood function. It has been experimentally observed that choosing a too small neighborhood size at the beginning of the analysis may prevent inducing a global ordering. To overcome this problem Kohonen [16] suggests starting with a fairly large $\sigma(0)$ (e.g. more than half the diameter of the network). Other alternative neighborhood functions used in the literature are the bubble Eq. (14.5), cutgauss Eq. (14.6) and Epanechikov Eq. (14.7) [17]:

$$h_{ci}(t) = \mathbf{1}(\sigma(t) - d_{ci}) \tag{14.5}$$

$$h_{ci}(t) = \exp\left(-\frac{d_{ci}^2}{2\sigma^2(t)}\right)\mathbf{1}(\sigma(t) - d_{ci}) \tag{14.6}$$

$$h_{ci}(t) = \max\left\{0,\ 1 - (\sigma(t) - d_{ci})^2\right\} \tag{14.7}$$

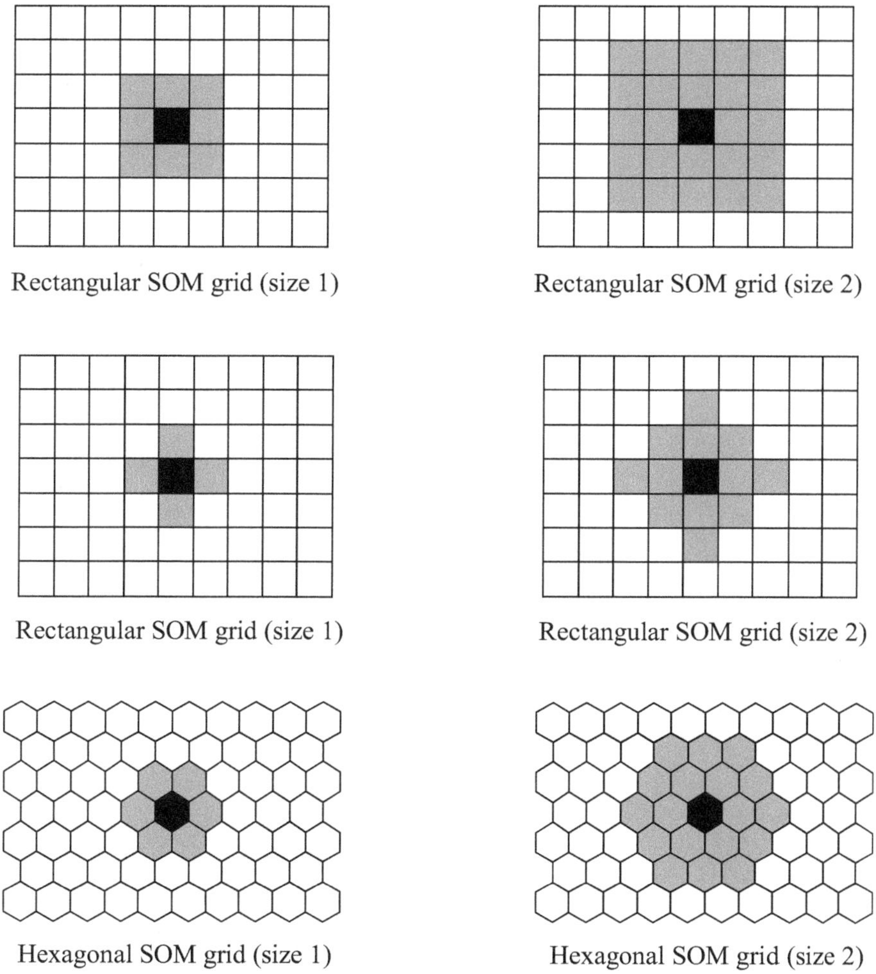

Fig. 14.4 Discrete neighborhoods (with size 1 and 2) for rectangular and hexagonal topologies

where $\mathbf{1}(x)$ is the step function: $\mathbf{1}(x) = 0$ if $x < 0$ and $\mathbf{1}(x) = 1$ if $x \geqslant 0$.

Finally, after defining the neighborhood set or kernel around the winning node c [16], a neighborhood function is chosen to update the weight vectors of the respective nodes. The following formula presents the update rule for the weight vector of unit i [4]:

$$w_i(t+1) = w_i(t) + \alpha(t)h_{ci}(t)[x(t) - w_i(t)] \quad (14.8)$$

All three processes – competition, cooperation and adaptation – described here are repeated for the remaining training data until the weights converge and no noticeable changes in the low-dimensional output layer are observed. The entire set of the resulting weight vectors reflects the distribution of the input data.

Example (continued) Recall that for the first sample record selected, the winning node was the first output node ($c = 1$) that had the minimum Euclidean distance. Now the weight vectors of this best matching node and its neighborhood will be adjusted so that they become more similar to the current input sample $x(0)$. Setting the initial learning rate close to one ($\alpha(0) = 0.9$) and applying the adaptation formula given in Eq. (14.3), the updated weights for the winning node ($c = 1$) at $t = 0$ are calculated as shown below:

$$w_1(1) = w_1(0) + \alpha(0)[x(0) - w_1(0)]$$

$$\begin{bmatrix} w_{11}(1) \\ w_{12}(1) \\ w_{13}(1) \\ w_{14}(1) \\ w_{15}(1) \\ w_{16}(1) \end{bmatrix} = \begin{bmatrix} 0.15 \\ 0.97 \\ 0.65 \\ 0.27 \\ 0.09 \\ 0.46 \end{bmatrix} + 0.9 \left(\begin{bmatrix} 0.25 \\ 0.75 \\ 0.75 \\ 0.50 \\ 0.25 \\ 0.50 \end{bmatrix} - \begin{bmatrix} 0.15 \\ 0.97 \\ 0.65 \\ 0.27 \\ 0.09 \\ 0.46 \end{bmatrix} \right) = \begin{bmatrix} 0.24 \\ 0.77 \\ 0.74 \\ 0.48 \\ 0.23 \\ 0.50 \end{bmatrix}$$

Note that if the learning rate was chosen equal to zero, the weights would not change, i.e. $w_1(1) = w_1(0)$. Alternatively, if the learning rate was chosen equal to one, the new weights would equal to the input data, i.e.$w_1(1) = x(0)$. Next, applying the Gaussian neighborhood function with $\sigma(0) = 1$ (Eq. (14.4)) and the adaptation formula given in Eq. (14.8), the updated weights for the neighboring units at $t = 0$ are calculated as shown below:

$$h_{12}(0) = \exp\left(-\frac{d_{12}^2}{2\sigma^2(0)}\right) = \exp\left(-\frac{0.544^2}{2 \cdot 1^2}\right) = 0.862$$

$$h_{13}(0) = \exp\left(-\frac{d_{13}^2}{2\sigma^2(0)}\right) = \exp\left(-\frac{0.515^2}{2 \cdot 1^2}\right) = 0.876$$

$$h_{14}(0) = \exp\left(-\frac{d_{14}^2}{2\sigma^2(0)}\right) = \exp\left(-\frac{0.676^2}{2 \cdot 1^2}\right) = 0.796$$

where

$$d_{12} = \sqrt{(0.15-0.06)^2+(0.97-0.84)^2+(0.65-0.50)^2+(0.27-0.35)^2+(0.09-0.39)^2+(0.46-0.85)^2} = 0.544$$

$$d_{13} = \sqrt{(0.15-0.11)^2+(0.97-1.00)^2+(0.65-0.99)^2+(0.27-0.64)^2+(0.09-0.19)^2+(0.46-0.47)^2} = 0.515$$

$$d_{14} = \sqrt{(0.15-0.38)^2+(0.97-0.78)^2+(0.65-0.79)^2+(0.27-0.68)^2+(0.09-0.44)^2+(0.46-0.22)^2} = 0.676$$

hence, we obtain for output node 2:

$$w_2(1) = w_2(0) + \alpha(0)h_{12}(0)[x(0) - w_2(0)]$$

$$\begin{bmatrix} w_{21}(1) \\ w_{22}(1) \\ w_{23}(1) \\ w_{24}(1) \\ w_{25}(1) \\ w_{26}(1) \end{bmatrix} = \begin{bmatrix} 0.06 \\ 0.84 \\ 0.50 \\ 0.35 \\ 0.39 \\ 0.85 \end{bmatrix} + (0.9)(0.862) \left(\begin{bmatrix} 0.25 \\ 0.75 \\ 0.75 \\ 0.50 \\ 0.25 \\ 0.50 \end{bmatrix} - \begin{bmatrix} 0.06 \\ 0.84 \\ 0.50 \\ 0.35 \\ 0.39 \\ 0.85 \end{bmatrix} \right) = \begin{bmatrix} 0.21 \\ 0.77 \\ 0.69 \\ 0.47 \\ 0.28 \\ 0.58 \end{bmatrix}$$

The weights of the output nodes 3 and 4 are calculated similarly. The adjusted weight vectors of the winning node and its neighborhood are given below:

$$w_1(1) = (0.24\ 0.77\ 0.74\ 0.48\ 0.23\ 0.50)'$$

$$w_2(1) = (0.21\ 0.77\ 0.69\ 0.47\ 0.28\ 0.58)'$$

$$w_3(1) = (0.22\ 0.80\ 0.80\ 0.53\ 0.24\ 0.49)'$$

$$w_4(1) = (0.29\ 0.76\ 0.76\ 0.55\ 0.30\ 0.42)'$$

14.2.2.4 *Summary of the SOM Algorithm*

The basic steps of Kohonen's SOM algorithm can be summarized by the following iterative procedure:

(i) *Initialization*. Choose the dimension and size of the output space. Assign random values or alternatively sample vectors drawn randomly from the training set to the initial weight vectors $w_i(0)$. Specify the neighborhood function as well as the functional form of the learning rate $\alpha(0)$ and radius of the neighborhood $\sigma(0)$. Assign starting values for $\alpha(0)$ and $\sigma(0)$. Normalize training data. Define a critical threshold value (T) for the maximum number of iterations. Set iteration index $t = 1$.

(ii) *Sampling*. Randomly select an input vector $x(t)$ from the training data set.

(iii) *Similarity Matching*. Compute the Euclidean distances between the input vector and each output node's weight vector and find the best matching node $c(t)$ at iteration t by applying the minimum distance criterion:

$$c(t) = \arg\min_i \{ \|x(t) - w_i(t)\| \} \quad i = 1, 2, \ldots, n$$

(iv) *Weight Updating*. Adjust the weights of the winning node and its neighborhood according to their distances to the winning node by using the update formula:

$$w_i(t+1) = w_i(t) + \alpha(t) h_{ci}(t) [x(t) - w_i(t)]$$

For the winning node the neighborhood function $h_{ci}(t)$ will be equal to 1.

(v) *Parameter Adjustment.* Set $t = t + 1$. Adjust the neighborhood size and the learning rate.

(vi) *Continuation.* Keep returning to Step 2 until the change of the weights is less than a prespecified threshold value or the maximum number T of iterations is reached. Otherwise stop.

14.3 Quality of SOM

SOM has the ability to produce simpler low-dimensional representations of complex high-dimensional data at the expense of information loss during the projection. This representation of topological relations and abstract structures is an approximation of the local, rather than the global, probability density of a given sample data set [9]. This explains why the outputs of an SOM can vary highly depending on the initial setting of parameters such as the number of output nodes, the learning rate, and the update speed of the neighborhood [4, 9]. It is, therefore, necessary to examine the quality of the results in order to determine the parameter values which give the most appropriate representation of the data set. However, it should be noted that there are no generally accepted rules for the determination of the parameters a priori. To address these issues different measurement methods have been developed in the literature to evaluate and compare the quality of SOM outputs. In this respect, the most studied properties of SOM projections are learning quality and projection quality, where the first one is sometimes referred to as vector quantization and the latter one as topology preservation. Technically, there is a tradeoff between these two, the topology preservation quality usually decreases with increasing data representation accuracy [27].

The methods suggested to measure the mentioned properties will be discussed in the remainder of this subsection. Note that not all the descriptions of the methods summarized contain formulas; for more detail on the concepts, please refer to the referenced publications.

Quantization Error

The Quantization Error (QE) provides a means to assess the quality of learning and shows how good the map fits to data. It is computed by determining the average distance of the sample vectors to its best matching unit [4]. The formula of the average quantization error is as follows:

$$QE = \frac{\sum_{t=1}^{T} \|x(t) - w_c(t)\|}{T} \tag{14.9}$$

where $w_c(t)$ is the reference vector associated to the BMU of $x(t)$.

By increasing the number of output nodes for any given dataset, the QE can be reduced as the data samples are distributed more sparsely on the map [27]. Here, the QE measure completely disregards the topological relations and because of the tradeoff between vector quantization and projection quality of the SOM, reducing the QE may lead to distortions of the map's topology [27]. Besides, Sun [28] suggests using a small neighborhood size in order to achieve a small quantization error and a good approximation of input distribution. Note that a low QE may also imply an over fitted model [29].

Topographic Error

Topology preservation, i.e. projection quality, is a property that is more complex to define and hard to measure. The most common quality measure for topology preservation is the Topographic Error (TE), which can be expressed as the percentage of data vectors for which the first and second BMUs are not adjacent units [30]. The topographic error is computed as follows [31]:

$$TE = \frac{1}{T} \sum_{t=1}^{T} u(x(t)) \tag{14.10}$$

where T is the number of input samples, and $u(x(t)) = 1$ if the best- and second best-matching units of $x(t)$ are non-adjacent, otherwise $u(x(t)) = 0$. So, the less the violation of the neighborhood relations on the grid is, the better the SOM preserves the topology. According to Pölzlbauer [27] this measure is unreliable for small maps, since the TE is almost too simplistic and suffers from the discrete nature of the output space.

Topographic Product

Introduced by Bauer and Pawelzik [32] the Topographic Product (TP) is a measure of the preservation of neighborhood relations in maps. According to TP, for each node on the map, the weight vectors of its k-nearest neighbor units are compared whether they coincide or not. If folds are observed on the map, this will be taken as an indicator that the map is trying to approximate a higher-dimensional input space, and thus producing topographic error [31]. Therefore, the result of the computation of the TP can be used to optimize the map size for a given input space. However, it was shown by Villmann *et al.* [33] that the topographic product fails to distinguish between the results caused by a folded input space, and the folds that are actually erroneous. Therefore, the TP presents reliable results only for nearly linear datasets.

Distortion Measure

One mathematical difficulty of the SOM algorithm is to prove that the algorithm converges to a good solution. Typically, in neural network learning algorithms, the weight update rule corresponds to some kind of gradient descent on an energy (or cost) function However, it has been shown that there exists no energy function for the Kohonen's original learning rule [34]. "Hence, strictly speaking, we cannot judge the degree of optimality achieved by the algorithm during learning" [23]. Only for a finite set of training patterns and an algorithm with a fixed neighborhood kernel radius, it is possible to come up with an energy function the SOM optimizes [35]. One such energy function used for measuring the overall quality of a given SOM is the Distortion Measure (E_d) [36], which is computed as follows:

$$E_d = \sum_{t=1}^{T} \sum_{i=1}^{n} h_{ci}(t) \left\| x(t) - w_i(t) \right\| \tag{14.11}$$

One important advantage of the Distortion Measure is that the error can be decomposed into three component terms [37]:

- Local Data Variance: it gives the quantization quality of the SOM,
- Neighborhood Variance: it analyzes the trustworthiness [38] (i.e. closeness of prototype vectors close to each other on the map grid) of the map topology,
- Neighborhood Bias: it links the other two properties together and can be regarded as the stress between them.

The Distortion Measure can be used to compare several maps trained with the same dataset according to their fit to the data.

Recall that finding the right size for the low-dimensional map is a quite important task in SOM. A too small map will be too rough to reveal some significant differences in data whereas a too large map will detect more detailed patterns where differences are too small. All the measures described above are critical in choosing the optimum size of the output space. As a result, for the same dataset, the map size where the values of QE and TE become minimum should be selected [29].

14.4 Applications of SOM

This section focuses on the applications of the SOM algorithm reported in the literature. Major practical application areas include process analysis, geoscience and remote control, image analysis and pattern recognition, information technology, intelligent systems design, signal processing, ecosystem modeling, robotics, optimization, expert systems and data mining (an exhaustive listing of the detailed applications can be found in Ref.39). Here, rather than attempting for an extensive overview, only the applications relevant to industrial engineering are presented.

Location-allocation (LA) problem. This problem consist of determining the locations of a set of facilities and the flows of products from these facilities to customers, such that the sum of location-specific fixed costs and operational as well as transportation costs are minimized. In this respect, the LA problem is a type of clustering problem based on minimum distances which can be solved by using an SOM algorithm [40, 41]. Here, the locations of K customers (i.e. K demand centers) taken as input patterns are clustered into N supply centers represented as output nodes in SOM. While the weights represent the location of supply centers, the input values denote the demands of customers. The learning algorithm adjusts the locations of the closest supply center (winning node) and its neighbors until no change with further training is observed.

Traveling-salesman problem (TSP). Given a list of cities and their pairwise distances, this problem searches the shortest (cheapest) possible route that visits each city exactly once and returns to the starting city. The neighborhood preserving property of the SOM and the convex-hull property of the TSP makes SOM a suitable algorithm to find the optimal route [42]. When the location of each city is represented by a point in a two-dimensional input space, and a tour by a sequence of N output nodes (i.e. closed chain), then a solution to the TSP can be viewed as a mapping from the input-space onto the nodes of the chain [23]. After a finite number of training cycles the weights of the output nodes are expected to coincide with the locations of the cities.

Cell designing problem. In cellular manufacturing machines are grouped together according to the families of parts produced. The cell formation problem is primarily concerned with machine–part clustering. As one of the clustering methods the SOM can successfully solve this problem by grouping machines and parts based on operations sequence [29]. Product parts are defined as the input data while machines as the output nodes. The SOM algorithm not only suggests a clustering solution but also provides a visualization of the solution.

Product positioning problem. Product positioning aims at exploring the perceptual differences between the brands of a product class and the way they are linked to consumer preferences or choice [43]. To this end, various sophisticated multivariate techniques (such as principle components, multidimensional scaling and unfolding, correspondence analysis) have been used to extract the competitive dimensions, product positions and consumer preferences [44]. The SOM approximates an unknown multivariate probability distribution by fitting topologically ordered units (product and customer positions) in a predefined low-dimensional perceptual space and links these positions to stated or revealed brand choice data. It allows performing segmentation (clustering) and positioning (mapping) analysis simultaneously. In comparison to the multivariate techniques, the SOM procedures are easy to implement, do not impose rigorous distributional assumptions and nor do they require particular scaling properties of the raw data. They effectively cope with very small dimensionality and samples of unlimited size [12]. For more detail on market segmentation using SOM the reader should refer to Kuo *et al.* [45], Kiang *et al.* [46], Hung and Tsai [47].

Multiple objective just-in-time (JIT) sequencing problem. This problem considers a JIT production-scheduling system where two inversely related objectives are present – minimization of setups between differing products and optimization of schedule flexibility. To address this problem, Kohonen's SOM has been used to search for sequences which are desirable in terms of both the number of setups and flexibility [48]. For this particular effort, each input sample associated with a particular item is represented by a two-dimensional vector denoting its coordinates (on a circle). The total number of inputs will be the total demand of all items needing sequencing. The weight vectors in this problem represent the sequence position of each of the individual items. During the learning process the weight vectors tend to organize themselves toward the (circular) patterns having minimal setups [48]. At the end of the training, each input vector will have been associated with a unique weight vector.

Product concept development, demand forecasting and approximation of probability density functions are further examples of industrial engineering relevant SOM applications, which, however, will not be discussed here.

14.5 Conclusion

The number of industrial engineering related SOM applications reported in the literature has shown a considerable increase in the past decade. Its ability to explore and visualize high-dimensional data with no rigorous assumptions makes SOM an appealing technique in this field. This chapter is intended to contribute to this evolving body of knowledge by providing a gentle introduction to self-organizing maps and discussing its properties, limitations and applications. It can be concluded that combining the SOM with appropriate statistical techniques, heuristic methods or decision making tools appears to be a promising approach for widening further the application field towards industrial engineering problems. However understanding the mathematics of self-organizing processes will remain a highly active field of research.

Bibliography

[1] P. Engelbrecht, *Computational Intelligence: An Introduction*, 2nd ed. John Wiley & Sons, England, (2007).

[2] N.R. Pal and S. Pal, Computational intelligence for pattern recognition, *International Journal of Pattern Recognition and Artificial Intelligence*, **16** (7), 773–779 (2002).

[3] T. Kohonen, Self-organized formation of topologically correct feature maps, *Biological Cybernetics*, **43**, 59–69 (1982).

[4] T. Kohonen, *Self-Organizing Maps*, Springer, Berlin, (2001).

[5] A.H. Holmbom, T. Eklund, and B. Back, Customer portfolio analysis using the SOM, *International Journal of Business Information Systems*, **8** (4), 396–412 (2011).

[6] D.T. Larose, *Discovering Knowledge in Data*, John Wiley & Sons, New Jersey, (2005).

[7] S. Haykin, *Neural Networks: A Comprehensive Foundation*, 2nd ed. Prentice-Hall, Englewood Cliffs, New Jersey, (1999).

[8] J.A. Mazanec, Positioning analysis with self-organizing maps: an exploratory study on luxury hotels, *Cornell Hotel and Restaurant Administration Quarterly*, **36** (6), 80–95 (1995).

[9] S. Wang and H. Wang, Knowledge discovery through self-organizing maps: data visualization and query processing, *Knowledge and Information Systems*, **4**, 31–45 (2002).

[10] J.E. Dayhoff, *Neural network architectures—An introduction*, New York: Van Nostrand Reinhold, (1990).

[11] A.J. Richardson, C. Risien, and F.A. Shillington, Using self-organizing maps to identify patterns in satellite imagery, *Progress in Oceanography*, **59**, 223–239 (2003).

[12] J.A. Mazanec, Neural market structure analysis: Novel topology-sensitive methodology, *European Journal of Marketing*, **35** (7/8), 894–914 (2001).

[13] J. Han and M. Kamber, *Data Mining: Concepts and Techniques*, Morgan Kaufmann Publishers, San Francisco, (2001).

[14] J.B. Greenwald, *Optical Character Categorization: Clustering as it Applies to OCR*, PhD thesis, Rochester Institute of Technology (1997).

[15] M. Zontul, O. Kaynar, and H. Bircan, Som tipinde yapay sinir ağlarını kullanarak Türkiye'nin ithalat yaptığı ülkelerin kümelenmesi üzerine bir çalışma, *C.Ü. İktisadi ve İdari Bilimler Dergisi*, **5** (2004).

[16] T. Kohonen, The self-organizing map, *Proceedings of the IEEE*, **78**, 1464–1480 (1990).

[17] J. Vesanto, J. Himberg, E. Alhoniemi, and J. Parhankangas, SOM Toolbox for Matlab 5, *Report A57*, Helsinki University of Technology Finland, (2000).

[18] T. Reutterer, Competitive market structure and segmentation analysis with self-organizing feature maps, in *Proceedings of the 27th EMAC Conference*, Ed. P. Andersson, Marketing Re-

search Stockholm, (1998).
[19] J.A. Lee and M. Verleysen, Self-organizing maps with recursive neighborhood adaptation, *Neural Networks*, **15**, 993–1003 (2002).
[20] C. Budayan, *Strategic group analysis: strategic perspective, differentiation and performance in construction*, PhD thesis, Middle East Technical University (2008).
[21] Y.M. Kiang, Extending the Kohonen self-organizing map networks for clustering analysis, *Computational Statistics & Data Analysis*, **38**, 161–180 (2001).
[22] C.R. Schmidt, S.J. Rey, and A. Skupin, Effects of irregular topology on self- organizing maps, *International Regional Science Review*, **34**, 215–229 (2010).
[23] M.M. Van Hulle, Self-Organizing Maps, in *Handbook of Natural Computing: Theory, Experiments, and Applications*, Ed. G. Rozenberg, T. Baeck, J. Kok, Springer, (2011).
[24] D. Alahakoon and S.K. Halgamuge, Dynamic self-organizing maps with controlled growth for knowledge discovery, *IEEE Transactions on Neural Networks*, **11**, 601–614 (2000).
[25] S. Aly, N. Tsuruta, and R.-I. Taniguchi, Face recognition under varying illumination using Mahalanobis self-organizing map, *Artificial Life and Robotics*, **13** (1), 298–301 (2008).
[26] R. Rojas, *Neural Networks*, Springer-Verlag, Berlin, (1996).
[27] G. Pölzlbauer, Survey and comparison of quality measures for self-organizing maps, in *Proceedings of the Fifth Workshop on Data Analysis* (*WDA-04*), Elfa Academic Press, Sliezsky dom, Vysoké Tatry, Slovakia, 67–82 (2004).
[28] Y. Sun, On quantization error of self-organizing map network, *Neurocomputing*, **34** (1-4), 169–193 (2000).
[29] M. Chattopadhyay, P.K. Dan, and S. Mazumdar, Application of visual clustering properties of self organizing map in machine–part cell formation, *Applied Soft Computing*, **12**, 600–610 (2012).
[30] J.I. Mwasiagi, H. XiuBao, W. XinHou, and C. Qing-Dong, The Use of K-Means and Kohonen Self Organizing Maps to Classify Cotton Bales, *Beltwide Cotton Conferences*, New Orleans, Louisiana, January 9-12 (2007).
[31] K. Kiviluoto, topology preservation in self-organizing maps, in *Proceedings of ICNN'96, IEE International Conference on Neural Networks*, IEEF, Service Center, Piscataway, pp. 294–299 (1996).
[32] H.U. Bauer and K.R. Pawelzik, Quantifying the neighborhood preservation of selforganizing feature maps, *IEEE Transactions on Neural Networks*, **3** (4), 570–579, (1992).
[33] T. Villmann, R. Der, and T. Martinez, A new quantitative measure of topology preservation in Kohonen's feature maps, *Proceedings of the IEEE International Conference on Neural Networks 94*, Orlando, Florida, USA, 645–648 (1994).
[34] E. Erwin, K. Obermayer, and K. Schulten. Selforganizing maps: ordering, convergence properties and energy functions, *Biological Cybernetics*, **67** (1), 47–55 (1992).
[35] T. Heskes, Energy functions for self-organizing maps, in *Kohonen Maps*, Ed. E. Oja and S. Kaski, pp. 303–316, Elsevier, Amsterdam, (1999).
[36] J. Lampinen and E. Oja, Clustering properties of hierarchical self-organizing maps, *Journal of Mathematical Imaging and Vision*, **2** (2-3), 261–272 (1992).
[37] J. Vesanto, M. Sulkava, and J. Hollmen, On the decomposition of the self-organizing map distortion measure, in *Proceedings of the Workshop on Self-organizing Maps*, WSOM'03, pp. 11–16 (2003).
[38] J. Venna and S. Kaski, Neighborhood preservation in nonlinear projection methods: an experimental study, in *ICANN 2001*, Ed. G. Dorffner, H. Bischof, and K. Hornik, LNCS, **vol. 2130**, pp. 485–491, Springer, Heidelberg, (2001).
[39] URL `http://www.cis.hut.fi/research/refs/`
[40] S. Lozano, F. Guerrero, L. Onieva and J. Larrafieta, Kohonen maps for solving a class of location-allocation problems, *European Journal of Operational Research*, **108**, 106–117

(1998).
[41] K.-H. Hsieh and F.-C. Tien, Self-organizing feature maps for solving location–allocation problems with rectilinear distances, *Computers & Operations Research*, **31**, 1017–1031 (2004).
[42] K.-S. Leung, H.-D. Jinb, and Z.-B. Xuc, An expanding self-organizing neural network for the traveling salesman problem, *Neurocomputing*, **62**, 267–292 (2004).
[43] P. Kotler, *Marketing Management* – International Millennium Edition, Prentice Hall, New Jersey, (2000).
[44] V. Trommsdorff, U. Asan, and J. Becker, "Produkt- und Markenpositionierung", in *Handbuch Markenführung*, Ed. Bruhn M., Volume 1, Second Edition, Gabler Verlag, Wiesbaden (2004).
[45] R.J. Kuo, Y.L. An, H.S. Wang, and W.J. Chung, Integration of self-organizing feature maps neural network and genetic K-means algorithm for market segmentation, *Expert Systems with Applications*, **30** (2), 313–324 (2006).
[46] M.Y. Kiang, M.Y. Hu, and D.M. Fisher, An extended self-organizing map network for market segmentation—a telecommunication example, *Decision Support Systems*, **42**, 36–47 (2006).
[47] C. Hung and C.-F. Tsai, Market segmentation based on hierarchical self-organizing map for markets of multimedia on demand, *Expert Systems with Applications*, **34** (1), 780–787 (2008).
[48] P.R. McMullen, A Kohonen self-organizing map approach to addressing a multiple objective, mixed-model JIT sequencing problem, *International Journal of Production Economics*, **72**, 59–71 (2001).

Chapter 15

Fuzzy Multi-Objective Programming for Energy Resource Planning

Adesh Jinturkar[1] and Sandip Deshmukh[2]

[1] *Anuradha Engineering College, Chikhli, Maharashtra, India*
[2] *Centre for Environmental Strategy, University of Surrey, Guildford, UK*

E-mail: ajinturkar@gmail.com, s.deshmukh@surrey.ac.uk

The energy resource planning takes into account various resources available and their demand. The energy planning endeavor involves finding a set of sources and conversion devices, so as to meet the energy requirements in an optimal manner. There are several exogenous and endogenous parameters of economy and environment causing huge uncertainty that is inferior in the longer term.

In this chapter, development and application of a fuzzy mixed integer goal programming (FMIGP) model has been presented for rural cooking and heating end-uses. The developed model considers various scenarios such as economical, environmental, social acceptance and local resources to trade off between socio-economical and environmental issues. Due to uncertainty involved in real world energy planning, exact input data is impossible to acquire. Hence, FMIGP model is used to consider four fuzzy objectives. The solutions provide energy resource allocations at micro-level with minimized cost, minimized emission, maximized social acceptance and use of local resources.

15.1 Introduction

The optimal allocation of rural energy resources to various energy end users is an important aspect for bridging the energy supply and demand gap in India. It has been observed that allocation should be channelized by multiple criteria. One important strategy towards bridging the gap is to optimally allocate various energy resources to the energy needs arising out of wide variety of end-uses for a given geographical region.

Decision Analysis is a set of systematic procedures for analyzing complex problems of decision making. Multiple Criteria Decision Making Decision Making (MCDM) was introduced as an important and promising tool in early 1970'es and has become systematic

C. Kahraman (ed.), *Computational Intelligence Systems in Industrial Engineering*,
Atlantis Computational Intelligence Systems 6, DOI: 10.2991/978-94-91216-77-0_15,

and rational way of decision making with multiple criteria. Multiple criteria include both multiple attributes and multiple objectives. Multi-criteria decision making (MCDM) methods are classified into two categories: multi-objective decision making (MODM) approach and multi-attribute decision making (MADM) approach. The decision problem in MODM is solved using multi-objective linear or nonlinear mathematical programming models in which several objective functions are integrated and optimized, subject to a set of constraints such as compromising programming, constraint method, goal programming, and fuzzy multi-objective programming. In MADM, each planning or design strategy is associated with a set of attributes whereby various planning or design strategies can be compared such as utility function method, tradeoff analysis method and analytical hierarchy process.

The decision problem in MODM is solved using multi-objective linear or nonlinear mathematical programming models in which several objective functions are integrated and optimized, subject to a set of constraints. In the present study, a fuzzy mixed integer goal programming model has been developed for optimizing the energy supply. The model takes into account the resource constraints with respect to supply, efficiency and utilization. The main purpose of the study is to analyze the technical options and resource utilization option for different energy end-uses for the development of the region. The purpose of imposing constraints on certain activities in the model is to ensure that the end-use demands energy is met. The model has been employed for rural household sector of Buldhana district located in Maharashtra, India. The data used in the study are based on detailed household energy survey of four villages.

The construction of scenario finds alternative future developments under a set of assumed conditions dates far back. The scenarios have now become the main tools used to consider the uncertainty in the fuel price and complexity in the supply and demand of energy resource to meet the future challenges. Energy scenarios provide a framework for exploring future energy perspectives in the process of research, development and deployment of renewable energy technologies. Four energy scenarios have been developed depending on the priorities by using fuzzy mixed integer goal programming approach.

15.2 Fuzzy Logic Approach

The concept of goal programming (GP) was first introduced by Charnes and Cooper [1]. Thereafter, GP has been widely discussed and applied for solving decision problems with multiple criteria [2, 3]. In classical GP models, unwanted deviations from target values defined by the decision maker are minimized in order to reach an acceptable solution. The unwanted deviations are measured using positive and negative deviation variables that are defined for each goal and they represent over achievement and underachievement of the goal, respectively. Recently, some extensions, which are formed by hybridizing approaches in order to obtain more general achievement function structures, have been proposed. However, from the modeling goal priorities point of view, weighted GP and preemptive priorities GP have been the two most widely used approaches [4].

The parameters of problems need to be defined precisely for using GP approach. Values of some parameters in most agricultural planning problems may not be known precisely. They are rather defined in fuzzy sense. Fuzzy set theory has initially been proposed by Zadeh that has been conventionally applied and used in Multi Criteria Decision Making problems [5]. The natural, social and human systems always have uncertainty and subjectivity causing different conclusion and systems could not be formulated perfectly by mathematical and stochastic methods. Accurate approximation and complex social and human systems lack exact mathematical model and, situations result fuzzy environment. Therefore, fuzziness in development and utility play important role in human behaviors.

The concept of fuzzy sets was first introduced to deal with the issue of uncertainty in systems modeling by Zadeh [6] who defined fuzzy sets as sets with boundaries that are not precise. The membership in a fuzzy set is not a matter of affirmation or denial, but rather a matter of degree. The concept of fuzzy set theory challenged conventional two-valued logic as when A is a fuzzy set and x is a relevant object, the proposition x is a member of A, is not necessarily either true or false, as required by two-valued logic, but it may be true only to some degree – the degree to which x is actually a member of A.

Application of fuzzy theories has been considered to various decision making processes by Bellman and Zadeh [7]. Fuzzy optimization technique to linear programming problems with multi-objective function has been introduced by Zimmermann [8]. Methods for all integer, mixed–integer and zero-one multi-objective problems have been presented by Lee and Morris [9]. Application of concept of fuzzy subsets to goal programming in fuzzy environment to specify imprecise aspiration levels of fuzzy goals has been introduced and also used combinations of linguistically defined weights and achievement degrees of the goals [10].

FGP approach has been widely used for a farm planning problem [11], agriculture planning problems [12, 13], land use problem in an agricultural system [14], vendor selection problems in a supply chain [15] and channel allocation problem in steel industry [16]. Energy planning problems are complex problems with multiple decision makers and multiple criteria. There are several exogenous and endogenous parameters of economy and environment causing huge uncertainly in the longer term in energy planning. Such uncertainly can easily be handled by fuzzy goal programming by introducing membership function. Fuzzy goal programming model structure can result in a wide range of outcomes. However, even when restricting oneself to a single model specification, uncertainty in some of the data or parameter values can yield a wide range of likely outcomes. We believe that a method such as fuzzy goal programming should be capable of producing a much more useful set of inputs for use in energy policy analysis and decision making.

15.3 Model Development

The problem of mixed integer goal programming approach is modeled by identifying six energy resources such as biogas ($i = 1$), biomass ($i = 2$), LPG ($i = 3$), kerosene ($i = 4$), dung cake ($i = 5$) and solarsolar thermal ($i = 6$) for cooking and four energy resources such as biomass ($i = 7$), biogas ($i = 8$), dung cake ($i = 9$) and solar thermal ($i = 10$) for heating energy end use. Four objectives and six constraints have been considered for rural cooking and heating energy planning in four villages namely *Malshemba, Muradpur, Savargaon Dukre* and *Yewta* of *Buldhana* district based on the availability and feasibility.

15.3.1 *Objective Functions*

15.3.1.1 *Objective 1: Minimization of cost*

The cost refers to the actual cost of energy delivered which is an important factor affecting the choice of energy systems. The objective function is represented as

$$\min \sum C_i x_i \tag{15.1}$$

where C_i is the cost of energy at end-use point for energy sources x_i ($i = 1, 2, 3, \ldots, 10$).

15.3.1.2 *Objective 2: Minimization of Emissions*

Three major pollutants are investigated to study the impact of human activities on environment. These are carbon oxides (CO and CO_2).
The objective functions are represented as:

$$\min \sum CO_i x_i \tag{15.2}$$

where CO_i are the emission factors of energy sources x_i ($i = 1, 2, 3, \ldots, 10$).

15.3.1.3 *Objective 3: Maximization of Social Acceptance of Energy System*

The energy sources should be accepted socially to reduce cost, emission and use of petroleum products and to promote use of local energy sources. The objective function is represented as

$$\max \sum S_i x_i \tag{15.3}$$

where S_i is the social acceptance of energy sources x_i ($i = 1, 2, 3, \ldots, 10$).

15.3.1.4 *Objective 4: Maximization of Local Resources Utilization*

To achieve self dependence in energy sources utilization and to be independent from commercial energy sources, the local resources should be maximally utilized. The resources except LPG and kerosene are considered as local resources and hence maximized. The objective function is represented as

$$\max_{i=1,2,5,6,7,8,9,10} \sum x_i \tag{15.4}$$

15.3.2 *Constraints*

The proposed energy allocation model is also subjected to following constraints.

15.3.2.1 *Constraint 1: Cooking Energy Requirement*

$$\sum x_i \geqslant \text{Total cooking energy requirement}, \ i = 1, 2, 3, \ldots, 6. \tag{15.5}$$

15.3.2.2 *Constraint 2: Heating Energy Requirement*

$$\sum x_i \geqslant \text{Total heating energy requirement}, \ i = 7, 8, 9, 10. \tag{15.6}$$

15.3.2.3 *Constraint 3: Limit Solar Thermal Usage for Cooking*

The total cooking requirement cannot be met by solar cooker and can be used for low temperature cooking purposes only which are approximately 20% of total cooking requirement. Therefore potential limit for the use of solar thermal cookers is assumed to be 20% of total cooking energy requirement. The constraint is

$$\sum x_6 \leqslant (0.2 \times \text{Total cooking energy requirement}). \tag{15.7}$$

15.3.2.4 *Constraint 4: Limit for Use of Dung Cake for Cooking and Heating*

Energy consumption pattern for cooking of the region shows that the dung cakes are not fully consumed for the cooking and heating applications. It is also observed during the survey, that most of the families use 10–25% dung available by making dung cakes. Therefore, it is assumed that 15% of the dung cakes produced are used for cooking and heating applications. Therefore, constraint function is

$$\sum\left(\frac{x_i}{\eta_i}\right) \leqslant \text{Dung energy available} \tag{15.8}$$

$i = 5,\ 9$ and η_i is efficiency of end- use devices for cooking and heating.

15.3.2.5 *Constraint 5: Potential Limit for Biogas Energy*

$$\sum\left(\frac{x_i}{\eta_i'}\right) \leqslant \text{Biogas energy available} \tag{15.9}$$

x_i $(i = 1)$ and η_i is efficiency of end- use devices for cooking and heating.

15.3.2.6 *Constraint 6: Potential Limit for Biomass Energy*

$$\sum\left(\frac{x_i}{\eta_i'}\right) \leqslant \text{Biomass energy available} \tag{15.10}$$

x_i $(i = 2,\ 8)$ and η_i is efficiency of end- use devices for cooking and heating

In practice, determining the goal value of each objective is difficult for decision makers due to uncertain and dynamic environment of fuel use pattern in rural region, supply and demand of fuel and prices of fuel. Such vague and uncertain information related to the objectives can be modeled as a FGP problem. A set of desirable achievement degrees and preemptive priorities can also be obtained in model to represent the relative importance and priority of the goals.

15.4 Fuzzy Goal Programming Model

The model allows consideration many objectives simultaneously. In reality, the decision-maker generally chooses the achievement of certain goals at the expense of others. Therefore, GP requires an ordinal ranking of the goals in order of importance by the decision-maker. The solution process then satisfies goals beginning with the goal with the highest priority. Lower order goals are considered only after higher order goal shave been satisfied. The solutions to the higher order goals then become constraints for the lower order goals. As such it may not be possible to satisfy all goals to the required extent. GP can then be used to find a satisfactory level of achievement of the goals.

The general formulation of goal programming model is as follows.

Objective

$$\min = \sum w_i \left(d_i^+ + d_i^-\right) \tag{15.11}$$

Subject to

$$\sum a_{ij}x_j + d_i^- - d_i^+ = b_i \ \forall I \tag{15.12}$$

$$x_j,\ d_i^-,\ d_i^+ \geqslant 0 \ \forall i, j \tag{15.13}$$

where x_j represents a decision variable, w_i represents the weights attached to goal i, and d_i^-, d_i^+ represent the under achievement and overachievement of a goal i respectively, a_{ij} represents an $i \times j$ matrix which expresses the relationship between goals and sub goals and b_i represents component column vector expressing goals $(b_1, b_2, \ldots, b_i)$.

The basic approach of fuzzy goal programming is to establish a specific numeric goal for each objective, and then seek a solution that corresponds to minimum weighted sum of deviation of the objective functions from their respective goals. In FGP formulation, the membership functions are transformed into membership goals by assigning the highest degree (Unity) as the aspiration level and introducing under and over-deviational variables to each of them. Then in the goal achievement function, the under-deviational variables are minimized on the basis of importance of achieving the aspired goal levels in the decision making context.

For the maximisation objectives, the membership functions $\mu_A(x)$ of the fuzzy sets characterizing the objective functions $Z_i(x_i)$ rise linearly from 0 at worst possible values (l_i) to 1 at the best possible values (b_i) of the objectives (Fig. 15.1). For the minimisation objectives, the membership functions decrease linearly from 1 at the best possible values (b_i) to 0 at worst possible values (u_i) (Fig. 15.2). The best possible values (b_i) and worst possible values (u_i) are obtained by maximizing and minimizing each objective function separately. This type of membership function is adopted due to ease in defining the aspiration levels that indicate the best possible solution as these values are obtained after ignoring other goals.

For the maximisation objectives

$$\mu_{[Z_i(x_i)]} = \begin{cases} 0 & \text{for } Z_i(x_i) \leqslant l_i \\ (Z_i(x_i) - l_i)/(b_i - l_i) & \text{for } l_i \leqslant Z_i(x_i) \leqslant b_i \\ 1 & \text{for } Z_i(x_i) \geqslant b_i \end{cases} \tag{15.14}$$

For the minimisation of objectives

$$\mu_{[Z_i(x_i)]} = \begin{cases} 1 & \text{for } Z_i(x_i) \leqslant b_i \\ (u_i - Z_i(x_i))/(u_i - b_i) & \text{for } b_i \leqslant Z_i(x_i) \leqslant u_i \\ 0 & \text{for } Z_i(x_i) \geqslant u_i \end{cases} \tag{15.15}$$

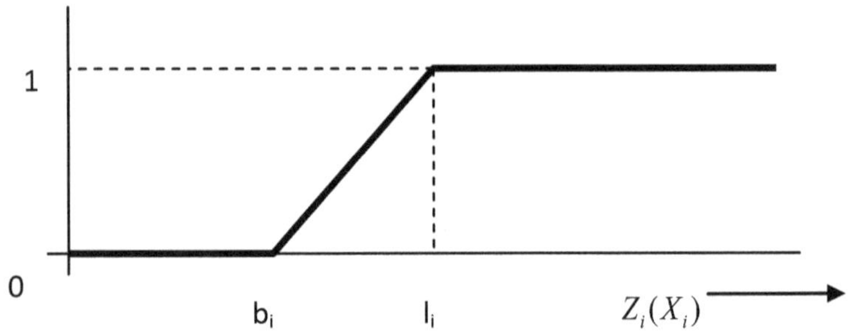

Fig. 15.1 Fuzzification of the maximization objective function $Z_i(x_i)$

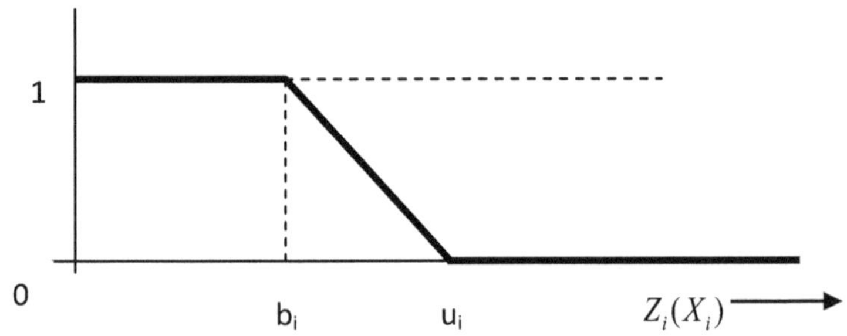

Fig. 15.2 Fuzzification of the minimization objective function $Z_i(x_i)$

The FGP model of the problem under a pre-emptive priority structure can be presented as:

Objective

$$\text{Minimise } Z = [d_k^+ + d_k^-] \tag{15.16}$$

Subject to

$$(Z_i(x_i) - w_i)/(b_i - w_i) - d_k^+ + d_k^- = 1 \tag{15.17}$$

$$(w_i - Z_i(x_i))/(w_i - b_i) - d_k^+ + d_k^- = 1 \tag{15.18}$$

$$d_k^+,\ d_k^- \geqslant 0 \quad k = 1,2,\ldots,K \tag{15.19}$$

where Z represents the vector of achievement function and d_k^+, d_k^- are the under and over-deviational variables of the kth goal. Integer programming is a special mathematical programming in which all of the variables are assumed to be only integer values. When there are not only integer variables but also conventional continuous variables, it is called mixed integer programming.

15.4.1 *Fuzzy Mixed Integer Goal Programming Model*

A mixed-integer programming (MIP) problem is one where some of the decision variables are constrained to have only integer values (i.e. whole numbers such as -1, 0, 1, 2, etc.) at the optimal solution. The use of integer variables greatly expands the scope of useful optimization problems that can be defined and solved. The four objectives of cost, emission, social acceptance and local resources are considered simultaneously but are allowed to have trade-off values in the final decision, the ambiguity of the decision making is modeled and problem is formulated as a fuzzy mixed integer goal programming as follows:
Objectives

$$\min \sum C_i x_i \cong \tilde{Z}_1 \qquad i = 1,\ 2,\ 3,\ 4 \tag{15.20}$$

$$\min \sum CO_i x_i \cong \tilde{Z}_2 \qquad i = 1,\ 2,\ 3,\ 4 \tag{15.21}$$

$$\max \sum S_i x_i \cong \tilde{Z}_3 \qquad i = 1,\ 2,\ 3,\ 4 \tag{15.22}$$

$$\max \sum x_i \cong \tilde{Z}_4 \qquad i = 1,\ 2,\ 3,\ 4 \tag{15.23}$$

Subject to

$$\sum x_i \geqslant \text{ Total cooking energy requirement,} \qquad i = 1,\ 2,\ 3,\ldots,\ 6 \tag{15.24}$$

$$\sum x_i \geqslant \text{ Total heating energy requirement} \qquad i = 7,\ 8,\ 9,\ 10 \tag{15.25}$$

$$\sum x_6 \leqslant (0.2\times \text{ Total cooking energy requirement}) \tag{15.26}$$

$$\sum \left(\frac{x_i}{\eta_i}\right) \leqslant \text{ Dung energy available} \qquad i = 5,\ 9 \tag{15.27}$$

$$\sum \left(\frac{x_1}{\eta_1'}\right) \leqslant \text{ Biogas energy available} \tag{15.28}$$

$$\sum \left(\frac{x_i}{\eta_i'}\right) \leqslant \text{ Biomass energy available} \qquad i = 2,\ 8 \tag{15.29}$$

$$x_i \geqslant 0 \text{ and are integer for } i = 1, 2, 3, \ldots, 10 \tag{15.30}$$

In the formulation of the objectives, the symbol '$\cong$' indicates the fuzziness of the goal. It represents the linguistic term of 'about' and means that the value of the objective should be in the vicinity of the aspiration level $(\tilde{Z}_i)$ within the deviations signified by tolerance limit. The constraints of the model are assumed to be crisp.

15.4.2 *Crisp Mathematical Formulation*

Using the approach formulated by Tiwari, Dharmar, and Rao [17], a linear membership function μ_k for the ith goal in the FMIGP formulation is considered. This membership function is used due to ease in defining the aspiration levels that indicate the best possible solution as these values are obtained after ignoring other goals, and lower (upper) tolerance limit for the ith fuzzy goal.

Chen and Tsai [18] formulated fuzzy goal programming incorporating different importance and preemptive priorities by using an additive model to maximize the sum of achievement degrees of all fuzzy goals. In contrast to previous works, the proposed approach allows the decision-maker to determine a desirable achievement degree for each fuzzy goal to reflect explicitly the relative importance of these goals. This approach can generate a set of achievement degrees consistent with the decision-maker's expectations, even though the relative importance of the goals may change. Furthermore, in this paper we incorporate the decision-maker's preemptive priority structure into a single formulation. The resulting solutions satisfy both the preemptive priority structure and have the maximum achievement degrees in sum. Gharehgozli *et al.* [19] presented mixed integer goal programming model for a parallel machine scheduling problem. Two fuzzy objectives have been considered in the model to minimize the total weight flow time. El-Wahed and Lee [20] discussed an interactive fuzzy goal programming approach to determine the preferred compromise solution for the multi objective integer programming. Kumar *et al.* [21] formulated vendor selection problem as a fuzzy multi-objective integer goal programming. Tsai *et al.* [16] presented fuzzy mixed integer goal programming model to address a steel supplier's channel allocation problem.

Using the approach presented by Yang *et al.* [22], which is based on a piecewise linear approximation with min-operator for aggregating fuzzy objectives, and the approach of Chen and Tsai [18], which allows decision makers to determine a desired achievement degree and importance (or weight) of each of the fuzzy goals, the energy planning FMIGP formulation is solved by maximizing the summation of achievement degrees. Once the membership functions of the fuzzy objectives are determined, the fuzzy optimization prob-

lem formulation is transferred into an equivalent crisp formulation for optimization with the variable of the achievement degree. An equivalent crisp additive model for energy allocation by fuzzy mixed integer goal programming approach is formulated as follows:
Objective function

$$\max f(u) = \sum_{k=1}^{4} \mu_k \tag{15.31}$$

Subject to

$$\mu_1 \leqslant \mu_{[Z_1(x_1)]} \quad \text{(minimization of cost)} \tag{15.32}$$

$$\mu_2 \leqslant \mu_{[Z_2(x_2)]} \quad \text{(minimization of emissions)} \tag{15.33}$$

$$\mu_3 \leqslant \mu_{[Z_3(x_3)]} \quad \text{(maximization of social acceptance)} \tag{15.34}$$

$$\mu_4 \leqslant \mu_{[Z_4(x_4)]} \quad \text{(maximization of local resources utilization)} \tag{15.35}$$

$$\sum x_i \geqslant \text{Total cooking energy requirement}, \; i = 1, 2, 3 \ldots, 6 \tag{15.36}$$

$$\sum x_i \geqslant \text{Total heating energy requirement}, \; i = 7, 8, 9, 10 \tag{15.37}$$

$$\sum x_6 \leqslant (0.2\times \text{ Total cooking energy requirement}), \tag{15.38}$$

$$\sum \left(\frac{x_i}{\eta_i}\right) \leqslant \text{Dung energy available}, \; i = 5, 9 \tag{15.39}$$

$$\sum \left(\frac{x_1}{\eta_1'}\right) \leqslant \text{Biogas energy available}, \tag{15.40}$$

$$\sum \left(\frac{x_i}{\eta_i'}\right) \leqslant \text{Biomass energy available}, \; i = 2, 8 \tag{15.41}$$

$$\mu_i \geqslant \alpha_i, \quad i = 1, 2, 3, 4 \tag{15.42}$$

where α_i is the desirable achievement degree for the ith fuzzy goal.

15.4.3 *Model Formulation*

15.4.3.1 *Objective Functions*

In this case, linear membership function as given in Eq. (15.14) or Eq. (15.15) is used. The aspiration level of each of the fuzzy goal is obtained as the best possible solution by having the same constraints defined in mixed integer goal programming formulation and ignoring other goals. Therefore, we have four membership functions of objective functions for *Malshemba*:

$$\mu_{[Z_1(x_1)]} \leqslant \frac{2.94 \times 10^6 - Z_1(x_1)}{1.68 \times 10^6} \tag{15.43}$$

$$\mu_{[Z_2(x_2)]} \leqslant \frac{1.15 \times 10^6 - Z_2(x_2)}{0.84 \times 10^6} \tag{15.44}$$

$$\mu_{[Z_3(x_3)]} \leqslant \frac{z_3(x_3) - 21.4 \times 10^6}{91.6 \times 10^6} \tag{15.45}$$

$$\mu_{[Z_4(x_4)]} \leqslant \frac{Z_4(x_4) - 0.77 \times 10^6}{1.41 \times 10^6} \tag{15.46}$$

15.4.3.2 *Crisp Formulation*

The fuzzy mixed integer goal programming is thus formulated as follows:

Objective

$$\text{Maximize } f(u) = \mu_1 + \mu_2 + \mu_3 + \mu_4 \tag{15.47}$$

Subject to

$$\mu_{[Z_1(x_1)]} \leqslant \frac{2.94 \times 10^6 - Z_1(x_1)}{1.68 \times 10^6} \tag{15.48}$$

$$Z_1(x_1) = 0.429x_1 + 0.72x_2 + 0.82x_3 + 1.71x_4 + 0.31x_5 + 0.5x_6$$
$$+ 0.429x_7 + 0.72x_8 + 0.31x_9 + 0.5x_{10}$$

$$\mu_{[Z_2(x_2)]} \leqslant \frac{1.15 \times 10^6 - Z_2(x_2)}{0.84 \times 10^6} \tag{15.49}$$

$$Z_2(x_2) = 0.189x_1 + 0.742x_2 + 0.217x_3 + 0.179x_4 + 0.633x_5$$
$$+ 0.189x_7 + 0.743x_8 + 0.633x_9$$

$$\mu_{[Z_3(x_3)]} \leqslant \frac{z_3(x_3) - 21.4 \times 10^6}{91.6 \times 10^6} \tag{15.50}$$

$$Z_3(x_3) = 10.49x_1 + 10.49x_2 + 74.49x_3 + 74.49x_4 + 10.49x_5 + 7.12x_6$$
$$+ 10.49x_7 + 10.49x_8 + 10.49x_9 + 7.12x_{10}$$

$$\mu_{[Z_4(x_4)]} \leqslant \frac{Z_4(x_4) - 0.77 \times 10^6}{1.41 \times 10^6} \tag{15.51}$$

$$Z_4(x_4) = x_1 + x_2 + x_5 + x_6 + 1x_7 + 1x_8 + x_9 + x_{10}$$

$$x_1 + x_2 + x_3 + x_4 + x_5 + x_6 \geqslant 1.69 \times 10^6 \tag{15.52}$$

$$5.88x_5 + 5.88x_9 \leqslant 2.76 \times 10^5 \tag{15.53}$$

$$5.88x_2 + 5.88x_8 \leqslant 7.61 \times 10^6 \tag{15.54}$$

$$2x_1 \leqslant 1.63 \times 10^5 \tag{15.55}$$

$$x_6 \leqslant 2.99 \times 10^5 \tag{15.56}$$

$$x_{10} \leqslant 1.39 \times 10^5 \tag{15.57}$$

$$x_3 \leqslant 2.22 \times 1.38 \times 10^5 \tag{15.58}$$

$$x_1, x_2, x_3, x_4, x_5, x_6, x_7, x_8, x_9, x_{10} \geqslant 0 \tag{15.59}$$

15.5 Result and Discussion

The analytical tool used in the study is scenario analysis which is selected to explore the range of possible outcomes from alternative policy options to achieve socio-economic goals and environmental sustainability. The objective of working with scenarios is not to predict the future to understand the uncertainties in order to take decision which is robust under a wide range of possible futures. The vast uncertainties exist in the price of fuel and its responses to climate change, necessitating the use of possible scenarios of the future to explore the potential consequences of different alternative options for the sustainable development.

To explore the consequences of different alternatives possible and their quantitative significance, four different scenarios have been developed using mixed integer fuzzy goal programming model. They are developed to assess the importance of critical policy options for meeting the energy requirements of the rural region. These scenarios are designed to map out the feasible options.

15.5.1 *Scenario 1: Economic Scenario*

The rural poor, who are large part of rural population, do not have the purchasing power to adopt the technology as per market demand. As a result, most of the households in rural region depend on biomass for cooking. The collecting firewood, which is women's daily household responsibilities, is tiring as the women often have to walk for long distances in search of firewood. Therefore, attention must be made to that section of population to find the areas of resource that meet their energy needs. Even most of them do not have access to modern energy carriers. The economic scenario refers to the minimization of actual cost of bringing the energy resource to the end-use point. In this scenario, cost is assigned a higher priority as compared to other objectives.

In this scenario, cost is assigned a higher priority as compared to other objectives. The scenario aims at reducing the cost of fuel and end use devises to increase acceptance of clean fuel. This scenario shows that biogas use should be increased from present 1.41% of total cooking need to 5.47% in *Malshemba*, 1.05% to 5.71% in *Muradpur*, 1.17% to 4.55% in *S. Dukre* and 1.63% to 6.32% in *Yewta*. Solar cookers need to be promoted to 20.07%

(*Malshemba*), 19.91% (*Muradpur*), 13.90% (*S. Dukre*) and 17.33% (*Yewta*) of total cooking energy need to replace biomass. For heating, optimal solution shows that the biogas should be promoted to use its untapped potential. The main hurdle in dissemination of these technologies is their high initial cost. Massive efforts are clearly required to increase the use of biogas and solar cookers in this region.

15.5.2 *Scenario 2: Low Carbon Scenario*

There has been increased concern over environmental issues in last decade. So, the relation between energy, economy, and environment has become an important factor for the decision makers. Therefore, in this scenario, emission is assigned higher priority as compared to other objectives. The scenario aims at reducing the emission to carry out energy planning for achieving sustainable energy development of the region considering associated environmental impacts. The cooking energy requirement in environmental scenario can be met by promoting energy sources like biogas and LPG for cooking and biogas for heating.

The cooking energy requirement in low carbon scenario can be met by promoting energy sources like LPG, kerosene and solar thermal. LPG has lowest carbon emission and burns cleanly with no soot. Kerosene (by using efficient stove designs with over 30% of thermal efficiency) seems to be the obvious choice as a transition cooking fuel in rural, particularly for populations with low and middle incomes. The equipment is less costly as compared with LPG stoves. For water heating, biomass fuel is to be replaced by biogas and solar thermal.

15.5.3 *Scenario 3: Sustainable Energy Scenario*

This scenario includes safe environmentally sound and sustainable technologies. A considerable change is expected in the fuel consumption pattern compared to other scenarios. Increase in research, development and deployment efforts are needed for the realization of this scenario which can lead to the sustainable rural energy development

The social acceptance is the main constraining factor in achieving the ambitious government targets to increase the share of renewable energy and to reduce the green house gas emissions. Socio-political and community acceptance are the two important influencing factors. Therefore, cost is assigned a higher priority as compared to other objectives in this scenario.

The scenario aims at increasing the acceptance of clean fuel. Kerosene and LPG are likely to be more acceptable in rural region due to comparatively cleaner fuel while solar thermal heating and cooking are not easily accepted due to high initial cost and lack of awareness. In this scenario, the optimal allocations are kerosene (15.06% of total cooking energy needs in *Malshemba*, 9.63% in *Muradpur*, 7.77% in *S. Dukre* and 10.02% in *Yewta*) and LPG (73.97% of total cooking energy needs in *Malshemba*, 84.66% in *Muradpur*, 84.84% in *S. Dukre* and 80% in *Yewta*). The use of LPG for cooking in rural region mainly relates to their standard of living. The real impact on the end use pattern is an income level. The people in the region rely mostly on biomass as it is freely available. The traditional cookstoves using fuelwood have low thermal efficiencies of about 14%. Efficient stove designs with over 30% of thermal efficiency are available and are being intensively implemented in India.

The biomass energy for the domestic use is essentially managed by women without technology or investment, and involves unsustainable practices, backbreaking drudgery, health problems especially for women and the girl child and likely environmental damage. Gasification or biogas can be used for using the biomass efficiently and conveniently to make energy this resource more sustainable. People living in the rural region aspire to have clean and convenient fuel just like their urban counterparts. Though falling in its share of the total energy mix, biomass dependence shall continue to rise in absolute terms in future. As such, a technology mission for enhanced and efficient use of biomass is highly desirable.

15.5.4 *Scenario 4: Local Resources Scenario*

The allocation process should prefer locally available resources for cooking and heating to reduce the vulnerability of the use of other expensive energy sources. In this scenario, local resources are assigned a higher priority as compared to other objectives. The scenario aims at increasing the use of locally available fuels and reducing the dependency on commercial fuels. Rural household energy supply mainly relies on the available quantities of local natural energy resources. Biomass is being increasingly substituted by commercial fuels; the shift to commercial fuels has been negligible in rural areas.

In this scenario, local resources are assigned a higher priority as compared to other objectives. The scenario aims at increasing the use of locally available fuels and reducing the dependency on commercial fuels. Rural household energy supply mainly relies on the available quantities of local natural energy resources.

Biomass is being increasingly substituted by commercial fuels; the shift to commercial fuels has been negligible in rural areas. In this scenario, biomass should be continued as fuel due to availability at reasonably low cost and use of solar cookers should be promoted to average of 20% of total cooking energy requirements in four villages. The dung cake for cooking is also to be promoted. The optimal allocations for cooking are biogas (5.47% of total cooking energy needs in *Malshemba*, 5.71% in *Muradpur*, 4.55% in *S. Dukre* and 6.32% in *Yewta*), biomass (71.31% of total cooking energy needs in *Malshemba*, 71% in *Muradpur*, 77.24% in *S. Dukre* and 73% in *Yewta)* and dung cake (3.15% of total cooking energy needs in *Malshemba*, 3.39% in *Muradpur*, 2.71% in *S. Dukre* and 3.32% in *Yewta*). Biogas is also allocated in optimal solution for heating (80.06% of total heating energy needs in *Malshemba*, 83% in *Muradpur*, 81.3% in *S. Dukre* and 82.7% in *Yewta*).

15.6 Conclusions

Fuzzy mixed integer goal programming model has been presented in this chapter for the rural cooking and heating energy planning. Four fuzzy objective and six constraints based on availability are considered for ten energy sources in four villages of Buldhana district, Maharashtra. Field studies haves shown that use of efficient stove results in only about 20% saving of fuelwood compared with traditional stove due to variation in cooking practices, inadequate user education; lack of repair and improper construction. So by shifting to best available technological options can reduce the dependence on costly fuels. Living standards in rural areas can be significantly improved by promoting a shift from direct combustion of biomass fuels (dung, crop residues, and fuelwood) in inefficient and polluting stoves to clean, efficient liquid or gaseous fuels.

In economic scenario, biogas and solar thermal have to be promoted for cooking and heating energy end use. In environmental scenario, LPG use has to be increased substantially to reduce consumption of biomass for cooking and biogas and solar thermal end uses for heating. Socially accepted scenario can be achieved by increasing the share of LPG along with kerosene. The biomass should be continued as a fuel with improved cook stoves for cooking and biogas and solar thermal may be extensively used for heating to get local resource scenario. Cost associated is more in social acceptance scenario due to LPG promotion while emission is highest in economical scenario indicating a need of technology improvement in cooking devices. As a future scope, the optimized result of various scenarios can further be integrated with behavioral analysis of fuel choice for the possible fuel use shift for cooking and heating in rural region and with policy evaluation framework for the effective implementation strategy of required scenario.

Bibliography

[1] A. Charnes and W.W. Cooper, *Operations Research,* **18** (1963).
[2] E.L. Hannan, *Fuzzy Sets and Systems*, **235** (1981).
[3] D. Jones, S. Mirrazavi, and M. Tamiz, *European Journal of Operational Research*, **1** (2002).
[4] R. Slowinski, *Fuzzy Sets and Systems*, **217** (1986).
[5] L.A. Zadeh, *IEEE Transactions on Automatic Control*, **59** (1963).
[6] L.A. Zadeh, *Fuzzy sets*, **338** (1965).
[7] R.E. Bellman and L.A. Zadeh, *Management science*, **141** (1970).
[8] H.J. Zimmermann, *International Journal of General Systems*, **209** (1975).
[9] S.M. Lee and R.L. Morris, *TIMS Stud. Mngt. Sci.*, **273** (1977).
[10] R. Narasimhan, *Decision Sciences*, **325** (1980).
[11] S.B. Sinha, K.A. Rao, and B.K. *Mangaraj, Socio-economic planning sciences*, **93** (1988).
[12] B.B. Pal, B.N. Moitra, and U. Maulik, *Fuzzy Sets and Systems*, **395** (2003).
[13] D.K. Sharma, D. Ghosh, and J.A. Alade, *Applied Mathematics and Computation*, **141** (2006).
[14] A. Biswas and B.B. Pal, *Omega*, **391** (2005).
[15] M. Kumar, P. Vrat, and R. Shankar, *Computers and Industrial Engineering*, **69** (2004).
[16] K.M. Tsai, S.Y. You, Y.H. Lin, and C.H. Tsai, *Expert Systems with Applications*, **1870** (2008).
[17] R.N. Tiwari, S. Dharmar, and J.R. Rao, *Fuzzy Sets and Systems*, **27** (1987).
[18] L. Chen and F. Tsai, *European Journal of Operational Research*, **548** (2001).
[19] A.H. Gharehgozli, R. Tavakkoli-Moghaddam, and N. Zaerpour, *Robotics and Computer Integrated Manufacturing*, **853** (2009).
[20] W.F.A. El-Wahed and S.M. Lee, *Omega*, **158** (2006).
[21] M. Kumar, P. Vrat, and R. Shankar, *International Journal of Production Economics*, **273** (2006).
[22] T. Yang, J.P. Ignizio, and H.J. Kim, *Fuzzy Sets and Systems*, **39** (1991).

Chapter 16

Fuzzy Simulation Based Chance Constrained Programming

Erhan Bozdag, Ceren Buket Kayı, Pınar Dursun

Industrial Engineering Department, ITU34367 Istanbul, Turkey

E-mail: {bozdagc, cakircer, dursunpi}@itu.edu.tr

This chapter describes chance constrained programming (CCP) to deal with optimization problems in fuzzy environment. Since in fuzzy CCP (FCCP) possibility or credibility of constraints must be considered, it is unavoidable to apply to the concepts of fuzzy variable and fuzzy measure which are based on possibility (credibility) theory. At first, the definitions of possibility (credibility) space and fuzzy variables are given, then several forms of FCCP models and their applications in the literature are presented. CCP models may be solved either by transforming them to their crisp equivalents or by means of fuzzy simulation. This chapter addresses fuzzy simulation which is used for estimating the possibility (credibility) of the constraints, and genetic algorithm (GA) which is applied as an optimization heuristic. Four numerical examples are given at the end of the chapter to illustrate the utilization of the specified methods.

16.1 Introduction

Many problems that decision makers encounter during in their managerial decision making process is considered in optimization theory. During optimization processes, decision maker aims to select appropriate values for a number of interrelated variables, by focusing on an objective which measures the system performance and the decision quality. This objective function is maximized (or minimized) while being subjected to a number of constraints that may limit the selection of values available for decision variables. Linear programming and non-linear programming are two most common optimization structures. Both of them may be classified in two parts: constrained and unconstrained. In unconstrained programming, there are no restrictions on the decision variable values.

C. Kahraman (ed.), *Computational Intelligence Systems in Industrial Engineering*,
Atlantis Computational Intelligence Systems 6, DOI: 10.2991/978-94-91216-77-0_16,

In their management decisions, decision makers usually come face to face with different kinds of uncertainties due to an uncertain environment that makes description of parameters difficult or even impossible. The uncertainties are two-fold: probability and possibility. If the problem uncertainties are probabilistic by nature, stochastic programming is the appropriate technique for reaching optimal decision. In stochastic programming, at least one parameter of a constraint is a random variable. One method that deals with stochastic parameters is chance constrained programming (CCP) which was proposed by Charnes and Cooper [1]. In this method uncertainty is handled by specifying a confidence level at which the stochastic constraint holds. The term chance signifies the probability that the constraints are satisfied.

However, randomness is not the unique uncertainty in real world. It is likely that the probability distribution of a random variable may be difficult to acquire due to lack of precise information and historical data. In such cases, parameters are approximately estimated based on decision makers' subjective judgment, and fuzzy theory introduced by Zadeh [2] offers a powerful tool with these cases. In 1978, Zadeh [3] puts forward the possibility measure and necessity measure respectively to deal with fuzzy events. In possibility theory two numbers, possibility and necessity are used to describe how likely a fuzzy event occurs. Analogous to CCP with stochastic parameters, when there are fuzzy parameters in the environment, possibility based CCP models can be used. In this case the chance is represented by possibility and a minimum acceptance level is specified for the fuzzy constraints.

Analogous to probability measure, Liu and Liu [4] proposed credibility measure in order to measure a fuzzy event. As the credibility measure has a self-dual property, it is advantageous over possibility measure. For example, in possibility measure, when a possibility value reaches 1, the fuzzy event may still fail to happen. However, in credibility measure if a fuzzy event achieves the credibility value 1, the fuzzy event will surely happen. After their study, credibility-based CCP has become prevalent. Another approach utilizes Hurwicz criterion which is a linear combination of possibility and necessity [5]. Credibility measure can be reached by taking equal weights in Hurwicz criterion.

The outline of this chapter is as follows, the next section gives preliminary knowledge of fuzzy measures on fuzzy events. Section 16.3 introduces FCCP models and provides several examples in the literature. In this section a hybrid algorithm which combines fuzzy simulation and genetic algorithm (GA), is also presented. Numerical examples are formulated and solutions are given in Section 16.4. Finally, the conclusions are presented in Section 16.5.

16.2 Preliminaries

16.2.1 *Fuzzy Variable*

In probability theory a random variable is defined as a function from probability space to the real line $\mathbb{R}$ and an event is a set of outcomes to which a probability assigned (for finite outcomes). However as mentioned in previous section, probability is one of the uncertainty types. Fuzzy variables were first introduced by Kaufmann [6], then it was re-analyzed by Zadeh [7]. Zadeh [3] proposed the possibility theory similar to the probability theory and developed a possibility measure to measure a fuzzy event. The possibility measure is not self-dual even though a self-dual measure is absolutely needed in both theory and practice [8]. Liu and Liu [4] presented the concept of credibility measure which has a self-dual property.

In possibility theory, a fuzzy variable is defined as a function from possibility space to the real line $\mathbb{R}$ and its membership function is derived from the possibility measure which is the degree of fuzzy event happening. Even though a fuzzy event has a possibility measure 1, it will not surely be happening. If the necessity measure of the fuzzy event is also 1, then this fuzzy event will definitely occur. Credibility measure overcomes the unidirectional property of possibility measure. The fuzzy event will certainly occur when its credibility is 1 and will certainly not occur if its credibility is 0. Liu [9] proposed credibility theory which is based on credibility measure and credibility space.

We will give definition of possibility space and axioms of possibility measure and credibility measure below.

16.2.2 *Spaces and Measures*

Before defining measures of a fuzzy variable, possibility space, necessity space and their axioms are given below.

Definition of Possibility. Let Θ be a non-empty set, and $\mathscr{P}(\Theta)$ the power set of Θ. The set function *Pos* is called a possibility measure and triplet $(\Theta, \mathscr{P}(\Theta), \text{Pos})$ is called a possibility space, if they satisfy following conditions:

(1) $\text{Pos}(\Theta) = 1, \ \ \text{Pos}(\varnothing) = 0$

(2) $\text{Pos}\{\bigcup_i A_i\} = \sup_i \text{Pos}(A_i)$ for any arbitrary collection $\{A_i\} \subseteq \mathscr{P}(\Theta)$

Let Pos be the measure of possibility then we have following axioms:

(1) $\forall A \in \mathscr{P}(\Theta),\ \ 0 \leqslant Pos(A) \leqslant 1$.
(2) If $A \subset B$, then $\mathrm{Pos}(A) \leqslant \mathrm{Pos}(B)$.
(3) $\forall A, B \in \mathscr{P}(\Theta),\ \ \mathrm{Pos}(A \cup B) \leqslant \mathrm{Pos}(A) + \mathrm{Pos}(B)$.

Definition of Necessity. Let $(\Theta, \mathscr{P}(\Theta), \mathrm{Pos})$ be a possibility space, and $A \in \mathscr{P}(\Theta)$, then $\mathrm{Nec}(A) = 1 - \mathrm{Pos}(A^c)$ is a necessity measure of each event A, where A^c is the complementary event of A.

Definition of Credibility. Let $(\Theta, \mathscr{P}(\Theta), \mathrm{Pos})$ be a possibility space, and $A \in \mathscr{P}(\Theta)$, then $\mathrm{Cr}(A) = [\mathrm{Pos}(A) + \mathrm{Nec}(A)]/2$ is called a credibility measure of each event A, i.e. credibility.

Axioms of credibility are:

Axiom 1. (Normality) $\mathrm{Cr}\{\Theta\} = 1$.
Axiom 2. (Monotonicity) $\mathrm{Cr}\{A\} \leqslant \mathrm{Cr}\{B\}$, whenever $A \subset B$.
Axiom 3. (Self-Duality) $\mathrm{Cr}\{A\} + \mathrm{Cr}\{A^c\} = 1$, for any event A.
Axiom 4. (Maximality) $\mathrm{Cr}\left\{\bigcup_i A_i\right\} = \sup_i \mathrm{Cr}\{A_i\}$,
for any events $\{A_i\}$ with $\sup_i \mathrm{Cr}\{A_i\} < 0.5$.

Now possibility, necessity and credibility measures can be defined on aforementioned spaces. A fuzzy number is a fuzzy subset of $\mathbb{R}$ with membership function $\mu : \mathbb{R} \to 0, 1$. Let $\tilde{a}$ be a fuzzy number with the membership function $\mu_{\tilde{a}}$, and b be a crisp number. Based on the concepts and techniques of possibility theory found by Zadeh [3], the possibility of $\tilde{a} \leqslant b$ is defined as follows:

$$\mathrm{Pos}\{\tilde{a} \leqslant b\} = \sup\{\mu_{\tilde{a}}(x) \mid x \in \mathbb{R},\ x \leqslant b\} = \sup_{x \leqslant b} \mu_{\tilde{a}}(x)$$

And similarly possibility of $\tilde{a} < b$ and $\tilde{a} = b$ are defined as follows:

$$\mathrm{Pos}\{\tilde{a} < b\} = \sup\{\mu_{\tilde{a}}(x) \mid x \in \mathbb{R},\ x < b\} = \sup_{x < b} \mu_{\tilde{a}}(x)$$

$$\mathrm{Pos}\{\tilde{a} = b\} = \sup\{\mu_{\tilde{a}}(x) \mid x \in \mathbb{R},\ x = b\} = \sup_{x = b} \mu_{\tilde{a}}(x) = \mu_{\tilde{a}}(b)$$

The necessity of $\tilde{a} \leqslant b$ is defined as follows:

$$\mathrm{Nec}\{\tilde{a} \leqslant b\} = 1 - \sup\{\mu_{\tilde{a}}(x) \mid x \in \mathbb{R},\ x > b\} = 1 - \sup_{x > b} \mu_{\tilde{a}}(x)$$

And similarly necessity of $\tilde{a} < b$ and $\tilde{a} = b$ are defined as follows:

$$\mathrm{Nec}\{\tilde{a} < b\} = \sup\{\mu_{\tilde{a}}(x) \mid x \in \mathbb{R},\ x < b\} = 1 - \sup_{x \geqslant b} \mu_{\tilde{a}}(x)$$

Table 16.1 Calculated fuzzy measures for $a \leqslant b$

$a \leqslant b$	$b=-2$	$b=0$	$b=3$	$b=4$	$b=6$
Possibility	0	0.25	1	1	1
Necessity	0	0	0	0.5	1
Credibility	0	0.125	0.5	0.75	1

$$\text{Nec}\{\tilde{a}=b\} = \sup\{\mu_{\tilde{a}}(x) \mid x \in \mathbb{R},\, x=b\} = 1 - \sup_{x \neq b} \mu_{\tilde{a}}(x)$$

Based on the concepts and techniques of credibility theory proposed by Liu and Liu [4], the credibility of any fuzzy event is calculated as the average of Possibility and Necessity of the event. For example, credibility of $\tilde{a} \leqslant b$ is defined as follows:

$$\text{Cr}\{\tilde{a} \leqslant b\} = \frac{1}{2}\left[\text{Pos}\{\tilde{a} \leqslant b\} + \text{Nec}\{\tilde{a} \leqslant b\}\right]$$

Example 16.2.1. Consider the triangular fuzzy number $\tilde{a} = (-1, 3, 5)$ given in Figure 16.1 and crisp number b with different values of $\{-2, 0, 3, 4, 6\}$.

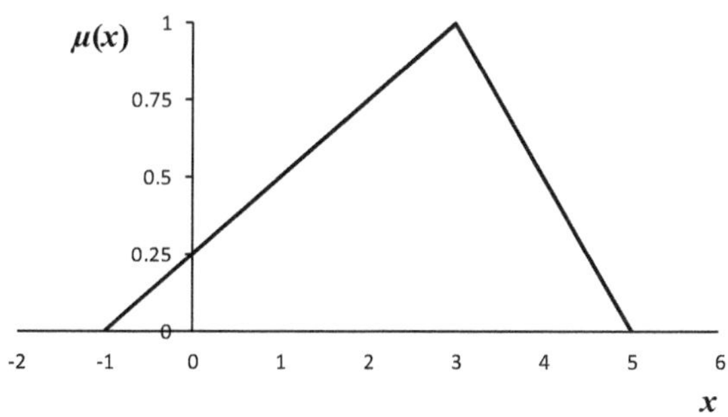

Fig. 16.1 Fuzzy number $\tilde{\mathbf{a}}$ given in example

Let us find all fuzzy measures for the comparison $\tilde{a} \leqslant b = 0$. In terms of possibility measure, considering portion of fuzzy number $\tilde{a}$ that is smaller than 0, the greatest membership value is seen as 0.25, which is possibility of $\tilde{a} \leqslant 0$. In terms of necessity measure, considering the portion of fuzzy number $\tilde{a}$ that is greater than 0, the greatest membership value is 1, then subtracting from 1, necessity measure is calculated as 0. Therefore, as being the average of the other two measures, credibility of $\tilde{a} \leqslant 0$ is calculated as 0.125.

Possibility, necessity and credibility measures of $\tilde{a} \leqslant b$ calculated for all b values are given in Table 16.1.

We can utilize similar calculations for $\tilde{a} \geqslant 0$ comparison. In terms of possibility measure, considering portion of fuzzy number $\tilde{a}$ that is greater than 0, the greatest membership

Table 16.2 Calculated fuzzy measures for $a \geqslant b$

$a \geqslant b$	$b = -2$	$b = 0$	$b = 3$	$b = 4$	$b = 6$
Possibility	1	1	1	0.5	0
Necessity	1	0.75	0	0	0
Credibility	1	0.875	0.5	0.25	0

value is seen as 1, which is possibility of $\tilde{a} \geqslant 0$. In terms of necessity measure, considering the portion of fuzzy number $\tilde{a}$ that is smaller than 0, the greatest membership value is 0.25, then subtracting from 1, necessity measure is calculated as 0.75. Therefore, as being the average of the other two measures, credibility of $\tilde{a} \geqslant 0$ is calculated as 0.875.

Similarly, possibility, necessity and credibility measures of $\tilde{a} \geqslant b$ calculated for all b values are given in Table 16.2.

In summary, considering the trapezoidal fuzzy number $\tilde{a} = (a_1, a_2, a_3, a_4)$, and the crisp number b, all three fuzzy measures are calculated as follows:

$$\text{Pos}\{\tilde{a} \leqslant b\} = \begin{cases} 0 & b \leqslant a_1 \\ \dfrac{b - a_1}{a_2 - a_1} & a_1 < b \leqslant a_2 \\ 1 & b > a_2 \end{cases}$$

$$\text{Nec}\{\tilde{a} \leqslant b\} = \begin{cases} 0 & b \leqslant a_3 \\ \dfrac{b - a_3}{a_4 - a_3} & a_3 < b \leqslant a_4 \\ 1 & b > a_4 \end{cases}$$

$$\text{Cr}\{\tilde{a} \leqslant b\} = \begin{cases} 0 & b \leqslant a_1 \\ \dfrac{b - a_1}{a_2 - a_1} & a_1 < b \leqslant a_2 \\ 0.5 & a_2 < b \leqslant a_3 \\ 0.5 + \dfrac{b - a_3}{a_4 - a_3} & a_3 < b \leqslant a_4 \\ 1 & b > a_4 \end{cases}$$

In Figure 16.2, the change in the three measures with respect to b is shown, considering comparison $\tilde{a} \leqslant b$. In the upper figures $\tilde{a}$ is considered as a triangular fuzzy number where $a_2 = a_3$ and in the lower figures it is considered as a trapezoidal fuzzy number. The first observation is that all 3 measures increase with the increasing values of b. Yet analyzing the figures in detail, it is seen that the possibility is the most optimistic one, while necessity is the most pessimistic, and in comparison to the other two measures, credibility is the most realistic one.

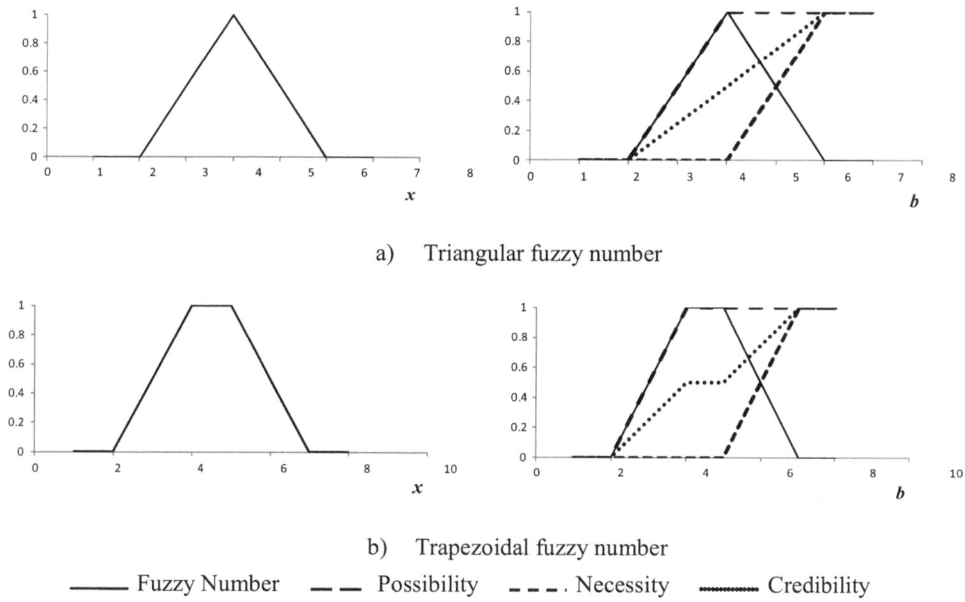

Fig. 16.2 Possibility, Necessity and Credibility measures for triangular and trapezoidal fuzzy numbers

16.3 Chance Constrained Programming

Stochastic programming offers a framework for modeling optimization problems with random variables. CCP which is developed by Charnes and Cooper [1] as a kind of stochastic programming, models stochastic decision systems using the assumption that the constraints with stochastic parameters must be maintained at least at the prescribed levels of probability.

CCP with fuzzy parameters is an extension of modeling from stochastic decision systems to fuzzy environment. In FCCP it is assumed that the constraints will hold for at least prescribed confidence levels of possibility. First studies on FCCP is based on the possibility theory and developed by Liu and Liu [4]. After they proposed credibility measure and Liu [9] developed the credibility theory, FCCP based on credibility theory have pervaded. In this section, firstly we will present several forms of CCP based on the possibility theory. But one can attain similar CCP models by using the credibility measure instead of possibility measure in constraints. Expected value method will be presented at the end of the section. We will give several applications that are seen in literature for all form of FCCP.

The general form of CCP with fuzzy parameters is as follows [10]:

$$\begin{aligned} &\max f(\mathbf{x}) \\ &\text{subject to:} \\ &\text{Pos}\{\boldsymbol{\xi} \mid g_i(\mathbf{x},\boldsymbol{\xi}) \leqslant 0, \quad i=1,2,\ldots,p\} \geqslant \alpha \end{aligned} \tag{16.1}$$

A decision vector $\mathbf{x}$ is accepted to be feasible if the possibility of the constraints is satisfied with the possibility α, that is, the predetermined confidence level.

Some instances of studies using programming form given in Eq. (16.1) are as follows: Qin *et al.* [11] formulated the FCCP model to solve problems in municipal solid waste management and considered the predetermined confidence levels of two facility capacity constraints are equal. He *et al.* [12] took the environmental problems into consideration to determine the optimal groundwater remediation strategies under uncertainty. Vehicle routing problems (VRP) have various uncertainties and may be considered by FCCP. When travel times are fuzzy, there is credibility constraints that provide customers are visited within their time windows with a confidence level emerge [13]. If demands are fuzzy in VRP, the objective is to minimize total distance together with additional distance caused by uncertainty of the demands [14]. In Open VRP, after serving all customers on the route, the vehicles do not have to return to the depot. Cao and Lai [15] proposed a FCCP model for Open VRP.

It is possible to separate chance constraints, so the model will be as follows [10]:

$$\begin{aligned} &\max f(\mathbf{x}) \\ &\text{subject to:} \\ &\text{Pos}\{\boldsymbol{\xi} \mid g_i(\mathbf{x}, \boldsymbol{\xi}) \leqslant 0\} \geqslant \alpha_i, \quad i=1,2,\ldots,p \end{aligned} \tag{16.2}$$

For example, Zheng and Liu [13] assumed travel times to be fuzzy variable in a VRP. Since the arrival times are fuzzy variables, they proposed a model to minimize the total distance of vehicles under fuzzy credibility constraints. Cao *et al.* [16] developed a model to solve crude oil scheduling considering both stochastic and fuzzy demand of crude.

Besides Eq. (16.1) and Eq. (16.2), the multiobjective programming model with chance constraints is proposed [10]:

$$\begin{aligned} &\max\,[f_1(\mathbf{x}),\ f_2(\mathbf{x}),\ \ldots,\ f_m(\mathbf{x})] \\ &\text{subject to:} \\ &\text{Pos}\{\boldsymbol{\xi} \mid g_i(\mathbf{x}, \boldsymbol{\xi}) \leqslant 0\} \geqslant \alpha_i, \quad i=1,2,\ldots,p \end{aligned} \tag{16.3}$$

The chance constrained goal programming (CCGP) is also proposed to the literature, the more general form of the model is [17]:

$$\begin{aligned}
&\min \sum_{j=1}^{l} P_j \sum_{i=1}^{m} (u_{ij}d_i^+ + v_{ij}d_i^-) \\
&\text{subject to:} \\
&\text{Pos}\{\boldsymbol{\xi} \mid f_i(\mathbf{x}, \boldsymbol{\xi}) + d_i^- - d_i^+ = b_i\} \geqslant \beta_i, \quad i = 1, 2, \ldots, m \\
&\text{Pos}\{\boldsymbol{\xi} \mid g_j(\mathbf{x}, \boldsymbol{\xi}) \leqslant 0\} \geqslant \alpha_j, \quad j = 1, 2, \ldots, p \\
&d_i^-, d_i^+ \geqslant 0, \quad i = 1, 2, \ldots, m
\end{aligned} \tag{16.4}$$

The P_j is the preemptive priority factor of the jth goal. $P_j \gg P_{j+1}$ for all j and u_{ij} and v_{ij} are the weighting factors of the positive and the negative deviation for goal i with priority j, respectively. b_i is the target value for goal i. There are m goals and l priorities in the model.

Hou and Wu [18] used CCGP for the series parallel redundancy optimization problem. Pal and Gupta [19] proposed a model integrates stochastic simulation and fuzzy goal programming. Using stochastic programming they get the best and the worst solutions for maximization of each objective, then put the solutions into fuzzy goal programming model to obtain the membership values of the goals. Liang and Gao [20] proposed a dependent-chance multi-objective programming model for the capital budgeting problem with conflicting objectives. To balance these objectives, the dependent-chance goal programming model was formulated and solved by fuzzy simulation-based GA.

A CCP with fuzzy coefficients occurring in both constraints and objective may be written as follows [10]:

$$\begin{aligned}
&\max \overline{f} \\
&\text{subject to:} \\
&\text{Pos}\{\boldsymbol{\xi} \mid f(\mathbf{x}, \boldsymbol{\xi}) \geqslant \overline{f}\} \geqslant \beta, \\
&\text{Pos}\{\boldsymbol{\xi} \mid g_j(\mathbf{x}, \boldsymbol{\xi}) \leqslant 0, \quad j = 1, 2, \ldots, p\} \geqslant \alpha.
\end{aligned} \tag{16.5}$$

For each fixed feasible solution x, the objective value $\overline{f}$ should be maximum that the objective function $f(x, \xi)$ is achieved with at least possibility β. It is also possible to separate the constraints of this problem, and formulate the multiobjective version of the model. For the rest of the models and detailed information, interested reader may read the study by Liu and Iwamura [17]. There are many applications in the literature for this model type. For example, Lan *et al.* [21] developed a mathematical programming model which has a credibility measure in the objective function and used approximation method to estimate the credibility value of the objective function. Wen and Iwamura [5] presented a model for the capacitated facility location-allocation problem with fuzzy demands. Since the demands are considered as fuzzy variables, the transportation cost becomes fuzzy too. In order to balance the pessimistic criterion and optimistic criterion, they utilized Hurwicz criterion. Since quality, budget, and demand may be given as fuzzy variables in a vendor selection problem, the objective function and some constraints are represented by fuzzy variables. Wang *et al.* [22] developed expected value model and FCCP model for the vendor selection problem.

Fuzzy expected value models were formulated by Liu and Liu [4]. The expected value of a fuzzy variable ξ is defined by

$$E[\xi] = \int_0^{\infty} \mathrm{Cr}\{\boldsymbol{\xi} \geqslant r\}dr - \int_{-\infty}^{0} \mathrm{Cr}\{\boldsymbol{\xi} \leqslant r\}dr$$

provided that at least one of the two integrals is finite. In order to obtain the decision with maximum expected return, we can employ the following single-objective fuzzy expected value model [4]:

$$\begin{aligned} &\max E[f(\mathbf{x}, \boldsymbol{\xi})] \\ &\text{subject to:} \\ &E[g_i(\mathbf{x}, \boldsymbol{\xi})] \leqslant 0, \quad i = 1, 2, \ldots, p. \end{aligned} \tag{16.6}$$

where $\mathbf{x}$ and $\boldsymbol{\xi}$ are decision vector and fuzzy vector, respectively, f is the objective function, and g_i are constraint functions for $i = 1, 2, \ldots, p$.

Zhou *et al.* [23] considered two different game structures for a supply chain. They estimated parameters of pricing strategies and expected value of profits using fuzzy simulation. Davari *et al.* [24] integrated fuzzy simulation with simulated annealing method to solve maximal covering location problem with fuzzy travel times. In order to solve the fuzzy programming model, they estimated expected value of the object function with fuzzy simulation.

The objective of a process planning problem may be to minimize the expected value of the project task path when completion times are fuzzy variables. In order to solve such a problem, Tian *et al.* [25] developed an expected value model with the expected completion time constraint of energy analysis for process planning. In the reverse logistic field, the design of the collection centers which collect used products from consumers and transfer to manufactory is an important problem. Qin and Ji [26] developed logistic network models in which returned products volume, cost of opening collection center and unit cost of handling in the collection center are independent fuzzy variables. They presented expected cost model, α-cost minimization model, credibility maximization model and their crisp equivalents. Lan *et al.* [27] presented a multi-period production planning and sourcing problem with a credibility service level model, in which demands, inventory and production costs are defined as fuzzy variables. They transformed the credibility constraint into its crisp equivalent form where demands are independent normal fuzzy variables. They also suggested an approximation algorithm to evaluate the objective function. In their hybrid particle swarm optimization algorithm, generated input-output set is evaluated by approximation approach and is used to train neural network to approximate the expected value function.

16.3.1 *Solution Methods of CCP*

In one of the ways of solving FCCP models, fuzzy chance constraints can be transformed to their deterministic versions (crisp equivalents) under predetermined confidence levels and the obtained crisp model can be solved by using general optimization techniques. For example this method is used in environmental management problems [28–31], portfolio selection problems [32–35], scheduling problems [16, 36], and green logistics [37].

However, this is possible if the membership functions of the fuzzy parameters encountered in the problem satisfy some necessary conditions. They are needed to be in simple forms (unimodal) such as triangular and trapezoidal fuzzy sets, so this leads to limitation of application of this approach [38]. However, in many real-world applications, some parameters may be suitable to be expressed as more complex forms, such as exponential or Gaussian. This would lead to difficulties in generating deterministic equivalents [11].

In other solution method, fuzzy simulation is used to handle this limitation when the fuzzy parameters are more complex. In this group of studies, the heuristic search methods as GA [11, 13], simulated annealing (SA) [12, 24, 39], particle swarm optimization (PSO) [21, 27], differential evaluation algorithm [14, 15] etc. are generally utilized to obtain optimal solutions by integrating with fuzzy simulation.

Some authors emphasize time consuming calculations in the fuzzy simulation and recommend a proxy simulator based on artificial neural network (ANN). In this approach, the fuzzy simulation is used at the initialization step of the algorithm to provide enough data for the training of ANN. When there is a need to calculate expected value or any fuzzy measure of a fuzzy variable, we apply to the trained network [12, 21, 25, 27].

In the context of this study, integration of fuzzy simulation and GA will be explained in detail in the next sub sections. Several illustrative examples and applications will be given in section 16.4.

16.3.1.1 *Fuzzy Simulation*

As proposed by Liu and Iwamura [10], fuzzy simulation technique is used to check the feasibility of fuzzy system's constraints without converting them to the crisp equivalents. Fuzzy simulation technique is basically an application of Monte Carlo simulation, and used in many studies to deal with chance constraints. It is important to be aware of the fact that the fuzzy simulation does not calculate the exact results of measures, but approximates to the exact results.

The calculation of credibility involves the calculation of possibility and necessity. Therefore, we will give only an algorithm for credibility measure. The following formulas approximately calculate the fuzzy measures (possibility, necessity, and credibility) of the event $F(\mathbf{x},\xi) = F_0$, where $\mathbf{x}$ is a decision vector, $\mathbf{u}_k = (u_{1k}, u_{2k}, \ldots, u_{nk})$ is a real number vector for k^{th} iteration, $\boldsymbol{\xi} = (\xi_1, \xi_2, \ldots, \xi_n)$ is a fuzzy parameter vector, $\mu(\mathbf{u_k})$ is the membership function of fuzzy parameter vector $\boldsymbol{\xi}$, and N is the number of iteration of fuzzy simulation algorithm.

$$\mathrm{Pos}\{F(\mathbf{x},\mathbf{u}_k) \leqslant F_0\} = \max_{1\leqslant k\leqslant N}\{\mu(\mathbf{u}_k) \mid F(\mathbf{x},\mathbf{u}_k) \leqslant F_0\}$$

$$\mathrm{Nec}\{F(\mathbf{x},\mathbf{u}_k) \leqslant F_0\} = \min_{1\leqslant k\leqslant N}\{1-\mu(\mathbf{u}_k) \mid F(\mathbf{x},\mathbf{u}_k) > F_0\}$$

$$\mathrm{Cr}\{F(\mathbf{x},\mathbf{u}_k) \leqslant F_0\} = \frac{1}{2}\left(\max_{1\leqslant k\leqslant N}\{\mu(\mathbf{u}_k) \mid F(\mathbf{x},\mathbf{u}_k) \leqslant F_0\} + \min_{1\leqslant k\leqslant N}\{1-\mu(\mathbf{u}_k) \mid F(\mathbf{x},\mathbf{u}_k) > F_0\}\right)$$

In the initialization step of a fuzzy simulation, possibility measure is set to 0 and necessity measure is set to 1. In each iteration, a crisp number u_i is randomly generated for each fuzzy variable ξ_i from its ε-level set (i.e. $\mu_i(u_i) > \varepsilon$) where ε is sufficiently small positive number. The membership function of fuzzy parameter vector $\boldsymbol{\xi} = (\xi_1, \xi_2, \ldots, \xi_n)$ in kth iteration can be obtained by $\mu(\mathbf{u_k}) = \mu(u_{1k}) \wedge \mu(u_{2k}) \wedge \cdots \wedge \mu(u_{nk})$. If constraint function $F(\mathbf{x},\mathbf{u}_k)$ is equal to or less than F_0 and $\mu(\mathbf{u_k})$ is greater than possibility calculated through previous $k-1$ iterations, possibility is replaced by $\mu(\mathbf{u_k})$. On the other hand, if $F(\mathbf{x},\mathbf{u}_k)$ is greater than F_0 and $1-\mu(\mathbf{u_k})$ is less than necessity calculated through the last $k-1$ iterations, necessity is replaced by $1-\mu(\mathbf{u_k})$. Credibility is the average of possibility and necessity. While possibility and necessity are recalculated in each iteration of fuzzy simulation, credibility is calculated only after the last iteration.

Fuzzy simulation algorithm is explained in the Figure 16.3.

Example 16.3.1. Consider a triangular fuzzy number $\tilde{a} = (0, 2, 5)$ given in the Figure 16.4 and events of $\tilde{a} > 3$ and $\tilde{a} < 3$. The exact value of credibility, possibility, and necessity of

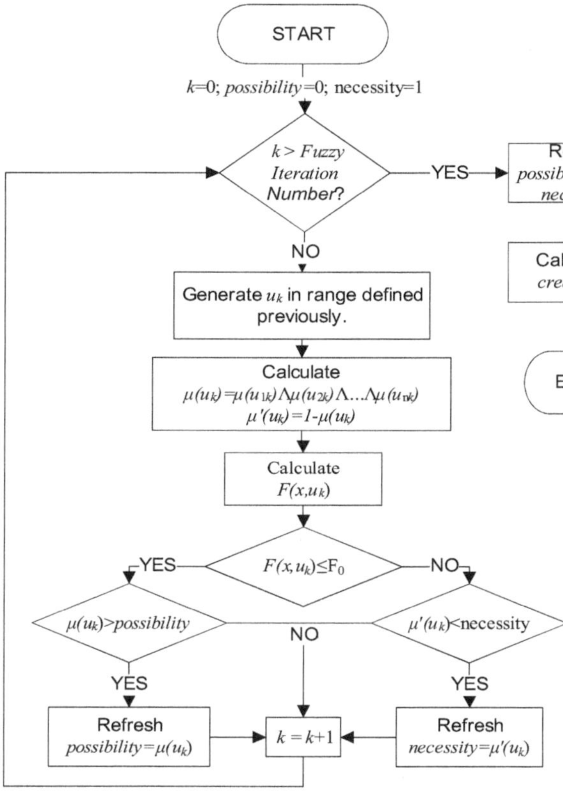

Fig. 16.3 Fuzzy simulation algorithm

event $\tilde{a} > 3$ is easily calculated as 0.333, 0.666, and 0.000, respectively. Similarly we can compute credibility, possibility and necessity for event $\tilde{a} < 3$ as 0.666, 1.000, and 0.333, respectively.

Fuzzy simulation was coded in C in order to calculate fuzzy measures for the example. We take the interval $[0, 5]$ as ε-level set of fuzzy number. In each iteration, a crisp number u is generated uniformly on $[0, 5]$ and its membership value is computed. If u is equal to or greater than 3, we compare its membership value with previous possibility. If previous possibility is less than the membership value, we replace it with the membership value. On the contrary, if u is less than 3, possibility of fuzzy event is zero. In this case if one minus the membership value is less than necessity, we assign it as necessity. Once all iterations are completed, credibility is calculated as the average of possibility and necessity.

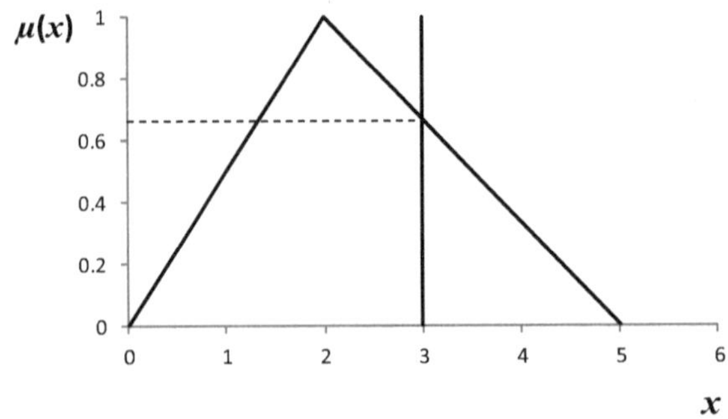

Fig. 16.4 Triangular fuzzy number $\tilde{a} = (0, 2, 5)$ and crisp number 3

When we perform computer program with $N = 200$ iterations, credibility, possibility and necessity values of fuzzy event $\tilde{a} > 3$ are obtained as 0.341, 0.664 and 0.019 respectively. We can accomplish to measure the event $\tilde{a} < 3$ by changing the direction of inequalities in the computer program for the event $\tilde{a} > 3$. For fuzzy event $\tilde{a} < 3$, credibility, possibility and necessity are 0.659, 0.981 and 0.336 respectively. Changes in possibility and necessity measures with respect to iteration number are given in Figure 16.5.

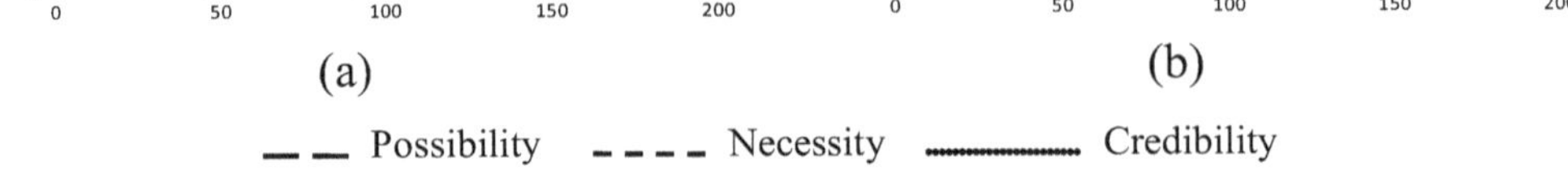

Fig. 16.5 Changes in measures with respect to iteration number (a) $\tilde{a} > 3$, (b) $\tilde{a} < 3$

16.3.1.2 *Genetic Algorithm*

This section attempts to present the implementation of GA in CCP. GA which is used with fuzzy simulation is explained in Fig. 16.6.

Some parameters used in the explanation of the GA are summarized in the Table 16.3. We classify these parameters in three groups: 1) parameters which should be defined before performing GA and set by decision maker, 2) parameters that affect the performance of GA, 3) parameters randomly generated in GA.

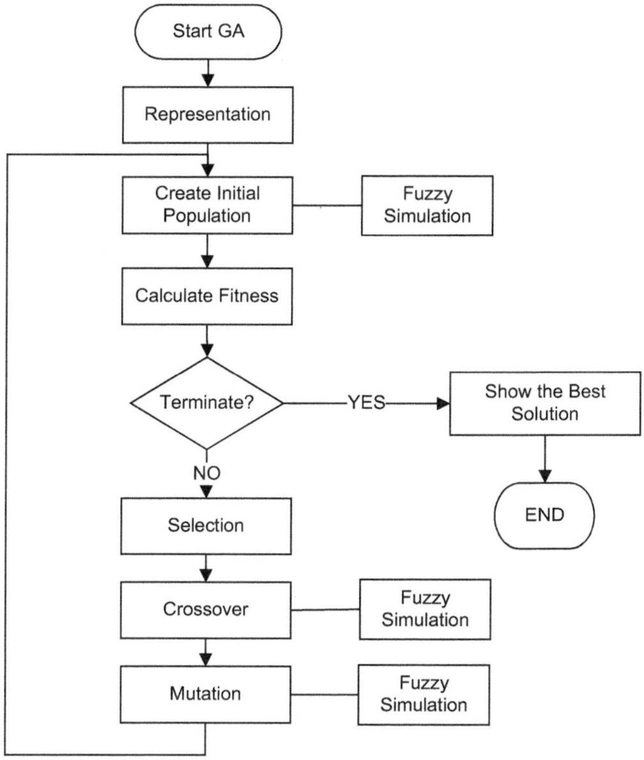

Fig. 16.6 GA for CCP

Terminating condition is also one of the important parameters that affect the performance of the GA. There are several ways to define the termination condition. It can easily be defined as a maximum number of iterations, and when this number is reached, the algorithm stops. It may also be defined as the number of iterations that the result does not improve. For example, if the number of iteration is 100 and an individual with better performance is generated in 99th iteration, then iteration number is set to zero and restart to iterations. Terminating condition may also be defined as a minimum level of improvement achieved in the last iteration. If an improvement that is less than the predetermined level is obtained, algorithm stops.

Table 16.3 Parameters used in GA

Symbol	Explanation	Needed in phase	Determination way
pop_size	Population size	All phases	Parameter
p_m	Mutation probability	Mutation	Parameter
p_c	Crossover probability	Crossover	Parameter
v_0	Interior point	Initialization, Mutation	Decision maker
M	Large number	Initialization, Mutation	Decision maker
d	Direction vector	Initialization, Mutation	In each iteration, randomly in unit sphere
a	$a \in (0,1)$	Natural Selection	Parameter
r	Random number	Mutation, Crossover	In each iteration, for each chromosome, randomly in $(0,1)$
c	Crossover ratio	Crossover	For each crossover, randomly in $(0,1)$

***Representation*:** Before implementing GA, it is essential to decide how to represent the chromosomes. A chromosome consists of decision variables of the model in question, so the number of genes must be equal to or greater than the number of decision variables. Binary or floating points may be chosen to represent genes.

***Initialization Process*:** We define *pop_size* as an integer that stands for the number of chromosomes in the population. It is possible to generate initial population in many different ways, but especially for complex optimization problems, one should use the rigorous method to produce feasible chromosomes. In this chapter, we explain such two methods which have been used in CCP. One very frequently used method is based on randomly generating feasible chromosomes [25, 34, 35]. If generated chromosome is not feasible, a new random chromosome is generated until a feasible one obtained. One another method used for generating chromosomes needs an interior point, denoted by v_0, in the feasible region [10]. A chromosome is defined as $x_j = v_0 + M * d$, where M is a large positive number, and d is a direction generated in $\mathbb{R}^n$ randomly. If chromosome does not hold for all of the constraints, we set M to a random number between 0 and M, and recalculate chromosome. This repetation is carried on until chromosome is feasible. Since v_0 is in feasible region, a feasible chromosome can be found in finite times. Repeating this method *pop_size* times, we get initial population.

***Fitness Value Calculation*:** Fuzzy objective function value of each chromosome is calculated by fuzzy simulation as fitness value of that chromosome. In this step, the chromosome which has the maximum fitness value among all chromosomes is selected and its fitness value is compared with the best solution reached. If it is better, we keep it as the best solution.

***Natural Selection*:** There are several natural selection procedures. In this chapter one of these methods will be explained which is frequently used in CCP [25, 34, 35, 40]. Chromosomes, whose fitness values have been calculated previously, are ranked with respect to their fitness values and their rank-probabilities are calculated by Eq. (16.7) where j is the rank of the chromosome x_j and a is a parameter in the genetic system. Then the cumulative probabilities are calculated by Eq. (16.8). Subsequently using the roulette wheel method new population is generated as follows: generate a random number r in $(0,1)$, if $p_{i-1} < r \leqslant p_i$, then, select i^{th} chromosome for the next generation. Repeat that procedure for *pop_size* times.

$$\text{Rank\_Probability}(x_j) = a(1-a)^{j-1}, \qquad j = 1, 2, \ldots, \text{ pop\_size} \tag{16.7}$$

$$p_0 = 0, \quad p_i = \frac{\sum\limits_{j=1}^{i} \text{Rank\_Probability}(x_j)}{\sum\limits_{j=1}^{\text{pop\_size}} \text{Rank\_Probability}(x_j)}, \qquad i = 1, 2, \ldots, \text{ pop\_size} \tag{16.8}$$

***Crossover Operation*:** Set the crossover probability as p_c. For each chromosome a random number r is generated to decide if this chromosome will be a parent or not. If $r < p_c$, then chromosome x_j is selected as a parent. Subsequently parents divided into pairs as $(\grave{x}_1, \grave{x}_2), (\grave{x}_3, \grave{x}_4), \ldots$, and crossover operation is done among pairs. For the pair $(\grave{x}_1, \grave{x}_2)$, a random number c is generated in $(0,1)$ and two children are produced as

$$x_{\text{child}\,1} = c\grave{x}_1 + (1-c)\grave{x}_2 \quad \text{and} \quad x_{\text{child}\,2} = (1-c)\grave{x}_1 + c\grave{x}_2.$$

After the production of children, feasibility of new chromosomes should be checked. If they are not feasible, new crossover ratio is generated in $(0,1)$ until the feasible chromosomes are obtained.

***Mutation Operation*:** Set the mutation probability as p_m. Similar to the crossover procedure, for each chromosome, a random number r is generated to decide if the chromosome will undergo to the mutation process or not. If $r < p_m$, then chromosome x_j is selected for the mutation operation. Mutation operation given in this chapter has been frequently used in CCP [13, 18, 25, 34, 40]. Similar to the initialization, a random mutation direction is selected as d, and new chromosome is generated as $x_{\text{new}\,j} = x_j + Md$. If the newly generated chromosome is not feasible then we confine M by replacing it with a random number between zero and M.

16.4 Numerical Examples

16.4.1 *Modified Blending Problem*

This problem is modified version of the problem in the study of Pagnoncelli *et al.* [41]. Some quantities in the original problem are considered as fuzzy numbers. The modified problem is as follows: Suppose a farmer has some crop and wants to use fertilizers to increase the production. He hires an agronomy engineer who recommends 7 g of nutrient A and 4 g of nutrient B. He has two kinds of fertilizers available: the first has w_1 g of nutrient A and w_2 g of nutrient B per kilogram. The second has 1 g of each nutrient per kilogram. The quantities w_1 and w_2 are uncertain: we will assume that they are fuzzy number with trapezoidal membership functions, $(1,2,3,4)$ and $(1/3,5/9,7/9,1)$ respectively. Furthermore, each fertilizer has a unitary cost per kilogram.

$$
\begin{aligned}
&\min\ x_1 + x_2 \\
&\text{subject to:} \\
&\text{Pos}\{w_1 x_1 + x_2 \geqslant 7\} \geqslant \alpha \\
&\text{Pos}\{w_2 x_1 + x_2 \geqslant 4\} \geqslant \alpha \\
&x_1,\ x_2 \geqslant 0
\end{aligned}
$$

x_1 and x_2 are decision variables and defines the amounts of the first type of fertilizer and the second type of fertilizer respectively. α is the confidence level and the objective of the model is to find the minimum total amount of fertilizers (total cost) where the possibilities of having nutrient A and nutrient B more than desired level are greater than α.

We have modeled our problem using fuzzy simulation and GA. Iteration numbers are selected 10 000 and 5 000 for the GA and the fuzzy simulation algorithms respectively. Number of individuals in population is taken as 30. Fuzzy parameter a, crossover probability P_c and mutation probability P_m are assigned 0.1, 0.2 and 0.2 respectively. The solution is encoded as floating numbers and genetic operators as explained in sub section 16.3.1.2 are utilized. The algorithm is modeled in C programming language and the best solution is obtained with objective function 4.261. The farmer needs to buy 1.304 kilograms of the first type of fertilizer and 2.957 kilograms of the second type of fertilizer. In this case, the chance constraints will be satisfied with the possibility 0.9. In Fig. 16.7, change in total cost with respect to GA iteration number is given. Total cost approaches to the original problem's result in the paper by Pagnoncelli *et al.* [41] only in 50 iterations. We only change the characteristics of the parameters from probabilistic to fuzzy by endeavoring to select membership functions which are close to original distributions.

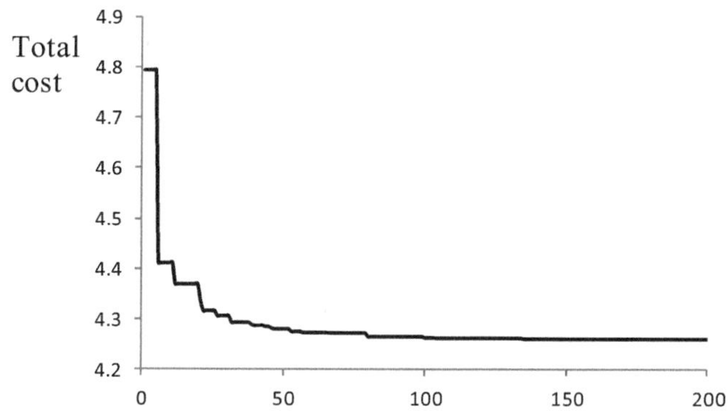

Fig. 16.7 The results of the modified blending problem

16.4.2 *Modified Burnit Goal Programming*

The next problem explained below is the modified version of the Burnit Goal Programming problem [42]. As presented in the reference book, Leon Burnit Advertising Agency is trying to determine a TV advertising schedule for Priceler Auto Company and the company has three goals: the ads of the company should be seen by at least 40 million high-income men (HIM), by at least 60 million low-income people (LIP) and by at least 35 million high-income women (HIW), respectively. Leon Burnit can purchase two types of ads: those shown during football games and those shown during soap operas. At most, \$600,000 can be spent on ads. The advertising costs are \$100,000 per one minute ads shown in football game and \$60,000 per one minute ads shown in soap opera.

We have modified the problem by assuming that the number of potential audiences of the ads is fuzzy numbers for any type of ads. Namely, the number of millions potential HIM audiences watch one minute ads is shown in football game by the fuzzy triangular number $(5, 7, 7.5)$ and in soap opera by the fuzzy triangular number $(1, 3, 4)$ million. Other triangular fuzzy numbers are $(9, 10, 11)$, $(3, 5, 6)$, $(3, 5, 5.4)$ and $(2, 4, 4.5)$ for the numbers of millions of audiences LIP for football game ads and soup opera ads, and HIW for football game ads and soup opera ads. The company tries to find the best advertisement schedule with minimum deviations from the goals under the budget constraint. It is accepted that the first goal is the most important goal and the weight of this goal is much larger than the other goal's weights. Also the weight of the second goal is much larger than weight of the third goal. The possibility of the fuzzy goal constraints must be at least 0.90, 0.85 and 0.85

respectively. FCCP model of the modified problem is as follows.

$$\begin{aligned}
&\text{lexmin}\,\{d_1^-, d_2^-, d_3^-\}\\
&\text{subject to:}\\
&\text{Pos}\,\{\tilde{a}_{11}x_1 + \tilde{a}_{12}x_2 + d_1^- - d_1^+ = 40\} \geqslant 0,90\\
&\text{Pos}\,\{\tilde{a}_{21}x_1 + \tilde{a}_{22}x_2 + d_2^- - d_2^+ = 60\} \geqslant 0,85\\
&\text{Pos}\,\{\tilde{a}_{31}x_1 + \tilde{a}_{32}x_2 + d_3^- - d_4^+ = 35\} \geqslant 0,85\\
&100x_1 + 60x_2 \leqslant 600\\
&x_1,\ x_2 \geqslant 0
\end{aligned}$$

x_1 and x_2 are the decision variables and state the number of minutes of ads shown during football games and soap opera respectively. $\tilde{a}_{ij}$ is the fuzzy number for the number of potential audiences of type i that watch the ads shown in the activity of type j ($i = 1, 2, 3$, $j = 1, 2$). d_i^+ is amount by which we numerically exceed the ith goal (the goal of ith type of audience) and d_i^- is the amount by which we are numerically undershoot the ith goal.

We have used the fuzzy simulation based GA to solve the problem by following the steps explained in subsection 16.3.1. Iteration numbers are selected 20 000 and 2 000 for the GA and the fuzzy simulation algorithms respectively. Number of individuals in population is taken as 30. Fuzzy parameter a, crossover probability P_c and mutation probability P_m are assigned 0.1, 0.3 and 0.2 respectively. The best solution to the problem is 4.264 with $x_1 = 5.71$ and $x_2 = 0.484$ with the possibilities 0.952, 0.915 and 0.9 respectively for the fuzzy chance constraints in the problem. The negative deviations from the first two goals are zero which means the solution does not undershoot the first and second goal. However, the last goal is not achieved; the negative deviation from this goal is 4.264 which is also objective function.

16.4.3 *Warehouse Location Problem*

Given a set of m warehouses and n stores, the incapacitated warehouse location problem (WLP) consists of choosing a subset of warehouses that minimizes the fixed costs of the warehouses and the transportation costs from the warehouses to the stores [43]. WLP has been solved with linear-programming, branch and bound, GA, and local search methods in the literature. For example, Michel and Hentenryck [43] developed a simple and efficient tabu search algorithm which found optimal warehouse location very quickly and with very high frequencies in deterministic conditions. Since deterministic conditions are very seldom in real world, methods that cope with uncertainties have been investigated in the literature. Some WLPs involve uncertain parameters that are characterized by imprecise or vague information, and are dealt with fuzzy sets theory. Wen and Kang [44] have presented the expected cost minimization model and the chance maximization model with random fuzzy demands. Wang and Watada [45] have proposed a facility location model with fuzzy parameters and a swarm intelligence approach.

We now present a FCCP model based on credibility measure for warehouse location selection with fuzzy demands. To simplify the problem, we will ignore the capacity of warehouses and the constraints on stores' demands.

Let s_i be the coordinates of store i for $i = 1,2,\ldots,n$ and w_j be the coordinates of warehouse j for $j = 1,2,\ldots,m$. The distance between store i and warehouse j is denoted with $d_{ij}(s_i, w_j)$. A store is supplied by a warehouse which has minimum distance among all warehouses to that specific store. The object is to minimize distance-demand function by changing the locations of warehouses. We denote distance-demand function as $F(\mathbf{s}, \boldsymbol{\xi}, \mathbf{w})$ where $\mathbf{s}$ is the vector of coordinates (x_i, y_j) of stores, $\boldsymbol{\xi}$ is the vector of demands of stores, and $\mathbf{w}$ is the vector of coordinates (x'_j, y'_j) of warehouses. The formula of distance-demand function is given below.

$$d_{ij}(s_i, w_j) = \sqrt{(x_i - x'_j)^2 + (y_i - y'_j)^2} \quad i = 1,2,\ldots,n \text{ and } j = 1,2,\ldots,m$$

$$f_i(s_i, \xi_i, \mathbf{w}) = \min(d_{i1}(s_i, w_1), \ldots, d_{im}(s_i, w_m)) * \xi_i$$

$$F(\mathbf{s}, \boldsymbol{\xi}, \mathbf{w}) = \sum_{i=1}^{n} f_i(s_i, \xi_i, \mathbf{w})$$

A CCP model which maximizes credibility of that distance-demand value is less than F_0 is given below.

$$\max \operatorname{Cr}\{F(\mathbf{s}, \boldsymbol{\xi}, \mathbf{w}) \leqslant F_0\}$$

We give a numerical example to illustrate the utilization of the model. A company with 7 stores aims to decide the location of 2 warehouses, with a minimum distance among the stores and the warehouses to those which supply to them. In the European side of Istanbul, the coordinates of 7 stores are obtained from map, and rescaled to $[1, 100]$ and $[1, 45]$ in x and y coordinates respectively. Fuzzy demands are given by decision makers as fuzzy normal numbers. Decision maker prefers distance-demand value to be less than $F_0 = 30,000$. Locations and fuzzy demands of stores are given in Table 16.4.

We represent a chromosome with four genes which keep x-coordinate and y-coordinate of two warehouses. Initial positions of warehouses is uniformly generated between $[1, 100]$ for the abscissa and $[1, 45]$ for the ordinate. Credibility of distance-demand function is found by employing fuzzy simulation. GA parameters are tuned with trial and error. Iteration numbers are selected 10 000 and 500 for the GA and the fuzzy simulation algorithms

Table 16.4 Coordinates and demands for stores

Stores	Rescaled Coordinates		Demands	
	x	y	μ	σ
Beylikduzu	5.32	10.88	300	25
Bakırkoy	50.47	2.44	390	40
Fatih	65.78	12.17	420	30
Cevahir	75.84	25.33	360	45
Profilo	77.75	27.12	240	20
Akmerkez	82.91	29.23	285	30
Istinye Park	84.43	38.23	405	50

respectively. Number of individuals in population is taken as 30. Fuzzy parameter a, crossover probability P_c and mutation probability P_m are assigned 0.2, 0.4 and 0.2 respectively.

Locations of two warehouses are $(52.09, 7.88)$ and $(80.00, 30.67)$. While first warehouse supplies to Beylikduzu, Bakırköy and Fatih stores, the second one supplies the rest of stores. Maximum credibility is 0.86 and distance-demand value is 27,473.

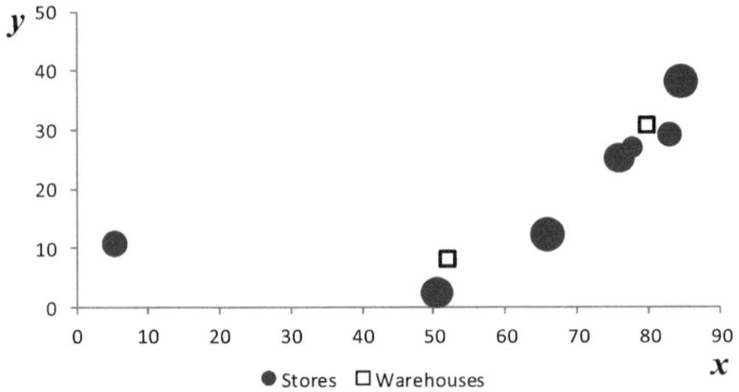

Fig. 16.8 Locations for stores and warehouses

16.4.4 *Shortest Path Problem with Fuzzy Travel Times*

Shortest path problem, which is finding the minimum total travel distance from an origin to a destination, is widely studied in the literature. The distances are deterministic but the travel time of these distances are assumed to be uncertain in this modified problem. There may be a time window for the vehicle that travels and may be obliged to reach to the destination within this time window. FCCP model for this problem is as follows:

$$\min \sum_{i=1}^{N} \sum_{j=1}^{N} x_{ij} \cdot c_{ij}$$

subject to:

$$\text{Pos}\left\{\left(\sum_{i=1}^{N} \sum_{j=1}^{N} x_{ij} \cdot \tilde{t}_{ij} + T\right) \in [a,b]\right\} \geqslant \alpha$$

$$\sum_{j=1, j\neq i}^{N} x_{ij} - \sum_{j=1, j\neq i}^{N} x_{ji} = 1 \text{ for } i = 1$$

$$\sum_{j=1, j\neq i}^{N} x_{ij} - \sum_{j=1, j\neq i}^{N} x_{ji} = -1 \text{ for } i = N$$

$$\sum_{j=1, j\neq i}^{N} x_{ij} - \sum_{j=1, j\neq i}^{N} x_{ji} = 0 \text{ for } i \neq 1 \text{ and } i \neq N$$

$$\sum_{j=1, j\neq i}^{N} x_{ij} \leqslant 1 \text{ if } i \neq N$$

$$\sum_{j=1, j\neq i}^{N} x_{ij} = 0 \text{ if } i = N$$

$$x_{ij} \in \{0, 1\} \ \forall i, \ \forall j$$

The binary decision variable x_{ij} is 1 if the link from i^{th} node to j^{th} node is on the path and 0 otherwise. The crisp parameter c_{ij} denotes the travel cost and fuzzy parameter $\tilde{t}_{ij}$ denotes the travel time on link (i, j). The vehicle starts trip at time T. To solve this problem fuzzy simulation based GA is utilized. But because the nature of the problem is different from previous examples, another encoding approach is used to represent the individuals in GA. A chromosome is formed as a sequence of positive integers representing the nodes. The length of a chromosome is variable and the first locus is reserved for the source node as denoted in the study of Ahn and Ramakrishna [46]. The number of nodes that a route includes can be at most N which is the total number of nodes in the network and the last locus signifies the destination node. Any node cannot be shown in a chromosome more than once. For instance, $V = (s, n_2, n_6, d)$ is a chromosome with four nodes. The crossover and mutation operations are also achieved as in the study of Ahn and Ramakrishna [46].

The numerical experiment given in the study of Hui *et al.* [47] is solved by using fuzzy simulation and GA. There are 14 nodes in the network, node 1 is the source and node 14 is the destination. The network is shown in Figure 16.9. The distance (as cost) of the links are crisp and shown in the figure.

The travel times of the links are trapezoidal fuzzy numbers (u_1, u_2, u_3, u_4) as given in Table 16.5. We assumed that there is only one vehicle that starts travelling from node 1 and try to reach node 14 at a minimum cost. But there is a time window that the vehicle must arrive to its destination.

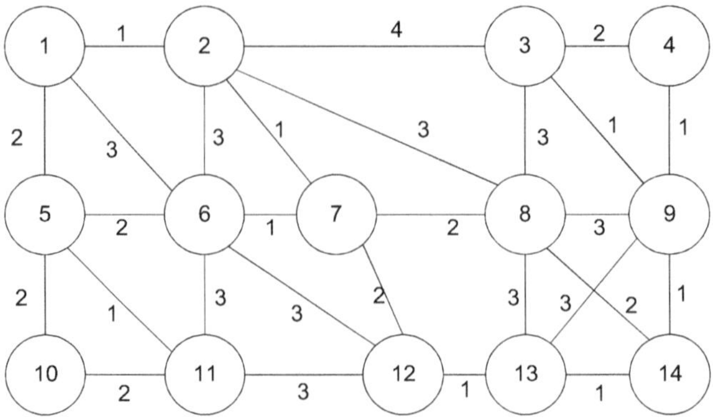

Fig. 16.9 Network

Table 16.5 Fuzzy travel times

Links	u_1	u_2	u_3	u_4
(1,2)	8.7	8.85	9.15	9.3
(1,5)	1.7	1.85	2.15	2.3
(1,6)	2.7	2.85	3.15	3.3
Others	0.7	0.85	1.15	1.3

The starting time of the travelling is zero and the arriving time should be less than 9. Target possibility of fuzzy constraint is 0.90. So the fuzzy constraint will be $\text{Pos}\left\{\sum_{i=1}^{N}\sum_{j=1}^{N} x_{ij}\cdot\tilde{t}_{ij} \leqslant 9\right\} \geqslant 0.90$. If there is no time window, the optimum solution can be found by classical methods and the optimal route is 1-2-7-12-13-14 with the distance 6. A computer program is developed for the proposed method in C programming language. Iteration numbers are selected 100 and 1000 for the GA and the fuzzy simulation algorithms respectively. Crossover probability P_c and mutation probability P_m are assigned 0.4 and 0.1 respectively. The algorithm finds the solution in average 5 iterations with quite small population sizes as 6. According to the results, the best path is 1-6-12-13-14 with the distance 8 and possibility of being in time windows is 0.99.

16.5 Conclusions

Optimization has been a powerful tool in exploration and solution of managerial problems. If conditions of the system or environment are probabilistic in nature, stochastic programming may provide adequate solution to the problems. CCP has been proposed and utilized to solve such complex stochastic problems. When uncertainties in an optimization problem emerge from vagueness, imprecision or subjectivity, CCP model with fuzzy parameters can be used. In this chapter, several forms of FCCP model are taken into consideration. Since one applies to iterative solution methods like fuzzy simulation and heuristics in these forms, length of solution time might be exhaustive for comprehensive models. Nevertheless, FCCP is a helpful technique for optimization problems which are impossible or difficult to solve with transforming model to the crisp equivalent.

Bibliography

[1] Charnes A., Cooper W.W., Chance-constrained programming. *Management Science.* 1959; **6** (1), 73–79.

[2] Zadeh L.A., Fuzzy sets. *Information and Control.* 1965; **8** (3), 338–353.

[3] Zadeh L.A., Fuzzy sets as a basis for a theory of possibility. *Fuzzy Sets and Systems.* 1978; **1** (1), 3–28.

[4] Liu B., Liu Y.K., Expected value of fuzzy variable and fuzzy expected value models. *IEEE Transactions on Fuzzy Systems.* 2002; **10** (4), 445–450.

[5] Wen M., Iwamura K., Fuzzy facility location-allocation problem under the Hurwicz criterion. *European Journal of Operational Research.* 2008; **184** (2), 627–635.

[6] Kaufmann A., *Introduction to the theory of fuzzy subsets*, 1975. New York: Academic Press.

[7] Zadeh L.A., The concept of a linguistic variable and its application to approximate reasoning. *Information Sciences.* 1975; **8** (3), 199–249.

[8] Liu B., *Theory and practice of uncertain programming*, 3 edn, 2002. Physica Verlag.

[9] Liu B., *Uncertainty theory*, 2010. Springer, Berlin.

[10] Liu B., Iwamura K., Chance constrained programming with fuzzy parameters. *Fuzzy Sets and Systems.* 1998; **94** (2), 227–237.

[11] Qin X., Xu Y., Xu T., Solving fuzzy waste management models through heuristic searching. In *3rd IASTED International Conference on Environmental Management and Engineering.* 2011: Calgary, AB, Canada.

[12] He L., Huang G.H., Lu H.W., A simulation-based fuzzy chance-constrained programming model for optimal groundwater remediation under uncertainty. *Advances in Water Resources.* 2008; **31** (12), 1622–1635.

[13] Zheng Y., Liu B., Fuzzy vehicle routing model with credibility measure and its hybrid intelligent algorithm. *Applied Mathematics and Computation.* 2006; **176** (2), 673–683.

[14] Erbao C., Mingyong L., A hybrid differential evolution algorithm to vehicle routing problem with fuzzy demands. *Journal of Computational and Applied Mathematics.* 2009; **231** (1), 302–310.

[15] Cao E., Lai M., The open vehicle routing problem with fuzzy demands. *Expert Systems with Applications.* 2010; **37** (3), 2405–2411.

[16] Cao C., Gu X., Xin Z., Chance constrained programming models for refinery short-term crude oil scheduling problem. *Applied Mathematical Modelling.* 2009; **33** (3), 1696–1707.

[17] Liu B., Iwamura K., A note on chance constrained programming with fuzzy coefficients. *Fuzzy Sets and Systems.* 1998; **100** (1-3), 229–233.

[18] Hou F., Wu Q., Genetic algorithm-based redundancy optimization problems in fuzzy framework. *Communications in Statistics – Theory and Methods.* 2006; **35** (10), 1931–1941.

[19] Pal B.B., Gupta S., A genetic algorithm approach for fuzzy goal programming formulation of chance constrained problems using stochastic simulation. In *Fourth International Conference*

on Industrial and Information Systems (ICIIS). 2009; pp. 187–192: Sri Lanka.

[20] Liang R., Gao J., Dependent-chance programming models for capital budgeting in fuzzy environments. *Tsinghua Science & Technology.* 2008; **13** (1), 117–120.

[21] Lan Y., Liu Y., Sun G., An approximation-based approach for fuzzy multi-period production planning problem with credibility objective. *Applied Mathematical Modelling.* 2010; **34** (11), 3202–3215.

[22] Wang J., Zhao R., Tang W., Fuzzy programming models for vendor selection problem in a supply chain. *Tsinghua Science & Technology.* 2008; **13** (1), 106–111.

[23] Zhou C., Zhao R., Tang W., Two-echelon supply chain games in a fuzzy environment. *Computers & Industrial Engineering.* 2008; **55** (2), 390–405.

[24] Davari S., Zarandi M.H.F., Hemmati A., Maximal covering location problem (MCLP) with fuzzy travel times. *Expert Systems with Applications.* 2011; **38** (12), 14535–14541.

[25] Tian G., Chu J., Liu Y., Ke H., Zhao X., Xu G., Expected energy analysis for industrial process planning problem with fuzzy time parameters. *Computers & Chemical Engineering.* 2011; **35** (12), 2905–2912.

[26] Qin Z., Ji X., Logistics network design for product recovery in fuzzy environment. *European Journal of Operational Research.* 2010; **202** (2), 479–490.

[27] Lan Y.-F., Liu Y.-K., Sun G.-J., Modeling fuzzy multi-period production planning and sourcing problem with credibility service levels. *Journal of Computational and Applied Mathematics.* 2009; **231** (1), 208–221.

[28] Xu Y., Qin X.S., Agricultural effluent control under uncertainty: An inexact double-sided fuzzy chance-constrained model. *Advances in Water Resources.* 2010; **33** (9), 997–1014.

[29] Guo P., Huang G.H., Li Y.P., An inexact fuzzy-chance-constrained two-stage mixed-integer linear programming approach for flood diversion planning under multiple uncertainties. *Advances in Water Resources.* 2010; **33** (1), 81–91.

[30] Zhang Y.M., Huang G.H., Inexact credibility constrained programming for environmental system management. *Resources, Conservation and Recycling.* 2011; **55** (4), 441–447.

[31] Zhang Y.M., Huang G.H., Lin Q.G., Lu H.W., Integer fuzzy credibility constrained programming for power system management. *Energy.* 2012; **38** (1), 398–405.

[32] Li X., Shou B., Qin Z., An expected regret minimization portfolio selection model. *European Journal of Operational Research.* 2012; **218** (2), 484–492.

[33] Wu C.-L., Wang Y.-Q., Huang C.-Y., Portfolio selection with credibility criterion. In *International Conference on Machine Learning and Cybernetics.* 2009; pp. 2620–2624.

[34] Huang X., Fuzzy chance-constrained portfolio selection. *Applied Mathematics and Computation.* 2006; **177** (2), 500–507.

[35] Qin Z., Li X., Ji X., Portfolio selection based on fuzzy cross-entropy. *Journal of Computational and Applied Mathematics.* 2009; **228** (1), 139–149.

[36] Huang C.-S., Huang Y.-C., Lai P.-J., Modified genetic algorithms for solving fuzzy flow shop scheduling problems and their implementation with CUDA. *Expert Systems with Applications.* 2012; **39** (5), 4999–5005.

[37] Pishvaee M.S., Torabi S.A., Razmi J., Credibility-based fuzzy mathematical programming model for green logistics design under uncertainty. *Computers & Industrial Engineering.* 2012; **62** (2), 624–632.

[38] Xu Y., Huang G.H., Qin X.S., Cao M.F., Sun Y., An interval-parameter stochastic robust optimization model for supporting municipal solid waste management under uncertainty. *Waste Management.* 2010; **30** (2), 316–327.

[39] Zarandi M.H.F., Hemmati A., Davari S., The multi-depot capacitated location-routing problem with fuzzy travel times. *Expert Systems with Applications.* 2011; **38** (8), 10075–10084.

[40] Shao Z., Ji X., Fuzzy multi-product constraint newsboy problem. *Applied Mathematics and Computation.* 2006; **180** (1), 7–15.

[41] Pagnoncelli B.K., Ahmed S., Shapiro A., Sample average approximation method for chance constrained programming: Theory and applications. *Journal of Optimization Theory and Applications.* 2009; **142** (2), 399–416.

[42] Winston W.L., *Operations research: Applications and algorithms.* 1991. Belmont: Duxbury Press.

[43] Michel L., Hentenryck P.V., A simple tabu search for warehouse location. *European Journal of Operational Research.* 2004; **157** (3), 576–591.

[44] Wen M., Kang R., Some optimal models for facility location-allocation problem with random fuzzy demands. *Applied Soft Computing.* 2011; **11** (1), 1202–1207.

[45] Wang S., Watada J., A hybrid modified PSO approach to VaR-based facility location problems with variable capacity in fuzzy random uncertainty. *Information Sciences.* 2012; **192** (in press), 3–18.

[46] Ahn C.W., Ramakrishna R.S., A genetic algorithm for shortest path routing problem and the sizing of populations. *IEEE Transactions on Evolutionary Computation.* 2002; **6** (6), 566–579.

[47] Hui F., Zi Z., Gang H., Developing a model for the emergency rescue routing problem using stochastic programming theory. In *Logistics Engineering and Intelligent Transportation Systems (LEITS)*, 2010 International Conference on. 2010; pp. 1–4.

Chapter 17

A Cooperative Intelligent Approach for Job-shop Scheduling Based on Bacterial Foraging Strategy and Particle Swarm Optimization

Hongwei Ge*, Guozhen Tan

College of Computer Science and Technology, Dalian University of Technology, Dalian 116024, China

*E-mail: * hwge@dlut.edu.cn*

The optimization of job-shop scheduling is very important because of its theoretical and practical significance. In this chapter, a computationally effective approach of combining bacterial foraging strategy with particle swarm optimization for solving the minimum makespan problem of job shop scheduling is proposed. In the artificial bacterial foraging system, a novel chemotactic model is designed to address the job shop scheduling problem and a mechanism of quorum sensing and communication are presented to improve the foraging performance. In the particle swarm system, a novel concept for the distance and velocity of a particle is presented to pave the way for the job-shop scheduling problem. The proposed cooperative algorithm effectively exploits the capabilities of distributed and parallel computing of swarm intelligence approaches. The algorithm is examined using a set of benchmark instances with various sizes and levels of hardness and compared with other approaches reported in some existing literatures. The computational results validate the effectiveness of the proposed algorithm.

17.1 Introduction

The job shop scheduling problem (JSSP) is a very important practical problem in the both fields of production management and combinatorial optimization. Efficient methods for solving the JSSP have significant effects on profitability and product quality. The JSSP has drawn the attention of researchers for the last three decades, but because scheduling problems vary widely according to specific production tasks and most of them are strongly NP-hardNP-hard problems such that some test problems of moderate size are still unsolved. As a matter of fact, only small size instances of the problems can be solved within a reasonable computational time by exact optimization algorithms such as branch and bound [1, 2],

C. Kahraman (ed.), *Computational Intelligence Systems in Industrial Engineering*, Atlantis Computational Intelligence Systems 6, DOI: 10.2991/978-94-91216-77-0_17,

and dynamic programming (DP) [3, 4], including the notorious 10×10 instance of Fisher and Thompson, which was proposed in 1963 and only solved 20 years later. Problems of dimension 15×15 are still considered to be beyond the reach of today's exact methods. By contrast, approximate and heuristic methods make a tradeoff between solution quality and computational cost. These methods mainly include dispatching priority rules [5–7], shifting bottleneck approach [8, 9], Lagrangian relaxation [10, 11], tabu search [12–14] and have made considerable achievement.

In recent years, much attention has been devoted to meta-heuristics with the emergence of new techniques from the field of artificial intelligence such as genetic algorithm (GA) [15–18], simulated annealing (SA) [19–22], ant colony optimization (ACO) [23], particle swarm optimization particle swarm optimization (PSO) [24, 25], artificial neural network (ANN) [26–28], bacterial foraging bacterial foraging algorithm (BFA) [29], and so on. These meta-heuristics can be regarded as problem-independent approaches and are well suited to solve complex problems that may be difficult to solve by traditional techniques, moreover, they are capable of producing high-quality solutions with a reasonable computational effort. Among the meta-heuristic algorithms, GA has been used with increasing frequency to address scheduling problems and may not remain much room for the improvement, however, the use of the BFA and PSO for the solution of scheduling problems has been scarce, especially the BFA. Besides, many research results of the JSSP show that it is difficult to obtain a good-enough solution only by single search scheme.

Motivated by these perspectives, we propose an effective cooperative intelligent algorithm for the job shop scheduling problem based on bacterial foraging strategy and particle swarm optimization in this chapter. Both BFA and PSO are evolutionary computation techniques based on swarm intelligence. They exhibit implicit parallelism and contain certain redundancy and historical information of past solutions. So the proposed hybrid algorithm also effectively exploits the capabilities of distributed and parallel computing of swarm intelligence approaches.

The remainder of the chapter is organized as follows. Section 17.2 introduces and formulates the job-shop scheduling problem. Section 17.3 describes a proper representation for the problem and proposes a schedule initialization algorithm based on the Giffler and Thompson (G&T) algorithm. A BFA-based scheduling algorithm and a PSO-based scheduling algorithm are proposed respectively in Section 17.4 and Section 17.5, which are followed in Section 17.6 showing the framework of the cooperative intelligence algorithm based on the above two sections. Section 17.7 reports the corresponding computational and comparative results. Finally, the conclusions are made in Section 17.8.

17.2 Job-Shop Scheduling Problem

The job shop problem studied in the chapter consists of scheduling a set of jobs on a set of machines with the objective of minimization the makespan, i.e., the maximum of completion times needed for processing all jobs. Each machine can handle at most one job at a time. Each job consists of a chain of operations to be processed in a specified sequence, on specified machines, and during an uninterrupted time period of given length.

Explaining the problem more specifically, let $J = \{1, 2, \ldots, n\}$ denote the set of jobs, $M = \{1, 2, \ldots, m\}$ denote the set of machines, and $O = \{0, 1, \ldots, n \times m, n \times m + 1\}$ denote the set of operations to be scheduled, where 0 and $n \times m + 1$ represent the dummy initial and final operations respectively. The operations are interrelated by the precedence constraints, which force each operation j to be scheduled after all predecessor operations P_j are completed. Besides, operation j can only be scheduled if the machine what it requires is idle. Further, let T_j and F_j denote the fixed processing time and the finish time of operation j respectively, let $A(t)$ be the set of operations being processed at time t, and let $e_{jm} = 1$ if operation j is required to process on machine m and $e_{jm} = 0$ otherwise.

The conceptual model of the JSSP can be stated as [30]:

$$\text{minimize} \qquad F_{n\times m+1}, \tag{17.1}$$

$$\text{subject to:} \qquad F_k \leqslant F_j - T_j, \qquad (j = 1, 2, \ldots, n \times m + 1;\; k \in P_j) \tag{17.2}$$

$$\sum_{j\in A(t)} e_{jm} \leqslant 1, \qquad (m \in M;\; t \geqslant 0) \tag{17.3}$$

$$F_j \geqslant 0, \qquad (j = 1, 2, \ldots, n \times m + 1) \tag{17.4}$$

The objective function (17.1) minimizes the finish time of the last operation, namely the makespan. Constraints (17.2) impose the precedence relations between operations. Constrains (17.3) represents that one machine can only process one operation at a time, and Constrains (17.4) force the finish times to be non-negative.

The solution to the JSSP can be represented as the operations permutation of the jobs on each machine. The total number of all possible schedules (both feasible and infeasible) is $(n!)^m$ for the problems with n jobs and m machines. Obviously, it is impossible to exhaust all the alternatives for finding the optimal solution even if the values of n and m are small. For example, for the Fisher-Thompson benchmark problem of the 10 jobs to 10 machines, it has a search space with a size at about 3.96×10^{65}. So it is necessary to restrict the search space and guide the search process. There are four types of feasible schedules in JSSP, namely inadmissible, semi-active, active and non-delay. Inadmissible schedules contain excess idle time, and they can be improved by forward-shifting operations until no excess idle time exists. Semi-active schedules contain no excess idle time, but they can be improved by shifting some operations to the front without delaying others. Active schedules contain no idle time and no operation can be finished earlier without delaying other operations. Non-delay schedules are active schedules, in which operations are placed into the schedule such that the machine idle time is minimized and no machine is kept idle if some operation can be processed. Their relationships [31] are shown in Figure 17.1. The optimal schedule is guaranteed to be an active schedule. So we only need to find the optimum solution in the set of active schedules.

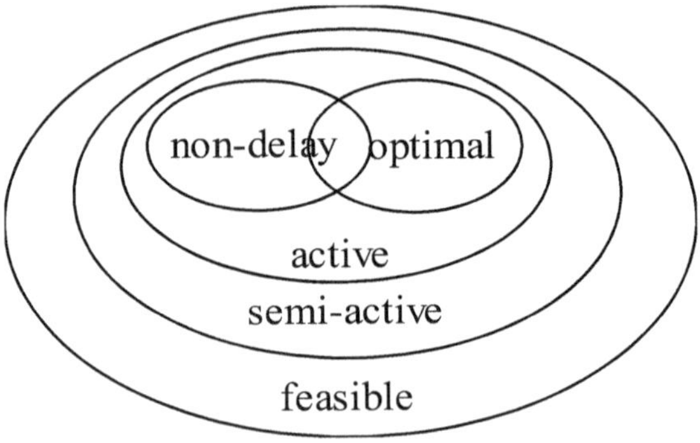

Fig. 17.1 Relationship of schedule properties

17.3 Representation and Initial Schedule Generation

17.3.1 *Representation*

Before solve the JSSP, we describe a proper representation for the solution of the problem, namely a scheduling, which is used in the later algorithms. In this chapter, an operation-based representation is adopted, which uses an unpartitioned permutation with m-repetitions of job numbers for the problems with n jobs and m machines. A job represents a set of operations that has to be scheduled on m machines. In this formulation, each job number occurs m times in the permutation. By scanning the permutation from left to right, the k-th occurrence of a job number refers to the k-th operation in the technological sequence of this job. A permutation with repetition of job numbers only expresses the order in which the operations of jobs are scheduled. For example, suppose a sequence is given as (2 1 3 4 2 4 3 2 1 2 3 3 1 1 4 4) in a 4 jobs $\times$ 4 machines problem, in which each gene stands for one operation. Each job consists of four operations, and the job number is thereby repeated four times. The fifth gene of the permutation implies the second operation of job 2 because number 2 has been repeated twice, similarly, the eighth gene represents the third operation of job 2, and so on. The prominent advantage is that the permutation is always feasible, moreover, it eliminates the deadlock schedules which are incompatible with the technological constraints and can never be finished. But it will produce redundancy in the search space and cause the search space size to expand to $(n \times m)!/(m!)^n$, that is, the mapping relation between sequence and schedule is many-to-one.

17.3.2 *Initial schedule generation*

A good initial solution is fundamental for the computational performance of the search algorithm. If the initial population is generated randomly, the generation of the optimal schedule not only requires longer time but also decreases the possibility to obtain an optimal solution. In this chapter, a new scheduling initialization algorithm is proposed based on the well-known algorithm of Giffler and Thompson [32] to produce the active schedules. The brief outline of the improved algorithm is as follows:

Step 1: Let the set A contain the first schedulable operation of each job. Let $s_{jm} = 0$, for all the operation (j,m) in the set A, where s_{jm} is the earliest time at which the operation (j,m) can start.

Step 2: Calculate $t(A) = \min(s_{jm} + p_{jm})$, for all the operation $(j,m) \in A$, where p_{jm} is the processing time of operation (j,m).

Step 3: Set up the set M, for all operation $(j,m) \in A$, if $s_{jm} < t(A)$, then put the machine m into set M.

Step 4: Set up the set G, randomly select a machine m^* from set M, for all the operations $(j,m) \in A$, put the operation (j,m) into set G if it is processed on the machine m^*.

Step 5: For all the operations $(j,m^*) \in G$, calculate the earliest completed time of job j if the operation (j,m^*) is currently processed, let j^* denote the job which is completed lastly.

Step 6: Select the operation $(j^*,m^*)$from set G and append it to the schedule.

Step 7: Delete the operation (j^*,m^*) from set A, then subjoin its immediate successor into set A.

Step 8: Update s_{jm} in set A and then return to step 2 until A is empty, namely all the operations are scheduled.

17.4 BFA-based Scheduling Algorithm

17.4.1 *Standard bacterial foraging algorithm*

Bacterial foraging algorithm (BFA), originally developed by K.M. Passino [33], is a new natural heuristic optimization method based on foraging behavior of *E. coli* bacterium. It is also an evolutionary computation technique based on swarm intelligence. BFA has attracted broad attention in the fields of evolutionary computing, optimization and many others [34–36]. The system of the *E. coli* bacterium enables it to achieve a complex type of search and avoidance behavior. Evolution has designed this control system. It is robust and clearly very successful at meeting its goals of survival when viewed from a population perspective. In addition the control system dictates that bacterial foraging process can be subdivided into four motile behaviors namely chemotaxis, swarming, reproduction, and elimination and dispersal. Below we briefly describe each of these processes.

(i) *Chemotaxis*: This process simulates the movement of an *E. coli* cell through swimming and tumbling via flagella. Biologically an *Escherichia coli* bacterium can move in two different ways. It can swim for a period of time in the same direction or it may tumble, and alternate between these two modes of operation for the entire lifetime. Suppose $\theta^i(j,k,l)$ represents the position of the i-th bacterium at j-th chemotactic, k-th reproductive and l-th elimination-dispersal step. $C(i)$ is the size of the step taken in the random direction specified by the tumble. Then in computational chemotaxis the movement of the bacterium may be represented by

$$\theta^i(j+1,k,l) = \theta^i(j,k,l) + C(i)\frac{\Delta(i)}{\sqrt{\Delta^T(i)\Delta(i)}} \tag{17.5}$$

Where Δ indicates a vector in the random direction whose elements lie in $[-1,1]$.

(ii) *Swarming*: An interesting group behavior has been observed for several motile species of bacteria including *E. coli* and *Salmonella typhimurium*, where intricate and stable spatio-temporal patterns (swarms) are formed in semisolid nutrient medium [37]. A group of *E. coli* cells arrange themselves in a traveling ring by moving up the nutrient gradient when placed amidst a semisolid matrix with a single nutrient chemo-effecter. The cells when stimulated by a high level of *succinate*, release an attractant *aspertate*, which helps them to aggregate into groups and thus move as concentric patterns of swarms with high bacterial density. The cell-to-cell signaling in *E. coli* swarm may be represented by the following function.

$$
\begin{aligned}
J_{cc}(\theta, P(j,k,l)) &= \sum_{i=1}^{S} J_{cc}(\theta, \theta^i(j,k,l)) \\
&= \sum_{i=1}^{S}\left[-d_{\text{attrac}\tan t}\exp\left(-w_{\text{attrac}\tan t}\sum_{m=1}^{p}(\theta_m-\theta_m^i)^2\right)\right] \\
&+\sum_{i=1}^{S}\left[h_{\text{repellant}}\exp\left(-w_{\text{repellant}}\sum_{m=1}^{p}(\theta_m-\theta_m^i)^2\right)\right] \qquad (17.6)
\end{aligned}
$$

Where $J_{cc}(\theta, P(j,k,l))$ is the cost function value to be added to the actual fitness function. $P(j,k,l)$ represents the population of bacteria, namely $P(j,k,l) = \{\theta^i(j,k,l) \mid i = 1,2,\ldots,S\}$. S is the total number of bacteria and p is the number of parameters to be optimized which are presented in each bacterium and $\theta = [\theta_1, \theta_2, \ldots, \theta_p]^T$ is a point in the p-dimensional search domain. $d_{\text{attrac}\tan t}, w_{\text{attrac}\tan t}, h_{\text{repellant}}$ and $w_{\text{repellant}}$ are different coefficients that should be chosen properly.

(iii) *Reproduction*: The least healthy bacteria eventually die while each of the healthier bacteria (those yielding lower value of the objective function) asexually split into two bacteria, which are then placed in the same location. This keeps the swarm size constant.

(iv) *Elimination and dispersal*: Gradual or sudden changes in the local environment where a bacterium population lives may occur due to various reasons, e.g. a significant local rise of temperature may kill a group of bacteria that are currently in a region with a high concentration of nutrient gradients. Events can take place in such a fashion that all the bacteria in a region are killed or a group is dispersed into a new location. To simulate this phenomenon in BFA some bacteria are liquidated at random with a very small probability while the new replacements are randomly initialized over the search space.

The step outline of the standard BFA for solving optimization problem is shown in Figure 17.2.

17.4.2 *Improved bacterial foraging algorithm for job shop Scheduling*

To tackle this job shop scheduling problem, some improvements are introduced to the basic BFA. In the chemotactic step, the vector Δ of length $j \times m$ for the problem of j jobs and m machines in the random direction is randomly filled with elements of the set $\{0,1\}$. The number of the element "1" is decided by the value of $C(i)$, namely the step size specified by the tumble, and $C(i)$ is a non-negative integer. The value of $C(i)$ is inversely proportional to the fitness of the i-th bacterium. The higher the fitness a bacterium has the

Pseudocode of bacterial foraging algorithm	
Step 1.	Initialize the population
Step 2.	Evaluate the individual using the evaluation function
Step 3.	Three loops of optimization;
	Inner loop: chemotactic event;
	Outer loop: elimination-dispersal event
Step 4.	Decode the optimal individual to obtain the final solution

Fig. 17.2 The step outline of the standard BFA

smaller the value. The $C(i)$ is decided by the following formulation

$$C(i) = int\left[\frac{(f_{\max} - f_i)(\rho - \eta)}{f_{\max} - f_{\min}}\right] + \eta \tag{17.7}$$

where $int[\cdot]$ is the truncation function, f_i is the fitness of the i-th bacterium, $f_{\max}$ and $f_{\min}$ are respectively the maximal and the minimal fitness in the bacterial colony, ρ and η are respectively the upper bound and the lower bound of the step size, which are adjustable integers. In the process of the bacterium tumble, the $C(i)$ positions are selected from the position vector of the bacterium, in which the values of the elements in these same positions of the corresponding vector Δ are equal to 1, and then the operation of the tumble is completed by permuting the elements on these positions randomly. After the bacterium tumbles, the swim mechanism is performed. In the process, randomly select an element in the position vector of the bacterium, respectively, make it move backwards or forwards in turn and once shift one position until it moves λ times, and then 2λ new bacterial are produced. We replace the original bacterium by the new one with the highest fitness among the produced bacterial if it is better than the original.

The experimentation with complex problems reveals that the BFO algorithm possesses a poor convergence behavior. By our analysis, the emergence of the poor convergence behavior is brought by the "swarming" mechanism. On the one hand, when the bacteria quickly congregate round the local optimal position on a large scale, a high level of attractant is formed in the region. It maybe induces that the bacteria near to the global optimal position depart from the original region and close to the congregated region; on the other hand, when a large numbers of bacteria close to the global optimal position, a high level of repellent maybe dispel the current optimal bacteria away from the global optimal position. Actually, to perform a social foraging, an animal needs communication capabilities and it gains advantages to exploit essentially the sensing capabilities of the group. So a novel mechanism of quorum sensing and communication (QUSAC) is presented instead of the "swarming" mechanism in this chapter. Besides, the QUSAC mechanism reduces the computation complexity largely in swarming stage.

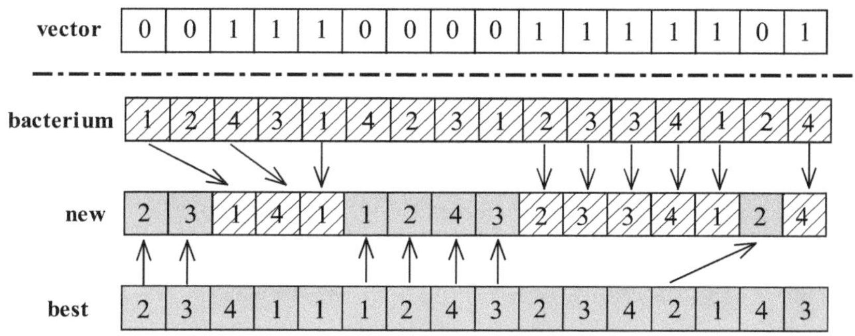

Fig. 17.3 An example of the QUSAC mechanism

The process of quorum sensing and communication is designed as follows. First, the bacterium with the highest fitness is recognized, and then a bacterium is selected for quorum sensing. Produce a vector of length $j \times m$ for the problem of j jobs and m machines which is randomly filled with elements of the set $\{0,1\}$. This vector defines the order in which the elements of the newly produced bacterium vector are drawn from the best bacterium and the selected bacterium, respectively. After an element is drawn from one bacterium and deleted from the other one, it is appended to the newly produced bacterium vector. This step is repeated until both the best and the selected bacterium vector are empty and the produced bacterium contains all the elements involved. A new bacterium is created based on quorum sensing and communication and accepted into the population if its fitness is higher than the original. Figure 17.3 gives an example of the QUSAC mechanism.

17.5 PSO-based Scheduling Algorithm

17.5.1 *Standard particle swarm optimization algorithm*

The particle swarm optimization (PSO), originally developed by Kennedy and Elberhart [38], is a method for optimizing hard numerical functions on metaphor of social behaviors of flocks of birds and schools of fish. It is an evolutionary computation technique based on swarm intelligence. A swarm consists of individuals, called particles, which change their positions over time. Each particle represents a potential solution to the problem. In a PSO system, particles fly around in a multi-dimensional search space. During its flight, each particle adjusts its position according to its own experience and the experience of its neighboring particles, making use of the best position encountered by itself and its neighbors. The effect is that particles move towards the better solution areas, while still having the ability to search a wide area around the better solution areas. The performance of each particle is measured according to a pre-defined fitness function, which is related to the problem being solved. The PSO has been found to be robust and fast in solving non-linear, non-differentiable and multi-modal problems. The mathematical description and executive steps of the PSO are as follows.

Let the i-th particle in a D-dimensional space be represented as $\vec{x}_i = (x_{i1},\ldots,x_{id},\ldots,x_{iD})$. The best previous position of the i-th particle is recorded and represented as $\vec{p}_i = (p_{i1},\ldots,p_{id},\ldots,piD)$, which gives the best fitness value and is also called *pbest*. The index of the best *pbest* among all the particles is represented by the symbol g. The location $P_{\vec{i}g}$ is also called *gbest*. The velocity for the i-th particle is represented as $\vec{v}_i = (v_{i1},\ldots,v_{id},\ldots,v_{iD})$. The concept of the particle swarm optimization consists of changing the velocity and location of each particle towards its *pbest* and *gbest* locations according to Eqs. (17.8) and (17.9) at each time step:

$$v_{id} = wv_{id} + c_1 r_1 (p_{id} - x_{id}) + c_2 r_2 (p_{gd} - x_{id}), \tag{17.8}$$

$$x_{id} = x_{id} + v_{id}, \tag{17.9}$$

where w is the inertia coefficient which is a constant in the interval $[0,1]$; c_1 and c_2 are learning rates which are nonnegative constants; r_1 and r_2 are generated randomly in the interval $[0,1]$; $v_{id} \in [-v_{\max}, v_{\max}]$, and $v_{\max}$ is a designated maximum velocity.

17.5.2 *Improved particle swarm optimization for job shop Scheduling*

In this section, we describe the formulations of the proposed PSO algorithm for the JSSP. Firstly, the concept of the difference of the locations is extended. Let us present the similarity measure of two particles. Denote the i-th and the j-th particles in a D-dimensional space as $X_i = (x_{i1},\ldots,x_{id},\ldots,x_{iD})$ and $X_j = (x_{j1},\ldots,x_{jd},\ldots,x_{jD})$, respectively. Define functions

$$s(k) = \begin{cases} 1 & \text{if } x_{ik} = x_{jk} \\ 0 & \text{if } x_{ik} \neq x_{jk} \end{cases} \quad \text{and} \quad S(X_i,X_j) = \sum_{k=1}^{D} s(k), \tag{17.10}$$

where $S(X_i,X_j)$ is called the similarity measure between particle X_i and particle X_j. Let $f(X_i)$ denote the fitness of particle i and assume that $0 \leqslant f(X_i) \leqslant C$, $\forall i \in \{1,2,\ldots,n\}$, where n is the population size. The difference of the locations between two particles, namely "distance", is redefined by the following equation:

$$\operatorname{dis}(X_i - X_j) = k \cdot \left[\alpha \cdot \frac{|f(X_i) - f(X_j)|}{C} + \beta \cdot \frac{D - S(X_i, X_j)}{D} \right], \tag{17.11}$$

where D is the dimension of a particle, k is a positive integer called acceleration coefficient, and α and β are two positive weights, which can be ascertained by trial and error and satisfy the relation that the sum of α and β is equal to 1. In this chapter, α and β is taken as 0.7 and 0.3, respectively. It is obvious that

$$0 \leqslant \frac{|f(X_i) - f(X_j)|}{C} \leqslant 1 \text{ and } 0 \leqslant \frac{D - S(X_i, X_j)}{D} \leqslant 1$$

Correspondingly, the concept of the velocity is also extended. The velocity is defined as the times of "adjustment operation". The brief outline of the adjustment algorithm is as follows.

Let $X = (x_1, \ldots, x_k, \ldots, x_D)$ and $Y = (y_1, \ldots, y_k, \ldots, y_D)$ be two particles in a D-dimensional space respectively.

Step 1: Select an index k of location randomly and set an indicator $m = 0$, if $x_k = y_k = s$, where s is the number on location k, go to Step 2, else if $x_k \neq y_k$, namely $x_k = s$ and $y_k \neq s$, go to Step 3.

Step 2: Scan set $X' = \{x_i \mid 1 \leqslant i < k\}$ and set $Y' = \{y_i \mid 1 \leqslant i < k\}$ from left to right, respectively, to find out the times that s appears, and denote them as $X'\_times(s) = t_x$ and $Y'\_times(s) = t_y$, respectively. If $t_x > t_y$, go to Step 4; if $t_x < t_y$, go to Step 5; if $t_x = t_y$, go to Step 6.

Step 3: Select a location index j in the particle Y randomly, which satisfies $y_j = s$ and $y_j \neq x_j$. Swap y_j and y_k, and set the indicator $m = 1$, then go to Step 2.

Step 4: Swap s which appears after location k and y_j which satisfies that $y_j \neq x_j$, $y_j \neq s$ and $j < k$ until $t_x = t_y$, then go to Step 7.

Step 5: Swap s which appears in the front of location k and y_j which satisfies that $y_j \neq x_j$, $y_j \neq s$ and $j > k$ until $t_x = t_y$, then go to Step 7.

Step 6: If $m = 0$, reselect an index k randomly; if $m = 1$, go to Step 7.

Step 7: Terminate.

Using "$\oplus$" to denote the adjustment operation, the proposed PSO algorithm could be performed by the following equations:

$$v_{id} = int\left[w v_{id} + c_1 r_1 \operatorname{dis}(p_{id} - x_{id}) + c_2 r_2 \operatorname{dis}(p_{gd} - x_{id}) \right], \tag{17.12}$$

$$x_{id} = x_{id} \oplus v_{id}, \tag{17.13}$$

where $int[\cdot]$ represents the truncation function and $\operatorname{dis}(\cdot)$ can be calculated using Eq. (17.11). In this chapter, the parameters are taken as $w = 0.5$, $c_1 = 1.8$, and $c_2 = 1.0$.

17.6 Cooperative Intelligence Algorithm Based on the BFA and PSO

Based on the models of the proposed BFA and PSO, a cooperative intelligence algorithm (CIA) is integrated to solve the JSSP. The objective is to search a scheduling such that

$$\min(T(JM)) = \min\{\max[T(1), T(2), \ldots, T(i), \ldots, T(m)]\} \tag{17.14}$$

where $T(i)$ is the final completion time of machining on the i-th machine, $T(JM)$ is the final completion time of all the jobs.

The fitness and the affinity are evaluated using the following equation

$$f_i = 100 \times opt/T_i(JM) \tag{17.15}$$

where f_i is the fitness value of the i-th bacterium in the BFA, and is also the fitness value of the i-th particle in the PSO, opt is the theoretical optimal makespan for a given problem and the value f_i is in the interval $(0, 100]$.

The value of the objective function $T(JM)$ can be obtained by the process of decoding. The makespan is produced by the process that assigns operations to the machines at their earliest possible starting time by the technological order of each job, scanning the permutation from left to right. It should be noticed that the scheduling acquired by this way is only a semi-active. Yet, the optimal solution must be an active schedule. Then the active decoding is applied, which checks the possible blank time interval before appending an operation at the last position, and fills the first blank interval before the last operation to convert the semi-active schedule to an active one so that the makespan can be shorter. For example, consider the 3 jobs $\times$ 2 machines problem shown in Table 17.1. In the example, job 1 must go to machine 2 for 1 unit of time, then to machine 1 for 2 units of time, and so on. Suppose an operation-based solution is the sequence $a = (1,\ 3,\ 2,\ 2,\ 1,\ 3)$ and its Gantt chart by the semi-active decoding scheme is drawn in Figure 17.4, whose makespan is 7 units. Whereas the makespan is 6 units by the active decoding and the corresponding Gantt chart is shown in Figure 17.5.

Table 17.1 Example for 3 jobs $\times$ 2 machines

	(m,t)	(m,t)
Job 1	(2, 1)	(1, 2)
Job 2	(1, 3)	(2, 2)
Job 3	(1, 1)	(2, 1)

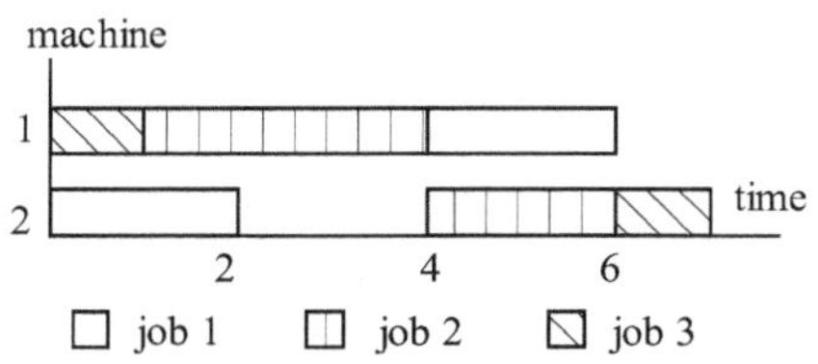

Fig. 17.4 The Gantt chart of semi-active scheduling

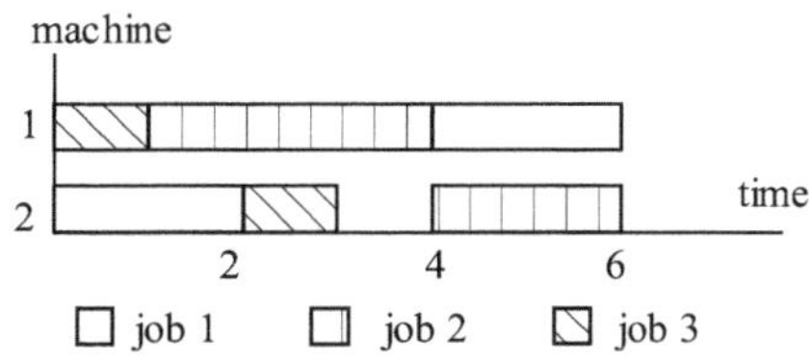

Fig. 17.5 The Gantt chart of active scheduling

It is worth mentioning that the fitness is as pivotal to the guidance of the search because the chance of a solution being chosen is determined solely by its objective function value. So their values are linearly adjusted to avoid early convergence and ensure the variety of the population. The adjustment process can be summarized as follows.

Step 1. Calculate the average fitness and affinity f_{ave}, the maximal $f_{\max}$ and the minimal $f_{\min}$ in the population;

Step 2. If $f_{\min} > 2f_{\text{ave}} - f_{\max}$, go to Step 3, else go to Step 4;

Step 3. Calculate coefficients $\alpha = \dfrac{f_{\text{ave}}}{f_{\max} - f_{\text{ave}}}$ and $\beta = \dfrac{(f_{\max} - 2f_{\text{ave}})f_{\text{ave}}}{f_{\max} - f_{\text{ave}}}$. Go to Step 5;

Step 4. Calculate coefficients $\alpha = \dfrac{f_{\text{ave}}}{f_{\text{ave}} - f_{\min}}$ and $\beta = \dfrac{-f_{\min} f_{\text{ave}}}{f_{\text{ave}} - f_{\min}}$. Go to Step 5;

Step 5. Calculate the new fitness value using the equation $f' = \alpha f + \beta$.

Furthermore, the proposed algorithm stops if the best obtained solution equals to the best known makespan or if for a given number *gens* of generations there has been no improvement of the best solution found so far, where $gens = 20$. The advantage of the proposed cooperative intelligence algorithm is that it can produce quite satisfactory solutions of the test problems in a reasonable time. The detailed pseudo-code of the improved algorithm is given below.

The cooperative intelligence algorithm for JSSP based on BFA and PSO

Step 1. Initialize parameters D, S, N_c, N_s, N_{re}, N_{ed}, P_{ed}, cnt, w, c_1, c_2,θ^i. where:

D: Dimension of the search space ($j \times m$ dimension for the problem of j jobs and m machines),

S: Total number of bacteria in the population,

N_c: The number of chemotactic steps,

N_s: The swimming length,

N_{re}: The number of reproduction steps,

N_{ed}: The number of elimination-dispersal events,

P_{ed}: Elimination-dispersal probability,

cnt: The iterative number of the PSO algorithm,

w: the inertia coefficient in PSO,

c_1, c_2: learning rates in PSO.

Step 2. Elimination-dispersal loop: $l = l + 1$.

Step 3. Reproduction loop: $k = k + 1$.

Step 4. Chemotaxis loop:$j = j + 1$.

[a] For $i = 1, 2, \ldots, S$, take a chemotactic step for bacterium i as follows.

[b] Compute the fitness value $J(i, j, k, l)$.

[c] Let $J_{\text{last}} = J(i, j, k, l)$ to save this value since we may find a better cost via a run.

[d] Update and save the best fitness value J^{best} found by far in the population and the corresponding best position θ^{best}.

[e] Compute the step size $C(i)$ using Eq. (17.7), generate a random vector $\Delta(i)$ with each element $\Delta_m(i)$, $m = 1, 2, \ldots, D$, a random number in set $\{0, 1\}$.

[f] Perform the tumble operation in Section 17.4.2.

[g] Perform the swim operation in Section 17.4.2.

(i) Let $m = 0$ (counter for swim length).

(ii) While $m < Ns$ (if the bacterium has not climbed down too long).

[h] Perform the operation of the quorum sensing and communication in Section 17.4.2.

[i] Go to next bacterium $(i + 1)$ if $i \neq S$ (i.e., go to [b] to process the next bacterium).

[j] Perform the PSO operation according to Eqs. (17.12)-(17.13) in Section 17.5 for a given iterative number cnt.

Step 5. If $j < N_c$, go to step 4. In this case, continue chemotaxis, since the life of the bacteria is not over.

Step 6. Reproduction:

[a] For the given k, and for each $i = 1, 2, \ldots, S$, let J^i_{health} be the health of the i-th bacterium. Sort bacteria in descending order of cost J_{health} (lower cost means lower health).

[b] The S_r bacteria with the lowest J_{health} values die and the other S_r bacteria with the best values split and the copies that are made are placed at the same location as their parent.

Step 7. If $k < N_{re}$, go to step 3. In this case, we have not reached the number of specified reproduction steps, so we start the next generation of the chemotactic loop.

Step 8. Elimination-dispersal: For $i = 1, 2, \ldots, S$ with probability P_{ed}, eliminate and disperse each bacterium, and this results in keeps the number of bacteria in the population constant. To do this, if a bacterium is eliminated, simply disperse another one to a random location on the optimization domain. If $l < N_{ed}$, then go to step 2; otherwise end.

17.7 Numerical simulation results and comparisons

The performance of the proposed cooperative intelligence algorithm (CIA) for the JSSP is examined by using some test problems taken from the OR-Library [39]. We consider 43 instances from two classes of standard JSSP test problems: instances FT06, FT10, FT20 designed by Fisher and Thompson (1963), and instances LA01 to LA40 designed by Lawrence (1984). Numerical experiments are performed in C++ language on a PC with Pentium IV 2.93 GHz processor and 2GB memory. The numerical results are compared with those reported in some existing literatures using other approaches [30, 40–44], including some heuristic and meta-heuristic algorithms. The selection of parameter settings in CIA is listed in Table 17.2. The proposed algorithm stops if the best obtained solution equals to the best known makespan or if for a given number *gens* (outer loop in CIA) of generations there has been no improvement of the best solution found so far, where $gens = 20$.

Table 17.2 CIA parameters

Parameters	value
Number of bacteria and particles (S)	80
Chemotactic steps (N_c)	100
Swim steps (N_s)	5
Reproduction steps (N_{re})	3
Elimination and dispersal steps (N_{ed})	$\ll 500$
Probability of elimination (P_{ed})	0.15
Iterative number of the PSO (cnt)	10
Inertia coefficient in PSO (w)	0.5
learning rates in PSO (c_1)	1.8
learning rates in PSO (c_2)	1.0

Table 17.3 summarizes the results of the experiments. The contents of the table include the name of each test problem (Instance), the scale of the problem (Size), the value of the best known solution for each problem (BKS), the value of the best solution found by using the proposed algorithm (HIA), the number of running generations (Iterations), the

percentage of the deviation with respect to the best known solution (RD%), and the best results reported in other literatures. It is worth mentioning that the makespan listed in the table is the best obtained from five executions by inching the defined fitness function (17.15), which is changed according to the following equation:

$$f_i = 100 \times (opt \pm \delta)/T_i(JM) \tag{17.16}$$

where δ is the tuning parameter and its value is taken as 0, 1 and 2 respectively. The proposed algorithm produces quite satisfactory solutions in reasonable amount of iterations by tuning only the parameter δ.

From the table it can be seen that the proposed algorithm is able to find the best known solution for 35 instances, i.e. in about 81% of the instances, and the deviation of the minimum found makespan from the best known solution is only on average 0.16%. The proposed algorithm yields a significant improvement in solution quality with respect to other algorithms. The superior results indicate the successful incorporation of the improved BFA and PSO, which facilitates the escape from local minimum points and increase the possibility of finding a better solution. So it could be concluded that the proposed cooperative intelligence algorithm solves the JSSP fairly efficiently.

As above-mentioned, the algorithm is performed five times for every test problem by taking the parameter δ as 0, 1 and 2 respectively. Table 17.4 lists the mean solutions (Mean), the relative deviation of the mean solution (MRD), the standard deviation of the solutions (SD), the best solution (Best), and the relative deviation of the best solution (BRD). The MRD is commonly zero for small size problem, and is not more than 1.5% for most other problems.

To illustrate the simulated results more intuitively, the notorious FT10 problem is specially described as an example, in which there are numerous local optima so that the problem is challenging enough. To investigate the effect of δ-tuning on the solution quality and convergence, Figure 17.6 shows a group of convergence curves for different values of the parameter δ. Figure 17.7 shows the Gantt chart of a best solution.

Define the utilization percentage of the i-th machine

$$U_i = \frac{P_i}{I_i + P_i} \times 100\%, \tag{17.17}$$

where P_i is the total processing time of machine i, I_i is the total idle time of machine i before the last operation in the machine is processed. Figure 17.8 shows the percentage of utilization of each machine.

Let D_i denote the blank interval between makespan and the finish time of the last operation completed in the machine i, here, called the dormancy time of machine i. Then

$$D_i + I_i + P_i = \text{makespan}, \quad \forall i \in \{1, 2, \ldots, M\} \tag{17.18}$$

where M is the number of machines.

The percentage of the total processing time, idle time and dormancy time for all the machines is respectively calculated by

$$P\% = \frac{\sum_i P_i}{\sum_i (D_i + I_i + P_i)} \times 100\% \tag{17.19}$$

$$I\% = \frac{\sum_i I_i}{\sum_i (D_i + I_i + P_i)} \times 100\% \tag{17.20}$$

$$D\% = \frac{\sum_i D_i}{\sum_i (D_i + I_i + P_i)} \times 100\% \tag{17.21}$$

Figure 17.9 shows the percentage of the processing time, idle time and dormancy time.

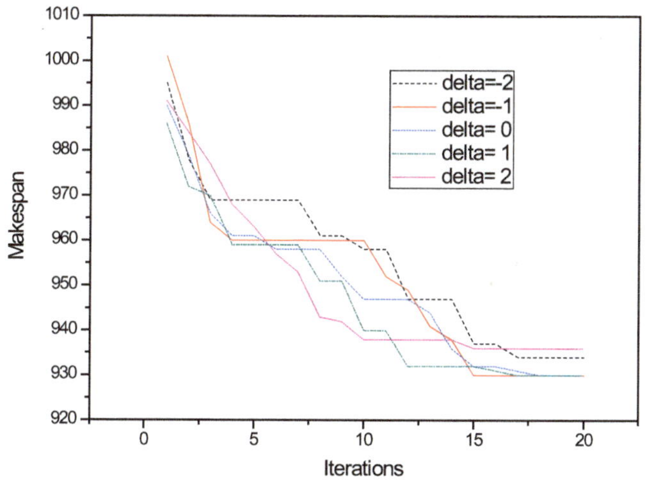

Fig. 17.6 Convergence curves for different values of the parameter δ

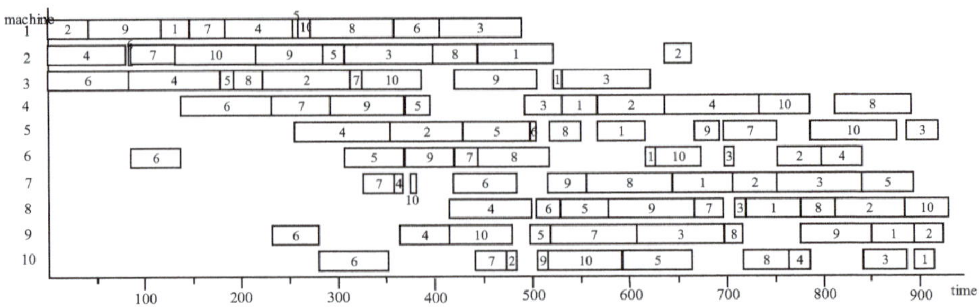

Fig. 17.7 The Gantt chart of an optimal schedule for FT10

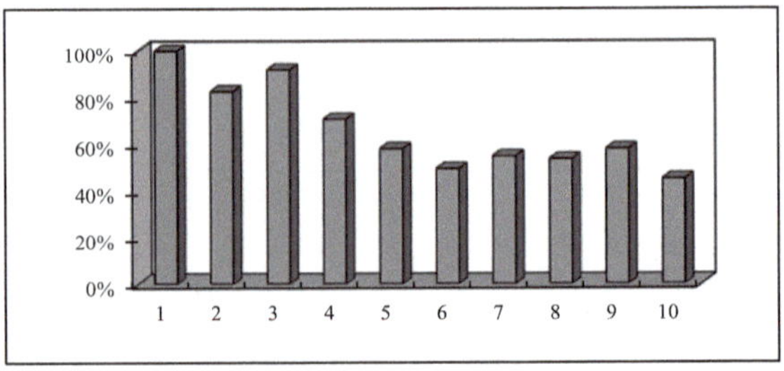

Fig. 17.8 The column chart of percentage of utilization

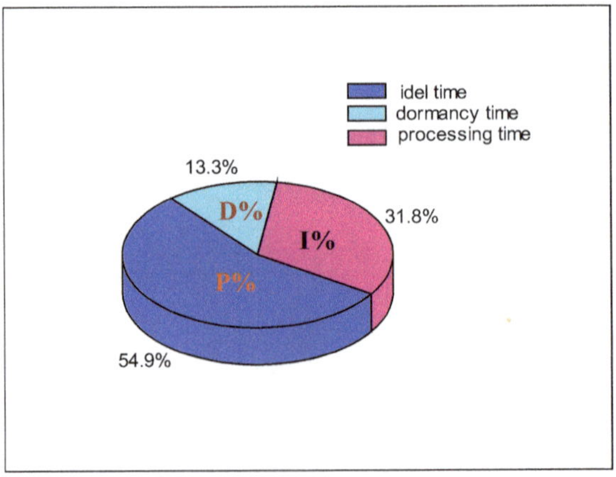

Fig. 17.9 The pie chart of percentage of different time

Table 17.3 Comparisons of the results between CIA and other approaches

Instance	Size	BKS	HEA	RD (%)	Lin [40]	Goncalves [30]			Ombuki [41]	Coello [42]	Aiex [43]	Binato [44]
						Param.A	HGA	Active	LSGA	AIS	GP+PR	MGA
Ft06	6×6	55	**55**	0.00	55	55	55	55	-	-	55	55
Ft10	10×10	930	**930**	0.00	930	930	951	945	-	941	930	930
Ft20	20×5	1165	**1165**	0.00	1165	1165	1178	1173	-	-	1165	1165
La01	10×5	666	**666**	0.00	666	666	666	666	-	666	666	666
La02	10×5	655	**655**	0.00	655	655	665	655	-	655	655	-
La03	10×5	597	**597**	0.00	597	597	604	603	-	597	597	-
La04	10×5	590	**590**	0.00	590	590	590	598	-	590	590	-
La05	10×5	593	**593**	0.00	593	593	593	593	-	593	593	-
La06	15×5	926	**926**	0.00	926	926	926	926	-	926	926	926
La07	15×5	890	**890**	0.00	890	890	890	890	-	890	890	-
La08	15×5	863	**863**	0.00	863	863	863	863	-	863	863	-
La09	15×5	951	**951**	0.00	951	951	951	951	-	951	951	-
La10	15×5	958	**958**	0.00	958	958	958	958	-	958	958	-
La11	20×5	1222	**1222**	0.00	1222	1222	1222	1222	-	-	1222	1222
La12	20×5	1039	**1039**	0.00	1039	1039	1039	1039	-	-	1039	-
La13	20×5	1150	**1150**	0.00	1150	1150	1150	1150	-	-	1150	-
La14	20×5	1292	**1292**	0.00	1292	1292	1292	1292	-	-	1292	-
La15	20×5	1207	**1207**	0.00	1207	1207	1207	1207	-	-	1207	-
La16	10×10	945	**945**	0.00	945	945	973	947	959	945	945	945
La17	10×10	784	**784**	0.00	784	784	792	784	792	785	784	-
La18	10×10	848	**848**	0.00	848	848	855	848	857	848	848	-
La19	10×10	842	**842**	0.00	842							
La20	10×10	902	**902**	0.00	902							
La21	15×10	1046	**1046**	0.00	1046	1046	1079	1074	1114	-	1057	1058
La22	15×10	927	**932**	0.54	932	935	950	962	989	-	927	-
La23	15×10	1032	**1032**	0.00	1032	1032	1032	1032	1035	-	1032	-
La24	15×10	935	**939**	0.43	941							
La25	15×10	977	**977**	0.00	977							
La26	20×10	1218	**1218**	0.00	1218	1218	1218	1237	1307	-	1218	1218
La27	20×10	1235	**1239**	0.32	1239	1256	1282	1280	1350	-	1269	-
La28	20×10	1216	**1216**	0.00	1216	1232	1250	1250	1312	1277	1225	-
La29	20×10	1152	**1179**	2.34	1173	1196	1206	1226	1311	1248	1203	-
La30	20×10	1355	**1355**	0.00	1355	1355	1355	1355	1451	-	1355	-
La31	30×10	1784	**1784**	0.00	1784	1784	1784	1784	1784	-	1784	1784
La32	30×10	1850	**1850**	0.00	1850	1850	1850	1850	1850	-	1850	-
La33	30×10	1719	**1719**	0.00	1719	1719	1719	1719	1745	-	1719	-
La34	30×10	1721	**1721**	0.00	1721	1721	1721	1721	1784	-	1721	-
La35	30×10	1888	**1888**	0.00	1888	1888	1888	1888	1958	1903	1888	-
La36	15×15	1268	**1274**	0.47	1278	1279	1303	1313	1358	1323	1287	1291
La37	15×15	1397	**1411**	1.00	1411	1408	1437	1444	1517	-	1410	-
La38	15×15	1196	**1213**	1.42	1208	1219	1252	1228	1362	1274	1218	-
La39	15×15	1233	**1233**	0.00	1233	1246	1250	1265	1391	1270	1248	-
La40	15×15	1222	**1229**	0.57	1225	1241	1252	1246	1323	1258	1244	-

Table 17.4 Average experimental results using the CIA

Instance	Size	BKS	Mean	MRD (%)	SD	Best	BRD (%)
Ft06	6×6	55	55	0.00	0.00	55	0.00
Ft10	10×10	930	932	0.22	2.19	930	0.00
Ft20	20×5	1165	1170	0.43	4.52	1165	0.00
La01	10×5	666	666	0.00	0.00	666	0.00
La06	15×5	926	926	0.00	0.00	926	0.00
La11	20×5	1222	1222	0.00	0.00	1222	0.00
La16	10×10	945	945	0.00	0.00	945	0.00
La21	15×10	1046	1050	0.38	4.20	1046	0.00

17.8 Conclusions

A promising cooperative intelligence algorithm combing an improved BFA and PSO is proposed to solve job shop problems with minimizing the makespan. In the artificial bacterial foraging system, a novel chemotactic model is designed to address the job shop scheduling problem and a mechanism of quorum sensing and communication are presented to improve the foraging performance. In the particle swarm system, a novel concept for the distance and velocity of a particle is presented for the job shop scheduling problem. Besides, a new scheduling initialization algorithm based on G&T algorithm is proposed to produce the initial population in the BFA. The proposed cooperative algorithm effectively exploits the capabilities of distributed and parallel computing of swarm intelligence approaches. The algorithm is tested on a set of 43 standard instances taken from the OR-Library. Computational results are compared with those obtained using other existing approaches and the proposed approach yields significant improvement in solution quality. The superior results indicate the successful incorporation of the two improved swarm intelligence approaches. The investigation on further study for the theoretical research as well as the performance of the technique is in progress. One of the important issues is to extend the proposed algorithm in order that it could be applied to more practical and integrated manufacturing problems such as dynamic arrivals, machine breakdown, or other factors that affect job status over time.

Acknowledgments

The authors are grateful to the support of the national natural science foundation of China (61103146, 60873256), the fundamental research funds for central universities (DUT11SX03), and the key technologies R&D program of China (2011BAK02B02-01).

Bibliography

[1] R. Nessah and I. Kacem, *Comp. Oper. Res.*, **471** (2012).
[2] C. Akkan and S. Karabati, *Eur. J. Oper. Res.*, **420** (2004).
[3] W. Jiang, K. Chen, X.Q. Zhong, C.E. Wang, and C.A. Zhu, *Compu. Eng.*, **19** (2008).
[4] C.N. Potts and L.N. Van Wassonhove, *Eur. J. Oper. Res.*, **405** (1987).
[5] Y.B. Canbolat and E. Gundogar, *J. Intell. Manuf.*, **527** (2004).
[6] R. Klein, *Eur. J. Oper. Res.*, **619** (2000).
[7] M.X. Weng and H.Y. Ren, *IIE Trans.*, **789** (2006).
[8] S. Topaloglu and G. Kilincli, *Int. J. Adv. Manuf. Tech.*, **781** (2009).
[9] S.Q. Liu and E. Kozan, *J. Oper. Res. Soc.*, **168** (2012).
[10] H.X. Chen and P.B. Luh, *Eur. J. Oper. Res.*, **499** (2003).
[11] I.J. Jeong and S.B. Yim, *Int. J. Prod. Res.*, **6783** (2009).
[12] F. Geyik and I.H. Cedimoglu, *J. Intell. Manuf.*, **439** (2004).
[13] J.P. Watson, J.C. Beck, and A.E. Howe, *Artif. Intell.*, **189** (2003).
[14] J.Q. Li, Q.K. Pan, P.N. Suganthan, and T.J. Chua, *Int. J. Adv. Manuf. Tech.*, **683** (2011).
[15] V. Kachitvichyanukul and S. Sitthitham, *J. Intell. Manuf.*, **355** (2011).
[16] R. Yusof, M. Khalid, G.T. Hui, S.M. Yusof, and M.F. Othman, *Appl. Soft. Comp.*, **5782** (2011).
[17] X. Gang and Z.M. Wu, *IEEE T. Sys. Man. Cy. A.*, **113** (2004).
[18] C. Fayad and S. Petrovic, *Lect. Notes. Comp. Sc.*, **524** (2005).
[19] R. Zhang and C. Wu, *Comp. Oper. Res.*, **854** (2011).
[20] R. Zhang and C. Wu, *Appl. Soft. Comp.*, **79** (2010).
[21] M. Kolonko, *Eur. J. Oper. Res.*, **123** (1999).
[22] R.K. Suresh and K.M. Mohanasundaram, *Int. J. Adv. Manuf. Tech.*, **184** (2006).
[23] L.N. Xing, Y.W. Chen, P. Wang, Q.S. Zhao, and J. Xiong, *Appl. Soft. Comp.*, **888** (2010).
[24] H.W. Ge, L. Sun, Y.C. Liang, and F. Qian, *IEEE T. Sys. Man. Cy. A.*, **358** (2008).
[25] W.J. Xia and Z.M. Wu, *Int. J. Adv. Manuf. Tech.*, **360** (2006).
[26] S.X. Yang, D.W. Wang, T.Y. Chai, and G. Kendall, *J. Scheduling*, **17** (2010).
[27] Yahyaoui, N. Fnaiech, and F. Fnaiech, *IEEE T. Ind. Electron.*, **1052** (2011).
[28] S. Yang and D. Wang, *IEEE T. Neural. Networ.*, **474** (2000).
[29] C.G. Wu, N. Zhang, J.Q. Jiang, and Y.C. Liang, *Lect. Notes. Comp. Sc.*, **562** (2007).
[30] J.F. Goncalves, J.J.D.M. Mendes, and M.G.C. Resende, *Eur. J. Oper. Res.*, **77** (2005).
[31] C.Bierwirth and D.C. Mattfeld, *Evol. Comput.*, **1** (1999).
[32] J. Giffler and G.L. Thompson, *Oper. Res.*, **487** (1960).
[33] K.M. Passino, *IEEE Contr. Syst. Mag.*, **52** (2002).
[34] B. Bhushan and M. Singh, *Appl. Soft. Comp.*, **4913** (2011).
[35] M.S. Kumar and P. Renuga, *Int. Rev. Electr. Engi.*, **1905** (2011).
[36] P.D. Sathya and R. Kayalvizhi, *Neurocomputing*, **2299** (2011).
[37] E. Budrene and H. Berg, *Nature*, **49** (1995).
[38] J. Kennedy and R. Eberhart, in *Proceedings of the IEEE International Conference on Neural Networks* (1995), Perth, Australia, IEEE Service Center, Piscataway, NJ, **4**, pp. 1942–1948.
[39] J.E. Beasley, *J. Oper. Res. Soc.*, **1069** (1990).
[40] T.L. Lin, S.J. Horng, T.W. Kao, Y.H. Chen, R.S. Run, R.J. Chen, J.L. Lai, and I.H. Kuo, *Expert. Syst. Appl.*, **2629** (2010).
[41] B.M. Ombuki and M. Ventresca, *Appl. Intell.*, **99** (2004).
[42] C.A.C. Coello, D.C. Rivera, and N.C. Cortes, *Lect. Notes. Comp. Sc.*, **2787** (2003).
[43] R.M. Aiex, S. Binato and M.G.C. Resende, *Parallel Comput.*, **393** (2003).
[44] S.Binato, W.J. Hery, D.M. Loewenstern, and M.G.C. Resende, in *Essays and Surveys in Metaheuristics* (2001), Kluwer Academic Publishers, Boston, pp. 59–80.

Chapter 18

Using Fuzzy Cognitive Maps as a Modeling Tool for Traveler Satisfaction in Public Transit Systems

Seda Ugurlu and Y. Ilker Topcu

Industrial Engineering Department, Istanbul Technical University, Macka Campus, 34367, Istanbu, Turkey

E-mail: sedayanik@itu.edu.tr, topcuil@itu.edu.tr

In order to increase the patronage of urban transit systems, improvement of customer satisfaction is a key element. As the factors affecting urban transit ridership are several, conflicting, and incommensurable; a multi criteria decision modeling technique is needed for the analysis of these factors. Fuzzy Cognitive Mapping (FCM) is a decision-support tool which gives the researcher opportunity to model complex systems especially in network shape. A FCM can be represented as a signed digraph which consists of factors (nodes) and relations between factors (edges between nodes). value are assigned to relations to represent the strength of impact. The value of factors are updated periodically by formulating the system in an iterative process. This is helpful in giving us the ability to see the system changes caused by the changes in factors.

18.1 Introduction

In modern societies; the quality of life, the economic productivity as well as the safety and the security are all affected by the transportation mobility. Nevertheless, our mobility in urban road networks is being interrupted because of traffic congestion due to increase in population, in income level, in car ownership, and in commuting.

Increasing population in urban areas also exerts pressure on infrastructure and the natural resources [1]. The key issues for sustainable growth are considered to be achieving better land use, lower noise, emission rates and energy use. This is mainly related to the considerable economic efficiency and environmental consciousness [2, 3]. For these reasons, transit is recognized as a key component in the management and planning of urban regions.

C. Kahraman (ed.), *Computational Intelligence Systems in Industrial Engineering*, Atlantis Computational Intelligence Systems 6, DOI: 10.2991/978-94-91216-77-0_18,

To achieve a sustainable and environmentally sensitive transportation form, the key is less private vehicle travel and higher levels of transit use in highly populated areas. Several modern cities take measures into consideration including increased investment in public transport and construction of new urban mass transit systems ranging from buses operating on streets to fully-grade-separated metro systems.

Transit represents a means by which people can efficiently move throughout a region with the least amount of impact on the environment. However, automobile travel offers individuals more freedom and flexibility in movement [1]. The advantages of private vehicles with the affordable vehicle ownership costs caused a significant increase in the use of private vehicles in the last decades. Despite the sustainable and environment-friendly mobility solution it presents, public transit ridership is not seen as convenient, available or comfortable compared to other private modes. People have become reluctant to travel by transit. And initiatives aimed at increasing the use of public transport have failed in the past with declining market shares [3, 4]. To overcome this resistance, urban planners need to identify strategies. A challenge for urban planners is to induce travelers and offer services that lead to use of public transportation.

It has been realized that transit should become a viable alternative competing with private vehicles. Customer satisfaction is one of the key elements for achieving higher demand for transit. Unless it is a necessity, public transportation ridership is determined largely by the match between the customer preferences and the service offered. On the other side, it is known that increasing levels of car ownership and income decrease the tendency to use public transportation. Thus, socio-economic variables constitute one of the major aspects for traveler satisfaction and mode choice. Besides, transit ridership is also mainly influenced by the urban form in terms of density and land use mix. In various studies, in effort to model the transit ridership, it has been concluded that origins in central business districts create higher ridership levels [5, 6].

In this study, we aim to introduce a decision support approach for transportation planning efforts to increase transit satisfaction and transit ridership in urban transportation. We employ fuzzy cognitive maps (FCMs) in order to observe the relationships among concepts clearly and facilitate impact analysis of alternatives. FCM is proposed as a tool which allows to model complex systems such as transportation. It enables to deal with various variables which may not be well-defined and are not known with certainty such as service quality described as intangible, heterogeneous.

18.2 Public Transit and Traveler Satisfaction

In the literature review made, transit service was characterized by some quality dimensions which are commonly used to determine service quality in many other service industries. The pioneering study of Parasuraman *et al.* [7] in which service quality dimensions have been systematically measured inspired the service quality research also in transportation industry. The features that define the transit service quality have been agreed as convenience (including ease of access and ease of use), performance/reliability (time, vehicles, schedules, and seamless connections), comfort (overcrowded vehicles without air conditioning), safety (feeling secure), value (cost savings) and facilities.

18.2.1 *Components of Public Transit System*

Public Transit system elements are well studied in the literature. We have concluded with four groups of elements to achieve better transit planning and higher transit ridership by reviewing the literature. In the following sections, we introduce the transit components which are related to quality of service and customer satisfaction, urban form, traveler characteristics and transit image.

18.2.1.1 *Transit and Service Quality*

Customer satisfaction research literature traditionally agrees that service quality is a measure of how well the service level delivered matches customer expectations. Delivering quality service means conforming to customer expectations on a consistent basis [8]. The minimum requirements from the transit system to achieve higher transit ridership are listed as [1]: more effective price structures; enhanced travel comfort; better suitability and convenience of service quality; reductions in travel time efficiency, increased service access.

In many practices in the service industry, a customer-defined service quality attribute list is developed by an exhaustive qualitative research conducted with customers [7–9]. The previous research shows that there are 40 to 50 attributes that can impact transit customer satisfaction. Investigating how transit customers aggregate attributes of service into collapsed quality dimensions is important to understand how customer satisfaction should be measured. In the literature, aggregated quality dimensions are developed from the customer-defined services attributes by statistical methods such as analyzing correlations, factor analysis and multiple regression analysis. As a result of the literature review, service quality dimensions related to the transit system can be expressed as follows:

Accessibility/Convenience: It represents the ease of access and ease of use of the transit system. Some of the underlying service attributes of the convenience dimension are availability of stations/bus stops close to home, close to work or shopping and availability of parking at stations/bus stops [10, 11].

Performance: It includes the issues of transit service, schedules, and vehicles. Performance of the transit system is mainly measured with the frequency of service, travel time by train/bus, wait time when transferring [12].

Reliability: It shows the consistency of the system to the declared performance. The most important indicator is the on-time performance. Some other indicators are the breakdown frequency, time takes to fix a breakdown and the delay a breakdown causes [13].

Comfort: It is concerned mainly with the overcrowded vehicles without air conditioning. The underlying service attributes are availability of seating at the station/bus stop, availability of seats on the vehicle, smoothness of the vehicle ride, comfort of the seats, degree of crowding on the vehicle, comfortable temperatures on the vehicle and availability of handrails/grab bars.

Safety: It includes the concerns about personal safety on transit vehicles as well as ability to avoid traffic accidents. Some indicators are safety from crime while riding, safety at stations/bus stops, safety related to the behavior of other persons and the rail/bus operation.

Value: It is focused on cost savings through employer subsidy programs and reduced transportation costs. value is mainly related to the cost of a one-way ride and a transfer, availability of discounted fares for senior citizens, students and availability of volume discounts such as monthly passes, as well as cost of parking at stations/bus stops.

18.2.1.2 *Transit and Urban Form*

The existing body of knowledge support that transit and urban form are strongly related to each other. Examined separately, each direction of the transit and urban form interaction is significant [6]. In Ref. [6], the term "urban form" is broken down into several elements to examine the ways in which urban form affects transit patronage.

Densities: In the literature, the density or compactness of an urban area, and its influence on transit use has been well examined. The relationship between residential densities, employment center densities, and transit patronage is proven to be robust.

Mix of Land Uses: The mix of land uses and urban design features in transit corridors also contributes to transit's attractiveness as a mode of travel. The characteristics of areas around stations also have an influence on the transit patronage.

Pedestrian Amenities: Though the bundle of attributes that makes for a successful, pedestrian and transit-friendly station area or neighborhood is difficult to break apart through statistical means, in many studies it has been shown that the presence of these attributes clearly makes transit a more attractive choice [14, 15, 9]. For example, a four-faceted measure of "pedestrian friendliness" was developed for Portland's travel demand forecasting model structure with the attributes of street connectivity, sidewalk connectivity, use of street crossing on principal arterials, and absence of topographic constraints to pedestrian mobility [15].

18.2.1.3 *Transit and Traveler Characteristics*

A number of factors including socioeconomic and demographic characteristics of the transit users were reviewed in the literature that may influence mode choice.

Car Ownership: Among the traveler socioeconomic and demographic characteristics, vehicle availability certainly seems to be the most powerful factor [5]. For a person or household having no vehicle, transit becomes almost a necessity issue. Contrastingly, as the vehicle availability increases, the tendency to use transit mode decreases.

Income: In modeling the traveler mode choice, income is also one of the most significant factors. The expectations, behaviors and perceptions of the individual/households from different income levels are diverse also in terms of travel.

Some of the other traveler characteristics which influence the mode choice are household type (size of the household, child in the household), age, gender and occupation [16].

18.2.1.4 *Transit and Its Image*

In transportation industry, it has been recently recognized that identifying perceptions, value and decision-making processes of people leverages the efforts to motivate people towards transit use [17]. In this perspective, successful marketing programs are essential to support transit use. The social marketing practice in transit systems aims to change the behavior of people in benefit of public. Named as social marketing, it uses the traditional marketing techniques to identify the actions for improving the image of the transit systems. The road map starts with understanding the personal attitudes towards transit, specifying the drivers to use transit for different groups, segmenting, targeting and communicating effectively. It is now widely believed that not only the operations but maintaining the visibility and image of the transit services makes transit more significant to people [18].

18.3 Methodology of Fuzzy Cognitive Maps

The first use of cognitive maps was designed to represent knowledge in social sciences, capture the casual relationships and make decisions of policies in social systems [19]. Cognitive mapping is a decision-making tool to model complex systems [20]. The signed diagraphs are developed to capture the causal assertion between elements of a system. They describe the behavior of a system showing the dynamics of the system. The graphs are generated using concepts representing an entity, a state, a variable or a characteristic of the system. The concepts are shown by nodes and the relations among the concepts are represented by weighted arcs.

FCMs are developed by fuzzy value of nodes and weighted arcs. The first use of FCMs was by Ref. [21] augmenting the power of CMs. After this pioneering work, FCMs attracted the attention of scientists in many fields and have been used in a variety of different scientific problems.

FCMs have been used for decision making in social sciences, for analyzing graph theoretic behavior. Moreover, FCMs also have been used in the control related themes such as Failure Modes and Effects Analysis for a system model and to model the supervisor of control systems.

In general, a FCM could be regarded as associative neural networks. A FCM describes a system in a one-layer network which is used in unsupervised mode, whose neurons are assigned concept meanings and the interconnection weights represent relationships between these concepts. Specifically, a FCM is a fuzzy signed oriented graph with feedback. The value of the nodes in a FCM are represented as fuzzy sets. They are expressed on a normalized range rather than an exact quantitative value, showing how active a concept is at a specific time. Thus, it provides a basis for comparing nodes such as "on or off", "active or inactive". This mapping is a variation of the "fuzzification" process in fuzzy logic [22].

The other constituent of the fuzzy signed graph, the causal relations between the concepts are to be fuzzy implications, too. A directed arc, W_{ij} from concept C_i to concept C_j measures how much C_i causes C_j. In simple FCMs, directional influences take on trivalent value $[-1,0,+1]$ where -1 indicates a negative relationship, 0 no causality relationship, and $+1$ a positive relationship. In general, the arcs W_{ij} can take value in the fuzzy causal interval $[-1;+1]$ allowing degrees of causality to be represented.

In Figure 18.1, a typical fuzzy graph is illustrated with nodes standing for concepts and causal relationships shown by weighted arcs. The graphical representation provides a clear understanding of the concepts influencing each other with the interconnections between each other. In this way, it is possible to figure out needed changes or updates for completing the construction of the system by adding or deleting concepts or links between them.

A mathematical model is formulated beyond the graphical representation. This model consists of an adjacency matrix which is a $n \times n$ weight matrix displaying all of the weights W_{ij} of the direct relationships from concept i to j for all n concepts. And the degree of activation of the n concepts (A_i) is shown by the $1 \times n$ state vector S. The initial state vector (S^0) consists of the concept value resulted from the fuzzification process which is the process of converting a crisp input value to a fuzzy value [23].

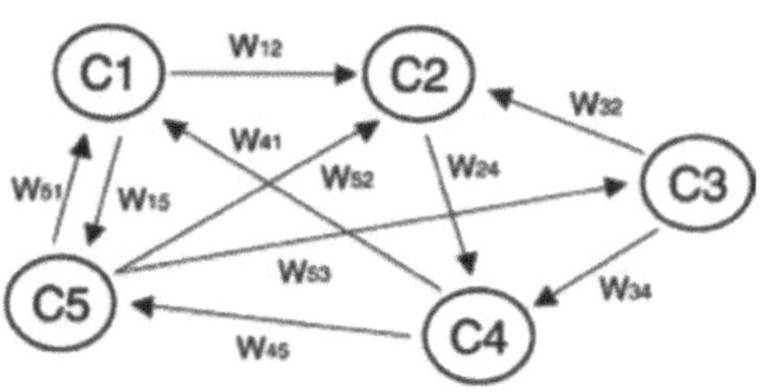

Fig. 18.1 A typical fuzzy cognitive map

The value of each concept is influenced by the value of the connected concepts with the appropriate weights and by its previous value. So, the initial value ($A_i^{(0)}$) and the arc value W_{ij} from the adjacency matrix are used to generate new concept value iteratively by equation (18.1):

$$A_i^{(k)} = f\left(\sum_{\substack{j=1 \\ j\neq i}}^{n} A_j^{(k-1)} W_{ji} + A_i^{(k-1)}\right) \tag{18.1}$$

where $A_i^{(k)}$ is the activation level of concept C_i at iteration step k, $A_i^{(k-1)}$ the activation level of the interconnected concept C_j at iteration step $k-1$, and W_{ji} is the weight of the interconnection between C_j and C_i, and f is a threshold function.

The threshold functions applied to the weighted sums are generally fuzzy in nature, reducing the unbounded inputs to a strict range and providing a basis for comparing nodes.

Some of the most common threshold functions used are bivalent, trivalent, logistic and sigmoid functions which are given respectively by equation (18.2), (18.3), (18.4), and (18.5).

$$f(x) = \begin{cases} 0, & x \leqslant 0 \\ 1, & x > 0 \end{cases} \tag{18.2}$$

$$f(x) = \begin{cases} -1, & -0.5 < x < 0.5 \\ 0, & x \geqslant 0.5 \\ 1, & x > 0 \end{cases} \tag{18.3}$$

$$f(x) = \frac{1}{1+e^{-cx}} \tag{18.4}$$

$$f(x) = \tanh(x) \tag{18.5}$$

In the study of Ref. [4], the capabilities of the functions are discussed. It is concluded that for specific situations, using specific functions are more suitable. For example, by using bivalent functions, only an increase at the activation level of the concepts can be achieved. However Trivalent and Sigmoid functions have the capabilities to reach increasing or decreasing levels. Using a sigmoid function, it is also possible to follow the changes in the degree of a concept.

After constructing the graph and the mathematical formulations with the initial state value and the weights, different scenariosscenarios can be simulated to observe the behavior of the complex systems dynamically. Simulations are run to a specific iteration number. The iteration number can be decided as a fix number or the value of the nodes could be observed until they converge to an expected form.

18.4 Applications of Fuzzy Cognitive Maps

The FCM methodology, Ref. [21] presented, have been used for decision making in various research and application areas. It provides an approach for causal reasoning, prediction and identification of improvement areas of complex situation. FCMs are argued to be successful when used with expert knowledge or other learning procedures in administrative sciences, game theory, information analysis, popular political developments, electrical circuits analysis, cooperative man machines, distributed group-decision support and adaptation and learning, etc. [22].

FCM has been applied used in many real-world problems for making societal, governmental, and corporate decisions. Among them, there are studies which analyze factors for the performance of a business unit or an electronic commerce web site: To analyze business performance, FCM was employed for identifying areas for improvement as well as benchmarking with the competition [25]. They first identified the metrics for the assessment of business performance. They used both internal and external factors such as information systems planning, business process reengineering and new entrants. Developing the FCM model for business performance, they identified the links of business strategies with business metrics and allowed for the simulation and evaluation of alternative ways of improving business performance.

Ref. [26] used FCM to analyze the design factors for the performance of electronic commerce web sites. They constructed the causal relationships among design factors of the electronic sites, and investigated the simulation to find the optimal strategy of adjusting the design factors. They used linear structural relationship to prove the original technology acceptance model. The degree of relationships in FCM was concluded based on the linear structural relationship. They performed cognitive map simulations for 15 hypothetical market situations to identify the most relevant design factors and thereby allocate limited resources.

Though FCM presents an effective methodology to analyze complex situation, its use in transportation is limited. In 1997, Ref. [27] proposed a path diagram to represent the linkages between factors affecting trip generation and mode choice which are the second and third steps of four-step transportation planning model. This diagram included factors such as income, car ownership, settlement density, trip length, perceptions, household characteristics, public transportation use, service quality, congestion, fares, costs, etc.

The FCM approach was used by Ref. [20] in the airline service to reveal the causal relationships among relationship management factors such as conflict, cooperation, communication, etc. They have used structural equation models to identify the relevant relationships among the factors and to determine their direction and strength. Then, they used standardized causal coefficients to create a cognitive map illustrating the effect of the status of one component on the status of another component.

Ref. [28] proposed an innovative use of FCM in the area of urban design. This paper proposed a tool for modeling of expert decision support. This methodology is based on a FCM-driven approach where the analogies between selective linear definite clause resolution trees and FCM are used. Selective linear definite clause resolution is a basic inference rule used in logic programming. In this study, they interpreted an existing logic-based knowledge to imprecise causal relationships among concepts within the FCM framework. Finally, the IF-THEN logic based-rules are generated into multi-branch map hierarchies.

Then, they showed the suitability of the tool by an application of the expert domain of esthetic analysis for urban design.

To our knowledge, the FCM methodology has not been used in analyzing the causal relationships between the service quality and other factors affecting the public transit use. This study proposes to model and analyze factors affecting customer satisfaction in public transit systems with utilizing FCM approach. The factors that motivate the use of public transportation are interlinked with each other complicatedly via direct and indirect causal relationships. A systematic approach analyzing the causality among these factors would help decision makers to estimate and understand the outcomes of the policies they would use for transportation planning of urban areas. Various models are presented to plan, develop and operate the public transportation services efficiently. However, FCM serves as an alternative approach to model interrelationships between travelers' demands, perceptions, urban form and the transportation service dimensions. We propose to use Fuzzy Cognitive Map (FCM) method for analyzing the causality and various scenarios under the cause and effect relations of the complex and dynamic nature of urban public transportation planning and traveler behavior.

18.5 FCM Development for Traveler Satisfaction and Transit Ridership

In diverse applications of FCMs, the goal is, commonly, to model the interrelations among the huge number of decision variables and the uncontrollable variables with each other as well as to analyze the hidden causal relationships in complex and dynamical systems. Moreover, FCMs allow researcher or practitioners to simulate the chain effect of activation of a specific node on the other nodes, make what-if sensitivity analysis to ensure the intended results and observe the indirect effects caused by activating specific nodes in a system where factors are highly interrelated [26].

In accordance with these applications, the aim in this study is to show the relationships among a variety of factors determining the level of transit ridership and the customer satisfaction from transit systems. The factors related to the transit domain are typically transit service dimensions such as performance, value, convenience, etc., urban form related factors, traveler characteristics as well as the visibility and image of transit systems. Using FCMs to model transit ridership will provide the ability to analyze all relevant factors simultaneously. FCM here offers a dynamic feedback system to observe all the direct and indirect effects of a change in the service quality in terms of traveler satisfaction and behavior to use transit for different traveler groups having diverse characteristics under various environmental conditions defined by the urban form.

The service quality dimensions, travels characteristics and the urban form related factors used to construct the FCM were concluded through the literature review. The list of the related concepts is given in Table 18.1 with brief explanations.

The service quality dimensions discussed in the literature review were represented by the concepts that describe the system best and that could be evaluated or measured easily. The concept "accessibility" was used to represent directly the accessibility dimension of quality which consists of availability of stations close to home or work, availability of parking places at stations, as well as pedestrian amenities that makes walking convenient for pedestrians. "travel time" concept was included as the main variable showing the performance of the travel service provided. Travel time was used as a general time including the in-vehicle time and wait time. In many studies, transfer that includes a walking between two stations and a wait time was treated as a specific determinant of quality related to both convenience and performance [29, 10]. Transfer is commonly perceived as a disutility element of travel. However, it provides the travelers a flexible multimodal travel service when transfers are managed such that it incurs minimum wait times and access problems. So that, "ease of transfer" was specified as an important concept of the transit system. "cost of travel" was included as a variable of the transit travel having a major impact on the value of the service. The service quality dimensions of "comfort" and "reliability" were also taken into consideration as major variables of the transit service. Other uncontrollable variables related to the urban form were represented by the concept of "residential/business density" whereas traveler characteristics related concepts were used as "income" and "car ownership". Income and car ownership were well studied by many researchers as the most significant factors influencing transit patronage. In addition, "image of the transit system" which was shown to have a robust influence on the preferences of travelers towards transit was considered as a concept in the constructed FCM. The common practice to construct the cognitive maps is to conduct questionnaires.

In this study, we first identified the concepts by reviewing the literature. Then, we asked four experts to evaluate which concepts influence other concepts and to determine the negative or positive effect for each relationship by assessing the degree of influence. A questionnaire of about 14 pages and 156 questions was developed to assess the influence of every concept on the other concepts. A part of the questionnaire form is presented at Table 18.2.

Table 18.1 The FCM concepts of public transit satisfaction

Concepts	Explanation of the concept
Service Quality	It is a measure of how well the customer needs are met with a certain level of performance.
Accessibility	It represents the ease of access to the transit system. Some underlying service attributes: availability of stations/bus stops close to home, close to work or shopping and availability of parking at stations/bus stops
Ease of Transfer	It indicates how well the transit system is integrated with each other in terms of scheduling, design, etc. Some underlying service attributes: wait time, walk distance while transferring.
Reliability	It shows the consistency of the system to the declared performance. The most important indicator is the on-time performance. Some other indicators are the breakdown frequency, time takes to fix a breakdown and the delay a breakdown causes
Travel Time	It is the main indicator of the performance of the transit system. In transit; the walk time, wait time and in-vehicle time are valued differently. The total journey time is also compared to the travel time of private vehicles and other modes of transit.
Comfort of Travel	It is concerned mainly with the overcrowded vehicles without air conditioning. The underlying service attributes are availability of seating at the station/bus stop, availability of seats on the vehicle, smoothness of the vehicle ride, comfort of the seats, degree of crowding on the vehicle, comfortable temperatures on the vehicle and availability of handrails/grab bars.
Cost of Travel	It is mainly related to the cost of a one-way ride and a transfer, availability of discounted fares for senior citizens, students and availability of volume discounts such as monthly passes, as well as cost of parking at stations/bus stops
Income Level	It is counted as the income of all residents in a household and generally expressed in monetary terms
Car Ownership	It is a percentage showing the ownership of a private vehicle for the people under consideration
Residential/ Business Density	High employment and population densities commonly serve as the major hub for transit lines. A primary incentive for transit use and other alternatives to the private vehicles is the cost of parking in these areas.
Image of Transit Agency	It includes understanding the personal attitudes towards transit, specifying the drivers to use transit for different groups, segmenting, targeting and communicating effectively
Customer Satisfaction	It is a measure of how products and services provided matches or surpass customer expectation
Level of Transit Ridership	It shows the share of all trips on public transit systems. It can be considered as the market share of transit versus private vehicles

We provided the explanations of the concepts to the experts in order to ensure that each concept is perceived in the same perspective among experts. A pairwise assessment of each concept with the other concepts was asked from the experts on the questionnaire form.

Table 18.2 Example of the questionnaire

Increase in "Comfort" will cause the following to:	+	0	-	*To what degree* Very Low 1	2	3	4	5	Very High 6
Travel time				1	2	3	4	5	6
Cost of Travel of Transit				1	2	3	4	5	6
Ease of Transfer				1	2	3	4	5	6
Reliability				1	2	3	4	5	6
Image of the Transit Agency				1	2	3	4	5	6
Car Ownership				1	2	3	4	5	6
Accessibility				1	2	3	4	5	6
Income				1	2	3	4	5	6
Service Quality				1	2	3	4	5	6
Residential/Business Density				1	2	3	4	5	6
Customer Satisfaction				1	2	3	4	5	6
Level of Transit Ridership				1	2	3	4	5	6

Two of the four experts are from the different Transit Agencies who have responsibilities in transport planning. One of them has taken role in the last Transportation Master Plan of Istanbul city. The other one is on the board of the Ispark which is responsible for managing the parking areas and policies within the city. The other two experts are academicians whose main research areas are transportation and they have been consulting for Transportation Agencies for the last 15 years. Three of the experts filled the questionnaire in the presence of one of the authors. It was completed in approximately one and a quarter hour. The other expert chose to fill the questionnaire in his leisure time and contacted the authors for asking some questions in order to understand the process.

In order to obtain an objective basis, the subjective evaluations of the experts were combined into one collective map. There are applications of FCM in which experts are weighted differently when it is believed that the level of expertise of each expert in the area is diverse. In this study, the evaluations of each expert were combined with equal weights. In other words, the combined adjacency matrices were specified using equation (18.6):

$$W_{ij}^{C} = \frac{1}{4}\left(W_{ij}^{1} + W_{ij}^{2} + W_{ij}^{3} + W_{ij}^{4}\right) \qquad (18.6)$$

The concluding adjacency matrix which shows the complex causal relationships with the highly interrelated elements of the transit system is given in Table 18.3.

Examining the adjacency matrix, it could be concluded that the output variables of the FCM, level of transit ridership, customer satisfaction and the service quality are the most affected variables. Besides, the most active variables which influence the other factors and the output variables strongly are found to be the car ownership, travel time, ease of transfer and comfort.

Table 18.3 Combined adjacency matrix

	Comfort	Travel Time	Cost of Travel	Ease of Transfer	Reliability	Image of Transit Agency	Car Ownership	Accessibility	Income	Service Quality	Residential/ Business Density	Customer Satisfaction	Level of Transit Ridership
Comfort	0.00	0.00	0.00	0.25	0.17	0.00	-0.17	0.17	0.00	0.17	-0.08	0.08	-0.25
Travel Time	0.00	0.00	0.00	0.00	0.00	0.00	-0.50	-0.17	0.00	-0.08	0.08	-0.08	0.00
Cost of Travel	0.08	0.08	0.00	-0.08	-0.08	0.00	0.00	0.00	0.00	0.17	0.00	0.08	0.00
Ease of Transfer	0.17	0.00	0.00	0.00	0.08	0.00	0.00	0.25	0.00	0.17	-0.08	0.08	0.00
Reliability	0.00	0.08	0.00	0.17	0.00	0.00	-0.17	0.00	0.00	0.17	-0.08	0.08	-0.08
Image of the Transit Agency	0.42	-0.33	-0.08	0.25	0.33	0.00	0.00	0.17	0.00	0.42	0.00	0.50	0.17
Car Ownership	-0.25	0.25	0.33	-0.17	-0.08	-0.17	0.00	-0.17	0.67	-0.25	0.08	-0.33	-0.42
Accessibility	0.17	-0.17	0.00	0.42	0.00	0.08	-0.25	0.00	0.00	0.08	0.08	0.08	0.00
Income	0.00	0.00	0.00	0.00	0.00	0.00	0.00	0.00	0.00	0.00	0.00	0.00	0.00
Service Quality	0.33	-0.50	-0.17	0.33	0.33	0.00	-0.08	0.25	0.00	0.00	0.25	0.00	-0.17
Residential/ Business Density	0.00	-0.08	0.00	0.08	0.00	0.08	-0.17	0.08	0.08	0.33	0.00	0.08	0.17
Customer Satisfaction	0.42	-0.50	-0.50	0.42	0.33	0.17	-0.08	0.33	-0.08	0.58	0.08	0.00	-0.17
Level of Transit Ridership	0.33	-0.50	-0.58	0.25	0.25	0.25	-0.50	0.17	-0.08	0.58	0.33	0.42	0.00

The adjacency matrix was formulated with a sigmoid threshold function for simulating various scenarios and obtaining the results of the scenarios. A sigmoid threshold function was employed because it allows to follow the smooth changes in the increasing or decreasing levels of the concepts.

18.5.1 *Scenario Analysis with Simulations*

In the simulation process, a concept was activated by initial value such as -1, 0 or 1. Iterations were run using equation (18.1). Accordingly, the state vector was calculated at each iteration. Iterations were terminated until the value of the concepts converge. For different scenarios, simulations were initialized with different initial state vectors. So the variations in the system behavior could be observed under different levels of service dimensions. Scenario analysis offers a tool for transport agencies to observe the chain effect caused by the improvement of a quality dimension and/or a change in the state of a sys-

Table 18.4 Iteration results after a decrease in the cost of transit travel

Iteration Nr.	Comfort	Travel Time	Cost of Travel	Ease of Transfer	Reliability	Image of the Transit Agency	Car Ownership	Accessibility	Income	Service Quality	Residential/ Business Density	Customer Satisfaction	Level of Transit Ridership
1	0.0	0.0	-1.0	0.0	0.0	0.0	0.0	0.0	0.0	0.0	0.0	0.0	0.0
2	0.0	0.0	-0.8	0.0	0.0	0.1	-0.3	0.0	0.0	0.2	0.0	0.5	0.5
3	0.0	0.1	-0.6	0.1	0.1	0.5	-0.8	0.1	0.0	0.2	0.2	0.7	0.9
4	0.0	0.4	-0.5	0.2	0.2	0.8	-0.9	0.4	0.0	0.3	0.6	0.8	1.0
5	0.2	0.6	-0.3	0.4	0.4	0.9	-1.0	0.7	0.0	0.5	0.8	0.9	1.0
6	0.4	0.7	-0.2	0.6	0.6	1.0	-1.0	0.9	0.0	0.7	0.9	0.9	1.0
7	0.7	0.8	0.0	0.8	0.7	1.0	-1.0	0.9	0.0	0.8	0.9	1.0	1.0
8	0.8	0.8	0.2	0.9	0.8	1.0	-1.0	0.9	0.0	0.9	0.9	1.0	1.0
9	0.9	0.8	0.4	0.9	0.9	1.0	-1.0	1.0	0.0	0.9	0.9	1.0	1.0
10	0.9	0.8	0.6	0.9	0.9	1.0	-1.0	1.0	0.0	0.9	1.0	1.0	1.0
11	0.9	0.8	0.7	0.9	0.9	1.0	-1.0	1.0	0.0	0.9	1.0	1.0	1.0
12	0.9	0.8	0.7	0.9	0.9	1.0	-1.0	1.0	0.0	0.9	1.0	1.0	1.0

tem variable. Since the system is represented as a network, the interrelationships cause the chain effect. When the state value converge, the results of the change can be identified. Using this tool, agencies may incorporate the complex interrelationships among the variables of the system and target to the most effective improvement areas in order to obtain most desirable and efficient outcome in the transportation system.

Some of the scenarios and the results corresponding to these scenarios are discussed in this section.

In the first scenario, the policy of offering the society a lower cost of travel by transit is investigated. In practice, such a policy is commonly used to give incentives to a group of the society for using transit through employer subsidy programs, discounted fares for senior citizens, students, etc. In Table 18.4, simulation results of the first scenario were given.

The state value of the concepts were given for the each iteration in the table. It is clearly seen that a decrease in the cost of travel by transit increases the level of transit ridership powerfully. The cost of travel also affects the car ownership negatively. On top of that, through indirect effects lowering cost of transit travel will also end up with higher levels of service quality dimensions which will be facilitated by the increasing volume in transit.

In the second scenario, the resulting levels of the concepts after a decrease in the travel time were analyzed. The results given at Table 18.5 are parallel to the results obtained after decreasing the cost. The level of transit ridership and the customer satisfaction levels are

Table 18.5 Iteration results after a decrease in the travel time of transit travel

Iteration Nr.	Comfort	Travel Time	Cost of Travel	Ease of Transfer	Reliability	Image of the Transit Agency	Car Ownership	Accessibility	Income	Service Quality	Residential/ Business Density	Customer Satisfaction	Level of Transit Ridership
1	0.0	-1.0	0.0	0.0	0.0	0.0	0.0	0.0	0.0	0.0	0.0	0.0	0.0
2	0.0	-0.8	-0.1	0.0	-0.1	0.3	-0.2	0.2	0.0	0.5	0.1	0.5	0.5
3	0.0	-0.6	0.0	0.1	0.0	0.8	-0.8	0.4	0.0	0.7	0.5	0.8	0.9
4	0.2	-0.4	0.1	0.4	0.1	1.0	-1.0	0.8	0.0	0.8	0.8	1.0	1.0
5	0.5	-0.1	0.3	0.6	0.4	1.0	-1.0	0.9	0.0	0.9	0.9	1.0	1.0
6	0.7	0.1	0.4	0.8	0.6	1.0	-1.0	1.0	0.0	0.9	1.0	1.0	1.0
7	0.8	0.4	0.5	0.9	0.8	1.0	-1.0	1.0	0.0	1.0	1.0	1.0	1.0
8	0.9	0.6	0.6	0.9	0.8	1.0	-1.0	1.0	0.0	1.0	1.0	1.0	1.0
9	0.9	0.7	0.7	0.9	0.9	1.0	-1.0	1.0	0.0	1.0	1.0	1.0	1.0
10	0.9	0.7	0.7	0.9	0.9	1.0	-1.0	1.0	0.0	0.9	1.0	1.0	1.0

Table 18.6 Iteration results after an increase in the car ownership

Iteration Nr.	Comfort	Travel Time	Cost of Travel	Ease of Transfer	Reliability	Image of the Transit Agency	Car Ownership	Accessibility	Income	Service Quality	Residential/ Business Density	Customer Satisfaction	Level of Transit Ridership
1	0.0	0.0	0.0	0.0	0.0	0.0	1.0	0.0	0.0	0.0	0.0	0.0	0.0
2	-0.2	-0.5	0.0	0.0	-0.2	0.0	0.8	-0.2	0.0	-0.1	-0.2	-0.1	-0.5
3	-0.2	-0.7	-0.1	-0.1	-0.3	-0.2	0.7	-0.4	0.0	-0.1	-0.4	-0.1	-0.7
4	-0.3	-0.8	-0.1	-0.3	-0.4	-0.4	0.8	-0.5	0.0	-0.1	-0.6	-0.1	-0.8
5	-0.4	-0.8	-0.2	-0.4	-0.5	-0.7	0.8	-0.7	0.0	-0.2	-0.7	-0.3	-0.9
6	-0.6	-0.8	-0.2	-0.6	-0.6	-0.9	0.9	-0.8	0.0	-0.4	-0.8	-0.6	-1.0
7	-0.7	-0.8	-0.4	-0.8	-0.7	-1.0	1.0	-0.9	0.0	-0.7	-0.9	-0.9	-1.0
8	-0.8	-0.8	-0.5	-0.9	-0.8	-1.0	1.0	-0.9	0.0	-0.9	-0.9	-1.0	-1.0
9	-0.9	-0.8	-0.6	-0.9	-0.8	-1.0	1.0	-1.0	0.0	-0.9	-0.9	-1.0	-1.0
10	-0.9	-0.8	-0.7	-0.9	-0.9	-1.0	1.0	-1.0	0.0	-0.9	-1.0	-1.0	-1.0
11	-0.9	-0.8	-0.7	-0.9	-0.9	-1.0	1.0	-1.0	0.0	-0.9	-1.0	-1.0	-1.0

observed to rise rapidly. On the other side, car ownership is decreasing whereas service quality dimensions are improving through the effect of the volume created in transit. However, due to high transit use, a feedback is observed in travel time. As a result, the travel time tends to increase again.

In the third scenario, we simulated a reality of increased car ownership. In real-life, we also experienced the higher levels of car ownership over the last decades. In the simulation results, it could be observed that, on the first hand, higher levels of car ownership cause an increase in the travel time of transit due to right of way shared with automobiles for modes such as buses.

Table 18.7 Iteration results after a decrease in the image of transit agency

Iteration Nr.	Comfort	Travel Time	Cost of Travel	Ease of Transfer	Reliability	Image of the Transit Agency	Car Ownership	Accessibility	Income	Service Quality	Residential/ Business Density	Customer Satisfaction	Level of Transit Ridership
1	0.0	0.0	0.0	0.0	0.0	1.0	0.0	0.0	0.0	0.0	0.0	0.0	0.0
2	0.0	0.0	0.0	0.0	0.0	0.8	-0.2	0.1	0.0	0.0	0.1	0.2	0.2
3	0.0	0.1	0.0	0.0	0.0	0.7	-0.4	0.2	0.0	0.0	0.2	0.3	0.6
4	0.0	0.2	0.0	0.1	0.1	0.8	-0.7	0.4	0.0	0.0	0.5	0.4	0.8
5	0.0	0.5	0.1	0.2	0.1	0.8	-0.9	0.6	0.0	0.1	0.7	0.5	0.9
6	0.1	0.7	0.2	0.3	0.3	0.9	-0.9	0.8	0.0	0.2	0.8	0.6	1.0
7	0.2	0.8	0.2	0.5	0.4	0.9	-0.9	0.8	0.0	0.3	0.9	0.7	1.0
8	0.5	0.8	0.3	0.7	0.6	0.9	-0.9	0.9	0.0	0.6	0.9	0.9	1.0
9	0.7	0.8	0.5	0.8	0.7	1.0	-1.0	0.9	0.0	0.8	0.9	0.9	1.0
10	0.8	0.8	0.6	0.9	0.8	1.0	-1.0	0.9	0.0	0.9	0.9	1.0	1.0
11	0.9	0.8	0.7	0.9	0.9	1.0	-1.0	1.0	0.0	0.9	0.9	1.0	1.0
12	0.9	0.8	0.7	0.9	0.9	1.0	-1.0	1.0	0.0	0.9	1.0	1.0	1.0
13	0.9	0.8	0.7	0.9	0.9	1.0	-1.0	1.0	0.0	0.9	1.0	1.0	1.0

As it could be seen in Table 18.6, the level of transit ridership quickly decreases. However the feedback of increase in car ownership on itself, causes a little decrease first, but due to the steep decrease in transit use and the further effects of low service quality levels through transit volume decrease, car ownership again starts to increase. Finally the value converge to lower service quality, lower customer satisfaction and lower ridership levels of transit.

The last example of the scenario analyses was investigating the results of a transit agency investing to increase its image. In Table 18.7, it is observed that, at the first iterations, even though the same service quality levels are achieved, the increase in customer satisfaction due to image improvement causes level of transit ridership to improve. Thus, the volume achieved in transit affects the service quality dimensions positively.

After examining the simulation results, it could be concluded that the improvements made in the transit service quickly results in higher levels of transit ridership. The actions simulated in the scenarios result with the expected impact over the system and it is observed that improving the service quality dimensions as well as investing to improve the visibility and image of transit has a powerful effect on increasing the transit use levels.

18.6 Conclusions and Further Research

In this study, a decision analysis approach which will help the Transit Agencies of Istanbul Metropolitan Municipality in strategic planning of transit ridership is introduced. Mostly, it is difficult to describe the complex systems such as transportation by a precise mathematical model. Thus, the presented FCM method which offers to observe the relationships among concepts clearly and gives an opportunity to facilitate impact analysis of alternatives is believed to contribute to the research in this area.

FCMs are flexible tools which allow decision makers to deal with many variables which may not be well-defined and are not known with certainty but can be described in degrees. The well-documented characteristics of quality and customer satisfaction research in service industry are intangibility, heterogeneity, and inseparability which cause a challenge for modeling. Thus, the FCM approach which can tackle with such characteristics is presented as a useful tool.

The feedback mechanism of the approach provides to represent the dynamic nature of the transportation system. When we examine the matrix representing the relations among factors, the level of transit ridership is directly related to two factors only, namely image of transit agencies and residential/business density positively as well as the factor of car ownership negatively. However, the dynamic nature of the transportation system can be analyzed using the simulations. Scenario analysis with simulations shows that the level of transit ridership is also affected by the cost and travel time of transit provided by the feedback mechanism of FCMs.

The notion of 'time' is very important for dynamic systems. In recent studies, the introduction of the time in FCMs is also investigated which could be also introduced for transport. The same research can be conducted for future state estimations of the transit system. On the other side, a similar research for the other cities of Turkey and in other countries may be performed. The comparison of the results with those in this study would be a fruitful research avenue.

Our approach for identifying the weights of the FCM may be compared with various other approaches in the literature. As one of those, a multivariate application will be made and data from passengers will be collected. The data will be analyzed with a method such as structural equation modeling and the output of the structural equation model will be used as the weights of the FCM.

A more comprehensive system comprising the interactions between different modes and the policies which are used by the transit agencies as control tools affecting the transportation system can also be studied.

Bibliography

[1] Murray A.T. (2001). Strategic analysis of public transport coverage. *Socio-Economic Planning Sciences*, **35**, p. 175–188.
[2] Anderson W., Kanaroglou P., and Miller E. (1996). Urban form, energy and the environment: a review of the issues, evidence and policy. *Urban Studies*, **33**, p. 7–35.
[3] Newman P. and Kenworthy J. (1999). *Sustainability and Cities: Overcoming Automobile Dependence*, Island Press, Washington, DC.
[4] Stern E. and Tretvik T. (1993). Public transport in Europe: requiem or revival? In: Salomon, I., Bovy, P.H.L., Orfeuil, J.P. (Eds.), *A Billion Trips a Day, Kluwer Academic Publishers*, The Netherlands, p. 129–148.
[5] Racca D. and Ratledge E. (2004). *Project Report for Factors That Affect and/or Can Alter Mode Choice*, University of Delaware, Newark.
[6] *Transit Cooperative Research Program* (1996). TCRP Report 16: Transit and Urban Form, Part1: Transit, Urban Form, and the Built Environment: A Summary of Knowledge, Transportation Research Board, National Research Council, Washington D.C.
[7] Parasuraman A., Zeithaml V., and Berry L. (1985). Conceptual Model of Service Quality and Its Implications for Future Research. *Journal of Marketing*, **49** (4), p. 41–50.
[8] *Transit Cooperative Research Program* (1999). TCRP Report 47: A Handbook for Measuring Customer Satisfaction and Service Quality, Transportation Research Board, National Research Council, Washington D.C.
[9] Dowling R., Reinke D., Flannery A., Ryus P., Vandehey M., Petritsch T., Landis B., Rouphail N., and Bonneson J. (2008). *Multimodal Level of Service Analysis for Urban Streets, Final Report*, Transportation Research Board, Washington, DC.
[10] Debrezion G., Pels E., and Rietveld P. (2009). Modelling the joint access mode and railway station choice. *Transportation Research Part E*, **45**, p. 270–283.
[11] Brons M., Givoni M., and Rietveld P. (2009). Access to railway stations and its potential in increasing rail use. *Transportation Research Part A*, **43**, p. 136–149.
[12] Krygsman S., Dijst M., and Arentze T. (2004). Multimodal public transport: an analysis of travel time elements and the interconnectivity ratio. *Transport Policy*, **11**, p. 265–275.
[13] Rietveld P., Bruinsma F., and Vuuren D.J. (2001). Coping with unreliability in public transport chains: A case study for Netherland. *Transportation Research Part A*, **35**, p. 539–559.
[14] Estupinan N. and Rodriguez D. (2008). The relationship between urban form and station boardings for Bogota's BRT. *Transportation Research Part A*, **42**, p. 296–306.
[15] Parsons, Brinckerhoff and Douglas, (1993). *LUTRAQ: The Pedestrian Environment Model Modifications*. Vol. **4A**.
[16] Arentze T. and Timmermans H. (2003). Measuring impacts of condition variables in rule-based models of space–time choice behavior: method of empirical illustration. *Geographical Analysis*, **35**, p. 24–45.
[17] Transit Cooperative Research Program, (2007). *TCRP Report 122 : Understanding How to Motivate Communities to Support and Ride Public Transportation.* Transportation Research Board, National Research Council, Washington D.C.
[18] Transit Cooperative Research Program (1999). *TCRP Report 63 :Enhancing the Visibility and Image of Transit in the United States and Canada,* Transportation Research Board, National

Research Council, Washington D.C.

[19] Axelrod R. (1976). *Structure of Decision: The Cognitive Maps of Political Elites,* Princeton University Press, Princeton, NJ.

[20] Kang I., Lee S., Choi J., 2004. Using fuzzy cognitive map for the relationship management in airline service. *Expert Systems with Applications*, **26**, 545–555.

[21] Kosko B. (1986). Fuzzy Cognitive Maps. *International Journal of Man-Machine Studies*, **24**, p. 65–75.

[22] Aguilar J. (2005). A Survey about Fuzzy Cognitive Maps Papers. *International Journal of Computational Cognition*, **3** (2), p. 27–33.

[23] Peña, A., Sossa H., and Gutiérrez A. (2007). Cognitive Maps: an Overview and their Application for Student Modeling. *Computación y Sistemas*, **10** (3), p. 230–250.

[24] Tsadiras A.K. (2007). Inference using Binary, Trivalent and Sigmoid Fuzzy Cognitive Maps, *Proceedings of The 10th International Conference on Engineering Applications of Neural Networks*, Thessaloniki, Greece.

[25] Kardaras D. and Gregory M. (1999). Using Fuzzy Cognitive Maps to Model and Analyze Business Performance Assessment. In: *Jacob Chen and Anil Mital* (Eds.), *Advances in Industrial Engineering Applications and Practice II*, p. 63–68.

[26] Lee K.C. and Lee S. (2003). A cognitive map simulation approach to adjusting the design factors of the electronic commerce web sites. *Expert Systems with Applications*, **24**, p. 1–11.

[27] Potter S., Enoch M., and Smith M. (1997). *Vital transportation statistics*, Landor Publishing.

[28] Xirogiannisa G., Stefanoua J., and Glykas M. (2004). A fuzzy cognitive map approach to support urban design. *Expert Systems with Applications*, **26**, p. 257–268.

[29] Hsu C.S. (2010). Determinants of passenger transfer waiting time at multi-modal connecting stations. *Transportation Research Part E*, **(46)**, p. 404–413.

Chapter 19

Genetic Algorithm to Solve Container Storage Problem for a Single and Various Types

Imen Ayachi[1,2], Ryan Kammarti[2], and Pierre Borne[1]

[1] *LAGIS, École Centrale de Lille, Villeneuve d'Ascq, France*
[2] *LACS, École Nationale des Ingénieurs de Tunis, Tunis - Belvédère, Tunisie*

E-mail: ayachiimen@gmail.com, kammarti.ryan@planet.tn, pierre.borne@ec-lille.fr

The loading and unloading of container and their temporary storage in the container terminal are the most important and complex operation in seaport terminals. It is highly inter-related with the routing of yard crane and truck and their costs increased significantly especially without an efficient terminal management. This paper focuses on optimizing the way of allocating inbound and outbound containers in storage locations, known as the Container Storage Problem (CSP). It consists on finding the most suitable storage location for incoming containers that minimises rehandling operations of containers during their transfer to the ship, truck or train. We propose a genetic algorithm to solve the CSP for a single and various container types (refrigerated, open side, empty, dry, open top and tank). The main objective of this approach is to find an optimal container arrangement which minimize the re-handle operations of containers at their departure dates (unloading time). In fact, the wait time of customer trucks, the transfer time of yard crane and the Ship turnaround time are advantageously reduced. The performances of our proposed approaches are verified comparing to the results generated by the Last In First Out algorithm. Genetic approaches generated good results and experimental study confirms these and shows their effectiveness.

19.1 Introduction

Containerization is the use of containers for goods transport, especially in the maritime domain. This process began in the 1960s and generalized in 1980s. It is a container logistics chain, which was put in place around the world. In fact, major ports have been adapted to this new transport mode by creating dedicated terminals for loading and unloading container ships, storing containers and their transfer to trains or trucks.

C. Kahraman (ed.), *Computational Intelligence Systems in Industrial Engineering*,
Atlantis Computational Intelligence Systems 6, DOI: 10.2991/978-94-91216-77-0_19,

At each port of destination, some containers are unloaded from the container ship and loaded in the port to be delivered to their customers. In fact, a container is either in the vessel waiting her unloading in the destination port or in at the harbor port waiting to be loaded in the vessel or to be delivered to his customer.

Most of previous works studied the container storage problem in the ship. There objective was to determine an optimal container arrangement that satisfy their criteria such the ship stability, the reduction of the re-handle operations of containers, the minimization of the total loading time...

Many approaches have been developed to solve this problem, rule based, mathematical model, simulation based, multi-agent model and heuristic methods.

Nevertheless, the temporary storage of containers at the container terminal is one of the most important and complex services. It influences the productivity, either for inbound or outbound containers and it is highly inter-related with the routing of yard crane and truck [11].

Indeed, the determination of an effective container organization reduces material handling costs (i.e., the costs associated with loading, unloading and transporting cargo) and minimizes the time of loading and unloading the containers.

In this paper, we studied the container storage problem (CSP). This problem can be defined as a bin packing problem in three dimensions where containers are items and storage spaces in the port or in the ship are bins used. It consists on affecting each container to the most suitable storage area in accordance with the problem objective.

Our aim is to determine a valid containers arrangement in the storage area, in order to minimize re-handle operations, the throughput of crane and Truck turnaround time and therefore the loading/unloading times of these containers.

Generally, this problem is studied considering a single container type. However, this does not stand the problem under its real-life statement as there are multiple container types that should be considered, (refrigerated, open side, empty, dry, open top and tank). This lets the problem more complicated, as various constraints appeared, related to the container type's requirements (e.g. refrigerated containers must be allocated to the blocks equipped by the power point, on an open top container, we cannot place a container at the top, tank container must be placed on each other, etc.)

Making a storage space allocation decision for different types of containers is too complicated especially for large scale instances and it is hard, even impossible, to solve it optimally. Therefore, most of the proposed solution approaches are based on metaheuristics.

In this paper, a genetic algorithm is proposed to solve the container storage space allocation for a single and various container types. This approach is chosen since his facility and quick achieve to the feasible solutions even for models having numerous equality constraints. Also, the genetic algorithm reported good results in many works on this problem [2, 4].

The rest of this paper is organized as follows: In Section 19.2, the literature review is given. Then, the mathematical formulation of the PSC for a single and multiple container type is presented. In Section 19.4, an adaptation of algorithm genetic to CSP is described. Then, some results and a comparative study with LIFO algorithm are presented. Finally, Section 19.6 covers our conclusion.

19.2 Literature review

The container storage is the most difficult task in container terminals since inbound and outbound containers are stacked together in the same storage area. After arrival at the terminal, each container picked up by transportation equipment and affected to one of the storage blocks. When the designated ship arrived, containers are unloaded from yard block, transported to the berth and loaded onto the vessel. The chain of operations for import containers are performed in the reverse order [6].

The container storage space allocation problem (SSAP) consists on affecting each container to the most suitable place in the storage area. The containers are often arranged with the objective of reducing the number of handling operations required later on to load/unload containers.

In the literature, various papers were proposed, treating different variants of the problem. Some of them will be presented in this section.

Kim and Park proposed a heuristic decision rule and a sub-gradient optimization technique to solve the storage space allocation for outbound containers [7]. Their objective was to find an arrangement of the containers that exploits efficiently the storage space and loading operations.

In the paper [15], authors modeled the seaport system as a container location model, with an objective function designed to minimize the distance between the vessel berthing locations and the storage zone. To solve this problem, they proposed a hybrid algorithm based on genetic (GA) and simulated annealing (SA) algorithm.

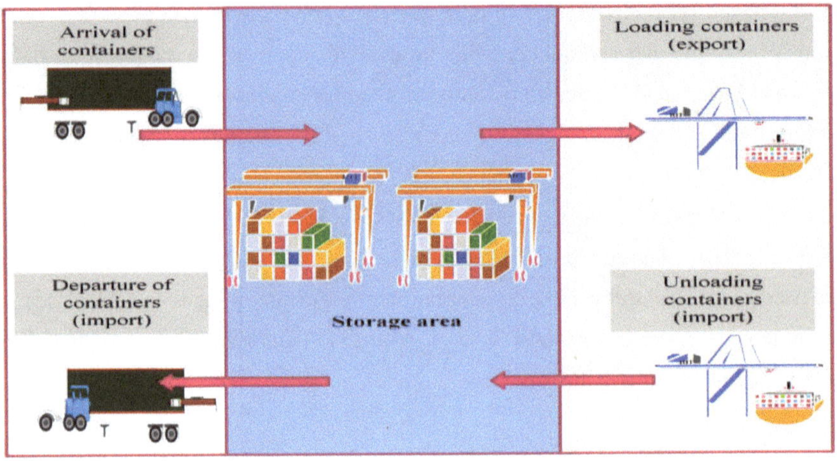

Fig. 19.1 Loading and unloading process

Park and his colleagues proposed a method of optimizing a stacking policy using a multi-objective evolutionary algorithm for the containers in the stacking yard of an automated container terminal (ACT). Their stacking policy takes account of both the locations for the incoming containers and those for temporary movement within the yard [17].

Preston and Kozan proposed a genetic algorithm to solve the container location model at seaport terminals. Their objective was to reduce the transfer and the handling time of containers. This approach took the Brisbane port as a case study and generated good results in comparison to the process already used in this port [12].

Kim presented a technique to estimate the rehandlings number for the next pick-up and the total number of rehandles to pick up all inbound containers in a bay [8].

In the paper [16], authors describe a randomized algorithm with Tabu Search for multi-objective optimization of large containership stowage plans. A group of non-dominated solutions is generated based on objectives such as the number of re-handles, the completion time of the longest crane, the number of stacks that exceed the weight limit, the number of idle slots, horizontal moment difference and cross moment difference.

Kim and Kim proposed a cost model to estimate various cost components related to the import container handling and to determine subsequently the storage space and the number of transfer cranes required [9].

Also, in the paper of Kumar and Vlacic (2008), a prediction model of unloading containers times and equipment utilization is presented.

Chen and his colleagues [3] combined diverse meta-heuristics (tabu search, simulated annealing and genetic algorithms) to solve the port yard storage optimization problem. It aims to minimize the space allocated to the cargo within a time interval.

Lee and his colleagues developed a heuristic algorithm to solve the yard truck scheduling and the storage allocation problems. Their objective is to minimize the weighted sum of total delay of requests and the cost of total travel time of yard trucks [10].

In 2003, Zhang solved the storage space allocation problem using a rolling-horizon approach. Both outbound and inbound containers are considered .Their aim was to minimize the total transportation distance of containers between blocks and vessel berthing locations [14].

Bazzazi and his colleagues [2] extended this approach in the case that different containers types and sizes are considered simultaneously. The authors proposed a genetic algorithm to solve this problem and they supposed that the allowable blocks to which a container type can be allocated are known in advance.

In the paper [1], a harmony search algorithm is proposed to solve the CSP *where a single container type was considered.* Its aim was to reduce the re-handle operations of containers. The results were compared to a genetic algorithm previously applied to the same problem in [5] and recorded good results.

Due to encouraging results reported in this paper and the complexity of CSP, a genetic algorithm is developed in this work. Our objective is to determine a storage plan, for the containers arriving to the port, which reduces the rehandling operations when leaving the storage space. The CSP is solved for a single and multiple container types (dry, refrigerated, open side, empty, open top and tank).

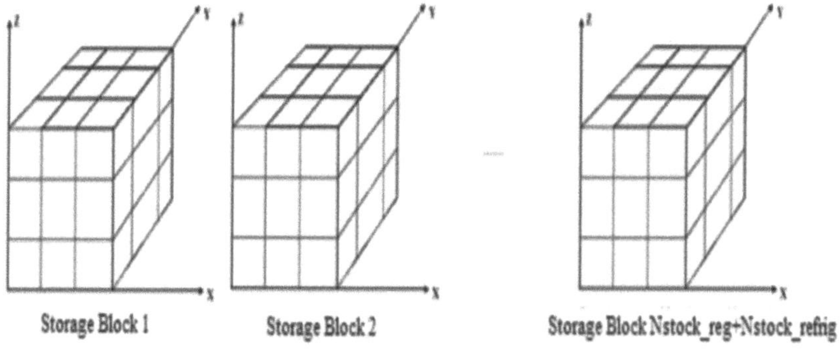

Fig. 19.2 Storage area

19.3 Problem Formulation

In this section, the Container Storage Problem is formulated for a single and multiple container type. It is based on the following assumptions.

19.3.1 *Assumptions*

- To unload a container, all containers above it in the same stack must be removed.
- Containers have the same size.
- The container's departure time are known.
- The initial state of the storage spaces is defined (available locations in storage blocks).
- A block is represented by a three dimensions plan.
- The storage area in the port is composed of several blocks that can be equipped by a power point to store reefer containers or regular blocks for the other containers types. Figure 19.1 shows an example of storage area.

19.3.2 *Inputs parameters*

i: container index, $i = 1,\ldots,N_c$

b: storage block index; $b = 1,\ldots,N_{\text{Block}}$

N_c: number of containers in the storage area

$N_{\text{Block}} = N_{\text{stock\_reg}} + N_{\text{stock\_refrig}}$: number of blocks

$N_{\text{stock\_reg}}$: number of standard blocks

$N_{\text{stock\_refrig}}$: number of refrigerated blocks

d_i: departure date of the container i

n_1: maximum containers number on the axis X

n_2: maximum containers number on the axis Y

n_3: maximum containers number on the axis Z

N_D: the number of container types

$N_c(D)$: the number of containers of type D, where $D = 1,\ldots,6$. denotes the type of container, Such as:

$$D = \begin{cases} 1 & \text{if it is a dry container} \\ 2 & \text{if it is an empty container} \\ 3 & \text{if it is an open top container} \\ 4 & \text{if it is a platform container} \\ 5 & \text{if it is a tank container} \\ 6 & \text{it is a refrigerated container} \end{cases}$$

Fig. 19.3 Container types

19.3.3 *Mathematical formulation*

The main objective of this paper is to determine a storage plan of arrival containers that reduce the rehandling operations at their departure times. We studied this problem for a single and various container types. The first objective function is described as follow.

19.3.3.1 *Single container type*

$$\text{Min} \sum_{i=1}^{N_c} \sum_{b=1}^{N_{\text{Block}}} m_{i,b} C_i(x,y,z,b); \quad \forall x = 1,\ldots,n_1; \ \forall y = 1,\ldots,n_2; \ \forall z = 1,\ldots,n_3. \tag{19.1}$$

where:

- m_i: the minimum number of container rehandlings to extract the container i.
- A re-handling is a container movement made in order to permit access to another, or to improve the overall stowage arrangement, and is considered a product of poor planning [13].

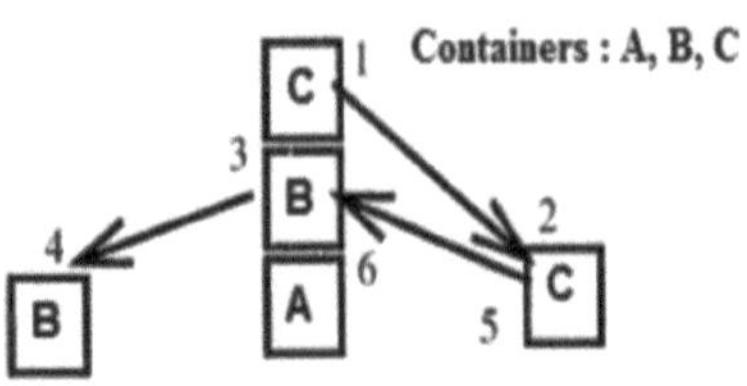

Fig. 19.4 Rehandling operations

- $C_i(x,y,z,b)$ designates the decision variable.

$$C_i(x,y,z,b) = \begin{cases} 1, & \text{the container } i \text{ is at position } x,\ y,\ z \text{ of block } b, \\ 0, & \text{otherwise.} \end{cases} \tag{19.2}$$

Subject to:

$$C_i(x,y,z,b) - C_i(x,y,z+1,b) \geqslant 0, \quad \forall i \in \{1,\ldots,N_c\}. \tag{19.3}$$

The constraint equation (19.3) ensures that a container must be placed on the ground or on another container.

19.3.3.2 *Multiple container types*

We extend the previous objective function in the case where the type of containers is different such as dry, tank, open top, platform, empty and refrigerated containers.

$$\text{Min} \sum_{D=1}^{N_D} \sum_{i=1}^{N_c(D)} \sum_{b=1}^{N_{\text{Block}}} m_{i,b} C_{i,D}(x,y,z,b). \tag{19.4}$$

Equation (19.4) describes the extended fitness function where D indicates the container type.

The difference between types of containers often results from the difference between goods and items inside them such as foods, chemical substances, liquids. Therefore, many storage constraints appeared, related to the container type's requirements. For instance, refrigerated containers must be allocated to the blocks equipped by the power point, empty containers must not be settled under a full container, on an open top container, we cannot place a container at the top; tank container must be placed on each other, etc.

The following equations described these storage constraints:

$$C_{i,r}(x,y,z,b) - C_{j,D}(x,y,z+1,b) = 1,$$
$$\forall i \in \{1,\ldots,N_c(r)\},\ \forall r = 3,4,\ \forall j \in \{1,\ldots,N_c(D)\} \tag{19.5}$$

$$C_{i,4}(x,y,z,b) - \sum_{m=x+1}^{n_3} C_{j,r}(m,y,z,b) = 1,$$
$$\forall i \in \{1,\ldots,N_c(4)\},\ \forall j \in \{1,\ldots,N_c(D)\},\ \forall r \in \{1,\ldots,N_D\} \tag{19.6}$$

$$C_{i,5}(x,y,z,b) - C_{j,r}(x,y,z+1,b) = 1,$$
$$\forall i \in \{1,\ldots,N_c(5)\},\ \forall j \in \{1,\ldots,N_c(r)\},\ \forall r \in \{1,2,3,4,6\} \tag{19.7}$$

- The constraint 5 ensures that an open top container or an open side container can not have another container above.
- The constraint 6 indicates that there aren't any containers at the open side of container type 4 (open side container)
- The constraint 7 indicates that tank containers must be placed on each other
- The constraint 8 suggests that a reefer container must be allocated to the blocks equipped by the power point.

19.4 Evolution procedure

In this section, the genetic algorithm implementation is detailed. Firstly, N different solutions are randomly generated to form the initial population. Next, two parents are selected using roulette-wheel method to which a crossover and a mutation operators are applied to get two new children. This process is reiterated until getting N new solutions. These solutions will be added to the current initial population to form an intermediate population noted P_{inter} of size $2N$. The N best solutions of Pinter, according to their fitness, will form the initial population for iteration $(t+1)$.

This procedure will be repeated until the termination criterion (number of iterations without improvement of the best solution) is satisfied. The principle of this procedure is illustrated in Figure 19.5.

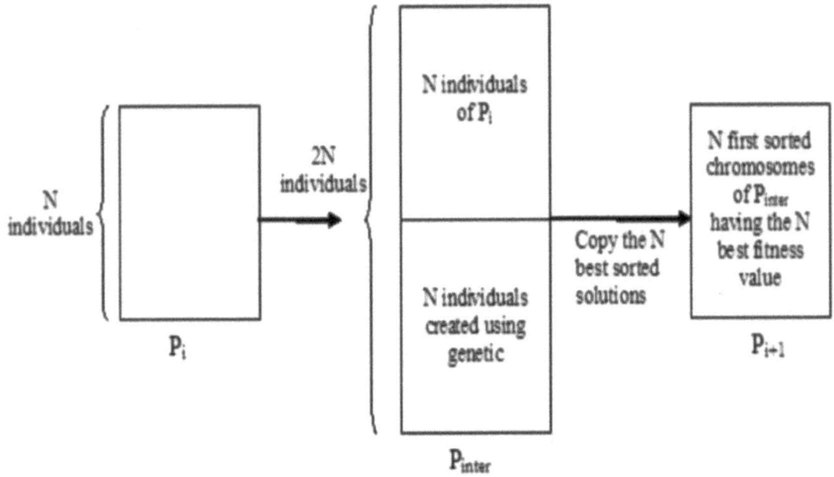

Fig. 19.5 Evolution procedure

19.4.1 *GA to solve CSP for Single container type*

19.4.1.1 *Chromosome representation*

According to the decision variable $C_i(x,y,z,b)$, we use a four dimension structure representation to describe the containers storage area.

These dimensions indicate respectively the container coordinates x, y, z in the plan (X,Y,Z) and b the number of the block to which it is allocated.

The chromosome represents a storage plan. It consists in assigning arriving containers to available storage spaces in order to minimize rehandling operations. Figure 19.6 shows an example of solution representation

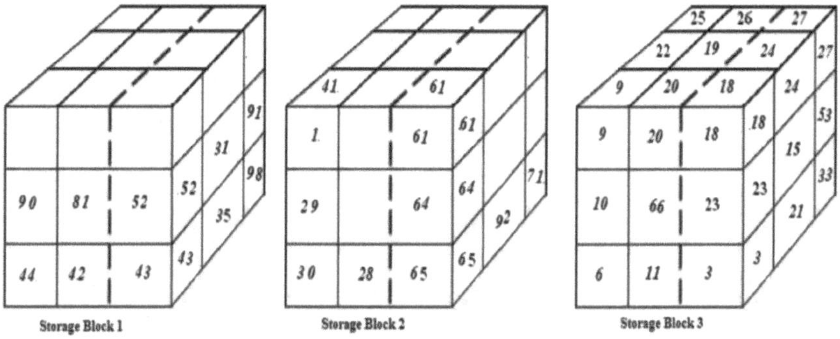

Fig. 19.6 Solution representation

19.4.1.2 *Initial population generation procedure*

This technique consists on affecting a randomly chosen container to an arbitrary available position satisfying constraint (19.3) in order to get a valid solution. This solution will be added to the initial population if it does not already exist.

```
Initial population generation procedure
Begin
 // Nc : Container number
 //Nbrpop : Population size
 pop=0
 do
     for i = 1 to Nc do
        Begin
          • Select a container c, randomly
          • Select an arbitrary block b with available places
          • Select randomly a position x, y z in block b satisfying the different
            constraints
          • Affect container c to position (x, y, z, b)
        End
 If the generated solution does not already exist then
 //Add the new solution to the initial population
         pop++
 while  (pop < Nbrpop)
End
```

Fig. 19.7 Initial population generation procedure for a single container type

19.4.1.3 *Crossover*

The crossover operator is applied to two solution (the parents) I_1 and I_2, selected using the roulette-wheel approach, to generate two new solutions (the children) E_1 and E_2. In this method, we adopted the one-point crossover operator.

It consists on selecting, according to X, Y and Z axes, three intersection points *p-crois-x*, *p-crois-y* and *p-crois-z*, which corresponds to the crossover plan for each block. Then, Child E_1 receives genes (containers) from parent I_1 located within the crossover plan and the remaining containers will be added according to their places in I_2. That is, for each container c that remains to add, if its coordinates (x,y,z,b) in I_2 is available in E_1, according to constraint 1.3, it will be added to it. Otherwise, it will be added the first presented place. Similarly, child E_2 will be generated according to the containers of parent E_2 and the remaining ones will be added considering their places in E_1.

Figure 19.8 illustrates the crossover operator applied to two parents I_1 and I_2 to generate two children E_1 and E_2. In this example, where a single bloc is presented, the crossover plan is defined by:

$$\text{p-crois-}x = 3,\ \text{p-crois-}y = 3 \text{ and p-crois-}z = 2.$$

It should be noted that the crossover operator is executed with a probability, generally between 75% and 90%, to be defined.

Figure 19.9 illustrates the crossover algorithm proposed in this adaptation where:

- Position: is a function giving the position of container r in parent I_2.
- Possible: is a function indicating if position Rpos in E_1 is available for affecting container r according to the storage constraint.
- Affect: affects container r at position Rpos in E_1.

19.4.1.4 *Mutation*

After crossover, the two children might undergo a tiny modification using a mutation operator. Its main objectives are to promote the diversification of solutions by exploring several areas of the search space, to maintain the diversity of the population in the successive generations and to escape from local optima.

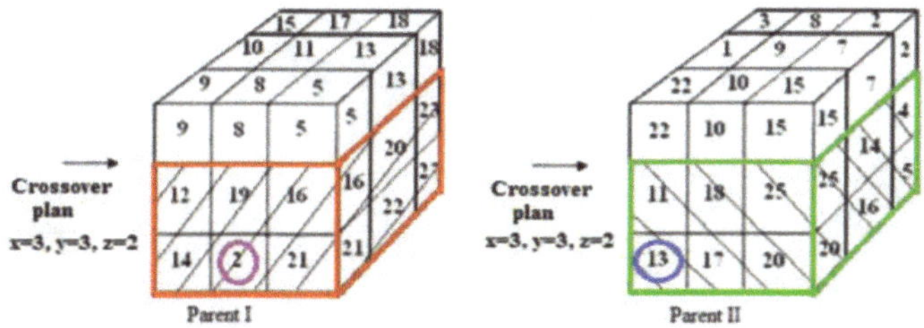

Parent 1

Floor 1(z=1)			Floor 2 (z=2)			Floor 3 (z=3)		
1	3	27	24	6	23	15	17	18
25	4	22	7	26	20	10	11	13
14	2	21	12	19	16	9	8	5

Parent 2

Floor 1(z=1)			Floor 2 (z=2)			Floor 3 (z=3)		
27	21	5	23	19	4	3	8	2
12	24	16	6	26	14	1	9	7
13	17	20	11	18	25	22	10	15

Child 1

Floor 1(z=1)			Floor 2 (z=2)			Floor 3 (z=3)		
1	3	27	24	6	23	15	9	8
25	4	22	7	26	20	11	18	10
14	2	21	12	19	16	13	17	5

Child 2

Floor 1(z=1)			Floor 2 (z=2)			Floor 3 (z=3)		
27	21	5	23	19	4	8	10	15
12	24	16	6	26	14	3	7	9
13	17	20	11	18	25	2	22	1

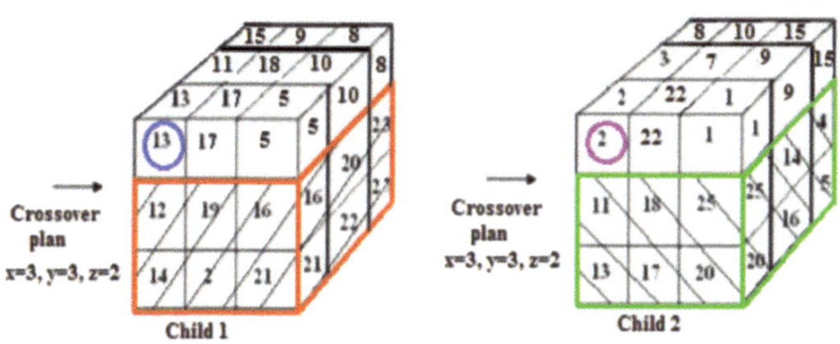

Fig. 19.8 Crossover operator

```
Crossover algorithm for a single type
Begin
    • Select randomly, 3 coordinates defining the crossover plan (p-crois-x, p-crois-y
      et p-crois-z) within [0, ..., n1-1], [0,...,n2-1]et [0,...,n3-1], respectively.
    • Copy to child E1the containers located in the intervals [0, p-crois-x],[0, p-crois-
      y] et [0, p-crois-z] of parent I1.
    • Determine the list of the remaining containersof size Nc_rem
      For r= 1 to Nc_rem do
          Begin
              Rpos=position(r,I2)
              If possible(Rpos,r,E1) then
                Affect(r,E1,Rpos)
              Else
                Rpos=Firstpos(r,E1)
                Affect(r,E1,Rpos)
              End if
          End
End
```

Fig. 19.9 Crossover algorithm for a single container type

The mutation operator adopted in this approach consists on switching two randomly selected containers in each block. This technique is presented in Figure 19.10.

```
Mutation algorithm for a single type
Begin
        For each block c do
          Begin
            ✓ Randomly select2 container from block c
            ✓ Switch the 2 containers
          End
End
```

Fig. 19.10 Mutation algorithm for a single container type

Figure 19.11 describes the containers arrangement in a block before and after mutation. In this figure, the containers with coordinates $(1,1,3)$ and $(2,1,1)$ are switched.

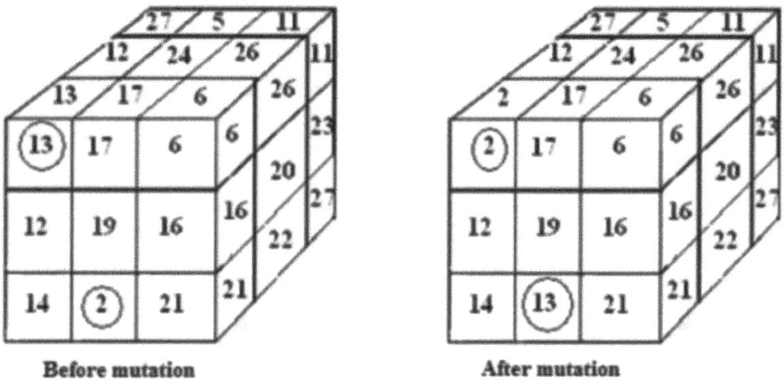

Fig. 19.11 Mutation operator

19.4.1.5 *Stopping criteria*

This criterion indicates when the optimization process stops. It can be either a maximum number of iterations or maximum number of iterations without improvement of the best solution. For this adaptation, the process of selection, crossover and mutation stops when there is no improvement of the best solution after a maximum number of iterations to be defined.

19.4.2 *Multiple container types*

19.4.2.1 *Solution coding*

To present a solution to the container storage problem where multiple containers types are considered, a four dimensions plan (X,Y,Z,B) is used. It represents for each container, identified by a number and a type (i,D), its coordinates in plan (X,Y,Z) at block B. The arrangement process of the different containers need to take into consideration the initial state of the storage areas and to satisfy the storage constraints given by equations (19.3) to (19.7) described above.

Figures 19.12 and 19.13 present the state of 3 blocks initially and after the affectation of 5 new dry containers (type 1), respectively.

19.4.2.2 *Initial population generation*

The initial population generation process is similar to the one applied when a single container type is considered. However, in order to get a valid solution, a container is affected to the available place at the block randomly selected, only if it verifies the constraint related to its type.

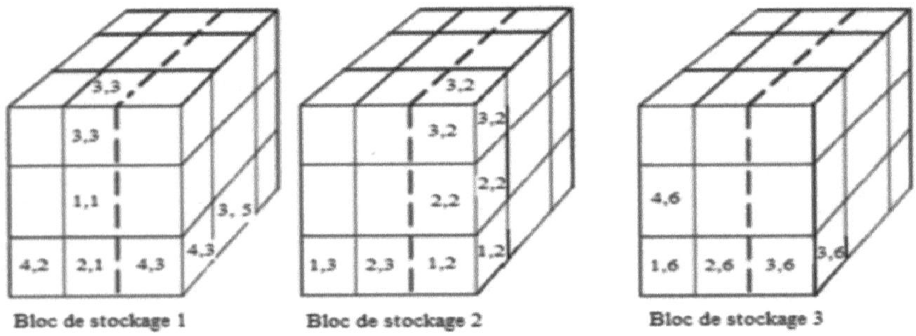

Fig. 19.12 Blocks state initially

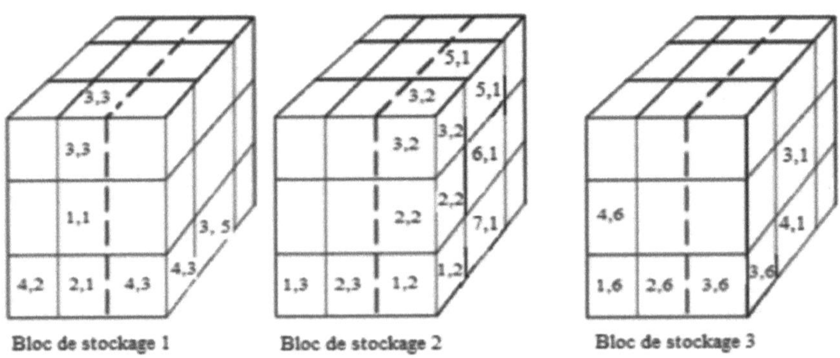

Fig. 19.13 Blocks state after affectation

19.4.2.3 *Crossover operator*

The crossover operation allows the creation of two children E_1 and E_2. Its adaptation in the proposed approach can be described as follows. Firstly, the blocks initial state is copied to E_1 and E_2. Next, two parents I_1 and I_2 are selected randomly from the initial population of the current iteration and a one point crossover plan defined by the coordinates p-crois-x, p-crois-y and p-crois-z is chosen.

Then, child E_1 (respectively E_2) receives genes (containers) from parent I_1 (respectively I_2) according to the crossover plan and the remaining containers are affected considering their locations in I_2 (respectively I_1). In fact, for each remaining container c, if its coordinates (x, y, z, b) in I_2 (respectively I_1) is available in E_1 (respectively E_2) and satisfies its storage constraints, it will be added to it. However, this operation may be impossible either because that it is already used or that it does not satisfies the type's constraints. In this case, the container will be affected to the first available position.

Begin

- Affect the refrigerated containers to the available places in isothermal blocks constraint 8.
- Affect tank containers to the available places satisfying constraints 3, 5, 6 and 7
- Affect Dry containers to the available places satisfying constraints 3, 5 and 6.
- Affect platform containers to the available places satisfying constraints 3, 5 and 6
- Affect open top containers to the available places satisfying constraints 3, 5 and 6
- Affect empty containers to the available places satisfying constraints 3, 5 and 6

End

Fig. 19.14 Initial population creation

```
Mutation algorithm for the CSP with multiple container types
Bégin
        For i=1 to NbBlock do
           begin
               Select randomly a container type, D
               Select randomly 2 container of type D
               Switch the two containers
           End
End
```

Fig. 19.15 Mutation algorithm for the CSP with multiple container types

19.4.2.4 *Mutation*

The mutation operator adopted in this adaptation consists on switching two containers having a similar type, in each block. It is illustrated by Figure 19.15.

19.5 Computational results

In this paper, a genetic algorithm is adopted to solve the container storage problem for a single and various types of containers. In order to evaluate the results generated by our proposed approach, a comparative study with a LIFO algorithm is presented for each case.

The LIFO algorithm (Last In First Out) consists on storing in first time the last arrived container at the storage area. This principle is always used by terminal planning to assign each container to an available storage place.

19.5.1 *Experimental results of CSP (single type)*

Table 19.1 describes six instances of the CSP for identical containers.

Table 19.1 Test instances

Problem size	Block number	Block size
$N_c = 50$	1	$n_1 = n_2 = n_3 = 4$
$N_c = 150$	3	$n1 = n2 = n3 = 4$
$N_c = 250$	5	$n_1 = n_2 = n_3 = 4$
$N_c = 500$	10	$n_1 = n_2 = n_3 = 4$
$N_c = 700$	12	$n_1 = n_2 = n_3 = 4$
$N_c = 900$	16	$n_1 = n_2 = n_3 = 4$

Each example is solved by GA 15 times and the mean of the fitness value and CPU time are reported.

As it can be seen in Figure 19.16, the solutions generated by the GA are largely better than those provided by the LIFO algorithm for all problem sizes and especially for medium and large sizes.

LIFO used the arrival date of container as an organization criterion which is not always correct since the last arriving container is not necessarily the first departing.

Genetic algorithm generated good results for all problem sizes in a tolerable time even if it is slower than LIFO, as it is reported in Figure 19.17. This can be explained that LIFO does not required a computation time and GA need more time to create a new solution (selection, crossover and mutation).

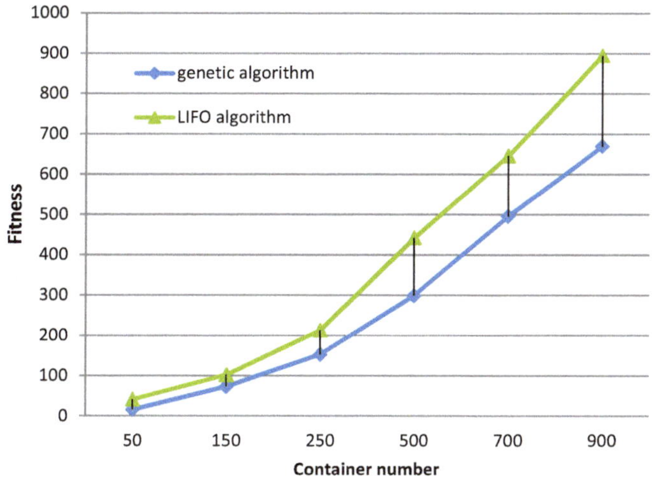

Fig. 19.16 Comparison between LIFO, GA's fitness values

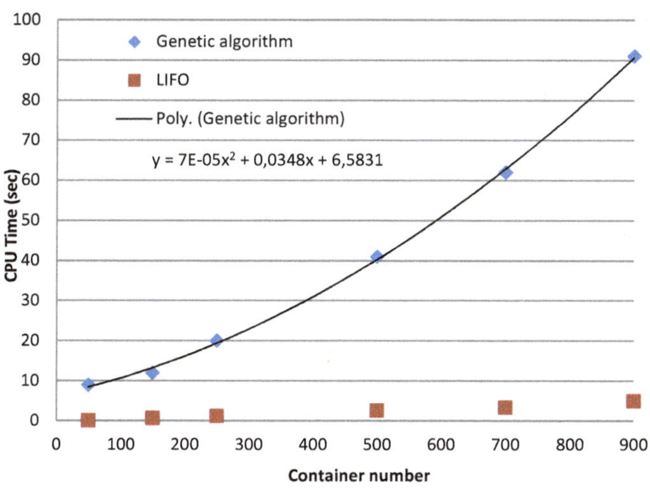

Fig. 19.17 Comparison between LIFO, GA's CPU times

GA's CPU time shows a polynomial behavior when the size of the problem increases and Figure 19.18 indicates a typical convergence of GA during 125 generations.

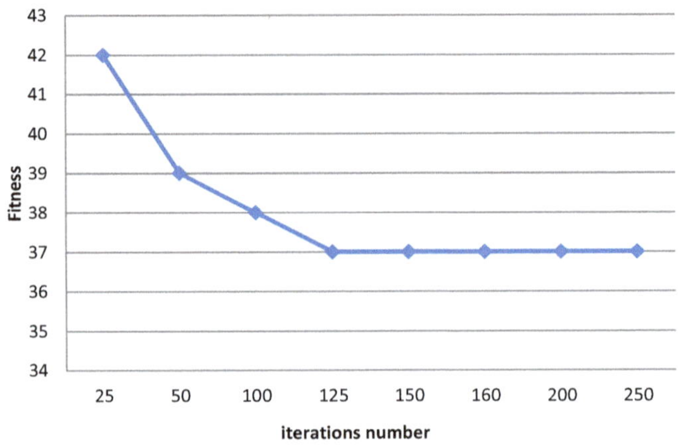

Fig. 19.18 Convergence of GA during 125 iterations

19.5.2 *Experimental results of CSP (multiples types)*

In this section we present the simulation results of the CSP for various container type. The storage area is composed of 5 blocks for refrigerated containers and 5 general blocks with $n_1 = n_2 = n_3 = 5$. Initially, there are in the storage area:

- 20 dry containers,
- 15 open top containers,
- 20 refrigerated containers,
- 15 tanks,
- 20 empty containers.

The performance of our genetic approach and the LIFO algorithm is verified according to the 5 case studies described in Table 19.2, by varying the containers numbers and types. For example, instance 1 indicates that 20 Dry container, 20 refrigerated, 15 Tank, 20 empty and 10 Platform will be stored in the storage blocks.

Table 19.2 Test instances for CSP (various container types)

Container type	Instance 1	Instance 2	Instance 3	Instance 4	Instance 5
Dry	20	50	70	100	120
Open top	0	10	20	20	25
refrigerated	20	30	50	100	100
Tank	15	20	50	70	75
empty	20	30	50	100	100
Platform	10	10	20	20	25

The results showed in Figure 19.19 indicate that the fitness value generated by the GA is largely better for all studied cases for an execution time tolerant.

Genetic algorithm has proven its effectiveness and its superiority to the LIFO especially for the last 3 instances where a high number of containers of various types is considered, which makes the problem more complicated. In fact, the storage of an open top container makes unavailable all locations above it. The same situation occurs for platform containers with an unavailability of positions that are on the whole space of the open side, etc.

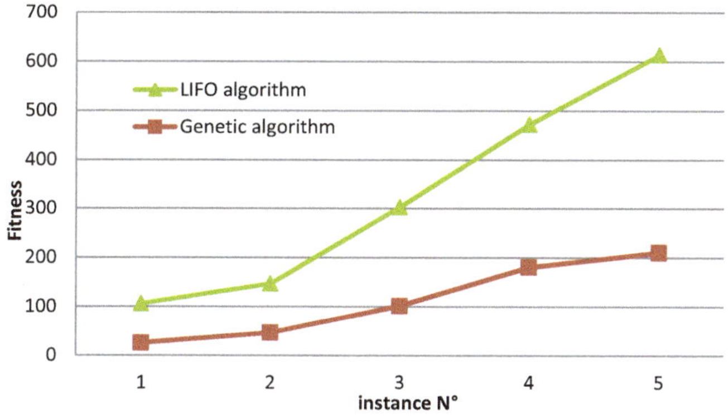

Fig. 19.19 Comparison between LIFO, GA's fitness values (various type)

The curve shown in the following figure confirms the complexity of the problem since GA's CPU time shows an exponential behavior for a various container types.

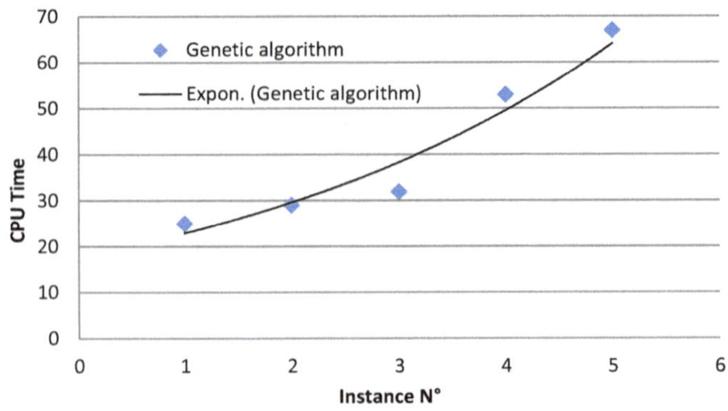

Fig. 19.20 Exponential trend of the GA's CPU time for CSP (various container types)

19.6 Conclusion

This paper proposed a genetic algorithm (GA) to solve the container storage problem for a single and multiple container type. In real world case, there are various types of container such as refrigerated, open side, empty, dry, open top, tank... Each container type has storage constraints that must be respected in the allocation process of the storage areas, which let the problem more difficult. That is refrigerated containers must be allocated to the blocks equipped by the power point, tank containers need to be placed on each others, etc.

Despite this difficulty, the proposed approach generated good results in a reasonable execution time. This can be confirmed by the experimental study that shows the effectiveness of the application of genetic algorithm in the resolution of the container storage problem.

An important extension of this research would be to formulate the problem as a dynamic storage space allocation in order to solve and to make decision in real time.

Bibliography

[1] Ayachi I., Kammarti R., Ksouri M., and Borne P., 2010, Harmony search algorithm for the container storage problem, *8th International Conference of Modeling and Simulation - MOSIM'10*, Tunisia.

[2] Bazzazi M., Safaei N., and Javadian N., A genetic algorithm to solve the storage space allocation problem in a container terminal, *Computers & Industrial Engineering*, 2009.

[3] Chen P., Fu Z., Lim A., and Rodrigues B., 2004, Port yard storage optimization, *IEEE Transactions on Automation Science and Engineering*, **1**, p. 26–37.

[4] Dubrovsky O., Levitin G., and Penn M., A genetic algorithm with compact solution encoding for the container ship stowage problem, *Journal of Heuristics*, **8**, 585–599, 2002.

[5] Kammarti R., Ayachi I., Ksouri M., and Borne P., 2009, Evolutionary Approach for the Containers Bin-Packing Problem, *Studies in Informatics and Control*, **18** (4).

[6] Kim Kap Hwan and Günther Hans-Otto, *Container Terminals and Cargo Systems, Design, Operations Management and Logistics Control Issues*, Springer-Verlag Berlin Heidelberg, 2007, pp. 3–14.

[7] Kim K.-H. and Park K.-T., 2003, A note on a dynamic space-allocation method for outbound containers, *European Journal of Operational Research*, p. 92–101.

[8] Kim K.-H., 1997, Evaluation of the number of rehandles in container yards, *Computers & Industrial Engineering*, **32** (4).

[9] Kim K.-H. and Kim, H.-B., 1998, The optimal determination of the space requirement and the number of transfer cranes for import containers, *Computers ind. Engng.*, **35**, p. 427–430.

[10] Lee D.-H., Cao J.-X., Shi Q., and Chen, J.-H., 2009, A heuristic algorithm for yard truck scheduling and storage allocation problems, *Transportation Research Part E*, **45**, p. 810–820.

[11] Murty K.G., Liu J., Wan Y.W., Zhang C., Tsang M.C.L., and Linn R., 2005, A decision support system for operations in a container terminal, *Journal Decision Support Systems*, **39** (3).

[12] Preston P. and Kozan E., 2001, An approach to determine storage locations of containers at seaport terminals, *Computers & Operations Research*, p. 983–995.

[13] Wilson I.D. and Roach P.A., 2000. Container stowage planning: a methodology for generating computerised solutions. *Journal of the Operational Research Society*, **51**, 1248–1255.

[14] Zhang C., Liu J., Wan Y.-W., Murty K.-G., and Linn R.-J., 2003, Storage space allocation in container terminals, *Transportation Research Part B*, **37**, p. 883–903.

[15] Moussi R., Fatma Ndiaye N., and Yassine A., Hybrid Genetic Simulated Annealing Algorithm (HGSAA) to Solve Storage Container Problem in Port, Intelligent Information and Database Systems, *Lecture Notes in Computer Science*, Vol. **7197**, 2012, pp 301–31.

[16] Liu F., Low M.Y.H., Hsu W.J., Huang S.Y., Zeng M., and Win C.A., Randomized algorithm with tabu search for multi-objective optimization of large containership stowage plans, in: Proc. 2nd Intl. Conference on Computational Intelligence, *Lecture Notes in Computer Science*, Vol. **6971**, pp. 256–272, 2011.

[17] Park T., Sohn M., and Rya K.R., Optimizing stacking policies using a MOEA for an automated container terminal, *40th Intl. Conference on Computers and Industrial Engineering*, 2010.

Chapter 20

Fuzzy Geometric Programming Model with Applications in Industrial Engineering

Esra Bas

Department of Industrial Engineering, Istanbul Technical University Macka, 34367 Istanbul, Turkey

E-mail: atace@itu.edu.tr

In this chapter, first, several fuzzy formulations are proposed for fuzzy geometric programming (FGP) by analogy to fuzzy linear programming (FLP). Then, the basic GP applications in industrial engineering (IE) are reviewed, and the proposed FGP formulations are adapted to two basic applications of FGP in IE, i.e. to inventory problems and entropy based transportation problems. In the last part of the chapter, a solution approach is proposed for fuzzy inventory problem formulated as a GP model.

20.1 Introduction

Geometric programming (GP) is a non-linear programming model with a special form of objective function and constraints. Many engineering problems can be represented as GP models. Once a problem is represented as a GP model, it can easily be solved by means of efficient and reliable solution methods [1]. Typical engineering applications of GP models are power control, truss design, gate sizing, wire sizing, and chemical equilibrium problems [2]. Particularly, GP applications in IE include optimal location problems, probability and statistics, inventory control, production system optimization, transportation engineering, revenue management and pricing, and systems and control theory [1].

Although GP models in deterministic case are already worthwhile, the uncertainty consideration in GP models is expected to yield additional value as a result of more robust solutions. With this respect, some optimization methods including fuzzy, robust, and stochastic optimization can be adapted to GP models for uncertainty consideration. Fuzzy optimization is more appropriate in cases where the vague or verbal information is inherent, and the

C. Kahraman (ed.), *Computational Intelligence Systems in Industrial Engineering*,
Atlantis Computational Intelligence Systems 6, DOI: 10.2991/978-94-91216-77-0_20,

parameters or relations can be represented as fuzzy numbers. Although fuzzy optimization is appropriate for modeling uncertainty, several possible formulations hinder the clarity of the approach. The purpose of this chapter is to systematically propose several formulations as fuzzy geometric programming models for general case, and extend the formulations to two basic applications of FGP in IE, i.e., to inventory problems and entropy based transportation problems. The crisp models are introduced prior to fuzzy formulations in general case and for IE applications. This chapter is an extension of Bas [3], where the author proposed a standard form of FGP defuzzified by t-norm and t-conorm fuzzy relations. In the final section of the chapter, a solution approach is proposed for fuzzy inventory problem formulated as a GP model.

20.2 A Geometric Programming Model

In this section, a basic GP model is provided for deterministic case, and three different optimization approaches are considered for uncertainty.

20.2.1 *A Geometric Programming Model in Deterministic Case*

20.2.1.1 *Standard form of Basic Geometric Programming Model*

A GP model is a non-linear programming model with a special form of objective function and constraints as follows [1]:

$$\begin{aligned} \min \quad & f_0(x) \\ \text{s.t.} \quad & f_i(x) \leqslant 1, \quad i = 1,2,\ldots,m \\ & g_i(x) = 1, \quad i = 1,2,\ldots,p \\ & x > 0 \end{aligned} \tag{20.1}$$

In model (20.1), $f_i(x)$ are posynomial functions (posynomials) for $i = 0, 1, 2, \ldots, m$ and $g_i(x)$ are monomial functions (monomials) for $i = 1, 2, \ldots, p$.

Definition 20.1 (Monomial function [1] (monomial)). A real-valued function $g(x)$ is a monomial function if

$$g(x) = c x_1^{a_1} x_2^{a_2} \cdots x_n^{a_n} \tag{20.2}$$

is defined where,

$x_j > 0$ are real decision variables for $j = 1, 2, \ldots, n$

$c > 0$ is the coefficient of the monomial

$a_j \in \mathbb{R}$ are exponents of the monomial for $j = 1, 2, \ldots, n$.

Definition 20.2 (Posynomial function [1] (posynomial)). A real-valued function $f(x)$ is a posynomial function if

$$f(x) = \sum_{k=1}^{K} c_k x_1^{a_{1k}} x_2^{a_{2k}} \cdots x_n^{a_{nk}} \tag{20.3}$$

is defined, where $f(x)$ is the sum of K monomial functions.

20.2.1.2 *Some Extensions to Standard Form of Basic Geometric Programming Model*

Two examples are *generalized geometric programming* model (GGP) and *signomial geometric programming* (SGP).

Generalized geometric programming model A GGP model is a GP model with generalized posynomial functions.

Definition 20.3 (Generalized posynomial function [1]). A generalized posynomial function is obtained by applying the operations addition, multiplication, positive (fractional) power, and maximum to the posynomial functions.

A GGP model can be converted into a standard form of GP model by using some transformations given in the literature [1].

Signomial geometric programming A SGP model is a standard form of GP model with objective function and constraint functions allowed to be signomial functions [1].

Definition 20.4 (Signomial function [1]). A signomial function is a function obtained from a posynomial function with coefficients of the monomials allowed to be negative.

20.2.2 *A Geometric Programming Model in Uncertain Case*

Although several insights are already drawn from a GP model in deterministic case, even small perturbations in the parameters of the deterministic model may invalidate the results of the problem. Thus, uncertainty consideration in a GP model should be studied.

The uncertainty in a GP model can be modeled by using different techniques such as fuzzy, robust or stochastic optimization. Fuzzy optimization is appropriate for the cases where the vague or verbal information is inherent, while stochastic optimization is more applicable to cases where stochastic nature of the problem dominates. Robust optimization has been proposed for providing robust solutions against possible perturbations from the expected values of parameters, and is more relevant to cases where the solutions are expected to be robust for some degrees of worst-case.

20.2.2.1 *Fuzzy Geometric Programming Model*

FGP has a well-developed theory including primal algorithm, dual algorithm, and dual theorem [4]. FGP formulations with different types of fuzzy coefficients and fuzzy decision variables, and several extensions to the standard form of GP model such as fuzzy reverse programming and fuzzy multi-objective programming are discussed in the literature [4]. Yang and Cao provided a literature review of theoretical research areas and various applications of FGP [5].

In this chapter, we propose several formulations of FGP problems based on the FLP formulations proposed by Bector and Chandra [6]. However, before proceeding with the formulations, we propose to convert model (20.1) into a convex optimization model [1]:

$$\begin{array}{lll} \min & \log\left(\sum_{k=1}^{K_0} e^{a_{0k}^T y + b_{0k}}\right) & \\ \text{s.t.} & \log\left(\sum_{k=1}^{K_i} e^{a_{ik}^T y + b_{ik}}\right) \leqslant 0 & i = 1,2,\ldots,m \\ & a_i^T y + b_i = 0 & i = 1,2,\ldots,p \end{array} \tag{20.4}$$

In model (20.4), $\log c_{0k} = b_{0k}$, $\log c_{ik} = b_{ik}$, $\log c_i = b_i$, and a_{0k}, a_{ik}, a_i, y are n-tuple vectors. Note that as a result of logarithmic change of variables by $\log x = y$ and logarithmic transformation of functions, model (20.1) is converted into model (20.4).

Fuzzy geometric programming model with fuzzy inequalities The following model is a FGP model with *fuzzy inequalities*:

$$\begin{array}{lll} \min & \log\left(\sum_{k=1}^{K_0} e^{a_{0k}^T y + b_{0k}}\right) & \\ \text{s.t.} & \log\left(\sum_{k=1}^{K_i} e^{a_{ik}^T y + b_{ik}}\right) \tilde{\leqslant} 0 & i = 1,2,\ldots,m \\ & a_i^T y + b_i \tilde{=} 0 & i = 1,2,\ldots,p \end{array} \tag{20.5}$$

In model (20.5), $\tilde{\leqslant}$ and $\tilde{=}$ are the fuzzy extensions of the classical binary relations $\leqslant$ and $=$, respectively. Based on Zimmermann's approach [7], the satisfaction of each fuzzy inequality can be measured with the following membership functions:

$$\mu_i(C_i) = \alpha = \begin{cases} 1, & \log\left(\sum_{k=1}^{K_i} e^{a_{ik}^T y + b_{ik}}\right) \leqslant 0 \\ 1 - \dfrac{\log\left(\sum_{k=1}^{K_i} e^{a_{ik}^T y + b_{ik}}\right)}{p_i}, & 0 < \log\left(\sum_{k=1}^{K_i} e^{a_{ik}^T y + b_{ik}}\right) \leqslant p_i \\ 0, & \log\left(\sum_{k=1}^{K_i} e^{a_{ik}^T y + b_{ik}}\right) > p_i \\ & i = 1, 2, \ldots, m \end{cases} \tag{20.6}$$

$$\mu_i(C_i) = \alpha = \begin{cases} 1, & \varepsilon_i^L \leqslant \left|a_i^T y + b_i\right| \leqslant \varepsilon_i^R \\ 1 - \dfrac{\varepsilon_i^L - (a_i^T y + b_i)}{\varepsilon_i^L - p_i^L}, & \varepsilon_i^L < \left|a_i^T y + b_i\right| \leqslant p_i^L \\ 1 - \dfrac{a_i^T y + b_i - \varepsilon_i^R}{p_i^R - \varepsilon_i^R}, & \varepsilon_i^R < a_i^T y + b_i \leqslant p_i^R \\ 0, & \left|a_i^T y + b_i\right| > p_i^L \text{ or } a_i^T y + b_i > p_i^R \\ & i = 1, 2, \ldots, p \end{cases} \tag{20.7}$$

In model (20.6) and model (20.7), $\mu_i(C_i)$ is the membership function for the satisfaction of the constraint C_i, p_i is the tolerance level for the satisfaction of the ith constraint for $i = 1, 2, \ldots, m$, ε_i^L and ε_i^R are very small positive real numbers representing the allowable deviations from the left side and right side of zero for $i = 1, 2, \ldots, p$, and p_i^L and p_i^R are the tolerance levels representing the deviations from the left side and right side of zero. By assuming the membership function (20.7) to be symmetric, Verdegay's approach [8] is proposed to be extended to FGP as follows:

$$\begin{array}{lll} \min & \log\left(\sum_{k=1}^{K_0} e^{a_{0k}^T y + b_{0k}}\right) & \\ \text{s.t.} & \log\left(\sum_{k=1}^{K_i} e^{a_{ik}^T y + b_{ik}}\right) \leqslant (1-\alpha) p_i & i = 1, 2, \ldots, m \\ & \left|a_i^T y + b_i\right| \leqslant \varepsilon_i + (1-\alpha)(p_i - \varepsilon_i) & i = 1, 2, \ldots, p \end{array} \tag{20.8}$$

In model (20.8), it is assumed that $\varepsilon_i = \varepsilon_i^L = \varepsilon_i^R$ and $p_i = p_i^L = p_i^R$ for $i = 1, 2, \ldots, p$. Model (20.8) can be solved for selected $\alpha \in [0, 1]$ values.

Fuzzy geometric programming model with fuzzy objective function The following model is a FGP model with *fuzzy objective function*:

$$\begin{array}{lll} \widetilde{\min} & \log\left(\sum_{k=1}^{K_0} e^{a_{0k}^T y + b_{0k}}\right) & \\ \text{s.t.} & \log\left(\sum_{k=1}^{K_i} e^{a_{ik}^T y + b_{ik}}\right) \leqslant 0 & i = 1, 2, \ldots, m \\ & a_i^T y + b_i = 0 & i = 1, 2, \ldots, p \end{array} \tag{20.9}$$

We propose to assign an aspiration level to the objective function as in Zimmermann [7] and Inuiguchi, Ramik, Tanino, and Vlach [9]:

$$\begin{aligned}
&\log\left(\sum_{k=1}^{K_0} e^{a_{0k}^T y+b_{0k}}\right) \tilde{\leqslant} Z_0 & \\
&\log\left(\sum_{k=1}^{K_i} e^{a_{ik}^T y+b_{ik}}\right) \leqslant 0 & i=1,2,\ldots,m \\
&a_i^T y+b_i=0 & i=1,2,\ldots,p
\end{aligned} \tag{20.10}$$

The satisfaction of the objective function inequality can be measured with the following membership function (adapted from Zimmermann [7]):

$$\mu_0(C_0)=\alpha=\begin{cases} 1, & \log\left(\sum_{k=1}^{K_0} e^{a_{0k}^T y+b_{0k}}\right) \leqslant Z_0 \\ 1-\dfrac{\log\left(\sum_{k=1}^{K_0} e^{a_{0k}^T y+b_{0k}}\right)-Z_0}{p_0}, & Z_0<\log\left(\sum_{k=1}^{K_0} e^{a_{0k}^T y+b_{0k}}\right) \leqslant Z_0+p_0 \\ 0, & \log\left(\sum_{k=1}^{K_0} e^{a_{0k}^T y+b_{0k}}\right) > Z_0+p_0 \end{cases} \tag{20.11}$$

In model (20.11), $\mu_0(C_0)$ is the membership function for the satisfaction of objective function inequality, Z_0 is the aspiration level, and p_0 is the tolerance level for the satisfaction of the objective function. Verdegay's approach [8] is proposed to be extended to FGP as follows:

$$\begin{aligned}
\max\quad & \alpha & \\
\text{s.t.}\quad & \log\left(\sum_{k=1}^{K_0} e^{a_{0k}^T y+b_{0k}}\right) \leqslant Z_0+(1-\alpha)p_0 & \\
& \log\left(\sum_{k=1}^{K_i} e^{a_{ik}^T y+b_{ik}}\right) \leqslant 0 & i=1,2,\ldots,m \\
& a_i^T y+b_i=0 & i=1,2,\ldots,p \\
& 0 \leqslant \alpha \leqslant 1 &
\end{aligned} \tag{20.12}$$

Fuzzy geometric programming model with fuzzy objective function and fuzzy inequalities The following model is a FGP model with *fuzzy objective function* and *fuzzy inequalities*:

$$\begin{aligned}
\widetilde{\min}\quad & \log\left(\sum_{k=1}^{K_0} e^{a_{0k}^T y+b_{0k}}\right) & \\
\text{s.t.}\quad & \log\left(\sum_{k=1}^{K_i} e^{a_{ik}^T y+b_{ik}}\right) \tilde{\leqslant} 0 & i=1,2,\ldots,m \\
& a_i^T y+b_i \tilde{=} 0 & i=1,2,\ldots,p
\end{aligned} \tag{20.13}$$

Fuzzy geometric programming with *fuzzy objective function* and *fuzzy inequalities* is a combination of model (20.5) and model (20.9), and can be reformulated as follows:

$$\begin{aligned}
\max \quad & \alpha \\
\text{s.t.} \quad & \log\left(\sum_{k=1}^{K_0} e^{a_{0k}^T y + b_{0k}}\right) \leqslant Z_0 + (1-\alpha)p_0 \\
& \log\left(\sum_{k=1}^{K_i} e^{a_{ik}^T y + b_{ik}}\right) \leqslant (1-\alpha)p_i \qquad i = 1, 2, \ldots, m \\
& \left|a_i^T y + b_i\right| \leqslant \varepsilon_i + (1-\alpha)(p_i - \varepsilon_i) \qquad i = 1, 2, \ldots, p \\
& 0 \leqslant \alpha \leqslant 1
\end{aligned} \tag{20.14}$$

Fuzzy geometric programming model with fuzzy parameters and fuzzy decision variables The following model is a FGP model with *all parameters* and *decision variables* assumed to be *fuzzy*:

$$\begin{aligned}
\min \quad & \log\left(\sum_{k=1}^{K_0} e^{\widetilde{a}_{0k}^T \widetilde{y} + \widetilde{b}_{0k}}\right) \\
\text{s.t.} \quad & \log\left(\sum_{k=1}^{K_i} e^{\widetilde{a}_{ik}^T \widetilde{y} + \widetilde{b}_{ik}}\right) \leqslant 0 \quad i = 1, 2, \ldots, m \\
& \widetilde{a}_i^T \widetilde{y} + \widetilde{b}_i = 0 \qquad i = 1, 2, \ldots, p
\end{aligned} \tag{20.15}$$

Several characterizations for fuzzy parameters and fuzzy decision variables are possible. For instance, fuzzy parameters can be assumed to be triangular, trapezoidal, or L-R fuzzy numbers [4]. Special cases of model (20.15) or a combination of models (20.13) and (20.15) are also possible. As an instance, only some parameters and some inequality constraints can be assumed to be fuzzy.

20.2.2.2 *Robust Geometric Programming Model*

To the best of our knowledge, the literature in robust geometric programming (RGP) is relatively scarce. Hsiung, Kim, and Boyd proposed a tractable approximate RGP model [10]. They assumed that the parameters of posynomial functions belong to a convex set, and formulated the RGP model as follows:

$$\begin{aligned}
\min \quad & \log\left(\sum_{k=1}^{K_0} e^{a_{0k}^T y + b_{0k}}\right) \\
\text{s.t.} \quad & \log\left(\sum_{k=1}^{K_i} e^{\widetilde{a}_{ik}^T(u) y + \widetilde{b}_{ik}(u)}\right) \leqslant 0 \quad i = 1, 2, \ldots, m \\
& a_i^T y + b_i = 0 \qquad i = 1, 2, \ldots, p
\end{aligned} \tag{20.16}$$

where

$$\left(\widetilde{a}_{ik}(u), \widetilde{b}_{ik}(u)\right) = \left(a_{ik}^0 + \sum_{j=1}^{L} u_j a_{ik}^j,\, b_{ik}^0 + \sum_{j=1}^{L} u_j b_{ik}^j\right), \quad u \in U \subseteq \mathbb{R}^L$$

is defined with U being the uncertainty set. As in robust linear programming (RLP), uncertainty set can be polyhedral or ellipsoidal.

Hsiung, Kim, and Boyd defined the worst-case RGP formulation of model (20.16) as follows [10]:

$$\begin{array}{lll} \min & \log\left(\sum_{k=1}^{K_0} e^{a_{0k}^T y+b_{0k}}\right) & \\ \text{s.t.} & \sup_{u\in U} \log\left(\sum_{k=1}^{K_i} e^{\tilde{a}_{ik}^T(u)y+\tilde{b}_{ik}(u)}\right) \leqslant 0 & i=1,2,\ldots,m \\ & a_i^T y+b_i=0 & i=1,2,\ldots,p \end{array} \quad (20.17)$$

Hsiung, Kim, and Boyd proposed to solve RGP model by piecewise-linear (PWL) convex approximation of each nonlinear constraint function [10].

As another RGP formulation, Shen, Ma, and Chen proposed a robust solution algorithm for generalized geometric programming [11].

20.2.2.3 *Stochastic Geometric Programming Model*

To the best of our knowledge, the literature in stochastic geometric programming is also scarce. Paul proposed the application of stochastic GP to heat exchanger design [12]. Jagannathan proposed a stochastic GP model with multiplicative recourse [13]. Abou-El-Ata, Fergany, and El-Wakeel proposed a probabilistic multi-item inventory model with GP approach [14].

20.3 Applications of Geometric Programming Model in Industrial Engineering

The main GP applications in IE are inventory problems, transportation problems, supply chain problems, and revenue management and pricing. Table 1 provides the main characteristics of deterministic GP models proposed in the literature with selected references.

In this section, the first inventory problem formulated as a GP model and a basic entropy based transportation problem as a GP model are provided as the pioneers of GP applications in industrial engineering.

20.3.1 *A Basic Inventory Model as a GP*

Cheng proposed the first inventory problem as a GP model [15]. Thereafter, several extensions to the basic inventory problem as a GP model have been made as summarized in Table 20.1. In this section, we provide a summary of the model by Cheng [15].

Table 20.1 Classification of selected literature with GP applications in industrial engineering

IE problem	Main characteristic	References
Inventory problems		
Optimal inventory policy	Demand dependent unit cost	Cheng [15]
	Flexibility and reliability considerations	Cheng [16], Leung [17]
	Optimal order quantity and selling price	Lee [18], Lee [19]
	Storage space limitations	Lee [19]
	Inventory investment limitations	Lee [19]
	Multi-item models	Abou-el-ata and Kotb [20] Hariri and Abou-el-ata [21]
	Varying holding cost	Abou-el-ata and Kotb [20]
	Varying order cost	Hariri and Abou-el-ata [21]
	Production lot-size model with trade-credit policy	Shawky and Abou-El-Ata [22]
	Various cost functions	Jung and Klein [23], Jung and Klein [24]
	Deteriorated items	Mandal, Roy and Maiti [25]
	Multi-objective inventory model	Islam [26]
Optimal shelf allocation	Retail space allocation	Corstjens and Doyle [27] Bookbinder and Zarour [28]
	Strategic retail space allocation	Corstjens and Doyle [29]
Transportation and supply chain problems		
Transportation problems	Trip distribution models	Kádas and Klafszky [30] Wong [31]
	Multimodal combined model	Marin [32]
	Entropy based transportation problem	Samanta and Majumder [33]
Supply chain problems	Optimal configuration of supply chains	Huang and Qu [34]
Revenue management and pricing	Predictive pricing and revenue sharing	Mungamuru and Garcia-Molina [35]

Notations [15]

S	Setup cost per batch
H	Inventory carrying cost per item per unit time
V	Unit production item cost
D	Demand rate (Decision variable)
q	Production (order) quantity per batch (Decision variable)
$C(D,q)$	Average annual total cost

$$\min C(D,q) = \left(S + Vq + \frac{Hq^2}{2D}\right)\frac{D}{q} \tag{20.18}$$

where

$$V = \alpha D^{-\beta} \tag{20.19}$$

is defined as unit production cost with $\alpha > 0$, $\beta > 1$

By plugging eq. (20.19) into eq. (20.18), Cheng obtained the following model [15]:

$$\min C(D,q) = SDq^{-1} + \alpha D^{1-\beta} + \frac{Hq}{2} \tag{20.20}$$

Then, Cheng obtained the following model [15]:

$$\begin{aligned} \min \quad & C(D,q) = u_1 + u_2 + u_3 \geqslant \left(\frac{u_1}{a_1}\right)^{a_1}\left(\frac{u_2}{a_2}\right)^{a_2}\left(\frac{u_3}{a_3}\right)^{a_3} = f(a) \\ \text{s.t.} \quad & a_1 + a_2 + a_3 = 1 \\ & a_1,\ a_2,\ a_3 > 0 \end{aligned} \tag{20.21}$$

by letting

$$u_1 = SDq^{-1} \tag{20.22}$$

$$u_2 = \alpha D^{1-\beta} \tag{20.23}$$

$$u_3 = \frac{Hq}{2} \tag{20.24}$$

Note that model (20.21) follows, since the arithmetic mean of non-negative real numbers is greater than or equal to their geometric mean [15].

By arbitrarily choosing the values of a_1, a_2, a_3 such that the right-hand side of objective function inequality of model (20.21) will be independent of decision variables and the constraint of model (20.21) will be satisfied, Cheng obtained the optimal values a_1^*, a_2^*, a_3^* [15]. Then, Cheng obtained the optimal values D^* and q^* in model (20.21) by considering the equality of objective function inequality and using [15]:

$$\frac{u_1}{a_1^*} = \frac{u_1}{a_2^*} = \frac{u_1}{a_3^*} = k \tag{20.25}$$

The reader is referred to Cheng [15] for the details of the model.

20.3.2 *An Entropy Based Transportation Model as a GP Model*

Samanta and Majumder proposed several entropy based transportation problems formulated as GP models [33]. Based on Islam and Roy [36], a classical transportation model can be extended with an entropy function as follows:

$$\begin{array}{lll} \max & -\sum_{i=1}^{m}\sum_{j=1}^{n}\frac{x_{ij}}{x}\ln\left(\frac{x_{ij}}{x}\right) & \\ \min & \sum_{i=1}^{m}\sum_{j=1}^{n} c_{ij}x_{ij} & \\ \text{s.t.} & \sum_{j=1}^{n} x_{ij}=s_i & i=1,2,\ldots,m \\ & \sum_{i=1}^{m} x_{ij}=d_j & j=1,2,\ldots,n \\ & x_{ij}\geqslant 0 & i=1,2,\ldots,m,\ \ j=1,2,\ldots,n \end{array} \tag{20.26}$$

where

$$-\sum_{i=1}^{m}\sum_{j=1}^{n}\frac{x_{ij}}{x}\ln\left(\frac{x_{ij}}{x}\right) \tag{20.27}$$

is the entropy function to be maximized, x_{ij} is the number of products to be shipped from supply point i to demand point j, s_i is the total number of products available at supply point i, d_j is the number of products ordered at demand point j, and

$$x=\sum_{i=1}^{m}\sum_{j=1}^{n} x_{ij}$$

is the total number of products to be shipped from supply points to demand points.

20.4 Applications of Fuzzy Geometric Programming Model in Industrial Engineering

Since the first inventory problem has been formulated and solved as a GP model, several extensions to the basic model have been made in deterministic case. Papers with fuzzy formulations of extensions to the deterministic case have also appeared. Table 20.2 provides selected papers including FGP formulations of inventory problems and one fuzzy entropy based transportation problem formulated as a GP model.

In this section, several FGP formulations for the basic inventory problem and entropy based transportation problem will be developed based on the FGP models proposed in Section 20.2.

20.4.1 *Several Fuzzy Formulations for Basic Inventory Model as a GP Model*

Roy and Maiti proposed the first FGP formulation of the basic inventory problem suggested by Cheng [15, 37]. They also considered limited storage capacity, and assumed

Table 20.2 Classification of selected literature with FGP applications in industrial engineering

IE problem	Main characteristic	References
Inventory problems		
Economic order quantity (EOQ) model	Demand-dependent unit cost	Roy and Maiti [37] Islam and Roy [38] Chou, Julian and Hung [39]
	Flexibility and reliability considerations	Islam and Roy [38] Bag, Chakraborty and Roy [40] Panda and Maiti [41]
	Storage space limitations	Roy and Maiti [37] Islam and Roy [38] Islam and Roy [42] Chou, Julian and Hung [39]
	Imprecise space constraint	Panda and Maiti [41]
	Budget constraints	Chou, Julian and Hung [39]
	Price dependent demand	Panda and Maiti [41]
	Uncertain demand	Bag, Chakraborty and Roy [40]
	Hybrid number cost parameters	Mandal and Roy [43] Panda, Kar and Maiti [44]
	Imperfect production lot size model	Mandal and Roy [43]
	Multi-item EOQ model	Mandal and Roy [43] Islam and Roy [42] Panda, Kar and Maiti [44] Panda and Maiti [41]
	Multi-objective model	Mandal, Roy and Maiti [45]
	Two warehouses	Maiti [46]
Transportation and supply chain problems		
Transportation problems	Entropy based transportation problem	Islam and Roy [36]

fuzzy objective function and fuzzy resources in their model as follows:

$$\begin{aligned} \widetilde{\min} \quad & C(D,q) = \left(S + Vq + \frac{Hq^2}{2D}\right)\frac{D}{q} \\ \text{s.t.} \quad & Aq \leqslant \widetilde{B} \\ & D,\ q > 0 \end{aligned} \tag{20.28}$$

In model (20.28), A is the batch size and B is the storage capacity. However, several fuzzy formulations of crisp model of model (20.28) are possible. In this chapter, analogous to FLP formulations proposed in Bector and Chandra [6], we give the possible fuzzy formulations of the crisp model of model (20.28). These fuzzy formulations can be reflected to several variations of basic inventory model outlined in Table 20.2.

20.4.1.1 *Fuzzy Inventory Model with Fuzzy Inequalities*

A fuzzy inventory model with *fuzzy inequalities* will be as follows:

$$\begin{aligned} \min \quad & C(D,q) = \left(S + Vq + \frac{Hq^2}{2D}\right)\frac{D}{q} \\ \text{s.t.} \quad & Aq \tilde{\leqslant} B \\ & D,\ q > 0 \end{aligned} \tag{20.29}$$

In model (20.29), storage space constraint is assumed to be a soft constraint, and $\tilde{\leqslant}$ is the fuzzy extension of the classical binary relation $\leqslant$. As in eq. (20.6), Zimmermann's approach [7] to the satisfaction of fuzzy inequalities can be extended to the fuzzy inventory model (20.29) with

$$\mu(Aq) = \alpha = \begin{cases} 1, & Aq \leqslant B \\ 1 - \dfrac{Aq - B}{P}, & B < Aq \leqslant B + P \\ 0, & Aq > B + P \end{cases} \tag{20.30}$$

defined as the membership function for the satisfaction of storage space constraint, where P is the tolerance level. As in model (20.8), Verdegay's approach [8] to solution of FLP can be extended to the solution of model (20.29) as follows:

$$\begin{aligned} \min \quad & [C]_\alpha(D,q) = \left(S + Vq + \frac{Hq^2}{2D}\right)\frac{D}{q} \\ \text{s.t.} \quad & Aq \leqslant B + (1-\alpha)P \\ & D,\ q > 0 \end{aligned} \tag{20.31}$$

20.4.1.2 *Fuzzy Inventory Model with Fuzzy Objective Function*

A fuzzy inventory model with *fuzzy objective function* will be as follows:

$$\begin{aligned} \widetilde{\min} \quad & C(D,q) = \left(S + Vq + \frac{Hq^2}{2D}\right)\frac{D}{q} \\ \text{s.t.} \quad & Aq \leqslant B \\ & D,\ q > 0 \end{aligned} \tag{20.32}$$

As in model (20.10), assigning an aspiration level to the satisfaction of fuzzy objective function is proposed as follows:

$$\begin{aligned} & C(D,q) = \left(S + Vq + \frac{Hq^2}{2D}\right)\frac{D}{q} \tilde{\leqslant} C_0 \\ & Aq \leqslant B \\ & D,\ q > 0 \end{aligned} \tag{20.33}$$

where, C_0 is the aspiration level of the objective function. Analogous to model (20.12), model (20.33) reduces to the following model:

$$\begin{aligned} \max \quad & \alpha \\ \text{s.t.} \quad & \left(S+Vq+\frac{Hq^2}{2D}\right)\frac{D}{q} \leqslant C_0+(1-\alpha)P_0 \\ & Aq \leqslant B \\ & 0 \leqslant \alpha \leqslant 1 \\ & D,\ q > 0 \end{aligned} \tag{20.34}$$

where P_0 is the tolerance level for the satisfaction of the objective function.

20.4.1.3 *Fuzzy Inventory Model with Fuzzy Objective Function and Fuzzy Inequalities*

A fuzzy inventory model with *fuzzy objective function* and *fuzzy inequalities* is a combination of model (20.29) and model (20.32) as follows:

$$\begin{aligned} \widetilde{\min} \quad & C(D,q) = \left(S+Vq+\frac{Hq^2}{2D}\right)\frac{D}{q} \\ \text{s.t.} \quad & Aq \lesssim B \\ & D,\ q > 0 \end{aligned} \tag{20.35}$$

By combining model (20.31) and model (20.34), model (20.35) reduces to the following model:

$$\begin{aligned} \max \quad & \alpha \\ \text{s.t.} \quad & \left(S+Vq+\frac{Hq^2}{2D}\right)\frac{D}{q} \leqslant C_0+(1-\alpha)P_0 \\ & Aq \leqslant B+(1-\alpha)P \\ & 0 \leqslant \alpha \leqslant 1 \\ & D,\ q > 0 \end{aligned} \tag{20.36}$$

20.4.1.4 *Fuzzy Inventory Model with Fuzzy Parameters and Fuzzy Variables*

A fuzzy inventory model with *all parameters* and *decision variables* as *fuzzy* numbers will be as follows:

$$\begin{aligned} \min \quad & C(\widetilde{D},\widetilde{q}) = \left(\widetilde{S}+\widetilde{V}\widetilde{q}+\frac{\widetilde{H}\widetilde{q}^2}{2\widetilde{D}}\right)\frac{\widetilde{D}}{\widetilde{q}} \\ \text{s.t.} \quad & \widetilde{A}\widetilde{q} \leqslant \widetilde{B} \\ & \widetilde{D},\ \widetilde{q} > 0 \end{aligned} \tag{20.37}$$

As a special case of model (20.37), some parameters and some decision variables can be allowed to be fuzzy.

20.4.1.5 *A Fully Fuzzified Inventory Model*

A fully fuzzified inventory model involves *fuzzy objective function*, *fuzzy inequalities*, *fuzzy parameters* and *fuzzy decision variables* as follows:

$$\begin{aligned}
\widetilde{\min} \quad & C(\widetilde{D},\widetilde{q}) = \left(\widetilde{S} + \widetilde{V}\widetilde{q} + \frac{\widetilde{H}\widetilde{q}^2}{2\widetilde{D}}\right)\frac{\widetilde{D}}{\widetilde{q}} \\
\text{s.t.} \quad & \widetilde{A}\widetilde{q} \widetilde{\leqslant} \widetilde{B} \\
& \widetilde{D},\ \widetilde{q} > 0
\end{aligned} \tag{20.38}$$

Several variations of model (20.38) are possible. As an example, Roy and Maiti [37] model is a special case of model (20.38) with fuzzy objective function and fuzzy storage capacity $\widetilde{B}$.

20.4.2 *Several Fuzzy Formulations for Basic Entropy Based Transportation Model as a GP*

Islam and Roy [36] proposed an entropy based transportation model with fuzzy cost parameters. However, several fuzzy formulations for the basic entropy based transportation model (20.26) can be developed analogous to the models developed for inventory model:

- Fuzzy entropy based transportation model with *fuzzy inequalities*
- Fuzzy entropy based transportation model with *fuzzy entropy function*
- Fuzzy entropy based transportation model with *fuzzy cost function*
- Fuzzy entropy based transportation model with *fuzzy entropy function* and *fuzzy cost function*
- Fuzzy entropy based transportation model with *fuzzy inequalities*, *fuzzy entropy function* and *fuzzy cost function*
- Fuzzy entropy based transportation model with *fuzzy parameters* and *fuzzy decision variables*
- *Fully fuzzified* entropy based transportation model

In this section, we provide some of the possible fuzzy formulations for model (20.26).

20.4.2.1 *An Entropy Based Transportation Model with Fuzzy Entropy Function*

An entropy based transportation model with *fuzzy entropy function* will be as follows:

$$\begin{array}{lll}
\widetilde{\max} & -\sum\limits_{i=1}^{m}\sum\limits_{j=1}^{n}\frac{x_{ij}}{x}\ln\left(\frac{x_{ij}}{x}\right) & \\
\min & \sum\limits_{i=1}^{m}\sum\limits_{j=1}^{n}c_{ij}x_{ij} & \\
\text{s.t.} & \sum\limits_{j=1}^{n}x_{ij}=s_i & i=1,2,\ldots,m \\
 & \sum\limits_{i=1}^{m}x_{ij}=d_j & j=1,2,\ldots,n \\
 & x_{ij}\geqslant 0 & i=1,2,\ldots,m,\ j=1,2,\ldots,n
\end{array} \tag{20.39}$$

Assigning an aspiration level E_0 to the satisfaction of fuzzy entropy function is proposed as in model (20.33):

$$\begin{array}{lll}
\min & \sum\limits_{i=1}^{m}\sum\limits_{j=1}^{n}c_{ij}x_{ij} & \\
\text{s.t.} & -\sum\limits_{i=1}^{m}\sum\limits_{j=1}^{n}\frac{x_{ij}}{x}\ln\left(\frac{x_{ij}}{x}\right)\ \widetilde{\geqslant}\ E_0 & \\
 & \sum\limits_{j=1}^{n}x_{ij}=s_i & i=1,2,\ldots,m \\
 & \sum\limits_{i=1}^{m}x_{ij}=d_j & j=1,2,\ldots,n \\
 & x_{ij}\geqslant 0 & i=1,2,\ldots,m,\ j=1,2,\ldots,n
\end{array} \tag{20.40}$$

Then, model (20.40) reduces to the following bi-objective model:

$$\begin{array}{lll}
\max & \alpha & \\
\min & \sum\limits_{i=1}^{m}\sum\limits_{j=1}^{n}c_{ij}x_{ij} & \\
\text{s.t.} & -\sum\limits_{i=1}^{m}\sum\limits_{j=1}^{n}\frac{x_{ij}}{x}\ln\left(\frac{x_{ij}}{x}\right)\geqslant E_0+(\alpha-1)P_{E_0} & \\
 & \sum\limits_{j=1}^{n}x_{ij}=s_i & i=1,2,\ldots,m \\
 & \sum\limits_{i=1}^{m}x_{ij}=d_j & j=1,2,\ldots,n \\
 & 0\leqslant\alpha\leqslant 1 & \\
 & x_{ij}\geqslant 0 & i=1,2,\ldots,m,\ j=1,2,\ldots,n
\end{array} \tag{20.41}$$

In model (20.41), P_{E_0} is the tolerance level for the satisfaction of the aspiration level of entropy function.

20.4.2.2 *An Entropy Based Transportation Model with Fuzzy Entropy Function and Fuzzy Cost Function*

An entropy based transportation model with *fuzzy entropy function* and *fuzzy cost function* will be as follows:

$$
\begin{aligned}
\widetilde{\max} \quad & -\sum_{i=1}^{m}\sum_{j=1}^{n}\frac{x_{ij}}{x}\ln\left(\frac{x_{ij}}{x}\right) \\
\widetilde{\min} \quad & \sum_{i=1}^{m}\sum_{j=1}^{n}c_{ij}x_{ij} \\
\text{s.t.} \quad & \sum_{j=1}^{n}x_{ij}=s_i \qquad i=1,2,\ldots,m \\
& \sum_{i=1}^{m}x_{ij}=d_j \qquad j=1,2,\ldots,n \\
& x_{ij}\geqslant 0 \qquad i=1,2,\ldots,m,\ j=1,2,\ldots,n
\end{aligned}
\tag{20.42}
$$

Assigning aspiration levels E_0 and C_0 to the entropy function and cost function are proposed as follows:

$$
\begin{aligned}
& -\sum_{i=1}^{m}\sum_{j=1}^{n}\frac{x_{ij}}{x}\ln\left(\frac{x_{ij}}{x}\right)\,\widetilde{\geqslant}\, E_0 \\
& \sum_{i=1}^{m}\sum_{j=1}^{n}c_{ij}x_{ij}\,\widetilde{\leqslant}\, C_0 \\
& \sum_{j=1}^{n}x_{ij}=s_i \qquad i=1,2,\ldots,m \\
& \sum_{i=1}^{m}x_{ij}=d_j \qquad j=1,2,\ldots,n \\
& x_{ij}\geqslant 0 \qquad i=1,2,\ldots,m,\ j=1,2,\ldots,n
\end{aligned}
\tag{20.43}
$$

Then, model (20.43) reduces to the following model analogous to model (20.41):

$$
\begin{aligned}
\max \quad & \alpha \\
\text{s.t.} \quad & -\sum_{i=1}^{m}\sum_{j=1}^{n}\frac{x_{ij}}{x}\ln\left(\frac{x_{ij}}{x}\right)\geqslant E_0+(\alpha-1)P_{E_0} \\
& \sum_{i=1}^{m}\sum_{j=1}^{n}c_{ij}x_{ij}\leqslant C_0+(1-\alpha)P_{C_0} \\
& \sum_{j=1}^{n}x_{ij}=s_i \qquad i=1,2,\ldots,m \\
& \sum_{i=1}^{m}x_{ij}=d_j \qquad j=1,2,\ldots,n \\
& 0\leqslant\alpha\leqslant 1 \\
& x_{ij}\geqslant 0 \qquad i=1,2,\ldots,m,\ j=1,2,\ldots,n
\end{aligned}
\tag{20.44}
$$

In model (20.44), P_{E_0} and P_{C_0} are the tolerance levels for the satisfaction of the aspiration level of entropy function and cost function, respectively.

20.4.3 *Solution of Fuzzy Inventory Model as a GP Model*

Several solution algorithms for solving a GP model are possible. Kuhn-Tucker conditions or solving dual GP model are some approaches. In this section, we propose a solution approach for the fuzzy inventory model (20.36). We consider the solution of model (20.36). By considering eq. (20.22), eq. (20.23), and eq. (20.24), we propose the following model by analogy to model (20.21):

$$\begin{aligned}
\max \quad & \alpha \\
\text{s.t.} \quad & u_1+u_2+u_3 \geqslant \left(\frac{u_1}{a_1}\right)^{a_1}\left(\frac{u_2}{a_2}\right)^{a_2}\left(\frac{u_3}{a_3}\right)^{a_3} \\
& u_1+u_2+u_3 \leqslant C_0+(1-\alpha)P_0 \\
& Aq \leqslant B+(1-\alpha)P \\
& a_1+a_2+a_3=1 \\
& 0 \leqslant \alpha \leqslant 1 \\
& a_1,\ a_2,\ a_3 > 0
\end{aligned} \tag{20.45}$$

Proposition 20.1. *The solution of model* (20.45) *is equivalent to the solution of model* (20.21) *by considering* (20.25) *and the following model:*

$$\begin{aligned}
\max \quad & \alpha \\
s.t. \quad & u_1^*+u_2^*+u_3^* \leqslant C_0+(1-\alpha)P_0 \\
& Aq^* \leqslant B+(1-\alpha)P_0 \\
& 0 \leqslant \alpha \leqslant 1
\end{aligned} \tag{20.46}$$

where q^*, u_1^*, u_2^*, u_3^* *are the optimal solutions from model* (20.21) *and eq.* (20.25).

20.5 Conclusions

In this chapter, we give the basic applications of geometric programming and fuzzy geometric programming in industrial engineering. We give several formulations for FGP by analogy to FLP in general case, and particularly for inventory and entropy based transportation problems. Although the literature in FGP formulations of several extensions to the basic inventory problem is relatively rich, the papers emphasize only special cases. In this chapter, we also propose a solution approach to solve a fuzzy inventory problem as a GP model.

Bibliography

[1] S. Boyd, S.-J. Kim, L. Vandenberghe and A. Hassibi. A tutorial on geometric programming. *Optimization Engineering*, **8**: 67–127 (2007).

[2] B.-Y. Cao and J.-H. Yang. Advances in fuzzy geometric programming, Fuzzy Information and Engineering, **40**: 497–502 (2007).

[3] E. Bas. Fuzzy geometric programming by t-norm and t-conorm fuzzy relations, Computational Intelligence Foundations and Applications, Proceedings of the 9th International FLINS Conference, World Scientific Publishing Co. Pte. Ltd., pp. 236–242 (2010).

[4] B.-Y. Cao. *Fuzzy Geometric Programming*, Kluwer Academic Publishers. Dordrecht. The Netherlands (2002).

[5] J.-H. Yang and B.-Y. Cao. Fuzzy geometric programming and its application, *Fuzzy Information and Engineering*, **2**: 101–112 (2010).

[6] C.R. Bector and S. Chandra. *Fuzzy Mathematical Programming and Fuzzy Matrix Games,* Springer, Berlin Heidelberg New York (2005).

[7] H.-J. Zimmermann. Fuzzy programming and linear programming with several objective functions, *Fuzzy Sets and Systems*, **1**: 45–55 (1978).

[8] J.L. Verdegay. A dual approach to solve the fuzzy linear programming problem, *Fuzzy Sets and Systems*, **14**: 131–141 (1984).

[9] M. Inuiguchi, J. Ramik, T. Tanino and M. Vlach. Satisficing solutions and duality in interval and fuzzy linear programming, *Fuzzy Sets and Systems*, **135**: 151–177 (2003).

[10] K.-L. Hsiung, S.-J. Kim and S. Boyd. Tractable approximate robust geometric programming, *Optimization and Engineering*, **9**: 95–118 (2008).

[11] P. Shen, Y. Ma and Y. Chen. A robust algorithm for generalized geometric programming, *Journal of Global Optimization*, **41**: 4593–4612 (2008).

[12] H. Paul. An application of stochastic geometric programming to heat exchanger design, *Computers & Industrial Engineering*, **10**: 253–262 (1986).

[13] R. Jagannathan. A stochastic geometric programming problem with multiplicative recourse, *Operations Research Letters*, **9**: 99–104 (1990).

[14] M. O. Abou-El-Ata, H.A.Fergany and M.F. El-Wakeel. Probabilistic multi-item inventory model with varying order cost under two restrictions: A geometric programming approach, *International Journal of Production Economics*, **83**: 223–231 (2003).

[15] T.C.E. Cheng. An economic order quantity model with demand-dependent unit cost, *European Journal of Operational Research*, **40**: 252–256 (1989).

[16] T.C.E. Cheng. An economic production quantity model with flexibility and reliability considerations, *European Journal of Operational Research*, **39**: 174–179 (1989).

[17] K.-N.F. Leung. A generalized geometric-programming solution to "An economic production quantity model with flexibility and reliability considerations", *European Journal of Operational Research*, **176**: 240–251 (2007).

[18] W.J. Lee. Determining order quantity and selling price by geometric programming: Optimal solution, bounds and sensitivity, *Decision Sciences*, **24**: 76–87 (1993).

[19] W.J. Lee. Optimal order quantities and prices with storage space and inventory investment limitations, *Computers & Industrial Engineering*, **26**: 481–488 (1994).

[20] M.O. Abou-el-ata and K.A.M. Kotb. Multi-item EOQ inventory model with varying holding cost under two restrictions: a geometric programming approach, *Production Planning & Control*, **8**: 608–611 (1997).

[21] A.M.A. Hariri and M.O. Abou-el-ata. Multi-item production lot-size inventory model with varying order cost under a restriction: a geometric programming approach. *Production Planning & Control*, **8**: 179–182 (1997).

[22] A.I. Shawky and M.O. Abou-El-Ata. Constrained production lot-size model with trade-credit policy: A comparison geometric programming approach via Lagrange, *Production Planning & Control*, **12**: 654–659 (2001).
[23] H. Jung and C.M. Klein. Optimal inventory policies under decreasing cost functions via geometric programming, *European Journal of Operational Research*, **132**: 628–642 (2001).
[24] H. Jung and C.M. Klein. Optimal inventory policies for profit maximizing EOQ models under various cost functions, *European Journal of Operational Research*, **174**: 689–705 (2006).
[25] N.K. Mandal, T.K. Roy and M. Maiti. Inventory model of deteriorated items with a constraint: A geometric programming approach, *European Journal of Operational Research*, **173**: 199–210 (2006).
[26] S. Islam. Multi-objective marketing planning inventory model: A geometric programming approach, *Applied Mathematics and Computation*, **205**: 238–246 (2008).
[27] M. Corstjens and P. Doyle. A model for optimizing retail space allocations, *Management Science*, **27**: 822–833 (1981).
[28] J.H. Bookbinder and F.H. Zarour. Direct product profitability and retail shelf-space allocation models, *Journal of Business Logistics*, **22**: 183–208 (2001).
[29] M. Corstjens and P. Doyle. A dynamic model for strategically allocating retail space, *Journal of the Operational Research Society*, **34**: 943–951 (1983).
[30] S.A. Kádas and E. Klafszky. Estimation of the parameters in the gravity model for trip distribution: A new method and solution algorithm, *Regional Science and Urban Economics*, **6**: 439–457 (1976).
[31] D.S. Wong. Maximum likelihood, entropy maximization, and the geometric programming approaches to the calibration of trip distribution models, *Transportation Research Part B: Methodological*, **15**: 329–343 (1981).
[32] A.G. Marï£¡n. A multimodal combined model analyzed via generalized geometric programming, *Transportation Research Part B: Methodological*, **22**: 391–397 (1988).
[33] B. Samanta and S.K. Majumder. Entropy based transportation problem use geometric programming method, *Applied Mathematics and Computation*, **179**: 431–439 (2006).
[34] G.Q. Huang and T. Qu. Extending analytical target cascading for optimal configuration of supply chains with alternative autonomous suppliers, *International Journal of Production Economics*, **115**: 39–54 (2008).
[35] B. Mungamuru and H. Garcia-Molina. Predictive pricing and revenue sharing, *Lecture Notes in Computer Science, Internet and Network Economics*, **5385**: 53–60 (2008).
[36] S. Islam and T.K. Roy. A new fuzzy multi-objective programming: Entropy based geometric programming and its application of transportation problems, *European Journal of Operational Research*, **173**: 387–404 (2006).
[37] T.K. Roy and M. Maiti. A fuzzy EOQ model with demand-dependent unit cost under limited storage capacity, *European Journal of Operational Research*, **99**: 425–432 (1997).
[38] S. Islam and T.K. Roy. A fuzzy EPQ model with flexibility and reliability consideration and demand dependent unit production cost under a space constraint: A fuzzy geometric programming approach, *Applied Mathematics and Computation*, **176**: 531–544 (2006).
[39] S.-Y. Chou, P.C. Julian and K.-C. Hung. A note on fuzzy inventory model with storage space and budget constraints, *Applied Mathematical Modelling*, **33**: 4069–4077 (2009).
[40] S.Bag, D.Chakraborty and A.R Roy. A production inventory model with fuzzy random demand and with flexibility and reliability considerations, *Computers & Industrial Engineering*, **56**: 411–416 (2009).
[41] D. Panda and M. Maiti. Multi-item inventory models with price dependent demand under flexibility and reliability consideration and imprecise space constraint: A geometric programming approach, *Mathematical and Computer Modelling*, **49**: 1733–1749 (2009).

[42] S. Islam and T.K. Roy. Fuzzy multi-item economic production quantity model under space constraint: A geometric programming approach, *Applied Mathematics and Computation*, **184**: 326–335 (2007).

[43] N.K. Mandal and T.K. Roy. Multi-item imperfect production lot size model with hybrid number cost parameters, *Applied Mathematics and Computation*, **182**: 1219–1230 (2006).

[44] D. Panda, S. Kar and M. Maiti. Multi-item EOQ model with hybrid cost parameters under fuzzy/fuzzy-stochastic resource constraints: A geometric programming approach, *Computers & Mathematics with Applications*, **56**: 2970–2985 (2008).

[45] N.K. Mandal, T.K. Roy and M. Maiti. Multi-objective fuzzy inventory model with three constraints: a geometric programming approach, *Fuzzy Sets and Systems*, **150**: 87–106 (2005).

[46] M.K. Maiti. Fuzzy inventory model with two warehouses under possibility measure on fuzzy goal, *European Journal of Operational Research*, **188**: 746–774 (2008).

Chapter 21

An Artificial Intelligent Classification System for Thermoplastic Recycling Using Sensor Fusion

S. Satorres Martínez[1], J.M. López Paniza[2], M. Cobo Ramírez[1], J. Gómez Ortega[1], and J. Gámez García[1]

[1] *System Engineering and Automation Department. University of Jaen, Spain*

[2] *Plastic Technological Center of Andalusia, Spain*

The first step in thermoplastic recycling is the classification of plastic wastes into categories. Nowadays, it is a manual task characterized by its high cost and inefficiency. This chapter presents a prototype of an automated classification system for thermoplastics that can be integrated on an industrial recycling plant. The main contribution of this new prototype is the use of a novel artificial intelligent classification system based on a multi-sensor strategy.

21.1 Introduction

The consumption of petroleum resources for plastics production and the accumulation of plastic wastes present well-known challenges to the sustainability of contemporary manufacturing [1]. One of these challenges is to increase the amount of recyclable plastic worldwide [1], and particularly, the recycled thermoplastic.

Thermoplastics are high molecular height polymers that turn to a liquid when heated and freezes to a very glassy state when cooled sufficiently. When solid, however, a thermoplastic becomes glass-like and subject to fracture. This characteristic, which lends the material its name, is reversible. That is, it can be reheated, reshaped, and frozen repeatedly. This quality also makes thermoplastics recyclable [2].

The interest in thermoplastic recycling is mainly because of its amount, since, nowadays, it is almost a component of everyday life. Nevertheless, it is a very complicated task because of the variety of thermoplastics, and also because different thermoplastics may have similar physical properties (color, shape, roughness, etc.). Furthermore, a plastic part may be

C. Kahraman (ed.), *Computational Intelligence Systems in Industrial Engineering*,
Atlantis Computational Intelligence Systems 6, DOI: 10.2991/978-94-91216-77-0_21,

composed of different kind of them. These reasons make that, at this moment, most sorting for mechanical recycling is done by trained operators who manually classify the plastics into polymer type and/or color [3].

Manual sorting is, as it is well known, expensive and ineffective; existing thus a high demand for new technologies that could quickly sort these materials with maximum purity while maintaining high throughput [4, 5]. Nowadays, the technologies to recycle thermoplastics are classified into four categories [2]:

- *Disassembly tools and systems*: disassembling the thermoplastic components before recycling increases purity and value of the recovered materials. This task is usually done manually, resulting highly expensive and inefficiency.
- *Thermoplastic identification*: this task can be carried out on single parts or using a mixture of thermoplastics. Different technologies have been applied such as spectroscopy (Raman, infrared, etc.) or computer vision-based solutions.
- *Mechanical plastic separation*: the success of the mechanical separation basically depends on the difference of the physical properties of the materials. The properties normally used are density, electrostatic energy, surface texture, etc. For the case of thermoplastics, this option is not possible because of their similitude.
- *Chemical plastic separation*: it works converting a mixture a different kind of thermoplastics into energy or monomers. Obviously, the material obtained cannot be reused for similar types of application.

In this work we propose a new classification system which contributes mainly to point 2 and also to point 3 (thermoplastic identification and mechanical separation, respectively). For the identification problem, we have proposed a sensor fusion methodology that integrates the information of three different sensors: inductive, computer vision computer vision in the visible spectra, and a NIR hyperspectral system. This methodology is based on a supervised learning classifier implemented by Artificial Neural Networks and where the NIR data were pretreated by the aid of PCA.

The remainder of this paper is as follows. Section 21.2 shows the State of the Art of this application, while Section 21.3 develops the classification task requirements and the hardware setup. The supervised learning controller is described in Section 21.4 and, in Section 21.5, the experimental result, obtained with this setup, is commented. Finally, the conclusions and future work are included in Section 21.6.

21.2 State of the Art

As commented in the previous section, thermoplastic recycling is currently done manually in the industry, although there also exist different unsuccessful attempts in this research field. Both are described in the following.

21.2.1 *Manual sorting*

To avoid the disassembly process of thermoplastic parts into their elements, in industry, these parts are chopped into small pieces (Fig. 21.1) and later, if is economically interesting, they are sorted by hand using a conveyor.

Fig. 21.1 Thermoplastics for manual sorting

In the manual process, operators usually select the small parts by hand using criteria like color, texture and appearance. The accuracy has to be extremely high; in fact, with only 10ppm of contamination a recycled plastic can reduce its price considerably. Another factor to take into consideration is the cadence of the process, since one expert worker can select 40kg/h of one kind of resin.

21.2.2 *Automatic sorting*

Usually, the recycling process development investigation is focused on automatic sorting for massive plastics. We can find several reviews for this issue [6, 7]. These plastics are used to make bottles, packaging and bags and are easy to identify in many ways: using machine vision: (1) shape recognition [8], (2) linear discriminant analysis (LDA) applied to gray level histogram on images [9], (3) X-ray transmission [10], (4) using other physical properties such us density, near infrared spectroscopy (NIR) [11, 12], (5) studying the difference of in the transmittance (or absorption) between different plastics to an infrared InGaAsP laser [13], or (6) adding tracers to the virgin material to allow an easier identification [14].

For thermoplastics, a compilation of technologies applied on the recovery of plastics waste from electronic and automotive sectors can be found in [2]. Other works explain the use of a classifier based on the measurement of the temperature after heating the waste with a laser diode [15], where the temperature depended on the material and its additives; or using medium infrared (MIR) spectrometry for identifying dark plastics (using the absorbance in this range for plastics and additives is lower than in NIR range) [16]. None of them demonstrated their validity for an efficient classification system.

21.3 Specifying the automatic system

A successful design of an automated inspection system is based on a detailed specification. The task and, as well, requirements of speed and accuracy need to be defined. Furthermore, the ambient influences such as the environmental light, mechanical tolerances, etc., have to be precisely described. Besides the specification, it is essential to provide a set of samples parts that covers all the different types of thermoplastics. These samples are required to design the hardware set-up and to get an impression of the problem complexity. Next subsections deal with these issues.

21.3.1 *Task specification and performance requirements*

Certainly, the task is one of the most important topics of the specification. For our case of study, the automatic classification system for thermoplastics has to properly sort the typical plastic resins that are used for automotive components, and these can be: PMMA (polymethylmethacrylate), PP (polypropylene), ABS (acrylonitrilebutadiene-styrene), PC (polycarbonate), PA (polyamide), EPDM (ethylene propylene diene monomer rubber), and blends of PC/ABS (Fig. 21.2).

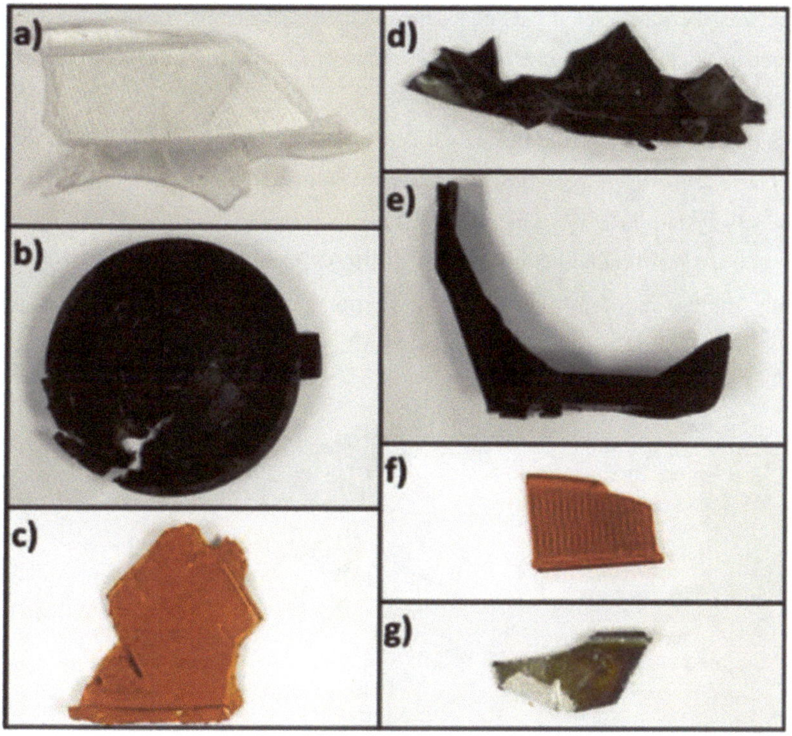

Fig. 21.2 Different types of thermoplastics. (a) Transparent PC with decorative pattern; (b) Black EPDM; (c) Red ABS-PC; (d) PP and talc; (e) PA and glass fiber; (f) Red PMMA with decorative pattern; (g) Metallized ABS

The same type of plastic could be colored or enhanced with a decorative pattern and also the overlapping and touching samples has to be avoided during the inspection. Despite these additional difficulties the system accuracy has to be next to 100% for obtaining usable resins from wastes.

From the economical point of view, the system has to sort 80 kg/hour for being profitable. Considering the size of the wastes, once they are milled, the cycle time has to be lower than 3.17 plastic unit/second.

21.3.2 *Hardware set-up*

The automatic system is mainly divided into three parts: the first, principally mechanical, separates the plastic elements and deposits them on a conveyor belt. This device is represented as "Feeder" in Figure 21.3, although in practice is a system more complex than a simple hopper.
The second part is composed of the sensors utilized to obtain information from the plastic wastes. These sensors are located over a conveyor belt and defined from now on as sensory system. This system is composed of: a CCD area scan camera, that provide data in the visible spectra, a set of inductive sensors and a NIR with a fast InGaAs-diode array detector.

Finally, the last part is composed of the ejection elements responsible for dispatching the plastic wastes into the corresponding container.

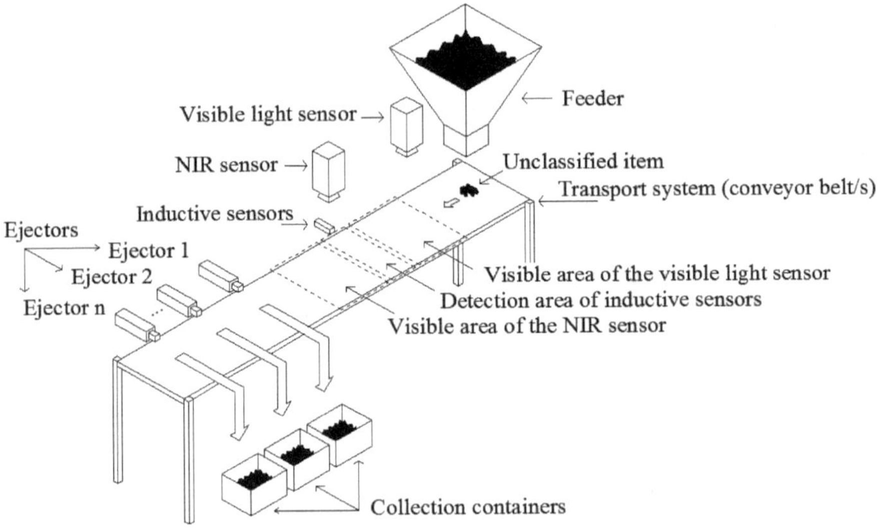

Fig. 21.3 The automatic classification system

21.3.3 *Electronic Exchange of Information*

Basically, management for feeding, movement and ejection of plastic is controlled by a PLC (Programmable Logic Controller). This PLC is connected via Ethernet with an industrial PC responsible for implementing the classification algorithm which uses the information provided by sensors included in the sensory system.

There are different types of data connections between the sensory system elements and the industrial PC. The visible spectra camera transfers the information via IEEE 1394. For the hyperspectral system the Camera Link interface is used and finally the Bus AS-Interface for the set of inductive sensors. All this electronic exchange of information is presented in Figure 21.4.

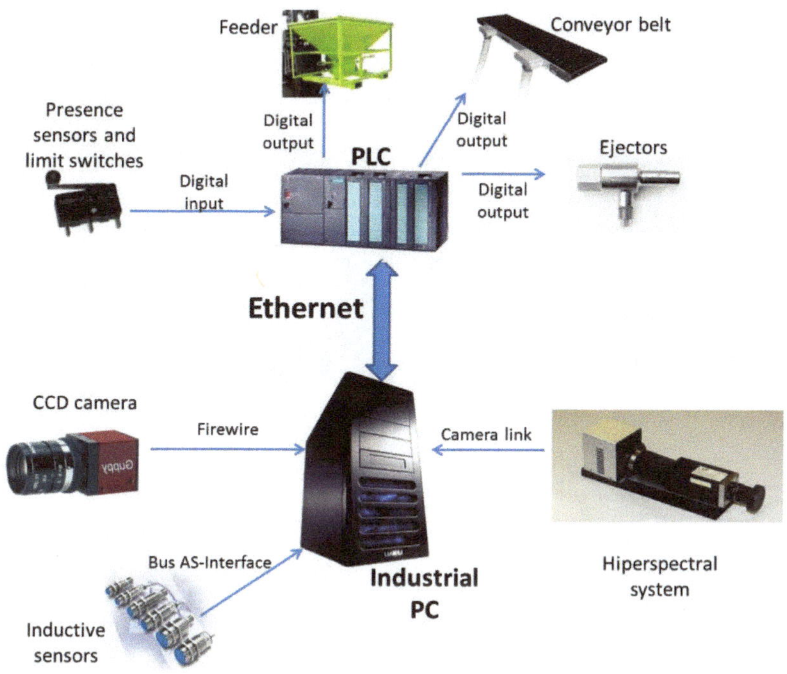

Fig. 21.4 Block diagram of the hardware setup communications

21.3.4 *The sensory system*

The identification problem for thermoplastics requires taking into account information collected from different devices. Utilizing one single device data is not useful for properly assigning a plastic to a class, so we need a conjunction of different sensors with different properties (known as complementary sensor fusion [17]. Every sensor has a predefined task within the automated classification system and their descriptions are given stated bellow.

21.3.4.1 *CCD camera*

The information in the visible spectra is collected from a color CCD area scan camera. The resolution of this sensor is 1034x778 square pixels of 4,65microns and its acquisition rate is 30 fps.
The purpose of this sensor is two-fold. Firstly,the center of gravity (c.o.g.) of the sample is computed (Fig. 21.5) and, secondly, data in the visible spectra such as color and texture, are obtained. The c.o.g.is useful for locating the sample for its further sorting, in the proper container, and also for reducing the number of acquisitions in the hyperspectral system.

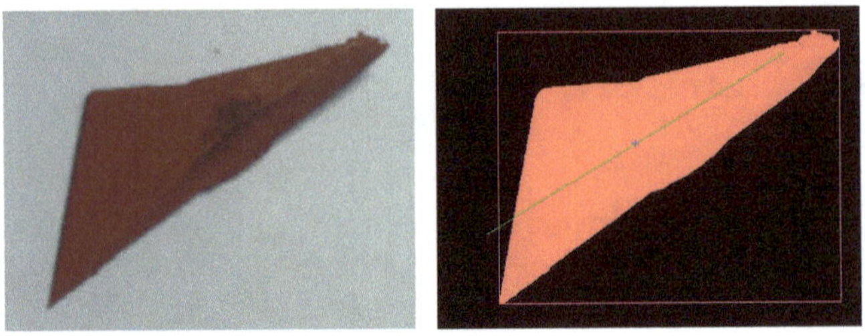

Fig. 21.5 Computing the center of gravity of the sample

As it is shown in the hyperspectral system description, thermoplastics that contain black master batch present a featureless spectra in the NIR-region. Hence, characteristics related to their surface finishing should be utilized for distinguishing these plastics. Color and texture are discriminant attributes from plastics like EPDM, PP and PA. Some thermoplastics such as PMMA may present a decorative pattern and, in this case, the texture information is useful for a reliable classification.

21.3.4.2 *Inductive sensors*

A set of inductive sensors have been attached to the conveyor belt for identifying, without physical contact, plastic components that have been metallized. They are also useful for avoiding metallic wastes in the plastic containers.

21.3.4.3 *Hyperspectral system*

The hyperspectral system is an integrated combination of an imaging spectrograph, an area monochrome camera and a hyperspectral lens (Fig. 21.6).

Fig. 21.6 The hyperspectral system

All the elements, included in this system, have been selected to work in the NIR-regionNIR-region.To be precise, the area scan camera is infrared and incorporates a sensor InGaAs of 320×256 pixels, the spectrograph has a spectral range between 900 and 1700 nm and the optical is corrected in the spectral range between 900 and 2500 nm.

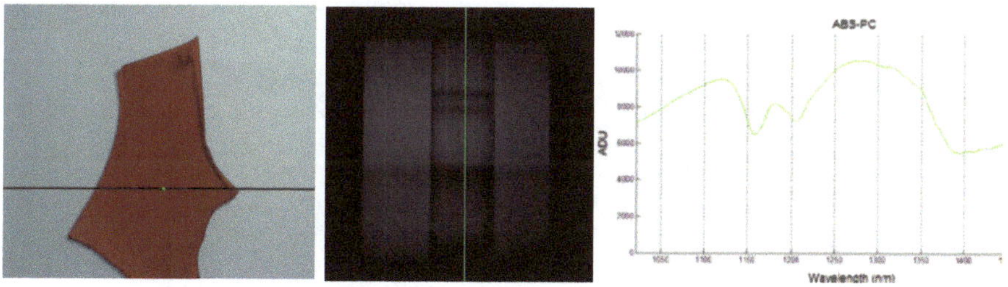

Fig. 21.7 The hyperspectral system data. From left to right: red ABS-PC sample with a pixel labeled in green; the hyperspectral system image that represents the full spectral information for the green pixel; the spectral features in the green pixel

Regarding its operation, it works as a line scan camera, providing full spectral information for each pixel on the imaged line in one image (Fig. 21.7). For acquiring the entire sample surface, it is necessary a movement mechanism. That is, the sample has to be scanned line by line in one spatial direction.

Irradiated by near infrared radiation, as produced by halogens lamps, thermoplastics absorb parts of the radiation. The recorded NIR spectra show minimums that correspond to the absorption peaks. Figure 21.8 shows the spectra in the wavelength area between 1000 and 1500 nm for five different types of thermoplastics. It is worth noticing that the PP spectra correspond to a PP sample with black master batch presenting a strong absorption in these wavelengths and thus resulting a plain spectra. Moreover, in occasions, there are subtle differences between thermoplastic spectra (e.g. PC and PC+ABS) and an adequate data pretreatment is mandatory.This issue is described in detail in the next section.

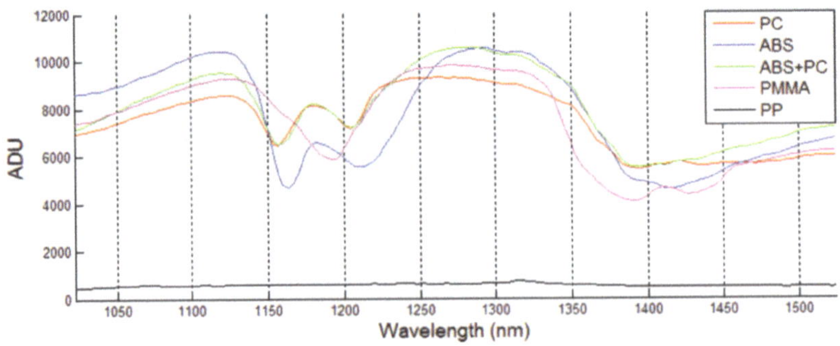

Fig. 21.8 Spectra in the NIR- region for different thermoplastics

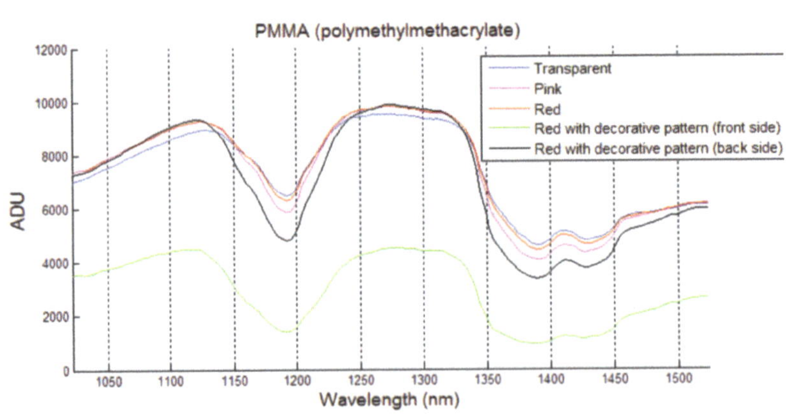

Fig. 21.9 Spectra in the NIR-region for PMMA with different color additives

21.4 The supervised learning classifier

The classifier task uses a feature vector, provided by the sensory system components, to assign a technical plastic to a certain category. Because the degree of difficulty in the classifier the design depends on the feature space definition, the feature vector has to be properly detailed. The feature values, for plastics in the same category, have to be similar and a great variability is expected for plastics included in different categories. Likewise, the classifier complexity increases with the length of the feature vector.

In selecting features, it is obviously desired that they were simple to extract, invariant to transformations, insensitive to noise, and useful for discriminating plastics in different categories. In occasions, combining prior knowledge and empirical data, relevant features could be found. But when data require an adequate pretreatment, for revealing small differences, or the number of them is huge, the former alternatives cannot be utilized.

That is the case of the information provided by the hyperspectral system and we have made use of the Principal Component Analysis method (PCA) for this data pretreatment. Once the feature space is determined from the sensory system main data, a supervised learning classifier based on Artificial Neural Networks (ANN) is used for classifying plastics into categories. A general overview of the classifier design is presented in Figure 21.10.

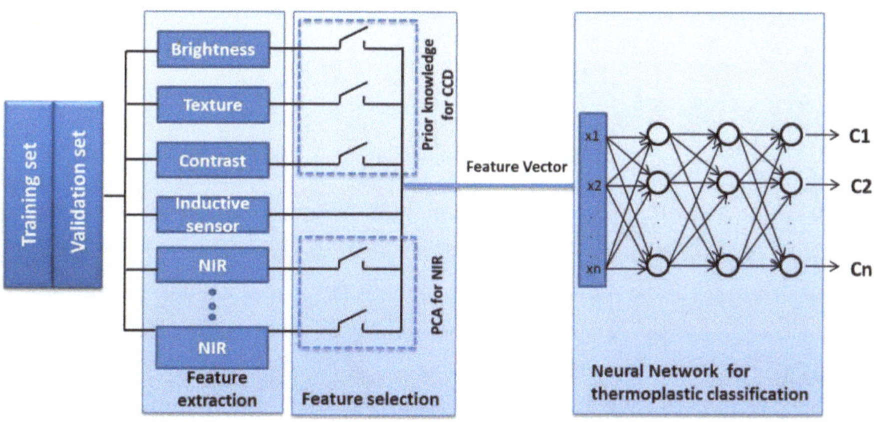

Fig. 21.10 The classifier design. It includes three main steps: Firstly, data are collected from the sensory system; Secondly, main features are selected; Thirdly, a neural network is designed for thermoplastic classification

21.4.1 *Reducing the dimensionality of data: the feature selection*

The choice of the distinguishing features is a critical design step and depends on the characteristics of the sensory system and the specification of the task.As thermoplastics that contain black masterbatch cannot be identified by the NIR system, features in the visible spectra have to be applied for identification. Prior knowledge is incorporated in this step by analyzing black samples and deciding which magnitudes discriminate between plastics. Since the NIR-spectroscopy is not dependent of the sample color, this sensor is utilized for the rest of thermoplastics.

However, NIR data require an adequate pretreatment to differentiate between plastics and, in this sense, the following methods are used before achieving our goal: the feature vector definition.

21.4.1.1 *Smoothing the spectra*

Previous to reduce the NIR data dimensionality, the smoothing of the spectra collected from this sensor is necessary. A simple form of smoothing a signal is the "moving average", which replaces each data value, $x(i)$, with the average of neighboring values, $\overline{x}(i)$. To avoid shifting the data, it is best to average the same number of values, before and after, where the average is being calculated. This value is defined by M. In equation form, the moving average is calculated by:

$$\overline{x}(i) = \frac{1}{2\cdot M+1}\sum_{j=-M}^{M} x(i+j) \qquad (21.1)$$

21.4.1.2 *Principal Component Analysis (PCA)*

PCA is a method for identifying patterns in data highlighting their similarities and differences. Another advantage of PCA is that, once patterns have been found in data, the dimensionality can be reduced without much loss of information. Starting with an original dataset, thePCA is computed by determining the eigenvectors and eigenvalues of the covariance matrix.

The original data is a set of 256-dimensional vectors, $\mathbf{x} = \{\mathbf{x}_1^T, \mathbf{x}_2^T, \ldots, \mathbf{x}_n^T\}$, where each vector consists of all ADU values in the 256 wavelengths from the same type of thermoplastic.The number of vectors, n, in this set corresponds to the samples used in for the analysis.

For PCA to work properly, the mean has to be subtracted from each of the data dimension. This produces a new data set, the matrix A, that it is utilized to calculate the covariance matrix, C. As the data is 256-dimensional, the covariance matrix is 256×256

whose values are:

$$C = \begin{pmatrix} cov(\mathbf{A}_1,\mathbf{A}_1) & cov(\mathbf{A}_1,\mathbf{A}_2) & \cdots & cov(\mathbf{A}_1,\mathbf{A}_{256}) \\ cov(\mathbf{A}_2,\mathbf{A}_1) & cov(\mathbf{A}_2,\mathbf{A}_2) & \cdots & cov(\mathbf{A}_2,\mathbf{A}_{256}) \\ \vdots & \vdots & \ddots & \vdots \\ cov(\mathbf{A}_{256},\mathbf{A}_1) & cov(\mathbf{A}_7,\mathbf{A}_2) & \cdots & cov(\mathbf{A}_7,\mathbf{A}_{256}) \end{pmatrix} \tag{21.2}$$

Note that $cov(\mathbf{A}_1,\mathbf{A}_1) = var(\mathbf{A}_1)$, and since $cov(a,b) = cov(b,a)$, the matrix is symmetrical about the main diagonal.

To compress data, 3 eigenvectors with the largest eigenvalues eigenvalues are calculated, from the matrix C, and chosen to derive the new data set in which the original dimensionality has been reduced in a 98% remaining its essential information. As it is shown in Figure 21.11 three principal components are sufficient for separation plastics without black masterbatch. For this reason EPDM and PP, that contain this additive, are closely placed in this space.

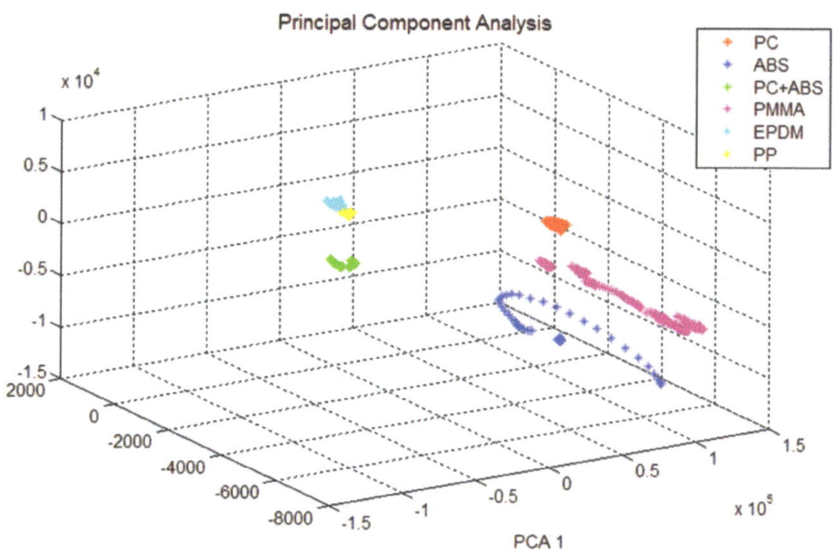

Fig. 21.11 Data from thermoplastic spectra after applying the PCA using three eigenvectors

21.4.2 *The classifier*

Since the decision process is based on the information obtained from three different devices, it is necessary a sensor fusion methodology to integrate the measurements collected from them.

Multisensor fusion refers to any stage in the integration process where there is an actual combination of different sources of sensory information in a representation format [17]. The idea is to merge two or more separate items, obtained fusing the sensory information, into a single entity with a multidimensional feature space. This space can be partitioned into distinct regions, each one representing a class or, for this case, a kind of thermoplastic. Neural networks have been extensively used in classification problems because have the capability of constructing the separating hypersurfaces between classes. In order to get a classifier able to construct general separating hypersurfaces a multi-layer perceptron (MLP) is designed, trained and tested.

21.4.2.1 *MLP design*

The first step in the design of a MLP neural network is the choice of its inputs and outputs. In our classifier, the input nodes, or neurons, are the values of the feature vector. The classes to be identified are the output nodes and, as it has been underlined before, seven types of thermoplastics have to be classified. Hence, in the MPL neural network, the output value of each of these neurons indicates the membership of the input to the corresponding class.

Other parameters to be considered are the number of hidden layers and the number of nodes in each hidden layer and, to define them, trial and error procedures have been needed. However, in order to minimize trials, the number of hidden layers has been limited to one or two. Previous work has demonstrated that increasing the number of layers increases the complexity of the network but its better performance is not guaranteed.

Consequently, the number of nodes in the hidden layer is approached by the Yei's formula [18]: $a+b/2$, where a is the number of input nodes and b is the number of output nodes. Based on these considerations, the MPL topology is the following:

- **Input layer**: it is composed of 9 neurons that present the following data from the sensory system components:

CCD Camera. Since the color and texture information is extracted using this sensor, five neurons are included in the input layer. By transferring the captured color image into the

HSV color space, the color feature of a plastic can be detected. The HSV color system is based on cylindrical coordinates so, in the feature vector, two neurons store the color data. For extracting texture information from images, the gray level co-occurrence matrix, P, is used. Three features, computed from the co-ocurrence matrix, are used and can be written as:

$$\begin{aligned} \text{Energy} &= \sum_i \sum_j P^2(i,j) \\ \text{Entropy} &= \sum_i \sum_j P(i,j)\log_2[P(i,j)] \\ \text{Contrast} &= \sum_i \sum_j (i-j)^2 P(i,j) \end{aligned}$$

Inductive sensor. This device presents a binary value and when "1" is assigned the plastic contains metal trace.

NIR sensor. The near infrared reflectance measurement are preprocessed and only 3 neurons are utilized to characterize every spectra.

- **Output layer:** This layer contains seven neurons that correspond with the seven classes.
- **Hidden layer(s):** The network training has been implemented using the back-propagation algorithm in its version with gradient descent, Gd, variable learning rate gradient descendent, Gdx, and Polak-Ribière conjugate gradient, Gcp. In addition, the activation function has been sigmoidal with value between $[0,1]$. Finally, two different criterions were applied to stop the training: in one case it was stopped when the Mean Squared Error, MSE, reached 0.01 and, in other, after performing 1000 epochs.

21.4.2.2 *MPL structure selection*

Different net structures have been tested to select the MPL topology for the classifier. The training set contains 330 samples that are representative of the different types of thermoplastics. Form these samples the 70% has been used for training, while the validation has been performed on the 30%. The results are summarized in Table 21.1, where the net structure is indicated in the second column (e.g., 30-15, two layers with 30 nodes, in the first, and 15 in the second). The training algorithm, epochs and the percentage of correct classified samples are reported in columns 1, 3 and 4, respectively.

According to the results of Table 21.1, net 2 shown in, the second row,has been chosen for being implemented in the real MLP network. In fact, it combines good performance, next to 100%,with the most reduced training.

Table 21.1 Training results for different network structures

Training	Neurons	Epochs	Accuracy
Gcp	30-15	380	89 %
Gcp	15	88	99%
Gdx	45	190	90 %
Gdx	65	145	93 %

Table 21.2 Results obtained in the classification of seven types of thermoplastics

Class	PMMA	PP	PC	PA	EPDM	ABS	PC+ABS	Accuracy
PMMA	106	0	0	0	0	0	0	100%
PP	11	6	0	0	0	0	0	35%
PC	15	0	40	0	0	0	0	72%
PA	0	0	0	0	11	0	5	0%
EPDM	0	13	0	0	14	0	0	52%
ABS	0	0	0	0	0	50	0	100%
PC+ABS	0	0	0	0	0	0	41	100%

21.5 Results and discussions

To validate theproposed classification system, we have analyzed a total of 312 samples with 7 different types of thermoplastics.

The result of the classifier is displayed by a confusion matrix (Table 21.2). In the resulting confusion matrix, rows show the number of every type of sample and the classification results are presented in columns. As can be seen, the classifier accuracy is 100% for three types of thermoplastics.For thermoplastics that contain black master batch the accuracy decreases. This is the case of EPDM, PA and PP. The worst results are presented for the PA. As these samples had a low width, properly data were difficult to collect. Although further analysis, with a higher number of samples should be done, this first attempt is useful to limit the types of thermoplastics that the automated system can sort fulfilling the accuracy requirement.

21.6 Conclusions

Recycling thermoplastic is a technologic problem that is not solved yet. The origin of the task complexity is that different thermoplastic parts, which can be found in a commercial device, present very similar physical properties. It makes highly difficult their recognition and later classification with standard sensor-based solutions.

In this work we have proposed a new classification system for thermoplastics that are in the recyclable stage. It is based on an artificial intelligent algorithm which fuses the information of three different sensors: a CCD of visible spectra, a NIR hyper-spectral sensor, and an inductive sensor. The algorithm developed uses a neural network whose inputs are the three sensors information but where the NIR data were pretreated by the aid of PCA.

The new classifier is integrated on an industrial prototype, showing initial results that validate partially this approach. Future research will focus on the use of other artificial intelligent algorithms and the realization of a commercial system specially adapted to classify thermoplastics from the automotive industry.

Acknowledgments

This work was supported bythe projects: DPI2008-05798/DPI, TEP2009-5363 and FUSISENSin which the Plastic Technological Center of Andalusia is included.

Bibliography

[1] T. Graedel, The evolution of industrial ecology, *Environmental Science Technology*, vol. **34** (1), pp. 28–31, 2000.

[2] X. Qu, J. Williams, and E. Grant, "Viable plastics recycling from end-of-life electronics," *IEEE Trans. Electronics Packaging Manufacturing,* vol. **29** (1), pp. 25–31, 2006.

[3] A. Iuga, A. Samuila, V. Neamtu, R. Morar, R. Beleca, S. Das, and L. Dascalescu, "Removal of Metallic Particles From Acrylonitrile Butadiene Styrene Wastes Using Electrostatic Separation Methods," *IEEE Transactions on Industry Applications,* vol. **47** (1), pp. 322–330, 2011.

[4] B. House, D. Capson, and D. Schuurman, "Towards real-time sorting of recyclable goods using support vector machines," *2011 IEEE International Symposium on Sustainable Systems and Technology* (ISSST), pp. 1–6, 2011.

[5] C. Murphy, P. Dillon, and G. Pitts, "Economic and logistical modeling for regional processing and recovery of engineering thermoplastics," Denver, 2001.

[6] Clean Washington Center, "Automated sorting systems for Bottles and Regrind in Sorting," 2012. [En línea]. Available: http://www.cwc.org/hdpe/hdpe4.htm.

[7] E. Bruno, "Automated Sorting of Plastics for Recycling," 2000. [En línea]. Available: http://www.p2pays.org. [Último acceso: 2012].

[8] D. Wahab, A. Hussain, E. Scabino, M. Mustafa and H. Basri, "Development of a Prototype Automated Sorting System for Plastic Recycling," *American Journal of Applied Sciences,* vol. **3** (7), pp. 1924–1928, 2006.

[9] R. Suzaimah, M. Mohd, H. Aini, and A. Dzuraidah, "Histogram of intensity feature extraction for automatic plastic bottle," *American Journal of Environmental Sciences,* vol. **4** (6), pp. 583–588, 2008.

[10] T. Jong y W. Dalmijn, "DEXTR Imaging for Automatic Particle Sorting," 2007.

[11] N. Eisenrich y T. Rohe, "Infrared spectroscopy in analysis of plastic recycling," de *Encyclopedia of analytical chemistry*, John Wiley & Sons Ltd, 2000.

[12] A. Kulcke y P. Kerschhaggl, "High Speed NIR Spectral Imaging in Recycling Applications," 2006.

[13] K. Inada, R. Matsuda, C. Fujiwara, M. Nomura, T. Tamon, I. Nishiwara, T. Takao, and T. Fujita, "Identification of plastics by infrared absorption using InGaAsP laser diode," *Resources, Conservation and recycling,* vol. **33**, pp. 131–146, 2001.

[14] E. Maris, A. Aoussat, E. Naffrechoux y D. Froelich, "Polymer tracer detection systems with UV fluorescence spectrometry to improve product recyclability," *Minerals Engineering,* 2011.

[15] H. Haferkamp and I. Burmester, "Fast and automatic thermographic material-identification for the recycling process," 1998.

[16] F. Kowol, M. Oleimeulen, H. Freitag, and T. Huth-Fehre, "Mid infrared acousto-optical tunable filter-spectrometer for rapid identification of black plastics from automotive construction," *Journal of Near Infrared Spectroscopy,* vol. **6**, 149–151, 1998.

[17] R. Luo, C. Yih y K. Su, "Multisensor fusion and integration: approaches, applications, and future research directions," *IEEE Sensors Journal,* vol. **2** (2), pp. 107–119, 2002.

[18] R. Duda, P. Hart y D. Stork, *Pattern Classification*, 2nd ed., John Wiley & Sons, 2001.

[19] R. Suzaimah, M. Mohd, H. Aini, and A. Dzuraidah, "Histogram of intensity feaute extraction for automatic plastic bottle recycling using machine vision," *American Journal of Environmental Science,* pp. 583–588, 2008.

Chapter 22

Linear Programming for Distribution Network Design under Fuzzy Environment

Alp Ustundag and Emre Cevikcan

Department of Industrial Engineering, Istanbul Technical University, P.O. Box 34367, Istanbul, Turkey

E-mail: ustundaga@itu.edu.tr

A key issue in contemporary supply chain management, Distribution Network Design (DND) addresses location and capacity of warehouses and retail outlets, together with the manufacturing plant and supply sources. An effective DND should not only provide the minimization of transportation and inventory costs, but also increase the level of service available to the customers. When designing a distribution network, parameters can not be frequently determined as crisp values. Fuzzy logic is utilized in many engineering applications so as to handle imprecise data. This chapter provides a mathematical programming basis for distribution network design problem under imprecise data environment. After providing background information about DND problem as well as fuzzy linear programming, fuzzy linear programming model of DND problem is solved for different α-cut values.

22.1 Introduction

There has been a trend towards supply chain management concept, as companies have become more global. The supply chain comprises of the manufacturer and suppliers as well as transporters, warehouses, retailers, and customers. The above mentioned chain is represented by the flow of materials and information both within and between business entities. As far as increasingly complicated supply chains are concerned, distribution network design has a significant role in controlling the cost of performing business activities. Distribution network design (DND) is an important strategic decision that companies must make to ensure that required raw materials and components can be distributed efficiently from their suppliers to their manufacturing plants and warehouses as well as the final products to their customers [1]. DND is a well recognized and widely studied problem in both supply

C. Kahraman (ed.), *Computational Intelligence Systems in Industrial Engineering*,
Atlantis Computational Intelligence Systems 6, DOI: 10.2991/978-94-91216-77-0_22,

chain management (SCM) and logistics. The problem consists of selecting sites to install plants, warehouses, and distribution centres; assigning customers to serving facilities; and interconnecting facilities by flow assignment [2]. Distribution network of a company should meet service goals at the lowest possible cost and at the maximum level of costumer satisfaction. To achieve this, an ideal network must have the optimum number, capacity, and location of distribution centres. Distribution, as a marketing function, can be defined as the steps taken to move and store a product from the supplier stage to the customer stage in a supply chain. Distribution directly affects cost and the customer experience and therefore drives profitability. Supply chain objectives ranging from low cost to high responsiveness are considered in distribution network design.

Fuzzy set theory has been widely applied in different disciplines, such as operations research, management science, control theory and artificial intelligence. Fuzzy mathematical programming is one of the most popular decision making approaches based on the fuzzy set theory [3]. In this chapter, imprecise parameters and fuzzy situation in distribution network design is addressed. The inspiration of this chapter is to provide a mathematical programming basis for DND under imprecise data.

The rest of this chapter is organized as follows. In Section 22.2, relevant literature is reviewed. In Section 22.3, background information about DND is given. Fuzzy linear programming techniques are included in Section 22.4. Application of fuzzy linear programming to DND problem is given in Section 22.5. Finally, conclusions are provided in Section 22.6.

22.2 Relevant Literature

In the literature, most of the previous studies focusing on the uncertainty in logistics network design use stochastic programming methods [4–6]. Nonetheless, this approach is not appropriate for real life cases because of the insufficient historical data of uncertain parameters and the high computational complexity of stochastic programming models. To handle this issue, the fuzzy mathematical programming method which is an efficient tool for handling uncertainty is recently used in the logistics network design problems [7, 8]. In fuzzy mathematical programming, models have been proposed to incorporate fuzziness of objective and constraint functions. In this chapter, imprecise parameters and fuzzy situation in distribution network design is addressed. As application, a distribution network design problem in ceramics industry is solved with the consideration of vagueness in customer demand, opening costs of distribution centres and production plants.

22.3 DND Model

In this chapter, a four-tier DND model is described considering a supply chain composed of suppliers, production plants, distribution centres and customers. Suppliers send materials to production plants which are sending products to the distribution centres. Finally, customer demands are met by the distribution centres. In this model, location and capacity allocation decisions are made for both production plants and distribution centres. Multiple distribution centres may be used to satisfy demand at a customer and multiple production plants may be used to replenish distribution centres. It is also assumed that units have been appropriately adjusted such that one unit of input from a supply source produces one unit of the finished product [9]. As customer demand, fixed costs of opening plants and distribution centres have fuzzy values, the other parameters have crisp values in the model. The parameters of the model are given below:

m	=	number of customer demand points
n	=	number of potential production plants
l	=	number of suppliers
t	=	number of potential distribution centre locations
$\tilde{D}_j$	=	annual demand from customer j
K_i	=	potential capacity of production plant at site i
S_h	=	supply capacity at supplier h
W_e	=	potential distribution centre capacity at site e
$\tilde{F}_i$	=	fixed cost of opening a plant at site i
$\tilde{f}_e$	=	fixed cost of opening a distribution centre at site e
c_{hi}	=	cost of shipping one unit from supplier h to factory i
c_{ie}	=	cost of producing and shipping one unit from production plant i to distribution centre e
c_{ej}	=	cost of shipping one unit from distribution centre e to customer j

The goal of the DND model is to identify production plant and distribution centre locations as well as quantities shipped between various points that minimize the total fixed and variable costs. The decision variables are given below:

$plop_i$ = 1 if production plant is located at site i, 0 otherwise

$dcop_e$ = 1 if distribution centre is located at site e, 0 otherwise

$suplquant_{hi}$ = quantity shipped from supplier h to production plant at site i $pldcquant_{ie}$ = quantity shipped from production plant at site i to distribution centre e

$dccuquant_{ej}$ = quantity shipped from distribution centre e to customer j.

The problem is formulated as a mixed integer linear program.

$$\text{Min} \sum_{i=1}^{n} \tilde{F}_i y_i + \sum_{e=1}^{t} \tilde{f}_e y_e + \sum_{h=1}^{l} \sum_{i=1}^{n} c_{hi} suplquant_{hi} + \sum_{i=1}^{n} \sum_{e=1}^{t} c_{ie} pldcquant_{ie} + \sum_{e=1}^{t} \sum_{j=1}^{m} c_{ej} dccuquant_{ej} \tag{22.1}$$

$$\sum_{i=1}^{n} suplquant_{hi} \leqslant S_h \text{ for } h = 1, \ldots, l \tag{22.2}$$

$$\sum_{h=1}^{l} suplquant_{hi} - \sum_{e=1}^{t} pldcquant_{ie} \geqslant 0 \text{ for } i = 1, \ldots, n \tag{22.3}$$

$$\sum_{e=1}^{t} pldcquant_{ie} \leqslant K_i y_i \text{ for } i = 1, \ldots, n \tag{22.4}$$

$$\sum_{i=1}^{n} pldcquant_{ie} - \sum_{j=1}^{m} dccuquant_{ej} \geqslant 0 \text{ for } e = 1, \ldots, t \tag{22.5}$$

$$\sum_{j=1}^{m} dccuquant_{ej} \leqslant W_e y_e \text{ for } e = 1, \ldots, t \tag{22.6}$$

$$\sum_{e=1}^{t} dccuquant_{ej} = \widetilde{D}_j \text{ for } j = 1, \ldots, m \tag{22.7}$$

$$plop_i,\ dcop_e \in \{0,1\},\ suplquant_{hi},\ pldcquant_{ie},\ dccuquant_{ej} \geqslant 0 \tag{22.8}$$

22.4 Fuzzy Linear Programming Techniques

Fuzzy linear programming can be derived by using fuzzy sets as coefficient values in objective function, constraints or right handsides of the constraints. Providing solutions for fuzzy linear programming models, three different techniques are discussed in this section.

22.4.1 *Method of Zimmermann*

Both fuzzy objective and constraint functions are considered in Zimmermann's approach [10].

$$\begin{aligned} &\widetilde{\text{Max}} \quad cx \\ &\text{s.t.} \quad a_i x \lesssim b_i, \quad i = 1, \ldots, m \\ &\qquad\quad x \geqslant 0 \end{aligned} \tag{22.9}$$

An aspiration level and a tolerance interval are proposed for the fuzziness of the objective function. In the fuzzy constraints, a fuzzy inequality can be considered as fuzzy right-hand sides. On the condition that an aspiration level of objective value is denoted as b_0, the fuzzy mathematical model is called a symmetric fuzzy model and can be written as follows [3]:

$$\begin{aligned} &cx \gtrsim b_0 \\ &a_i x \lesssim b_i, \quad i = 1, \ldots, m \\ &x \geqslant 0 \end{aligned} \tag{22.10}$$

The following matrix notation is used for defining symmetric fuzzy model:

$$\begin{aligned} &Ax \lesssim b \\ &x \geqslant 0 \\ &A = \begin{bmatrix} -c \\ a_i \end{bmatrix}, \quad b = \begin{bmatrix} -b_0 \\ b_i \end{bmatrix} \end{aligned} \tag{22.11}$$

Having an interval, the fuzzy inequality violation of right-hand-side values, b_i, is equivalent to the fuzzy inequality. The membership function of the degree of violation of the fuzzy inequality is expressed as follows where p_i is a tolerance level in the fuzzy relationship:

$$\mu_i(a_i x) = \left\{ \begin{array}{cc} 1 & a_i x \leqslant b_i \\ 1 - \dfrac{a_i x - b_i}{p_i} & b_i \leqslant a_i x \leqslant b_i + p_i \\ 0 & b_i + p_i \leqslant a_i x \end{array} \right\} \tag{22.12}$$

The problem is now to obtain the maximum value of the membership grade in expression (22.4) which can be expressed as the following model which presents the final solution.

$$\begin{aligned} &\text{Max} \quad \lambda \\ &\text{s.t.} \quad \lambda \leqslant 1 - \frac{a_i x - b_i}{p_i} \\ &\qquad\quad \lambda \leqslant 1 \\ &\qquad\quad x \geqslant 0 \end{aligned} \tag{22.13}$$

22.4.2 *Method of Chanas*

The approach of Chanas can be considered as parametric programming method for fuzzy linear programming. A parameter is included to Eq. (22.2) as follows [11]:

$$\begin{aligned} & cx \geqslant b_0 - \theta p_0 \\ & a_i x \leqslant b_i + \theta p_i, \quad i = 1, \ldots, m \\ & x \geqslant 0 \end{aligned} \tag{22.14}$$

The optimization of the membership of constraints can be given as the following max-min operation oriented equation:

$$\mu_c(A_x(\theta)) = \min_{i=1,\ldots,m} [\mu_i(a_i x(\theta))] = \alpha = 1 - \theta \tag{22.15}$$

α: the minimum membership grade of all constraints,
θ: the complementary term of α.
In this approach, the objective value has a piecewise-linear membership function in which aspiration and tolerance levels are utilized.

$$\mu_0(x(\theta)) = \left\{ \begin{array}{cc} 1 & cx^* > b_0 \\ 1 - \dfrac{b_0 - cx^*(\theta)}{p_0} & b_0 - p_0 \leqslant cx^*(\theta) \leqslant b_0 \\ 0 & cx^*(\theta) < b_0 - p_0 \end{array} \right\} \tag{22.16}$$

$x^*(\theta_0)$: the admissible solution with a fixed parameter θ.
Finally, a decision is made via max-min operation oriented membership function given as the following equation where μ_0 and μ_c are the membership functions of objective and constraints, respectively [3].

$$\max \mu_D(\theta) = \max\{\min[\mu_0(\theta), \mu_0(\theta)]\} \tag{22.17}$$

22.4.3 *Method of Julien*

Julien [12] integrates the a-cut concept with Buckley's [13] possibility programming to resolve the problem including fuzzy objective and fuzzy right hand side by solving pairs of crisp linear programming problems in Eq. (22.18) and (22.19). As previously mentioned Eq. (22.9) is a general form of fuzzy linear programming model [14]. The superscript represents an α-cut of the fuzzy parameters, and the subscripts L and U are the corresponding lower and upper cuts.

$$\begin{aligned} \text{Max} \quad & c_L^\alpha x \\ \text{s.t.} \quad & A_U^\alpha x \leqslant b_L^\alpha, \quad i = 1, \ldots, m \end{aligned} \tag{22.18}$$

$$\begin{aligned} \text{Max} \quad & c_U^\alpha x \\ \text{s.t.} \quad & A_L^\alpha x \leqslant b_U^\alpha, \quad i=1,\dots,m \\ & x \geqslant 0 \end{aligned} \tag{22.19}$$

When method of Julien is applied to a minimization problem given in (22.20), the following formulation should be considered. Then, the objective function has the interval between the solutions of (22.22) and (22.21).

$$\begin{aligned} \widetilde{\text{Min}} \quad & cx \\ \text{s.t.} \quad & a_i x \underset{\sim}{\geqslant} b_i, \quad i=1,\dots,m \\ & x \geqslant 0 \end{aligned} \tag{22.20}$$

$$\begin{aligned} \text{Min} \quad & c_U^\alpha x \\ \text{s.t.} \quad & A_L^\alpha x \geqslant b_U^\alpha, \quad i=1,\dots,m \\ & x \geqslant 0 \end{aligned} \tag{22.21}$$

$$\begin{aligned} \text{Min} \quad & c_L^\alpha x \\ \text{s.t.} \quad & A_U^\alpha x \geqslant b_L^\alpha, \quad i=1,\dots,m \\ & x \geqslant 0 \end{aligned} \tag{22.22}$$

22.5 Application – Distribution Network Design with Fuzzy Parameters

A company in ceramics industry plans to design its distribution network with the consideration of vagueness in customer demand and opening costs. The goal is to obtain the optimal transhipment amounts among suppliers, production plants, distribution centres and customers and making opening decision for alternative production plants and distribution centres by using fuzzy linear programming method of Julien [11]. Figure 22.1 displays the transportation network of the company. Related capacity parameters and required transportation costs for the problem are given in Table 22.1 and Table 22.2, respectively.

Demand amounts and opening costs for production plants and distribution centres are expressed as fuzzy triangular numbers, since they can not be determined precisely. Their upper and lower values of opening costs and demand forecasts for different α-cuts are given in Table 22.3 and Table 22.4, respectively.

The model is coded by using X-press IVE optimization software. Results are obtained for the α-cut values of 0, 0.25, 0.50, 0.75, 1. Lower and upper bounds of optimum distribution network design and their related non-zero variables are given in Table 22.5.

As can be seen from Table 22.5, upper bound of optimal distribution network design increases with higher level of fuzziness (highest value: 62,797,000, α-cut$=0$). Inversely,

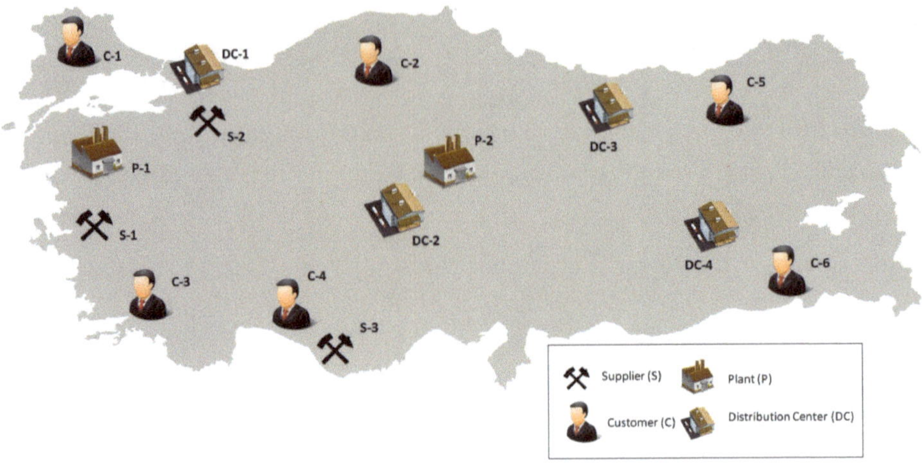

Fig. 22.1 Transportation network of the company

Table 22.1 Capacity parameters

Location	Product Equivalent Capacity
S-1	20,000
S-2	24,000
S-3	25,000
P-1	120,000
P-2	90,000
DC-1	30,000
DC-2	35,000
DC-3	33,000
DC-4	28,000

lower α-cut values yields decreasing lower bounds (lowest value: 33,544,000, α-cut= 0). In fact, α-cut value can be considered as the level of certainty. Range between lower and upper bounds is inversely related to α-cut value. The underlying reason of this fact is that range of fuzzy parameters (demand and activating costs) gets wider with higher level of vagueness. Lower bound has the range of $[33,544,000; 44,326,0000]$. Similarly, upper bounds are computed as the values between 51,514,000 and 62,797,000. In addition, activating decisions for plants and distribution centres are not sensitive to fuzzy parameters. All the plants and distribution centres are opened, except distribution centre 4 for each situation. As for the distribution quantity aspect, upper bounds provide more non-zero variables of distribution quantity. Since upper bound models has higher demand values as right hand

Table 22.2 Transportation Costs ($/unit)

	P-1	P-2	DC-1	DC-2	DC-3	DC-4	C-1	C-2	C-3	C-4	C-5	C-6
S-1	140	420										
S-2	160	400										
S-3	500	450										
P-1			210	420	570	840						
P-2			400	110	220	410						
DC-1							180	230	430	420	710	820
DC-2							510	230	360	220	470	520
DC-3							730	320	650	490	140	360
DC-4							890	550	760	580	170	110

Table 22.3 Opening Costs for production plants and distribution centres ($)

	Upper/Lower Cut	P-1	P-2	DC-1	DC-2	DC-3	DC-4
1	-	1,500,000	1,000,000	440,000	320,000	260,000	280,000
0.75	Lower	1,387,500	925,000	407,000	296,000	240,500	259,000
	Upper	1,612,500	1,075,000	473,000	344,000	279,500	301,000
0.5	Lower	1,275,000	850,000	374,000	272,000	221,000	238,000
	Upper	1,725,000	1,150,000	506,000	368,000	299,000	322,000
0.25	Lower	1,162,500	775,000	341,000	248,000	201,500	217,000
	Upper	1,837,500	1,225,000	539,000	392,000	318,500	343,000
0	Lower	1,050,000	700,000	308,000	224,000	182,000	196,000
	Upper	1,950,000	1,300,000	572,000	416,000	338,000	364,000

Table 22.4 Demand forecasts (product units)

α	Upper/Lower Cut	C-1	C-2	C-3	C-4	C-5	C-6
1	-	14,000	9,000	9,000	12,000	9,000	8,000
0.75	Lower	12,950	8,325	8,325	11,100	8,325	7,400
	Upper	15,050	9,675	9,675	12,900	9,675	8,600
0.5	Lower	11,900	7,650	7,650	10,200	7,650	6,800
	Upper	16,100	10,350	10,350	13,800	10,350	9,200
0.25	Lower	10,850	6,975	6,975	9,300	6,975	6,200
	Upper	17,150	11,025	11,025	14,700	11,025	9,800
0	Lower	9,800	6,300	6,300	8,400	6,300	5,600
	Upper	18,200	11,700	11,700	15,600	11,700	10,400

side values in constraints. For instance, an additional distribution from distribution centre 2 to customer 5 occurs for lower bound scenarios with the α-cuts of 0.5; 0.25; 0.

Table 22.5 Lower and upper bounds of optimum distribution network design

α-cut				
1	0.75		0.5	
	Lower	Upper	Lower	Upper
Total Cost:47,920,000$	Total Cost:44,326,000$	Total Cost:51,514,000$	Total Cost:40,732,000$	Total Cost:55,150,000$
suplquant(1,1)=32,000	suplquant(1,1)=29,600	suplquant(1,1)=34,400	suplquant(1,1)=27,200	suplquant(1,1)=36,800
suplquant(2,2)=29,000	suplquant(2,2)=26,825	suplquant(2,2)=31,175	suplquant(2,2)=24,650	suplquant(2,2)=33,350
pldcquant(1,1)=32,000	pldcquant(1,1)=29600	pldcquant(1,1)=34,400	pldcquant(1,1)=27,200	pldcquant(1,1)=36,800
pldcquant(2,2)=20,000	pldcquant(2,2)=18,500	pldcquant(2,2)=21,500	pldcquant(2,2)=17,000	pldcquant(2,2)=23,350
pldcquant(2,3)=9,000	pldcquant(2,3)=8,325	pldcquant(2,3)=9675	pldcquant(2,3)=7,650	pldcquant(2,3)=10,000
dccuquant(1,1)=14,000	dccuquant(1,1)=12,950	dccuquant(1,1)=15,050	dccuquant(1,1)=11,900	dccuquant(1,1)=16,100
dccuquant(2,4)=12,000	dccuquant(2,4)=11,100	dccuquant(2,4)=12,900	dccuquant(2,4)=10,200	dccuquant(2,4)=13,800
dccuquant(1,2)=9,000	dccuquant(1,2)=8,325	dccuquant(1,2)=9,675	dccuquant(1,2)=7,650	dccuquant(1,2)=10,350
dccuquant(1,3)=9,000	dccuquant(1,3)=8,325	dccuquant(1,3)=9,675	dccuquant(1,3)=7,650	dccuquant(1,3)=10,350
dccuquant(3,5)=9,000	dccuquant(3,5)=8,325	dccuquant(3,5)=9,675	dccuquant(3,5)=7,650	dccuquant(3,5)=10,000
dccuquant(2,6)=8,000	dccuquant(2,6)=7,400	dccuquant(2,6)=8,600	dccuquant(2,6)=6,800	dccuquant(2,6)=9,200
plop(1)=1	plop(1)=1	plop(1)=1	plop(1)=1	dccuquant(2,5)=350
plop(2)=1	plop(2)=1	plop(2)=1	plop(2)=1	plop(1)=1
dcop(1)=1	dcop(1)=1	dcop(1)=1	dcop(1)=1	plop(2)=1
dcop(2)=1	dcop(2)=1	dcop(2)=1	dcop(2)=1	dcop(1)=1
dcop(3)=1	dcop(3)=1	dcop(3)=1	dcop(3)=1	dcop(2)=1
				dcop(3)=1

α-cut			
0.25		0	
Lower	Upper	Lower	Upper
Total Cost:31,138,000$	Total Cost:58,877,000$	Total Cost:33,544,000$	Total Cost:62,797,000$
suplquant(1,1)=24,800	suplquant(1,1)=39,200	suplquant(1,1)=22,400	suplquant(1,1)=40,000
suplquant(2,2)=22,475	suplquant(2,2)=34,000	suplquant(2,2)=20,300	suplquant(2,2)=32,400
pldcquant(1,1)=24,800	suplquant(1,2)=800	pldcquant(1,1)=22,400	suplquant(3,2)=5,300
pldcquant(2,2)=15,550	suplquant(3,2)=725	pldcquant(2,2)=14,000	suplquant(2,1)=1,600
pldcquant(2,3)=6,975	pldcquant(1,1)=39,200	pldcquant(2,3)=6,300	pldcquant(1,1)=41,600
dccuquant(1,1)=10,850	pldcquant(2,2)=25,525	dccuquant(1,1)=9,800	pldcquant(2,2)=27,700
dccuquant(2,4)=9,300	pldcquant(2,3)=10,000	dccuquant(2,4)=8,400	pldcquant(2,3)=10,000
dccuquant(1,2)=6,975	dccuquant(1,1)=17,150	dccuquant(1,2)=6,300	dccuquant(1,1)=18,200
dccuquant(1,3)=6,975	dccuquant(2,4)=14,700	dccuquant(1,3)=6,300	dccuquant(2,4)=15,600
dccuquant(3,5)=6,975	dccuquant(1,2)=11,025	dccuquant(3,5)=6,300	dccuquant(1,2)=11,700
dccuquant(2,6)=6,200	dccuquant(1,3)=11,025	dccuquant(2,6)=5,600	dccuquant(1,3)=11,700
plop(1)=1	dccuquant(3,5)=10,000	plop(1)=1	dccuquant(3,5)=10,000
plop(2)=1	dccuquant(2,6)=9,800	plop(2)=1	dccuquant(2,6)=10,400
dcop(1)=1	dccuquant(2,5)=1,025	dcop(1)=1	dccuquant(2,5)=1,700
dcop(2)=1	plop(1)=1	dcop(2)=1	plop(1)=1
dcop(3)=1	plop(2)=1	dcop(3)=1	plop(2)=1
	dcop(1)=1		dcop(1)=1
	dcop(2)=1		dcop(2)=1
	dcop(3)=1		

22.6 Conclusion

In this chapter, fuzzy linear programming model is provided to solve the distribution network design problem. Then the proposed model is applied to a real life problem in Turkey. The advantage of using fuzzy linear programming is the incorporation of uncertainty of the costumer demands, and activating cost for plants and distribution centres. The

optimal transhipment amounts in the distribution network and the required facility and distribution centres have been determined for different fuzziness levels.
According to the uncertainties influencing the distribution network, fuzzy numbers are used to model the problem. Fuzzy logic helps distribution network planners to know the value of membership degree of development plan in the optimum set. In fact, for specific uncertainty, the planner can understand the range of optimum network planning costs. According to the results of this research, planners will be able to decide how to develop the distribution network under imprecise demand data.
As a future research topic, different fuzzy linear programming methods can be compared for distribution network design problem. Another research direction would be to analyze a distribution network problem with fuzzy left hand side values.

Bibliography

[1] C.L.F., Michelle, *New Models in Logistics Network Design and Implications for 3PL Companies*, PhD Dissertation, The SINGAPORE-MIT Alliance (SMA) Office (2005).
[2] P. A. Miranda, R. A. Garrido, Inventory service-level optimization within distribution network design problem, *Int. J. Prod. Econ.,* **122** (2009), p. 276.
[3] L.-H. Shih, Cement transportation planning via fuzzy linear programming, *Int. J. Prod. Econ.*, **58** (1999), p. 277.
[4] M. El-Sayed, N. Afia, and A. El-Kharbotly, A stochastic model for forward reverse logistics network design under risk, *Comp. & Ind. Eng.*, **58** (2010), p. 423–431.
[5] M. S. Pishvaee, F. Jolai, and J. Razmi, A stochastic optimization model for integrated forward/reverse logistics network design, *J. Manuf. Sys.*, **28** (2009), p. 107–114.
[6] P. Schütz, A. Tomasgard and S. Ahmed, Supply chain design under uncertainty using sample average approximation and dual decomposition, *Eur. J. Oper. Res.*, **199** (2009), p. 409–419.
[7] M.S. Pishvaee, F. Torabi, and J. Razmi, Credibility-based fuzzy mathematical programming model for green logistics design under uncertainty, *Comp. & Ind. Eng.*, **62** (2012), p. 624–632.
[8] M.S. Pishvaee, M. Rabbani, and S. Torabi, A robust optimization approach to closed-loop supply chain network design under uncertainty, *App. Math. Model.*, **35** (2011)., p. 637–649.
[9] S. Chopra and P. Meindl, *Supply Chain Management: Strategy, Planning and Operation*. Pearson Education Inc (2010).
[10] H.J. Zimmermann, *Fuzzy Set Theory and Its Applications*, Kluwer Academic Publishers (1991).
[11] S. Chanas, The Use of Parametric programming in Fuzzy Linear Programming, *Fuzzy Set. Syst.*, **11** (1983), p. 243.
[12] B. Julien, An Extension to Possibilistic Linear Programming, *Fuzzy Set. Syst.*, **64** (1994), p. 195.
[13] J.J. Buckley, Solving possibilistic linear programming, *Fuzzy Set. Syst.*, **31** (1989) p. 329.
[14] M. Allahviranloo and S. Afandizadeh, Investment Optimization on Port's Development by Fuzzy Integer Programming, *Eur. J. Oper. Res.*, **186** (2008), p. 423–434.

Chapter 23

GSA–Based Training of Convolutional Neural Networks for OCR Applications

Lucian-Ovidiu Fedorovici, Radu-Emil Precup, Radu-Codrut David, and Florin Drăgan

Department of Automation and Applied Informatics, "Politehnica" University of Timisoara, Bd. V. Parvan 2, RO-300223 Timisoara, Romania

E-mail: lucian.fedorovici@aut.upt.ro, radu.precup@aut.upt.ro, davidradu@gmail.com, florin.dragan@aut.upt.ro

The chapter proposes a new training algorithm to train Convolutional Neural Networks (CNNs) in Optical Character Recognition (OCR) applications. This algorithm is characterized by the use of Gravitational Search Algorithms (GSAs) in combination with the back-propagation algorithm (BP), and a new family of CNNs is thus suggested.

23.1 Introduction

Optical Character Recognition (OCR) is the technology that transforms an image that contains text into an editable document in electronic format. The development of the technology started in the early 30's and has known an effervescent development in the last decade due to advances in computing power, image processing algorithms, feature extraction, classification algorithms and improvements in the printing quality and image retrieval [1].

The main objectives for an OCR engine are to increase accuracy, to become more independent from the input, to perform less image preprocessing, to recognize handwriting characters and all of those objectives must be performed in a timely manner. Usually there is a balance between recognition rates and required computing power. Newly developed handheld devices offer possible platforms for running an OCR engine for various applications; on those platforms a part from a high recognition rate the execution time also becomes an important objective given the hardware constraints.

C. Kahraman (ed.), *Computational Intelligence Systems in Industrial Engineering*, Atlantis Computational Intelligence Systems 6, DOI: 10.2991/978-94-91216-77-0_23,

An OCR engine consists of several interconnected modules, depending on the architecture, in which one of them is the character recognition module. Character recognition implies recognition of a single character or symbol from a pre-segmented document; this module can be implemented using different approaches that can be grouped in two main classes: supervised classification [2] and unsupervised classification [3].

The supervised classification approach is based on a classifier which recognizes various types of features that can be extracted from the image of a symbol. These features can be grouped in: geometric moments [4, 3, 5, 6], features extracted from contour [7] or descriptors [8, 9].

Artificial neural networks can be used in OCR engines as simple classifiers or as combined feature extractors and classifiers [10–13]. This chapter focuses on the recognition module of an OCR application, and it proposes a novel approach to train Convolutional Neural Networks (CNNs) for character recognition. The architecture of the CNN is proposed by LeCun [14] and designed to recognize handwritten series of digits.

The back-propagation (BP) algorithm is classical in neural network training. The main disadvantage of BP is that it cannot guarantee that the neural network converges to a global minimum and not to a local one. Several techniques aim the improvement of BP such as applying distortions to characters or determining when the learning rate should be adjusted. In this chapter, a novel approach is proposed that uses heuristic algorithms combined with BP to force the neural network convergence to a global minimum. This new training approach consists of two stages and of a new algorithm: in the first stage the neural network is trained by means of Gravitational Search Algorithms (GSAs) until the objective function converges to a threshold value; the second stage represents the fine tuning of the best agent obtained in the first stage by means of BP.

Our combination between the BP and the GSAs has resulted in the modifications of GSAs. These modifications focus performance improvement and reduction of computational cost, and they concern the reduction of the number of active agents and of the gravitational constant using a sub-unitary factor. These modifications are expected to work and produce improved results in optimization problems which involve other models as well [15–21].

This chapter will show the improvement of the recognition accuracy offered by the proposed training algorithm compared to that ensured by the classical approach. This represents the main contribution with respect to the state-of-the-art.

The chapter is organized as follows: the detailed architecture of the CNN is presented in Sec. 23.2 and accompanied by detailed mathematical models and figures to depict the connections between different hidden layers; the classical and the new training algorithms are presented and backed up by mathematical justifications in Sec. 23.3; the modifications of the GSAs are pointed out; the case study presented in Sec. 23.4 includes an extended set of results and discussion; the conclusions are given in Sec. 23.5.

23.2 The architecture of the CNN

The LeNet 1 [14] architecture is applied in this chapter as the first CNN proposed by LeCun to demonstrate that complex neural networks can be used in order to recognize handwritten series of digits. Further on, LeCun proposed other architecture for CNNs: LeNet 4, Boosted LeNet 4 and LeNet 5 [22]. LeNet 1 is applied as follows because it has the smallest number of free parameters that need to be trained, so it represents the best candidate to use the GSA algorithm in order to get the global optimum after training.

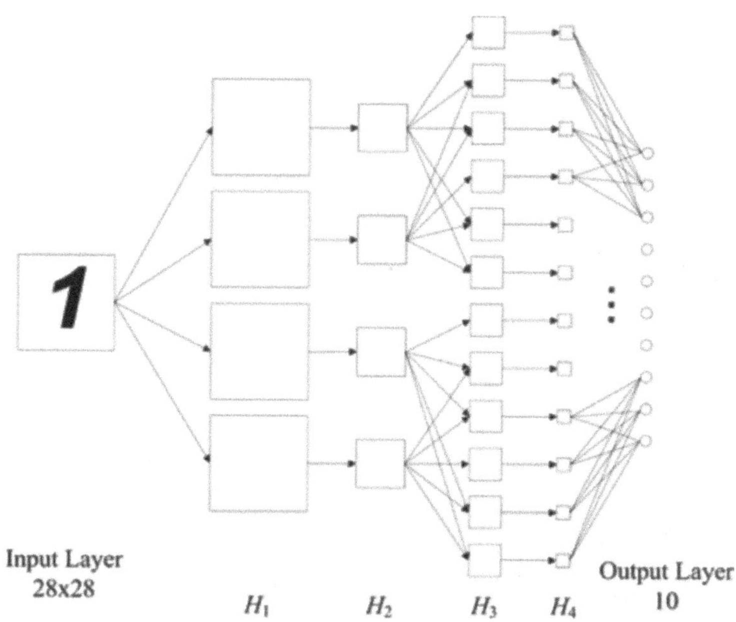

Fig. 23.1 The architecture of the convolutional neural network

The architecture of the CNN is depicted graphically in Fig. 23.1. The CNN is formed of four hidden layers, two convolutional layers, H_1 and H_3 and two averaging and sub-sampling layers, H_2 and H_4. The role of the convolutional layers is to extract features, like orientated edges, corners or end-points from the previous layer. Each convolutional layer is formed from a number of feature maps within each neuron takes its synaptic inputs from a local receptive field, called a convolutional kernel, from the previous layer. The role of the averaging and sub-sampling layers is to reduce the precision of the feature map. This will make the CNN invariant to slight distortions and/or translations that can occur in the input layer [10].

The input layer is formed of 28×28 neurons and receives the image of a character that is previously centered, normalized and converted to binary format. The value of the input layer belongs to the $[-1,1]$ interval.

The first hidden layer, H_1, is a convolutional layer with four feature maps, each feature map consisting of 24×24 neurons. Each neuron in each feature map is a 5×5 convolutional kernel of the input layer. The convolutional kernel is moved by one position for each neuron from the feature map resulting in 24 positions where the kernel will fit in each row and 24 positions where the kernel will fit in each column. Therefore, layer H_1 will consist of: $24 \times 24 \times 4 = 2034$ neurons, $(5 \times 5 + 1) \times 4 = 104$ weights and $(25 + 1) \times 2304 = 59904$ connections. The connections are depicted graphically in Fig. 23.2.

If the following notations are considered:

- $x_{H_1, f_k}^{i,j}$ represents the output of the neuron from H_1 in the f_k feature map at coordinates (i, j), where $k = 1, \ldots, 4$ and i, $j = 1, \ldots, 24$,
- $x_I^{i,j}$ represents the input value from (i, j) coordinates where i, $j = 1, \ldots, 28$,
- w_{H_1, f_k}^{l} represents the weight of the l connection from f_k feature map in H_1,
- F represents the activation function of the neuron,

the mathematical model that describes the connection of H_1 to the input layer is:

$$x_{H_1, f_k}^{i,j} = F\left(\sum_{u=1}^{5}\sum_{v=1}^{5}\left(w_{H_1, f_k}^{u+v} \cdot x_I^{i+(u \bmod 5), j+(v \bmod 5)}\right) + w_{H_1, f_k}^{26}\right). \tag{23.1}$$

The second hidden layer, H_2, is an averaging and sub-sampling layer consisting of four maps. Each map connects to a single feature map, as presented in Fig. 23.1 and is formed from 12×12 neurons. Each neuron takes its synaptic inputs from a receptive field of size 2×2. The receptive field is moved for each neuron so that it will not overlap, as presented in Fig. 23.3. The neurons in a map are constrained to share the same set of weights; moreover,

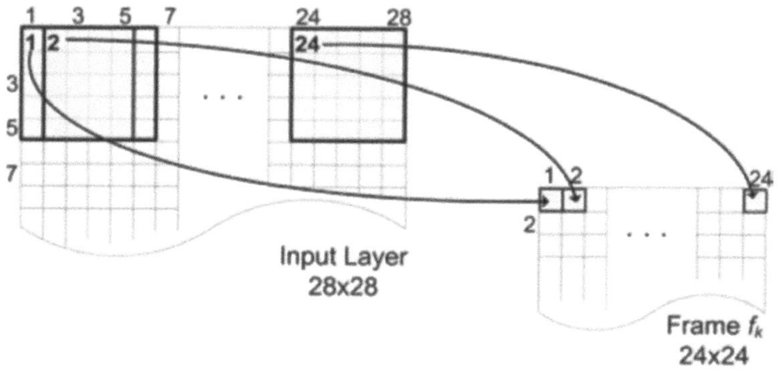

Fig. 23.2 Detailed representation of the synapses between input layer and H_1

all input connections of a neuron will share the same weight. Therefore, layer H_2 will consist of: $12 \times 12 \times 4 = 576$ neurons, $(1+1) \times 4 = 8$ weights and $(2 \times 2+1) \times 576 = 2880$ connections.

Considering the notations introduced before and:

- $x^{i,j}_{H_2,m_k}$ represents the output of the neuron from H_2 in the m_k averaging map at coordinates (i, j), where $k = 1,\ldots,4$ and $i, j = 1,\ldots,12$,
- $w^1_{H_2,m_k}$ represents the shared weight of the input synapses and $w^2_{H_2,m_k}$ represents the shared weight of the bias,

the mathematical model that describes the connection of H_2 to H_1 is:

$$x^{i,j}_{H_2,m_k} = F\left(\sum_{u=1}^{2}\sum_{v=1}^{2}\left(w^1_{H_2,m_k} \cdot x^{i+(u \bmod 2),j+(v \bmod 2)}_{H_1,f_k}\right) + w^2_{H_2,m_k}\right). \tag{23.2}$$

The third hidden layer, H_3, is a convolutional layer with 12 feature maps, each feature map consisting of 8×8 neurons. Each neuron takes its synaptic inputs from one or two 5×5 receptive fields that are at identical positions within each map in layer H_2. The receptive fields are moved with one position for each neuron, similar to the connections between H_1 and the input layer. Table 23.1 describes the connection between feature maps from H_3 and averaging maps from H_2. Layer H_3 consists of: $8 \times 8 \times 12 = 768$ neurons, $(5 \times 5+1) \times 4 + (2 \times 5 \times 5+1) \times 8 = 512$ weights and $768 \times 512 = 393216$ connections.

Besides the notations introduced before the following notations are used:

- $x^{i,j}_{H_3,f_k}$ represents the output of the neuron from H_3 in the f_k feature map at coordinates (i, j), where $k = 1,\ldots,12$ and $i, j = 1,\ldots,8$,

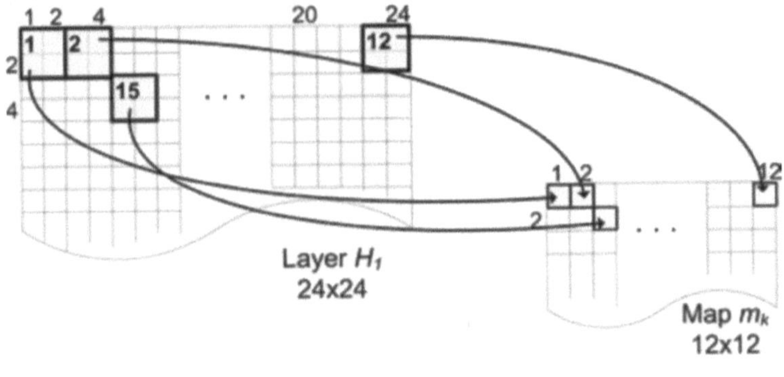

Fig. 23.3 Detailed representation of the synapses between H_1 and H_2

Table 23.1 Connections between averaging maps in H_2 to feature maps in H_3

c_{m_k,f_k}	f_1	f_2	f_3	f_4	f_5	f_6	f_7	f_8	f_9	f_{10}	f_{11}	f_{12}
m_1	1	1	1	0	1	1	0	0	0	0	0	0
m_2	0	1	1	1	1	1	0	0	0	0	0	0
m_3	0	0	0	0	0	0	1	1	1	0	1	1
m_4	0	0	0	0	0	0	0	1	1	1	1	1

- $w^l_{H_3,f_k}$ represents the weight of the l connection of a neuron from f_k feature map in H_3,
- c_{m_k,f_k} is a coefficient (shown in Table 23.1) which indicates the connections between the feature maps f_k from layer H_3 to averaging maps m_k from layer H_2.

In this context the mathematical model that describes the connections between the neurons from H_3 to H_2 is:

$$x^{i,j}_{H_3,f_k} = F\left(\sum_{m_k}^{4} c_{m_k,f_k}\left[\sum_{u=1}^{5}\sum_{v=1}^{5}\left(w^{(u+v)(m_k \bmod 2)}_{H_3,f_k},\right.\right.\right.$$

$$\left.\left.\left.\cdot\, x^{i+(u \bmod 5),\,j+(v \bmod 5)}_{H_2,m_k}\right)\right] + w^{1+25\sum^{4}_{m_k} c_{m_k,f_k}}_{H_3,f_k}\right). \qquad (23.3)$$

The last hidden layer, H_4, is an averaging and sub-sampling layer, similar to H_2. It consists of 12 averaging maps, each map connected to the corresponding feature map from the previous layer. An averaging map is formed from 4×4 neurons, where each neuron will have its receptive input a 2×2 kernel. All the neurons from an averaging map and all the input synapses of a neuron except for the bias are constrained to share the same weight.

Therefore, the layer H_4 will have: $4\times4\times12=192$ neurons, $(1+1)\times12=24$ weights and $(2\times2+1)\times192=960$ connections.

Introducing the additional notations:

- $x^{i,j}_{H4,m_k}$ represents the output of the neuron from H_4 in the m_k averaging map at coordinates (i,j), where $k=1,\dots,12$ and $i,j=1,\dots,4$,
- $w^1_{H_4,m_k}$ represents the shared weight of the input synapses and $w^2_{H_4,m_k}$ represents the shared weight of the bias,

the mathematical model that describes the connection from H_3 to H_4 is:

$$x^{i,j}_{H_4,m_k}=F\left(\sum_{u=1}^{2}\sum_{v=1}^{2}\left(w^1_{H_4,m_k}\cdot x^{i+(u\bmod 2),j+(v\bmod 2)}_{H_3,f_k}\right)+w^2_{H_4,m_k}\right). \tag{23.4}$$

The output layer is a fully connected layer and contains 10 neurons, each neuron representing a digit recognized by the neural network. Therefore, the output layer consists of: 10 neurons, $(192+1)\times10=1930$ weights and connections. Using the notations:

- x^i_O represents the output of the ith neuron from the output layer,
- $w^{i,j}_O$ represents the weight that the ith neuron from the output layer applies to the output of the jth neuron from the previous layer,
- C_n represents the number of neurons from the output layer, $C_n=192$,

the mathematical model that describes the connections of the output layer with H_4 is:

$$x^i_O=F\left(\sum_{j=1}^{C_n}w^{i,j}_O\cdot x^j_{H_4}+w^{i,C_n+1}_O\right). \tag{23.5}$$

The activation function F is selected as a scaled version of the hyperbolic tangent because it is symmetric and its derivative can be calculated easily:

$$x=F(y)=\tanh(y)=\frac{\sinh(y)}{\cosh(y)}, \tag{23.6}$$

$$\frac{dF}{dy}=\frac{d}{dy}\left(\frac{\sinh(y)}{\cosh(y)}\right)=\frac{\cosh^2(y)-\sinh^2(y).}{\cosh^2)}=1-\tanh^2(y)=1-x^2 \tag{23.7}$$

The function is scaled to belong to the $[-1.7,1.7]$ interval in order to allow the neural network to be trained in the $[-1,1]$ interval and to avoid saturating the output units [11].

23.3 Training algorithms

The training of the CNNs consists of getting the parameters of the CNNs which belong to the vector $\boldsymbol{\rho}$:

$$\boldsymbol{\rho}=[\rho_1\ \rho_2\ \dots\ \rho_q]^T, \tag{23.8}$$

with q – the number of trained parameters and T – matrix transposition, as solutions to the optimization problem:

$$\boldsymbol{\rho}^* = \arg\min_{\rho \in D_\rho} E(\boldsymbol{\rho}), \tag{23.9}$$

where $\boldsymbol{\rho}^*$ is the optimal value of the vector $\boldsymbol{\rho}$, and $D_{\boldsymbol{\rho}} \in \mathbb{R}^q$ is the feasible domain of $\boldsymbol{\rho}$. Accepting c classes of the classifier and M training examples, the expression of the objective function $E(\boldsymbol{\rho})$, whose minimization leads to trained CNNs, is [10]:

$$E(\boldsymbol{\rho}) = \frac{1}{M}\sum_{n=1}^{M}\sum_{j=1}^{c}(t_j^n - y_j^n(\boldsymbol{\rho}))^2, \tag{23.10}$$

where t_j^n is j^{th} dimension of the target that corresponds to n^{th} input pattern, and y_i^n is the value of jth output layer unit obtained as the response of the CNN to the nth input pattern.

23.3.1 *Back-propagation algorithm*

The back-propagation algorithm is a supervised training algorithm based on the error-correction learning rule [23] and widely used to train CNNs [10, 11, 14, 24].

The BP algorithm consists of two passes through the different layers of the network: a forward pass and a backward pass. In the forward pass a pattern is presented to the neural network and its effect propagates through the network layer by layer until it reaches the output layer. Finally, a set of outputs is produced as the actual response of the neural network. During the forward pass the synaptic weights of the network are all fixed. During the backward pass the synaptic weights are all adjusted in accordance with an error-correction rule. Specifically, the actual response of the network is subtracted from a desired response to produce the error signal. The error signal is then propagated backward through the network against the direction of the synaptic connection and the weights are adjusted to make the actual response of the network move closer to the desired response.

Considering the following notations:

- E_n^P is the error due to single pattern P at the last layer n,
- T_n^i is the desired output at the last layer,
- x_n^i is the actual value of the output at the last layer,
- F is the activation function,

the implementation of the BP algorithm consists of the following steps. The first step calculates the error due to a single pattern presented to the input of the neural network as follows:

$$E_n^P = \frac{1}{2}\sum(x_n^i - T_n^i)^2. \tag{23.11}$$

The partial derivative of Eq. (23.11) gives the starting value for the BP process. The numerical values of the derivatives are obtained in terms of:

$$\frac{\partial E_n^p}{\partial x_n^i} = x_n^i - T_n^i. \tag{23.12}$$

Using the numeric values of the derivatives, the delta change of the weights are calculated using Eqs. (23.13) and (23.14):

$$\frac{\partial E_n^p}{\partial y_n^i} = F'(x_n^i)\frac{\partial E_n^p}{\partial x_n^i}, \tag{23.13}$$

$$\frac{\partial E_n^p}{\partial w_n^{ij}} = x_{n-1}^j \frac{\partial E_n^p}{\partial y_n^i}. \tag{23.14}$$

Using Eqs. (23.12) and (23.13), the derivative of the error for the previous layer is calculated in terms of:

$$\frac{\partial E_{n-1}^p}{\partial x_{n-1}^k} = \sum_i w_n^{ij} \frac{\partial E_n^p}{\partial y_n^i}. \tag{23.15}$$

The values obtained with Eq. (23.15) are used as starting values for the calculations on the previous layer. In other words, the values obtained are used in a repetition of Eqs. (23.13), (23.14) and (23.15) for the immediately preceding layer until the first layer is reached.

The values of the weights in layer n are changed using Eq. (23.14) and:

$$\left(w_n^{ij}\right)_{\text{new}} = \left(w_n^{ij}\right)_{\text{old}} - eta \cdot \left(\frac{\partial E_n^p}{\partial w_n^{ij}}\right), \tag{23.16}$$

where $eta > 0$ represents the learning rate. The learning rate is chosen to be as large as possible without leading to oscillations, and it is decreased during the training process.

23.3.2 *Gravitational Search Algorithm*

GSA is a heuristic algorithm inspired by the Newtonian laws of gravity and mass interactions [25–29]. In GSA, agents are considered as objects and their performance is measured by masses. All these objects attract each other by the gravity force, and this force causes a global movement of all objects towards the ones with heavier masses. Each agent has four specifications: position, inertial mass, active gravitational mass, and passive gravitational mass. The position of the agent corresponds to a solution of the problem, and its gravitational and inertial masses are determined using an objective function.

Considering N agents and a q-dimensional search space, the position of ith agent is defined by the vector $\mathbf{X}_i$:

$$\mathbf{X}_i = [x_i^1 \ldots x_i^d \ldots x_i^q]^T, \quad i = 1, \ldots, N \tag{23.17}$$

where x_i^d is the position of ith agent in d^{th} dimension, $d = 1, \ldots, q$.

To ensure the stochastic characteristic of the GSA the total force acting on i^{th} agent in the dth dimension is a randomly weighted sum of all forces exerted from the other agents:

$$F_i^d(k) = \sum_{\substack{j=1 \\ j \neq i}}^{N} (\rho_j F_{ij}^d(k)), \tag{23.18}$$

with ρ_j, $0 \leqslant \rho_j \leqslant 1$ – a random generated number, the force acting on i^{th} agent from j^{th} agent is:

$$F_{ij}^d(k) = G(k) \frac{M_{Pi}(k) M_{Aj}(k)}{R_{ij}(k) + \varepsilon} \left(x_j^d(k) - x_i^d(k) \right), \tag{23.19}$$

where $M_{Pi}(k)$ is the active gravitational mass of the ith agent, $M_{Aj}(k)$ is the passive gravitational mass of the jth agent, ε is a small constant and $R_{ij}(k)$ is the Euclidean distance between i^{th} agent and j^{th} agent.

$$R_{ij}(k) = \left\| X_i(k) - X_j(k) \right\|. \tag{23.20}$$

The law of motion leads to the acceleration $a_i^d(k)$ of the ith agent at iteration k in d^{th} dimension:

$$a_i^d(k) = \frac{F_i^d(k)}{M_{Ii}(k)}, \tag{23.21}$$

where $M_{Ii}(k)$ is the inertial mass of the ith agent. The next velocity, $v_i^d(k+1)$ and the next position, $x_i^d(k+1)$ of an agent are obtained in terms of the state-space equations [25, 26]:

$$v_i^d(k+1) = \rho_i \cdot v_i^d(k) + a_i^d(k), \tag{23.22}$$

$$x_i^d(k+1) = x_i^d(k) + v_i^d(k+1), \tag{23.23}$$

where ρ_j, $0 \leqslant \rho_j \leqslant 1$, is a uniform random generated number.

The gravitational constant G is initialized at the beginning, and it will be reduced with time to control the search accuracy. Eq. (23.24) expresses the depreciation of the gravitational constant with the advance of GSA's iterations [28]:

$$G(k) = G_0 \cdot e^{-\frac{\zeta \cdot k}{k_{\max}}}. \tag{23.24}$$

The gravitational and inertial masses are calculated using the fitness evaluation. A heavier mass indicates a more efficient agent. In other words, agents with higher masses exercise a higher force over the rest of the agents and move slower. Assuming that the gravitational and inertial masses are equal, the mass of an agent is given by:

$$M_{Ai} = M_{Pi} = M_{Ii} = M_i, \tag{23.25}$$

$$m_i = \frac{f_i(k) - b(k)}{w(k) - b(k)}, \tag{23.26}$$

$$M_i = \frac{m_i}{\sum_{j=1}^{N} m_j}, \tag{23.27}$$

where $f_i(k)$ is the fitness value of ith agent at the iteration index k, $b(k)$ represents the fitness value of the best agent and $w(k)$ represents the fitness value of the worst agent calculated as follows:

$$b(k) = \min_{j=1,\dots,N} f_j(k), \tag{23.28}$$

$$w(k) = \max_{j=1,\dots,N} f_j(k). \tag{23.29}$$

The relation between the fitness function f, the fitness value $f_i(k)$ in the GSA and the objective function in Eq. (23.10) is:

$$f_i(k) = E(\boldsymbol{\rho}), \quad i = 1, \dots, N. \tag{23.30}$$

The relation between agents' position vector $\mathbf{X}_i$ in the GSA and the parameter vector $\boldsymbol{\rho}$ of CNN is:

$$\mathbf{X}_i = \boldsymbol{\rho}, \quad i = 1, \dots, N. \tag{23.31}$$

To improve the performance of the GSA described above, the following changes are applied and adapted to our CNN [30]:

- Allow the configuration of the number of first best agents that will be used to update the acceleration of one agent.
- From Eqs. (23.19) and (23.21) the acceleration of ith agent in d^{th} dimension can be calculated as follows:

$$a_i^d(k) = \frac{F_i^d(k)}{M_{Ii}(k)} = \sum_{\substack{j=1 \\ j \neq i}}^{N_K(k)} \left(\rho_j G(k) \frac{M_j(k)}{R_{ij}(k) + \varepsilon} \left(x_j^d(k) - x_i^d(k) \right) \right), \tag{23.32}$$

  where $N_K(k)$ represents the number of first best agents that are taken into consideration.
- Calculate the Euclidian distance between i^{th} agent and j^{th} agent only once as follows:

$$R_{ij}(k) = R_{ji}(k) = \sqrt{\sum_{d=1}^{q} \left(x_j^d(k) - x_i^d(k) \right)^2}. \tag{23.33}$$

The value of $N_K(k)$ can be calculated using one of the following equations:

$$N_K(k) = N, \tag{23.34}$$

$$N_K(k) = K, \tag{23.35}$$

$$N_K(k) = N \cdot e^{-\frac{\zeta \cdot k}{k_{\max}}}, \tag{23.36}$$

where $k = 1,\ldots,k_{\max}$, K represents a configurable constant, and ζ is a constant that determines how fast the number of first best agents decreases with the advance of GSA's iterations.

The implementation of our GSA which trains the CNN consists of the following steps:

(i) Initialize the q-dimensional search space, the number of N agents, randomly generate the agents' position vector $\mathbf{X}_i$ and determine the number of agents that will be used in Eq. (23.32) using one of the following Eqs. (23.34), (23.35) or (23.36). The q-dimensional search space consists of the trainable parameters of the CNN, the synaptic weights.

(ii) Evaluate the agents' fitness using Eqs. (23.10) and (23.30). This step is related to a forward propagation of the training set to all agents.

(iii) Update $G(k)$, $b(k)$, $w(k)$ and $M_i(k)$ using Eqs. (23.24), (23.28), (23.29) and (23.27) for all agents, $i = 1,\ldots,N$.

(iv) Calculate the acceleration in different directions using Eq. (23.32).

(v) Update agents' velocities and positions using Eqs. (23.22) and (23.23).

(vi) Increment k and go to step (ii) until the maximum number of iterations is reached, i.e. $k = k_{\max}$.

23.4 Case study and discussion of results

The benchmark used to train and test our CNN and training algorithm contains printed digits extracted from black and white scanned documents at 300 DPI. The benchmark is created from documents selected such that to contain both fixed size font types and variable size font types with the size of text ranges from 8 to 40 pixels. The benchmark is divided into two sets, one composed of around 3500 characters used in the training phase and the other one composed of around 7000 characters used to test the CNN accuracy.

The image of the characters from the training and testing set needs to be first normalized to a size of 28×28 pixels, which represents the size of CNN's input layer. The image is scaled to a size of 26×26 and centered on the final 28×28 image to avoid problems when a kernel overlaps a boundary [14].

To improve the CNN's generalization capability, the following techniques were used in the training process: the training set was passed to the CNN in a randomized order for each epoch and distortions were applied to the digits [12]. The first technique forces the CNN to not settle on weights that emphasize attributes from a single digit and causes the weights to change for each input pattern. The second technique was used to extend the training set by applying the following type of distortions to the input digit: elastic, scale and rotation. The three types of distortions are generated by applying affine displacement fields to the image of the digit. This is performed by calculating for every pixel a new target location with respect to the original location. For example, if $\Delta x(x,y) = 1$ and $\Delta y(x,y) = 0$, the new location of every pixel is shifted by 1 to the right. For printed digits distortions helped simulating low quality documents combined with text printed using small font sizes. The following coefficients were used: elastic factor, $e_f = 0.5$, scale factor, $s_f = 2$, rotation factor, $r_f = 5$ and severity factor, $S_f = 0.5$. The values of the coefficients were determined experimental. If the values are too high, the CNN is trained to recognize only highly degraded images, while if the values are too small the generalization capability of the CNN does not improve.

The comparison between the classical BP-based trained CNN and the proposed GSA-BP trained CNN uses the character recognition accuracy quality index defined as:

$$A_p = \frac{D_c}{D_t} 100\,[\%], \tag{23.37}$$

where D_c is the number of digits that were recognized correctly, and D_t is the total number of digits contained in the testing set.

The case study follows two scenarios:

- The first scenario presents the results obtained by using only BP to train the CNN.
- The second scenario focuses on the proposed approach which consists of two stages: course training by means of GSA and fine tune by means of BP.

The first scenario starts with a CNN for which the synaptic weights are initialized randomly in the $[-0.05, 0.05]$ interval. The algorithm is run for 20 epochs starting with a learning rate of 0.005. An epoch represents the backward propagation and weight adjustment for all the digits from the training set. The learning rate is adjusted after each epoch in accordance with:

$$eta(k+1) = 0.99\, eta(k). \tag{23.38}$$

The evolution of the objective function $E(\boldsymbol{\rho})$ and of the accuracy percentage A_p after each epoch is presented in Figures 23.4 and 23.5.

The second scenario of the case study starts by training the CNNs using GSA and fine tune the best agent by means of BP. The first stage is applied until the objective function $E(\boldsymbol{\rho})$ reaches a preset threshold value. Due to the computational power required by the GSA when used with a large number of agents the training set was filtered out so that it would contain only artificially rendered digits with high quality. The number of digits in the training set was reduced to 600. The number of agents employed in CNN training was also relative small in comparison with the number of weights contained by a CNN due to the limitations in computational power and ranges from 100 to 250.

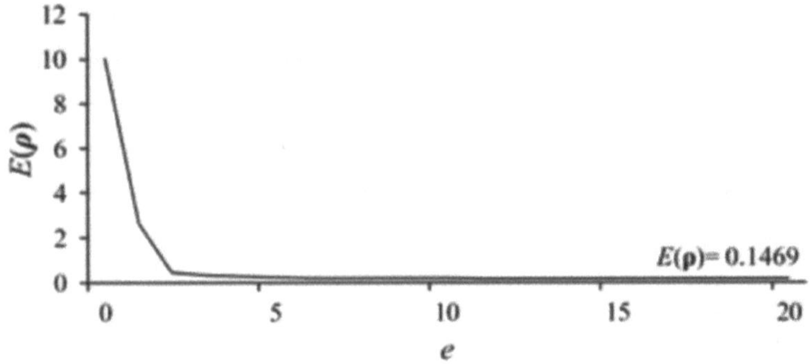

Fig. 23.4 Objective function $E(\boldsymbol{\rho})$ versus epoch number obtained by the CNN when trained with BP

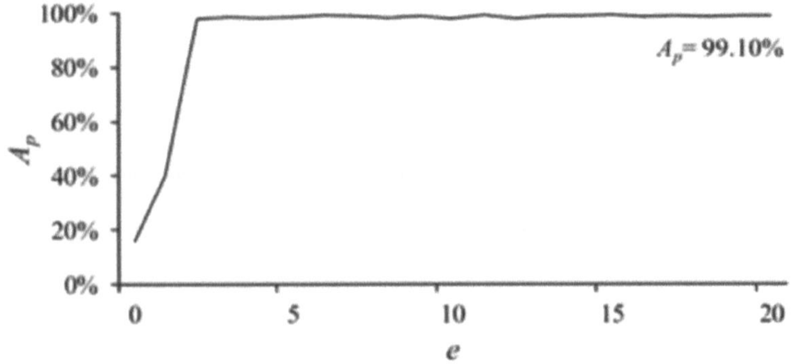

Fig. 23.5 Accuracy percentage A_p versus epoch number obtained by the CNN when trained with BP

The first results were obtained using Eq. (23.34) to calculate the number of active agents that are taken into consideration when calculating the next position of an agent. The following values were chosen for the GSA parameters: $N = 100$, $G_0 = 1$, $\varepsilon = 0.001$, $k_{\max} = 50$, the initial velocity was randomly generated such that to belong to the $[-0.05, 0.05]$ interval.

Considering the average fitness value defined as:

$$f_{\text{avg}}(k) = \frac{\sum_{i=1}^{N} f_i(k)}{N}, \tag{23.39}$$

Fig. 23.6 presents the value of the best agent after each iteration and Fig. 23.7 presents the value of the worst agent and the average fitness.

The best fitness value reaches only 6.89, which does not represent a good candidate for the fine tuning process in the second phase. These results are caused by the small number of agents compared to the number of trainable parameters of the GSA; in addition, the heuristic nature of the GSA algorithm does not take into account the information contained by the patterns that are presented at CNN input; since all agents actively apply forces to each other, it is shown that the value of f_{avg} increases versus the iteration index. This indicates that the agents with relatively low fitness value cannot compensate for the ones with high fitness value.

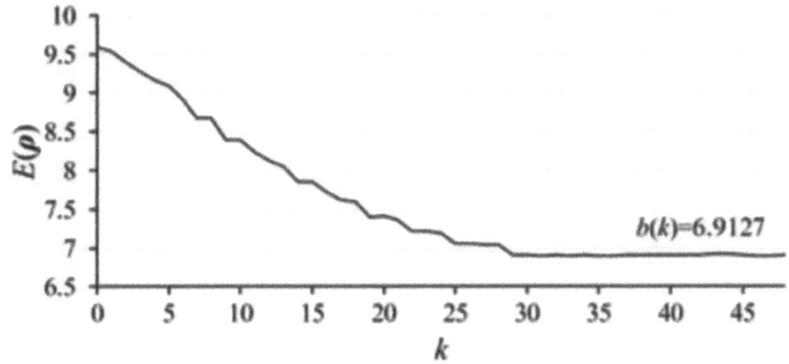

Fig. 23.6 Fitness value of best agent $b(k)$ versus iteration index obtained by the CNN when trained with GSA and the number of active agents is calculated using Eq. (23.34)

The second set of results, presented in Figs. 23.8, 23.9 and 23.10, was obtained by calculating the number of active agents using Eq. (23.35). The parameters of the GSA were chosen as follows: $N = 200$, $G_0 = 10$, $\varepsilon = 0.00001$, $k_{\max} = 100$, the initial velocity was randomly generated such that to belong to the $[-0.1, 0.1]$ interval, the number of active agents $K = N/2 = 100$ and the gravitational constant decrease factor $\zeta = 5$.

As shown in Fig. 23.8, the fitness value of the best agent decreases in the first 35 iterations while the gravitational constant is still high. This is due to the higher initial

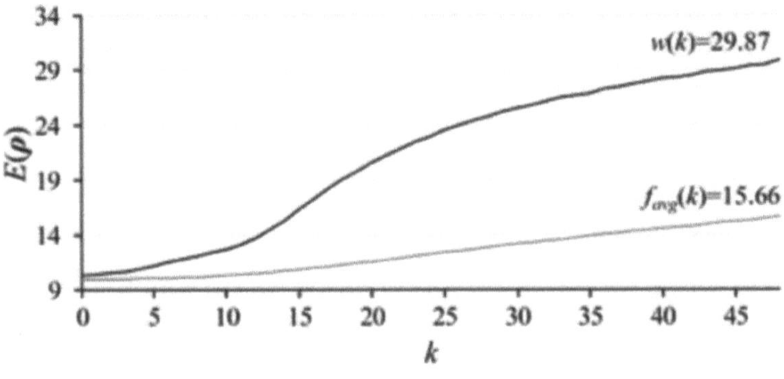

Fig. 23.7 Fitness value of worst agent $w(k)$ and average fitness $f_{avg}(k)$ versus iteration index obtained by the CNN when trained with GSA and the number of active agents is calculated using Eq. (23.34)

velocity and also due to the fact that only the first half agents are actively applying forces to the other agents. Fig. 23.9 also enforces this statement by depicting graphically the average fitness value after each iteration which settles to an average value of 4.3. The fitness value of the worst agent varies a lot due to the heuristic nature of the algorithm where the value in one dimension can accelerate and pass the desired value.

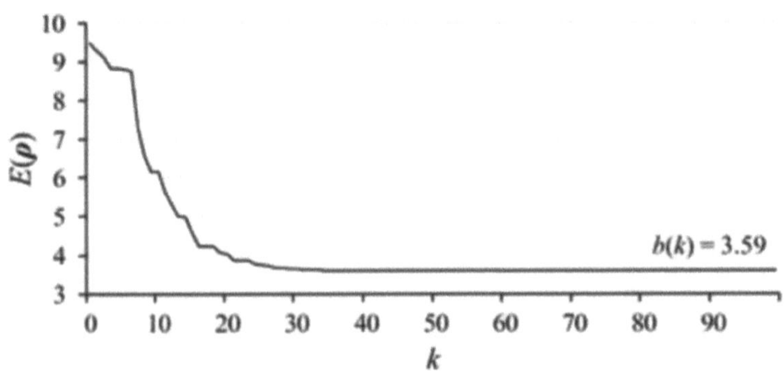

Fig. 23.8 Fitness value of best agent $b(k)$ versus iteration index obtained by the CNN when trained with GSA and the number of active agents is calculated using Eq. (23.35)

Figure 23.10 presents the evolution of the gravitational constant G during the k_{max} iterations. The best agent that will be used in phase two was obtained during this second set of results. Even though the fitness value of the agent is relatively high and the CNN cannot be used in the recognition process at this stage, it should have been settled near the

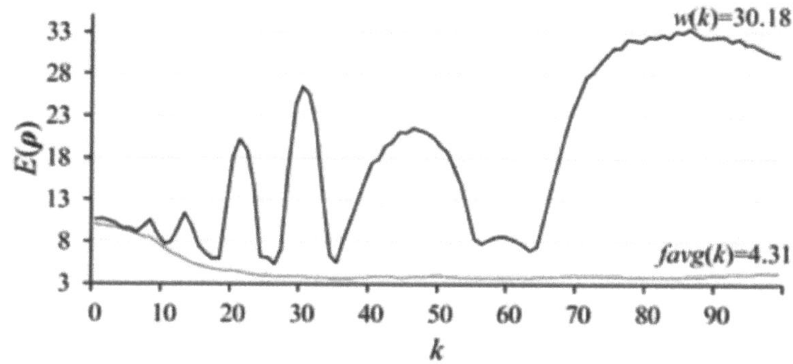

Fig. 23.9 Fitness value of worst agent $w(k)$ and average fitness $f_{avg}(k)$ versus iteration index obtained by the CNN when trained with GSA and the number of active agents is calculated using Eq. (23.35)

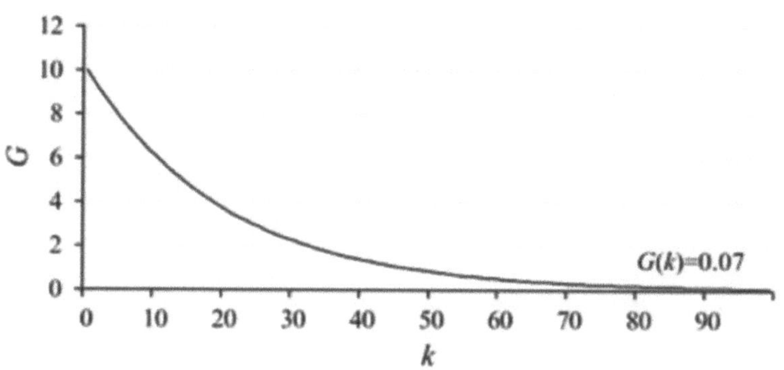

Fig. 23.10 Gravitational constant $G(k)$ versus iteration index when the CNN is trained with GSA and the number of active agents is calculated using Eq. (23.35)

global minimum and by applying the BP algorithm it can be fine-tuned to converge towards it.

The third set of results is obtained using Eq. (23.36) for calculating the number of agents that actively apply forces. The GSA is configured using the following parameters: $N = 250$, $G_0 = 20$, $\varepsilon = 0.00001$, $k_{\max} = 100$, the initial velocity was randomly generated such that to belong to the $[-0.25, 0.25]$ interval, the number of active agents and the gravitational constant are reduced using a factor $\zeta = 5$.

The fitness value of the best agent converges to approximately the same value as in the previous set of results as presented in Fig. 23.11. It does not converge as fast as in the previous case because the number of active agents is higher at the beginning allowing agents with high fitness value to contribute to the computation of the acceleration of other agents.

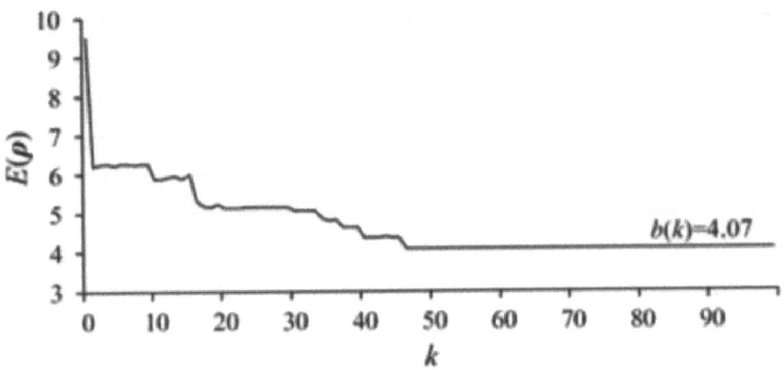

Fig. 23.11 Fitness value of best agent $b(k)$ versus iteration index obtained by the CNN when trained with GSA and the number of active agents is calculated using Eq. (23.36)

Even though the number of active agents decreases with the iteration index, so does the gravitational constant and the forces applied by the remaining active agents are not high enough to attract the agents with high fitness values as depicted in Fig. 23.12.

Figures 23.13 and 23.14 represent graphically the evolution of the gravitational constant, G, and of the number of active agents, N_K, during the $k_{\max}$ iterations performed with the GSA.

The relatively small number of agents used in the GSA training is due to the exponential increase of the cost as depicted graphically in Fig. 23.15. The cost is measured in time required to run a single iteration for the GSA algorithm. Increasing the training set will also reflect in the time required to run the algorithm, but this applies to both BP and GSA, and it will determine only a linear increase due to the time required to calculate the objective function.

The agent selected to perform the second step, viz. the fine tuning of the CNN with BP, is the best agent obtained in the second set. The parameters of the BP are selected to be the same as in the first scenario, $eta = 0.005$, decreasing after each epoch by Eq. (23.38), and the same distortions are applied to the digits from the training set. The initial and complete training and testing sets are used in the second stage training. The final results, obtained after fine tuning the CNN resulted from the best agent for 20 epochs with BP, are presented in Fig. 23.16 and Fig. 23.17.

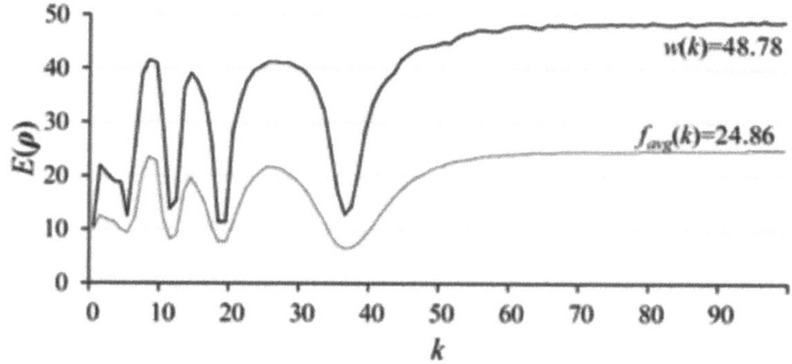

Fig. 23.12 Fitness value of worst agent $w(k)$ and average fitness $f_{\text{avg}}(k)$ versus iteration index obtained by the CNN when trained with GSA and the number of active agents is calculated using Eq. (23.36)

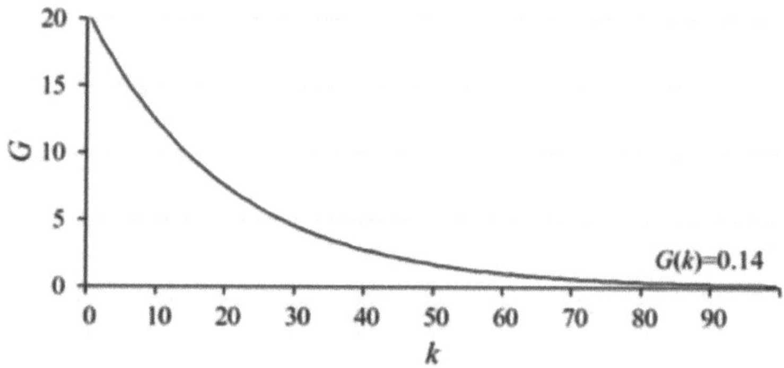

Fig. 23.13 Gravitational constant $G(k)$ versus iteration index obtained by the CNN when trained with GSA and the number of active agents is calculated using Eq. (23.36)

Figure 23.16 represents the evolution of the objective function $E(\boldsymbol{\rho})$, which decreases to a minimum of 0.1330 compared to the minimum value obtained using only BP, 0.1460. While $E(\boldsymbol{\rho})$ settled after 10 epoch when only BP is used, Fig. 23.16 shows that using our new algorithm the value of $E(\boldsymbol{\rho})$ is still converging indicating that the CNN avoids trapping in a local minimum, which is the main problem when training a neural network. The parameters for BP were chosen the same for the two scenarios to clearly illustrate the benefit of using the GSA for training the CNN before fine tuning the network.

Figure 23.17 depicts graphically the evolution of the accuracy percentage, A_p obtained on the testing set after each epoch of fine-tuning with BP the CNN produced by the best

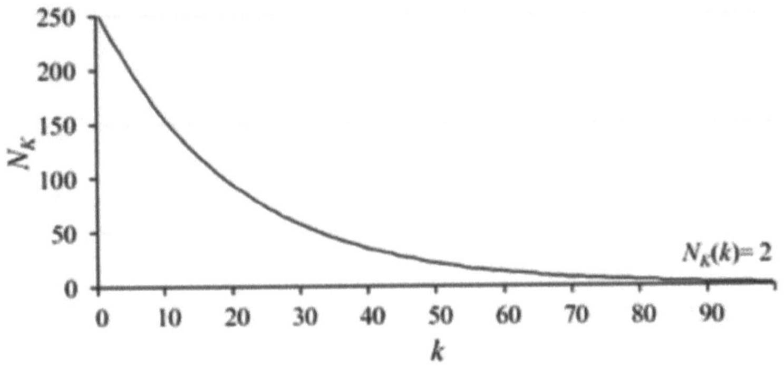

Fig. 23.14 Number of agents that apply forces $N_K(k)$ versus iteration index obtained by the CNN when trained with GSA and the number of active agents is calculated using Eq. (23.36)

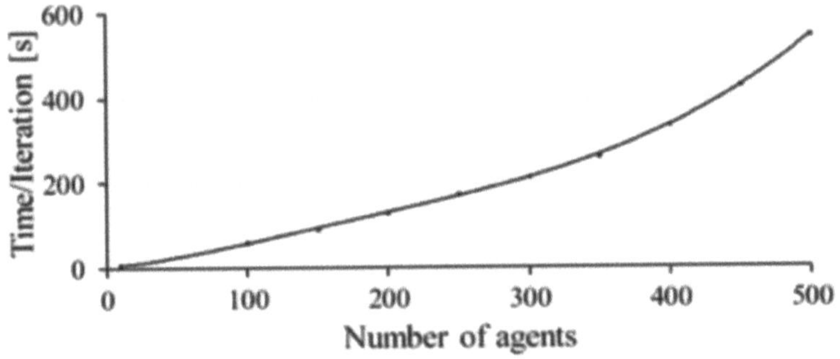

Fig. 23.15 Computational cost versus the number of active agents for the GSA training algorithm

agent. A comparison of A_p obtained using only BP and GSA-BP is presented in Fig. 23.18. It shows that the CNN trained with GSA-BP obtains higher accuracy percentage by at least 0.5% with less variation caused by distortions applied to the input patterns.

Applying the BP for more than 20 epochs may result in a CNN with a better A_p especially in the second scenario, where it was still converging, but it may also result in over-training the CNN which will reduce the generalization capability of the CNN on real-world data. The results clearly indicate that by using a heuristic algorithm as a first step in the training process, the CNNs can be forced to avoid settling to a local minimum and to converge to the global minimum when the classical BP algorithm is involved.

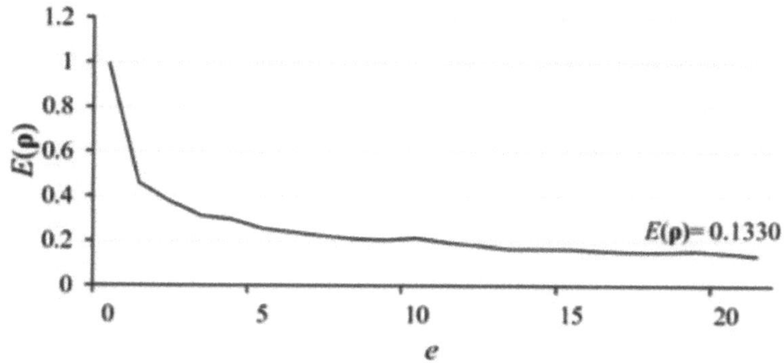

Fig. 23.16 Objective function $E(\boldsymbol{\rho})$ versus epoch number for CNN trained with GSA and fine-tuned with BP

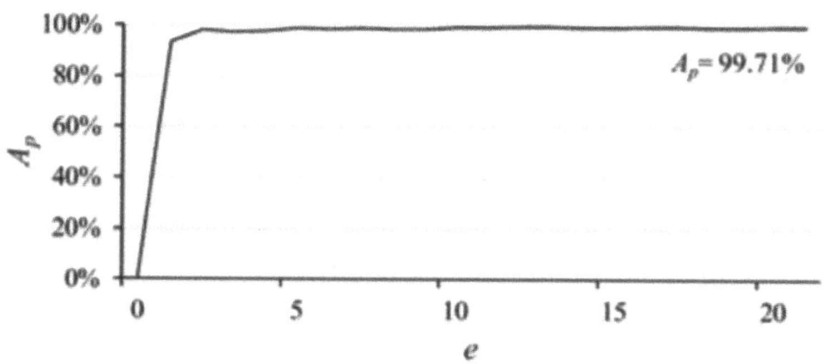

Fig. 23.17 Accuracy percentage A_p versus epoch number for CNN trained with GSA and fine-tuned with BP

23.5 Conclusions

This chapter has focused on the recognition module of an OCR application, and proposed a novel approach on training CNNs for character recognition based on the combination of the classical BP and of GSAs. An improved GSA has been suggested in this context.

The results obtained by the proposed algorithm clearly show that the trained CNN offers better recognition accuracy with respect to that obtained by using only the classical backpropagation algorithm.

Even though the results shows the superiority of the new algorithm, there are some limitations imposed by the computational power required by the GSA algorithm when used

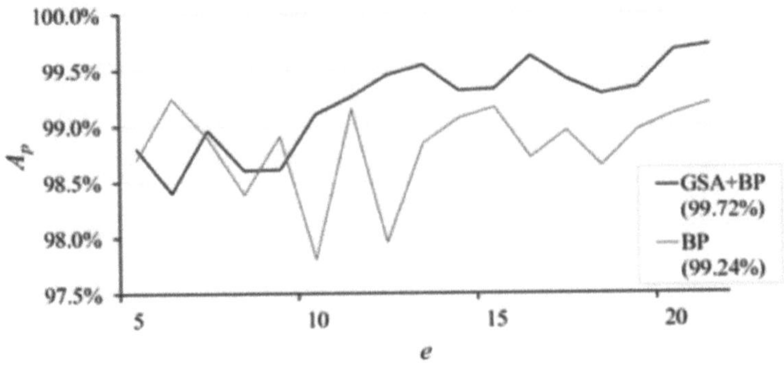

Fig. 23.18 Comparison between A_p obtained on by the CNN on the testing set when trained using BP and GSA with BP

on problems with a high number of free parameters and large sets of patterns, like CNNs. Our approach to use GSAs in CNN training has required a reduced number of patterns in the training set and also a reduced number of agents so that the algorithm would run in a reasonable amount of time. To take advantage of the full capability of the algorithm, the desired number of agents would have been at least the number of free parameters.

The results presented by the case study are very important because they validate the proposed algorithm, and they represent a direction of future process modeling applications that are not restricted to OCR applications or to a specific architecture of the neural network [31–35].

Acknowledgments

This work was supported by a grant of the Romanian National Authority for Scientific Research, CNCS – UEFISCDI, project number PN-II-ID-PCE-2011-3-0109.

Bibliography

[1] S.V. Rice, F.R. Jenkins, and T.A. Nartker. *The fifth annual test of OCR accuracy*, Tech. Rep., Information Science Research Institute, University of Nevada, Las Vegas, NV, USA (1996).
[2] V.N. Vapnik. An overview of statistical learning theory, *IEEE Trans. Neural Netw.*, **10**, pp. 988–999 (1999).
[3] O.R. Duda, P.E. Hart, and D.G. Storck. *Pattern Classification*, 2nd Edition, John Wiley & Sons, New York (2001).
[4] S.O. Belkasim, M. Shridhar, and M. Ahmadi. Pattern recognition with moment invariants: A comparative study and new results, *Pattern Recognit.*, **14**, pp. 1117–1138 (1991).

[5] J. Flusser and T. Suk. Pattern recognition by affine moment invariants, *Pattern Recognit.*, **26**, pp. 167–174 (1993).

[6] J. Flusser and T. Suk. Affine moment invariants: a new tool for character recognition, *Pattern Recognit. Lett.*, **15**, pp. 433–436 (1994).

[7] T. Zhang, X. Wang, C. Chen, and J. Liu. Connected numeral strings segmentation based on the combination of characteristic position and contour detecting, *Proc. 2009 International Conference on Digital Image Processing* (*ICDIP 2009*), Bangkok, Thailand, pp. 81–84 (2009).

[8] K. Arbeter, W.E. Snyder, H. Burkhardt, and G. Hirzinger. Application of affine-invariant Fourier descriptors to recognition of 3-D objects, *IEEE Trans. Pattern Anal. Mach. Intell.*, **12**, pp. 640–647 (1990).

[9] D. Zhang and G. Lu. A comparative study of three region shape descriptors, *Proc. 6th International Conference on Digital Image Computing Techniques and Applications* (*DICTA 2002*), Melbourne, Australia, pp. 21–22 (2002).

[10] L. Bottou, Y. LeCun, and Y. Bengio. Global training of document processing systems using graph transformer networks, *Proc. 1997 IEEE Computer Society Conference on Computer Vision and Pattern Recognition*, San Juan, Puerto Rico, pp. 489–494 (1997).

[11] Y. LeCun, L. Bottou, Y. Bengio, and P. Haffner. Gradient-based learning applied to document recognition, *Proc. IEEE*, **86**, pp. 2278–2324 (1998).

[12] P.Y. Simard, D. Steinkraus, and J. Platt. Best practices for convolutional neural networks applied to visual document analysis, *Proc 7th International Conference on Document Analysis and Recognition* (*ICDAR 2003*), Edinburgh, UK, pp. 958–962 (2003).

[13] L.-O. Fedorovici, E. Voişan, F. Drăgan, and D. Iercan. Improved neural network OCR based on preprocessed blob classes, *Proc. International Joint Conference on Computational Cybernetics and Technical Informatics* (*ICCC-CONTI 2010*), Timisoara, Romania, pp. 559–564 (2010).

[14] Y. LeCun, B. Boser, J.S. Denker, D. Henderson, R.E. Howard, W. Hubbard, and L.D. Jackel. Handwritten digit recognition with a back-propagation network, *Adv. Neural Inf. Process. Syst.*, **2**, pp. 396–404 (1990).

[15] P. Baranyi, P. Korondi, H. Hashimoto, and M. Wada. Fuzzy inversion and rule base reduction, *Proc. IEEE International Conference on Intelligent Engineering Systems* (*INES'97*), Budapest, Hungary, pp. 301–306 (1997).

[16] P. Baranyi and Y. Yam. Singular value-based approximation with Takagi-Sugeno type fuzzy rule base, *Proc. 6th IEEE International Conference on Fuzzy Systems* (*FUZZ-IEEE'97*), Barcelona, Spain, **1**, pp. 265–270 (1997).

[17] C. Kahraman and S. Çebi. A new multi-attribute decision making method: Hierarchical fuzzy axiomatic design, *Expert Syst. Appl.*, **36**, pp. 4848–4861 (2009).

[18] D. Hladek, J. Vaščák, and P. Sinčák. Multi-robot control system for pursuit-evasion problem, *J. Electr. Eng.*, **60**, pp. 143–148 (2009).

[19] N. Kasabov and H.N.A. Hamed. Quantum-inspired particle swarm optimisation for integrated feature and parameter optimisation of evolving spiking neural networks, *Int. J. Artif. Intell.*, 7, pp. 114–124 (2011).

[20] E. Kayacan, Y. Oniz, A.C. Aras, O. Kaynak, and R.H. Abiye. A servo system control with time-varying and nonlinear load conditions using type-2 TSK fuzzy neural system, *Appl. Soft Comp.*, **11**, pp. 5735–5744 (2011).

[21] Y. Li and J. Liang. A hybrid multi-swarm co-evolutional particle swarm optimizer, *Int. J. Artif. Intell.*, **7**, pp. 274–287 (2011).

[22] Y. LeCun, L. Jackel, L. Bottou, A. Brunot, C. Cortes, J. Denker, H. Drucker, I. Guyon, U. Muller, E. Sackinger, P. Simard, and V. Vapnik. Comparison of learning algorithms for handwritten digit recognition, *Proc. 4th International Conference on Artificial Neural Networks,* Cambridge, UK, pp. 53–60 (1995).

[23] S. Haykin. *Neural Networks. A Comprehensive Foundation*, 2nd Edition., Prentice-Hall, Englewood Cliffs, NJ (1999).
[24] F.H.C. Tivive and A. Bouzerdoum. Efficient training algorithms for a class of shunting inhibitory convolutional neural networks, *IEEE Trans. Neural Netw.*, **16**, pp. 541–556 (2005).
[25] E. Rashedi, H. Nezamabadi-pour, and S. Saryazdi. GSA: A gravitational search algorithm, *Inf. Sci.*, **179**, pp. 2232–2248 (2009).
[26] E. Rashedi, H. Nezamabadi-pour and S. Saryazdi. BGSA: binary gravitational search algorithm, *Natural Comput.*, **9**, pp. 727–745 (2009).
[27] R.-C. David, R.E. Precup, S. Preitl, J.K. Tar, and J. Fodor. Parametric sensitivity reduction of PI-based control systems by means of evolutionary optimization algorithms, *Proc. 6th IEEE International Symposium on Applied Computational Intelligence and Informatics (SACI 2011)*, Timisoara, Romania, pp. 241–246 (2011).
[28] R.-E. Precup, R.-C. David, E.M. Petriu, S. Preitl, and A.S. Paul. Gravitational search algorithm-based tuning of fuzzy control systems with a reduced parametric sensitivity, *Soft Computing in Industrial Applications*, A. Gaspar-Cunha, R. Takahashi, G. Schaefer, L. Costa (Eds.), Springer-Verlag, Berlin, Heidelberg, pp. 141–150 (2011).
[29] R.-E. Precup, R.-C. David, E.M. Petriu, M.-B. Rădac, S. Preitl, and J. Fodor. Evolutionary optimization-based tuning of low-cost fuzzy controllers for servo systems, *Knowledge-Based Syst.*, DOI 10.1016/j.knosys (2011).
[30] M. Ghalambaz, A.R. Noghrehabadi, M.A. Behrang, E. Assareh, A. Ghanbarzadeh, and N. Hedayat. A hybrid neural network and gravitational search algorithm (HNNGSA) method to solve well known Wessinger's equation, *World Acad. Sci. Eng. Tech.*, **73**, pp. 803–807 (2011).
[31] S. Blažič and I. Škrjanc. Design and stability analysis of fuzzy model-based predictive control - a case study, *J. Intell. Robot. Syst.*, **49**, pp. 279–292 (2007).
[32] J.A. Iglesias, P. Angelov, A. Ledezma, and A. Sanchis. Evolving classification of agents' behaviors: a general approach, *Evolving Syst.*, **1**, pp. 161–171 (2010).
[33] Z.C. Johanyák. Student evaluation based on fuzzy rule interpolation, *Int. J. Artif. Intell.*, **5**, pp. 37–55 (2010).
[34] O. Linda and M. Manic. Uncertainty-robust design of interval type-2 fuzzy logic controller for delta parallel robot, *IEEE Trans. Ind. Informat.*, **7**, pp. 661–670 (2011).
[35] B. Öztayşi, T. Kaya, and C. Kahraman. Performance comparison based on customer relationship management using analytic network process, *Expert Syst. Appl.*, **38**, pp. 9788–9798 (2011).

Chapter 24

Fuzzy Logic Applications in Cellular Manufacturing System Design

Hulya Behret[1] and Sule Itir Satoglu[2]

[1] *Industrial Engineering Department, Erciyes University, 38039 Kayseri, Turkey*
[2] *Industrial Engineering Department, Istanbul Technical University, 34367 Macka, Sisli, Istanbul, Turkey*

E-mail: hbehret@erciyes.edu.tr, onbaslis@itu.edu.tr

Cellular manufacturing system (CMS) is based on the Group Technology philosophy, where similar products having operational similarities are manufactured within the same production unit called a "cell", to decrease the setup times, manufacturing lead times and work-in process. Fuzzy Set Theory was employed to consider the uncertainty in real life during the CMS design. The approaches used for this purpose include hard computing methods and soft computing methods. The former do not consider the uncertainty, and the parameters are deterministic. However, the latter (Fuzzy Set Theory) intends to reflect vagueness through the parameters, and it is associated with the soft computing methods. An effective CMS design requires concurrent consideration of real aspects of the system, which are usually uncertain and hard to measure. For instance, parts demands, cost of inter-cell flow, machine purchasing costs etc. are imprecise which stipulates use of soft computing techniques. The purpose of this study is basically to summarize the fuzzy techniques, especially the fuzzy clustering and fuzzy programming employed for the CMS design, and their past applications, and offer directions for future research.

24.1 Introduction

Group Technology (GT) is a concept that seeks to identify and group similar parts/products and to exploit their similarities in manufacturing and design [24]. In this context, Cellular Manufacturing is an application of the GT principles to manufacturing system design, reconfiguration and facility layout design. A cellular manufacturing system (CMS) entails arrangement of manufacturing cells that are comprised of all manufacturing resources including the machines, tools, vehicles and workers that are required to manu-

C. Kahraman (ed.), *Computational Intelligence Systems in Industrial Engineering*, Atlantis Computational Intelligence Systems 6, DOI: 10.2991/978-94-91216-77-0_24,

facture the components of a part family. Furthermore, a part family can be defined as a collection of parts which are similar either because of geometric shape and size or because of similar processing requirements [22]. Ideally, a manufacturing cell is dedicated to a single part family, and the components of the part family should be completed within that cell. In other words, manufacturing cells are preferred to have minimum interaction with each other [24].

There is an extensive literature about the CMS design issues. A detailed review of the CMS literature was provided by Selim *et al.* [50]. Moreover, Singh and Rajamani wrote a book [54] where design methods, layout and production planning issues, and production control of the CMSs were extensively explained.

The past approaches used for the CMS design problem can be categorized into two based on the type of computing: soft computing and traditional hard computing [62]. The primary requirements of the hard-computing approaches are precision, certainty and rigor. In contrast, soft computing approaches are tolerant of imprecision, uncertainty, and partial truth. Computation and decision making should make use of the tolerance for imprecision and uncertainty, since precision and certainty are not always possible in real life [62].

In particular, hard computing approaches to the CMS problem include mathematical programming, heuristics; similarity based clustering methods, and graph theoretic approaches [62]. Besides, main approaches to the CMS design problem based on soft computing include fuzzy models/fuzzy logic, genetic algorithms and neural network models [62]. Until recently, soft computing approaches to the CMS design problem has increased. The genetic algorithms, particle swarm optimization and scatter search techniques were developed which can be all categorized as evolutionary algorithms [42]. In addition, ant colony optimization is also another soft-computing technique recently employed for the CMS design. The evolution of cell formation problem methodologies was examined by Papaioannou and Wilson [42], recently.

As the design and planning processes of the CMSs include uncertainty and imprecision, fuzzy theory has been employed. The purpose of this chapter is to explain the fuzzy logic applications in CMS design, and offer directions for future research.

24.2 Fuzzy Techniques for the Cell Formation Problem

The most important decision in designing effective CMS is the Cell Formation (CF) which is defined as grouping parts with similar design features or processing requirements into part families and corresponding machines into machine cells. Cellular Manufacturing (CM) gives opportunity to small and batch-type production to get economic advantages of mass production while keeping the advantages of flexibility associated with job shop production.

During the past decades, much effort has been devoted to solve CF problems. Papaioannou and Wilson [42] presented a literature review of the CF problem focused on formulations proposed in the last decade. According to them a general CF process consists of the following steps; (i) part families are formed according to their processing requirements, (ii) machines are grouped into manufacturing cells, (iii) part families are assigned to cells. However, these steps are not necessarily operated in the above order or even sequentially. Selim *et al.* [50] categorized the solution strategies for CF problems into three categories depending upon the procedures/formulations employed to form manufacturing cells and part families; (i) part families are formed first and then machines are grouped into cells according to the part families. This solution strategy is referred to as part-family identification (PFI), (ii) manufacturing cells (grouped machines) are first created based on similarity in part routings and then the parts are allocated to cells. This solution strategy is referred to as machine groups' identification (MGI), (iii) part families and manufacturing cells are formed simultaneously. This is referred to as part families/machine grouping (PF/MG) solution strategy.

In the CF literature, there are various studies using these solution strategies. Among these studies, some of them used design features for part/machine similarity, while most used processing requirements such as routing data. However, most of these methods do not comprise the uncertainty or imprecision in measuring similarities among parts/machines [5].

Traditional clustering methods assume that there are well defined boundaries between part/machine groups and for that reason assign each part or machine to only one family. Nevertheless, sometimes a part/machine can have a high similarity to both families and the distinction cannot be so precise. Traditional models fail to fully reflect the complex nature of part features or routing data, when boundaries between groups are imprecise or fuzzy [31]. In addition, the part demands, costs of inter-cell flow, machine purchasing costs, intercell transfer cost, machine capacity are among the CMS parameters that are uncertain, and cannot be captured by determinism [53].

However, hard computing techniques such as mathematical programming do not consider the uncertainty within these parameters during the CF [53].

In order to cope with these deficiencies, fuzzy models/techniques are appropriate for the CF problems having part/machine information under uncertain or inexact environment. Among the fuzzy models/techniques for the CF problem, some studies consider design features for grouping parts while the others use processing requirements. There are fuzzy clustering, fuzzy classification-coding and other techniques, fuzzy programming, and fuzzy ART techniques. Classification of fuzzy models/techniques for the CF problem is represented in Table 24.1.

The most important issue in applying fuzzy models to CF problem is to quantify the fuzziness through membership function [62]. Since the determination of membership functions includes subjectivity, the accurate representation of the characteristics of each feature of the part of machining process is not easy to identify with these membership functions. To solve this difficulty Ben-Arieh and Triantaphyllou [8], Ben-Arieh *et al.* [7], and Narayanaswamy *et al.* [38] have proposed methods to quantify the fuzziness in part-features and machining processes. In another study, Gill and Bector [19] used linguistic variables to capture design features. Liao [33] used classification and coding approach for part family formation under a fuzzy environment. Gungor and Arikan [21] used a fuzzy decision making system in part-machine grouping considering both design and manufacturing attributes and operation sequences as factors.

In the following sections, these fuzzy techniques for the CF problem will be explained in detail.

24.2.1 *Fuzzy clustering*

Clustering techniques are generally unsupervised methods that can be used to arrange data into groups based on similarities among the individual data items. Most clustering algorithms do not rely on assumptions common to conventional statistical methods, such as the underlying statistical distribution of data, and therefore they are useful in situations where little prior knowledge exists. In real life applications there is very often no sharp boundary between clusters so that fuzzy clustering is often better suited for the data. Membership degrees between zero and one are used in fuzzy clustering instead of crisp assignments of the data to clusters.

The first studies in the fuzzy clustering applications of Group Technology (GT) literature have used fuzziness in part's features as inputs and applied the fuzzy models to find part-machine grouping. Batra and Rajagopalan [3] and Rajagopalan [45] introduced fuzzy clustering in the CF problem. They used an iterative algorithm to form fuzzy part families

Table 24.1 Classification of fuzzy techniques for CF

Fuzzy Clustering	
Rajagopalan (1975) [45]	Susanto *et al.* (1999) [58]
Batra and Rajagopalan (1977) [3]	Lozano *et al.* (2002) [34]
Li *et al.* (1988) [32]	Josien and Liao (2002) [25]
Xu and Wang (1989) [67]	Yang *et al.* (2006) [68]
Chu and Hayya (1991) [16]	Li *et al.* (2007) [31]
Zhang and Wang (1992) [69]	Torkul *et al.* (2006) [60]
Gindy *et al.* (1995) [20]	Behret and Kahraman (2008) [5]
Wen *et al.* (1996) [64]	Naadimuthu *et al.* (2010) [37]
Leem and Chen (1996) [30]	Susanto *et al.* (2010) [59]
Masnata and Settinery (1997) [36]	Hung *et al.* (2011) [23]
Fuzzy Classification, Coding, and Other Techniques	
Ben-Arieh and Triantaphyllou (1992) [8]	Gill and Bector (1997) [19]
Su (1995) [55]	Gungor and Arikan (2000) [21]
Ben-Arieh *et al.* (1996) [7]	Liao (2001) [33]
Narayanaswamy *et al.* (1996) [38]	
Fuzzy Programming	
Tsai *et al.* (1997) [61]	Paydar *et al.* (2008) [43]
Shanker and Vrat (1999) [53]	Safaei and Tavakkoli-Moghaddam (2009) [48]
Ravichandran and Rao (2001) [46]	Arıkan and Güngör (2009) [2]
Arikan and Gungor (2005) [1]	Feyzioglu and Pierreval (2009) [18]
Safaei *et al.* (2007) [47]	Mahdavi *et al.* (2011) [35]
Eski and Ozkarahan (2007) [17]	Nunkaew and Phruksaphanrat (2011) [39]
Fuzzy Neural Networks	
Kaparthi and Suresh (1994) [27]	Pai and Lee (2001)
Suresh and Kaparthy (1994) [56]	Park and Suresh (2003) [41]
Burke and Kamal (1992,1995) [10, 11]	Peker and Kara (2004) [44]
Kamal and Burke (1996) [26]	Kuo *et al.* (2006) [28]
Kuo *et al.* (1999)	Ozdemir *et al.* (2007) [40]
Suresh *et al.* (1999) [57]	Won and Currie (2007) [66]
Josien and Liao (2000)	Sengupta *et al.* (2011) [51]

according to the design features of parts. Xu and Wang [67] used fuzzy classification and fuzzy equivalence approaches for part-family formation and then they applied fuzzy pattern recognition to assign new parts to existing part-families. Another related work in the area of application of fuzzy cluster analysis and pattern recognition is presented by Li *et al.* [32]. More recent studies in the fuzzy CF literature have used processing requirements for part-machine grouping. Chu and Hayya [16] used fuzzy c-means (FCM) algorithm to solve CF problem. Based on the Chu and Hayya's approach [16], FCM algorithm was modified by several researchers including Gindy *et al.* [20], Wen *et al.* [64], Masnata and Settineri [36], Susanto *et al.* [58], Josien and Liao [33], Lozano *et al.* [34], Li *et al.* [31] and Behret and

Kahraman [5]. Yang *et al.* [68] used a mixed variable fuzzy c-means clustering algorithm to cell formation in group technology. Torkul *et al.* [60] studied the fuzzy logic approach in the design of part families and machine cells simultaneously through FCM clustering algorithm. Naadimuthu *et al.* [37] studied an approach based on fuzzy clustering and they proposed aggregation operators to design cell formation involving multiple criteria or multiple attributes. The study of Susanto *et al.* [59] focuses on stages selection of parts and part families' generation, and selection of machines and process and grouping of these into cells in the design of a new manufacturing system considering sequence of operation, alternative routing and part-volume.

Fuzzy clustering methods take not only non-binary data as input but also give non-binary classification as output. That is, in addition to specifying the cell or family a machine or part belongs to, they also reveal the degree to which a machine fits into a cell with the other elements. This allows more flexibility in the formation of cells, especially when the assignment is not clear. Although there have been several studies applying FCM clustering algorithm to CF problem, there are relatively few studies applying other clustering algorithms for part-machine grouping. Zhang and Wang [69] used a non-binary machine-part matrix as an input to conventional single linkage and rank order clustering methods. Leem and Chen [30] considered the non-binary matrix in a similarity-coefficient based fuzzy clustering algorithm. Hung *et al.* [23] applied a fuzzy relational data clustering algorithm to form part families and machine groups.

In designing an effective CMS, fuzzy c-means (FCM) clustering is the mostly used algorithm for CF problems. The FCM clustering is the fuzzified adaptation of the conventional K-means (C-means) clustering method [62]. Besides FCM clustering algorithm, fuzzy single linkage clustering, fuzzy rank order clustering and fuzzy relational clustering algorithms are also used in CF problems.

The FCM algorithm performs well with small and well-structured data sets. However in the usage of larger and especially ill-structured data sets the algorithm may result in clustering errors, infeasible solutions, and uneven distribution of parts/machines.

The fuzzy cell formation problem can be defined as: given the routing information of p parts and m machines, the goal of CF is to cluster the parts into c part families and the corresponding machines into machine cells where $2 \leqslant c < \min\{p,m\}$ [31]. The standard application of FCM to cell formation assumes a binary part-machine incidence matrix A

whose components are:

$$a_{ij} = \begin{cases} 1 & \text{if part type } i \text{ requires machine } j \\ 0 & \text{otherwise} \end{cases}$$

We define the membership value (μ_{ki}) which represents the membership degree of part i to part family k. The membership values must verify the following conditions:

$$\mu_{ki} \in [0,1], \quad i = 1,2,3,\ldots,p; \;\; k = 1,2,3,\ldots,c \tag{24.1}$$

The sum of all membership values for a single data point in all of the classes has to be unity:

$$\sum_{k=1}^{k=c} \mu_{ki} = 1, \quad i = 1,2,3,\ldots,p \tag{24.2}$$

There must be no empty classes and there must be no class that contains all the data points:

$$0 < \sum_{i=1}^{i=p} \mu_{ki} < p, \quad k = 1,2,3,\ldots,c \tag{24.3}$$

To determine the fuzzy c-partition matrix $U = [\mu_{ki}]_{c\times p}$ for grouping p parts to c part families, we minimize the objective function $J_m(U,V)$ which is defined as following:

$$J_m(U,V) = \sum_{i=1}^{p}\sum_{k=1}^{c} (\mu_{ki})^m (d_{ki})^2 \tag{24.4}$$

where $1 < m < \infty$ is a weighting exponent for the degree of fuzziness and d_{ki} is the distance between part i and part family k.

Steps of the FCM algorithm

The FCM algorithm is an iterative procedure which has been introduced by Bezdek [9]. Xu and Wang [67] and Chu and Hayya [16] adapted the FCM algortihm to solve CF problems. The steps of the algorithm can be summarized as follows:

1. *Parameter selection*: Fix the group number c $(2 \leqslant c < \min\{p,m\})$, select a value for m $(m > 1)$, for the degree of fuzziness and identify a value for the stopping criteria (ξ).
2. *Initialization*: Initialize the fuzzy partition matrix $(U^{(0)})$. Each iteration of the algorithm will be labeled as r, where $r = 0,1,2,\ldots$
3. *Center function calculation*: For iteration $r = 1,2,\ldots$ calculate the cluster center matrix $(V^{(r)})$ for each step.

$$V = [v_{kj}] = \frac{\sum_{i=1}^{p} (\mu_{ki})^m a_{ij}}{\sum_{i=1}^{p} (\mu_{ki})^m}, \quad k = 1,2,3,\ldots,c; \quad j = 1,2,3,\ldots,m \tag{24.5}$$

4. *Distance calculation*: Calculate the Euclidean distance (d_{ki}) between part i and part family k.

$$d_{ki} = \|\overline{a}_i - \overline{v}_k\| \tag{24.6}$$

Here, $\overline{a}_i$ is the column vector of matrix A corresponding to part type i and $\overline{v}_k$ is the cluster center corresponding to part family k.

5. *Membership degree calculation*: Update the partition matrix for the rth step, $U^{(r)}$ as follows:

$$U = [\mu_{ki}] = \frac{1}{\sum_{q=1}^{c}\{d_{ki}/d_{qi}\}^{2/(m-1)}}, \quad i = 1,2,3,\ldots,p \tag{24.7}$$

6. *Stopping decision*: Compare $U^{(r)}$ to $U^{(r+1)}$ in a convenient matrix norm i.e. if $\|U^{(r+1)} - U^{(r)}\| \leqslant \xi$, stop; otherwise return to step 3.
Form part families according to the partition matrix $U^{(r)}$ and machine families according to $V^{(r)}$. Part type i is assigned to part family k iff:

$$\mu_{ki} = \max\{\mu_{1i}, \mu_{2i}, \ldots, \mu_{ci}\}$$

Machine type j is assigned to machine family k iff:

$$v_{kj} = \max\{v_{1j}, \mu_{2j}, \ldots, v_{cj}\}$$

7. Rearrange the part-machine matrix $\acute{A}$ according to part families and machine cells to form manufacturing cell blocks in the diagonal sub-matrix.

Example 1. Let we have totally 6 parts and 5 machines. Part-machine matrix is formed by using the routing data, where if part type i requires machine j than 1, otherwise 0.

$$A = [a_{ij}] = \begin{array}{c} \\ 1 \\ 2 \\ 3 \\ 4 \\ 5 \\ 6 \end{array} \begin{array}{c} \begin{array}{ccccc} 1 & 2 & 3 & 4 & 5 \end{array} \\ \begin{bmatrix} 1 & 0 & 1 & 0 & 1 \\ 0 & 0 & 0 & 1 & 0 \\ 1 & 1 & 0 & 0 & 1 \\ 0 & 0 & 1 & 1 & 0 \\ 0 & 1 & 1 & 0 & 1 \\ 1 & 1 & 0 & 1 & 0 \end{bmatrix} \end{array}$$

Step 1. Parameter selection. $c = 2$, $m = 2$, $\xi = 0.01$ Step 2. Initialization.

$$U^{(0)} = \begin{bmatrix} 1 & 0 & 1 & 0 & 1 & 1 \\ 0 & 1 & 0 & 1 & 0 & 0 \end{bmatrix}$$

Step 3. Center function calculation.

$$V^{(1)} = \begin{bmatrix} 0.75 & 0.75 & 0.50 & 0.25 & 0.75 \\ 0 & 0 & 0.50 & 1 & 0 \end{bmatrix}$$

Step 4. Distance calculation.

$$d_{ki} = \begin{bmatrix} 1 & 1.85 & 0.71 & 1.58 & 1 & 1.22 \\ 1.80 & 0.50 & 2.06 & 0.50 & 1.80 & 1.50 \end{bmatrix}$$

Step 5. Membership degree calculation.

$$U^{(1)} = \begin{bmatrix} 0.76 & 0.09 & 0.89 & 0.09 & 0.76 & 0.60 \\ 0.24 & 0.91 & 0.11 & 0.91 & 0.24 & 0.40 \end{bmatrix}$$

Step 6. Stopping decision.

$\|U^{(1)} - U^{(0)}\| = 0.4 > 0.01$ Go to step 3.

$$V^{(2)} = \begin{bmatrix} 0.74 & 0.74 & 0.50 & 0.16 & 0.83 \\ 0.11 & 0.11 & 0.48 & 0.93 & 0.06 \end{bmatrix}$$

$$U^{(2)} = \begin{bmatrix} 0.75 & 0.09 & 0.89 & 0.10 & 0.75 & 0.50 \\ 0.25 & 0.91 & 0.11 & 0.90 & 0.25 & 0.50 \end{bmatrix}$$

We reach a value smaller than the stopping criteria at iteration 6.

$$V^{(6)} = \begin{bmatrix} 0.71 & 0.71 & 0.56 & 0.09 & 0.91 \\ 0.19 & 0.19 & 0.42 & 0.94 & 0.06 \end{bmatrix}$$

$$U^{(6)} = \begin{bmatrix} 0.78 & 0.08 & 0.87 & 0.13 & 0.78 & 0.41 \\ 0.22 & 0.92 & 0.13 & 0.87 & 0.22 & 0.59 \end{bmatrix}$$

$\|U^{(6)} - U^{(5)}\| = 0.0091 < 0.01$, stop.

According to $U^{(6)}$ and $V^{(6)}$ matrices, we form the part families and machine families respectively.

Part family: $\{(1,3,5);\ (2,4,6)\}$

Machine family: $\{(1,2,3,5);\ (4)\}$

Step 7. Rearrange the part-machine matrix according to part families and machine cells to form manufacturing cell blocks in the diagonal sub-matrix.

$$\acute{A} = \begin{array}{c} \\ 1 \\ 3 \\ 5 \\ \hline 2 \\ 4 \\ 6 \end{array} \begin{array}{c} \begin{array}{cccc|c} 1 & 2 & 3 & 5 & 4 \end{array} \\ \left[\begin{array}{cccc|c} 1 & 0 & 1 & 0 & 1 \\ 0 & 0 & 0 & 1 & 0 \\ 1 & 1 & 0 & 0 & 1 \\ \hline 0 & 0 & 1 & 1 & 0 \\ 0 & 1 & 1 & 0 & 1 \\ 1 & 1 & 0 & 1 & 0 \end{array}\right] \end{array}$$

Grouping efficiency

Grouping efficiency is defined as in the following [14]:

$$z = w * z_1 + (1 - w) * z_2 \tag{24.8}$$

where z_1 stands for intercellular density, z_2 represents the cost of exceptional elements, w represents the weighting factor, z represents the grouping efficiency, z_1 stands for the # of ones in the diagonal block/all the # of zeros and ones in the diagonal block and z_2 stands for the # of zeros in the nondiagonal blocks/ # of all zeros and ones in the nondiagonal block. The grouping efficiency of our partition in Example 1 will be found as following:

$z = w * z_1 + (1 - w) * z_2$

$z = 0.5 * 0.67 + 0.5 * 0.67$

$z = \mathbf{0.67}$

24.2.2 *Fuzzy Programming*

First, an overview to the Fuzzy Programming models will be presented. Next, the fuzzy programming applications in the CMS design will be discussed.

24.2.2.1 *Overview to the Fuzzy Programming Models*

The types of the fuzzy programming models depend on the type of fuzziness and the way it is prescribed by the decision maker [4]. Fuzzy programming techniques include fuzzy linear programming and fuzzy quadratic programming. Fuzzy linear programming models will be considered, but the fuzzy quadratic programming models are not included in the scope of this chapter.

In this chapter, two important models will be discussed, namely symmetric and non-symmetric models. According to the Bellman and Zadeh's fuzzy decision definition [6], a fuzzy linear programming problem (FLP) is *fully symmetric* if the objective(s) and constraints of this model can be represented by fuzzy sets, under imprecise conditions [6]. Then, a decision can be expressed as a confluence of objective(s) and constraints and can be defined by a max-min operator. Let's assume that the fuzzy objective set G and the fuzzy constraint set c is given in a space of alternatives X. Then, G and c combine to form a fuzzy decision set D that is the intersection of G and c, and the corresponding membership functions can be stated as $\mu_D = \mu_G \cap \mu_C$.

The non-symmetric models are based on two approaches [29]:

Determination of the fuzzy set decision

Determination of a crisp maximizing decision by aggregating the objective functions after suitable transformations, with the constraints.

After this basic introduction, the models will be classified. The FLP problems/models can be roughly classified as follows [29]:

1) Linear programming problems with fuzzy resources
2) Linear programming problems with fuzzy resources and objective
3) Linear programming problems with fuzzy parameters in the objective function
4) Linear programming problems with all fuzzy coefficients.

This section is based on the assumption that linear membership functions and max-min operators are employed. Now, these problems will be briefly explained.

1) Linear Programming Problems with Fuzzy Resources

In FLP, the fuzziness of the available resources is characterized by the membership function over the (ambiguous) tolerance range. Membership function of the optimal solutions can be achieved by the convolution of the membership functions of the constraints. The general model of this linear programming with fuzzy resources available is formulated as follows [29]:

$$\begin{aligned} &\text{Max } Z = cx \\ &\text{s.t.} \quad (Ax)_i \leqslant \tilde{b}_i, \quad i = 1, \dots, m \\ &x \geqslant 0 \end{aligned} \tag{24.9}$$

where $\tilde{b}_i$, $\forall i$ are in $[b_i, b_i + p_i]$ with given tolerance p_i. We may also consider the following fuzzy inequality constraints:

$$\begin{aligned} &\text{Max } Z = cx \\ &\text{s.t.} \quad (Ax)_i \,\tilde{\leqslant}\, \tilde{b}_i, \quad i = 1, \dots, m \\ &x \geqslant 0 \end{aligned}$$

where $\tilde{\leqslant}$ is called fuzzy less than or equal to. Assume that tolerance p_i for each constraint is known. $(Ax)_i \tilde{\leqslant} b_i$ will be equivalent to $(Ax)_i = (b_i + \theta p_i)$, $\forall i$ where θ is in $[0, 1]$. If the membership functions of these cases are the same, then the equations stated above will correspond to the same problem [29]. There are two solution approaches for this type of problem: Verdegay's approach [63] and Werners's approach [65]. Here, only the basic ideas of these approaches will be explained.

Verdegay [63] considered the linear, monotonic membership functions of the fuzzy constraints $\mu_i(x)$, where a tolerance value p_i is employed for each fuzzy constraint i. Then the problem will be equivalent to:

$$\begin{aligned}&\text{Max } Z = cx\\&\text{s.t.} \quad x \in X_\alpha\\&\text{where } X_\alpha = \{x \mid \mu_i(x) \geqslant \alpha,\ \forall i,\ x \geqslant 0\}, \text{ for each } \alpha \in [0,1].\end{aligned}$$

According to the Werners's approach [65], the objective must be also fuzzy. Similar to the Verdegay's approach, a tolerance value p_i is defined for each fuzzy constraint i. Then, the optimal solutions of the two models are achieved where the right hand side of the constraints are either b_i or $b_i + p_i$. These solutions are z^0 and z^1, respectively. Then, the membership function of the objective (μ_0) is constructed. Then, max-min operator is employed to achieve the optimal solution. The following model can be solved:

$$\begin{aligned}&\text{Max } \alpha\\&\text{s.t.} \quad \mu_0(x) \geqslant \alpha\\&\mu_i(x) \geqslant \alpha,\ \forall i\\&\alpha \in [0,1] \text{ and } x \geqslant 0.\end{aligned}$$

Example 2. This example is a modified version of the model proposed in [53] about fuzzy programming for the CMS design . This example will be used through this section. Decision variables' and parameters' explanations are presented below.

Decision variables X_i: Units of part i to be subcontracted
Y_{jf}: Number of machines of type j to be purchased for cell f
Z_{ij}: Number of intercellular transfers required by part i as a result of machine type j not being available within a part's manufacturing cell
M_{ij}: Number of machines of type j dedicated to production of part i.

Parameters A_j: Annual cost of acquiring a machine of type j
S_i: Incremental cost of subcontracting a unit part i
D_i: Annual forecasted demand for part i
C_j: Annual capacity of machine type j
P_{ij}: Processing time to produce part i on a machine type j
I_i: Incremental cost for moving part i out of a cell as opposed to moving it within that cell

G_f: Set of exceptional parts in cell f

H_f: Set of bottleneck machines required by parts in cell f

B: Upper budgetary limit on purchase of new machines

$$\text{Min } C_0 = \sum_f \left(\sum_{i \in G_f} S_i X_i + \sum_{i \in H_f} A_i Y_{if} + \sum_{i \in G_f} I_i Z_{ij} \right) \tag{24.10}$$

$$\text{s.t.} \quad Z_{ij} + X_i + (C_i M_{ij}/P_{ij}) \geqslant \widetilde{D}_i \text{(For all exceptional elements)} \tag{24.11}$$

$$\sum_{i \in G_f} M_{ij} \leqslant Y_{if} \quad \text{(for all } j,\ f) \tag{24.12}$$

$$X_i = 0; \quad \text{for all exceptional elements not to be subcontracted} \tag{24.13}$$

$$\sum_i \sum_f A_i Y_{if} \leqslant \widetilde{B} \tag{24.14}$$

$$A_i Y_{if} \text{ and } Z_{ij} \text{ are all integers}$$

In this example, in the crisp objective function minimization of total costs that incur due to the exceptional elements and the bottleneck machines, such as subcontracting cost, inter-cell transfer cost and discounted machine purchasing cost is aimed. However, constraint (24.11) is a fuzzy constraint where only the right hand size (demand) is fuzzy. Here, it is stipulated that demand of the exceptional elements are satisfied by inter-cell part processing, subcontracting and intra-cell part processing. In addition, constraint (24.14) is also a fuzzy constraint where the machine purchasing cost is limited to a budget value that is imprecise.

2) Linear Programming Problems with Fuzzy Resources and Objective

Both the objective and right hand sides of the constraints are fuzzy in this type of problem. Due to the same reason mentioned above, the following problems are equivalent [29]:

$$\widetilde{\text{Max}}\, Z = cx \tag{24.15}$$

$$\text{s.t.} \quad (Ax)_i \leqslant \tilde{b}_i, \quad i = 1, \ldots, m \tag{24.16}$$

$$x \geqslant 0 \tag{24.17}$$

and

$$\text{Max } \tilde{c}x \tag{24.18}$$

$$\text{s.t.} \quad (Ax)_i \widetilde{\leqslant} b_i, \quad i = 1, \ldots, m \tag{24.19}$$

$$x \geqslant 0 \tag{24.20}$$

For the solution of this type of problem, Zimmermans's approach [70] can be used. In this approach, the goal b_0 and the corresponding tolerance p_0 of the objective function are predetermined. In addition, the similar parameters b_0 and p_0 of the fuzzy resources are initially given. Based on these values, the membership functions of the objective and the fuzzy constraints are constructed. By using the Bellman and Zadeh's max-min operator [6], the membership function of the decision space can be obtained. For an α-level cut, the following model can be solved:

$$\begin{aligned}
&\text{Max } \alpha \\
&\text{s.t.} \quad cx \geqslant b_0 - (1-\alpha)p_0 \\
&(Ax)_i \leqslant b_i + (1+\alpha)p_i, \quad \forall i \\
&\alpha \in [0,1] \text{ and } x \geqslant 0
\end{aligned}$$

Example 3. The same example presented above is converted into an LP with fuzzy resources and objective, that is the original example given by Shanker and Vrat [53]. The model is as follows:

$$\widetilde{\text{Min}}\, C_0 = \sum_f \left(\sum_{i \in G_f} S_i X_i + \sum_{i \in H_f} A_i Y_{if} + \sum_{i \in G_f} I_i Z_{ij} \right) \tag{24.21}$$

$$\text{s.t.} \quad Z_{ij} + X_i + (C_i M_{ij}/P_{ij}) \geqslant \widetilde{D}_i \tag{24.22}$$

$$\sum_{i \in G_f} M_{ij} \leqslant Y_{if} \text{ for all } j,\ f \tag{24.23}$$

$$X_i = 0; \text{ for all exceptional elements not to be subcontracted} \tag{24.24}$$

$$\sum_i \sum_f A_i Y_{if} \leqslant \widetilde{B} \tag{24.25}$$

$$A_i Y_{if} \text{ and } Z_{ij} \text{ are all integers} \tag{24.26}$$

Both the demand and the budget and the corresponding constraints, and the goal corresponding to the objective are fuzzy.

3) Linear Programming Problems with Fuzzy Parameters in the Objective Function

The coefficients of the objective functions often cannot be precisely determined. Therefore, in these cases considering imprecise coefficients in the objective function is required. This problem can be formulated as follows [29]:

$$\text{Max } \tilde{c}x \tag{24.27}$$

$$\text{s.t.} \quad (Ax)_i \leqslant b_i, \quad i = 1, \ldots, m \tag{24.28}$$

$$x \geqslant 0 \tag{24.29}$$

For this problem, Verdegay proposed the following parametric linear programming model [63]:

$$\text{Max } cx \tag{24.30}$$

$$\text{s.t.} \quad \mu(c) \geqslant 1 - \alpha \tag{24.31}$$

$$(Ax)_i \leqslant b_i, \quad i = 1, \ldots, m \tag{24.32}$$

$$\alpha \in [0,1] \text{ and } x \geqslant 0 \tag{24.33}$$

where $c = (c_1, c_2, \ldots, c_n)$, $\mu(c) = \inf_j \mu_j(c_j)$ and $\mu(c)$ is the membership function of (cx) and $\mu_j(c_j)$, $\forall j$, are the membership functions of c_j.

Example 4. In this modified version of the model, the parameters in the objective function that are incremental cost of subcontracting a unit part i , and incremental cost for moving part i out of a cell are fuzzy.

$$\text{Min } \widetilde{C}_0 = \sum_f \left(\sum_{i \in G_f} \widetilde{S}_i X_i + \sum_{i \in G_f} \widetilde{I}_i Z_{ij} \right)$$

$$\text{s.t.} \quad Z_{ij} + X_i + (C_i M_{ij} / P_{ij}) \geqslant D_i$$

$$\sum_{i \in G_f} M_{ij} \leqslant Y_{if} \text{ for all } j \text{ and } f$$

$$X_i = 0; \text{ for all exceptional not to be subcontracted}$$

$$\sum_i \sum_f A_i Y_{if} \leqslant B$$

$$X_i,\ Y_{if} \text{ and } Z_{ij} \text{ are all integers}$$

4) Linear programming problems with all fuzzy coefficients

When all of the coefficients of the linear programming model are imprecise, the problem can be formulated as follows [29]:

$$\text{Max } \tilde{c}x \tag{24.34}$$

$$\text{s.t.} \quad \left(\widetilde{Ax}\right)_i \leqslant \tilde{b}_i, \quad i = 1, \ldots, m$$

$$x \geqslant 0$$

For this type of problem, Carlsson and Korhonen [12] proposed a parametric approach, where a full trade-off between the fuzzy parameters c, $-A$, and b is considered. A feasible solution will be found at $\mu = \mu_c = \mu_A = \mu_b$.

Example 5.

$$\text{Min}\,\widetilde{C}_0 \sum_f \left(\sum_{i\in G_f} \widetilde{S}_i X_i + \sum_{i\in H_f} \widetilde{A}_i Y_{if} + \sum_{i\in G_f} \widetilde{I}_i Z_{ij} \right)$$

$$\text{s.t.}\quad Z_{ij} + X_i + (C_j M_{ij}/P_{ij}) \geqslant \widetilde{D}_i$$

$$\sum_{i\in G_f} M_{ij} \leqslant Y_{if} \text{ for all } j,\, f$$

$$X_i = 0; \text{ for all exceptional elements not to be subcontracted}$$

$$\sum_i \sum_f \widetilde{A}_i Y_{if} \leqslant \widetilde{B}$$

X_i, Y_{if} and Z_{ij} are all integers.

After this brief explanation of the types of FLP models, past implementations of them in the CMS design problem will be discussed.

24.2.2.2 *Past Applications of Fuzzy Programming in CMS Design*

In former sections, fuzzy clustering techniques were explained. Fuzzy programming and fuzzy clustering methods differ in some ways during the CMS design. In the fuzzy clustering problem, the fuzzy membership of parts or machines is defined and hierarchical clustering of parts or machines is applied. However, in fuzzy programming, vagueness of the many other design parameters can be modeled and the solution can be obtained [53].
In CMS design, conventional mathematical programming techniques can be applied to maximize or minimize single/multiple objectives that satisfy the constraints. But, in many practical situations, the objectives and the constraints cannot be defined precisely, but these can be stated in fuzzy sense. In these cases, fuzzy programming techniques are useful for the CMS design.
The fuzzy programming techniques applied for the CMS design are fuzzy linear programming (FLP), fuzzy multi-objective linear programming (FMOLP), fuzzy parametric programming (FPP) and fuzzy goal programming (FGP). Table 24.2 summarizes the past fuzzy programming studies based on the fuzzy programming method used, objective functions and the fuzzy parameters.

Table 24.2. Classification of the Fuzzy Programming Studies for the CMS Design

		(Shanker and Vrat, 1999) [53]	(Tsai, Chu and Barta, 1997) [61]	(Safaei, Saidi-Mehrabad, Tavakkoli-Moghaddam and Sassani,2008) [49]	(Arikan and Gungor, 2005) [1]	(Safaei, Saidi-Mehrabad, Babakhani, 2007) [47]	(Papaioannou and Wilson, 2009) [42]	(Safaei and Tavakkoli-Moghaddam, 2009) [48]	(Arikan and Gungor, 2009) [2]	(Safaei and Tavakkoli-Moghaddam, 2009) [49]	(Nunkaew and Phruksaphanrat, 2011) [39]	(Eski and Ozkarahan, 2007) [17]	(Mahdavi, Aalaei, Paydar and Solimanpur, 2011) [35]	(Feyzioglu and Pierreval, 2009) [18]	(Paydar, Mahdavi, Solimanpur, and Tajdin, 2008) [43]
Fuzzy Programming Method	FPP			X			X								
	FMOPP							X							
	FLP	X	X		X	X									X
	FMOLP	X						X		X				X	X
	FGP	X								X	X	X	X		X
	FQP				X				X						
Objective Function(s)	Total Purchasing Cost	X		X				X						X	X
	Total Intercell Material Handling Cost	X		X	X		X	X	X				X		X
	Total Intra-cell Material Handling Cost				X								X		
	Total Subcontracting Cost	X		X				X							X
	Utilized Capacity in the System	X						X			X				X
	Number of Exceptional Operations	X				X		X		X		X			X
	Reconfiguration Cost				X		X		X						
	Setup Cost					X									
	Number of Intercell Movements due to Exceptional Operations					X					X			X	
	Mean Tardiness										X				
	Percentage of Tardy Jobs										X				
	Inventory Holding Cost											X			
	Backorder Cost											X			

Continued on next page

Table 24.2. Classification of the Fuzzy Programming Studies for the CMS Design (continued)

	Machine Constant/ Variable Cost				X		X		X	X	X	
Fuzzy Parameters	Purchasing Price of a Machine			X				X				
	Capacity of a Machine per Period			X	X		X	X	X			
	Operation Time										X	
	Demand per Period	X		X	X		X	X	X		X	X
	Intercell Material Handling Cost			X				X				
	Subcontracting Cost Unit/Piece of Part			X				X				
	Capacity Utilization of a Machine for a Part Type			X				X				
	Number of Machines Allowed at Each Cell		X			X					X	
	Budget Limit	X										X

Tsai, Chu and Barta [61] formulated a mixed integer FLP model where the total machine acquisition cost, total inter-cell material transfer cost and total part subcontracting cost are attempted to be minimized during the cell formation stage. The number of machine types allowed in each cell (NM) is a fuzzy parameter on the right hand side of the corresponding constraint. Except this parameter, all parameters of the objective function and those of constraints are crisp. Therefore, this is a LP model with fuzzy resources, as explained above. Through the solution of the model, the optimal solution achieved using the minimum NM value and the one achieved using the maximum NM value are determined. Later, according to the Werners' approach [65] the objective function was fuzzified and the membership function of the objective is attained. Linear non-increasing and triangular membership functions of the fuzzy objective and constraint are formulated, separately. Beside the well known Max-min operator, three more operators were employed to achieve the decision. The data set of Shafer *et al.* [52] was employed as a numerical example that consists of ten parts and nine machine types, and the computational results were illustrated. The authors concluded that choosing a suitable fuzzy NM parameter is vital to obtain a good solution of this model. The drawback of this study is that it neglects the imprecision through other parameters such as machine acquisition costs, subcontracting costs and the imprecision inherent within the forecasted demand and machine capacities.

Shanker and Vrat [53] proposed three FLP models for the design of CMSs at the post-clustering stage. The models developed in this study intend to eliminate the exceptional elements that arise after the clustering stage. Another utility of the paper is that it explains the reasons of imprecision within the parameters of the CMS design problem. The objective function of the first model attempts to minimize total subcontracting cost, inter-cell transfer cost and discounted machine acquisition cost. Although the parameters of the objective function are not fuzzy, the objective is fuzzy itself. Besides, the forecasted demands of the parts and the budget limit for machine acquisition are the fuzzy parameters at the right hand side of the two fuzzy constraints. Minimum and maximum aspiration levels were defined for each of the objective, and the fuzzy constraints. Based on Zimmerman's approach [70], the solution that maximizes the membership function of the fuzzy decision set of the model is sought. This model was illustrated based on the data of the [52]. In addition, a multi-objective fuzzy linear programming model (FMOLP) was proposed as an extension of this aforementioned model. Beside the objectives of the former model, minimization of the unutilized capacity of the bottleneck machines and minimization of total inter-cell transfer of parts are aimed as the objectives. However, this model was quadratic and it was converted into a linear model. Then, this model was solved by taking one objective function at a time and the membership functions of the objectives were constructed. The solution that maximizes the membership function of the fuzzy decision set of the model is sought. Besides, for this FMOLP, deviational variables were defined and the model was converted into a fuzzy goal programming problem (FGP). Utility of the FMOLP is that it saves decision maker's efforts in determining the priority of the goals and maximizes the aspirations of the decision maker in meeting all objectives [53].

Feyzioglu and Pierreval [18] developed a FMOLP model for the design of the hybrid CMS that consists of manufacturing cells and the functional layout of machines. The parts demand, operation times, number of machines allowed at each cell and part-cell assignment cost are the fuzzy parameters. The fuzzy objectives of the model are stability based assignment cost, intra-cellular transfer ratio and net present value of machine purchases and variable operations costs. This model consists of fuzzy coefficients in both the objectives and the constraints.

Safaei, Saidi-Mehrabad and Babakhani [47] proposed a mixed integer fuzzy quadratic programming (FQP) model for the design of CMSs under dynamic and uncertain conditions.

The authors emphasize the dynamic conditions and argued that one cellular configuration designed for a certain planning period may not be efficient for another period due to imprecise nature of the parts demand and availability of the manufacturing resources. Thus, the CMS under these dynamic conditions was called dynamic cellular manufacturing system (DCMS). In the proposed FLP model, minimization of the total variable/constant machine costs, inter and intra-cell movement costs and reconfiguration costs over the planning horizon was aimed. The objective function was a non-linear. The parts' demand per period is a fuzzy coefficient considered through both the objective and the constraints. Besides, the available capacity of a machine per period was also assumed as a fuzzy resource at the right hand side of the corresponding constraint. The membership functions of the objective function and the fuzzy constraint were constructed, and the FQP was converted to a crisp quadratic programming problem and it was solved. The proposed FQP was illustrated on a very small problem.

Papaioannou and Wilson [42] developed crisp integer program for the CMS design and then converted it to a FLP model. The FLP model intends to minimize total number of parts allocations to the cells, setup costs of the machines, number of returns of part batches to a cell that was already visited. The objective function of this model distinguishes this study from others. Moreover, this study considers fuzzy maximum number of machines as a fuzzy resource in the corresponding constraints. Moreover, in the FLP, the objective function is also fuzzy, but the coefficients of the objective are crisp. During the solution of the problem, combinations of the two membership functions and two operators were employed. Membership functions used are linear non-increasing and triangular functions. The operators are "fuzzy and" and max-min operators. The model was solved for three small size data sets.

Now, the fuzzy parametric programming (FPP) studies will be discussed. Arikan and Gungor have two FPP studies where the former one has a single objective [1] and latter has multiple objectives [2]. In the former one [1], the objective is the minimization of the total subcontracting, machine purchasing and inter-cell material handling costs and the coefficients of the objective are fuzzy. In addition, the resources and the parameters within the constraints are fuzzy. These are purchasing price of a machine, capacity of a machine per period, demand per period, inter-cell material handling cost subcontracting cost unit per piece of part capacity utilization of a machine for a part type. The authors used the data set of Shafer *et al.* [52] with some modifications for the numerical illustration. Fuzzy parameters' membership functions were constructed and their values for different membership levels (μ) between 0 and 1 such as 0.1, 0.2, etc. were calculated. Based on these values different crisp LP models were constructed and solved using commercial software.

In addition to the objective of [1], the FMOPP model of [2] intends to maximize utilized capacity of the system, and to minimize total number of exceptional operations. The fuzzy parameters are same with those of [1]. For the solution of the FMOPP, the authors employed the approach of Carlsson and Korhonen's approach [12] explained above.

Moreover, Safaei and Tavakkoli-Moghaddam [48] developed a FPP model for the dynamic CMS design. Total system reconfiguration, machines constant and variable, and inter-cell material handling costs are attempted to be minimized. However, the objective is non-linear. Machine capacities and part demands are the only fuzzy parameters. The model was solved for different levels of precision (α) of the parameters. The model and the solution were illustrated on a small example.

Finally, the studies that employed fuzzy goal programming will be discussed. Eski and Ozkarahan [17] developed a FGP model where maximization of the utilized system capacity, minimization of inter-cell movements due to the exceptional operations, and minimization of both mean tardiness and percentage of tardy jobs are aimed. These are the fuzzy goals. Minimum and maximum limits for the first goal were determined analytically. The limits of the other three goals were determined by means of a simulation model. Solution of the model was carried out by tabu search. With respect to the objectives of the problem and the means of determining the limits of the goals and the solution of the model differentiate this study from the other FGP studies.

In addition, Mahdavi *et al.* [35] proposed a FGP model for the CMS design and production planning in a dynamic virtual CMS. The difference between the Virtual CMS and the real CMS is that the machines of the same group are not necessarily brought into physical proximity [35]. The objectives of the proposed FGP model are minimization of the number of exceptional operations, and minimization of total inventory holding and back order costs. The model was illustrated using a small example.

Paydar *et al.* [43] also developed a FGP model where minimization of total inter-cell, intra-cell material handling cost and minimization of machines' constant/variable costs are aimed. The model was solved for a small problem. Moreover, Nunkaew and Phruksaphanrat [39] developed a preemptive FGP model where minimization of the exceptional operations and minimization of the void elements within the cells are aimed. In preemptive FGP, the goals have different priorities or weights. The satisfactory level for each goal must be satisfied consecutively. Solution of the proposed model was shown using two small-scale problems.

Based on this review of fuzzy programming studies, it can be deduced that the number of fuzzy programming models that consider multiple objectives simultaneously have increased, recently. This is due to the fact that optimizing a single objective does not satisfy the requirements of a CMS design. Moreover, it was observed that the models were attempted to be illustrated on very small examples. Although most of the fuzzy programming studies emphasized that real life problems consists of imprecision and this technique would be helpful in this setting, the authors could not encounter any study that solved a fuzzy model on a real life problem. This is a gap within the literature. Moreover, the tools or techniques that must be employed for the solution of the large problem sizes also need attention, in future studies.

24.2.3 *Fuzzy neural networks*

Neural networks are particularly parallel computer algorithms with an ability to learn from experience [51]. They can adapt and generalize the new information and present reliable classifications of data. Artificial neural networks can be classified as either supervised or unsupervised based on the learning type. To identify similar patterns, supervised neural network models can learn from previous experience or have to be trained, whereas unsupervised neural network models are able to self-organize with little or no training [66]. In the GT literature, unsupervised neural networks are applied to CF problems. One of the well known unsupervised neural networks is the Adaptive Resonance Theory (ART) neural networks. Adaptive Resonance Theory (ART) neural networks were introduced as a theory of human cognitive information processing in an evolving series of models [15]. Fuzzy Art which is proposed by Carpenter *et al.* [13] is an unsupervised category learning and pattern recognition network. The algorithm integrates computations from fuzzy set theory into the Adaptive Resonance Theory (ART) based neural network. Although Fuzzy ART does not require a completely binary representation of the parts to be grouped, it possesses the same desirable stability properties as ART1 and a simpler architecture than that of ART2.

Burke and Kamal [10, 11] showed the use of fuzzy ART and demonstrated that fuzzy ART is superior to other ART neural networks. Suresh and Kaparthi [56] and Kaparthi and Suresh [27] examined the performance of fuzzy ART on large scale data sets with different degrees of imperfection and showed that the performance of fuzzy ART is better and more consistent. Fuzzy ART is capable of rapid, stable clustering of analog or binary input pat-

terns. In another work, Kamal and Burke [26] introduced a new clustering algorithm which is an improved version of fuzzy ART called FACT (Fuzzy ART with Add Clustering Technique). Their algorithm can be used to group parts-machines under a multiple objective environment. Suresh *et al.* [57] developed a Fuzzy ART neural network to address the part-machine grouping problem considering operation sequences. Park and Suresh [41] compared the performance of Fuzzy ART with traditional hierarchical clustering techniques for grouping parts based on operation sequences with large size data sets. Peker and Kara [44] investigated the parameter setting concept for Fuzzy ART networks. Kuo *et al.* [28] demonstrated fuzzy ART2 neural network for clustering parts into several families. They showed that fuzzy ART2 neural network can correctly cluster the parts even under the shift and noise conditions and also fuzzy ART2 can have very promising results compared with fuzzy self-organizing feature map and fuzzy c-means. In the recent studies, Won and Currie [66] proposed Fuzzy ART/RRR-RSS (Fuzzy ART/ReaRRangement-ReaSSignment) algorithm which uses non-binary part machine incidence matrix which effectively reflects real manufacturing factors such as the operation sequences including multiple visits to the same machine and production volumes of parts. They show that their algorithm has robustness and recoverability to large-size ill-structured data sets by producing highly independent block diagonal solution close to the near-best one. Ozdemir *et al.* [40] presented a new pattern recognition neural network algorithm for clustering problems, and demonstrated its use for machine cell design in group technology. Their algorithm includes modifications of the learning procedure and resonance test of the Fuzzy ART neural network which enable the neural network to process integer values rather than binary valued inputs. Sengupta *et al.* [51] proposed an artificial neural network based hybrid clustering algorithm which incorporates Fuzzy ART and K-means clustering to solve the cell formation problem using the non-binary real valued work load data as an input matrix. Their algorithm gives better results compared to simple K-means algorithm and modified ART1 algorithm as found in the recent literature.
The advantages of fuzzy ART are

i) it is capable of rapid stable clustering of both binary and non-binary input patterns,
ii) it can overcome the problem of category proliferation in fast learn/slow recode option.

The disadvantage of the algorithm is that the quality of the solutions depends on the accurate choice of the parameters.

The main components of fuzzy ART network are displayed in Fig. 24.1. The network consists of two layers of neurons: the comparison (input) layer and the recognition (output) layer. The output layer is composed of neurons, each representing one class of input patterns. The neurons required in output layer are dynamically created when a new class is identified. Conversely, the input layer is

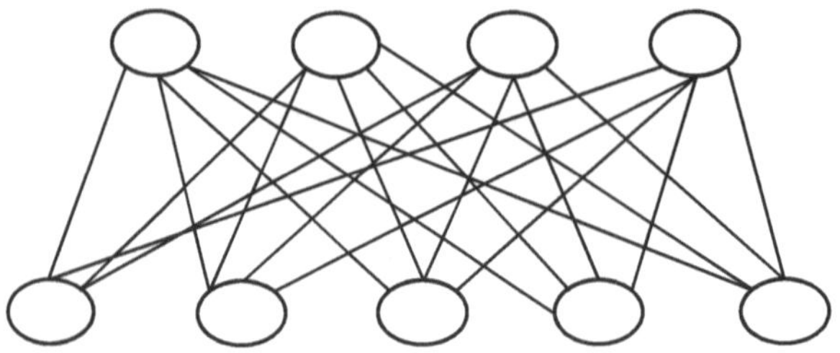

Fig. 24.1 Fuzzy ART network

composed of neurons, each representing an input pattern. The neurons of the input layer are connected to the output neurons of the network by weight vectors. The values held by the weight vectors are changed whenever a new input pattern is learned by the network. In CF applications, the neurons in the output layer corresponding to the classes of input patterns represent part families identified. The number of neurons in the input layer depends on the characteristics of the input pattern (such as routing data). Thus, the number of neurons in the input layer equals to the number of machine types.

The Fuzzy ART neural network involves three parameters; the choice parameter ($\alpha > 0$), the learning rate ($\beta \in [0,1]$) and the vigilance parameter ($\rho \in [0,1]$). The outputs of the network may vary when the selected values of these parameters are changed.

Steps of the Fuzzy ART network

The step-by-step illustration of Fuzzy ART network is as follows [56]:

1. Initialization.
   Connection weights: $w_{ij}(0) = 1$.
   $0 \leqslant I \leqslant N-1, 0 \leqslant j \leqslant M-1$.
   Select values for $\alpha > 0$, $\beta \in [0,1]$.
2. Read a new input vector I consisting of binary or analog elements.

3. For every output node j, compute choice function
   $T_j = \|I \wedge w_j\| / [\alpha + \|w_j\|]$, $0 \leqslant j \leqslant M-1$, where $\wedge$ is the fuzzy AND operator defined as: $x \wedge y = \min(x_i, y_i)$.
4. Select best matching exemplar (maximum output value): $T_\theta = \max_j\{T_j\}$.
5. Resonance test (i.e. degree of similarity with best matching exemplar).
   If $\|I \wedge w_\theta\| / \|I\| \geqslant \rho$ then go to step 7 (learning step); else go to step 6
6. Mismatch reset: set $T_\theta = -1$ and go to step 4.
7. Update best matching exemplar (learning law):
   $w_\theta^{\text{new}} = \left[\beta \times \left(I \wedge w_\theta^{\text{old}}\right)\right] + \left[(1-\beta) \times w_\theta^{\text{old}}\right]$
8. Repeat: go to step 2.

24.3 Conclusion

In this book chapter, fuzzy logic applications in CMS design were classified based on the technique used, a concise explanation of each technique was presented, and the techniques were tried to be exemplified. Then, the studies that employed each fuzzy technique for the CMS design were discussed. The major fuzzy techniques for the CMS design are fuzzy clustering, fuzzy programming, and fuzzy neural networks. These techniques can also be divided into sub categories that are explained in the corresponding subsection.
The most important decision in designing effective CMS is the Cell Formation (CF) which groups parts with similar design features or processing requirements into part families and corresponding machines into machine cells. In the past, most of the studies based on CF considered only design features or processing requirements for grouping. However, to reach the goal of creating efficient manufacturing cells it is very important to consider workers, tools, machine utilizations or other dimensions related to parts and machines respectively to meet industrial specifications through the grouping process. Hence, in future studies, these dimensions should also be considered for CF problems. Including additional dimensions to the fuzzy CMS design applications would increase the complexity and size of the problem. Hence, these large-size complex problems may require more complex solution algorithms such as hybrid methods or meta-heuristics. More attention should be given to using a combination of fuzzy logic with meta-heuristics for the solution of complex CF problems. The integration of fuzziness for a CF problem with a dynamic condition in CMS where the dynamic condition indicates a multi-period planning horizon, in which product mix and demand in each period are different, is another important research area

which has been studied by only a few researchers in the past literature. Thus, as a further study it would be useful to extend the solution methods of CF problems for dynamic conditions. Moreover, industrial examples of fuzzy CMS design that embrace the imprecision and dynamic nature will be beneficial to reveal the utility of the fuzzy logic for this group of problems.

Bibliography

[1] Arıkan, F. and Güngör, Z. (2005). A parametric model for cell formation and exceptional elements' problems with fuzzy parameters, *J. Intell. Manufacturing,* **16** (1), pp. 103–114.

[2] Arıkan, F. and Güngör, Z. (2009). Modeling of a manufacturing cell design problem with fuzzy multi-objective parametric programming. *Mathematical and Computer Modelling*, **50** (3-4), pp. 407–420.

[3] Batra, J.L. and Rajagopalan, R. (1977). A composite components through graphs and fuzzy clusters. *Proc. of the 18th International MTDR Conference*, London.

[4] Bector, C.R. and Chandra, S. (2005) Fuzzy mathematical programming and fuzzy matrix games, Springer-Verlag, Berlin Heidelberg.

[5] Behret, H. and Kahraman, C. (2008). A Fuzzy Clustering Application in a Cellular Manufacturing System, *Proc. of the 8th International FLINS Conference. Computational Intelligence in Decision and Control*, pp. 1117–1122.

[6] Bellman, R.E., Zadeh, L.A. (1970). Decision making in fuzzy environment, *Manage. Sci.*, **17** (4), pp. B141–B164.

[7] Ben-Arieh, D., Lee, S.E. and Chang, P.T. (1996). Fuzzy part coding for group technology. *Euro. J. Op. Res.*, **92**, pp. 637–648.

[8] Ben-Arieh, D. and Triantaphyllou, E. (1992). Quantifying data for group technology with weighted fuzzy features. *Int. J. Prod. Res.*, **30** (6), pp. 1285–1299.

[9] Bezdek, J.C. (1981). *Pattern Recognition with Fuzzy Objective Function and Algorithms,* New York: Plenum Press.

[10] Burke, L.I. and Kamal, S. (1992). Fuzzy ART and cellular manufacturing. *Proc. of the ANNIE 92 Conference*, pp. 779–784.

[11] Burke, L.I. and Kamal, S. (1995). Neural networks and the part family/machine group formation problem in cellular manufacturing: a framework using fuzzy ART. *Journal of Manufacturing Systems*, **14**, pp. 148–159.

[12] Carlsson, C. and Korhonen, P. (1987). A parametric approach to fuzzy linear programming. *Fuzzy sets and systems,* **20**, pp. 17–30.

[13] Carpenter, G.A., Grossberg, S., and Rosen, D.B. (1991). Fuzzy ART: fast stable learning and categorization of analog patterns by an adaptive resonance system, *Neural Networks*, 4, pp. 759–771.

[14] Chandrasekharan, M.P. and Rajagopalan, R. (1986). An ideal seed nonhierarchical clustering algorithm for cellular manufacturing. *Int. J. Prod. Res.*, **24**, pp. 451–464.

[15] Chen, C.H. (1995). *Fuzzy Logic and Neural Network Handbook,* IEEE Press, McGraw-Hill.

[16] Chu, C.H. and Hayya, J.C. (1991). A fuzzy clustering approach to manufacturing cell formation. *International Journal of Production Research,* **29**, pp. 1475–1487.

[17] Eski, O. and Ozkarahan, H. (2007). Design of Manufacturing Cells for Uncertain Production Requirements with Presence of Routing Flexibility, *Advanced Intelligent Computing Theories and Applications. With Aspects of Artificial Intelligence*, pp. 670–681.

[18] Feyzioglu O. and Pierreval, H. (2009). Hybrid organization of functional departments and manufacturing cells in the presence of imprecise data. *Int. J. Prod. Res.*, **47**, pp. 343–368.

[19] Gill, A. and Bector, C.R. (1997). A fuzzy linguistic approach to data quantification and construction of distance measures for the part family formation problem. *Int. J. Prod. Res.*, **35** (9), pp. 2565–2578.

[20] Gindy, N.N.Z., Ratchev, T.M., and Case, K. (1995). Component grouping for GT applications: a fuzzy clustering approach with validity measure. *International Journal of Production Research*, **33**, pp. 2493–2509.

[21] Gungor, Z. and Arikan, F. (2000). Application of fuzzy decision making in part–machine grouping. *International Journal of Production Economics*, **63**, pp. 181–193.

[22] Ham, I., Hitomi, K., and Yoshida, T. (1985). *Group Technology: Applications to Production Management.* MA: Kluwer-Nicho.

[23] Hung, W.L., Yang, M.S., and Lee, E.S. (2011). Cell formation using fuzzy relational clustering algorithm. *Mathematical and Computer Modelling*, **53**, pp. 1776–1787.

[24] Irani, S.A., Subramanian, S., and Allam Y.S. (1999). Introduction to cellular manufacturing systems. In *Hand-book of cellular manufacturing systems,* ed. S. Irani. New York: John Wiley & Sons, Inc.

[25] Josien, K. and Liao, T.W. (2002). Simultaneous grouping of parts and machines with an integrated fuzzy clustering method. *Fuzzy Sets and Systems,* **126**, pp. 1–21.

[26] Kamal, S. and Burke, L.I. (1996). FACT: A new neural network based clustering algorithm for group technology. *International Journal of Production*, **34**, pp. 919–946.

[27] Kaparthi, S. and Suresh, N.C. (1994). Performance of selected part-machine grouping techniques for data sets of wide ranging sizes and imperfection. *Decision Sciences*, **25**, pp. 515–539.

[28] Kuo, R.J., Su, Y.T., Chiu, C.Y., Chen, K.Y. and Tien, F.C. (2006). Part family formation through fuzzy ART2 neural network. *Decision Support Systems,* **42**, pp.89–103.

[29] Lai, Y.J. and Hwang, C.L. (1992). *Fuzzy mathematical programming: methods and applications.* Springer, Heidelberg.

[30] Leem, C.W. and Chen, J.J. (1996), Fuzzy-set based machine-cell formation in cellular manufacturing. *Journal of Intelligent Manufacturing*, **7**, pp. 355–364.

[31] Li, J., Chu, C.H., Wang, Y., and Yan, W. (2007). An improved fuzzy clustering method for cellular manufacturing, *International Journal of Production Research*, **45** (5), pp. 1049–1062.

[32] Li, J., Ding, Z., and Wei, W. (1988). Fuzzy cluster analysis and fuzzy pattern recognition methods for formation of part families (NAMRC). *Society of Manufacturing Engineers*, pp. 558–563.

[33] Liao, T.W. (2001). Classification and coding approaches to part family formation under a fuzzy environment. *Fuzzy Sets and Systems,* **122** (3), pp. 425–441.

[34] Lozano, S., Dobado, D., Larrañeta, J., and Onieva, L. (2002). Modified fuzzy C-means algorithm for cellular manufacturing. *Fuzzy Sets and Systems,* **126**, pp. 23–32.

[35] Mahdavi, I., Mahdi Paydar, M.M., and Solimanpur, M. (2011). Solving a new mathematical model for cellular manufacturing system: A fuzzy goal programming approach. *Iranian Journal of Operations Research,* **2** (2), pp. 35–47.

[36] Masnata, A. and Settineri, L. (1997). An application of fuzzy clustering to cellular manufacturing. *International Journal of Production Research*, **35** (4), pp. 1077–1094.

[37] Naadimuthu, G., Gultom, P., and Lee, E.S. (2010). Fuzzy clustering in cell formation with multiple attributes, *Computers and Mathematics with Applications*, **59**, pp. 3137–3147.

[38] Narayanaswamy, P., Bector, C.R., and Rajamani, D. (1996) Fuzzy logic concepts applied to machine-componentmatrix formation in cellularmanufacturing. *European Journal of Operational Research*, **93**, pp. 88–97.

[39] Nunkaew, W. and Phruksaphanrat, B. (2011). Concurrent Cell Formation for Cellular Manufacturing System by Preemptive Fuzzy Goal Programming. *Proc. of the International MultiConference of Engineers and Computer Scientists*, Hong Kong.

[40] Ozdemir, R.F., Gencyilmaz, G., and Aktin, T. (2007). The modified fuzzy art and a two-stage clustering approach to cell design. *Information Sciences*, **177** (23), pp. 5219–5236.

[41] Park, S. and Suresh, N.C. (2003). Performance of Fuzzy ART neural network and hierarchical clustering for part-machine grouping based on operation sequences. *International Journal of Production Research*, **41** (14), pp. 3185–3216.

[42] Papaioannou, G. and Wilson, J.M. (2010). The evolution of cell formation problem methodologies based on recent studies (1997-2008): Review and directions for future research. *Eur. J. Oper. Res.*, **206**, pp. 509–521.

[43] Paydar, M.M., Mahdavi, I., Solimanpur, M., and Tajdin, A. (2008). Solving a New Mathematical Model for Cellular Manufacturing System: Fuzzy goal Programming, *Proc. of the 2008 IEEE IEEM*, pp. 1224–1228.

[44] Peker, A. and Kara, Y. (2004). Parameter setting of the Fuzzy ART neural network to part–machine cell formation problem. *International Journal of Production Research*, **42** (6), pp. 1257–1278.

[45] Rajagopalan, R. (1975). *Design retrieval and group layout through fuzzy sets and graphs*. Unpublished PhD thesis, Indian Institute of Technology, Kanpur, India.

[46] Ravichandran, K.S. and Rao, K.C.S. (2001) . A new approach to fuzzy part-family formation in cellular manufacturing systems. *International Journal of Advanced Manufacturing Technology*, **18**, pp. 591–597.

[47] Safaei, N., Saidi-Mehrabad, M., and Babakhani, M. (2007) Designing cellular manufacturing systems under dynamic and uncertain conditions. *J Intell Manuf*, **18**, pp. 383–399.

[48] Safaei, N. and Tavakkoli-Moghaddam, R. (2009). An extended fuzzy parametric programming-based approach for designing cellular manufacturing systems under uncertainty and dynamic conditions, *International Journal of Computer Integrated Manufacturing*, **22** (6), pp. 538–548.

[49] Safaei, N., Saidi-Mehrabad, M., Tavakkoli-Moghaddam, R., and Sassani, M. (2008). A fuzzy programming approach for a cell formation problem with dynamic and uncertain conditions, *Fuzzy Sets and Systems*, **159**, pp. 215–236.

[50] Selim, H.M., Askin, R.G., and Vakharia, A.J. (1998). Cell formation in group technology: Evaluation and directions for future research, *Computers and Industrial Engineering*, **34** (1), pp. 3–20.

[51] Sengupta, S., Ghosh, T., and Dan, P.K. (2011). Fuzzy ART K-Means Clustering Technique: a hybrid neural network approach to cellular manufacturing systems. *International Journal of Computer Integrated Manufacturing*, **24** (10), pp. 927–938.

[52] Shafer, S.M., Kern, G.M., and Wei, J.C. (1992). A mathematical programming approach for dealing with exceptional elements in cellular manufacturing. *International Journal of Production Research*, **30**, 1029–1036.

[53] Shanker, R. and Vrat, P. (1999). Some design issues in cellular manufacturing using the fuzzy programming approach, *International Journal of Production Research*, **37**, pp. 2545–2563.

[54] Singh, N. and Rajamani, D. (1996). *Cellular Manufacturing Systems*. New York: Chapman & Hall.

[55] Su, C.T. (1995). A fuzzy approach for part family formation. *Proc. of The IEEE Industry Application Society 30th Annual Meeting*, Orlando, Florida.

[56] Suresh, N.C. and Kaparthi, S. (1994). Performance of fuzzy ART neural network for group technology cell formation. *International Journal of Production Research*, **32**, pp. 1693–1713.

[57] Suresh, N.C., Slomp, J., and Kaparthi, S. (1999). Sequence dependent clustering of parts and machines: a Fuzzy ART neural network approach. *International Journal of Production Research*, **37** (12), pp. 2793–2816.

[58] Susanto, S., Kennedy, R.D. and Price, J.W. (1999). A new fuzzy-c-means and assignment technique-based cell formation algorithm to perform part-type clusters and machine-type clusters. *Production Planning and Control,* **10** (4), pp. 375–388.

[59] Susanto, S., Bhattacharya, A. and Al-Dabass, D. (2010). Fuzzy Clustering-Based Optimised Cell Formation Algorithm Considering Sequence of Operations, Alternative Routing and Part-Volume. *Proc. of the 12th International Conference on Computer Modelling and Simulation,* Cambridge, UK.

[60] Torkul O., Cedimoglu I.H., and Geyik A.K. (2006). An application of fuzzy clustering to manufacturing cell design. *Journal of Intelligent and fuzzy systems,* **17** (2), pp. 173–181.

[61] Tsai, C.C., Chu, C.H., and Arnold Barta, T. (1997). Modeling and analysis of manufacturing cell formation problem with fuzzy mixed-integer programming. *IIE Transactions*, **29**, pp. 533–547.

[62] Venugopal, V. (1999). Soft-computing-based approaches to the group technology problem: A state-of-the-art review, *International Journal of Production Research,* **37**, pp. 3335–3357.

[63] Verdegay, J.L., (1982). Fuzzy mathematical programming. In *M.M. Gupta and E. Sanchez, Eds., Fuzzy Information and Decision,* Processes North Holland, Amsterdam, 231–237.

[64] Wen, H.J., Smith, C.H., and Minor, E.D. (1996). Formation and dynamic routing of part families among flexible manufacturing cells. *International Journal of Production Research*, **34**, pp. 2229–2245.

[65] Werners, B., (1987). Interactive multi objective programming subject to flexible constraints. *European Journal of Operational Research,* **31**, pp. 342–349.

[66] Won, Y. and Currie, K.R. (2007). Fuzzy ART/RRR-RSS: a two-phase neural network algorithm for part-machine grouping in cellular manufacturing. *International Journal of Production Research*, **45** (9), pp. 2073–2104.

[67] Xu, H. and Wang, H. (1989). Part family formation for GT applications based on fuzzy mathematics. *International Journal of Production Research*, **27**, pp. 1637–1651.

[68] Yang, M.S., Hung, W.L., and Cheng, F.C. (2006). Mixed-variable fuzzy clustering approach to part family and machine cell formation for GT applications, *International Journal of Production Economics,* **103** (1), pp. 185–198.

[69] Zhang, C. and Wang, H. (1992). Concurrent formation of part families and machine cells based on the fuzzy set theory. *Journal of Manufacturing Systems*, **11**, pp. 61–67.

[70] Zimmerman, H.J., (1976). Description and optimization of fuzzy system. *International Journal of General System,* **2**, pp. 209–216.

Chapter 25

The Shape of Crisis Lessons from Self Organizing Maps

Marina Resta

Department of Economics, University of Genova 16126 Genova, Italy

E-mail: resta@economia.unige.it

Self Organizing Maps are computational tools whose engagement in various research fields has grown faster and wider in latest year, with the notable exception of macroeconomics, where contributions are somewhat lacking. However, we are going to provide evidence that joining Self Organizing Maps together with some graphs theory tools (namely: the Minimum Spanning Tree), they can be successfully employed to develop macroeconomic models thus taking both static and dynamic (i.e. over a moving period of time) snapshots of countries financial situations. In this way it is possible to generate useful information for policy makers, in order to realize more efficient interventions in periods of either higher instability or full-blown crisis situation.

25.1 Problem Statement

Self Organizing Maps (SOMs) are nowadays a landmark among soft computing techniques, with applications which virtually span over all fields of knowledge. However, while the use of SOMs in robotics, medical imaging, characters recognition, to cite most important examples, is celebrated by a consistent literature corpus, economics and financial markets seem niche wise fields, relatively lesser explored, with some notable exceptions. Such lack is quite nonsense, provided the great potential that relies on this kind of technique.

The rationale of this contribution is to offer some insights into the use of SOMs to explore both how financial markets organize during critical periods (deflation, recession and so on), and in the exploration of countries exposure to financial crisis risk. One could straightforwardly think that something similar has been already discussed. Nevertheless, although there is an extensive literature dealing with so-called Early Warning Systems

C. Kahraman (ed.), *Computational Intelligence Systems in Industrial Engineering*,
Atlantis Computational Intelligence Systems 6, DOI: 10.2991/978-94-91216-77-0_25,

(EWS), i.e. pure econometric tools (at least in their original release) that are able to develop alarms to alert policy makers in case of critical economic situations before they effectively happen. Attempts to develop SOM-based EWS are documented in more recent years, as well. Besides, the European Central Bank (ECB) Financial Stability Review (December, 2011, p. 149) highlighted the importance of SOM as tools to allow early identification of sources of vulnerabilities, making it possible to take preventive targeted policy actions.

This contribution, however, goes one step forward under various points of view. The first element of innovation relies on the examined data. We studied the situation in markets characterized by different levels of (in)stability; to such purpose, instead of using financial indicators as it is generally done, we employed historical time-series of price levels for every company quoted in the related stock exchanges. This choice may be easily justified by observing that in recent years critical situations waved from financial markets to flood throughout other sectors of the country economy. We then trained a SOM for data of each market.

Another innovative item is of technical nature (i.e. related to the features of SOM algorithm), since it relied on a peculiar use of SOMs units. We extracted from SOMs their best matching units and we employed them to build for each market the related Minimum Spanning Tree (MST) structure. In this way we were able to get a kind of market "skeleton" frame, and hence to capture both clusters structure of every market, and to analyze the impact of emerging patterns over the economic situation of the country. This second target was pursued both in a static and in a dynamic way. In the earlier case, we examined the markets topology with data referring to a fixed period with length one year (from December 2010 to December 2011). In the latter case we acted by comparison of up to ten MST that we obtained going back in time for a maximum of 3000 days, with a 300 day long moving window.

What we are going to show throughout the paper may be easily summarized as follows: (i) using SOMs we got an original representation of financial markets topology; (ii) building from SOMs winning nodes the corresponding MST structure made us possible both to emphasize the relations among various quoted enterprises, and to check for the emergence of critical patterns; (iii) we provided a global representation of countries financial situation thus generating information that can be of help to policy makers, in order to realize more efficient interventions in periods of higher instability.

What remains of the chapter is organized as follows. Section 25.2 is made of three parts. In the first and the second ones we will provide a brief review of SOM and MST respectively. In the third part we will illustrate how the two concepts can cope together in order to obtain a skeleton frame of the examined data domain. In Section 25.3 we will present some technicalities concerning both the examined markets and the data we studied for each of them. Section 25.4 will detail the simulation experiments and the results. Section 25.5 will conclude.

25.2 Technical Background

The simulations that we are going to describe and discuss rely on the joined use of both SOM and graphs formalism to analyze the available data. To make clearer the framework we are merging into, it is then straightforward to introduce some basic definitions and notational conventions for both the aforementioned tools.

25.2.1 *Self Organizing Maps*

Self Organizing Maps (SOMs), also known as Kohonen maps, are very powerful unsupervised methods to project data from the original higher dimensional space into a lower dimensional one (usually bi-dimensional), maintaining unchanged existing topological relationships.

The way SOMs work may be easily described. Consider first a finite set $X = \{x(t)\}_{t=1,\ldots,T}$ of r-dimensions input data items:

$$\mathbf{x}(t) = [x_1\ x_2\ \cdots\ x_r]. \tag{25.1}$$

Besides, let us assume that M is the $m \times k$ bi-dimensional projection grid whose elements (units, nodes) are arranged into m rows and k columns: each unit of M is associated with an array w_{ij} ($i = 1,\ldots,m$, $j = 1,\ldots,k$) whose number of components is the same as that of input data (i.e.: w_{ij} has r elements).

After initializing the map nodes at random, the Kohonen algorithm in its simplest (and more commonly used) version assumes to iteratively modify the map nodes by the following rules. For each $t = 1,\ldots,T$, we will have:

$$\mathbf{w}^*_{ij}(t) = \min_{\mathbf{w}_{ij}(t)\in\mathbf{M}} \{d[\mathbf{x}(t), \mathbf{w}_{ij}(t)]\} \tag{25.2}$$

and

$$\mathbf{w}_{ij}(t+) = \mathbf{w}_{ij}(t) + \alpha(t)h\left[t, \mathbf{x}(t), \mathbf{w}^*_{ij}(t)\right]\left[\mathbf{x}(t) - \mathbf{w}_{ij}(t)\right]. \tag{25.3}$$

Where $d[\mathbf{x}(t)-\mathbf{w}_{ij}(t)]$ is the function that computes the distance among input patterns and each node in M. Despite very sophisticate functions may be used, most common choice is the Euclidean distance:

$$d_E[\mathbf{x}(t)-\mathbf{w}_{ij}(t)] = \sqrt{([d[\mathbf{x}(t)-\mathbf{w}_{ij}(t)]]' \cdot d[\mathbf{x}(t)-\mathbf{w}_{ij}(t)])} \tag{25.4}$$

the symbol $'$ being the standard notation for transposition. The node satisfying (25.2) is called winner or Best Matching Unit (BMU). The notation $\alpha(t)$ indicates a scalar decreasing factor in the range $[0,1]$ depending on time t that defines the size of the correction: starting from values closer to one (that means maximum correction), as time goes on $\alpha(t)$ decreases to values nearer to zero (no correction at all). Finally, h is the neighborhood function; it models the distance between map nodes and BMU. Function h may assume various shapes, but here we refer to the simplest one:

$$h(p_{\text{BMU}}, p_{\mathbf{w}_{ij}}, t) = e^{-\sum |p_{\text{BMU}} - p_{\mathbf{w}_{ij}}|} \tag{25.5}$$

where p_{BMU}, $p_{\mathbf{w}_{ij}}$ are the grid coordinates of both BMU and generic map nodes respectively.

25.2.2 *Some brief remarks about the Minimum Spanning Tree*

From the formal viewpoint, a graph can be fully described as: $G=(V,E)$, where V denotes a (finite) set of nodes (vertexes, points), and $E \subseteq V \times V$ indicates edges, i.e. a collection of unordered pairs $\{u,v\}$ of distinct elements from V. Graphs representation passes through the building of the adjacency matrix, i.e. the matrix that marks neighbor vertexes with one, and labels with a zero not adjacent nodes. Figure 25.1 provides an explanatory example.

The way linkages among nodes organize may origin different structures. This is the case of the Minimum Spanning Tree (MST), which is nothing but a particular graph with no cycles, where all nodes are connected and edges are selected in order to minimize the sum of distances. Here, we will provide some pseudo-code of the procedure to build it.

Assume C being a square symmetric matrix: by this assumption we can focus only on its lower (upper) triangular part $L(U)$. Then:

1. sort the elements of $L(U)$ in decreasing order, thus moving from L to the list Lord (from U to $Uord$).
2. Set the coordinates in C of the first element of *Lord* ($Uord$) as those of the first two nodes of the MST.

$$\begin{pmatrix} 1 & 1 & 0 & 0 & 1 & 0 \\ 1 & 0 & 1 & 0 & 1 & 0 \\ 0 & 1 & 0 & \mathbf{1} & 0 & 0 \\ 0 & 0 & 1 & 0 & 1 & 1 \\ 1 & 1 & 0 & \mathbf{1} & 0 & 0 \\ 0 & 0 & 0 & \mathbf{1} & 0 & 0 \end{pmatrix}$$

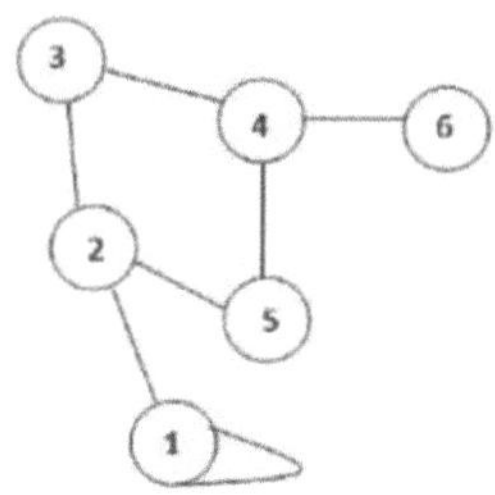

Fig. 25.1 From left to right: the adjacency matrix for an undirected graph, and the corresponding graph. The ones in the matrix indicate the existence of connection among nodes, while zeroes mean no connection.

3. For each element in *Lord* (*Uord*) add the corresponding couple from C to MST; in particular, if the graph still maintains acyclic (i.e. no loops are added to MST), then hold the inserted link, otherwise discard it.
4. Repeat step 3 until all the elements in *Lord* (*Uord*) have been examined, and then stop the procedure.

25.2.3 *Coping SOM to MST*

SOM achieves a bi-dimensional representation of the input domain, maintaining unchanged the basic relations among neighbor patterns: more simpler, closer points in their r-dimensions ($r \gg 2$) initial space are still nearer one to each other in the SOM grid; in addition, they are projected into a space where relations can be easily visualized and understood. However, sometimes this cannot be enough.

Consider the issue to represent basic relations among quoted societies in a market (for example: in the German market). Figure 25.2 shows the appearance of the SOM, once existing relations among German quoted companies have been learned.

The example clearly shows an intrinsic limitation of the SOM procedure: projecting points onto lower dimensional spaces may lead to clusters whose components (in our case: the enterprises) are even too much closer than desired, thus making difficult to understand the effective significance of their closeness. Besides, if we take a look to the existing connections among nodes, we could obtain a representation like the one in Figure 25.3.

One can immediately note the huge number of connections: almost all nodes seem to be reciprocally connected, and it is almost impossible to separate strongly correlated nodes from those with weaker connections. Besides, SOM declared aim, i.e. to unveil that part of information that traditional techniques hardly disclose, seemed completely missing.

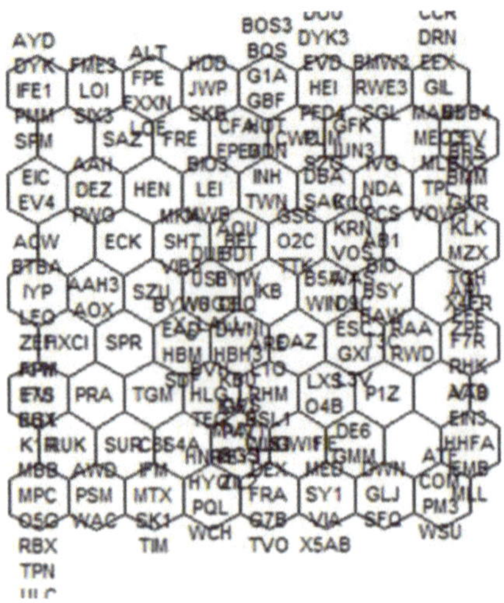

Fig. 25.2 SOM representation of German quoted enterprises

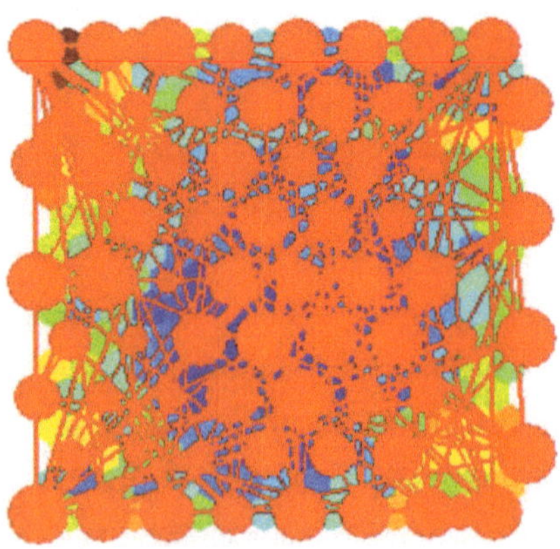

Fig. 25.3 Connections among SOM nodes in the German market

We then suggested a procedure that copes the SOM algorithm to the MST to extract essential features from data. Basic steps are reported on following.

Step 1. Run SOM on data.

Step 2. Extract BMUs from SOM.

Step 3. Build the correlation matrix C among BMU nodes.

Step 4. Use C as starting point to compute the MST.

Clearly, in this way the available information is filtered, and basically this is our idea: to use a filtering procedure and hence to let more significant patterns to emerge.

25.3 Data and Methodology

Our goal is to investigate to what extent SOMs can be used as macroeconomic tools, to develop alerts, hence allowing policy makers to take preventive targeted policy actions.

Classical approaches to the topic assume to describe the country's situation by means of a bunch of financial ratios like (but not limited to): Debt to equity, Capital reserve to asset, Capital growth, Debt growth, Assets turnover, Return on asset, Asset growth, Business growth, Capital reserve growth, and so on. However, we argue that over the past decade most serious crises arose from financial markets to flood throughout other sectors. We hence decided to monitor countries financial wealth by analyzing the basic relations that are formed on stock exchanges among quoted enterprises.

This was done in two ways, both statically and dynamically.

By applying the static procedure we took a picture of markets referring to the latest year. Once selected a market (say for instance the German Stock Exchange) for each quoted enterprise we performed the following actions. Firstly, we selected all available price levels from December 2010 to December 2011. We were then able to manage for the generic i-th stock ($i = 1,\dots,N$, where N is the overall number of quoted enterprises) the time-series $S^{(i)} = \{pv_k^{(i)}\}$ with length $T-1$, being:

$$pv_k^{(i)} = \log \frac{pl_k^{(i)}}{pl_{k-1}^{(i)}}, \quad k = 2,\dots,T \tag{25.6}$$

The data transformation described in (25.6) turns price levels into price variations (also known as log-returns). This is a perfectly straightforward task to eliminate any trend effect from data.

Assuming to use each $S^{(i)}$ time-series as input pattern, we got a matrix Σ of dimensions $N \times T-1$, where N is, as previously said, the number of quoted enterprises, and $T-1$ is the length of $S^{(i)}$, for every i.

We then run on Σ the technique we explained in Section 25.2.3, joining together both SOM and MST approaches.

The dynamical procedure is similar to the static one, but instead of considering last year sample for each stock, we examined a number of fixed length samples, going back in time (when possible) up to ten years. In practice, assuming as starting point $t = 3000$ the day

Table 25.1 Markets main features

Country Mkt	Ref.Idx	NrPS	MatrixDim
Germany	CDAX	207	207×245
Italy	Mibtel	166	166×245

Dec. 30 2011, we build for each stock a block B1 going from $t = 2701$ to $t = 3000$; a block B2 with data from $t = 2401$ to $t = 2700$; and so on towards the block B10 that goes from $t = 1$ to $t = 300$. In this way we were able to draw how market evolved over the latest ten years.

The two approaches were tested on two markets (German, and Italian) each of them characterized by a different level of (in)stability: at the end of 2011 Italian financial equilibrium seemed heavily compromised, while Germany still maintained its leading role in Europe.

25.4 Results

As said in previous section, we examined German and Italian markets in both a static and dynamic approach. Before going to separately discuss such cases, we will spend a few words about some common features shared by our simulation in the static case.

As first step we examined the aforementioned markets over the time frame from 30 December 2010 to 30 December 2011. For each market we considered data of quoted enterprises, transforming them according to the formula given in (25.6). Data main features are reported in Table 25.1.

The column *Country Mkt* details the countries whose assets have been examined, while *Ref.Idx* shows national market index we have employed to find out quoted stocks that we have included into the study; NrPS is the number of stocks we picked up for every market; finally *MatrixDim* highlights input matrix dimensions in our simulation study.

It is noteworthy to observe that in order to pick up assets, we referred to the CDAX (*Composite Deutscher Aktienindex*) index for the German market, and to Mibtel (*Milano Borsa Telematica*) index for Italy. Using stocks featuring such indexes assured us to consider "robust" enterprises: index components, in fact, must comply with international transparency standards.

One could argue that in general the overall number of quoted enterprises is higher than the number we have reported in the third column of Table 25.1. However, in order to make comparisons among obtained graphs, we needed to eliminate from each market those stocks for which it was not possible to go back in time enough (at least 600 days, approximately corresponding to two years and half of market tradings).

25.4.1 *The case of Germany*

The market topology of Germany, as resulting by our procedure is given in Figure 25.4.

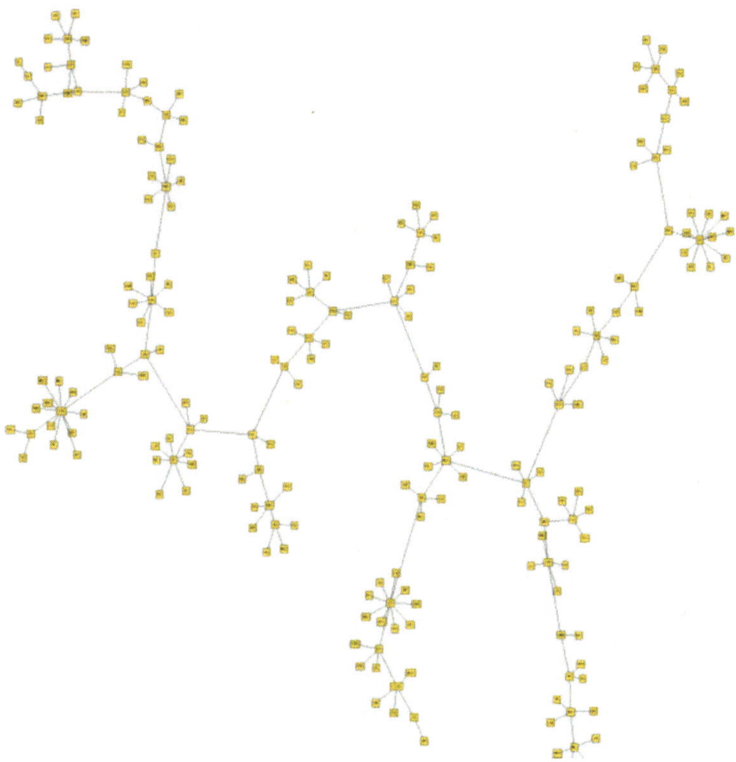

Fig. 25.4 Skeleton framework of German companies in the period: 30 December 2010 – 30 December 2011.

As general remark, the procedure of market skeletonization found 14 clusters. The situation captured by our procedure is quite heterogeneous and it is noteworthy to examine it in deeper detail. To such purpose, Table 25.2 shows clusters' main statistics.

For each cluster we examined basic statistics: mean (mu), standard deviation (std), skewness (sk), and kurtosis (ku). We also evaluated the Sharpe Ratio of every cluster (SR), since this is a performance index which can be easily interpreted, giving an idea about the

Table 25.2 Markets main features

CL. ID.	mu	std	sk	ku	SR
CL01	0.0001270	0.045235	1.651532	411.3881	0.281%
CL02	0.0002923	0.055717	10.48408	2521.463	0.525%
CL03	0.0001208	0.143701	9.605624	906.0551	0.084%
CL04	0.0002904	0.056795	10.3774	2448.29	0.511%
CL05	0.0003361	0.056718	10.14951	2386.374	0.593%
CL06	0.0004759	0.057566	11.4447	2608.218	0.827%
CL07	0.0003254	0.056369	10.4055	2471.74	0.577%
CL08	0.0003683	0.05673	10.54512	2485.83	0.649%
CL09	0.0003223	0.056334	10.14833	2402.134	0.572%
CL10	0.0004811	0.057746	11.11191	2515.727	0.833%
CL11	0.0002920	0.05616	10.06927	2395.238	0.520%
CL12	0.0005635	0.058396	11.85686	2655.044	0.965%
CL13	0.0002770	0.058349	11.88501	2663.531	0.475%
CL14	0.0003343	0.056866	10.46831	2462.213	0.588%

general attractiveness/profitability of the companies included in each group. The analysis of the results evidenced that all clusters have positive mean, relatively low variability, and good profitability (with the exception of CL01 and CL03 whose Sharpe ratio is the lowest over all examined cases). Besides, companies returns are positively skewed. We then analyzed clusters composition (Table 25.3), in order to verify if companies tend to aggregate according to the sector they belong to, as well as if clusters composition may have affected the results that we have shown in Table 25.2.

We found out that, in general, clusters did not have an exclusive, but rather a dominant composition. Looking at Table 25.3, in fact, CL01 exhibits a dominant percentage of companies from both Services (Serv) and Logistics (Log) sector (20%), CL02 is equally divided into firms belonging to Banking and Finance (B&F), Health-Care (HC), Logistics and Components (Comp) sectors which share the same 14.29% percentage. B&F dominates (30%) cluster CL03 as well as CL05 (42.86%), CL11 (21.43%), and CL13 (23.08%). Hi-Tech companies (HT) are preferably grouped into clusters CL04 (25%) and CL10 (28.57%). Companies working in the Health Care sector (HC) are more numerous in clusters CL07 and CL08 (17.39% and 27.28% respectively). Finally, clusters CL06 and CL09 have their most representative elements in societies of energy sector (En) (20% and 14.29%), while CL14 is dominated by Heavy Industry (HI) companies.

The data we have examined in previous rows make us possible to draw some remarks: on the one hand, despite the variety of sectors German quoted companies belong to, only a reduced number of them (i.e. clusters dominant sectors) may be considered the very

Table 25.3 Markets main features

Sector	CL01	CL02	CL03	CL04	CL05	CL06	CL07	CL08	CL09	CL10	CL11	CL12	CL13	CL14
Mkt	0.00	7.14	0.00	0.00	0.00	0.00	0.00	0.00	0.00	0.00	7.14	7.69	0.00	0.00
B&F	6.67	14.29	30.77	16.67	42.86	6.67	13.04	21.43	4.76	7.14	21.43	7.69	23.08	9.09
HI	0.00	7.14	7.69	0.00	0.00	0.00	4.35	0.00	9.52	14.29	0.00	15.38	0.00	4.35
Serv	20.00	0.00	0.00	0.00	0.00	6.67	4.35	15.00	9.52	0.00	14.29	0.00	15.38	9.09
Fash	13.33	0.00	0.00	16.67	0.00	13.33	5.88	0.00	0.00	7.14	0.00	7.69	15.38	0.00
HT	0.00	0.00	7.69	25.00	14.29	0.00	9.52	0.00	9.52	28.57	7.14	0.00	15.38	27.27
HC	13.33	14.29	0.00	8.33	14.29	6.67	17.39	27.28	9.52	14.29	0.00	0.00	0.00	0.00
Log	20.00	14.29	15.38	8.33	0.00	0.00	4.35	0.00	0.00	0.00	21.43	0.00	4.35	4.52
Lux	0.00	0.00	0.00	0.00	0.00	0.00	0.00	0.00	5.00	0.00	0.00	0.00	0.00	4.52
Hou	13.33	7.14	7.69	8.33	14.29	13.33	8.70	7.14	4.76	0.00	7.14	0.00	7.69	0.00
Comp	0.00	14.29	0.00	8.33	0.00	0.00	4.52	0.00	0.00	7.14	14.29	7.14	0.00	9.09
Re. Serv	0.00	0.00	0.00	0.00	0.00	6.67	0.00	0.00	0.00	0.00	0.00	0.00	0.00	0.00
F&D	0.00	7.14	7.69	0.00	0.00	0.00	9.52	0.00	0.00	0.00	0.00	0.00	7.69	0.00
En	0.00	7.14	7.69	0.00	0.00	20.00	0.00	7.14	14.29	7.14	0.00	0.00	0.00	9.09
Ent	0.00	0.00	0.00	0.00	7.14	6.67	0.00	0.00	2.50	7.14	0.00	30.77	0.00	0.00
Press	0.00	0.00	7.14	0.00	7.14	6.67	0.00	0.00	10.00	0.00	0.00	7.69	4.14	0.00
Imp/Exp	0.00	0.00	0.00	0.00	0.00	0.00	4.35	0.00	0.00	0.00	0.00	0.00	0.00	9.09
PU	0.00	0.00	0.00	0.00	0.00	0.00	0.00	0.00	0.00	0.00	0.00	4.52	0.00	4.79
TCom	0.00	7.14	8.24	0.00	0.00	0.00	0.00	0.00	13.33	0.00	7.14	0.00	6.89	0.00
Auto	6.67	0.00	0.00	0.00	0.00	0.00	9.52	0.00	2.50	0.00	0.00	7.14	0.00	9.09
Gard	0.00	0.00	0.00	0.00	0.00	0.00	0.00	0.00	0.00	0.00	0.00	0.00	0.00	0.00
Man	6.67	0.00	0.00	8.33	0.00	13.33	4.52	7.14	4.76	7.14	0.00	4.28	0.00	0.00

driving engine of the German economy. Such information together with the one retrieved by looking at the Sharpe Ratio scores has strengthened the belief that Hi-Tech and Energy are, at present, the most challenging areas for investors in German market.

As a counterpart, we observed that there is a plenty of sectors[1] whose incidence on clusters composition is lower, or better, they did not seem to cluster anyway. If this sounds reasonably for some niche-wise sectors (Luxury and Gardening, to make some examples), this is more surprising for other sectors (mainly Automotive and Telecommunications) that are worldwide known as strengths of German economy. This evidence, however, is somewhat aligned to the policy strategy that the German government has adopted in most recent times.

Earlier conclusions we can draw suggest that over the past year Germany did not particularly suffered for the critical situation common to greater part of European countries. The role played by both Hi-Tech and Energy sectors has been probably a key issue. However, from now on Germany should carefully monitor the state of B&F companies that are those

[1]Fashion (Fash), Luxury Goods (Lux), Housing (Hou), Retail Services (Ret Serv.), Food and Drinking (F&D), Entertrainment (Ent), Press (Press), Import/Export (Imp/Exp), Public Utilities (PU), Telecommunications (TCom), Automotive (Auto), Gardening (Gard) and Manifacturing (Man).

that actually are performing worse. Other sectors like F&D, Hou, Press, and Auto need to be constantly checked as well since they seem to be in a stage whose evolution (towards either better or worse phases) is uncertain.

At this point it make sense to test whether or not the actual snapshot we have captured for Germany is the result of either a strategic issue, or a kind of natural evolution from previous situations. To do this we performed a dynamical analysis going back in time from December 2011 to December 2001. As explained in Section 25.3, we scanned data by means of a moving window, thus obtaining 10 matrices of dimensions 207×300, where 207 is the number of companies included into the simulation and 300 is the number of log-returns we took for each of them.

In order to make the discussion as clear as possible we focused on the analysis of the results by comparison of the static situation (i.e. last year market skeleton) with those referring to 2004-2005 and 2007-2008 periods. This choice may be easily motivated: on the one hand the period 2004-2005 is a starting point of some symptoms anticipating the world financial crisis; on the other hand the period 2007-2008 is generally acknowledged as the one where deepest effects of the crisis were felt. In such a mood, actual time could be interpreted to either confirm or neglect the path traced in past times.

Figure 25.5 shows the market skeleton frame for the German Stock Exchange in the periods 2007-2008 and 2004-2005 respectively. Tables 25.4–25.7 detail basic statistics and clusters composition for the aforementioned periods.

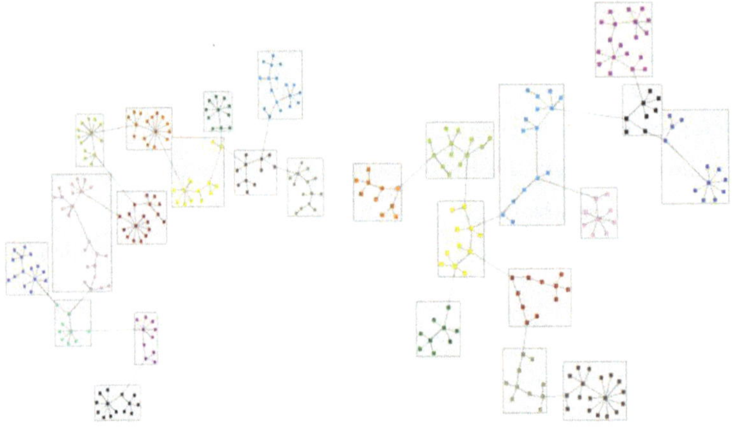

Fig. 25.5 From left to right, skeleton framework of German companies in the periods: December 2007 – December 2008 and December 2004 – December 2005

Table 25.4 Clusters main features: the case of Germany in the period 2007-2008

CL.ID	mu	std	sk	ku	SR
CL01	0.0003094	0.057005	10.44936	2451.638	0.543%
CL02	0.0003942	0.057219	10.25497	2390.428	0.689%
CL03	0.0002678	0.05579	9.94812	638.8266	0.480%
CL04	0.0002824	0.056436	10.02534	2373.059	0.500%
CL05	0.0003270	0.056945	10.9984	2587.711	0.574%
CL06	0.0002756	0.055918	10.00477	2389.595	0.493%
CL07	0.0002644	0.055723	10.02986	2411.497	0.475%
CL08	0.0004753	0.058852	11.32831	2525.394	0.808%
CL09	0.0003157	0.056422	10.36997	2460.894	0.559%
CL10	0.0005666	0.059179	11.77725	2602.168	0.957%
CL11	0.0005439	0.058084	11.26676	2532.204	0.936%
CL12	0.0003427	0.056806	10.10251	2371.656	0.603%
CL13	0.0004168	0.057255	10.66783	2490.513	0.728%

Table 25.5 Clusters main features: the case of Germany in the period 2004-2005

CL.ID	mu	std	sk	ku	SR
CL01	0.0003849	0.056505	10.48629	2485.513	0.681%
CL02	0.0004037	0.057215	1.653466	2378.495	0.706%
CL03	0.0008787	0.061519	12.6148	2713.425	1.428%
CL04	0.0002657	0.055689	9.951796	2394.033	0.477%
CL05	0.0003660	0.056965	10.57433	2485.67	0.642%
CL06	0.0003745	0.056659	10.46166	2471.775	0.661%
CL07	0.0003209	0.056475	10.12393	2390.405	0.568%
CL08	0.0001570	0.045558	1.733983	413.5084	0.345%
CL09	0.0003237	0.056749	10.47651	2469.166	0.570%
CL10	0.0003062	0.056178	10.10382	2404.7	0.545%
CL11	0.0005619	0.058529	11.85308	2648.211	0.960%

In both cases cluster statistics evidence (once again) positive mean and skewness, and lower variability. The Sharpe Ratio is generally higher than that evidenced in the static analysis. Looking at clusters composition, we primarily observe that, moving from one period to another, it did not maintain unchanged. However, it has been possible to isolate dominant sectors. In particular, in the period 2007-2008, B&F companies prevail in five over thirteen clusters (CL01, CL03, CL07, CL08 and CL13); HC and Imp/Exp firms share dominance in CL02; Hi-Tech is the dominant sector in cluster CL06, CL10, CL11 and CL12. Finally, Logistics and TCom societies are concentrated in CL04 and CL09 respectively. Coping such results to the values of Sharpe Ratio, it seems possible to claim that good performances are mainly due to the leading activity of the High-Tech sector. Besides,

Table 25.6 Clusters percentage composition during the period 2007-2008

Sec.	CL01	CL02	CL03	CL04	CL05	CL06	CL07	CL08	CL09	CL10	CL11	CL12	CL13
Mkt	0.00	0.00	0.00	8.33	0.00	0.00	4.35	0.00	11.76	0.00	0.00	8.33	0.00
B&F	23.81	8.33	15.38	0.00	22.22	0.00	17.39	37.50	11.76	7.69	14.29	8.33	36.36
HI	4.76	0.00	0.00	8.33	0.00	6.29	0.00	0.00	12.75	0.00	0.00	0.00	0.00
Serv	0.00	8.33	7.69	8.33	0.00	7.14	4.35	12.50	0.00	0.00	0.00	0.00	7.14
Fash	4.76	8.33	7.69	8.33	5.56	7.14	4.35	12.50	5.88	15.38	0.00	16.67	0.00
HT	4.76	12.00	7.69	0.00	0.00	21.43	4.35	0.00	11.76	23.08	21.43	25.00	9.09
HC	0.00	16.67	8.33	0.00	16.67	0.00	0.00	0.00	0.00	15.38	0.00	0.00	0.00
Log	4.76	0.00	8.33	16.67	0.00	7.14	4.35	12.50	0.00	0.00	0.00	0.00	4.75
Lux	0.00	8.33	0.00	0.00	0.00	0.00	0.00	0.00	0.00	0.00	0.00	7.14	0.00
Hou	19.05	0.00	7.69	0.00	5.56	7.14	8.70	12.50	5.88	0.00	14.29	8.33	9.09
Comp	4.76	0.00	0.00	8.33	0.00	0.00	4.00	0.00	0.00	7.69	0.00	0.00	9.09
Ret Serv	0.00	0.00	0.00	0.00	0.00	5.00	0.00	0.00	11.76	0.00	4.21	0.00	0.00
F&D	4.76	0.00	7.69	8.33	5.56	0.00	16.67	0.00	0.00	0.00	0.00	8.33	4.75
En	4.76	0.00	4.76	0.00	16.67	0.00	9.20	0.00	0.00	0.00	0.00	0.00	4.76
Ent	0.00	0.00	0.00	8.33	0.00	7.14	0.00	7.14	5.88	23.08	14.29	0.00	0.00
Press	4.76	0.00	8.33	8.33	0.00	0.00	0.00	0.00	0.00	0.00	0.00	0.00	7.14
Imp/Exp	0.00	16.67	3.94	8.33	0.00	7.14	0.00	0.00	0.00	0.00	7.14	0.00	0.00
PU	4.76	1.00	0.00	0.00	0.00	0.00	0.00	0.00	0.00	0.00	0.00	0.00	0.00
TCom	0.00	0.00	7.69	0.00	0.00	7.14	12.41	0.00	16.67	7.69	7.69	0.00	0.00
Auto	9.52	12.00	0.00	8.33	11.11	7.14	5.90	5.36	5.88	0.00	16.67	16.67	9.09
Gard	0.00	0.00	0.00	0.00	0.00	3.00	0.00	0.00	0.00	0.00	0.00	1.19	0.00
Man	4.76	8.33	4.76	0.00	16.67	7.14	4.00	0.00	0.00	0.00	0.00	0.00	0.00

by comparison with the performances discussed in the static analysis, Germany gave the impression to have suffered for the global crisis with some delay.

Most interesting results, in our opinion, come by the analysis of the period: December 2004-December 2005). The first element to highlight is that in this case we have only 11 clusters (versus 13 in the period: 2007-2008, and 14 in the period: 2010-2011). For what is concerning clusters composition now we have: B&F companies dominating clusters CL01, CL05, CL06, CL07, CL08, and CL10; fashion sector prevails in CL02, entertainment companies in CL03, housing societies in CL09, while Hi-Tech is the king of remaining clusters (CL04, CL05 and CL11). If we compare the results to those we have previously discussed, it is quite clear that during the observed period we have witnessed various companies reactions to the crisis: while Hi-Tech as well as financial companies maintained similar behaviors (and this is confirmed by the tendency to be clustered together), companies in other sectors did not grouped in any way. A possible explanation might stay in some policy action made by the national government, in order to address the economy, and to protect sectors with higher exposure.

Table 25.7 Clusters percentage composition during the period 2004-2005

Sec.	CL01	CL02	CL03	CL04	CL05	CL06	CL07	CL08	CL09	CL10	CL11
Mkt	0.00	6.25	12.50	6.67	0.00	6.25	0.00	0.00	0.00	0.00	0.00
B&F	20.00	12.50	0.00	0.00	18.18	25.00	41.67	35.00	0.00	40.00	14.29
HI	0.00	0.00	7.50	16.80	5.00	0.00	0.00	0.00	16.43	0.00	0.00
Serv	0.00	6.25	0.00	0.00	9.09	0.00	0.00	10.00	0.00	10.00	0.00
Fash.	10.00	25.00	0.00	6.67	0.00	6.25	0.00	5.00	0.00	0.00	14.29
HT	10.00	0.00	12.50	26.67	18.18	0.00	8.33	0.00	14.29	0.00	41.43
HC	5.00	0.00	0.00	0.00	0.00	12.50	0.00	0.00	0.00	0.00	0.00
Log	0.00	18.50	0.00	0.00	10.00	0.00	8.33	20.00	0.00	0.00	12.00
Lux	0.00	0.00	12.50	0.00	0.00	0.00	0.00	0.00	0.00	0.00	0.00
Hou	10.00	6.25	0.00	6.67	0.00	5.50	8.33	0.00	41.43	0.00	0.00
Comp	0.00	11.50	0.00	6.67	0.00	6.25	0.00	5.00	0.00	0.00	0.00
Ret Serv	0.00	0.00	10.00	0.00	0.00	0.00	0.00	0.00	0.00	15.00	0.00
F&D	0.00	0.00	0.00	6.67	0.00	6.25	0.00	5.00	0.00	0.00	0.00
En	10.00	0.00	7.50	0.00	0.00	0.00	0.00	0.00	0.00	0.00	0.00
Ent	0.00	6.25	25.00	6.67	0.00	12.50	0.00	0.00	0.00	10.00	7.14
Press	0.00	0.00	0.00	0.00	10.00	0.00	0.00	10.00	0.00	10.00	0.00
Imp/Exp	0.00	0.00	0.00	9.86	0.00	0.00	0.00	0.00	5.71	0.00	5.00
PU	0.00	0.00	0.00	0.00	0.00	0.00	0.00	0.00	0.00	5.00	0.00
TCom	0.00	7.50	0.00	0.00	0.00	0.00	16.67	0.00	0.00	0.00	0.00
Auto	30.00	0.00	12.50	6.67	9.09	12.50	16.67	10.00	15.00	0.00	0.00
Gard	5.00	0.00	0.00	0.00	10.00	0.00	0.00	0.00	0.00	0.00	0.00
Man	0.00	0.00	0.00	0.00	10.00	7.00	0.00	0.00	7.14	10.00	3.00

Table 25.8 Basic statistics for clusters in the Italian stock exchange. The reference period is: December 2010-December 2011

	mu	std	sk	ku	SR
CL01	−0.000215501	0.027887	−11.0446	1044.121	−0.773%
CL02	−0.000318218	0.028256	−12.7722	1103.929	−1.126%
CL03	−0.000303881	0.026743	0.933632	76.79378	−1.136%
CL04	−0.00031791	0.028483	−9.59839	884.5868	−1.116%
CL05	−0.000295205	0.028527	−9.87737	906.3181	−1.035%
CL06	−0.000208052	0.026202	1.480943	70.91261	−0.794%
CL07	−0.000260298	0.028065	−10.8747	1018.387	−0.927%
CL08	−0.000227621	0.028127	−10.3357	980.1618	−0.809%
CL09	−0.000265711	0.025861	0.599355	54.41501	−1.027%
CL10	−0.000266599	0.028507	−10.3303	936.4683	−0.935%
CL11	−0.000199087	0.026208	1.485088	70.85024	−0.760%

25.4.2 *The case of Italy*

As done for Germany, we begin by the static analysis during the period: December 2010-December 2011. Our procedure identified eleven clusters, as shown in Figure 25.6.

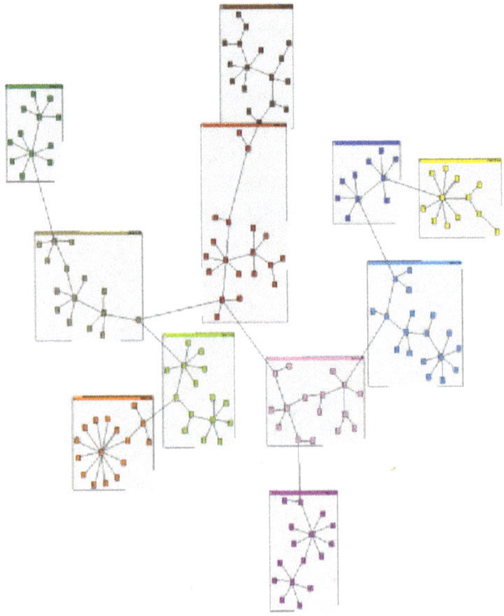

Fig. 25.6 Skeleton framework of the Italian stock exchange in the period: 30 December 2010 – 30 December 2011

Table 25.8 immediately captures the enormous difference from the situation we have discussed for Germany up to now: clusters mean is always negative, and the same holds for skewness. This obviously affects Sharpe Ratio values that are negative in turn.

The analysis of clusters composition (Table 25.9) evidenced the dominance of the B&F sector in five over eleven clusters (CL01, CL04, CL06, CL07, CL11), Heavy Industry in CL02, CL03, and CL10, companies in the Services sector (CL09), Public Utilities (CL05) and Housing (CL08).

Nevertheless, the results are interesting if we cope clusters composition together with Sharpe ratio values. In this way it is easy to see that worst performances are those in clusters where Heavy Industry is dominant: this is perfectly aligned with the long lasting crisis that affected this sector in Italy.

By replicating for Italy the analysis we have performed for Germany, Figure 25.7 shows the market organization in the periods: 2007-2008 and 2004-2005, while Tables 25.10–25.13 report the corresponding basic statistics and clusters composition.

In order understand the case of Italy, the key issue is that the present (critical) situation was anticipated in previous times by unheard warnings. Looking back to the periods 2004-2005 (Table 25.11) and 2007-2008 (Table 25.10), we can see that basic statistics (exhibiting

Table 25.9 Cluster percentage composition for Italy in the period December 2010-December 2011

	CL01	CL02	CL03	CL04	CL05	CL06	CL07	CL08	CL09	CL10	CL11
Mkt	0.00	0.00	8.33	0.00	0.00	0.00	0.00	0.00	0.00	0.00	0.00
B&F	26.32	11.11	0.00	31.25	12.50	60.00	17.65	12.50	20.00	13.33	25.00
HI	5.26	27.78	25.00	18.75	6.25	0.00	11.76	0.00	13.33	20.00	8.33
Serv	0.00	0.00	0.00	0.00	0.00	0.00	5.88	0.00	26.67	0.00	0.00
Fash	5.26	0.00	8.33	0.00	0.00	10.00	11.76	0.00	6.67	0.00	0.00
HT	5.26	0.00	33.33	6.25	12.50	0.00	5.88	6.25	13.33	20.00	0.00
HC	5.26	0.00	0.00	0.00	6.25	0.00	5.88	6.25	6.67	6.67	0.00
Log	5.26	0.00	11.11	0.00	6.25	0.00	0.00	6.25	0.00	6.67	0.00
Lux	5.26	0.00	0.00	6.25	6.25	10.00	5.88	0.00	0.00	0.00	8.33
Hou	0.00	5.56	0.00	0.00	0.00	0.00	11.76	25.00	0.00	6.67	0.00
Comp	5.26	0.00	0.00	6.25	6.25	0.00	0.00	18.75	6.67	6.67	8.33
Ret Serv	0.00	0.00	5.56	0.00	0.00	0.00	0.00	0.00	0.00	0.00	0.00
F&D	5.27	5.56	0.00	0.00	6.25	0.00	5.88	6.25	0.00	6.67	8.33
En	0.00	0.00	0.00	0.00	0.00	0.00	0.00	0.00	0.00	0.00	0.00
Ent	5.26	16.67	0.00	12.50	6.25	0.00	5.88	6.25	6.67	0.00	0.00
Press	5.26	0.00	0.00	0.00	0.00	0.00	0.00	6.25	0.00	6.67	8.33
Imp/Exp	0.00	0.00	0.00	0.00	6.25	10.00	11.76	0.00	0.00	0.00	8.33
PU	0.00	0.00	0.00	0.00	0.00	0.00	0.00	0.00	0.00	0.00	0.00
TCom	5.26	16.67	0.00	12.50	18.75	0.00	0.00	0.00	0.00	6.67	0.00
Auto	0.00	16.67	8.33	0.00	0.00	0.00	0.00	0.00	0.00	0.00	0.00
Gard	5.26	0.00	0.00	0.00	6.25	0.00	0.00	0.00	0.00	0.00	25.00
Man	5.26	0.00	0.00	0.00	0.00	0.00	0.00	6.25	0.00	0.00	0.00

Table 25.10 Basic statistics for clusters in the Italian stock exchange for the period: 2004-2005

	mu	std	sk	ku	SR
CL01	−0.000247	0.028264	−10.0774	955.6451	−0.874%
CL02	−0.000240	0.028246	−12.567	1085.316	−0.849%
CL03	−0.000322	0.025733	0.648867	39.43743	−1.252%
CL04	−0.000296	0.02845	−10.1097	920.5785	−1.042%
CL05	−0.000225	0.028156	−10.3772	982.2957	−0.798%
CL06	−0.000318	0.028483	−9.59839	884.5868	−1.116%
CL07	−0.000393	0.028098	−10.5085	989.4136	−1.398%
CL08	−0.000243	0.028126	−11.0344	1030.068	−0.863%
CL09	−0.000275	0.028378	−10.1387	930.7036	−0.969%
CL10	−0.000285	0.027973	−13.3502	1158.986	−1.020%
CL11	−0.000285	0.027986	−12.7585	1113.521	−1.018%

both mean and skewness negative values) could have provided very alerting signals, if correctly interpreted. Coping such results to those referring to clusters composition, it would be possible for Italian policy makers to have a direction for interventions: B&F, HI, Services and Hi-Tech were key sectors either to protect or to promote in order to improve

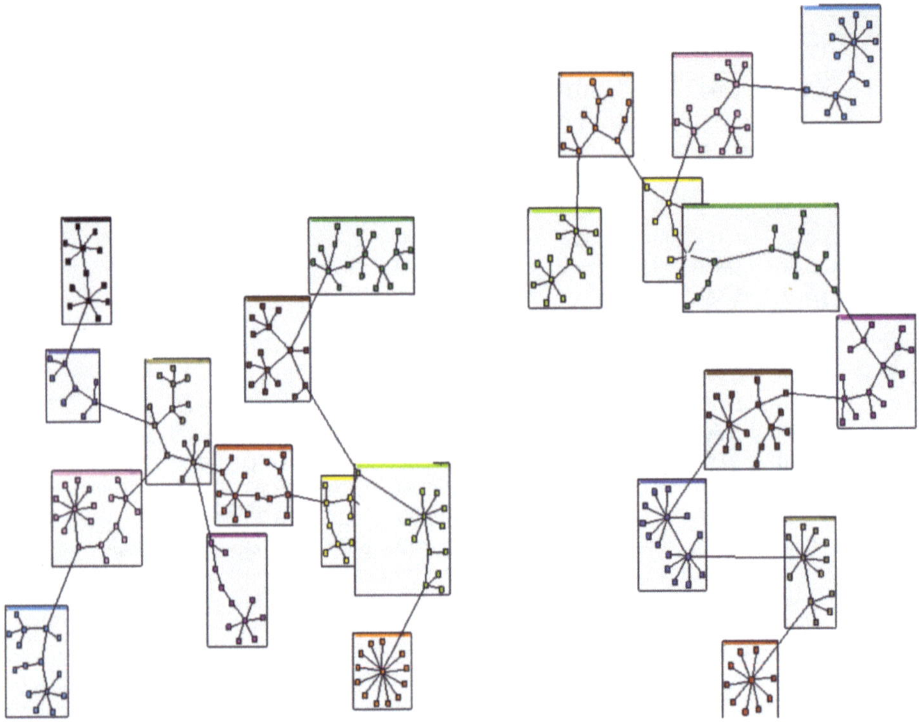

Fig. 25.7 From left to right, skeleton framework of Italian companies in the periods: 2007-2008 and 2004-2005

the wealth of the country. However, this message has been partially kept in the case of HI sector, while it has been completely missed in other cases.

25.5 Conclusion

In this chapter we provided evidence that soft computing techniques may be proficiently used as tool for macroeconomic analysis.

We moved from row data (price levels) of quoted enterprises to provide a snapshot of countries financial situation, based on a procedure that joins together Self Organizing Maps (SOMs) and Minimum Spanning Tree (MST).

Our technique was tested on two markets featuring different levels of (in)stability: the German and the Italian Stock Exchange.

Table 25.11 Basic statistics for clusters in the Italian stock exchange for the period: 2007-2008

	mu	std	sk	ku	SR
CL01	−0.000256	0.028378	−10.829	982.8535	−0.901%
CL02	−0.000322	0.028588	−9.67184	887.7461	−1.127%
CL03	−0.000238	0.02816	−10.0597	957.5814	−0.844%
CL04	−0.000318	0.028483	−9.59839	884.5868	−1.116%
CL05	−0.000361	0.026546	0.861259	38.8684	−1.361%
CL06	−0.000267	0.028131	−12.4423	1079.345	−0.950%
CL07	−0.000226	0.028098	−10.5085	989.4136	−0.804%
CL08	−0.000274	0.02868	−11.3805	1043.52	−0.956%
CL09	−0.000223	0.028133	−11.3811	1064.015	−0.794%
CL10	−0.000258	0.028292	−10.4348	982.5524	−0.911%
CL11	−0.000276	0.028135	−11.813	1085.863	−0.981%
CL12	−0.000263	0.028087	−10.4526	984.4849	−0.935%

Table 25.12 Cluster percentage composition for Italy in the period 2007-2008

	CL01	CL02	CL03	CL04	CL05	CL06	CL07	CL08	CL09	CL10	CL11	CL12
Mkt	0.00	0.00	0.00	0.00	0.00	0.00	0.00	7.69	0.00	0.00	0.00	0.00
B&F	20.00	0.00	27.78	40.00	25.00	12.50	50.00	23.08	32.00	20.00	0.00	15.38
HI	20.00	0.00	16.67	0.00	6.25	0.00	16.67	7.69	0.00	6.67	30.00	7.69
Serv.	6.67	21.43	0.00	0.00	0.00	0.00	0.00	7.69	0.00	0.00	0.00	0.00
Fash.	5.26	0.00	0.00	10.00	0.00	12.50	8.33	0.00	4.00	6.67	10.00	0.00
HT	6.67	21.43	5.56	0.00	0.00	25.00	0.00	1.44	8.00	6.67	0.00	30.77
HC	0.00	0.00	0.00	10.00	0.00	0.00	0.00	0.00	0.00	13.33	0.00	15.38
Log	0.00	0.00	11.11	0.00	8.33	25.00	0.00	7.69	0.00	6.67	0.00	0.00
Lux	0.00	7.14	5.56	0.00	0.00	0.00	8.33	0.00	4.00	13.33	0.00	0.00
Hou	6.67	0.00	5.56	0.00	8.33	0.00	8.33	7.69	8.00	0.00	0.00	0.00
Comp	6.67	7.14	11.11	10.00	0.00	0.00	8.33	7.69	0.00	0.00	0.00	0.00
Ret. Serv	0.00	0.00	5.56	0.00	0.00	0.00	0.00	0.00	5.00	0.00	0.00	0.00
F&D	5.27	7.14	5.56	10.00	10.75	0.00	0.00	0.00	0.00	0.00	20.00	0.00
En.	5.26	0.00	0.00	0.00	8.33	0.00	0.00	15.38	8.00	0.00	0.00	0.00
Ent	5.26	7.14	0.00	0.00	0.00	0.00	0.00	7.69	4.00	6.67	0.00	0.00
Press	0.00	0.00	0.00	0.00	0.00	12.50	11.76	0.00	4.00	0.00	10.00	7.69
Imp/Exp	0.00	0.00	0.00	0.00	0.00	0.00	0.00	0.00	0.00	0.00	0.00	0.00
PU	6.67	4.76	5.56	10.00	0.00	12.50	0.00	0.00	4.00	13.33	20.00	0.00
TCom	0.00	16.67	0.00	0.00	8.00	0.00	0.00	0.00	0.00	0.00	0.00	7.69
Auto	0.00	7.14	0.00	0.00	25.00	0.00	0.00	0.00	0.00	6.67	0.00	0.00
Gard	0.00	0.00	0.00	0.00	0.00	0.00	0.00	6.25	5.00	0.00	10.00	0.00
Man	0.00	0.00	0.00	0.00	0.00	0.00	0.00	00.00	4.00	0.00	0.00	0.00

Our study made us possible to highlight most important relations among quoted societies, as well as the natural clusters that tend to be created into those markets.

In particular, in the case of Germany we captured the country situation in three periods (2004-2005, 2007-2008 and 2010-2011). The study suggested that the German government

Table 25.13 Cluster percentage composition for Italy in the period 2004-2005

	CL01	CL02	CL03	CL04	CL05	CL06	CL07	CL08	CL09	CL10	CL11	CL12
Mkt	0.00	0.00	0.00	0.00	0.00	0.00	0.00	4.39	0.00	6.67	0.00	0.00
B&F	6.67	0.00	16.67	30.00	33.33	50.00	33.33	30.77	32.00	6.67	40.00	0.00
HI	23.33	0.00	16.67	30.00	6.25	25.00	8.33	30.77	8.00	6.67	0.00	0.00
Serv.	6.67	0.00	5.56	15.00	0.00	0.00	0.00	0.00	0.00	5.00	0.00	0.00
Fash.	5.26	14.29	0.00	0.00	0.00	0.00	8.33	7.69	0.00	0.00	0.00	0.00
HT	6.67	0.00	0.00	2.50	8.33	25.00	33.33	1.44	4.00	6.67	0.00	0.00
HC	6.67	7.14	0.00	2.50	0.00	0.00	0.00	0.00	8.00	6.67	0.00	0.00
Log	6.67	0.00	11.11	0.00	8.33	0.00	0.00	0.00	4.00	6.67	0.00	0.00
Lux	5.26	0.00	5.56	0.00	0.00	0.00	0.00	0.00	0.00	0.00	10.00	0.00
Hou	0.00	28.57	0.00	5.00	8.33	0.00	0.00	0.00	4.00	5.00	10.00	0.00
Comp	5.26	7.14	0.00	0.00	8.33	0.00	0.00	0.00	18.00	5.00	0.00	0.00
Ret. Serv	0.00	0.00	5.56	0.00	0.00	0.00	0.00	0.00	5.00	0.00	5.00	0.00
F&D	5.27	0.00	5.56	0.00	10.75	0.00	8.33	7.69	0.00	6.67	0.00	0.00
En.	5.26	0.00	11.11	5.00	0.00	0.00	8.33	0.00	2.00	0.00	10.00	0.00
Ent	5.26	0.00	0.00	0.00	0.00	0.00	0.00	0.00	4.00	6.67	0.00	0.00
Press	0.00	7.14	0.00	0.00	0.00	0.00	0.00	0.00	4.00	6.67	10.00	0.00
Imp/Exp	0.00	0.00	0.00	0.00	0.00	0.00	0.00	0.00	0.00	5.00	0.00	0.00
PU	6.67	4.76	11.11	5.00	8.33	0.00	0.00	0.00	0.00	0.00	10.00	0.00
TCom	0.00	16.67	0.00	0.00	8.00	0.00	0.00	0.00	0.00	0.00	0.00	0.00
Auto	0.00	0.00	11.11	5.00	0.00	0.00	0.00	11.00	0.00	6.67	0.00	0.00
Gardv	0.00	7.14	0.00	0.00	0.00	0.00	0.00	6.25	5.00	6.67	0.00	0.00
Man	5.10	7.14	0.00	0.00	0.00	0.00	0.00	0.00	2.00	0.00	5.00	0.00

was able to pay attention to warning signals emerging from the market. In this way Germany applied measures that allowed it to face last year critical situation. Protecting sectors with a strength tradition and promoting the challenge in emerging sectors Germany played a game that seems to maintain the country at the marginal side of current global crisis.

On the other hand, in the case of Italy, local governors neither did take into account in the right way alerting signals or did apply correction/protection measures. As a consequence, Italy is presently facing against a higher unstable financial situation. Our procedure, however, highlighted some direction towards which policy makers could operate in order to reduce instability.

To conclude the joined SOM-MST approach seems able to suggest proper recipes that governments might consider in order to address their policy efforts.

Bibliography

[1] Cattinelli I., Bolzoni E., Barbieri C., Mari F., Martin-Guerrero J., Soria-Olivas E., Martinez-Martinez J., Gomez-Sanchis J., Amato C., Stopper A., and Gatti E., *Health Care Managem. Sci.*, (2011).

[2] Deboeck G. and Kohonen T., *Visual Explorations in Finance: with Self-Organizing Maps*, (Springer Finance, New York, 1998).
[3] Montefiori M. and Resta M., *Health Health Care Managem. Sci.*, **12**, (2009).
[4] Kaski S., Kangas J., and Kohonen T., *Neur. Comp. Surv.*, **1**, (1998).
[5] Kohonen T., *Self-Organizing Maps*, (Springer, 2001).
[6] Polla M., Honkela T., and Kohonen T., *TKK Reports in Information and Computer Science*, Helsinki University of Technology, (2009).
[7] Oja M., Kaski S. and Kohonen T., *Neur. Comp. Surv.*, **3**, (2003).
[8] Resta M., in *Proceedings of the 2009 conference on New Directions in Neural Networks: 18th Italian Workshop on Neural Networks: WIRN 2008*. (IOS Press Amsterdam, The Netherlands, 2010).
[9] Resta M., in: *J. Laaksonen, T. Honkela, Advances in f, LNCS* **6731**, (Springer Heidelberg, 2011), 30–39.
[10] Sarlin P. and Eklund T., in: J. Laaksonen, T. Honkela, *Advances in Self Organizing Maps, LNCS* **6731**, (Springer Heidelberg, 2011), 40–50.
[11] Martin B. and Serrano Cinca C., *Neur. Comp. & Appl.*, **1**, (1993).

Chapter 26

Evolutions in Particle Swarm Optimization: Benchmark on Continuous Cases

Ayca Altay and Gulgun Kayakutlu

Department of Industrial Engineering, Istanbul Technical University34357, Macka, Istanbul, Turkey

E-mail: altaya@itu.edu.tr

This chapter aims to present the results of developing a novel swarm optimization method that responds to the need of using multidimensional parameters. This meta-heuristic optimization approach is inspired by the ecological system of animals and their hierarchical relationship. Animal Food Chain in the nature is known to have three groups: herbivores (plant eaters), omnivores (both plant and meat eaters) and carnivores (meat eaters). In the food pyramid, the number of herbivores is higher than the number of omnivores which is higher than the number of carnivores in ratios depending on the environment. Furthermore, herbivores are known to be slowest and carnivores are known to be the fastest of the chain. These features are represented by conditional and multidimensional parameters in the Foraging Search algorithm. Tests are run on continuous and non-linear benchmark problems with various numbers of constraints. Cross validations of the results are realized by comparison of classical and Predator-Prey based Particle Swarm algorithms. Test results emphasize the power of the Foraging Search as an optimization tool.

26.1 Introduction

This chapter is reserved to review research on particle swarm optimization (PSO); a technique widely preferred in decision making and decision support applications. This metaheuristic optimization method is based on the population approach like ant colony or genetic algorithm, foundations of which are taken from the natural life. The richness of nature allows continuous improvements as a response to complex decisions.

The objective of this study is two-fold; first, to present improvements in particle swarm optimization with examples and, second, to introduce a new algorithm with the imperative causes and the improved results.

C. Kahraman (ed.), *Computational Intelligence Systems in Industrial Engineering*, Atlantis Computational Intelligence Systems 6, DOI: 10.2991/978-94-91216-77-0_26,

In this chapter, the first section will give the history of particle swarm optimization, along with the classical algorithm definition. Second section will be analyzing the most recent algorithms with sample problems to demonstrate the improvements. Third section will be introducing the Foraging Search algorithm, emphasizing the needs and the achievements. The fourth section will demonstrate the result of experiments on continuous nonlinear benchmark problems, and the final section will provide concluding remarks and recommendations for future studies.

26.2 Background on the PSO Method

Particle Swarm Optimization (PSO) is a population based optimization technique invented by Kennedy and Eberhart in 1995 influenced by the social behaviour of fish schooling and bird flocking [1]. It simulates the "collective behaviour" of animals, which socio-cognitively share information among the swarms [2].

Animals in nature, urge into swarms for different objectives: finding food, escaping predators, etc. These swarms, also called boids (a special name for bird swarms), have three vital principles for their collaborative movements: "collision avoidance – separation", "velocity matching – alignment" and "flock centring – cohesion" [3, 4]. Collision avoidance refers to not crashing with nearby elements of the swarm, whereas velocity matching refers to adapt velocities according to the velocities of neighbouring elements. The flock centring is a result of velocity matching. Since each flock-mate tends to adapt his velocity according to his neighbours, the flock tends to stay close to each other. Generalizing it to the whole swarm, they tend to stay close to a neighbourhood centre [5]. Apart from these three principles, environmental principles such as obstacle avoidance and following a desired path are also valid [5] (Veenhuis and Köppen, 2006; Reynolds, 1987).

Imitating the aforementioned principles, the Particle Swarm Optimization field has gained interest because of its applicability, simplicity, and efficiency. The algorithm is proven to be robust and effective in various types of problems from single objective to combinatorial [6].

In the PSO algorithm, each solution is presented with a particle, that is, an element of the swarm. The swarm consists of a number of particles, in other words, a population of solutions [7]. The particles move through the search space at a random degree of freedom limited by the parameters of each other [8]. The moving particles have two properties, position and velocity, which are updated at every iteration of the algorithm [9]. Each particle is intelligent in a way that it keeps the memory of best position of himself and the neighbourhood, which introduces the neighbourhood concept. In the ultimate condition, generally the neighbourhood denotes the whole swarm [6].

Let p_i be the position of the ith particle in the swarm which consists of N particles, and let each particle have n dimensions defined over a maximization objective function f. The steps of the algorithm is given below [10]:

Step 1. The particle velocity and position of each p_i are initiated such that

$$x_{i,j} = x_{\min} + r(x_{\max} - x_{\min}), \qquad i = 1,\ldots,N; \;\; j = 1,\ldots,n \tag{26.1}$$

$$v_{i,j} = \alpha \frac{x_{\min} + r(x_{\max} - x_{\min})}{\Delta t}, \qquad i = 1,\ldots,N; \;\; j = 1,\ldots,n \tag{26.2}$$

where x denotes the position, v denotes the velocity and α is constant in the range $[0,1]$.

Step 2. The objective value of each particle is calculated as $f(x_i)$.

Step 3. The best position for each particle and the global best position for the swarm is updated. For a problem

$$f(x_i) < f(x_i^{\text{pb}}) \longrightarrow x_i^{\text{pb}} = x_i \tag{26.3}$$

$$f(x_i) < f(x_i^{\text{sb}}) \longrightarrow x_i^{\text{sb}} = x_i \tag{26.4}$$

where pb denotes the particle best and sb denotes the swarm best.

Step 4. Particle velocity and particle position are updated, that is, the new velocities and positions are calculated for each particle.

$$v_{i,j} \longleftarrow wv_{i,j} + c_1 r_1 \left(\frac{x_{i,j}^{\text{pb}} - x_{i,j}}{\Delta t} \right) + c_2 r_2 \left(\frac{x_{i,j}^{\text{sb}} - x_{i,j}}{\Delta t} \right),$$
$$i = 1,\ldots,N; \;\; j = 1,\ldots,n \tag{26.5}$$

$$x_{i,j} \longleftarrow x_{i,j} + v_{i,j}\Delta t, \qquad i = 1,\ldots,N; \;\; j = 1,\ldots,n \tag{26.6}$$

where w is the inertia rate between $[0,1]$, and r_1 and r_2 are random numbers between $[0,1]$. In the velocity update formula, $v_{i,j}$ is the inertia term where the particle attempts to save its own velocity, $c_1 q((x_{i,j}^{\text{pb}} - x_{i,j})/\Delta t)$ is the cognitive term where the particle attempts to reach at least its best position, and $c_2 r((x_{i,j}^{\text{sb}} - x_{i,j})/\Delta t)$ is the social term where the particle attempts to keep up with the best position of the swarm.

Step 5. Step 2 is returned to until a termination criterion is satisfied. Various termination criteria include iteration number, convergence of the result, convergence of error in results, etc.

Particle Swarm Optimization algorithm provides a number of parameters to be tuned. The effects of these parameters are presented below:

The inertia coefficient – w

The inertia coefficient adapts Newton's first law of motion, that is, "if there are no forces acting on a particle the velocity will stay the same" [11]. If $w = 0$, then the velocity update formula is called selfless [10]. A higher value of the inertia parameter indicated a higher self-confidence in the particle and a lower value indicates a high trust in personal memory and the swarm. Furthermore, a higher self-confidence signifies a better exploitation and a higher self-distrust signifies a better exploration in terms of the algorithm [12]. The inertia parameter is also omitted in the earlier studies, that is, $w = 1$ [13].

Over time, various variants of the inertia parameter have been developed. These variants involve [14] the following:

Constant and random inertia: The inertia parameter is selected randomly before the utilization of the algorithm, and is not changed through the iterations.

Time varying inertia: The inertia parameter is changed through iterations. The value of the parameter is assigned a high value in the earlier phases of iterations to provide exploration and decreased through iterations to provide better exploitation. The decrease can be achieved linearly and nonlinearly.

Adaptive inertia: The instant solution of the algorithm is monitored and inertia is adjusted according to the quality of the solution.

The cognitive coefficient – c_1

The cognitive term is an indicator of the affinity of the particle to its own success. A high cognitive coefficient yields the particle to a cyclic trajectory around its best. If $c_1 = 0$, the formula becomes memoryless in terms of cognition [15].

The social component – c_2

The social term is an indicator of the affinity of the particle to the success of the whole swarm. A high social coefficient implies a greater degree of affinity. Moreover, the trajectory of the particle is defined as a resultant of both the cognitive and the social component. This resultant attempts to balance exploration and exploitation only at particle level.

26.3 Advanced PSO Algorithms

It is the basis of all improvements of metaheuristic algorithms, to provide more robust solutions as well as to avoid local optima. It is also known that even though existing algorithms are fast and efficiently convergent, there is always possibility to attain algorithms with better efficiency and robustness. Hence, since its foundation Particle Swarm Optimization algorithm is both hybridized with other metaheuristic operators and improved with modifications. The hybridizations of various metaheuristics and several modifications that can be considered evolutionary in the context of Particle Swarm Optimization are provided below.

26.3.1 *Hybridization of the algorithm*

Hybridizations of other metaheuristics involve another aim apart from modified algorithms, that is, hybrid algorithms aim balancing the conflicting objectives of exploration and exploitation. Exploration ensures that all search space is covered during the utilization of the algorithm, whereas, exploitation ensures that all neighbourhoods of good solutions are thoroughly analyzed. Hybridizations are achieved using two means: utilizing the algorithms one after another or embedding the operators of one metaheuristic into the other. Both means are applied for PSO algorithm and it is most hybridized with Genetic Algorithm in order to capture the advantages and eliminate the disadvantages of both algorithms [16]. Gnamabal *et al.* [17] utilize an elitist reproduction following the determination of particle best and swarm best procedure to ensure that the algorithm produces better solutions per iteration. Similarly, Shunmugalatta and Slochanal [18] and El-Dib *et al.* [19] use the reproduction operator and update velocity at a parental level. Gao and Xu [20] and Wu [21] embed mutation operator to the PSO algorithm. Both crossover and mutation operators are embedded in the PSO algorithm by Lian *et al.* [22]. All aforementioned studies provide an instance for embedded hybridization whereas Yin *et al.* [23] utilizes algorithms one after another where PSO is utilized first and then Hill Climbing is applied according to the results of the first algorithm.

Other hybridizations involve embedding Ant Colony Optimization sequentially [24] or Local Search Algorithm as a part of the PSO Algorithm [25, 26]. Hybridizations are also achieved with exact algorithms. Kayhan *et al.* [27] utilize a sequential hybrid PSO algorithm with Solver where Solver obtains its inputs from the PSO algorithm. Different metaheuristic hybridizations involve Tabu Search [28], Simulated Annealing [29], Fuzzy Pareto Optimal PSO [30], GRASP [31] and problem-specific heuristics [32, 33].

26.3.2 *The Predator-Prey Particle Swarm Optimization*

As the PSO algorithm has evolved into more complicated algorithms in order to provide efficient results and effective iterations, the algorithms remained devoted to the motion of nature. One of these algorithms is the Predator-Prey PSO algorithm which is a competitive PSO approach [10].

Three problems faced with the classical PSO are exploration overwhelming the exploitation, being blocked by the local optima and the early convergence. As a solution, the hunting scheme of nature is simulated by the introduction of a second swarm [34] in order to overcome the three difficulties. If a prey swarm meets a predator swarm, they diffuse just to regroup again after the predator is gone. Diffusion provides a better exploration, whereas regrouping provides a better exploitation. The steps of the Predator-Prey PSO algorithm are as follows:

Step 1. Initiating particle velocity and position of each p_i for two swarms of equal sizes as in Eq.s (26.1)-(26.2).

Step 2. The objective function values for each particle of both swarms are calculated.

Step 3. Particle bests, that is, $x_{i,j}^{\mathrm{pb}}$, are found four each particle of each swarm.

Step 4. Swarm bests, that is $x_{i,j}^{\mathrm{sb}}$ are found for each swarm.

Step 5. The overall best value is updated.

Step 6. Velocities are updated separately according to swarms.

For the prey swarm, the velocity update formula is determined by a uniform random number rn between 0 and 1.

If $rn < pf$

$$v_{i,j} = wv_{i,j} + c_1 r_1 \left(\frac{x_{i,j}^{\mathrm{pb}} - x_{i,j}}{\Delta t} \right) + c_2 r_2 \left(\frac{x_{i,j}^{\mathrm{sb}} - x_{i,j}}{\Delta t} \right) + c_3 r_3 \frac{D(d)}{\Delta t},$$

$$i = 1, \dots, N; \ \ j = 1, \dots, n \tag{26.7}$$

where d is the Euclidean distance to the nearest and $D(\cdot)$ is an exponential function of the related distance, that is, $D(d) = \alpha e^{-\beta d}$.

The velocity update formula for the predator swarm in case that $rn \geqslant pf$ is

$$x_{i,j} \longleftarrow x_{i,j} + v_{i,j} \Delta t, \ \ i = 1, \dots, N; \ \ j = 1, \dots, n \tag{26.8}$$

$$v_{i,j} \longleftarrow wv_{i,j} + c_1 r_1 \left(\frac{x_{i,j}^{\mathrm{pb}} - x_{i,j}}{\Delta t} \right) + c_2 r_2 \left(\frac{x_{i,j}^{\mathrm{sb}} - x_{i,j}}{\Delta t} \right),$$

$$i = 1, \ldots, N; \;\; j = 1, \ldots, n \qquad (26.9)$$

which is the velocity update formula of the PSO algorithm.

The last term of the formula, $D(d)/\Delta t$, is multiplied by a random number. The prey either totally fears or does not fear at all its predator. This decision is based upon a random number.

When it comes to the predator swarm, the velocity update formula is:

$$v_{i,j} \longleftarrow \frac{r(x_{i,j}^{\mathrm{sb}} - x_{i,j})}{\Delta t} \qquad (26.10)$$

In this formula, the swarm best $x_{i,j}^{\mathrm{sb}}$ is not the swarm best of the predator swarm, but the swarm best of the prey swarm. Any predator particle hunts individually after the best particle. This trajectory of the predator particle allows the prey swarm best to perturb the local optima.

Step 7. The positions are updated as in the PSO algorithm.

Step 8. Steps 2–7 are revisited if any finishing criterion is not reached.

The Predator-Prey PSO algorithm provides additional parameters to be tuned.

Fear probability –pf

If the fear probability is assigned 0 for all prey particles, then the particles are treated as an ordinary swarm as given in Section 26.2. A higher pf value provides a better exploration, whereas a lower pf value provides a better exploitation. Hence, this parameter is assigned a relatively high value in the first iterations of the algorithm to encourage exploration and is decreased through iterations for exploitation [10].

Prey coefficient – c_3

If the prey coefficient is assigned much greater than the cognitive coefficient $-c_1$ and the social coefficient $-c_2$, the prey group is expected to diverge and not to regroup which results in random search for the prey particles.

Distance coefficients $-\alpha$ and β

The coefficient α has the same effect as the fear probability and should be decreased over time. On the other hand, β has the counter effect of α and should be increased over time.

26.4 The Foraging Search

The classical Particle Swarm Optimization algorithm utilizes one swarm, which increases the probability of being attracted by a local optimum. As an advanced Particle Swarm Optimization algorithm, the Predator-Prey Particle Swarm Optimization algorithm introduces a predator swarm that enforces the prey swarm to move forward to skip the local optima, yet, this new approach causes focusing only on exploration and ignoring exploitation. The need for a new algorithm to handle both drawbacks led to the construction of the Foraging Search algorithm, which is explained below in detail.

26.4.1 *Biological foundations*

In the food chain, animals form three groups: herbivores, omnivores and carnivores. Herbivores are animals that eat plants, carnivores are animals that eat other animals and finally omnivores are animals that eat both animals and plants. In nature, herbivores are primary consumers that are below in the food pyramid. Omnivores are in secondary consumers the middle in the food pyramid and feed on both plants and specific herbivores. Lastly, carnivores are tertiary consumers which are at the top of the food pyramid and feed on specific herbivores and omnivores. This makes herbivores the ultimate hunts, the carnivores the ultimate hunters, and omnivores both hunters and hunts.

In nature, according to the transformation of energy, for a specific food chain, the number of herbivores is greater than omnivores and the number of omnivores is greater than the number of carnivores. In wild environments, the herbivore-omnivore-carnivore ratio can be $10:3:1$ whereas in calm environments the ratio can be $40:10:1$. Moreover, omnivores are the slowest of the food chain whereas carnivores are the fastest [35].

26.4.2 *The algorithm*

The Foraging Search is based on this food flow in nature. Unlike the PSO and the Predator-Prey PSO algorithm, the Foraging Search stimulates three swarms of different sizes. The first swarm is constituted by herbivores, the ultimate hunts and the third swarm is constituted by omnivores, the ultimate hunters. However, the second swarm is a balancing swarm between the hunters and the hunts, which is the omnivore swarm. Utilization of multiple swarms based on the PSO algorithm have been exploited in literature [34, 35], yet, these swarms provide avoiding the local optima merely by encouraging exploration. The introduction of a novel balancing swarm aims to balance the exploration and the exploitation of all swarms. This introduction was achieved in Altay and Kayakutlu [36].

Each swarm is initialized relative to an environment parameter. In relatively wild environments, the ratio between hunters and hunts become higher, whereas in relatively calm environments, this ratio decreases. This environment is modeled as an algorithm parameter. Hunting is modeled two-folds: first level hunters and second level hunters. First level hunters for a specific animal are the animals that are one level above the food chain. First level hunters of herbivores are omnivores and first level hunters of omnivores are carnivores. Second level hunters are animals that are two levels above the food chain. Only herbivores have their second level hunters, that is, carnivores. Thus, herbivores possess two swarms of hunters and omnivores possess one swarm of hunters to avoid. This change in fear factors affects the formula for velocity updates. The Foraging Search algorithm maintains an exploration-exploitation balance, while providing a realistic simulation of nature.

The steps of the algorithm are given below:

Step 1. Define the herbivore-omnivore-carnivore ratio according to the environment

IF the environment is harsh – wild $\longrightarrow 10 : 3 : 1$

IF the environment is regular $\longrightarrow 25 : 6 : 1$

IF the environment is calm $\longrightarrow 40 : 10 : 1$

Step 2. The swarms are initialized randomly according to the environment and herbivore-omnivore-carnivore ratio where n_h denotes the number of herbivores, no denotes the number of omnivores and n_c denotes the number of carnivores.

Step 3. The objective function is calculated for each particle in each swarm.

Step 4. Particle bests, that is $x_{i,j}^{\mathrm{pb}}$, are found four each particle of each swarm.

Step 5. Swarm bests, that is $x_{i,j}^{\mathrm{sb}}$, are found for each swarm.

Step 6. The overall best value is updated.

Step 7. The fear degrees of herbivores from omnivores (the first level hunters) and carnivores (the second level hunters) are calculated as

$$pfho_i = 1 - \frac{d_{fho,i}}{d_{fho}^{\min}} \tag{26.11}$$

and

$$pfhc_i = 1 - \frac{d_{fhc,i}}{d_{fhc}^{\min}} \tag{26.12}$$

where

$i = 1, \ldots, n_h$

$pfho_i$: fear degree from omnivores of the ith herbivore (in the interval $[0,1]$)

$pfhc_i$: fear degree from carnivores of the ith herbivore (in the interval $[0,1]$)

$d_{fho,i}$: the distance of the ith herbivore to the nearest omnivore

$d_{fhc,i}$: the distance of the ith herbivore to the nearest carnivore

$d_{fho}^{\min}$: the minimum distance for a herbivore to fear an omnivore

$d_{fhc}^{\min}$: the minimum distance for a herbivore to fear an carnivore

Step 8. The fear degree of omnivores from carnivores (the first level hunters) are calculated as

$$pfoc_i = 1 - \frac{d_{foc,i}}{d_{foc}^{\min}} \tag{26.13}$$

where

$i = 1, \ldots, n_0$

$pfhc_i$: fear degree from carnivores of the ith omnivore (in the interval $[0,1]$)

$d_{foc,i}$: the distance of the ith omnivore to the nearest carnivore

$d_{foc}^{\min}$: the minimum distance for an omnivore fear an omnivore

Step 9. For omnivores, the probability of being a hunt rather than a hunter is calculated as

$$pp_i = \frac{dc_i}{dc_i + dh_i} \tag{26.14}$$

where

$i = 1, \ldots, n_0$

pp_i: the probability of omnivores being a hunter

dh_i: the distance of ith omnivore to the nearest herbivore

dc_i: the distance of ith omnivore to the nearest carnivore

Step 10. The velocity update formulae are applied to each particle of each swarm, according to the swarm they belong to.

For herbivores, the velocity update formula is

$$v_{i,j} \longleftarrow \omega v_{i,j} + c_1 r_{1i}(y_{i,j} - x_{i,j} + c_2 r_{2i}(\widehat{y}_j - x_{i,j}) + pfho_i c_3 r_{3i} D(d_{fho,i}) + pfhc_i c_4 r_{4i} D(d_{fhc,i}) \tag{26.15}$$

where

$i = 1, \ldots, n_h$

$v_{i,j}$: velocity of jth dimension i^{th} particle of the swarm

w: the inertia coefficient

c_1 and c_2: cognitive and social coefficient

r_{1i}, r_{2i}, r_{3i}, r_{4i}: random numbers for the ith particle in the interval $[0, 1]$

$y_{i,j}$: the position of jth dimension of personal best for the ith particle of the swarm

$x_{i,j}$: the position of jth dimension of the ith particle of the swarm

y_j: j^{th} dimension of the best position of the swarm

c_3: distance based coefficient of herbivores from omnivores

c_4: distance based coefficient of herbivores from carnivores

and $D(\cdot)$ is a measure of the effect that the hunter has on the hunt and it is formulated as

$$D(x) = \alpha e^{-\beta x} \tag{26.16}$$

where d is the Euclidean distance between the prey particle and the nearest hunter particle. α and β are positive constants that define the effect of distance to velocity.

For carnivores, the velocity update formula is

$$v_{i,j} \longleftarrow r \cdot \left(\widehat{y}_j - x_{i,j}\right) \tag{26.17}$$

where

$i = 1, \ldots, n_c$

$v_{i,j}$: velocity of jth dimension of the ith particle of swarm

r: random number in the interval $[0, 1]$

y_j: j^{th} dimension of the best position of the nearest hunt swarm

$x_{i,j}$: the position of the jth dimension of the ith particle of the swarm

For omnivores, the velocity update formula is

$$\begin{aligned} v_{i,j} \longleftarrow (1 - pp_i) \cdot \Big(&\omega v_{i,j} + c_1 r_{1i}(y_{i,j} - x_{i,j}) \\ &+ c_2 r_{2i}(\widehat{y}_j - x_{i,j}) + pfoc_i c_3 r_{3i} D(d_{foc,i}) \Big) + pp_i \cdot \left(r \cdot (\widehat{y}_j - x_{i,j})\right) \end{aligned} \tag{26.18}$$

where

$i = 1, \ldots, n_0$

pp_i: the probability of omnivores being a hunter calculated in Step 8

$v_{i,j}$: velocity of jth dimension of the ith particle of swarm

w: the inertia coefficient

c_1 and c_2: cognitive and social coefficient

r_{1i}, r_{2i}, r_{3i}: random numbers for the ith particle in the interval $[0,1]$

$y_{i,j}$: j^{th} dimension of the personal best of the ith particle of the swarm

$x_{i,j}$: the position of the jth dimension of the ith particle of the swarm

y_j: best position of the jth dimension of the swarm

c_3: distance based coefficient of omnivores from carnivores

r: random number in the interval $[0,1]$

Step 11. The particle positions are updated for each particle of each swarm using Eq. (26.2), where

$x_{i,j}$: the position of the jth dimension of the ith particle

$v_{i,j}$: the velocity of jth dimension of the ith particle

Step 12. Steps 3 and 11 are repeated until the finishing criterion is satisfied.

The newly introduced algorithms are

$d_{fho}^{\min}$ – the threshold distance of a herbivore to an omnivore: This parameter determines the minimum closeness of an omnivore to a herbivore that the herbivore fears the omnivore.

$d_{fhc}^{\min}$ – the threshold distance of a herbivore to a carnivore: This parameter determines the minimum closeness of an carnivore to a herbivore that the herbivore fears the carnivore.

$d_{fhc}^{\min}$ – the threshold distance of an omnivore to a carnivore: This parameter determines the minimum closeness of an carnivore to an omnivore that the omnivore fears the carnivore.

In the Foraging Search algorithm, the fear elements are gradual, that is, as the hunter approaches more to the hunt, the hunt fears more. In terms of the algorithm, the difference lies in the speed of exploration and exploitation. Gradual increase in these parameters provides a better balance rather than a constant value of the parameter.

26.5 The Foraging Search on Continuous Nonlinear Unconstrained Problems

Metaheuristic algorithms may serve as the optimization method for certain problems. There is a need for problem-specific metaheuristics as in the studies of Nearchou [37] and Tsubakinati and Evans [38], while, there are some metaheuristics such as the classical PSO algorithm, that can be applied in any kind of optimization problem.

Nonlinear continuous functions are considered as widely accepted benchmarking cases for new algorithms. The Foraging Search is compared with the Classical and Predator-Prey PSO algorithms according to some performance measures including the robustness of results, computer time, and number of functions utilized. At every trial, random numbers are varied in order to test the robustness and efficiency of the algorithms, which leads to different initial swarm positions and velocities for same algorithm parameters. Two benchmark problems are analyzed in this section: a two dimensional unconstrained problem and the Griewangk function of multi dimensions.

26.5.1 *Two-dimensional problem*

The nonlinear, continuous, two-dimensional problem is minimizing

$$f(\vec{x}) = 100(x_2 - x_1)^2 + (1 - x_1)^2 \tag{26.19}$$

where

$$0 \leqslant x_1 \leqslant 10 \text{ and } 0 \leqslant x_2 \leqslant 10$$

The problem has a global optimum at $x_1 = 1$, $x_2 = 1$ and $f(\vec{x}) = 0$ at this optimum. For measuring success rates of algorithms, three success levels are assigned. This assignment is considered necessary for such problems since the closeness to the number 0 is relative and arguable in mathematical terms. Success Rate 1 (SR1) is assigned as 10^{-4}, which denotes the rate out of 500 trials where the solution is equal to or smaller than 10^{-4}. The second success level, Success Rate 2 (SR2) is assigned as 10^{-7}, which denotes the rate out of 500 trials where the solution is equal to or smaller than 10^{-7}. Finally, the third success rate, Success Rate 3 (SR3) is assigned the exact solution, or 0. SR1 is determined as the standard MATLAB Package precision and also noted to be appropriate for two dimensional problems. Other success rates are determined upon the results obtained in order to reveal the distinction between different algorithm results. The results of the algorithm are provided in Table 26.1 and Figure 26.1:

Comparing the Predator Prey PSO and the Foraging Search, it is obtained that in all criteria, the Foraging Search algorithm has produced better results and dominates the Predator Prey PSO algorithm. In terms of mean number of functions, it is observed that the Predator Prey PSO algorithm uses significantly big number of functions, meaning more number of iterations, than the Foraging Search.

Table 26.1 Results of the 2-dimensional unconstrained problem

	Classical PSO	Predator-Prey PSO	The Foraging Search
Mean of 500 trials	0.1144	4.3219×10^{-6}	1.5096×10^{-8}
Standard deviation of 500 trials	0.2325	6.7716×10^{-5}	2.7201×10^{-8}
Minimum of 500 trials	1.0454×10^{-7}	4.7293×10^{-22}	0
Maximum of 500 trials	1.3707	0.0150	9.9149×10^{-8}
Mean CPU time of 500 trials (seconds)	1.5018	16.9321	12.4584
Mean number of functions	4454.51	32054.86	9515.40
SR1 (%)	17.40	99.60	100.00
SR2 (%)	00.00	84.20	100.00
SR3 (%)	00.00	00.00	57.60

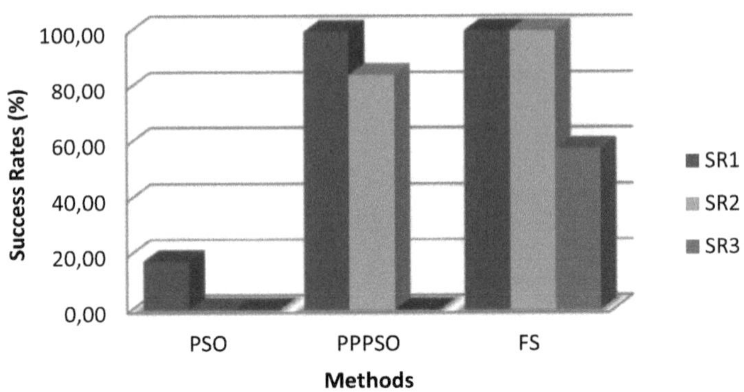

Fig. 26.1 Success Rates for Unconstrained Problem with Two Variables

26.5.2 *The Griewangk function*

The Griewangk Function is generalized as

$$f_{GR}(\overrightarrow{x}) = \sum_{i=1}^{d} \frac{x_i^2}{4000} - \prod_{i=1}^{d} \cos\left(\frac{x_i}{\sqrt{i}}\right) + 1 \tag{26.20}$$

where $-511 \leqslant x_i \leqslant 512$ for all i. In this study, this function is benchmarked for $d =$ 5, 20, 100. The function has a global minimum at $x_i = 0$ for all i and $f(\overrightarrow{x}) = 0$ at the optimum.

26.5.2.1 *The Griewangk function for 5 dimensions*

Results obtained for this or larger values of d indicate that the exact solution is not achieved by all three algorithms. However, according to the range of solutions, four success levels are attained regarding to the comparison of the algorithms. Success Rate 1 (SR1) is

Table 26.2 Results of the Griewangk Function for $d = 5$

	Classical PSO	Predator-Prey PSO	The Foraging Search
Mean of 500 trials	21.6020	0.9303	0.9189
Standard deviation of 500 trials	35.2344	0.0132	0.0063
Minimum of 500 trials	0.0019	0.9087	0.9087
Maximum of 500 trials	320.4325	0.9823	0.9431
Mean CPU time of 500 trials (seconds)	3.9652	16.3842	22.8372
Mean number of functions	4664.32	14644.64	13653.00
SR1 (%)	51.20	100.00	100.00
SR2 (%)	34.40	100.00	100.00
SR3 (%)	11.20	100.00	100.00

assigned as 10, which denotes the rate out of 500 trials where the solution is equal to or smaller than 10. The second success level, Success Rate 2 (SR2) is assigned as 5, which denotes the rate out of 500 trials where the solution is equal to or smaller than 5. Finally, the third success rate, Success Rate 3 (SR3) is assigned 1, where the rate of results are equal to or smaller than 1. Finally, the last success level (SR4) is assigned 0.1. The results are given in Table 26.2 and the success rates are modeled in Figure 26.2.

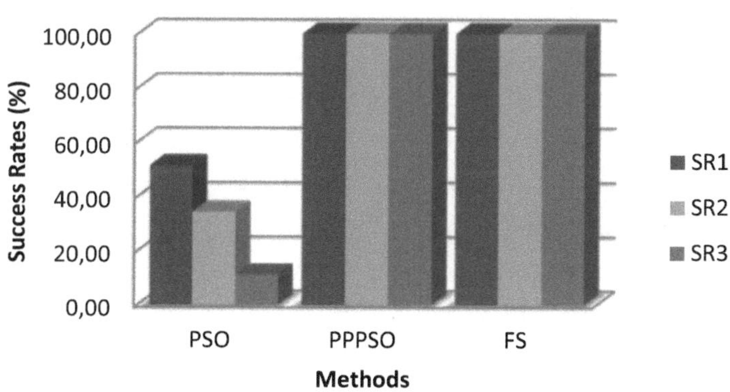

Fig. 26.2 Success Rates for Griewangk Function for $d = 5$

According to the results, the Classical PSO Algorithm has a very wide range of solutions. Out of 500 trials, 4 of them are better (smaller) than the other algorithm results. However, 96% of the results are worse than the other two algorithms and cannot be considered compensatory at a great extent.

The comparison between Predator Prey PSO and the Foraging Search results indicate that the Foraging Search produces slightly better results with less number of functions.

Table 26.3 Results of the Griewangk Function for $d = 20$

	Classical PSO	Predator-Prey PSO	The Foraging Search
Mean of 500 trials	3.9603	10.7054	1.0196
Standard deviation of 500 trials	2.5965	3.8943	0.0697
Minimum of 500 trials	1.8122	2.9039	1.0000
Maximum of 500 trials	27.5322	25.4187	1.3550
Mean CPU time of 500 trials (seconds)	17.9161	92.6173	100.9847
Mean number of functions	5214.00	20842.00	9210.97
SR1 (%)	99.40	97.80	100.00
SR2 (%)	97.00	47.60	100.00
SR3 (%)	82.80	04.20	100.00

A simple hypothesis testing under the assumptions for the true variances are not known and are not equal and the mass distributes normally, where $H_0 = \mu_{PPPSO} = \mu_{FS}$ and $H_1 = \mu_{PPPSO} > \mu_{FS}$ yields a z value of 12.26, implies that H_0 hypothesis to be rejected. Hence, under given assumptions, at 99% significance level, significant evidence has been found on that Predator Prey PSO solution values are greater than the Foraging Search Solutions.

26.5.2.2 *The Griewangk function for 20 dimensions*

The exact solution is not achieved by all three algorithms. However, according to the range of solutions, four success levels are attained. Success Rate 1 (SR1) is assigned as 20, which denotes the rate out of 500 trials where the solution is equal to or smaller than 20. The second success level, Success Rate 2 (SR2) is assigned as 10, which denotes the rate out of 500 trials where the solution is smaller than equal to or 10. Finally, the third success rate, Success Rate 3 (SR3) is assigned 5, where the rate of results are equal to or smaller than 5. Finally, the last success level (SR4) is assigned 1, where the rate of results are equal to or smaller than 1. The results are presented in Table 26.3 and the success rates are presented in Figure 26.3.

In case of $d = 20$, Classical PSO and PPPSO algorithm both have larger means and variations than the Foraging Search. Surprisingly, PPPSO results are worse than the Classical PSO algorithm in terms of all aspects: mean result of trials, time and number of function. On the other hand, the Foraging Search has produced better results in terms of trial means, minimums and maximums. However, the computational time is higher than other algorithms. Yet, the mean number of functions is smaller than of PPPSO, which indicates that the Foraging Search is a faster and more convergent algorithm in terms of number of iterations.

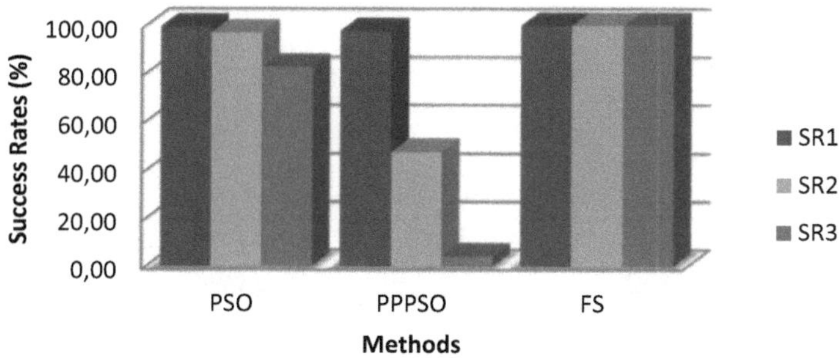

Fig. 26.3 Success Rates for Griewangk Function for $d = 20$

Table 26.4 Results of the Griewangk Function for $d = 100$

	Classical PSO	Predator-Prey PSO	The Foraging Search
Mean of 100 trials	106.7919	68.9474	63.0681
Standard deviation of 100 trials	13.9043	9.3685	8.6069
Minimum of 100 trials	79.9879	48.7990	48.1292
Maximum of 100 trials	140.8094	88.7055	84.0837
Mean CPU time of 100 trials (seconds)	12.0473	158.5205	95.9338
Mean number of functions	681.00	9835.6	2843.00
SR1 (%)	37.00	100.00	100.00
SR2 (%)	01.00	84.00	93.00
SR3 (%)	00.00	18.00	34.00
SR4 (%)	00.00	02.00	02.00

26.5.2.3 *The Griewangk function for 100 dimensions*

The exact solution is not achieved by all three algorithms. However, according to the range of solutions, four success levels are attained. Success Rate 1 (SR1) is assigned as 100, which denotes the rate out of 100 trials where the solution is equal to or smaller than 100. The second success level, Success Rate 2 (SR2) is assigned the rate out of 100 trials, where the solution is equal to or smaller than 80%. Finally, the third success rate, Success Rate 3 (SR3) is assigned 60, where the rate of results are equal to or smaller than 60. Finally, the last success level (SR4) is assigned 50, where the rate of results are equal to or smaller than 50. The results are presented in Table 26.4 and the success rates are presented in Figure 26.4.

For the case $d = 100$, the classical PSO algorithm has produced less successful result compared to the other two algorithms. On the other hand, in terms of computational time and mean number of functions, the least successful outcomes belong to Predator Prey PSO

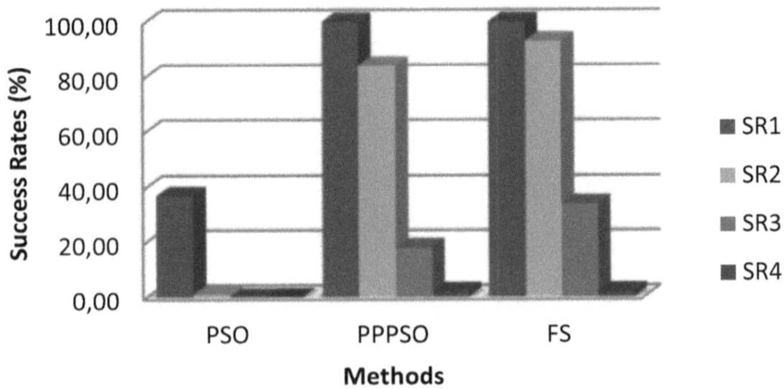

Fig. 26.4 Success Rates for Griewangk Function for $d = 100$

algorithm. According to the trial results and success rates, for case $d = 100$, the most successful and efficiently convergent algorithm is the Foraging Search.

26.6 Commentary and Recommendations

In literature, - metaheuristics have been developed for problems which are computationally expensive or unsolvable with exact methods [13]. Each metaheuristic attempt is to overcome the flaws of the formerly developed metaheuristics. One important weakness is convergence to the local optima [10]. Early convergence can be caused by either a poor exploration of the solution space or an early aborted exploitation within the neighbourhood of the best point found by the algorithm [39]. Nonetheless, exploration and exploitation are known to be two conflicting objectives and therefore, it has been the focus of the recent metaheuristic evolutions [40].

The Foraging Search algorithm imitates the ecological Animal Food Chain or the Animal Food Pyramid. The algorithm engages less mathematical computations and does not employ derivations. A two dimensional, unconstrained and nonlinear problem and Griewangk function are utilized to benchmark the Foraging Search with Particle Swarm Optimization and Predator-Prey PSO algorithm. The algorithm has been found to be effective and robust for all dimensions.

Further studies include an increase in the number of trials in order to generalize obtained results. Moreover, the comparison of the algorithm with other meta-heuristics such as Ant Colony Optimization, Harmony Search, Memetic Algorithms, etc is necessary as a proof of validation of the algorithm . The Foraging Search Algorithm can also be benchmarked with combinatorial optimization problems such as Travelling Salesman Problem, Assignment Problem, and so on. Moreover, this new metaheuristic can be tested on real world engineering problems.

Bibliography

[1] Kennedy J. and Eberhart R.C., Particle Swarm Optimization. In *IEEE International Conference on Neural Networks*, 1995; pp. 1942–1948, vol. 1994 Neural Networks, (1906 November 1995), doi:1910.1109/ICNN.1995.488968 Key: citeulike:379793.

[2] Hassan R., Cohanim B., de Weck O., and Venter G., A Comparison of Particle Swarm Optimization and The Genetic Algorithm. In 46th AIAA/ASME/ASCE/AHS/ASC Structures, Structural Dynamics, and Materials Conference (2005). 2005, pp. 1–13: Austin, Texas, USA.

[3] Dutot A., Olivier D., and Savin G., Collaboration and Competition in Boids Simulation with Predation. In *Emergent Properties in Natural and Artificial Complex Systems, EPNACS 2010*. 2010, pp. 1–3: Lisbon, Portugal.

[4] Lungu V. and Sofron A., Using Particle Swarm Optimization to Create Particle Systems. *CSMR*. 2011, In Print.

[5] Veenhuis C. and Köppen M., Data Swarm Clustering. *Swarm Intelligence in Data Mining*. 2006, 221–241.

[6] Léon-Javier A., Cruz-Cortez N., Morenao-Armendariz M.A., and Orantes-Jiménez S., Finding Minimal Addition Chains with a Particle Swarm Optimization Algorithm. In *Lecture Notes in Computer Science*, 2009, Volume **5845**, MICAI 2009: Advances in Artificial Intelligence. 2009; pp. 680–691.

[7] Engelbrecht A.P., *Computational Intelligence: An Introduction*, 2003. Wiley Publishing.

[8] Ciuprina G., Ioan D., and Monteanu I., Use of Intelligent-Particle Swarm Optimization in Electromagnetics. *IEEE Transactions on Magnetics*. 2002, **38** (2), 1037–1040.

[9] Castro E.G. and Tsuzuki M.S.G., Simulation Optimization Using Swarm Intelligence as Tool for Cooperation Strategy Design in 3D Predator-Prey Game. In *Swarm Intelligence, Focus on Ant and Particle Swarm Optimization*. (eds. Felix T.S., Tiwari M.K.), 2007. In Tech.

[10] Engelbrecht A.P., *Fundamentals of Computational Swarm Intelligence*, 2006. Wiley Publishing.

[11] Mikki S.M. and Kishk A.A., Physical Theory for Particle Swarm Optimization. *Progress in Electromagnetics Research*. 2007, **75**, 177–207.

[12] Liou C.D. and Hsieh Y.C., A PSO-based Algorithm For Two-Machine No-Wait Flowshop Scheduling With Setup Time. In *the International Conference on Business and Information*. 2009: Kuala Lumpur, Malaysia.

[13] Talbi E.G., *Metaheuristics from Design to Implementation*, 2011. Wiley Interscience: New York.

[14] Nickabadi A., Ebadzadeh M.M., and Bakhahsh S., A Novel Particle Swarm Optimization Algorithm With Adaptive Inertia Weight. *Applied Soft Computing*. 2011, **11** (4), 3658–3670.

[15] Weise T., *Global Optimization Algorithms - Theory and Applications*. 2009.

[16] Thangaraj R., Pan M., Abraham A., and Bouvry P., Particle Swarm Optimization: Hybridization Perspectives and Experimental Illustrations. *Applied Mathematics and Computation*. 2011, **217**, 5208–5226.

[17] Gnamabal K., Marimuthu N.S., and Babulal C.K., Three-Phase Power Flow Analysis in Sequence Component Frame Using Hybrid Particle Swarm Optimization. *Applied Soft Computing*. 2011, **11**, 1727–1734.

[18] Shunmugalatha A. and Slochanal S.M.R., Optimum Cost of Generation for Maximum Loadability Limit of Power System Using Hybrid Particle Swarm Optimization. *Electrical Power and Energy Systems*. 2008, **30**, 486–490.

[19] El-Dib A.A., Youssef H.K.M., El-Metwally M.M., and Osman Z., Maximum Loadability of Power Systems Using Hybrid Particle Swarm Optimization. *Electric Power System Research*. 2006, **76**, 485–492.

[20] Gao H. and Wu W., Particle Swarm Algorithm with Hybrid Mutation Strategy. *Applied Soft Computing*. 2011, **11**, 5129–5142.

[21] Wu Q., Power Load Forecasts Based on Hybrid PSO with Gaussian and Adaptive Mutation and Wv-SVM. *Expert Systems with Applications*. 2010, **37**, 194–201.

[22] Lian Z., Gu X., and Jiao B., A Similar Particle Swarm Optimization Algorithm for Permutation Flowshop Scheduling to Minimize Makespan. *Applied Mathematics and Computation*. 2006, **175**, 773–785.

[23] Yin P.Y., Yu S.S., and Wang Y.T., Task Allocation for Maximizing the Reliability of a Distributed System Using Hybrid Particle Swarm Optimization. *The Journal of Systems and Software*. 2007, **80**, 724–735.

[24] Shelokar O.S., Siarry P., Jayaraman V.K., and Kulkarni B.D., Particle Swarm and Ant Colony Algorithms Hybridized for Improved Continuous Optimization. *Applied Mathematics and Computation*. 2007, **188**, 129–142.

[25] Haibing G., Chi Z., and Liang G., Particle Swarm Optimization Based Algorithm for Economic Load Dispatch. *In International Symposium on Intelligence Computation & Applications* (**ISICA 2005**). 2006, pp. 594–599: Wuhan, China.

[26] Shiau D.F., A Hybrid Particle Swarm Optimization for a University Course Scheduling Problem with Flexible Preferences. *Expert Systems with Applications*. 2011, **38** (1), 235–248.

[27] Kayhan A.H., Ceylan H., Ayvaz M.T., and Gurarslan G., PSOLVER: A New Hybrid Particle Swarm Optimization Algorithm for Solving Continuous Optimization Problems. *Expert Systems with Applications*. 2010, **37**, 6798–6808.

[28] Zhang G., Zhao X., Li P., and Gao L., An Effective Hybrid Particle Swarm Optimization Algorithm for Multi-Objective Flexible Job-Shop Scheduling Problem. *Computers & Industrial Engineering*. 2009, **56**, 1309–1318.

[29] Wang X.H. and Li J.J., Hybrid Particle Swarm Optimization with Simulated Annealing. In 2004 International Conference on Machine Learning and Cybernetics. 2004, pp. 2402–2405: Shangai, China.

[30] Niknam T. and Firouzi B.B., A Practical Algorithm For Distribution State Estimation Including *Renewable Energy Sources*. 2009, **34** (11), 2309–2316.

[31] Marinakis Y. and Marinaki M., A Hybrid Multi-Swarm Particle Swarm Optimization Algorithm for the Probabilistic Traveling Salesman Problem. *Computers & Operations Research*. 2010, **37** (3), 432–442.

[32] Lin S.Y., Horng S.J., W K.T., *et al.*, An Efficient Bi-Objective Personnel Assignment Algorithm Based on a Hybrid Particle Swarm Optimization Model. *Expert Systems with Applications*. 2010, **37**, 7825–7830.

[33] Sha D.Y. and Hsu C.Y., A Hybrid Particle Swarm Optimization for Job Shop Scheduling Problem. *Computers & Industrial Engineering*. 2006, **51**, 791–808.

[34] Silva A., da Silva A.P.N.F., and Costa E., An Empirical Comparison of Particle Swarm and Predator Prey Optimisation. In *Lecture Notes in Computer Science*. 2002, pp. 1–45.

[35] Lenbury S., Rattanamongkonkul S., Tumravsin N., and Amornsamankul S., Predator-Prey Interaction Coupled by Parasitic Infection: Limit Cycles and Chaotic Behavior. *Mathematical & Computer Modeling*. 1999, **30**, 131–146.

[36] Altay A. and Kayakutlu G., Animal Food Chain Particle Swarm Optimization. In *The 2011 International Conference of Computational Intelligence and Intelligent Systems, WCE 2011*.

2011, pp. 1094–1099: London, England.

[37] Nearchou A.C., A Novel Meta-Heuristic Approach for the Flow Shop Scheduling Problem. *Engineering Applications of Artificial Intelligence*. 2004, **17**, 289–300.

[38] Tsubakitani S. and Evans J.R., An Empirical Study of a New Metaheuristic for the Traveling Salesman Problem. *European Journal of Operations Research*. 1998, **17**, 289–300.

[39] Dorigo M., Stützle T., *The Ant Colony Optimization Metaheuristic: Algorithms, Applications, and Advances. In Handbook of Metaheuristics*. (eds. Glover F., Kochenberger G.A.), 2003. Springer: New York.

[40] Zapfel G., Braune R., and Bögl M., *Metaheuristic Search Concepts: A Tutorial with Applications to Production and Logistics*, 1 edn, 1950. Springer: New York.

Chapter 27

Active Distribution Networks and Smart Grids: Optimal Allocation of Wind Turbines by using Hybrid GA and Multi-Period OPF

P. Siano[1], G. Rigatos[2], and A. Piccolo[1]

[1] *Department of Industrial Engineering, University of Salerno, Ponte don Melillo 1, Fisciano (SA), Italy*

[2] *Department of Engineering, Harper Adams University College, TF10 8NB, Shropshire, UK*

E-mail: psiano@unisa.it, apiccolo@unisa.it, grigat@ieee.org

The increasing penetration of Renewable Energy Sources (RES) in distribution electrical networks will require a deep revision of the planning and management methodologies for distribution electrical networks. Indeed, the current passive distribution networks, characterized by unidirectional power flows and a partial and centralized control, do not allow optimally managing of RES, therefore limiting their exploitation. Consequently, the gradual increase of Distributed Generation (DG) penetration, especially of renewable type, will determine a deep change in existing electrical distribution networks that will evolve from passive to active networks and smart grids. In order to overcome the integration problems of DG and RES, active networks and smart grids will be managed through systems based on Information and Communication Technology (ICT).

In this chapter a hybrid optimization method that aims at maximizing the Net Present Value related to the investment made by Wind Turbines developers in an active distribution network and smart grids is proposed. The proposed method combines a Genetic Algorithm with a multi-period optimal power flow. The method is demonstrated on a 69-bus 11 kV radial distribution network.

27.1 Introduction

The environmental and energetic problems lead the international politics of energy development towards the reduction of pollutant emissions, a rational use of energy resources and the use of renewable energy resources as strategic targets, thus an increasing spread of Distributed Generation (DG) based on renewable sources is expected.

C. Kahraman (ed.), *Computational Intelligence Systems in Industrial Engineering*, Atlantis Computational Intelligence Systems 6, DOI: 10.2991/978-94-91216-77-0_27,

The wide use of DG will be also supported by the continuous technological development, by the spread of automation in the control and management of electrical networks and by the liberalization of energy markets. The increasingly significant use of management systems based on automatic controls and advanced Information & Communications Technology (ICTs) for distribution networks management will lead to the realization of microgrids (MGs), active networks and smart grids (SGs).

The term smart grid refers to a fully automated electric power system, controlling and optimizing the operation of all its interconnected elements, in order to guarantee safe and efficient operations of energy generation, transmission and distribution [3, 4]. Today, many interesting examples of SGs are available in almost all countries [3, 10–12], often categorized in literature with different conceptual models such as internet model, active networks and microgrids (MGs).

According to the first one, i.e. the internet model, supported by Information and Communication Technology (ICT): “every node in the electrical network of the future will be awake, responsive, adaptive, price-smart, eco-sensitive, real-time, flexible, humming and interconnected with everything else” [3, 10–12]. The internet model implementation is based on the concept of active networks identified as facilitators to offer connectivity and interaction capability for both DGs and customers [3, 10–12].

The second model, i.e. the active network, employs two novel concepts: at first the primary role of the network is to provide connectivity between the points of power supply and demand, besides the distribution network must be capable of interacting with customers. Demand side management (DSM) and real-time power markets can permit end users to modify their energy consumption profiles, thus giving an important contribution to the balance between demand and supply [31, 35, 40].

The third model, i.e. the microgrid, will play a key role in the evolution of the SG [21, 24], it can be considered as a small-scenario version of a centralized electricity system that locally generates, distributes and regulates the flow of energy to consumers. MGs are connected to the bulk power grid allowing high reliability and energy cost reductions by encouraging the end consumers to locally purchase generated electric power at privileged tariffs. The smart grid can be considered a system of integrated intelligent MGs [12, 42], representing the main constituents of a SG [21].

At the distribution level, a smart grid can be achieved through the evolution of active distribution networks that employ various active network management (ANM) schemes. Such ANM schemes intend to make better use of available energy resources [1, 5, 6, 30].

ANM is expected to emerge as the preferred solution to the connection and operation of DG in the near-future. In the initial stage, ANM will allow monitoring and remote control of DG to facilitate its integration in the system. In the intermediate stage, ANM will permit accommodating significant amount of DG once local and global services and trading issues have been defined. Ultimately, the full active power management will rely on network management regime and use real-time communication and remote control to meet the majority of the network service requirements [10].

Among other issues, this chapter focuses on the technical challenges faced by both the Distribution Network Operators (DNOs) and DG developers when considering the grid integration of large amounts of DG [13–15, 17, 23, 32].

Since such network reinforcement usually calls for high capital investment, DNOs and WT developers would like to explore less costly means that can improve the capability of the network to accommodate new generation. One way is to make the best use of the existing network by encouraging development at the most suitable locations [13–15]. In order to do this, DNOs and WTs developers require a reliable and repeatable method of quantifying the capacity of new DG that may be connected to distribution networks without the need for reinforcement.

The challenge of identifying the best network location and capacity for DG has attracted significant research effort, albeit referred to by several terms: optimal 'capacity evaluation' [1, 13–15], 'DG placement' [28], or 'capacity allocation' [18, 19, 43]. These optimization problems apply different numerical algorithms with various objectives and constraints. Several other algorithms are adopted to handle optimization problems with discrete variables [9, 20, 34]. Other approaches require network locations of interest to be prespecified with algorithms guiding capacity growth within network constraints [14, 15, 32].

Nevertheless, as values associated with WTs are time and location dependent, methods that simply consider one specific power value at a specific moment are not able to account for time dependence. Therefore, WTs optimal allocation should consider their capability of delivering power at the right time and WTs should be located in the right place to be able to deliver energy while satisfying network constraints. Simulating load and generation variations during a year and computing the WTs delivered energy allows including the time dimension, when compared with methods that simply consider the power at one specific point in time [8]. In order to account for the time dependence of load and generation, some approaches focused on the concept of energy from DG [13, 30, 39].

This chapter proposes a hybrid optimization method that aims at maximizing the Net Present Value (NPV) related to the investment made by WTs developers in a distribution network endowed with active management schemes. The hybrid optimization method combines the Genetic Algorithm (GA) and the multi-period optimal power flow (OPF) algorithm.

The GA is suitable for selecting the optimal site and number of WTs among some selected WTs types allowing WTs developers to maximize the NPV. The evaluation of the wind energy production is based on a multi-period OPF algorithm, which takes into account distribution network constraints. Such a multi-period OPF considers the time-varying characteristics of the load demand and wind power generation. The method also integrates active management schemes such as coordinated voltage control, energy curtailment and power factor control. The analyses are demonstrated on a 69-bus 11 kV radial distribution network.

27.2 The Smart Grid infrastructure

Active power networks require new control methodologies and techniques needed for achieving a dynamical control of all the interconnected elements, thus optimizing network operations and ensuring regulatory status of the whole electrical system. A possible scenery for the actualization of this infrastructure is described in Figure 27.1.

The main elements of the system are:

- Remote Terminal Units (RTUs);
- Supervisory Control And Data Acquisition (SCADA);
- Energy Management System (EMS);
- Advanced Metering Infrastructure (AMI);
- State Estimation Algorithms (SEA);
- Generation and Load Forecast System (GLFS).

The SCADA transmits the measurement data provided by remote collecting data devices (RTUs) placed in strategic positions along the grid to the EMS.

In order to obtain a full control of the grid, real time information of the state of the network are needed to identify the current and voltage profiles at all its nodes. To achieve this goal the use of SEA, i.e. a set of methods and mathematical procedures for evaluating the state of a network by making use of remote measurements of apposite parameters is required. Moreover, as load flow and voltage profiles estimation suffers from a substantial inaccuracy and in order to obtain an accurate state estimation, adequate monitoring systems [3, 4, 39] as well as an AMI, based on wireless and/or wired automatic metering systems (smart meters) with bidirectional capabilities, are needed [11, 16, 25, 26, 29, 36].

Optimization, monitoring and control of the SG performances are entrusted to a suite of hardware/software applications that constitutes the so-called EMS [25]. After evaluating the data coming from SCADA using SEA it determines those actions required for achieving the optimum state of the grid, setting the optimal combination of control variables including the reference values for the active and reactive power dispatching and the position of the OLTC transformer in the substation.

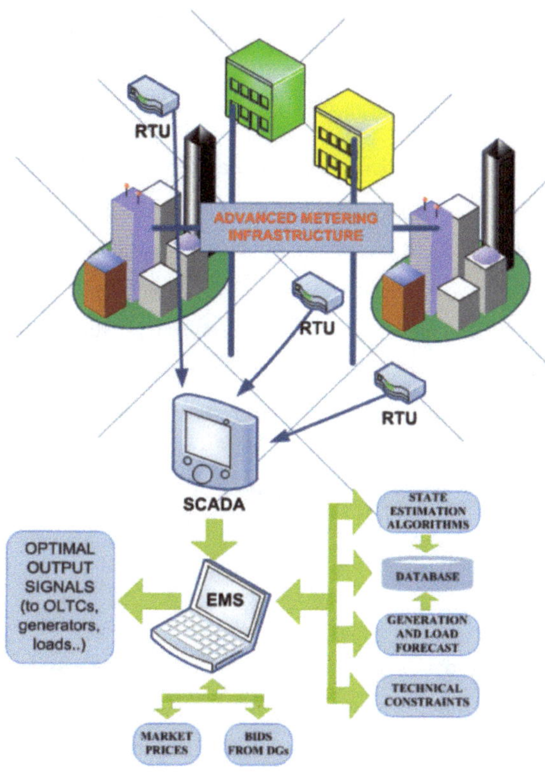

Fig. 27.1 Energy Management System in the Smart Grid infrastructure

27.3 Assessing maximum wind energy exploitation in active distribution networks by means of multi-period OPF

Active management represents an alternative approach to enable national targets for renewable energy and increase the penetration of WTs into the existing distribution networks [22, 33]. It has, indeed, the potential to maximize DG penetration level while minimizing DG-related network reinforcements [6, 7, 27, 37, 38, 40, 44].

Active management can be realized, for instance, through generation dispatch, transformer tap adjustment or reactive power compensators. WTs generation curtailment during low demand, reactive power management using a reactive compensator, and area-based on-load tap-changer coordinated voltage control have been used in the active management [22].

27.3.1 *Coordinated on-load tap-changer voltage control*

Traditional control strategies of on-load tap-changers (OLTCs) are either based on the voltage regulation at a single busbar or voltage drop compensation on a particular line [22]. Such voltage control strategies are based on local measurements and are suitable for traditional distribution systems with unidirectional power flow. However, these strategies may cause problems in distribution networks with bi-directional power flows. On the other hand, the area-based control strategy of OLTCs is based on measurements from various locations of the network. In this way, the voltage regulation of OLTCs can be based on the voltage information of the bus that has the most severe over voltage problem. Consequently, the maximum wind energy penetration level may be increased by the implementation of the control strategy.

27.3.2 *Energy Curtailment*

In order to alleviate the over voltage problem, it may be necessary to curtail a certain amount of wind energy injected into the network [22]. Although the output wind energy is reduced, the WT developer may still gain more profits due to the possibility of installing more WTs.

In the proposed method, wind energy may be curtailed during certain periods in order to alleviate any voltage or thermal constraint violation. For example, for a specific period, there are different possible combinations of load demand and wind power. Wind energy is curtailed during periods of minimum demand and high wind power generation. The same strategy is applied to each of the periods analyzed. In the method, energy curtailment is implemented in each period by introducing a negative generation variable to represent the curtailed energy from each WT. For a given period, the maximum energy that can be curtailed from a given WT is set to a fraction of the potential energy that the WT could have produced without energy curtailment.

27.3.3 *Coordinated generator reactive power control*

The recent grid codes of many countries, such as Denmark, Germany, Italy, Ireland and the UK, require that WTs should provide reactive power control capabilities and that network operators may specify power factor or reactive power generation requirement for grid-connected WTs [41].

In practice, a grid-connected WT needs to fulfill the specific requirement depending on the regulation of the country. For example, in the Danish grid code for grid-connected WTs, reactive power generation is confined to a control band with respect to active power generation (with a power factor between 1.00 and 0.995 lagging). The German grid code specifies different reactive power limits according to voltage value at interconnection (with a power factor ranging between 1.00 and 0.925 lagging). The Irish grid code requires a power factor between 0.835 leading and 0.835 lagging when the active power output level is below 50% of the rated capacity. In Italy and the UK, the power factor at a WT's terminal should be between 0.95 leading and 0.95 lagging. Although it is important to fulfill the grid code when connecting a WT, this paper intends to illustrate the concept of the proposed method, but not to design a WT that fulfills a specific requirement.

WTs, especially those with power electronic controllers, are able to provide necessary reactive power support to the grid. The reactive power generation can be dispatched centrally by the DNOs [41]. In other words, power factors of WTs can be controlled so that wind energy penetration level in the network is maximized. The proposed control scheme requires WTs to generate reactive power during load peak hours and low generation, and to absorb reactive power during load off-peak hours and high generation.

27.4 Multi-Period Optimal Power Flow

The optimization method aims to find the optimal locations and capacities of WTs so that the wind energy exploitation in the network is maximized. Such an objective is subject to a number of technical constraints imposed by regulations including bus voltage limits, line/transformer thermal limits, and system short-circuit levels. By fulfilling these constraints, the network reinforcement due to the connection of WTs may be avoided. In addition, such a method can be used to investigate the impact of the foregoing active management strategies on the maximum wind energy penetration level in the network.

The proposed approach, based on the nonlinear programming formulation of the multi-period OPF [39], has been modified in order to maximize the wind energy exploitation and to include active management schemes, the time-varying characteristics of the load demand and wind power generation, and the system short-circuit constraints.

The multi-period OPF is formulated as:

$$\begin{aligned} \text{maximize} \quad & E = \sum_{j=1}^{N_j} \sum_{g=1}^{N_G} E_g^j(P_g, \mathbf{x}_j) \\ \text{subject to:} \quad & h(\mathbf{x}_j) = 0, \\ & g(\mathbf{x}_j) \leqslant 0 \end{aligned} \tag{27.1}$$

where $E_g^j(P_g, \mathbf{x}_j)$ is the wind energy generated during the time period j by the gth WT with rated capacity P_g, N_j is the total number of periods in a year corresponding to different combinations of load demand and wind power generation; N_G is the number of WTs (indexed by g).

The vector $\mathbf{x}_j$ consists of a set of controllable quantities and dependent variables during each period j. The optimization variables include the capacity of each WT, and for each period j: the secondary voltage of the OLTC, the power factor angle and the curtailed energy of each WT, and the import/export power at the interconnection to the external network.

The equality constraints $h(\mathbf{x}_j)$ represent the static load flow equations such as Kirchhoff current law $\forall j \in J$ and $\forall b \in B$, where J is the set of periods (indexed by j), B is the set of buses (indexed by b) and Kirchhoff voltage law, $\forall j \in J$ and $\forall l \in L$, where L is the set of lines (indexed by l).

The inequality constraints $g(\mathbf{x}_j)$ are listed in the following.

– Capacity constraints for the interconnection to external network (slack bus) $\forall j \in J$, $\forall x \in X$:

$$\begin{aligned} P_x^- &\leqslant P_{x,j} \leqslant P_x^+ \\ Q_x^- &\leqslant Q_{x,j} \leqslant Q_x^+ \end{aligned} \tag{27.2}$$

where X is the set of external sources (indexed by x), $P_{x,j}$ and $Q_{x,j}$ are the active and reactive power outputs of x, respectively and P_x^-, P_x^+ and Q_x^-, Q_x^+ are the min/max active and reactive power outputs of x, respectively.

– Capacity constraints for the WTs: maximum capacity that may be installed at each site $\forall j \in J, \forall g \in G$:

$$\begin{aligned} P_g^- &\leqslant P_{g,j} &\leqslant P_g^+ \\ Q_g^- &\leqslant Q_{g,j} &\leqslant Q_g^+ \end{aligned} \tag{27.3}$$

where G is the set of WTs (indexed by g), $P_{g,j}$ and $Q_{g,j}$ are the active and reactive power outputs of g, respectively and P_g^-, P_g^+ and Q_g^-, Q_g^+ are the min/max active and reactive power output of g, respectively.

– Voltage level constraints $\forall j \in J, \forall b \in B$:

$$V_b^- \leqslant V_{b,j} \leqslant V_b^+ \tag{27.4}$$

where $V_{b,j}$ is the voltage at bus b, V_b^+ and V_b^- are the max/min voltage at bus b, respectively.

Flow constraints for lines and transformers $\forall j \in J, \forall l \in L$:

$$\sqrt{(f_{l,j}^P)^2 + (f_{l,j}^Q)^2} \leqslant f_l^+ \tag{27.5}$$

where $f_{l,j}^P$ and $f_{l,j}^Q$ represent the active and reactive power injection onto l respectively and f_l^+ the maximum power flow on l.

– Short-circuit level constraint: the requirement of not exceeding the design short-circuit capacity in typical radial networks, fed by a MV/LV substation and with wind generation, should be satisfied as it could constrain new generation capacity. WTs connected to the distribution network may contribute to the short-circuit level at the distribution substation. The upstream grid provides the dominant contribution to the short circuit capacity, which rapidly diminishes downstream the network due to the series impedance of the lines. The short-circuit requirement normally needs to be checked at the MV (or LV) busbars of the substation [2]. Therefore, given the typical radial arrangement of distribution networks, the maximum short-circuit level will be obtained when considering a three-phase short-circuit at the low-voltage side of the substation.

The magnitude of the expected short-circuit current $|I_{cc}|$ at the low-voltage side of the substation, calculated from the phasor sum of the maximum short-circuit currents from the upstream grid, through the step-down transformer, and from the WTs connected to the distribution network, is, therefore, limited by the design short-circuit capacity $I_{cc}^{\max}$.

$$|I_{cc}| \leqslant I_{cc}^{\max} \tag{27.6}$$

The grid contribution is calculated according to IEC 60909 and the contribution of WTs is computed [20, 39].

The additional constraints derived from the active management schemes are: coordinated on-load tap-changer voltage constraint, curtailed energy, WTs power factor angles.

-Curtailed energy constraint $\forall j \in J$:

$$CE_g^j \leqslant CE_{g\,\max}^j \tag{27.7}$$

where CE_g^j represents the amount of curtailed energy from generator g during period j and $CE_{g\,\max}^j = C_f^j \times E_g^{j\text{-max}}$ the maximum permitted curtailed energy from generator g during j, where C_f^j is the curtailment index, varying in the range $[0,1]$ and $E_g^{j\text{-max}}$ is the maximum energy that generator g could have produced during j without curtailment.

– Coordinated on-load tap-changer voltage constraint $\forall\, j \in J$:

$$V_{\text{OLTC}}^- < V_{j,\text{OLTC}} < V_{\text{OLTC}}^+ \tag{27.8}$$

where $V_{j,\text{OLTC}}$ is the secondary voltage of the OLTC during j, V_{OLTC}^- and V_{OLTC}^+ are the (max/min) voltage of the OLTC, respectively.

– Coordinated generator reactive power constraints, $\forall\, j \in J$, $\forall\, g \in G$:

$$\phi_g^- < \phi_{g,j} < \phi_g^+ \tag{27.9}$$

where $\phi_{g,j}$ is the power factor angle of g during j, ϕ_g^- and ϕ_g^+ are the (max/min) power factor angle of g, respectively.

27.5 Genetic Algorithm - Multi-Period OPF Hybrid Optimization Method for optimal allocation of wind turbines

The GA is used in order to select the types and number of WTs to be allocated at each candidate bus. The GA randomly generates the initial population of solutions (individuals) by defining a set of vectors. Each vector, or called a chromosome, has a size $N_e = N_C \times N_T$, where N_C is the number of candidate locations and N_T is the number of defined WT types.

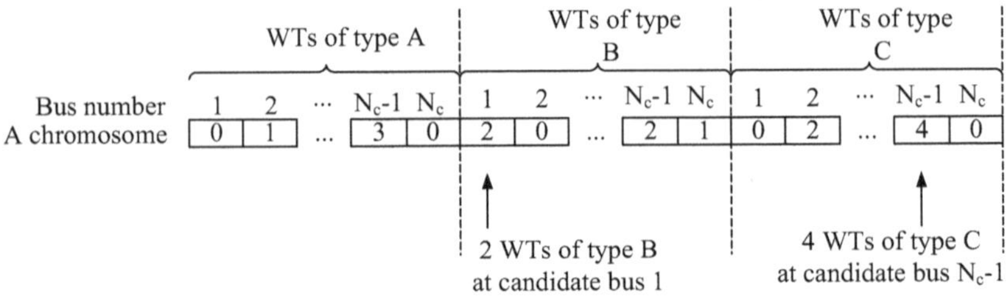

Fig. 27.2 Schematic of the GA chromosome

As shown in Fig. 27.2, a chromosome consists of a vector of integers, each of which represents the number of WTs of a given type to be allocated at a candidate bus. For instance, WTs of type A is associated with the first part of the vector with the size of N_C, which is the number of the candidate locations. Each element of this vector is an integer representing the number of WTs of type A connected to the corresponding bus. As such, the locations and types of WTs are expressed as a string of integers.

At each generation of the GA, a new set of improved individuals is created by selecting individuals according to their fitness; the selection mechanism used here is the normalized geometric ranking scheme. After the new population is selected, genetic operators are applied to selected individuals for a discrete number of times. These genetic operators are simple crossover and binary mutation. A simple crossover randomly selects a cut-point dividing each parent into two segments. Then, two segments from different parents are combined to form a new child (individual). A binary mutation changes each of the bits of the parent based on the probability of mutation. An elitism mechanism is also adopted to ensure the best member of the population is not lost. The iteration process continues until one of the stopping criteria is reached.

For each chromosome of the GA, specifying the number and location of WTs, the maximum wind energy generation over a year is evaluated by a multi-period OPF algorithm nested in the GA algorithm.

In order to analyze the profitability of the WTs investment, the GA objective function is the NPV, i.e. the difference between the sum of the discounted cash flows which are expected from the investment and the amount which is initially invested:

$$NPV = -C_0 + \sum_{k=1}^{n} \frac{C_k}{(1+i)^k} \tag{27.10}$$

where: k is the year of the cash flow, i the discount rate (the rate of return that could be earned on an investment in the financial markets with similar risk), C_k the net cash flow (the amount of cash, inflow minus outflow) at year k, C_0 is the investment, n is the lifetime of WTs.

Consequently, this hybrid method will deliver the best locations as well as the best WT types in the end. The flow chart of the foregoing hybrid optimization method is shown in Fig. 27.3. The proposed method has been implemented in Matlab® and is based on MATPOWER suite [45] and demonstrated through the study system described in the following section.

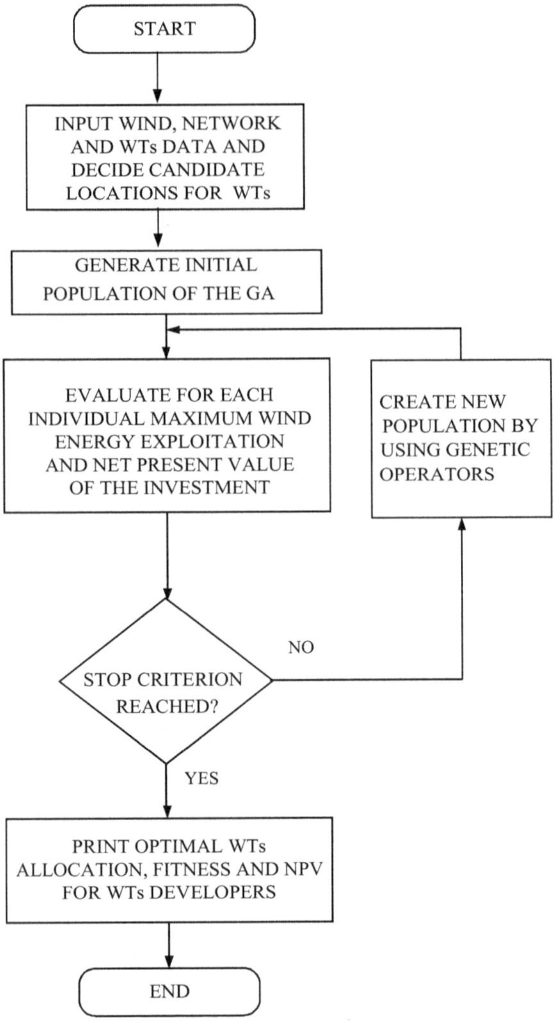

Fig. 27.3 Flow chart of the Hybrid optimization method

27.6 System Description

This section describes the distribution system and data that are used to demonstrate the hybrid optimization approach proposed in the previous section for the optimal allocation of WTs.

The following analyses are based on a 69-bus 11 kV radial distribution system [39]. The four feeders are supplied by two identical 6 MVA 33/11 kV transformers. Fig. 4 shows the distribution system and the potential WT locations, selected to demonstrate the capabilities of the method.

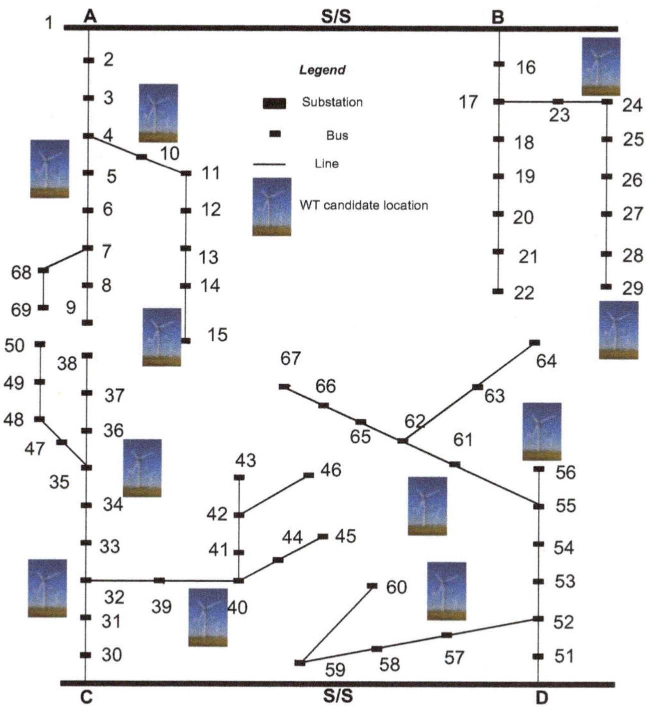

Fig. 27.4 69-bus network indicating potential locations for WTs

27.6.1 *Modelling of time-varying load and wind power generation*

For the modeling of time-varying load and wind power generation, real data from a local distribution network have been used and processed.

Based on their joint probability of occurrence, defining the number of coincident hours over the year, wind availability and demand have been aggregated into a number of wind/demand scenarios [4]. The set of scenarios obtained by combining wind availability and load demand real data for one year have been considered. Each scenario represents the combination between wind speed and load demand values, indicated in percentage terms,

Table 27.1 Maximum Network Loading

Active Power [MW]	Reactive Power [MVar]
2.90	1.99

and is characterized by a defined number of hours over the year. Such a number represents the time (number of hours) during which each combination wind/demand occurs in the course of the year [4].

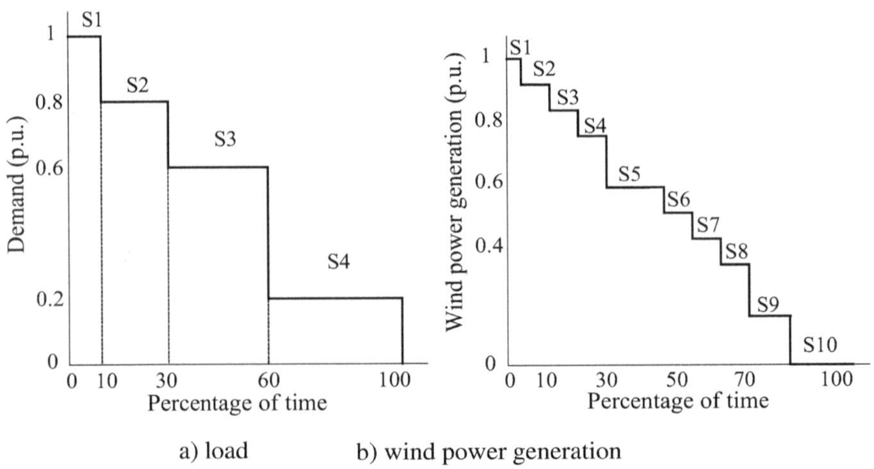

Fig. 27.5 Discrete states of a) load and b) wind power generation at a given hour

As shown in Fig. 27.5, each type of day consists of 24 hours each of which can have 400 ($10 \times 10 \times 4$) different combinations of load-generation considering as two different wind speed distributions at buses have been assumed. In order to create the multi-period interdependency, at each iteration of the MP-OPF, to a unique set of WTs capacity variables correspond 400 sets of power flow variables. The maximum load level of each bus given in Table 27.1.

27.6.2 *Short-circuit calculations*

The short-circuit capacity of the grid is assumed to be 1000 MVA. The grid is connected to the transformers through a series inductive impedance of 8%. Losses for the substation's transformers are not considered. The designed short-circuit capacity for the 11 kV network is assumed to be 200 MVA. As recommended by the IEC Standard, the R/X ratio for the corresponding equivalent impedance is assumed to be 0.1. The voltage factor used to consider the variations of the system voltage is assumed to be 1.1 [2]. The WTs are connected to the 11 kV network through a 0.69/11 kV transformer with a series impedance of 4% and a rated resistive component of the short-circuit voltage of 1.2% [2]. In order to evaluate the WTs' contribution to the short-circuit level at each bus, the equivalent short-circuit impedances of WTs and transformers have been computed for each bus according to the installed WTs capacities and to the parallel connections of WTs.

27.6.3 *Network operation constraints*

Voltage limits are taken to be $\pm 6\%$ of nominal and feeder thermal limits are 5.1 MVA (270 A/phase). The substation power exports to the upstream grid are limited to the capacity of the transformers (12 MVA).

In order to demonstrate the capabilities of the energy curtailment scheme, it has been assumed that wind energy can be curtailed only during the period in which the combination of minimum demand and high wind power occurs.

Power factor is assumed to vary between 0.9 leading and 0.9 lagging when the coordinated generators reactive power control option is considered. Otherwise, WTs are assumed to operate with a fixed power factor of 1.

The short-circuit limit constraint of 200 MVA has been assumed accordingly to the designed short-circuit capacity for the network.

27.7 Case Study

The hybrid optimization algorithm was applied on the above-described distribution system.

It is assumed that WTs of three different capacities are chosen by the WT developers. These capacities are 225 kW, 660 kW and 900 kW.

Maximum four WTs of can be allocated at a given location. This requirement may be set by the available land for building WTs.

Consequently, GA is used to search for the optimal number of WTs of each type at the candidate locations. It is also assumed that the power factor is the same for all WTs connected to the same bus.

Table 27.2 Weibull wind speed distributions parameters

Weibull distribution	Scale Parameter	Shape Parameter
WD1	6	3
WD2	10	3

The multi-period OPF has been applied for evaluating its annual maximum wind energy exploitation considering the following active management options simultaneously: coordinated OLTC voltage control, energy curtailment and WTs reactive power control.

The basic parameters of the GA are summarized as follows. The total control variables are 33 ($= 3 \times 11$), corresponding to the number of three types of WTs at the eleven candidate locations. The population size of each generation is 20. The initial population is generated at random between zero and three.

The GA stops if any of the following conditions is reached:

1) the maximum generation number exceeds 150,
2) there is no improvement in the objective function for 50 consecutive generations, and
3) the cumulative change in the fitness function value over 5 generations is less than $1e^{-6}$.

Sensitivity analyses have been carried out to consider different values for the GA parameters such as stop criteria, population size and genetic operators. From these analyses, it was shown that the used values guarantee the convergence of the algorithm to a satisfactory solution in this case.

In order to evaluate the effectiveness of the proposed hybrid method, different scenarios for the wind speed distributions in each bus of the network have been assumed. In particular, two different Weibull wind speed distributions have been considered, namely WD1 and WD2. The corresponding wind power generation have been derived and discretized on the basis of power curves of commercial WTs. The Weibull wind speed distributions parameters are shown in Table 27.2. Table 27.3 lists the capital costs (C_C) associated to the three candidate WTs.

Table 27.4 lists the optimal numbers of WTs at different locations found by the GA and the corresponding wind speed distribution assumed at each location. As shown in Table 27.5 a total WTs capacity of 13.380 MW is installed allowing delivering 36,166 MWh/year, with a total capital cost of the investment is of 14.976 M€ and the NPV equals 13.088 M€.

Table 27.3 Capital costs associated to the three candidate WTs

WT Type	Rated output electric power [kW]	Capital cost [€/kW]	Total capital cost [M€]
A	225	1400	0.315
B	660	1200	0.792
C	900	900	0.810

Table 27.4 Optimal numbers of WTs at different locations found by the GA and corresponding wind speed distribution

Bus no.	Weibull distribution	225 kW	660 kW	900 kW	Capacity (kW)
5	WD1	1	1	1	1785
10	WD1	2	0	1	1350
15	WD1	0	1	0	660
24	WD1	1	1	1	1785
29	WD1	1	1	0	885
32	WD2	2	1	0	1110
35	WD2	1	1	0	885
40	WD2	2	0	1	1350
56	WD2	0	1	0	660
57	WD2	1	1	2	2685
61	WD2	1	0	0	225
Total		12	8	6	13380

Table 27.5 Results of WTs obtained from the hybrid optimisation method

Total capacity [MW]	Delivered wind energy [MWh]	Total capital cost [M€]	NPV WTs developers [M€]
13.380	36166	14.976	13.088

27.8 Conclusion

This chapter has described a hybrid optimization method that can help WTs developers to plan investments in an active distribution network.

The proposed optimization method combines the GA and the multi-period OPF and considers the time-varying characteristics of the load demand and wind power generation. The method allows the WTs developers to optimally allocate a chosen number and types of WTs among a large number of potential combinations in an active distribution network.

Furthermore, the method can be used to evaluate the feasibility of a project in WTs before carrying out investments. Simulation results evidenced that the proposed hybrid method, which maximizes the NPV related to the investment in WTs, is suitable for selecting the optimal site and number of WTs among selected WTs types.

Different active management schemes have been considered in the proposed optimization formulation as they are able to enhance the total amount of wind energy exploitation and thus offer more economic benefits to both WTs developers and DNOs. In fact, active management is expected to provide higher profits to the WTs developers by allowing them connecting more WTs to the network. The active management is also an effective and indispensable strategy for DNOs to integrate and operate WTs in distribution networks and to defer network investments caused by annual load growth and/or DG connections.

Nevertheless, practical implementation of active management schemes requires additional commercial arrangements and financial evaluations. New market rules should be implemented to offer economic benefits to DNOs in order to drive them to provide the active management service to WTs developers. On the other hand, new revenue mechanisms should be developed so that WTs developers and DNOs share the benefits as well as the costs of active management.

Further simulations with larger networks, not presented here, have demonstrated the scalability of the proposed method and its applicability to larger networks. The method is also able to cope with a larger number of control variables. Although this will lead to an increase in the computing time, this is not a constraint as the method is intended for long-term planning studies. Different load profiles (by considering a mix of industrial, commercial and residential customers) for each node can be easily introduced in the method.

The main drawback of the proposed hybrid optimization algorithm is that the simulation time is very long. This is due to the evaluation of the fitness function which each time calls for a lengthy multi-period OPF. However, this drawback can be tolerated as simulation time is not a major concern for long-term system planning. In addition, a more powerful computer may improve the simulation speed to a certain extent.

Bibliography

[1] Ault G.W., F. Currie R.A., and McDonald J.R., Active Power Flow Management Solutions for Maximising DG Connection Capacity, *Proc. of IEEE Power Eng. Society General Meeting*, pp. 1–5, (2006).

[2] Boutsika T.N. and Papathanassiou S.A., Shortcircuit calculations in networks with distributed generation, *Electric Power Systems Research*, **78**, 1181–1191 (2008).

[3] Cecati C., Citro C., Piccolo A., and Siano P., Smart Operation of Wind Turbines and Diesel Generators According to Economic Criteria, *IEEE Transactions on Industrial Electronics*, **58** (10), pp. 4514–4525 (2011).

[4] Cecati C., Citro C., Siano P., Combined Operations of Renewable Energy Systems and Responsive Demand in a Smart Grid. *IEEE Transactions on Sustainable Energy*, **2** (4), pp. 468–476 (2011).

[5] Celli G., Pilo F., Pisano G., and Soma G.G., Optimal participation of a microgrid to the energy market with an intelligent EMS, *Proc. of the 7th International Power Engineering Conference*, **2**, pp. 663–668 (2005).

[6] Currie R.A.F., Ault G.W., Foote C.E.T., Burt G.M., and McDonald J.R., Fundamental research challenges for active management of distribution networks with high levels of renewable generation. *Proc. of 39th Int. Universities Power Eng. Conf.*, **2**, pp. 1024–1028 (2004).

[7] Currie R.A.F., Ault G.W., Foote C.E.T., and McDonald J.R., Active power-flow management utilising operating margins for the increased connection of distributed generation, *IET Generation, Transmission & Distribution*, **1** (1), pp. 197–202 (2007).

[8] Dugan R. and Wacławiak M., Using energy as a measure of risk in distribution planning, *Proc. of 19th Int. Conf. on Electricity Distribution*, pp. 1–4 (2007).

[9] El-Khaltam W., Bhattacharya K., Hegazy Y., and Salama M.M.A., Optimal investment planning for distributed generation in a competitive electricity market, *IEEE Trans. Power Systems*, **19** (3), pp. 1674–1684 (2004).

[10] European Commission, EUR 22040 European Technology Platform SmartGrids, *Office for Official Publications of the European Communities*, (2006).

[11] Fabro M., Roxey T., and Assante M., No Grid Left Behind, *IEEE Security & Privacy*, **8** (1), pp. 72–76 (2010).

[12] Farhangi H., The path of the smart grid, *in IEEE Pwr & En. Mag.*, **8** (1), pp. 18–28 (2010).

[13] Harrison G.P., Piccolo A., Siano P. and Wallace A.R., Exploring the Trade-offs Between Incentives for Distributed Generation Developers and DNOs, *IEEE Trans. on Power Systems*, **22**, pp. 821–828 (2007).

[14] Harrison G.P., Piccolo A., Siano P., and Wallace A.R., Hybrid GA and OPF evaluation of network capacity for distributed generation connections, *Electrical Power Systems Research*, **78**, pp. 392–398 (2008).

[15] Harrison G.P., Piccolo A., Siano P., and Wallace A.R., Distributed Generation Capacity Evaluation Using Combined Genetic Algorithm and OPF, *International Journal of Emerging Electric Power Systems*, **8**, pp. 1–13 (2007).

[16] Kamireddy S., Schulz N.N., and Srivastava A.K., Comparison of state estimation algorithms for extreme contingencies, *NAPS '08 Power Symposium*, 40th North American, pp. 1–5 (2008).

[17] Keane A. and O'Malley M., Optimal Utilization of Distribution Networks for Energy Harvesting, *IEEE Trans. on Power Systems*, **22** (1), 467–475 (2007).

[18] Keane A. and O'Malley M., Optimal Allocation of Embedded Generation on Distribution Networks, *IEEE Trans. Power Systems*, **20** (3), pp. 1640–1646 (2005).

[19] Kim K.H., Lee Y.J., Rhee S.B., Lee S.K., and You S.K., Dispersed generator placement using fuzzy-GA in distribution systems, *Proc. of IEEE PES Summer Meeting*, Chicago, USA, pp. 1148–1153 (2002).

[20] Kuri B., Redfern M., and Li F., Optimization of rating and positioning of dispersed generation with minimum network disruption, *Proc. of IEEE Power Eng. Soc. Gen. Meeting*, Denver, CO, pp. 2074–2078 (2004).

[21] Lidula N. and Rajapakse A.D., Microgrids research: A review of experimental microgrids and test systems, *Ren. and Sust. En. Rev.*, **15** (1), pp. 186–202 (2011).

[22] Liew S.N. and Strbac G., Maximising penetration of wind generation in existing distribution networks, *Proc. of IEE Proc. Generation, Transmission and Distribution*, **149**, pp. 256–262

(2002).

[23] Masters C.L., Voltage rise: The big issue when connecting embedded generation to long 11 kV overhead lines. *Power Eng. J.*, **16** (1), 5–12 (2002).

[24] Mehrizi-Sani A. and Iravani R., Potential-function based control of a microgrid in islanded and grid-connected modes, *IEEE Trans. Pwr Syst.*, vol. **25** (4), pp. 1883–1891 (2010).

[25] Muller H., Rudolf A., and Aumayr G., Studies of distributed energy supply systems using an innovative energy management system, *Power Industry Computer Applications*, PICA 2001, pp. 87–90 (2001).

[26] Muscas C., Pilo F., Pisano G., and Sulis S., Optimal allocation of multichannel measurement devices for Distribution State Estimation, *IEEE Trans. on Instrumentation and Measurement*, **58** (6), (2009).

[27] Mutale J., Benefits of Active Management of Distribution Networks with Distributed Generation, *Proc. of Power Systems Conf. and Exp.*, pp. 601–606 (2006).

[28] Nara K., Hayashi Y., Ikeda K., and Ashizawa T., Application of tabu search to optimal placement of distributed generators, *Proc. of IEEE PES Winter Meeting*, pp. 918–923, (2001).

[29] Neenan B. and Hemphill R.C., Societal Benefits of Smart Metering Investments, *The Electricity Journal*, **21** (8), pp. 32–45 (2008).

[30] Ochoa L.F., Keane A., Dent C., and Harrison G.P., Applying active network management schemes to an Irish distribution network for wind power maximization, *Proc. of Int. Conf. on Electricity Distribution*, pp. 1–4 (2009).

[31] Palensky P. and Dietrich D., Demand Side Management: Demand Response, Intelligent Energy Systems, and Smart Loads, *IEEE Trans. on Ind. Inf.s*, **7** (3), pp. 381–388 (2011).

[32] Piccolo A. and Siano P., Evaluating the Impact of Network Investment Deferral on Distributed Generation Expansion, *IEEE Trans. Power Systems*, **24** (3), 1559–1567 (2009).

[33] Pudjianto D., Castro M., Djapic P., Stojkovska B., Strbac G., and Allan R.N., Transmission Investment and Pricing in Systems with Significant Penetration of Wind Generation, *Proc. of IEEE Power Engineering Society General Meeting*, pp. 1–3 (2007).

[34] Rau N.S. and Wan Y.H., Optimum location of resources in distributed planning, *IEEE Trans. on Power Systems*, **9** (4), pp. 2014–2020 (1994).

[35] Roscoe A.J. and Ault G., Supporting high penetrations of renewable generation via implementation of real-time electricity pricing and demand response, *IET Ren.. Pow. Gener.*, **4** (4), pp. 369–382 (2010).

[36] Russell B.D. and Benner C.L., Intelligent Systems for Improved Reliability and Failure Diagnosis in Distribution Systems, *IEEE Trans. on Smart Grids*, **1** (1), pp. 48–56 (2010).

[37] Schwaegerl C., Bollen M.H.J., Karoui K., and Yagmur A., Voltage control in distribution systems as a limitation of the hosting capacity for distributed energy resources, *Proc. of 18th Int. Conf. on Electricity Distribution*, pp. 1–5 (2005).

[38] Shafiu A., Bopp T., Chilvers I., and Strbac G., Active management and protection of distribution networks with distributed generation, *Proc. of IEEE Power Eng. Society General Meeting*, **1**, pp. 1098–1103 (2004).

[39] Siano P., Chen P., Chen Z., and Piccolo A., Evaluating maximum wind energy exploitation in active distribution networks", *IET Gener. Transm. Distrib.*, **4** (5), pp. 598–608 (2010).

[40] Strbac G., Demand side management: benefits and challenges, *Energy Policy*, **36** (12), pp. 4419–4426 (2008).

[41] Tsili M. and Papathanassiou S., A review of grid code technical requirements for wind farms, *IET Renewable Power Generation*, **3** (3), pp. 308–332 (2009).

[42] Vasquez J.C., Guerrero J.M., Miret J., Castilla M., and De Vicuna L.G., (2010): Hierarchical control of intelligent microgrids, *IEEE Ind. El. Mag.*, **4** (4), pp. 23–29.

[43] Vovos P.N., Kiprakis A.E., Wallace A.R., and Harrison G.P., Centralised and Distributed Voltage Control: Impact on Distributed Generation Penetration, *IEEE Trans. Power Systems*, **22**

(1), pp. 476–483 (2007).

[44] Zhang J., Cheng H., Wang C., Xia Y., Shen X., and Yu J., Quantitive assessment of active management of distribution network with distributed generation, *Proc. of Third Int. Conf. on Electric Utility Deregulation and Restructuring and Power Technologies*, pp. 2519–2524 (2008).

[45] Zimmerman R.D. and Gan D. MATPOWER – A MATLAB Power System Simulation Package, User's Manual, *School of Electrical Engineering*, Cornell University (1997).

Chapter 28

Re-Mining Association Mining Results Through Visualization, Data Envelopment Analysis, and Decision Trees

Gurdal Ertek and Murat Mustafa Tunc

Faculty of Engineering and Natural Sciences, Sabanci University, 34956, Orhanli, Tuzla, Istanbul, Turkey

E-mail: ertekg@sabanciuniv.edu, muratmustafa@sabanciuniv.edu

Re-mining is a general framework which suggests the execution of additional data mining steps based on the results of an original data mining process. This study investigates the multi-faceted re-mining of association mining results, develops and presents a practical methodology, and shows the applicability of the developed methodology through real world data. The methodology suggests re-mining using data visualization, data envelopment analysis, and decision trees. Six hypotheses, regarding how re-mining can be carried out on association mining results, are answered in the case study through empirical analysis.

28.1 Introduction

This study investigates how the results of association mining analysis can be analyzed further using data visualization, data envelopment analysis, and the computational intelligence method of decision trees.

Association mining is a popular data mining technique for operations management and industrial engineering, and reveals associations between item sets within a large item set [1]. While extensive literature exists on how to carry out association mining in a more efficient way, research on the interpretation of association mining results is relatively narrow. The initial motivation for data mining and association mining, which is the discovery of actionable insights, has been mostly neglected in theoretical and algorithmic studies. What is therefore needed is a set of methodologies that enable further analysis of data mining results, including association mining results that will yield further discoveries.

C. Kahraman (ed.), *Computational Intelligence Systems in Industrial Engineering*, Atlantis Computational Intelligence Systems 6, DOI: 10.2991/978-94-91216-77-0_28,

Re-mining is a general framework which suggests the execution of additional data mining steps based on the results of a data mining analysis [2]. Re-mining extends post-mining: Post-mining directly analyzes the data mining results, whereas re-mining suggests the incorporation of new attributes into the initial data mining results, enabling a richer analysis and deeper insights. Even though re-mining suggests a general approach, there is considerable opportunity with respect to how it can be executed for various domains, data structures, and data mining goals.

This chapter suggests three types of analysis for re-mining of association mining outputs: As a pre-processing step, graph theoretic metrics are computed. Then, as the first analysis, graph visualization is used for the display of items and their associations and to discover actionable insights. Secondly, Data Envelopment Analysis (DEA) is adopted to formalize the insights from the first analysis, and to identify items that deserve most focus in planning. Finally, decision tree analysis is carried out to obtain additional insights regarding how domain attributes and graph metrics affect the tendency to get involved in only positive associations.

Ceteris paribus, given that every other factor remains the same, positive associations are highly sought after in business practice. A manager would be highly interested in not only identifying the positive and negative associations in items, but also the percentages of the positive and negative associations for an item. A frequent item that engages in positive associations can be considered as an attractor item, which increases the sales of the items it clusters with [3]. This is the reason that the presented research focuses on the tendency of an item to get involved in only positive associations.

The contribution of this chapter is three-folds: Firstly, this is the first study, to the best of our knowledge, that analyzes association mining results using graph theoretical metrics. Our study formalizes how this analysis can be carried out through visualization and machine learning. Secondly, data envelopment analysis (DEA) is adopted for the first time in literature as a part of re-mining. Thirdly, the applicability and the usefulness of the proposed methodology is demonstrated through real world data coming from a case study.

The research questions explored in this study are as follows:

Question Q1. How can visual re-mining that considers both positive and negative associations of an item be carried out visually?

Question Q2. How can DEA be incorporated into re-mining?

Question Q3. Can the tendency of an item to engage in only positive associations be related to its domain-specific attributes?

Question Q4. Can the tendency of an item to engage in only positive associations be related to its metric values computed from its association graphs?

Question Q5. Can the relation in Q3 be predicted?

Question Q6. Can the relation in Q4 be predicted?

The remainder of the paper is organized as follows. Section 28.2 provides a brief review of some relevant literature as the background for conceptual development. Section 28.3 discusses the methodology developed. Section 28.4 is devoted to the demonstration of the methodology through a case study with data from the real world. Finally, Section 28.5 draws some conclusive remarks from the research.

28.2 Literature

28.2.1 *Association Mining*

Association mining is a highly popular and useful data mining technique that is used in both academia and industry for both production and service systems [4]. The input to association mining is a transaction data, where each transaction contains a subset of items coming from a given set. The typical output of association mining, also referred to as association mining results, is the list of item sets that appear together frequently in transactions, and the rules that describe how these associations affect each other [1]. The first output is referred to as frequent itemsets, and the second output is referred to as association rules. The frequent itemsets and the association rules can be extremely numerous, and hence it is typical to specify a threshold support value while computing these results.

Support of an itemset (e.g.: $\{A,B\}$) or a rule (e.g. $A \Rightarrow B$) is the percentage of transactions that the items in the itemset or the rule appear in. Support is also the primary metric related with an itemset or rule that signifies importance. Another common metric in association mining is confidence, and is defined only for association rules (when only the above two outputs are generated). Confidence of a rule $A \Rightarrow B$ is the conditional probability of item B appearing in a transaction, given that item A readily appears in that transaction. A mathematical description of the aforementioned concepts can be found in Demiriz *et al.* [2].

The standard algorithm for association mining is the Apriori algorithm, which was first introduced by Agrawal *et al.* [5] and has gained increased popularity ever since (1991 citations for the original paper as of February 2012). Association mining is a standard module in almost every data mining platform [6–8] and can also be conducted through specialized software [9, 10].

The term association mining is commonly used to refer to only positive association mining. However, negative association rules are also essential in real world applications. Negative association does not simply refer to two items not appearing together. It refers to the situation of two items not appearing together in transactions as frequently as they should, when they are expected to do so, due to their independent associations with another mediator item. Two applicable methods for computing negative associations are introduced in Savasere *et al.* [11] and Tan *et al.* [12], and the latter is applied in our study due to its convenience of implementation.

28.2.2 *Graph Visualization*

Graph drawing refers to the drawing of graphs, which consist of nodes and arcs, through specialized algorithms so as to obtain actionable insights [13, 14]. Graph visualization is a sub-field of information visualization [15–17]. Many successful applications of graph visualization are listed in Ertek *et al.* [4].

28.2.3 *Association Graphs*

In this chapter, the associations between items are represented as graphs, and are visualized using *grid layout* algorithm. Since the graphs presented in this study exclusively depict associations, rather than anything else, they will be referred to as association graphs [3, 4], following the terminology in Ertek *et al.* [3]. In the presented association graphs, items are represented as nodes and the positive or negative associations between the items are represented as arcs. Grid layout algorithm snaps the items that are laid out according to a *force-directed algorithm* onto a grid, such that associated items are positioned closer to each other on the grid. The knowledge discovery from the grid layouts is enhanced through the mapping of additional attributes to the node colors and sizes.

28.2.4 *Re-mining*

Re-mining is a broadly applicable concept, which refers to the "mining of a newly formed data that is constructed upon the results of an original data mining process" [2]. In re-mining, the enriched data includes not only the results of the original data mining process, but also additional attributes. The goal is to obtain new insights that couldn't have been discovered otherwise, and to characterize, describe, and explain the results of the original data mining process. Re-mining is a generalized extension of *post-mining*, which only summarizes the results of the original process. Re-mining can involve any type of data mining analysis, including *exploratory*, *descriptive*, and *predictive* analysis. In Demiriz *et al.* [2], the original data mining process is association mining, and it is proven through complexity analysis that the running time for re-mining is polynomial in problem size, whereas it is exponential in the alternative quantitative association mining (QAM).

28.2.5 *Re-mining on Association Graphs*

While the concept of re-mining was first introduced in a journal paper by Demiriz *et al.* [2], Ertek *et al.* [3] is the first study that formalizes *visual re-mining* on association graphs. Ertek *et al.* [18] suggest a visual data re-mining process, based on association graphs, where values of additional attributes are mapped onto node colors. The authors demonstrate the applicability of the approach through a *market basket analysis* (*MBA*) case study. Specifically, the mentioned study maps statistics on the attributes of customers (such as gender, knowledge of French language, and hunger level) linearly to the node colors. The statistics displayed (through node color) within a node are computed only for the customers that select the item/itemset represented in that node. It is demonstrated that visual re-mining of association graphs enables the discovery of links between transactional patterns and customer attributes.

In this chapter, the node colors and sizes are not related to customer attributes, but are rather related to domain-specific item attributes and the tendency of the item to get involved in positive associations, rather than negative ones.

The visual re-mining on the association graphs answers research question Q1.

28.2.6 *Data Envelopment Analysis (DEA)*

Data Envelopment Analysis (*DEA*) [19] is a highly popular [20] and effective methodology that can be implemented to benchmark a group of entities through *efficiency scores*. Through the results of DEA, managers can obtain a multi-dimensional benchmark of the entities they are comparing, such as items, stores, suppliers, etc., and gain actionable insights on how to improve inefficient entities. These entities being compared are referred to as *DMU*s (Decision Making Units). In DEA, calculation of efficiency scores (the primary benchmark metric) for the DMUs is based on their input and output values. Inputs are the resources consumed by the DMUs, for generating desired outputs. Efficiency score, which takes a value between 0 and 1, increases as a DMU generates higher values of any of the outputs when the input values are the same, or when a DMU generates the same values of the outputs when any of the input values is lower.

Our methodology suggests the use of DEA to integrate the results from multiple visual analyses. In the presented case study, a DEA model was constructed to identify the most "efficient" items, where efficiency was defined as appearing in many transactions, engaging in mostly positive associations, and having a high price (specifically, a high end-of-season price), within a short lifetime.

The DEA model, which constitutes part of our methodology, answers research question Q2.

28.2.7 *Graph Metrics*

Graph theory is the field of mathematics that investigates problems related to graphs. A *graph* is composed of two types of entities, *nodes* that represent discrete, distinct entities and *arcs* that connect the nodes. The nodes in a graph and the graph itself can be characterized through *graph metrics*. In this chapter, several novel research questions are pursued as listed in introduction, where Q4 relates to graph theory. To this end, graph metrics are computed for each item (node on the association graph), and a classification model is used for prediction. The computations are done on two different association graphs, the one that displays only positive associations, and the one that displays only negative ones. The graph metric attributes are named with the suffix POS or NEG, based on which association graph the metric is computed for.

The graph metrics computed and used in this chapter are given as follows:

Degree shows the number of connections for each node. It is an integer value, and it is the summation of in degrees and out degrees of the node.

Betweenness centrality represents total number of shortest paths for each pair of nodes, if the node is on that path. It can take values between 0 and 1.

Closeness centrality shows the distance between the node and every other node. It only takes values between 0 and 1.

Eigenvector centrality shows the distance between the node and every other "special" node. It is a value between 0 and 1.

Page rank is the value which only increases if the node is closely related with "special" nodes, in other words, its closeness-centrality of "special" nodes is higher [21].

Clustering coefficient represents the tendency of aggregation for several nodes [22]. It can only take values between 0 and 1.

Detailed information about these and other graph metrics are explained in Opsahl *et al.* [23] and Christensen and Reka [24].

28.2.8 *Decision Trees*

Decision trees summarize rule-based information regarding classification as trees. In decision tree models, each node is split (branched) according to a criterion. Then, a tree is constructed with depth. At each level, the attribute that creates the most increase compared to the previous level is observed. The algorithms for decision tree analysis are explained in Chien and Chen [25]. In our study, ID3 algorithm [26] is used, and branches are created in Orange software [27]. In decision trees, identifying the nodes that differs considerably from its root node are our main focus. By observing the shares of slices and comparing with the parent nodes, one can discover classification rules.

The decision trees answer research questions Q3 and Q4.

28.2.9 *Classification*

In the classification models, the dataset is divided into two groups, namely *learning dataset* and *test dataset*. Algorithms that are referred as *classifiers* (or *learners*) use the learning dataset in order to learn from data and predict the class attributes in the test dataset. The prediction success of each learner is measured through *classification accuracy* (CA), which is the percentage of correct predictions among all [1]. The following classification algorithms are among the most well-known classifiers in the machine learning field: Naive Bayes, k-Nearest Neighbor (kNN), C4.5, Support Vector Machines (SVM), and Decision Trees [28].

The classification model searches for answers to two of the novel research questions posed in this chapter: Q5 and Q6.

28.3 Methodology

The methodologies described in the literature reviewed in section 2 have already been developed and shown to be applicable. Association mining, classification, and decision trees are especially well established methodologies within data mining [1, 29], and graph theory is a very extensive field of mathematics that has found its applications in almost every field of science [30–32]. DEA is a well-researched and vastly applied methodology for benchmarking [20]. Association graphs, re-mining, and re-mining on association graphs are relatively much newer methodologies. However, to the best of our knowledge, there does not exist a methodology that links these diverse methodologies in a unified frame. Specifically, our research shows that insights can be obtained for items that engage in only positive association (and no negative associations) based on their association graph metrics or domain-specific attributes. Our research also shows that exploratory and descriptive re-mining may be very successful, whereas predictive re-mining may not yield satisfactory results.

The methodology that we propose consists of the following steps:

Step 1. Perform positive association mining on the transactions database for obtaining frequent item pairs (2-itemsets).

Step 2. Find negatively associated item pairs, using the results of Step 1.

Step 3. For each item, compute the percentage of positive associations it gets involved in.

Step 4. Construct two association graphs; one that shows only positive associations, and the other that shows only negative associations. Items are represented as nodes, and associations (positive or negative) are represented as arcs.

Step 5. Compute the graph metrics for each node (each item) in each of the two association graphs.

Step 6. Construct the dataset for re-mining, where each row is an item (that got involved in at least one positive association), and the columns are support count (SupC), domain-specific item attributes (StartWeek, EndWeek,...), percentage of positive associations (PercOfPositiveAssoc), class labels that are defined based on percentage of positive associations (PositiveAssoc takes value of 1, if PercOfPositiveAssoc = 1) and graph metrics obtained from Step 4 (DegreePOS, BetweennessCentralityPOS,..., DegreeNEG, BetweennessCentralityNEG,...).

Step 7. Apply grid layout for the association graph, and map domain-specific attributes to node color and size. Visually analyze the graphs for discovering actionable insights.

Step 8. Construct a DEA model, to combine the insights from multiple visual analysis, and to find the most important items.

Step 9. Construct a classification model, to predict the items that engage in only positive associations (items that have PercOfPositiveAssoc = 1). Construct decision trees, using the classification model of Step 8.

Step 10. Apply multiple classifiers in the classification model and evaluate classification accuracy.

In the next section, the described methodology is applied to a real world dataset coming from an apparel retail company.

28.4 Case Study

28.4.1 *Problem Description*

The company that we work with in our research projects is one of the largest apparel retail chains in Eurasia. Headquartered in Istanbul, Turkey, the company has more than 300 retail stores in Turkey alone, as well as more than 30 stores in more than 10 other countries. The data analyzed comes from the men's clothes line (one of the merchandise groups) and belongs to the 2007 summer season. Earlier research with the data proved the applicability of the re-mining methodology [2], and focused on the analysis of item pairs. The findings in this chapter relate to the items as single entities.

The selection of retail industry for the case study is highly relevant, because this is an industry where the business is centered on customer transactions as the revenue stream. Therefore, novel and applicable data mining and machine learning methodologies for association mining can be incorporated into the retail operations. Yet, there is a bigger picture: As of November 2010, the retail industry in U.S. alone exceeded $ 377.5 billion [33]. This is a huge industry, and even the smallest percentage improvement in operations can correspond to significant financial gains.

28.4.2 *Analysis Process*

As **Step 1** of the proposed methodology, positive association mining was carried out on the transaction database for obtaining frequent item pairs. The minimum support count was taken as 100, which means that only the item pairs that appeared in at least 100 transactions were considered. A total of 3930 such frequent item pairs have been identified, involving 538 items.

Next, in **Step 2**, negatively associated item pairs were identified based on the items that formed the frequent item pairs. A total of 2433 negatively associated item pairs were identified, involving 537 items of Step 1 (all items except one of them).

In **Step 3**, the percentage of positive associations each item gets engaged in (PercOfPositiveAssoc) was computed. For this, all the associations for that item were filtered and the percentage was computed as the number of positive associations divided by the total number of associations. The distribution of PercOfPositiveAssoc is shown in Figure 28.1.

In **Step 4**, positive and negative association graphs were constructed using the Fruchterman-Reingold force-directed algorithm [34]. It was not possible to obtain any immediate insights from the association graphs.

In **Step 5**, graph metrics were computed for each node in the association graph, using NodeXL add-in [35, 36] for Microsoft Excel. The computed metrics were the ones listed earlier in the literature review.

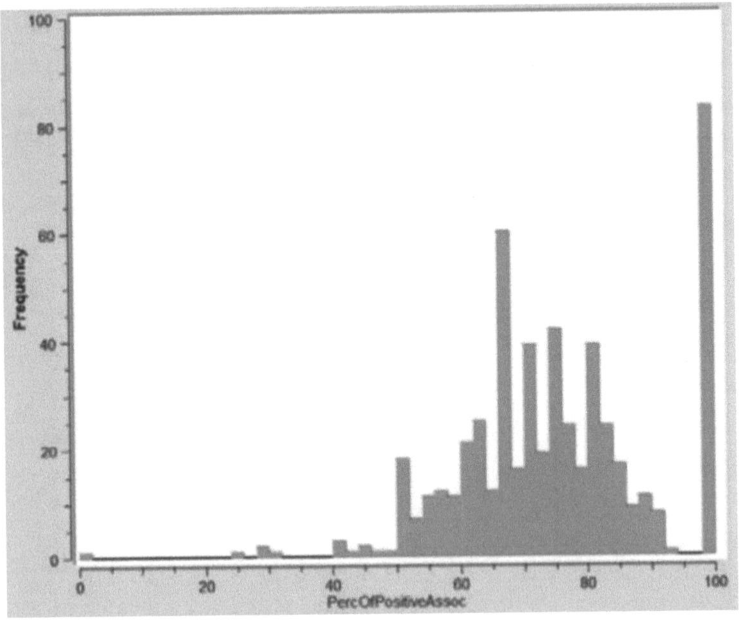

Fig. 28.1 The distribution of PercOfPositiveAssoc values for analyzed items

In **Step 6**, a dataset was formed for re-mining, where each row is an item that got involved in at least one positive association. The first column in the dataset (key attribute) is the unique item number. The second column in the dataset is the support count (SupC), which is the number of transactions the item appears in. The domain specific attributes are the following:

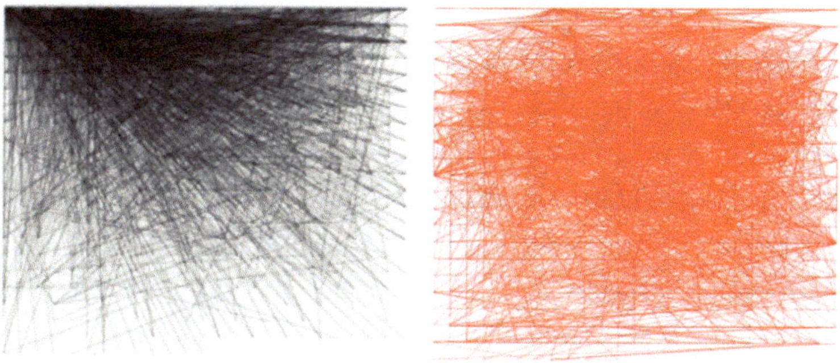

Fig. 28.2 The positive and negative association graphs for the transactions data set

- StartWeek: The week within the season when the sale of that item was started.
- EndWeek: The week within the season when the sale of that item was ended.
- LifeTime: The number of week the item stayed on sale.
- MaxPrice: The initial price of the item (in apparel retail industry, item prices are subject to continuous markdowns, without any increases throughout the season).
- MinPrice: The final price of the item.
- PriceDiff: The difference between MaxPrice and MinPrice.
- MerchSubGrp: The merchandise subgroup the item belongs to.
- Category: The category the item belongs to.

The next column in the dataset is PercOfPositiveAssoc which was computed in Step 5. This continuous value was then discretized into a class attribute (PositiveAssoc), where the target class is AllPositiveAssoc (meaning that the item engages only in positive association, and no negative associations). The remaining columns are the graph metric values for the item.

In **Step 7**, grid layout was applied in NodeXL for visualizing the association graph, based on snapping the nodes positioned by the Fruchterman-Reingold algorithm [34] on a grid. The insights obtained in here are described detail later.

In **Step 8**, a DEA model was constructed using the SmartDEA Solver software [37], which implements the basic CCR and BCC models, and enables the generation of DEA outputs in a database format.

In **Step 9**, decision trees were constructed for descriptive re-mining using C4.5 algorithm (Figure 28.3).

Finally in **Step 10**, a classification model was constructed (Figure 28.3) in Orange data mining software [27, 38], to predict the class attribute PositiveAssoc. In our model for classification accuracy, learning dataset was taken from 70% of overall dataset, with 20 repetitive sampling. This means that for each learner, 20 random sampling of 70% of the dataset were executed as the learning dataset, and the remaining was experimented as the test dataset. In the classification models, random sampling was applied with %70 of the rows being the training set (376 rows), and the remaining %30 being the test set (162 rows).

In the classification step, three models, namely Model A, B, C, have been developed. In all three models, PositiveAssoc is the class attribute, in Model A, only domain related attributes are used as predictors; in Model B, only graph metric attributes are used as predictors; in Model C, both domain-specific attributes and graph metric attributes are used as predictors.

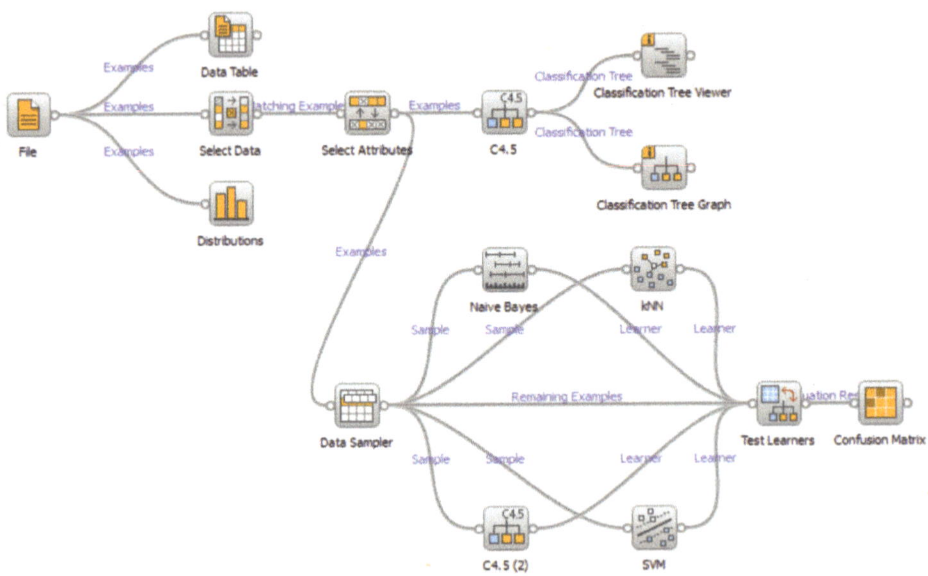

Fig. 28.3 The machine learning model for decision tree analysis and classification

28.4.3 *Graph Visualization*

The analysis of association graphs can yield significant insights [4]. This section presents the analysis of association graph visualizations, where each node's color, size, and shape represents attributes of that item. Actionable and practical insights and policies are obtained through the graph visualizations.

Each node in the visualizations shows a particular apparel item and each arc (connecting two nodes) shows an association (Figures 28.2, 28.4–28.6). Black arcs denote positive associations and red arcs denote negative associations. The graph visualization is automatically constructed through the NodeXL graph analysis software, with the constraint of snapping all the nodes a rectangular grid. The color of a node can denote any selected variable. Since the focus of this research is the investigation of percentage of positive associations, color of each node is based on the PercOfPositiveAssoc value for that item. Yellow node color denotes items that engage in mostly negative associations, and darker nodes denote items that engage in more positive associations (higher PercOfPositiveAssoc values). Similarly, the size of can denote any selected variable. In our visualizations (Figure 28.5 and 28.6), the size of that node denotes the minimum price of that item.

The first observation in Figure 28.4 is that the upper left quadrant of the graph consists mostly of nodes. A further examination of the node colors reveals that the node colors in general are lighter compared to the remaining regions of the graph, suggesting items that have many associations tend to get involved in more negative associations, as well. This is consistent with the definition and the computation method of negative associations.

Figure 28.4 also highlights very dark-colored outlier items within the items on the upper left region, which engage in not only many associations, but also mostly (or completely) positive ones. Everything else being the same, these items are very important, because they engage in many associations on the positive association graph, and do not engage in much negative associations. They complement many other items, and do not substitute many or even any items.

Figure 28.5 appends an additional dimension to Figure 28.4: The end-of-season sales prices (MinPrice) of the items, represented through the node sizes. Larger node sizes denote higher MaxPrice values. These high priced items also reside in the region that contains many positive associations, and are thus very important for the retailer. The suggestion for the retailer would be to understand the characteristics on these particular items and also sell these high-priced items throughout the season to maximize the revenue, both through their high-price and through the sales that they trigger as complements.

Finally, Figure 28.6 adds a new dimension to Figure 28.5: The categories of the items (category), represented through the node shapes. This analysis enables the identification of whether the important items in Figure 28.4 and 28.5 belong to a particular category

Fig. 28.4 Node color denotes the PercOfPositiveAssoc value of each item

or not. Highlighted nodes are in the upper left region of the positive association graph (involvement in positive associations with many items), have dark colors (high percentage of positive associations), have relatively large sizes (high end-of-season prices), and all come from only one of the three item categories (the category shown with a square). It is easy to see that almost all the items with the desired characteristics are from the same item category. The analysis in this section shows that exploratory visual re-mining brought significant insights and has thus been successful in this case study. The research question Q1 has been answered positively.

28.4.4 *Data Envelopment Analysis (DEA)*

The DEA model was constructed so that the insights obtained from visual re-mining can be integrated within a unified analytical method. Thus, DEA is introduced not as a data mining method, but as a method to formally and analytically integrate the insights

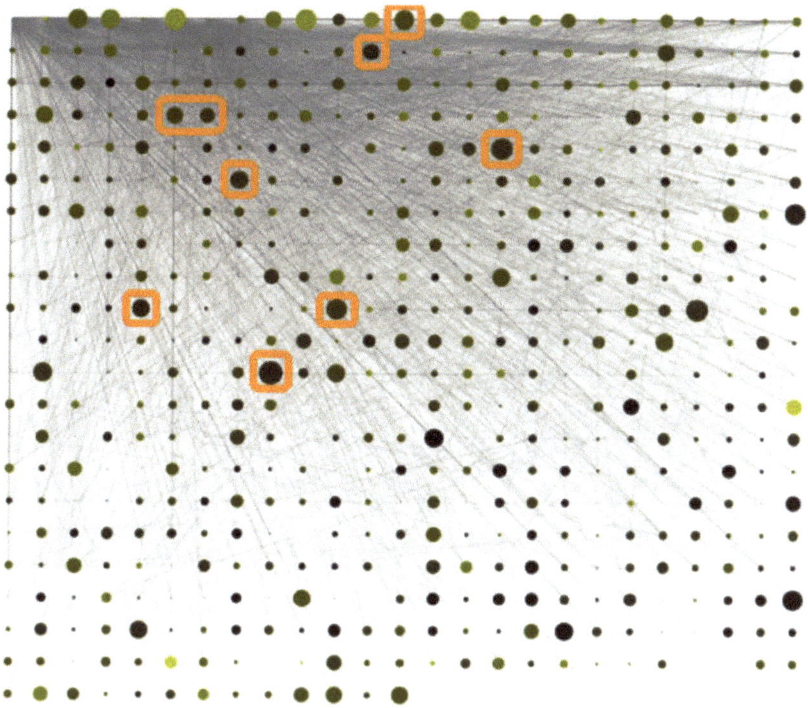

Fig. 28.5 Node size denotes the item's minimum price, and the node color denotes the item's PercOfPositiveAssoc value. The highlighted item triggers the sales of many items and also brings more revenue itself

that can be obtained through the above visual data analysis. The DEA model considers three criteria, and treats their values as outputs. Since there were no inputs discussed in the visual re-mining section, there exists only a single auxiliary input, which takes the value of 1 for all the DMUs. The outputs are the support count (SupC), percentage of positive associations the items gets engaged in (PercOfPositiveAssoc), and the end-of-season price (MinPrice) of the items. Since the values for all the outputs are non-negative, the BCC models, which were introduced by Banker *et al.* [39] were found most appropriate selected, due to their allowance for variable returns to scale. Since the input values are auxiliary, all equal to 1, and cannot be changed, the goal of a DMU is to increase the values of its outputs (the values for performance criteria). This perspective is reflected in the selected DEA model, which is the output-oriented BCC model (BCC-O).

The results of DEA for the described model generate 84 efficient items. However, all but one of these items are efficient because they have PercOfPositiveAssoc=100, the

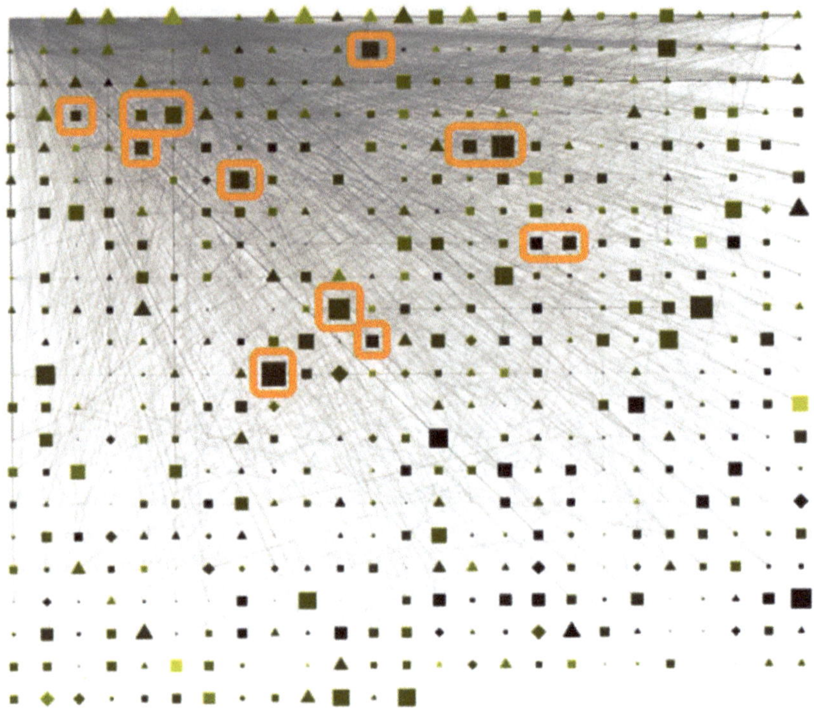

Fig. 28.6 The size of the nodes shows the minimum price of the items, the color of the nodes illustrates PercOfPositiveAssoc values, and node shape denotes the category

highest value possible. Therefore, a new DEA model was constructed, that excluded these 83 items that engage in only positive associations. This time, five items were found to be efficient (first five items of Table 28.1). Finally, a new, third DEA model was constructed, that treated LifeTime as the only input. This time, nine items were found to be efficient (Table 28.1). This third model considers the effect of life time of an item, and praises items that achieve more of the desired qualities in less time. The research question Q2 of how DEA can be integrated into re-mining is now answered.

28.4.5 *Decision Trees*

In both decision tree analysis (for Q3 and Q4) and classification analysis (for Q5 and Q6), three models, namely Model A, B, C, have been developed. In all three models, AllPositiveAssoc (whether the item engages only in positive associations) is the class attribute. In Model A, only domain related attributes are used as predictors; in Model B, only

Table 28.1 Results of the DEA model, together with the inputs and outputs

Item	Eff1*	Eff2**	Input_Auxiliary	Input_LifeTime	PercOfPositiveAssoc	SupC	MinPrice
059	Yes	Yes	1	16	91.67	4157	19.99
087	Yes	Yes	1	26	92.31	8947	14.90
094	Yes	Yes	1	11	75.00	4647	41.57
106	Yes	Yes	1	32	30.00	346933	9.25
169	No	Yes	1	7	75.00	4464	34.90
289	No	Yes	1	8	87.50	4317	23.06
412	Yes	Yes	1	13	88.89	2658	34.90
438	No	Yes	1	10	91.67	4999	14.90
513	No	Yes	1	4	80.00	5115	13.80

the graph metric attributes computed from the positive association graph are used as predictors; in Model C, only the graph metric attributes computed from the negative association graph are used as predictors. The list of predictors in each model is given below:

Model A: StartWeek, EndWeek, LifeTime, MaxPrice, MinPrice, PriceDiff;

Model B: DegreePOS, BetweennessCentralityPOS, ClosenessCentralityPOS, EigenvectorCentralityPOS, PageRankPOS, ClusteringCoefficientPOS

Model C: DegreeNEG, BetweennessCentralityNEG, ClosenessCentralityNEG, EigenvectorCentralityNEG, PageRankNEG, ClusteringCoefficientNEG

Figures 28.7–28.9 show the decision trees constructed for Models A, B, and C, respectively. In these decision trees, dark slices denote the share of items that engage in only positive associations (AllPositiveAssoc = AllPositive).

In Figure 28.7, it can be observed that items that have large price differences (> 26.490) through the season have a higher chance of having only positive associations, as denoted by the larger share of the dark-colored slice. This chance increases if the item is taken out of the market before week 37, and increases even more if the item is introduced before week 22. These three pieces of information print out strategies on when an item should be introduced, until when it should be sold in stores, and how much of a price difference (PriceDiff = MaxPrice − MinPrice), it should have between its initial price (MaxPrice) and final end-of-season price (MinPrice).

In Figure 28.8, it can be observed that items that have a BetweenCentralityPOS = 0 in the positive association graph have no chance of engaging in only positive associations. Among the other items, ClusteringCoefficientPOS = 5 and even more when EigenVectorCentralityPOS = 0.

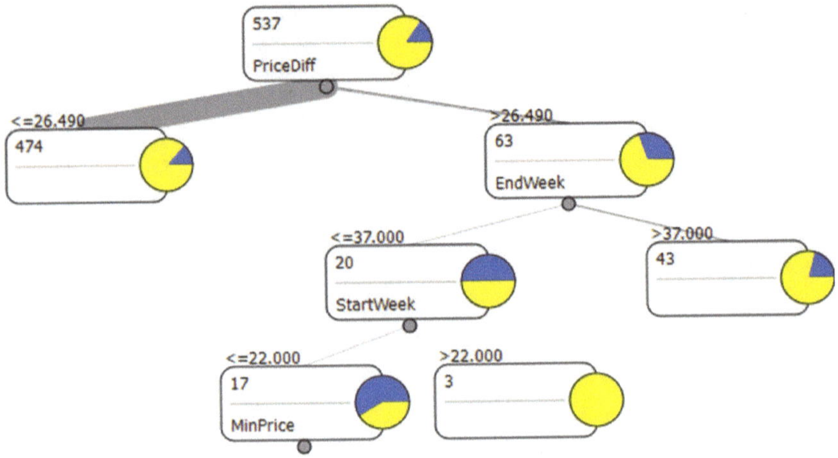

Fig. 28.7 Decision Tree for Model A

How can the findings of this decision tree be used in practice? The results obtained from decision tree help us indentify the subgroup of items that have high chance of engaging in only positive associations. So, without carrying negative association mining for new data sets (not a readily available analysis type in many commercial software), one can obtain idea on how to search for important items.

Figure 28.9 shows the decision tree for Model C. This decision tree summarizes the behavior of all-positively associated items with respect to negative associations. Where would one find all-positively associated items the most? The answer is "when DegreeNEG = 3 or 4 and ClusteringCoefficientNEG = 0.4". Research questions Q3 and Q4 have been answered in this section.

28.4.6 *Classification*

The classification accuracies obtained for the three models using various learners are presented in Table 28.2. In all three models, it is possible to achieve a classification accuracy of at most 83.57%. However, the percentage of items with only positive associations is 15.45%, meaning that if a classifier guessed none of the items to have all-positive associations, the average classifier accuracy would still be 84.55%. Clearly, the classification models predict worse than a blind guess, and thus predictive re-mining through classification was not successful in this case study, despite the success of exploratory and descriptive re-mining. Furthermore, a stepwise regression analysis, which is not described here, yielded a

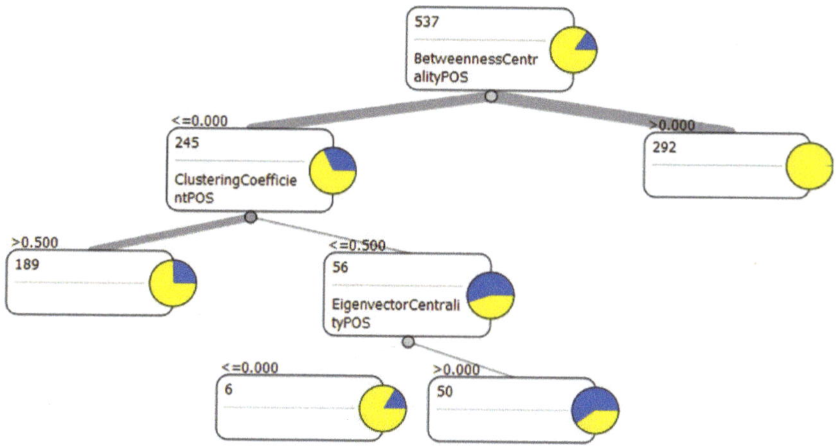

Fig. 28.8 Decision Tree for Model B

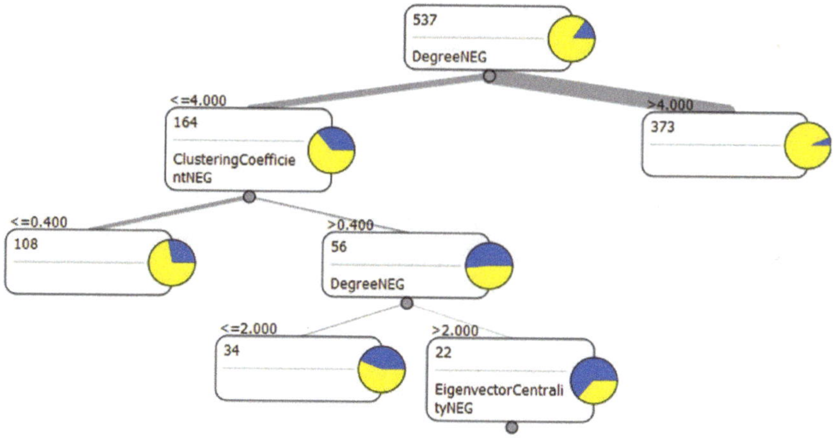

Fig. 28.9 Decision Tree for Model C

very low adjusted R2 value of 0.2180, suggesting that neither the domain attributes nor the graph metrics cannot be used in a linear regression model to predict PercOfPositiveAssoc. The answers to both Q5 and Q6 are "No".

Table 28.2 Classification Results (Classification Accuracies)

Classifier	Model A	Model B	Model C
C4.5	0.8204	0.8000	0.7562
SVM	0.8357	0.8063	0.8094
kNN	0.7592	0.7406	0.7094
Naïve Bayes	0.7622	0.6313	0.7281

28.5 Conclusion

In this chapter, a novel computational intelligence methodology was introduced for practical re-mining. The methodology combines association mining, graph theory, classification, DEA, and re-mining, for answering six novel research questions.

The application of the methodology in a case study demonstrates the applicability and usefulness of exploratory and descriptive re-mining. The tendency of an item to engage in positive associations is related to its graph metric values computed from the association graphs, and to its domain-specific attributes. However, predictive re-mining through classification and stepwise regression did not prove useful in the context of the case study.

Acknowledgement

The authors thank Ayhan Demiriz of Sakarya University for providing the data for this study. The authors also thank Ilhan Karabulut for her work that inspired the visual re-mining approach on association graphs.

Bibliography

[1] J. Han, M. Kamber, and J. Pei, *Data Mining: concepts and techniques*, (2011).

[2] A. Demiriz, G. Ertek, T. Atan, and U. Kula, Re-mining item associations: Methodology and a case study in apparel retailing, *Decision Support Systems*, **52** (1), pp. 284–293, (2011).

[3] G. Ertek and A. Demiriz, A framework for visualizing association mining results, *Lecture Notes in Computer Science* (*LNCS*), **4263**, pp. 593–602, (2006).

[4] G. Ertek, M. Kaya, C. Kefeli, O. Onur, and K. Uzer, Scoring and Predicting Risk Preferences,in *Behavior Computing: Modeling, Analysis, Mining and Decision*, Cao, L., Yu, P.S. (Eds.), Springer, (2012).

[5] R. Agrawal, T. Imielinski, and A.N. Swami, Mining association rules between sets of items in large databases,in *SIGMOD Conference*, P. Buneman and S. Jajodia, (Eds.), (1993).

[6] SAS. `http://www.sas.com/`

[7] RapidMiner. https://rapid-i.com/content/view/181/190/

[8] Weka. `http://www.cs.waikato.ac.nz/ml/weka/`

[9] C. Borgeltand and R. Kruse, *Graphical models: methods for data analysis and mining*, Wiley, (2002).

[10] E.N. Cinicioglu, G. Ertek, D. Demirer, and H.E. Yoruk, A framework for automated association mining over multiple databases, in *Innovations in Intelligent Systems and Applications (INISTA), International Symposium, IEEE*, (2011).

[11] A. Savasere, E. Omiecinski, and S. Navathe, Mining for strong negative associations in a large database of customer transactions, in *Data Engineering, Proceedings, 14th International Conference, IEEE*, (1998).

[12] P.N. Tan, V. Kumar, and H.Kuno, in *Western Users of SAS Software Conference*, (2001).

[13] I. Herman, G. Melancon, and M.S. Marshall, Graph visualization and navigation in information visualization: A survey, *Visualization and Comp. Graphics*, **6**, (2000).

[14] M. Van Kreveld and B. Speckmann, *Graph Drawing, Lecture Notes in Computer Science (LNCS)*, **7034**, (2012).

[15] R. Spence, *Information Visualization*, ACM Press, (2001).

[16] H. Ltifi, B. Ayed, A.M. Alimi, and S. Lepreux, Survey of information visualization techniques for exploitation in KDD, in *Int. Conf. Comp. Sys.and App.*, (2009).

[17] C. Chen, Information Visualization, *Wiley Interdisciplinary Reviews: Computational Statistics*, **2**, (2010).

[18] G. Ertek, A. Demiriz, and F. Çakmak, Linking Behavioral Patterns to Personal Attributes through Data Re-Mining, in *Behavior Computing: Modeling, Analysis, Mining and Decision*, Cao, L., Yu, P.S. (Eds.), Springer, (2012).

[19] W.W. Cooper, L.M. Seiford, and K. Tone, *Introduction to Data Envelopment Analysis and Its Uses: With DEA Solver Software and References*, Springer, (2006).

[20] S. Gattoufi, M. Oral, and A. Reisman, Data envelopment analysis literature: A bibliography update (1951–2001), *Journal of Socio-Econ. Planning Sci.*, **38**, pp. 159–229, (2004).

[21] L. Page, S. Brin, R. Motwani, and T. Winograd, The PageRank Citation Ranking: Bringing Order to the Web, *The PageRank Citation Ranking: Bringing Order to the Web*, (1999).

[22] D.J. Watts and S. Strogatz, Collective dynamics of 'small-world' networks, *Nature*, **393**, (1998).

[23] T. Opsahl, F. Agneessens, and J. Skvoretz, Node centrality in weighted networks: Generalizing degree and shortest paths, *Social Networks*, **32**, pp. 245, (2010).

[24] C. Christensen and A. Réka, Using graph concepts to understand the organization of complex systems, *International Journal of Bifurcation and Chaos*, **17**, pp. 2201–2214, (2007).

[25] C.F. Chien and L.F. Chen, Data mining to improve personnel selection and enhance human capital: A case study in high-technology industry, *Expert Systems with Applications*, **34**, pp. 280–290, (2008).

[26] J.R. Quinlan, Induction of decision trees, *Machine Learning*, **1** (1), pp. 81–106, (1986).

[27] Orange. `http://orange.biolab.si/`

[28] E. Alpaydin, *Introduction to Machine Learning*, The MIT Press, (2010).

[29] P.N. Tan, M. Steinbach, and V. Kumar, *Introduction to Data Mining*, Pearson Addison Wesley, (2006).

[30] B. Bollobas, *Modern Graph Theory*, Springer, (1998).

[31] J.L. Gross and J. Yellen, *Handbook of Graph Theory, CRC Press*, (2003).

[32] M.E.J. Newman, *Networks: An Introduction*, Oxford University Press, (2010).

[33] WolframAlpha. `http://www.wolframalpha.com/`

[34] T.M.J. Fruchterman and E.M. Reingold, Graph Drawing by Force-Directed Placement, *Software – Practice & Experience*, **21** (11), pp. 1129–1164, (1991).

[35] E.M. Bonsignore, C. Dunne, D. Rotman, M. Smith, T. Capone, D.L. Hansen, and B. Shneiderman, First Steps to NetViz Nirvana: Evaluating Social Network Analysis with NodeXL, in *International Symposium on Social Intelligence and Networking*, (2009).

[36] NodeXL. `http://nodexl.codeplex.com/`

[37] A.E. Akcay, G. Ertek, and G. Buyukozkan, Analyzing the solutions of DEA through information visualization and data mining techniques: SmartDEA framework, *Expert Systems with Applications*, **39**, pp. 7763–7775, (2012).
[38] T. Curk, J. Demsar, Q. Xu, G.Leban, U. Petrovic, I. Bratko, G. Shaulsky, and B. Zupan, Microarray data mining with visual programming, *Bioinformatics*, **21** (3), pp. 396–398, (2005).
[39] R.D. Banker, A. Charnes, and W.W. Cooper, Some models for estimating technical and scale inefficiencies in data envelopment analysis, *Management Science*. **30** (9), pp. 1078–1092, (1984).

Chapter 29

A Multi-Objective Evolutionary Algorithm for Finding Efficient Solutions to a Competitive Facility Location Problem

José Maria Pangilinan[1], Gerrit K. Janssens[2], and An Caris[2]

[1] *Department of Computer Science, St. Louis University Baguio, The Philippines*
[2] *Operations Management and Logistics, Hasselt University Diepenbeek, Belgium*

E-mail: joeyp@slu.edu.ph, {gerrit.janssens,an.caris}@uhasselt.be

This chapter investigates the performance of several Multi-Objective Evolutionary Algorithms (MOEAs) in discovering solutions to the Competitive Facility Location (CFL) problem and a sensitivity analysis of the solutions related to input parameters. In terms of the optimality, the MOEAs are able to find Pareto-optimal solutions in the discrete-valued objective (captured weights) search space but show inferior Pareto-optimal solutions to the real-valued objective (facility quality). In terms of runtime complexity, the MOEAs run in polynomial time but are computationally expensive in terms of repeated executions for sensitivity analysis. The sensitivity analyses show that the mutation rate causes much of the variation in the output and requires a high probability value in order to generate CFL solutions near the Pareto-optimal front. The section on robustness examined the effects of changing the area location in finding solutions to the competitive facility location problem while preserving a small set of previously effective GA-parameter combinations. The results show that there is no single GA-parameter combination that dictates the generation of best solutions. The MOEA in this case, is robust as long as the mutation probability and the bit-turn probability are set to a probability of more than 60% while keeping the crossover probability low.

29.1 Introduction

Facility location is the process of determining a geographic site for operations of a company or any organization in general. Managers of both service and manufacturing organizations must weigh several factors when assessing the desirability of a particular site, including proximity to customers and suppliers, labor costs, and transportation costs. Facility location is often determined by one critical factor. In the case of plant or warehouse location, economic factors usually are dominant.

C. Kahraman (ed.), *Computational Intelligence Systems in Industrial Engineering*, Atlantis Computational Intelligence Systems 6, DOI: 10.2991/978-94-91216-77-0_29,

A location model is said to be *competitive* when the problem of locating a new facility in the market incorporates the existence of other facilities and that the new facility has to compete for its market share (Plastria 2001). A number of facilities with known fixed locations exist in the market. That is, Competitive Facility Location (CFL) models describe how facilities capture their market share and where a new facility should be located to maximize its market share. The models mostly start from a measure of *attraction* that a customer feels for a facility. The attraction is determined by several factors, but most CFL models represent market share as a function of the *distance* between the customer and the facility on the one hand, and on the other hand on internal characteristics of the facility, which generally can be called the *quality* of a facility. Various types of attraction functions might be formulated, but the study will deal with a multiplicative type, which leads to a gravity type attraction, given by the quality divided by some strictly positive power of the distance. Anyway, the attraction function should be non-increasing with distance and non-decreasing with quality.

The models take care of two decisions: the location of the facility and its design. The location of the site influences the distance part of the objective and the design relates to the quality. Quality is determined by a mixture of attributes of the facility like floor area, number of check-counters, product mix and price level. Raising the level of these attributes of quality involves a higher cost. The main objective of the decision is to maximize profit, which can be expressed as a function of sales and cost, like sales minus cost or sales divided by cost. The link is made due to the results obtained by Carrizosa and Plastria (1995) who show that profit–maximization with respect to both the location and the quality of the new facility can be obtained by inspecting only a finite number of solutions. The solutions are obtained after solving a bi-objective optimization problem of finding efficient solutions that maximize the number of "captured" consumers, while minimizing the quality costs of the new facility. The type of problem is known as a maxcovering-minquantile location problem, which is discussed further on.

29.2 Problem Definition

The competitive location model involving bi-objective maximization of location and quality of a facility is due to Plastria and Carrizosa (2004). They present a general profit maximizing competitive location model with different attraction types on a consumer and limit the location of the new facility within a bounded area.

A finite set of *customers* is denoted by AF. Each customer $a \in AF$ has a known location $\boldsymbol{x}_a$ and a strictly positive weight w_a representing his buying power. A finite set of *competing facilities* with which the new facility is to compete is denoted by CF. A competing facility $f \in CF$ is located at site $\boldsymbol{x}_f$ and has a quality α_f considered to be known and fixed. Any customer $a \in AF$ feels an attraction $attr(\alpha, f)$ towards a facility f at $\boldsymbol{x}_f$, which depends on factors such as distance from $\boldsymbol{x}_a$ to $\boldsymbol{x}_f$ or other factors like tradition etc. Plastria and Carrizosa (2004) show that a maximal profit solution to the CFL problem through the determination of efficient solutions reduces to the bi-objective optimization of

Min α
Max $CW(\alpha, \boldsymbol{x})$; $\alpha = \alpha_0, \boldsymbol{x} \in \boldsymbol{S}$
where $\alpha =$ unknown quality cost of the new facility at an unknown site $\boldsymbol{x}$
$CW(\alpha, \boldsymbol{x}) =$ captured weight of the new facility
$\alpha_0 =$ minimal quality cost ($\alpha_0 > 0$)
$\boldsymbol{S} =$ closed set in the plane.

Consider a new facility with unknown site $\boldsymbol{x}$ and unknown quality α, of at least some minimal quality $\alpha_0 > 0$. Its attraction on a customer $a \in AF$ can be expressed by a function $A_a(\alpha, dist_a(\boldsymbol{x}))$, a function of the quality α and of the distance $dist_a(\boldsymbol{x})$ from the customer to the facility. The function A_a is defined on $[\alpha_0, +\infty[\times[0, +\infty[\rightarrow [0, +\infty]$. A typical example is attraction of *gravity type*, given by

$$A_a(\alpha, x) = \frac{k_a \alpha}{||x - x_a||^p} \tag{29.1}$$

where p is any strictly positive exponent, and $k_a > 0$ represent some proportionality constant depending on a. In pure gravity models $p = 2$, which may appear in a wide variety of physical contexts. But also linear markets may be assumed (Eiselt and Laporte, 1988). A common feature of all attraction functions is that a larger quality cost increases the attraction of a customer to a facility whereas a larger distance between them decreases it. With a deterministic customer choice rule, the facility captures those customers attracted more to the new facility than to any competing facility in CF. The set of captured customers is given by:

$$Capt(\alpha, x) = \{a \in AF \mid \forall f \in CF : A_a(\alpha, dist_a(x)) \geqslant attr(a, f)\} \tag{29.2}$$

The total weight captured by the new facility is given by

$$CW(\alpha, x) = \sum_{a \in Capt(\alpha, x)} w_a \tag{29.3}$$

Profit is expressed by a *profit-indicator* function π which is strictly increasing in sales income $\sigma(CW(\alpha,\boldsymbol{x}))$ and strictly decreasing in operating costs $\gamma(\alpha)$

$$\Pi(\alpha,x) = \pi(\sigma(CW(\alpha,x)),\gamma(\alpha)) \tag{29.4}$$

It is the profit-indicator Π that needs to be optimised by an adequate choice of both the quality $\alpha > \alpha_0$ and the site x within some set of feasible sites $\boldsymbol{S}$:

$$\max\{\Pi(\alpha,x) \mid x \in S,\ \alpha \geqslant \alpha_0\}. \tag{29.5}$$

Given a fixed quality α, profit maximisation is achieved by maximisation of the total captured weight. This optimisation problem is called a *maximal covering problem* as studied in the planar context by Drezner (1981). Drezner's objective is to maximise a weighted number of demand points (read: customers) within a given (Euclidean) distance from the facility. This constraint might be motivated as follows: the location of an emergency facility may be satisfying a bound on the distance between each demand point and the new facility, or when a shopping centre is planned, only customers within a given distance use its services. Drezner (1981) presents an algorithms of complexity $O(n^2 \log n)$ where n is the number of demand points. Since, in this problem, α is a decision variable, such a maximal covering problem needs to be solved for each possible value of α, given that only a finite number of feasible quality values are available.

When, however, optimisation needs to be done over the full range of positive α-values (with $\alpha = \alpha_0$), another approach is to be advised. In Carrizosa and Plastria (1995) it is shown how the optimisation problem can be solved by reducing it to a bi-objective problem. In some cases the optimisation problem can be solved by inspecting a finite number of points, polynomial in the cardinality of AF. They have proven that, if maximal profit solutions exist, they are all efficient (or nondominated, or Pareto-optimal) for the bi-objective problem. The authors call the problem a minquantile/maxcovering bi-objective problem.

The cases, in which the optimisation problem can be solved analytically, depend on (a) the type of attraction function; (b) the characteristics of the subset; and (c) the characteristics of the search space. In terms of attraction function a limited number of functions are feasible, including the gravity-type functions. In terms of the characteristics of the subset, the subset S needs to be a closed convex subset. In terms of the search space, solutions need to be found in the plane $\mathfrak{R}^2$, in which distances are measured by a norm. If the requirements are met then Carrizosa and Plastria (1995) state that efficient solutions to the bi-objective problem are optimal solutions to single-objective problems, i.e. single-facility minmax location problems. To explain this procedure in further detail, Theorem 8 from Plastria and Carrizosa (2004) is stated as follows:

"**Theorem 8.** *Let* $(\alpha^*, \boldsymbol{x}^*)$ *be an efficient solution for the bi-objective problem. Then, one has:*

1. *When* $Capt(\alpha^*, x^*) \neq \varnothing$ *there exists a nonempty subset* $T \subset AF$*, with cardinality at most* 3 *such that*

   a) $\boldsymbol{x}^*$ *solves the generalizes single-facility minmax location problem*

   $$\min_{x \in S} \max_{a \in T} D_{\alpha}(x)$$

   with $D_a(\boldsymbol{x}) : \mathfrak{R}^2 \to [0, +\infty]$ *defined as* $D_a : x \to B_a(dist_a(\boldsymbol{x}))$ *in which* B_a *indicates, as a function of the distance d, the quality threshold above which a customer a at distance d is captured.*

   b) α *is the optimal value of* (a), *which is finite;*

2. *In case* $Capt(\alpha, \boldsymbol{x}^*) = \varnothing$*, one must have* $\alpha^* = \alpha_0$*, and any other pair* $(\alpha_0, \boldsymbol{x})$ *is then also efficient for the bi-objective problem."*

After finding, in (b) the optimal value α_T of each of the $O(n^3)$ optimisation problems in (a), and finding the set S_T of optimal solutions for the problem in (a), a list of pairs $(\alpha_T, \boldsymbol{x}_T)$ is available which contains the set of efficient solutions for the bi-objective problem.

29.3 Literature review

A number of models exist that solve different competitive facility location problems. A survey of such models is discussed by Eiselt *et al.* (1993). The authors suggest fives major components in competitive facility location, which are space, number of players, pricing policy, rules of the game, and behavior of customers. The mostly used metric space in competitive facility location is the m-dimensional real space $\mathfrak{R}^m$ and, in most instances, is bounded by a convex polygon denoted by $|\mathfrak{R}^m|$. A pricing policy distinguishes between models with both price and location as decision variables and models that consider location as the only decision variable. Rules of the game define the general concept of equilibrium as applied to games but most competitive facility location studies in operations research do not fall under equilibrium models because competing facilities already exist. The last component, behavior of customers, defines deterministic and probabilistic models. In a deterministic model, customers always patronize a single facility that they are most attracted to. On the other hand, probabilistic models assign probabilities of customer attraction to each facility.

Drezner (1994) proposes a solution for the location of a new facility in a continuous planar space and in an environment where competing facilities have different levels of attraction on a consumer. Drezner's (1994) algorithm solves a single-objective optimization problem that first calculates a "break-even" distance, which is the maximum distance that a consumer is willing to travel to the new facility. Afterwards, the market share is computed for the new facility relative to the break-even distance. She concludes that her model guarantees a superior location based on information about consumer preferences and facility attributes. Plastria and Carrizosa (2004) observe that most algorithms solve competitive facility location problems by either setting a fixed site and attempt to maximize profit by an adequate choice of quality or setting the quality fixed and then find an optimal site. They show that profit maximization with respect to both location and the quality of the new facility may be obtained by solving efficient solutions (Pareto-optimal) in a bi-objective location problem. number of factors also conspire to make the process difficult to study experimentally.

29.4 Multi-objective evolutionary algorithm

The study uses the EA algorithm by Zitzler and Thiele (1999), the Strength Pareto Evolutionary Algorithm (SPEA). Zitzler *et al.* (2002) have shown that SPEA2, an improved version of SPEA, outperforms other evolutionary techniques and seems to have performance advantages in solving optimization problems with higher dimensional objective spaces. For such reasons, this section employs SPEA2 to find non-dominated solutions to the CFL problem.

29.4.1 *Genetic Algorithm for the CFL Problem*

Four MOEA methods are introduced in the study. The first is a simple MOEA (S-MOEA) that uses mutation. The second is a variant of the simple MOEA wherein a local search procedure is added to the algorithm (MOEA-LS). The third MOEA-AX uses average crossover and non-uniform mutation. The non-uniform mutation acts as a local search operator at the latter parts of the EA run. The last MOEA-BLX uses blend crossover (Eshelman and Schaffer, 1993) and non-uniform mutation. The blend crossover preserves the mean objective function values of the population. All four MOEAs use the SPEA2 selection algorithm.

Genetic Representation. A chromosome $\boldsymbol{x}$ for the CFL problem is represented by real-parameters, which define the location of a new facility in two dimensions using (x,y)-coordinates. The objective functions are quality α, and captured weight CW of a new facility. Variation and selection operators are applied directly to these real parameter values. The main difficulty here is how to create a new pair of offspring vectors or how to mutate a facility location to a new location in a meaningful manner.

Genetic Operators. Deb (2001) argues that binary crossover techniques hardly create meaningful offspring vectors in real-parameter EAs. Given this observation, implementing a binary crossover operator in a real-parameter search space is not very useful. The problem is then reduced to the design of a crossover and a mutation operator that are strong enough to create diversified solutions and at the same time ensure convergence to optimal solutions. A simple solution would be to implement random mutation per new generation to guarantee diversity but this operator does not ensure convergence. The next step is to generate solutions not from the entire search space but from a search space near the parent solution with a uniform probability distribution. It is also possible to create a non-uniform mutation where the probability of creating solutions closer to the parent increases as the number of generations increase. Our experiment adopts a non-uniform mutation operator, which is described as follows

$$x_i^{t+1} = x_i^t + p(x_i^l - x_i^u)h(t) \tag{29.6}$$

$$h(t) = 1 - r^{(1-t/t_{\max})b} \tag{29.7}$$

where p is either -1 or 1 with a probability of 0.5. The terms x_i^l and x_i^u express the lower and upper bounds that define the decision space and restrict the decision variable x_i to take a value within the decision space at generation t. The parameter r is a real random number in the interval $[0,1]$, t is the generation counter, and $t_{\max}$ is the maximum number of generations. The parameter b is a user-defined input that determines non-uniformity and has a value usually greater than 1. Equation 29.6 assures that mutation accomplishes uniform exploration at the first generations and that search becomes local at the last generations. Equation (29.7) makes it possible to create a non-uniform mutation where the probability of creating solutions closer to the parent gets higher as the number of generations increases.

Recombination is implemented in two ways, as the arithmetic average of both parent vectors or as a blend crossover of both parents. The arithmetic crossover is defined as a vector of elements

$$z_i = \frac{1}{2}(x_i + y_i) \tag{29.8}$$

where $\boldsymbol{x}$ and $\boldsymbol{y}$ are the parents, $\boldsymbol{z}$ is the offspring, $i = 1,\ldots,m$, and m is the number of dimensions of the chromosome. The offspring genes represent the arithmetic mean of the values of the parent vectors.

The blend crossover (BLX) is defined as

$$z_i = (1 - \gamma_i)x_i + \gamma_i y_i \tag{29.9}$$

where $\boldsymbol{x}$ and $\boldsymbol{y}$ are the parents, $\boldsymbol{z}$ is the offspring, $i = 1,\ldots,m$, and m is the number of dimensions of the chromosome, $\gamma_i = (1+2\beta)u_i - \beta$. The term u_i is a real random number between 0 and 1. If β is zero, the crossover creates a random solution in the range (x_i, y_i). Eshelman and Schaffer (1993) report that $\beta = 0.5$ performs better than any other β value for the BLX operator.

Fitness Function. The raw fitness function in this MOEA evaluates the real parameters α and $CW(\alpha,\boldsymbol{x})$ and is a multi-objective function that minimizes α and maximizes $CW(\alpha,\boldsymbol{x})$. Tournament selection as defined in SPEA2 is used for the selection of parent solutions.

Local Search. In addition to the basic genetic operators described above, the MOEA-LS uses a local search procedure that improves the convergence rate of the simple MOEA. The local search procedure iteratively searches for the lowest quality α of a facility (individual) in location $\boldsymbol{x}$. This value is applied to all new offspring individuals in every generation.

29.4.2 *Experiments and results*

The problem considered is taken from Plastria and Carrizosa (2004). Specifically, a new facility must be located in the planar region $\boldsymbol{S}$ bounded by a convex polygon with corner points (0,0), (50,0), (50,20), (25,45), (0,45) as shown in Figure 29.1.

A set of two facilities, $CF = \{f_1, f_2\}$ already exists. They are located at coordinates $(20,73)$ for f_1 and $(50,70)$ for f_2. The facilities have invested a quality cost of 1250 for f_1 and 1000 for f_2. They compete to attract a set of customers $AF = \{a_1, a_2, \ldots, a_{10}\}$. For each consumer a_i $(i = 110)$ the gravity-type attraction to any facility is defined in equation (29.1) with each $k_a = 1$ representing some proportionality constant depending on consumer a, some minimal quality $\alpha_0 \approx 0$, $\alpha_0 = 0.0000001$ and $p = 2$. The parameter p represents the sensitivity of attraction *attr* to distance. Consumer a is captured by the existing facility

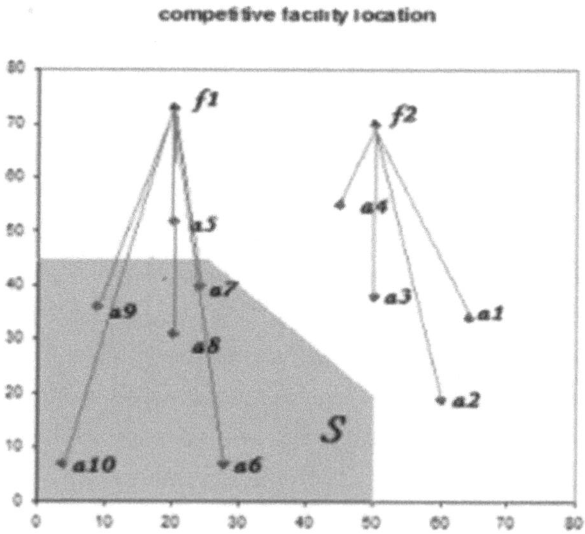

Fig. 29.1 Two facilities, ten customers and a bounded region S

yielding the highest attraction (before the event of the new facility). The attraction $attr_a$ is determined by:

$$attr_a = \max_{f \in CF} \left(\frac{\alpha_f}{\|x_a - x_f\|^2} \right) \quad (29.10)$$

An overview of the attraction parameters of ten potential consumers is presented in Table 29.1. The terms l_a and w_a represent the location and weight of each consumer a respectively.

Table 29.1 Attraction parameters

Consumer	Location l_a	Weight w_a	Attraction $attr_a$	Facility $f(a)$
a_1	(64.0, 34.0)	600	0.6702	f_2
a_2	(60.0, 19.0)	100	0.3702	f_2
a_3	(50.0, 38.0)	100	0.9766	f_2
a_4	(45.0, 55.0)	100	4.0000	f_2
a_5	(20.0, 52.0)	400	2.8345	f_1
a_6	(27.8, 7.0)	300	0.2830	f_1
a_7	(24.0, 40.0)	100	1.1312	f_1
a_8	(20.0, 31.0)	100	0.7086	f_1
a_9	(9.0, 36.0)	100	0.8389	f_1
a_{10}	(3.8, 7.0)	600	0.2707	f_1

One difficulty of using evolutionary algorithms is its sensitivity to the input parameters such as population size, random seed, number of generations, and probability values for crossover and mutation. Determining a balanced interaction between parameter settings to find efficient solutions is not an easy task. A number of simulation runs is usually required to come up with near optimal solutions. The set of parameters shown in Table 29.2 is one of the many combinations that may be used for the competitive facility location problem.

Parameter *maxgen* stands for the maximum number of generations. The maximum quality value is indicated as *quality_max*. Parameters *xy_range* and *quality_range* signify maximum perturbation for location and quality. *Mutation_probability* is the probability that a mutation is performed. *Bit_turn_probability* represents the probability of a gene to be mutated. The *initial population* is the number of individuals at the start of the MOEA run. From an *initial population, a parent population* is selected to reproduce a population of *offspring individuals*. The process of selecting parents and reproducing offspring is repeated until *maxgen* is reached.

Table 29.2 Input parameters

Parameter	S-MOEA	MOEA-LS	MOEA-AX	MOEA-BLX
initial population	25	25	25	25
parent population	25	25	25	25
offspring individuals	25	25	25	25
Maxgen	2000	250	250	250
quality_max	2000	2000	2000	2000
xy_range	5	5	4.28	7.19
quality_range	15	15	1.25	13.13
mutation_probability	1.00	1.00	0.69	0.47
crossover_probability	0.00	0.00	0.44	0.91
bit_turn_probability	0.50	0.50	0.81	0.53

Table 29.3 shows comparative results of Plastria and Carrizosa's algorithm (called Plastria in the column heading), S-MOEA, MOEA-LS, MOEA-AX, and MOEA-BLX from the given sample problem. Plastria and Carrizosa's results show the complete set of optimal solutions. Clearly, all MOEA results show inferior solutions to that of Plastria and that the MOEA-LS, the MOEA-BLX, and the MOEA-AX perform better than the S-MOEA. The worst solutions are generated by the S-MOEA (at 250 generations) because its search operator is weak. It requires 2000 generations to have comparable results with the other MOEAs. This shows that a non-uniform local search is more effective than a uniform search. The best solutions are obtained by the MOEA-AX, which uses arithmetic crossover and non-uniform mutation.

The solutions of all MOEA algorithms are near optimal because the CFL problem has a real-parameter search space. Deb (2001, p. 124) explains that after a few generations, a real-parameter genetic algorithm treats a continuous search space problem as a discrete search space problem and the global optimum is difficult to find.

Table 29.3 Comparative results of efficient solutions

CW	Plastria α	S-MOEA α	MOEA-LS α	MOEA-AX α	MOEA-BLX α
		2000 generations	250 generations		
600.00	0.00	0.00	0.00	0.00	0.08
900.00	39.85	40.45	39.92	40.36	40.01
1000.00	89.83	91.41	89.87	90.46	93.15
1100.00	135.27	136.69	135.84	135.51	144.72
1200.00	182.72	184.10	183.23	182.75	200.52
1300.00	359.56	362.34	360.86	361.92	363.88
1600.00	362.00	363.41	362.81	362.32	365.47
1800.00	440.48	447.06	443.73	441.31	441.46
1900.00	446.91	448.73	448.89	447.34	459.72
2000.00	566.04	568.70	566.40	566.11	589.47
2400.00	767.59	770.36	768.99	768.60	769.34
2500.00	1800.00	1812.14	1806.45	1805.07	1814.01

In terms of runtime complexity, Plastria and Carrizosa's algorithm is able to find the list of efficient solutions in $O(n^3 \log n)$ time where n is the number of consumers. On the other hand, the MOEAs takes $O(M^2)$ to evaluate the raw fitness of all candidate solutions per generation, where M is the sum of the list of candidate solutions n and the archive size. The fitness assignment in SPEA2 per generation takes $O(M^2 \log M)$ and the environmental selection per generation takes $O(M^2 \log M)$. Calculation of the quality α and captured weights CW per generation takes $O(nM)$. The runtime complexity of the MOEAs is dominated by its selection operator and is dependent on the initial population size N and not on the number of customers n. With regard to the MOEAs, the MOEA-LS and the MOEA-AX generates better solutions i.e. solutions nearer to the Pareto-front and finds these solutions in a much shorter time (250 generations against 2000 generations).

29.5 Sensitivity analysis

The purpose of *uncertainty* analysis is to determine the uncertainty in estimates for dependent variables of interest (Saltelli *et al.* 2000). The purpose of *sensitivity analysis* (SA) is to determine the relationships between the uncertainty in the independent variables and the uncertainty in the dependent variables. Uncertainty analysis typically precedes

SA since there is no reason to perform SA when the uncertainty in a dependent variable is below an acceptable bound or range. Sensitivity analysis is often defined as a local measure of the effect of a given input on a given output. This measure can be achieved most often by Monte Carlo methods in conjunction with a variety of sampling strategies (Saltelli *et al.* 2004). Monte Carlo sensitivity analysis is based on performing multiple model evaluations with probabilistically selected model inputs, and the results of such evaluations are used to 1) determine the uncertainty in model predictions and 2) identify the input variables that influence uncertainty.

SIMLAB (2004) is a program designed for global uncertainty and sensitivity analysis based on Monte Carlo methods. It offers several techniques for sample generation, sensitivity analysis, and a link to external model execution. The link allows execution of complex models that can hardly be coded as simple mathematical functions such as genetic algorithms.

29.5.1 *Experiments and results*

A sensitivity analysis is conducted of the MOEA-AX and the MOEA-BLX for the CFL problem to determine which input parameters affect the output and to avoid assigning arbitrary values to the input parameters. A robustness test for the MOEA-AX is added to show that the MOEA can find near-optimal solutions without repeating a sensitivity analysis when the position of $\boldsymbol{S}$ is altered.

The five input parameters for the genetic algorithm model and their configuration are shown in Table 29.4. The factor *xyrange* defines the upper and lower limit for searching a point in the polygon $\boldsymbol{S}$. The factor $\alpha range$ defines the limits for searching the quality α of the new facility. *Mrate*, *xover*, *bitrate* are the mutation, recombination, and bit-turn probabilities respectively.

Table 29.4 Probability distribution of input parameters

Input Factor	Description	Probability Distribution
xyrange	Range for searching location	Uniform(0, 10)
$\alpha range$	Range for searching quality	Uniform(0, 20)
Mrate	Mutation probability	Uniform(0, 1)
xover	Crossover probability	Uniform(0, 1)
Bitrate	Bit-turn probability	Uniform(0, 1)

The bit-turn probability is the probability that a gene will undergo mutation whereas the mutation rate defines the probability that a chromosome or an individual will undergo mutation. The MOEA runs for 250 generations for each configuration. The Sobol' method in SIMLAB is used for sensitivity analysis.

Table 29.5 shows the comparative results of captured weights and facility quality as computed from Plastria and Carrizosa (2004) and the best solution set from 384 different solution sets generated by the genetic algorithms. Plastria and Carrizosa's (2004) results show the Pareto-optimal set $\boldsymbol{P}^*$. The genetic algorithms are able to compute the exact values of captured weight *CW* of the new facilities but show inferior solutions of quality values α.

The performance of a solution set $\boldsymbol{Q}$ from the genetic algorithm is based on three parameters: (29.1) the ratio of *CW* Pareto-optimal solutions in $\boldsymbol{Q}$ over the number of solutions in $\boldsymbol{P}^*$, (29.2) the ratio of *CW* solutions in $\boldsymbol{Q}$ that are members of $\boldsymbol{P}^*$ over the number of solutions in $\boldsymbol{Q}$, and (29.3) the sum of differences in quality values of $\boldsymbol{Q}$ and $\boldsymbol{P}^*$. The best solution $\boldsymbol{Q}^*$, is the solution that has all the *CW* solutions in $\boldsymbol{P}^*$ and the smallest sum of the differences of quality values between $\boldsymbol{Q}^*$ and $\boldsymbol{P}^*$. The quality values of the best solutions from the MOEA-AX and MOEA-BLX are shown in Table 29.5. The quality values generated by the MOEA-AX are nearer to the optimal values than the values generated by the MOEA-BLX, which means that the arithmetic crossover performs better than the blend crossover in this CFL problem set.

Table 29.5 Comparative results of best solutions

CW	Plastria (α) $\boldsymbol{P}^*$	MOEA-AX ($\boldsymbol{Q}^*$)	MOEA-BLX ($\boldsymbol{Q}^*$)
600	0.0000	0.0003	0.0818
900	39.8488	40.3606	40.0118
1000	89.8289	90.4611	93.1452
1100	135.2698	135.5118	144.7237
1200	182.7161	182.7483	200.5216
1300	359.5603	361.9230	363.8839
1600	361.9952	362.3166	365.4698
1800	440.4785	441.3086	441.4561
1900	446.9055	447.3366	459.7156
2000	566.0434	566.1086	589.4663
2400	767.5907	768.5999	769.3388
2500	1800.0000	1805.0749	1814.0083

As regards the sensitivity analysis, Table 29.6 shows the Sobol first-order and total-order indices. The first-order sensitivity index shows the individual effect of an input factor on the output. More specifically a first-order index gives a measure of the direct effect of an input factor on the output variation. An input parameter having a first-order index with the least value means that it has the least influence on the output whereas a factor with the highest first-order value is most important for further investigation. If the sum of the first-order indices equals 1.0, then the model is linear. If the sum of the Sobol first-order indices does not equal 1.0 then the model is nonlinear and implies that some effects on the output are due to interactions among the input factors. First-order sensitivity measures do not capture interactions between factors whereas total-order sensitivity measures detect interactions among input factors. The total-order indices describe the share of the output variation that is related to each input factor. This includes the direct effect as well as interactions with other factors. Removal of a factor means removal of the amount of the total-order index from the output variation. Hence, only factors with very small total-order indices can be removed to avoid significant changes in the output.

Table 29.6 has three columns that represent the performance metrics for the MOEA-AX. The values represent the first-order and total-order sensitivity indices. The mutation rate has the highest first-order index over the three output metrics and causes nearly 30% of the variation in each of those metrics. Similarly, the factors *xyrange* and the *bitrate* share almost 50% of the variation as shown in the third column output metric-the sum of the differences in the quality α of a facility between solutions $\boldsymbol{Q}$ and $\boldsymbol{P}^*$ and can be interpreted as the proximity of an α solution in $\boldsymbol{Q}$ to $\boldsymbol{P}^*$. The parameters α range and *xover* have almost no direct influence on the output metrics.

Table 29.6 Sobol first-order and total-order indices for MOEA-AX

Sobol first order indices			
	Number of *CW* solutions in $\boldsymbol{Q}$ over $\lvert\boldsymbol{P}^*\rvert$	Number of *CW* solutions in $\boldsymbol{Q}$ that are in $\boldsymbol{P}^*$ over $\lvert\boldsymbol{Q}\rvert$	Sum of differences in quality α
xyrange	0.047	0.115	0.298
α *range*	−0.089	−0.077	0.039
Mrate	0.286	0.318	0.292
xover	−0.002	0.082	0.054
bitrate	0.065	−0.070	0.207
Sobol total order indices			
xyrange	0.602	0.642	0.407
α *range*	0.450	0.264	0.153
Mrate	0.611	0.637	0.528
xover	0.200	0.300	0.137
bitrate	0.405	0.243	0.372

The crossover probability factor shows small values and can be a candidate for removal from the list of parameters. The results also show that the mutation rate has the greatest influence on the output but specific mutation probabilities that generated near-optimal solutions are not known. Further investigation reveals that, out of the 384 input configurations, 21 samples have generated near-optimal solutions i.e. they have detected all Pareto-optimal solutions for the function "captured weight", *CW* and have small differences in values for the function "facility quality", α. Those 21 samples show high values for the mutation rate, specifically an average of 0.75 and a standard deviation of 0.16. Other averages for the input factors are $bitrate = 0.62$, $xyrange = 5.96$, $\alpha range = 6.46$, and $xover = 0.40$.

Table 29.7 shows the performance results from the MOEA-BLX. The crossover rate has the highest first-order index, followed by the *bitrate* and mutation rate. They account for 68% of the variation in the first column output metric and 53% of the third column output metric. The factors *xyrange* and the $\alpha range$ have almost no direct influence on all the output metrics. However, their total-order effects are significant. 23 near-optimal solutions are generated by the MOEA-BLX. The mutation rate averages at 0.63, the crossover rate at 0.60 and the bit rate at 0.53. The average of *xyrange* is 5.04 and $\alpha range$ is 9.60.

Table 29.7 Sobol first-order and total-order indices for MOEA-BLX

	Number of *CW* solutions in $\boldsymbol{Q}$ over $\lvert\boldsymbol{P}^*\rvert$	Number of *CW* solutions in $\boldsymbol{Q}^*$ that are in $\boldsymbol{P}^*$ over $\lvert\boldsymbol{Q}\rvert$	Sum of differences in quality α
Sobol first order indices			
xyrange	−0.023	−0.017	0.018
α *range*	0.117	0.041	−0.013
Mrate	0.222	−0.005	0.153
xover	0.240	0.035	0.229
bitrate	0.235	−0.108	0.168
Sobol total order indices			
xyrange	0.486	0.801	0.386
α *range*	0.849	0.742	0.708
Mrate	0.665	0.895	0.544
xover	0.631	0.694	0.687
bitrate	0.920	0.650	0.847

Tables 29.6 and 29.7 show the sensitivity indices of two MOEAs having different crossover operators. Their values are very different which means that the type of crossover operator, while having the same mutation and selection operator, changes the behavior of the MOEA in terms of the individual and interaction effects of each input parameter to the output. A

common result between both algorithms is that the mutation rate and the bit-turn rate are kept high (> 0.60) to get near-optimal solutions.

29.5.2 *Robustness*

The interaction among parameters with high-valued indices should be further investigated in order to improve the solutions in the real-valued objective (quality of the facility). A method that requires less input sample sizes is beneficial in the investigation of sensitivity analyses of evolutionary algorithms as applied to CFL problem. To test whether a smaller input sample set is beneficial, Table 29.8 shows a list of ten parameter combinations that produced the best CFL solutions. These are tested on different positions of the bounded region $\boldsymbol{S}$. The succeeding experiment explores and describes the solutions that are generated by these ten parameter combinations using MOEA-AX for nine different area locations as defined by the polygon $\boldsymbol{S}$. The MOEA-AX is chosen for the robustness test since it produces the better solutions than the other MOEAs.

Table 29.8 GA parameters that give best solutions to MOEA-AX

	xyrange	*qrange*	*mutation*	*xover*	*bitturn*
1	4.38	1.25	0.69	0.44	0.81
2	3.13	3.75	0.94	0.06	0.44
3	5.31	0.63	0.91	0.72	0.34
4	7.81	0.63	0.78	0.59	0.34
5	3.13	3.75	0.94	0.56	0.44
6	1.88	1.25	0.94	0.19	0.56
7	7.50	5.00	0.75	0.25	0.75
8	9.22	14.69	0.95	0.55	0.83
9	7.19	13.13	0.47	0.28	0.53
10	7.81	9.38	0.78	0.72	0.34

The experiment intends to determine whether the same parameter combinations are able to generate near-optimal CFLP solutions on differing areas as shown in Figures 29.2 till 29.9. Figure 29.2 is a translation of the original area by $+15$ units on the x-axis and Figure 29.3 is a translation of the original area by $+15$ units on the y-axis. Figure 29.4 is a translation of the original area by $+15$ units on both the x- and y-axes. Figures 29.5, 29.6, and 29.7 are reflections of the original area on the y-axis, x-axis, and the xy-axes respectively. Figure 29.8 is the area rotated clockwise by 90 degrees at location $(50,20)$ and Figure 29.9 is the original area rotated counter-clockwise by 90 degrees at location $(25,45)$.

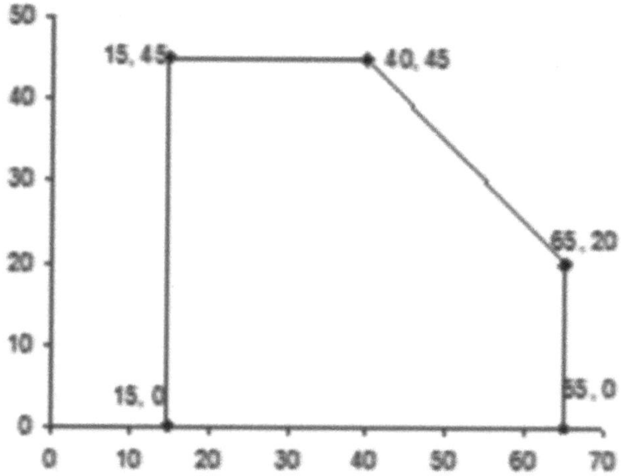

Fig. 29.2 Translation of S on x-axis

Table 29.9 shows the quality values and captured weights after translation of $\boldsymbol{S}$ on the x-axis (see Figure 29.2). The best solution is indicated by column $\boldsymbol{P}^*$, and the quality values are from non-dominated sets $\boldsymbol{Q}_1$ to $\boldsymbol{Q}_{10}$. The term $\boldsymbol{Q}_i$, where $i \in \{1,2,\ldots,10\}$ represents a non-dominated set of CFL problem solutions generated from an input combination listed in Table 29.7. Solutions $\boldsymbol{Q}_1$, $\boldsymbol{Q}_3$, $\boldsymbol{Q}_4$, $\boldsymbol{Q}_8$, and $\boldsymbol{Q}_{10}$ have the complete set of solutions and have mutation rates $= 0.69$ and crossover rates $= 0.44$. Whereas solution sets $\boldsymbol{Q}_2$, $\boldsymbol{Q}_5$, $\boldsymbol{Q}_6$, $\boldsymbol{Q}_7$, and $\boldsymbol{Q}_9$ miss one captured weight ($CW = 1800$) and its corresponding quality values. The incomplete solutions have mutation rates $= 0.75$ (except for $\boldsymbol{Q}_9$, which has 0.47) and crossover rates $\leqslant 0.56$. The nearest solution to $\boldsymbol{P}^*$, according to the performance criteria described above is $\boldsymbol{Q}_8$ with a mutation rate of 0.95, bit-turn probability of 0.83, and a crossover rate of 0.55 followed by $\boldsymbol{Q}_3$ with a mutation rate of 0.91, bit-turn probability of 0.72, and a crossover rate of 0.34. The worst vector is $\boldsymbol{Q}_6$ with a mutation rate of 0.94, bit-turn probability of 0.56, and a crossover rate of 0.19.

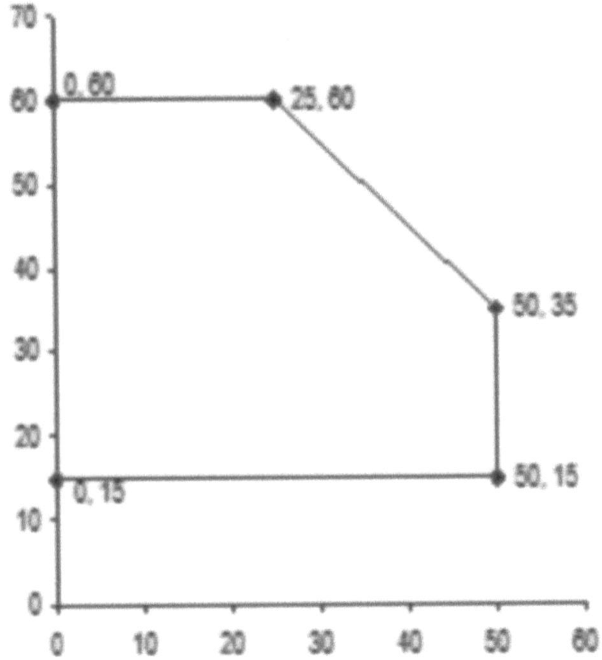

Fig. 29.3 Translation of S on y-axis

Table 29.9 Quality values after translation of S on the x-axis

$\boldsymbol{Q}_1$	$\boldsymbol{Q}_2$	$\boldsymbol{Q}_3$	$\boldsymbol{Q}_4$	$\boldsymbol{Q}_5$	$\boldsymbol{Q}_6$	$\boldsymbol{Q}_7$	$\boldsymbol{Q}_8$	$\boldsymbol{Q}_9$	$\boldsymbol{Q}_{10}$	CW	$\boldsymbol{P}^*$
0.077	1E-04	0.001	9E-04	2E-04	0.014	0.003	0.215	5E-04	0.016	300	1E-04
33.9	33.9	33.9	33.9	33.9	33.8	34.1	33.9	34.2	34.2	600	33.8
39.9	39.9	39.9	89.8	40.3	39.9	39.9	40.4	40.9	41.3	900	39.9
91.5	103.3	89.9	90.5	92.3	97.8	91.0	90.5	96.7	97.3	1000	89.9
135.8	139.6	135.7	135.3	137.1	135.4	135.7	135.9	135.6	140.0	1100	135.3
187.4	183.0	185.4	184.9	192.2	188.8	182.8	182.8	194.0	190.9	1200	182.8
359.8	368.7	359.9	359.6	355.7	370.6	359.8	359.8	369.9	358.7	1300	355.7
398.6	400.5	398.1	398.9	396.1	399.1	397.4	364.9	397.1	397.3	1400	364.9
441.9		441.2	441.0				442.6		443.0	1800	441.0
447.6	455.8	449.8	447.7	449.2	473.2	449.0	453.1	463.3	447.0	1900	447.0
566.3	566.8	566.5	566.5	567.0	566.4	566.3	566.4	566.6	568.1	2000	566.3
687.9	688.6	688.0	689.6	689.1	693.4	687.9	688.3	690.4	689.9	2400	687.9
789.1	789.0	789.1	791.0	789.0	788.8	789.2	789.2	790.1	790.0	2500	788.8

Table 29.10 shows the quality values and captured weights after translation of $\boldsymbol{S}$ on the y-axis (see Figure 29.3). Vectors $\boldsymbol{Q}_2$ and $\boldsymbol{Q}_6$, have the complete set of solutions and have

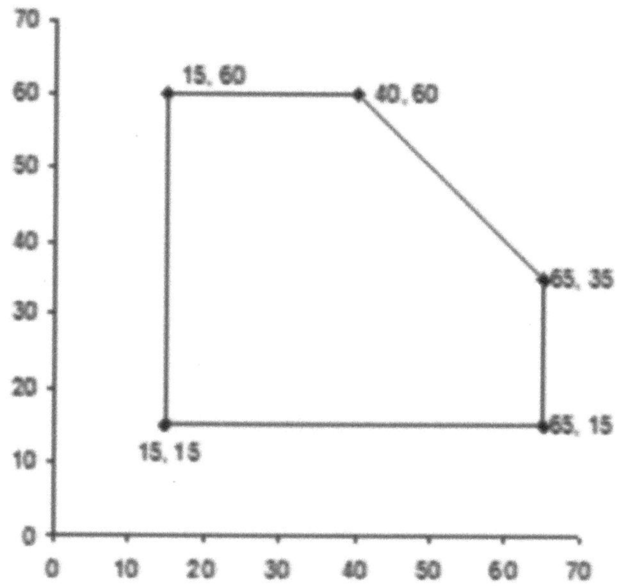

Fig. 29.4 Translation of S on xy-axis

mutation rates equal to 0.94 and crossover rates $\leqslant 0.44$. Vectors $\boldsymbol{Q}_1$, $\boldsymbol{Q}_4$ and $\boldsymbol{Q}_{10}$ miss one solution ($CW = 2100$). Solution sets $\boldsymbol{Q}_5$, $\boldsymbol{Q}_8$, and $\boldsymbol{Q}_9$ miss two solutions each whereas $\boldsymbol{Q}_3$ and $\boldsymbol{Q}_7$ miss three solutions. The nearest solution to $\boldsymbol{P}^*$ is $\boldsymbol{Q}_2$ with a mutation rate of 0.94, bit-turn probability of 0.44, and a crossover rate of 0.06 followed by $\boldsymbol{Q}_6$ with a mutation rate of 0.94, bit-turn probability of 0.56, and a crossover rate of 0.19. The worst vector is $\boldsymbol{Q}_3$ with a mutation rate of 0.91, bit-turn probability of 0.34, and a crossover rate of 0.72.

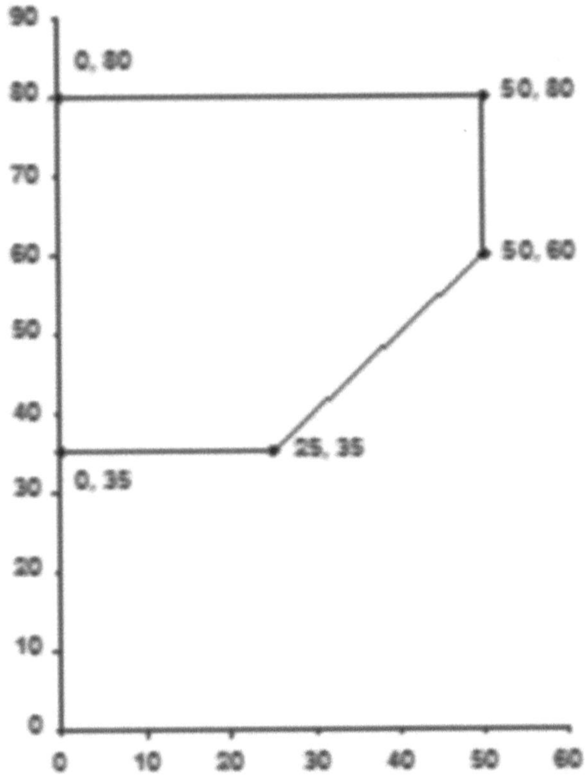

Fig. 29.5 Reflection of S at y-axis

Table 29.10 Quality values after translation of S on the y-axis

Q_1	Q_2	Q_3	Q_4	Q_5	Q_6	Q_7	Q_8	Q_9	Q_{10}	CW	P^*
0.002	0.0001	0.0001	0.0001	0.001	0.001	0.001	0.095	0.003	0.0003	100	0.0001
25.1	17.3	17.4	17.4	17.7	17.4	17.4	18.0	23.8	19.2	600	17.3
59.0	57.9	57.6	57.9	57.6	58.2	57.9	62.1	58.3	59.0	900	57.6
90.3	90.2	90.5	90.4	91.0	90.0	90.0	91.7	92.1	90.5	1000	90.0
136.0	135.3	136.5	136.6	136.2	135.6	135.6	135.6	137.2	139.9	1100	135.3
182.8	183.5	182.9	183.7	185.0	182.8	182.8	183.0	183.0	182.9	1200	182.8
361.9	361.8	366.7	361.5	361.8	362.0	360.1	361.0	359.8	361.7	1300	359.8
363.3	362.1		364.5	362.2	362.3		364.2	363.0	363.4	1600	362.1
446.6	444.7		440.9		440.6			443.7	442.6	1800	440.6
447.8	449.0	452.7	448.7	451.3	459.7	447.6	450.9	447.7	449.3	1900	447.6
566.4	567.3	566.2	566.3	566.4	566.4	566.9	566.4	567.7	566.4	2000	566.2
	684.4				685.5					2100	684.4
686.4	686.9	686.5	686.7	686.1	686.2	686.3	686.9		687.8	2400	686.1
688.7	689.2	688.8	688.7	689.3	688.8	689.0	689.2	689.4	690.1	2500	688.7

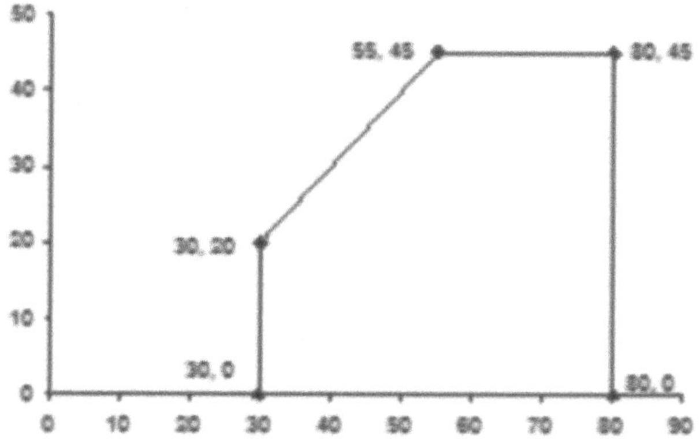

Fig. 29.6 Reflection of S on x-axis

Table 29.11 shows the quality values and captured weights after translation of $\boldsymbol{S}$ on both the x and y axes (see Figure 29.4). No single solution set captures all quality values and captured weights as listed in the column of $\boldsymbol{P}^*$. Solution sets $\boldsymbol{Q}_1$, $\boldsymbol{Q}_7$, and $\boldsymbol{Q}_{10}$ miss one solution ($CW = 2100$) and vectors $\boldsymbol{Q}_5$ and $\boldsymbol{Q}_8$ miss also one solution, $CW = 1800$ and $CW = 100$, respectively. The remaining vectors $\boldsymbol{Q}_2$, $\boldsymbol{Q}_3$, and $\boldsymbol{Q}_4$ miss two solutions whereas $\boldsymbol{Q}_6$ and $\boldsymbol{Q}_9$ miss three solutions. The nearest solution to $\boldsymbol{P}^*$ is $\boldsymbol{Q}_{10}$ with a mutation rate of 0.78, bit-turn probability of 0.34, and a crossover rate of 0.72 followed by $\boldsymbol{Q}_7$ with a mutation rate of 0.75, bit-turn probability of 0.75, and a crossover rate of 0.25. The worst vector is $\boldsymbol{Q}_8$ with a mutation rate of 0.95, bit-turn probability of 0.85, and a crossover rate of 0.55.

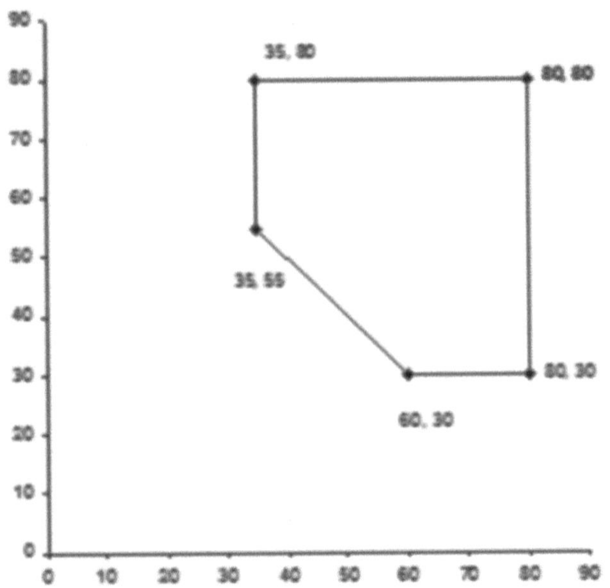

Fig. 29.7 Reflection of S at xy-axes, $y = 80 - x$

Table 29.11 Quality values after translation of S in both x and y-axes

Q_1	Q_2	Q_3	Q_4	Q_5	Q_6	Q_7	Q_8	Q_9	Q_{10}	CW	P^*
1.868	2E-05	0.0001	0.0001	0.0008	0.0002	0.124	0.011	0.022	0.024	600	2E-05
29.8	43.1	42.6	31.2	43.6	42.6	29.7	31.5	31.9	43.7	700	29.7
58.7	58.0	58.0	57.9	57.7	74.9	57.9	58.3	59.2	57.9	900	57.7
95.1	92.6	90.1	90.6	90.6	90.1	91.8	90.6	90.6	90.4	1000	90.1
135.7	135.5	135.4	135.4	135.5	135.4	135.5	135.5	137.3	136.5	1100	135.4
183.1	183.6	183.0	183.9	185.3	183.4	182.8	184.1	189.4	184.8	1200	182.8
350.5	359.9	360.1	359.9	363.4	350.2	362.6	350.6	361.2	372.6	1300	350.2
398.1	396.5	401.3	395.3	395.7	397.0	398.5	398.3		397.3	1400	395.3
444.8		440.7	443.7			447.9	447.3	445.7	440.8	1800	440.7
447.4	449.4	454.0	448.5	448.3	452.1	448.4	455.8	448.4	449.5	1900	447.4
566.8	566.8	566.5	566.2	566.3	566.1	566.2	566.1	566.2	569.7	2000	566.1
	683.9	686.6		685.2			689.2			2100	683.9
686.3	687.2	689.2	687.3	686.1	685.8	686.0	1344.6	687.0	687.2	2400	685.8
689.0	689.0		688.9	689.0	688.7	688.7	1462.5	689.3	691.3	2500	688.7

Table 29.12 shows the quality values and captured weights of the reflection of $\boldsymbol{S}$ on the y-axis (see Figure 29.5). Vectors $\boldsymbol{Q}_3$ and $\boldsymbol{Q}_6$, have the complete set of solutions. The nearest solution to $\boldsymbol{P}^*$ is $\boldsymbol{Q}_6$ with a mutation rate of 0.94, bit-turn probability of 0.56, and a crossover rate of 0.19 followed by $\boldsymbol{Q}_3$ with a mutation rate of 0.91, bit-turn probability

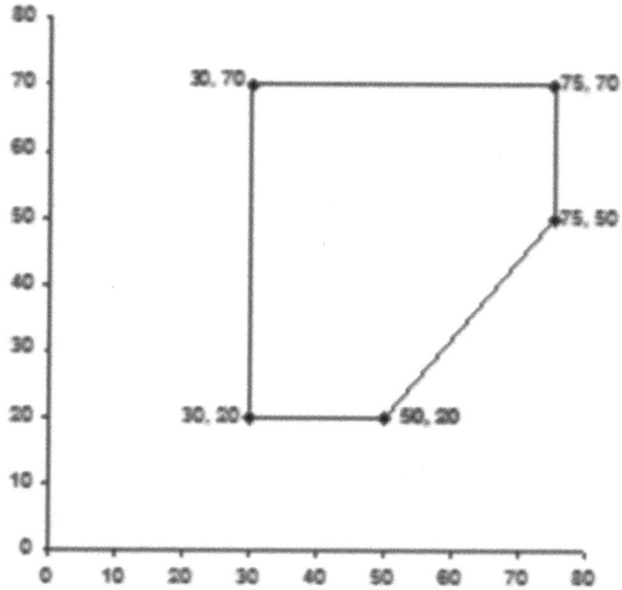

Fig. 29.8 Rotation of S at $(50, 20)$ clockwise 90°

of 0.72, and a crossover rate of 0.34. The worst vector is $\boldsymbol{Q}_8$ with a mutation rate of 0.95, bit-turn probability of 0.85, and a crossover rate of 0.55.

Table 29.12 Quality values of the reflection of S at y=40

$\boldsymbol{Q}_1$	$\boldsymbol{Q}_2$	$\boldsymbol{Q}_3$	$\boldsymbol{Q}_4$	$\boldsymbol{Q}_5$	$\boldsymbol{Q}_6$	$\boldsymbol{Q}_7$	$\boldsymbol{Q}_8$	$\boldsymbol{Q}_9$	$\boldsymbol{Q}_{10}$	CW	$\boldsymbol{P}^*$
0.087	0.0022	0.0039	0.0028	0.0003	5E-05	0.013	0.141	0.002	0.0002	400	5E-05
73.3	69.3	68.0	68.6	68.5	69.2	70.5	69.9	69.0	73.5	500	68.03
135.7	139.7	133.1	132.8	133.2	133.0	133.6	133.0	135.7	140.3	600	132.8
143.7	144.0	143.0	143.3	143.7	142.8	143.3	144.8	146.3	145.6	700	142.8
212.2	212.6	212.2	212.5	212.4	212.2	212.3	212.5	212.5	212.8	800	212.2
225.6	226.1	225.6	226.6	226.2	225.9	226.5	227.3	230.5	226.1	900	225.6
257.0	257.0	257.1	257.3	257.1	257.3	257.4	258.1	257.8	257.7	1200	257
		361.9	361.5		361.7				362.1	1300	361.5
362.6	364.7	362.5	362.1	362.2	362.1	362.4	362.8	362.7	363.5	1600	362.1
485.7	485.7	487.7	485.1	485.3	485.1	487.6	489.4	486.2	488.6	1700	485.1
538.5	537.9	537.7	537.6	538.1	537.5	537.6	542.0	538.9	539.7	1800	537.5
649.8	649.4	649.4	651.1	650.1	649.4	651.3	649.3	652.9	650.5	2000	649.3
685.6	684.0	685.2		683.5	683.6				687.2	2100	683.5
686.6	686.7	685.8	686.6	686.5	686.4	686.2		686.5		2400	685.8
688.7	688.8	689.0	689.0	689.0	688.7	689.0	689.1	689.0	690.8	2500	688.7

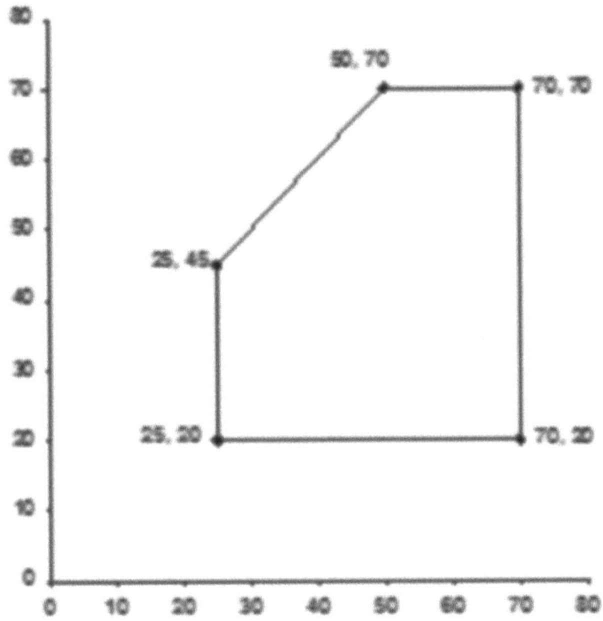

Fig. 29.9 Rotation of S at $(25,45)$ counter-clockwise 90°

Table 29.13 shows the quality values and captured weights of the reflection of $\boldsymbol{S}$ on the x-axis (see Figure 29.6). Vector $\boldsymbol{Q}_2$ has the complete set of solutions and all the other vectors miss two or more solutions. Following the performance criteria described previously the nearest solution to $\boldsymbol{P}^*$ is $\boldsymbol{Q}_2$ with a mutation rate of 0.94, bit-turn probability of 0.44, and a crossover rate of 0.06 followed by $\boldsymbol{Q}_7$ with a mutation rate of 0.75, bit-turn probability of 0.75, and a crossover rate of 0.25. The worst vector is $\boldsymbol{Q}_3$ with a mutation rate of 0.91, bit-turn probability of 0.34, and a crossover rate of 0.72.

Table 29.13 Quality values of the reflection of S at x=40

Q_1	Q_2	Q_3	Q_4	Q_5	Q_6	Q_7	Q_8	Q_9	Q_{10}	CW	P^*
	4E-05		0.0003		1E-06					100	1E-06
0.04	5.0	3E-04	12.9	0.0001	0.0001	0.001	0.04	0.0002	0.0003	600	1E-04
30.7	43.0	42.6	29.7	30.3	42.6	29.8	30.6	34.4	29.9	700	29.7
67.5	66.9	69.2	69.4	66.6	66.0	66.5	69.3	66.1	69.7	800	66.0
185.1	185.2				185.1	185.4	185.3			900	185.1
	212.1									1000	212.1
212.4	212.4	220.1	212.7	212.7	217.8	212.3	213.6	212.2	215.2	1100	212.2
335.5	335.8	335.5	337.2	337.5	336.3	335.8	338.4	336.3	336.6	1200	335.5
382.9	384.5	383.5	383.2	384.8	383.4	403.8	383.0	383.8	385.1	1300	382.9
440.5	440.9	440.6	441.0	440.7	440.6	441.4	441.4	440.7	442.4	1800	440.5
451.0	447.6	448.3	448.4	447.4	447.4	452.5	448.0	448.7	448.9	1900	447.4
681.2	655.9	682.0	674.0	652.0	652.0	678.4	653.4	652.9	673.1	2000	652.0
1245	1245	1246	1246	1246	1245	1246	1246	1256	1246	2100	1245

Table 29.14 shows the quality values and captured weights of the reflection of $\boldsymbol{S}$ on both the x and y axes (see Figure 29.7). Vectors $\boldsymbol{Q}_1$, $\boldsymbol{Q}_6$, $\boldsymbol{Q}_7$, and $\boldsymbol{Q}_9$ have the complete set of solutions. Vectors $\boldsymbol{Q}_2$, $\boldsymbol{Q}_3$, $\boldsymbol{Q}_5$, and $\boldsymbol{Q}_{10}$ miss one solution each. The nearest solution to $\boldsymbol{P}^*$ is $\boldsymbol{Q}_1$ with a mutation rate of 0.69, bit-turn probability of 0.81, and a crossover rate of 0.44 followed by $\boldsymbol{Q}_7$ with a mutation rate of 0.75, bit-turn probability of 0.75, and a crossover rate of 0.25.

Table 29.14 Quality values of the reflection of S at $y = 80 - x$

Q_1	Q_2	Q_3	Q_4	Q_5	Q_6	Q_7	Q_8	Q_9	Q_{10}	CW	P^*
1.307	4E-05	3E-04	0.0002	3E-05	0.003	0.001	0.022	0.0026	0.0011	600	3E-05
42.6	42.7	42.7	42.6	42.8	42.5	43.2	44.9	45.8	43.2	700	42.5
74.7	74.8	75.0	77.5	75.1	80.6	75.2	76.9	80.0	75.2	800	74.7
333.4	331.6	331.4	331.8	331.5	331.2	334.6	331.3	331.6	337.2	900	331.2
430.9	431.2	430.6	432.3	430.7	431.8	431.1	444.1	430.6	431.5	1100	430.6
482.0	481.7	482.0	483.7	482.3	482.5	482.1	482.9	482.1	483.1	1200	481.7
500.9	506.1	501.4	505.6	503.4	502.2	502.1	502.4	507.8	500.8	1300	500.8
542.8	547.5	543.5	546.8	542.8	549.0	543.0	546.2	543.1	546.0	1400	542.8
769.0			770.5		792.0	768.1		770.1	848.3	1800	768.1
848.3	848.1	848.1	848.1	848.0	848.4	848.4	848.1	848.1		2100	848.0
868.5	867.2	910.8	867.7	897.6	869.3	867.8	867.1	870.3	911.8	2400	867.1
882.0	908.5	917.6	879.5	914.8	893.0	887.6	907.4	887.0	917.9	2500	879.5

The worst vector is $\boldsymbol{Q}_{10}$ with a mutation rate of 0.78, bit-turn probability of 0.34, and a crossover rate of 0.72.

Table 29.15 shows the quality values and captured weights after $\boldsymbol{S}$ is rotated clockwise by 90 degrees at (50, 20) (see Figure 29.8). No single vector captures all quality values and captured weights as listed in the column of $\boldsymbol{P}^*$. Solution sets $\boldsymbol{Q}_1$, $\boldsymbol{Q}_3$, and $\boldsymbol{Q}_6$ miss only one solution ($CW = 1400$) and $\boldsymbol{Q}_{10}$ misses solution $CW = 2400$ respectively. Vectors $\boldsymbol{Q}_9$,

$\boldsymbol{Q}_4$ and $\boldsymbol{Q}_5$ miss solution $CW = 2100$. The remaining vectors $\boldsymbol{Q}_2$, $\boldsymbol{Q}_5$, $\boldsymbol{Q}_7$, and $\boldsymbol{Q}_8$ miss two solutions each. The nearest solution to $\boldsymbol{P}^*$ is $\boldsymbol{Q}_6$ with a mutation rate of 0.94, bit-turn probability of 0.56, and a crossover rate of 0.19 followed by $\boldsymbol{Q}_3$ with a mutation rate of 0.91, bit-turn probability of 0.34, and a crossover rate of 0.72 . The worst vector is $\boldsymbol{Q}_8$ with a mutation rate of 0.95, bit-turn probability of 0.83, and a crossover rate of 0.55.

Table 29.15 Quality values after a 90° clockwise rotation at (50,20)

$\boldsymbol{Q}_1$	$\boldsymbol{Q}_2$	$\boldsymbol{Q}_3$	$\boldsymbol{Q}_4$	$\boldsymbol{Q}_5$	$\boldsymbol{Q}_6$	$\boldsymbol{Q}_7$	$\boldsymbol{Q}_8$	$\boldsymbol{Q}_9$	$\boldsymbol{Q}_{10}$	CW	$\boldsymbol{P}^*$
0.001		0.0003	0.0053	0.0001	4E-05	0.024	0.039	0.0189	0.0025	100	4.E-05
4.9	2.4	3.8	2.5	4.2	3.1	2.6	2.4	2.3	5.4	600	2.3
42.9	42.7	43.2	41.6	43.7	43.6	42.8	44.5	47.8	43.1	700	41.6
70.0	67.4	66.6	65.9	66.1	66.1	66.4	66.6	67.2	67.5	800	65.9
214.7	212.5	220.0	215.5	218.0	219.1	219.7	213.9	226.1	212.5	1100	212.5
335.3	335.7	335.8	335.3	336.0	335.1	308.6	336.8	308.0	337.1	1200	308.0
350.8	350.4	350.2	358.8	350.5	354.7	359.6	353.1	361.4	351.4	1300	350.2
			395.5			395.4	395.6	398.8	397.3	1400	395.4
443.4	443.0	442.6	440.9	442.6	443.4			445.2	443.6	1800	440.9
447.4	449.7	447.8	450.2	448.9	447.4	448.1	448.6	447.4	448.8	1900	447.4
566.5	566.9	566.1	566.7	567.0	568.8	566.6	567.4	568.8	571.0	2000	566.1
683.4	684.6	684.4			683.6	683.9	683.3		688.2	2100	683.3
686.2	686.2	686.7	686.5	687.4	686.5			686.9		2400	686.2
688.7	688.7	689.0	688.9	688.9	689.1	689.0	689.6	689.6	691.0	2500	688.7

Table 29.16 shows the quality values and captured weights after $\boldsymbol{S}$ is rotated counter-clockwise by 90 degrees at (25,45) (see Figure 29.9). Vector $\boldsymbol{Q}_8$ has the complete set of solutions and all other vectors have one or more missed solutions. Solution sets $\boldsymbol{Q}_1$, $\boldsymbol{Q}_4$, $\boldsymbol{Q}_7$, and $\boldsymbol{Q}_{10}$ miss only one solution each and vectors $\boldsymbol{Q}_2$, $\boldsymbol{Q}_3$, $\boldsymbol{Q}_6$, and $\boldsymbol{Q}_9$ miss two solutions each. Vector $\boldsymbol{Q}_5$ misses three solutions. Following the performance criteria described previously, the nearest solution to $\boldsymbol{P}^*$ is $\boldsymbol{Q}_8$ with a mutation rate of 0.95, bit-turn probability of 0.83, and a crossover rate of 0.55 followed by $\boldsymbol{Q}_1$ with a mutation rate of 0.69, bit-turn probability of 0.81, and a crossover rate of 0.44. The worst vector is $\boldsymbol{Q}_3$ with a mutation rate of 0.91, bit-turn probability of 0.34, and a crossover rate of 0.72.

Table 29.16 Quality values after a 90° counter-clockwise rotation (25, 45)

$\boldsymbol{Q}_1$	$\boldsymbol{Q}_2$	$\boldsymbol{Q}_3$	$\boldsymbol{Q}_4$	$\boldsymbol{Q}_5$	$\boldsymbol{Q}_6$	$\boldsymbol{Q}_7$	$\boldsymbol{Q}_8$	$\boldsymbol{Q}_9$	$\boldsymbol{Q}_{10}$	CW	$\boldsymbol{P}^*$
0.017	0.0001	0.0195	0.0004	4E-05	0.0001	0.11	0.038	0.0014	0.0021	600	4.E-05
33.2	29.9	29.4	29.9	31.6	31.4	30.1	33.1	33.6	43.1	700	29.4
66.4	66.1	66.2	66.1	66.8	66.0	67.5	70.3	67.6	68.6	800	66.0
216.5	218.8	214.1	216.4	215.8	218.0	213.0	213.3	223.9	214.2	1100	213.0
253.7	307.8	335.3	307.8	307.7	307.7	307.9	253.7	312.4	309.6	1200	253.7
354.0	359.8	351.7	360.9	360.1	359.6	361.3	360.9	361.3	352.1	1300	351.7
401.2			396.2			395.9	399.0	400.6	401.8	1400	395.9
		443.9	443.1			444.7	443.6	442.3	441.1	1800	441.1
489.9	458.4	448.1	463.0	448.0	447.0	448.0	448.7	447.8	447.7	1900	447.0
566.2	567.1	566.2	566.4	566.1	566.1	566.6	566.6	566.2	566.7	2000	566.1
686.5	683.1				684.9		685.0			2100	683.1
686.7	685.8	686.4	687.1	686.4	686.4	686.5	686.9		686.9	2400	685.8
689.0	688.7	688.8	689.2	689.1	688.7	688.8	688.8	689.2	690.3	2500	688.7

29.6 Conclusions and Outlook

The results show that there is no single parameter combination that dictates the generation of the best non-dominated solutions. This means a parameter combination that finds the best solution in one case area does not necessarily find the best solution in another case. There exist also cases wherein not even one combination produced a complete set of solutions from all ten parameter combinations. This means that the ten best parameter combinations in the original configuration do not always generate the best solutions when changes to the location of the area are introduced. However, the multiobjective evolutionary algorithm can generate the best solution $\boldsymbol{Q}^*$ from an ensemble of only ten parameter combinations by combining the best solutions for each combination.

The tables also show that high mutation probabilities ($> 60\%$) and low crossover probabilities ($< 55\%$) produce the nearest nondominated solution to $\boldsymbol{P}^*$. On the contrary, high crossover rates ($> 50\%$) often produce the worst nondominated solutions among the parameter combinations when using arithmetic crossover and non-uniform mutation.

Bibliography

[1] Carrizosa E. and Plastria F., (1995). On minquantile and maxcovering optimization. *Mathematical Programming*, **71**: 101–112.

[2] Deb K., (2001). *MultiObjective Optimization Using Evolutionary Algorithms*. Chichester, UK: Wiley.

[3] Drezner T., (1994). Locating a single new facility among existing, unequally attractive facilities. *Journal of Regional Science*, **34** (2): 237–252.

[4] Drezner Z., (1981). On a modified one-center problem, *Management Science*, **27** (7): 848–851.

[5] Eiselt H.A. and Laporte G., (1988). Location of a new facility on a linear market in the presence of weights. *Asia-Pacific Journal of Operational Research*, **5**: 160–165.

[6] Eiselt H.A., Laporte G., and Thisse J.-F., (1993). Competitive Location Models: A Framework and Bibliography. *Transportation Science*, **27** (1): 44–54.

[7] Eshelman L.J. and Schaffer J.D., (1993). Crossover's niche. In *Proceedings of 5th International Conference on Genetic Algorithms*, Virginia, USA. pp. 9–194.

[8] Plastria F., (2001). Static competitive facility location: an overview of optimisation approaches. *European Journal of Operational Research*, **129**: 461–470.

[9] Plastria, F. and E. Carrizosa (2004). Optimal location and design of a competitive facility. *Mathematical Programming*, **100** (2): 247–265.

[10] Saltelli A., Chan K., and Scott M., (2000). *Sensitivity Analysis*. Probability and Statistics Series. New York, NY: Wiley.

[11] Saltelli A., Tarantola S., Campolongo F., and Ratto M., (2004). *Sensitivity Analysis in Practice: a Guide to assessing Scientific Models*. Chichester: Wiley.

[12] SIMLAB 2.2, (2004). *Reference Manual*. European Commission-ISPC.

[13] Zitzler E. and Thiele L., (1999). Multiobjective evolutionary algorithms: a comparative study and the strength Pareto approach. *IEEE Transactions on Evolutionary Computation*, **3** (4): 257–271.

[14] Zitzler E., Laumanns M., and Thiele L., (2002). SPEA2: Improving the strength Pareto evolutionary algorithm for multiobjective optimization. In Giannakoglou K., Tsahalis D., Periaux J., Papailou K., Fogarty T. (eds.), *Evolutionary Methods for Design, Optimization and Control*, pp. 19–26, CIMNE, Barcelona, Spain.

Chapter 30

Human Fatigue: Evaluation with the Usage of Bayesian Networks

Nikolaos P. Ventikos, Georgios V. Lykos, and Nikolaos A. Vagias

Laboratory for Maritime Transport, School of Naval Architecture and Marine Engineering, National Technical University of Athens, 9 Iroon Polytechneiou st, Zografou, 15773, Hellas

E-mail: niven@deslab.ntua.gr

In the environment and the day-to-day operations of modern high-risk industries much of the attention tends to be (or better should be) on the performance of people; this can be justified through the acknowledgment that the competitive advantage of organizations and/or companies can be generated by Human Factors (HF), and therefore the effective adaption of best practices of Human Resource Management (HRM). Fatigue amongst workers in such industries has proved to be the outmost reason affecting their performance and the overall rating of the process. In this context, the approach described hereafter is focused on the offset of the interdependencies between aspects of HRM and the phenomena of fatigue; specifically, the implementation of Bayesian Networks (BNs) with the use of evidence(s) leads in a systematic manner to the projection and evaluation of fatigue. To this end, the developed model yields the quality of sleep and other more specific factors such as the non-disturbance of circadian rhythms as important parameters for the reliable evaluation of fatigue; hence these can be considered as significant aspects in the effort of strengthening the efficiency and safety of operations. The chapter concludes with interesting insights of the aforementioned tasks.

30.1 Introduction

The foundations of a modern safety-oriented organization lie with the shifting of focus on the human element; many involved stakeholders turn their attention towards the role and capabilities of personnel. According to Kermally [1], "the twentieth century was the age of machine; the twenty-first century will be the age of people". This can be justified by the revolution and enhanced reliability of the technological equipment, fact that revealed the human element as the most vulnerable link in the operational chain.

C. Kahraman (ed.), *Computational Intelligence Systems in Industrial Engineering*,
Atlantis Computational Intelligence Systems 6, DOI: 10.2991/978-94-91216-77-0_30,

The purpose of this chapter is to highlight the significance of evaluating human fatigue in safety-critical industries and subsequently to outline the effects of fatigue on operations in high-risk environments; hence, aspects of fatigue and their correlation with topics such as the Human Resource Management (HRM) and Crew Resource Management (CRM) are presented. Particular emphasis is given on the evaluation of human fatigue with the usage of Bayesian Network (BNs). The objective of this chapter is to discuss the way in which BN models for studying human fatigue can be applied in the real working environment of safety-critical industries.

The remainder of this chapter is organized as follows. Section 30.2, focuses on the aspects of human fatigue with regards to HRM and CRM and Sec. 30.3 deals with fatigue uncertainties and BNs. Then, Sec. 30.4 presents the developed BN model for the study of fatigue in the maritime domain, followed by Sec. 30.5 in which this model is applied on a case study of a well-known marine casualty. Finally, this chapter concludes with the discussion of the findings from the evaluation of human fatigue with the usage of BNs and with interesting insights on various points of the aforementioned topics.

30.2 Human Fatigue

In safety-critical domains, increasing numbers of fatigue-related accidents are recorded despite the significant improvement of the technical reliability of equipment. In this context, the purpose of this Sec. is to highlight fatigue as the outmost feature of human factors in high-risk industries. Hence, the fatigue phenomenon is analyzed and the correlation of fatigue with HRM and CRM is presented.

30.2.1 *Human Factors in safety-critical industries*

The emphasis of HRM is given to the fact that human factor can make a major contribution into organizational structure; human factors is a dynamic multi-disciplinary concept developed on the condition that human element is strongly linked with the functioning of technology, people, design, and operation [2]. In consequence, the field of human factors is the discipline devoted to the optimization of human element, and of the human-system interaction.

A mutual relationship can be found between human element and technology, as there is a progressive interaction between frontline personnel and technical equipment. Thus, the aim of human factors is to optimize this mutual relationship through the implementation of social and applied sciences in combination to ergonomics and human physiology.

Human factors are mainly about the performance of workers as part of complex and dynamic systems. Complexity in the technical field includes parameters such as the number of components, and the interactions between various pieces of equipment. When the focus is upon the human element, complexity can be expressed in cognitive metrics, human limitations, or social skills [2]; the number of interface tasks to be performed, or the number of critical events to be monitored, are indicative examples of this complexity. In safety-critical industries the effort is to select and implement proactive techniques to reduce excessive complexity and subsequently unnecessary risk for the examined operations.

Recent research in the field of human factors shows that technological improvements and equipment reliability do not necessarily reduce the probability of human-related accidents [2]. Indeed, numerous technological evolutions often increase the system complexity and furthermore they require new demanding skills from the operators (to meet their new tasks). The impacts by the technological development and motivation have succeeded in sustaining the records of accidents in high-risk industries low, sometimes even decreasing. Nevertheless, catastrophic failures still occur and the number of "losses" has yet failed to reach the desired level for the society.

Hence, human factors have been involved in almost 80% of the causes of all accidents within safety-critical domains [3]. However this is a statistic that needs further consideration; on the one hand, human error has obvious implications with regards to safety and on the other hand, to err is human and therefore inevitable. Hence it must be stressed that it is the human failure that causes accidents and not in general the so called human factor.

30.2.1.1 *Fatigue as the Outmost Feature of Human Factors*

The feature of human factor that has been recorded in almost all accidents in safety-critical domains is fatigue. An important aspect of these industries is their need to operate efficiently 24 hours per day, 365 days per year; though equipment is designed to work uninterruptedly under these conditions, human are simply not. In effect, fatigue can be traced on the involved human factors. Moreover, fatigue has alarming impacts to the rest of the non-technical skills; it can be the most important variable of human under performance, since it inhibits the application of cognitive and social skills of workers.

It is also important to recognize that there is a significance difficulty in identifying and measuring (in a direct manner) fatigue [4]. Imminence consequence of that is the fact of underreporting/misreporting of fatigue both in normal operations and accidents investigations.

In the contemporary competitive environment with the increasing demands for effectiveness, and pressure for minimized manning levels the issues of fatigue are now as timely and critical as never before. Numerous studies have revealed lack of knowledge about the perception of workers regarding fatigue, as well as their inability to recognize its symptoms. Indicatively, Hethrighton *et al.* [5] pinpoint that there are potential disastrous outcomes from fatigue, in terms of health, performance and effectiveness.

An interesting test bed with regards to the existence of fatigue is the maritime domain. According to Cole-Davies survey [6], out of 1000 participating officers, 84% felt fatigue widely prevalent onboard ships and stated that the impact of fatigue has been significantly greater in the past 10 years than it used to be. An even more alarming finding is the result of NTUA's non-technical skills research (2012) in 813 Greek merchant officers, according to which although 85% of officers stressed out fatigue as the dominant factor affecting safety onboard ships, but only 5% of them finally suggested the increasing of hours of sleep as a solution [7].

The significance of fatigue can be also derived by the fact that even cognitive or social human impairments can be the outcome of a fatigued individual; a typical example is when a fatigued worker can follow a wrong decision making process in a non-routine case [8]. The likelihood of this wrong decision would be noticeable lower if the worker had enough sleep, since fatigue influences adversely decision making capability and situational awareness. Hence, such a human failure would be documented as decision making failure rather than one related to fatigue; due to the underreporting and misinterpretation of fatigue. In effect, this misunderstanding would have misled researchers to emphasize on the decision making skills rather than to the efforts to coping with fatigue.

30.2.1.2 *Fatigue-related accidents*

In most accidents there is this common belief: the accidents could have been avoided had the operators made different choices, than those they actually did. It is in human nature to avoid getting involved in accidents; however the human element is the major contributor/cause leading to accidents and disasters. In this context, fatigue is the human-related parameter which can be found in almost all accidents/disasters and can affect all other contributing factors [9, 10].

In safety-critical domains there is a complex interaction between operators and equipment. Although equipment is human-centered designed and the operators are trained to use it, in some cases operators simply fail to control it. Furthermore, careful observation in major catastrophes revealed that they generally occurred in "dangerous" periods, namely the hours that fatigue, low alertness, and circadian rhythm disruption were present.

Paradigms of severe catastrophes in high-risk industries or even military operations comprise the accidents of Three Mile Island (1979) [11] and Chernobyl (1986) from the nuclear power industry [12], the one at the Tenerife Airport (1977) from aviation, of the Herald of Free Enterprise (1987) and Exxon Valdez (1989) in maritime industry, the explosion of Challenger (1986) from the aerospace industry, the Piper Alpha (1988) in offshore oil & gas industry [13], and the USS Vincennes (1988) from military operations [14].

In all these cases, fatigue was the direct or indirect contributing human factor, which also affected numerous other causes. In many of these accidents the effects of fatigue and disruption of the circadian rhythm were the reasons that impaired the ability of the operators to deal with the situation; they failed to use their cognitive or social skills. Hence, fatigue appears to be the contributing factor in many disasters; it reduces the resilience ability of the operators and consequently the resilience capacity of the industry.

30.2.2 *Fatigue*

Fatigue can be outlined as a separate factor influencing adversely operational efficiency and performance. It must be stressed that fatigue can influence other non-technical skills such as situational awareness, decision making, or others under the umbrella of social skills like teamwork, communication, and leadership. It comes out naturally that fatigue does not affect everyone at the same level and manner. In effect, the impact of fatigue on the rest of the non-technical skills is up to personal variables such as age, physical condition, gender, experience, or even personality. Additionally, the effect of fatigue is further increased by the inability of human to manage stress adequately.

The rich international literature in definitions of fatigue reveals the difficulty to a commonly accepted definition. According to Akerstedt [15] fatigue is a synonym to drowsiness, sleepiness and tiredness. In this context, fatigue can be described as the feeling of lethargy, lassitude, exhaustion, languor and even languidness that diminishes the capability to accomplish routine or emergency tasks. Fatigue can surface as an intensive feeling of tiredness driving in a great urge to sleep; moreover the symptoms can gradually grow and even pain may appear [8]. An even more alarming observation is the fact that tests on guinea-pigs showed that mice that were forced to stay without any sleep for few days, finally died [16].

Using terms of physiology, fatigue can be described as a lack of energy, feelings of muscle weakness, and slowed movements or central nervous system reactions [8]. In addition, fatigue can also generate notable cognitive tiredness. It have been noticed in many cases that fatigue may affect a lack of cognitive clarity, difficulty in understanding and concentrating, and even difficulty in memory process; fatigue affects all cognitive processes which for non-technical skills can be translated into lack of situational awareness and decision making capability.

A correlation can be found between fatigued operators and the probability to experience deterioration of performance and physiological malfunctioning. Reporting of fatigue through the existence of its symptoms is the first step in dealing efficiently with these problems.

30.2.2.1 *Categories and types of fatigue*

Fatigue can be divided into the categories of physical and mental fatigue; and also divided into the types of central and peripheral fatigue, which will all be briefly analyzed hereafter [8].

Physical fatigue Physical fatigue is the classic fatigue everyone has witnessed as tiredness which comes after protracted periods of (intense) physical activities. In the case of safety-critical domains the progress of technology has minimized the amount of physical effort that operators have to apply to accomplish their tasks; nevertheless, sometimes this need still exists. In addition, physical fatigue can be described as muscle weakness and can be expressed as the failure to apply energy with muscles to the degree that would be expected in comparison with the presumable human levels [8].

Mental fatigue Many times in safety-critical organizations operators have to deal with complex tasks and to operate high tech equipment; these current trends, accompanied with the introduction of automation and decreased manning levels, impose excessive mental workload. The effects that operators witness, during and after prolonged periods of cognitive activity that require work efficiency can be described by the term of mental fatigue. Hence, this category of fatigue can be expressed as minimized levels of wakefulness, awareness, and even perception, with not necessarily including symptoms of sleepiness [8]. The symptoms of mental fatigue can lead to dangerous conditions and outcomes when performing complex tasks that require continuous awareness, e.g. in the transport domain, or in advanced control rooms.

Similarly, mental fatigue describes the state caused by extended period of challenging cognitive activity. The effect of mental fatigue on cognitive skills and mental performance is rather profound [4]. However, strong emphasis must be given to the impact of mental fatigue on physical performance which is often neglected and not sufficiently studied.

Central fatigue The reduction in the neural drive or nerve-based motor command to working muscles that result in a decline of force output can be described as central fatigue [8]. Additionally it has been suggested that the reduced neural drive during exercise may be a protective mechanism to prevent organ failure should the work continued at the same intensity.

Peripheral fatigue Fatigue during physical work can be described as the inability of the body to apply energy to the contracting muscles to meet the increased energy demand. Peripheral fatigue causes contractile malfunction; that is manifested by the eventual reduction, or lack of ability of a single muscle or local groups of muscles to do the assigned work [8].

The fundamental difference between the peripheral and central theories of fatigue is that the peripheral model of fatigue assumes failure at one or more sites in the chain that initiates muscle contraction. In particular, central models of fatigue imply a malfunction of the CNS (central nervous system), such as impaired transmission between the CNS and the peripheral nervous system, or malfunction of selected areas of CNS, such as the hypothalamic region. On the other hand, the peripheral models view fatigue as result from the malfunction of the peripheral nervous system, such as impaired neuromuscular transmission at the motored-plate. While dualistic approaches have proven to be popular, they fail to capture the multidimensional nature of fatigue [17].

30.2.2.2 *Effects of fatigue*

According to recent research in safety-critical domains, fatigue has been proved to cause severe effects in terms of physical and mental health, and work performance. It is usually accompanied with stress and it can affect in various ways almost all other non-technical skills. The strong impact of fatigue is in cognitive skills such as situational awareness and decision-making; however it also has significant impacts on social skills such as leadership, communication, and teamwork [4].

Effects of fatigue in work performance Individuals, organizations and authorities are usually aware of the consequences of alcohol on human performance, especially in the transportation domain. It is of utmost importance to recognize that fatigue has similar influence in safety and performance records; for example, it has been proved that staying awake for 21 hours has parallel effects on performance and overall safety as measuring a blood alcohol content of 0.1% [18]. High levels of fatigue usually yield diminished efficiency, and consequently higher risk for an incident to take place. It has been observed that performance levels tend to drop as working periods are extended and periods without sleep stretched [8].

Fatigued operators are unable to understand the level of their own fatigue and they think that they can perform effectively the assigned tasks. Indicatively, according to NTUA attitude research (2012) on 813 Greek merchant officers the responses on the statement "even when I am fatigued I can cope with emergency situations" showed that 60.8% of the interviewees answered positively; this is an indication of the alarming shortcoming of officers on the perception of fatigue in the maritime domain [7].

Cognitive impairment is one more aspect of fatigue that affects the levels of performance and efficiency. According to recent surveys increased fatigue levels decrease the ability of operators for decision-making and situational awareness [4]. With regards to situational awareness, fatigue diminishes the ability for concentration, impairs vigilance and visual perception, and reduces predictability. Similarly, decision making skills are impaired from poor judgment, inability of recollection of time and events, and shortage of short-term memory. Hence, fatigue reduces the capacity for efficient judgment, and increases the reaction times; fatigued operators tend to adopt more "safe" actions, affecting this way productivity and efficiency in a negative manner.

Chronic Fatigue Syndrom (CFS) The Chronic Fatigue Syndrome (CFS) is one of the notable symptoms that have been observed in the presence of fatigue for long periods; it may be followed by symptoms, such as headaches, muscle pain, sore throats, multi-joint pain, and tender lymph nodes. CFS may also present cognitive symptoms such as impaired memory or concentration. Moreover it can be described as a debilitating disorder, usually uncorrelated to exertion, accompanied with specific symptoms for a minimum of a six month period; these extra symptoms include muscle pain, flu-like illness, encephalomyelitis, that impairs ability of individuals to cope with the required tasks.

Micro-sleeps In safety-critical industries fatigue symptoms can be dangerous, in terms of both safety and performance. It has been observed that fatigued workers may fall asleep for a short period of time even when they deal with complex tasks. This phenomenon is common and it may have been witnessed by almost everyone in their lifetime; it is broadly known as micro-sleep. In effect, a micro-sleep is a brief nap that lasts for a limited period of time. Operators who perform tedious tasks are more vulnerable to the micro-sleep effect. It is alarming that individuals experiencing micro-sleeps cannot perceive this fact and therefore they cannot control it, in effect, micro-sleep is a risky and hazardous phenomenon, in safety-critical domains, since the operator actually may fall asleep and lose control for limited time periods, usually no more than fifteen seconds. High-risk organizations have zero tolerance to this phenomenon because it can lead to severe consequences.

Sleep inertia In specific safety-critical domains, such as maritime, offshore oil & gas industry, or military, operators may have to respond to emergency situations during their sleep breaks. In such cases the operator has to wake up suddenly and rush to undertake his/her task; however immediately after he/she wakes up (and must go to complete the task) he/she experiences the phenomenon of sleep inertia.

The experience of sleep inertia is rather common and depends on the stage of sleep at the time of the wakening. The feeling of sleep inertia is equivalent to the feeling of exhaustion (i.e. having no energy at all) and it can last for almost half an hour after waking up. Hence this period may be critical safety-wise, since operators can make wrong judgments and errors; moreover, operators usually have to go fast to undertake their emergency mission usually through narrow corridors and so they can hurt themselves. Tokens of sleep inertia are confusion, loss of orientation, and lack of situational awareness. In this context, operational managers should take into consideration this human limitation when designing tasks and performing drills.

Health issues Lack of sleep and subsequently fatigue has severe effects on human health (mainly correlated to the perturbation of the circadian rhythm). These meta-effects of fatigue are summarized in high blood pressure, heart diseases, mental illnesses, stomach disorders, or even decreased fertility. These effects are correlated to human age with fifty years being a critical value for this type of studies; this can be attributed to the fact that elderly people generally, experience lighter and more fragmented sleep which impedes their body to fully recuperate from fatigue. Moreover the disruption of circadian rhythm has significant impact on human health and it can be identified through depression, diabetes, and asthma.

Circadian Rhythm As mentioned earlier the disruption of circadian rhythm can have major impact on human health. Circadian process is a physiological process for every living being (animals and even plants) that oscillates on a 24 hour basis; hence the circadian rhythm can be simply explained as the internal biological clock of beings. This rhythm can be depicted with peaks and low points during the day and night that affect the alertness and the performance of the human body. The mechanism that regulates the energy cycles is the internal biological clock, which follows specific levels of alertness in correlation to day and night time. This way the optimum time periods for sleeping, waking-up, working and exposing to daylight can be identified [19].

The quality of human sleep is affected in a direct way from circadian rhythms. Jet lag which is the unpleasant feeling witnessed from people when traveling over global time zones is attributed to circadian disruption. In specific safety-critical domains such as the maritime, aviation and military operations shift schedules disrupt the biological clock and quality of sleep, making operators more prone to excessive fatigue [15, 20].

According to NASA [21], sleep loss and circadian disruption can lead to critical decrements in performance and alertness levels. Symptoms witnessing the problem of circadian rhythm disruption include perturbation of body temperature, abnormal cardiovascular function, poor digestion, lack of alertness, and experiencing of wakefulness.

30.2.2.3 *Fatigue measurement*

There is a general consensus among investigators that measuring fatigue in real operating environments is a complex process with sometimes uncertain results [22]. The best practices from literature review conclude that fatigue can be estimated with the simultaneous use of both objective and subjective methods [4]; these methods include behavioral, physiological, self-reporting, attitude and performance measurements which are conducted either in real working environments, or in simulators. Moreover, these different types of methods act supplementary one to each other, and many times researchers have to carry out a combination of them to identify and measure fatigue in a reliable manner.

Physiological measures Most methods that have been developed for measuring fatigue tend to fall into the category of physiological metrics. The most common physiological measure is the Electroencephalograph (EEG) which measures brain activity and correlates it to recorded fatigue levels. Another application of this type for determining fatigue is the Multiple Sleep Latency Test (MSLT) [8]; this effort is based on measuring of time needed for operators to fall asleep after exhaustive tasks (under sleep-friendly conditions). A particularly exciting development in measuring fatigue, with wide and ever increasing application in the automotive industry is the tracking of eye movements. The prevalence of this type of fatigue measurement is demonstrated by the "Percentage of Eye Closure" (PERCLOS) [8]. The great advantage of this method is that determines fatigue and consequently it can be applied as a warning system; it tracks the percentage of eye-lid closure and correlates it to fatigue in real-time.

Behavioral measures The level of fatigue and drowsiness with symptoms such as lethargy, decreased social participation, yawn, and micro-sleep can be also determined with the implementation of behavioral methods. The most common application of behavioral methods is actigraphs, which are high accuracy instruments, usually worn like a watch on a wrist that measure wrist motion and consequently the frequency of the body movement [23]. This body movement can be then directly correlated to alertness levels, tendency of sleepiness, or even sleep stage [24].

Self-report measures Another method used to determine fatigue levels that fall in the category of subjective methods is self-report measuring. The strong emphasis on these measurements is encountered on the symptoms of fatigue as these are evaluated from the operators. The main advantage of these methods is that they can be conducted efficiently on large samples; however humans are rather subjective in assessing themselves and therefore fatigue data may be underreported [25]. People in the context of these methods have to respond in simple attitude or mood statements using Likert scales, or just keep logs of their activities with details about their work/tasks, periods of sleep etc.

Performance measures Performance measurement is the simplest and most common practice to identify fatigue levels. Typical paradigms include tracking the performance of humans under fatigue in terms of time to accomplish specific tasks or testing the cognitive capabilities of operators. Such an application is the Walter Reed Performance Assessment Battery (PAB); this process measures the cognitive and behavioral performance in correlation to recorded reaction times.

30.2.2.4 *Fatigue countermeasures*

The existence of fatigue in high-risk domains and the adverse effect that it has on safety is a fact generally accepted from the majority of researchers [26]. Fatigue related disasters forced investigators to suggest techniques and practices to ensuring quality of sleep and acceptable circadian rhythms, aspects that will contribute to safe and efficient operations. Strong emphasis with regards to fatigue management is given to fatigue assessment. Safety organizations have suggested and adopted effective methods of reducing fatigue; these approaches include mainly educational programs for operators and risk assessment.

On the other hand, fatigue countermeasures may have a significant cost for the organization. In effect, there is a need to identify and calculate the probability of having fatigue along with the corresponding consequences in safety-critical domains, to determine the necessity of adopting the identified control measures.

30.2.3 *Human Resource Management (HRM)*

The Human Resource Management (HRM) is an integral function of an organization; it epitomizes a sequence of important activities directly related to human factor. HRM covers operational aspects with regards to safe operation and especially with a focus on frontline operators of safety-critical industries. Safety is of paramount importance in high-risk industries and as such attention must be given equally to equipment and the performance of human element [27].

Incident and accident investigations in all safety-critical domains revealed that the human element was always the most vulnerable link of the accident chain. On the other side human-oriented industries achieved to minimize accidents rates by simply adopting human-centered techniques; these techniques are part of HRM of the organizations.

The purpose of HRM in high-risk industries is to improve the ability of the organization to manage and develop its competence and abilities, particularly in emergency situations. Day-to-day practices in such organizations have been changing dramatically through the last years following the impacts from recorded catastrophes in all levels. Human inability, or simply put human error, has been proved to be the prime contributing factor in the majority of these accidents [3]; in effect, attempts to reduce the occurrence of accidents should concentrate on eliminating errors of the front line personnel. Hence, a more holistic approach is needed, which should be based on the development of positive safety culture to promote and maintain a safe working environment. In safety-critical industries this need is mainly expressed though Crew Resource Management.

30.2.4 *Crew Resource Management (CRM)*

Safety-critical industries, with aviation as pioneer, have recognized the significance of human error in the occurrence of accidents, the last thirty years [4]. The adoption/integration of human factors in high-risk industries is known as Crew Resource Management (CRM). CRM development is shown by the application of human factor awareness to frontline personnel, the interaction between the members of the team/group, and with regards to technical equipment.

CRM was the academic response to a series of catastrophic aviation accidents, precipitated by the ineffective management of available resources in the late 1970s [4]. The definition of CRM has taken several forms since its development; CRM is generally accepted as the effective utilization of all available human and informational resources towards the effective performance of a safe and efficient operation. There is a recorded consensus that CRM is about behaviors, attitudes and social interactions [4]. Thus, CRM is a strategic choice that has been proved to be extremely useful for an organizational approach in its effort to minimize and cope with human error.

Human factors are the theoretical knowledge of phycology and CRM is the practical use of that knowledge [28]. Although CRM was initially developed to minimize human error and improve performance in emergency situations, it actually enhanced the safety, efficiency, and productivity of safety-critical organizations; this can be attributed to the positive culture that CRM cultivates in the structure of the organization.

The notable success of CRM programs have proved their significance in most safety-critical domains and have contributed to the reduction of accidents and incident rates. A noticeable example of effective application of CRM in the maritime domain is the one of the Danish company "Maersk". In particular, the implementation of CRM training was the dominant factor for the significant decrease of maritime casualties from one per thirty ship-years in 1992, to one per ninety ship-years in 1996. Sequel consequence of this achievement was the reduction in insurance premiums at a level of 15% [29].

The strong emphasis on CRM is focused on non-technical skills, or otherwise known as soft skills. Non-technical skills are "the cognitive, social and personal resource skills that complement technical skills, and contribute to safe and efficient task performance" [30]. In this context, cognitive skills include situational awareness and decision-making, social skills comprise leadership, teamwork and communication, and finally personal limitations address the management of stress and how to cope with fatigue [30].

30.3 Fatigue uncertainty and Bayesian Networks (BNs)

In safety-critical industries fatigue is an issue that has to deal with uncertainty with regards to both its occurrence and potential outcomes. Bayesian Networks (BNs) have proved to be a reliable methodology to cope with uncertainty [31–33]. The purpose of this Sec is to present a limited theoretical review on BNs (or Belief Nets); a detailed analysis of the theory of BNs lies outside the scope of this publication.

30.3.1 *Fatigue uncertainties*

BNs are considered as a state-of-the-art methodology in dealing with fatigue, in terms of estimation, definition of consequences, and uncertainty analysis [14, 34, 35]. In effect, there are probabilistic models based on BNs that determine the probability of human fatigue by getting data from various visual cues and similar appropriate sources. The most demanding part of such efforts is the parameterization of the model which is based either on the extraction of data from the international literature, or on simulations, or on subjective knowledge (e.g. from experts).

30.3.2 *Bayesian Networks*

BNs are probabilistic graphical models that deal with uncertainty and combine the dependencies of random variables with the use of Directed Acyclic Graph (DAG). These DAGs are directed graphs with no directed cycles and they include vertices and directed edges [33]; hence, a cyclic route is impossible to appear. The nodes in BNs represent random variables; these may be observed quantities, latent variables, unknown parameters, or hypotheses. On the other hand, edges depict conditional dependencies between parent and child nodes. Each node is related to a probability function that takes as input a particular set of values coming from parent node(s) and gives the probability of the variable reflected on the node. The probabilities of the child nodes are inputs in the Conditional Probabilities Tables (CPTs).

30.3.2.1 *Probabilistic graphical models*

The presence of uncertainty is of outmost significance when dealing with fatigue, so this is a fact that should be taken into account when trying to estimate it. In effect, uncertainty is managed and represented by probabilistic graphical models; these are generated from the conjunction of graphs with probability theory.

In causal networks (such as BNs are) the managing of certainties is done through the usage of conditional probability. A conditional probability is defined as follows: "given event B, the probability of event A equals x"; in mathematical this is given below.

$$P(A/B) = x \tag{30.1}$$

The main types of connections in BNs include serial, diverging and converging connections between parent and child nodes [31].

30.3.2.2 *BN Structure process*

BNs present the following structure [31]:

- A set of variables and a set of directed edges between variables;
- Each variable has a finite set of mutually exclusive states;
- The variables together with the directed edges form a Directed Acyclic Graph (DAG);
- To each variable A with parents $B(1),\ldots,B(n)$ there is attached a conditional probability table $P(A \mid B(1),\ldots,B(n))$.

A Bayesian Network is depicted by the functioning of the chain rule [31]: If X is a BN over $U = \{A_1,\ldots,A_m\}$; then the joint probability distribution $P(U)$ is the product of all conditional probabilities specified in X:

$$P(U) = \prod P(A_i \mid p\alpha(A_i)) \tag{30.2}$$

where $p\alpha(A_i)$ are the parents of A_i.

30.3.2.3 *Inference process*

BNs have been proved to be the solution to the issue of probabilistic inference; in particular, Bayesian inference is a process of statistical inference, according to which evidence or observations are used to estimate the probability that a hypothesis may be true, or else to update its prior probability. In the context of Bayesian Inference, the simplified version of the Bayes Theorem is presented.

$$P(B/A) = \frac{P(A/B)P(B)}{P(A)} \tag{30.3}$$

- B stands for a specific hypothesis, which may be a null hypothesis;
- A stands for the evidence that has been observed;
- $P(B)$ is the prior probability of H (= hypothesis) that was inferred before new evidence is available;

- $P(A/B)$ is the conditional probability of seeing E (= evidence) if the hypothesis H turns out to be true;
- $P(A)$ is the marginal probability of A, which is the prior probability of witnessing the new evidence A under all possible hypotheses.
- $P(B/A)$ is the posterior probability of B given A, hence it is the new estimation for the probability that the hypothesis B is true, taking evidence A into account.

30.4 BN models for studying fatigue in the Maritime domain

In this Sec., the model for estimating the probability of fatigue in the maritime domain is presented; the categorization was made according to the guidelines from the International Maritime Organization (IMO) with regards to fatigue mitigation [36]. In Fig. 30.1 the described BN for studying fatigue is shown.

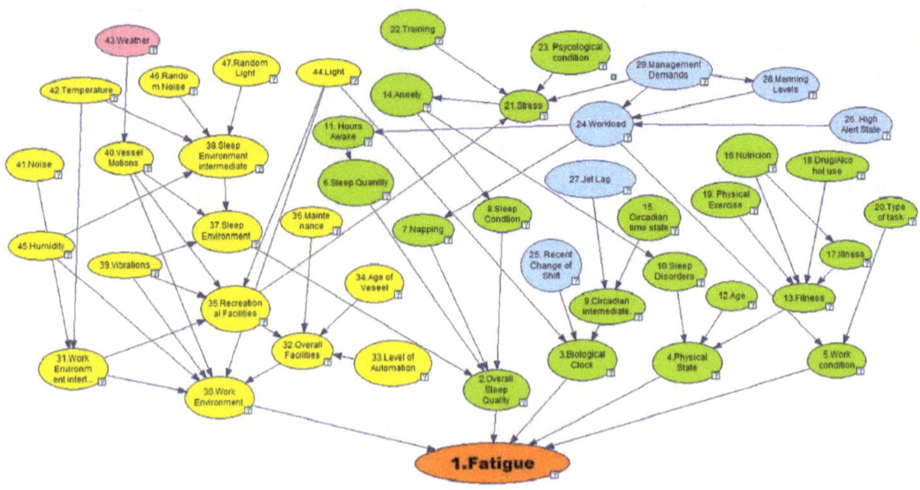

Fig. 30.1 NTUA Bayesian Network for studying fatigue

30.4.1 *BN models*

30.4.1.1 *Target Node (Child)*

Fatigue is the target node (in orange) of the BN developed by the National Technical University of Athens (NTUA) in the sense of estimating a metric of its posterior probability. The assumption that "fatigue is inevitable even if all factors favor its mitigation" was taken into consideration in the effort to enhance the reliability of the model [35]. Furthermore, the similarity of various factors of fatigue throughout many safety-critical domains facilitated the adoption of data from the international literature during the development of the model [8]. Finally, specific adjustments were made based on subjective judgments, experience records, log books, and conducted interviews. The state of fatigue is expressed initially in a binary manner, namely "Yes" or "No".

30.4.1.2 *Parent Nodes*

On the developed BN model parent nodes include crew specific factors, management factors, ship specific factors, and environmental factors. Starting with the crew specific factors, the "overall sleep quality" has a major role in introducing fatigue; "Sleep quality" is expressed with the states of "poor" and "fair". The circadian rhythm or simply put the "biological clock" is defined through the states of "drowsy" and "awake" for highlighting the time periods when body is awake. Similarly, the "physical state" in the model is characterized either as "poor" or "healthy" and it connects fatigue with human physical impairments. Indicatively, another contributing node is the "work condition" which employs the states of "poor" or "fair' and it represents the burdening and tediousness of the existing workload.

According to the BN model, "sleep quantity" is a major contributing factor with regards to fatigue and it can be expressed in hours of sleep; its employed states are "deprived" or "sufficient". Moreover, the chance for "napping" is addressed in relation to the nature of the workload; the possible states for the node "napping" are either "no" or "minimum 30 min". The ability of workers to sleep by putting aside distractive factors such as anxiety is also integrated in the presented BN through the "sleep condition" node; the states of this node are either "bad" to "fair".

The node of "circadian intermediate" is used in order to introduce in the model the influence from jet lag and time state. Fatigue related research in safety-critical domains has shown that both shift rotation and the existence of sunlight have great impact on the circadian rhythms [19, 34, 37]; in this context, the "circadian intermediate" is expressed either as "drowsy", or "awake".

"Sleep disorders" represent (medical) disorders of operators with regards to their sleeping patterns which can interfere with physical, mental and emotional functioning. The potential states of the node of "sleep disorders" are "regular" and "rare".

"Hours awake" was integrated to give to the model a dynamic constituent. The introduction of this node was based on the human need for certain hours of sleep; therefore it was selected to show the impact of sleep deprivation. The employed probabilities for the respective CPTs are mainly taken from [39].

Research on maritime operators has shown that the overall physical efficiency decreases with age [39]. Specifically, the age of 45 has been marked as the hub of the distribution of age against physical condition. The potential states of "age" are "45<" and "45>".

The analysis of all employed nodes in the BN of Fig. 30.1 is out of the scope of this publication, however a nominative review of them is provided so as the potential reader can have a high level picture of the complexity and completeness of the developed model. In this context, the rest contributing nodes in seafarers fatigue are anxiety, circadian time state, nutrition, illness, drug or alcohol use, physical exercise, type of task, stress, training and psychological condition.

The category covering management factors includes nodes such as workload, recent change of shift, high alert state, jet lag, manning levels, and management demands.

There are also the nodes covering ship specific and environmental factors; these are the total work environment, intermediate work environment, overall facilities, level of automation, age of vessel, recreational facilities, maintenance, sleep environment, intermediate sleep environment, vibrations, vessel motions, noise, temperature, weather, light, humidity, random noise and random light.

30.4.2 *Fatigue acceptance scale*

To determine and attribute the probability of human fatigue amongst seafarers the scale listing below is presented as an index of levels of fatigue prevalence. The influence of physiological and psychological parameters is determined for the estimation of the probability of fatigue. A fundamental assumption that was made was that seafarers will eventually suffer the effects of fatigue even if all factors, such as sleeping quality or work environment, are at a favorable state. The scale of Fig. 30.2 presents the posterior probability for fatigue according to the presented BN model; the red color represents the dangerous zone for the operators where the consequences of fatigue weaken human ability and efficiency, thus possibly leading to critical outcomes. These dangerous situations include impairments in cognitive, personal and social non-technical skills.

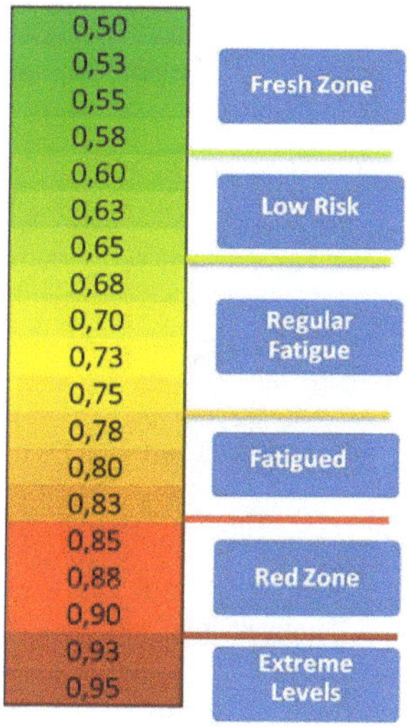

Fig. 30.2 Fatigue acceptance scale

30.4.3 *Fatigue and hours of sleep*

The overall sleeping quality is the most significant node of the BN with regards to the occurrence of fatigue. The results from research in numerous safety-critical domains show that the critical hub between sufficient sleep and its deprivation lies within the period from six to eight hours of uninterrupted and undisturbed sleep, in relation to parameters such as physical condition, or age.

Figure 30.3 shows the posterior probability for fatigue from the presented BN against the various states of the "hours awake" node that are imported as evidence in the model.

Figure 30.3 yields the 20 hours of being awake as the critical limit for the manifestation of fatigue on maritime crews; beyond that point the crew members/operators enter the red zone of Fig. 30.2 and consequently all of their non-technical skills are being gradually blocked. In Fig. 30.3 the abrupt raise recorded in the area from 16 to 20 hours being awake highlights the explicit distinction between enough sleep and sleep deprivation; it is reminded that this can translated in 6 to 8 of uninterrupted hours of sleep. When the workload is increased onboard vessels (e.g. during loading/unloading or entering into a port) there is rarely the chance for more than 6 hours of sleep. In effect, according to Fig. 30.3 if seafarers manage to sleep more than 8 hours per day then they will have a small probability to experience fatigue.

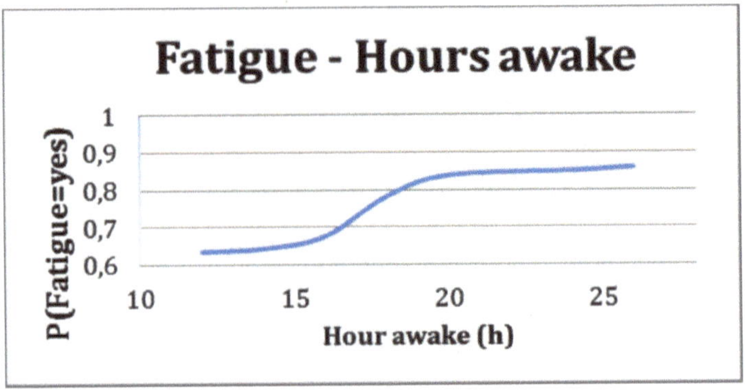

Fig. 30.3 Fatigue against Hours Awake

30.5 A case study from the maritime domain

In this Sec. the effectiveness of the presented BN model is testified against a fully investigated and well documented maritime accident.

30.5.1 *Exxon Valdez Catastrophe*

The Exxon Valdez oil spill occurred on March 24, 1989 and it has been considered as one of the most typical examples of human driven accidents. It is a fact that the disastrous environmental consequences forged ahead particularly significant changes in the way of transferring crude oil in the maritime domain, namely the phasing out of single hull tankers [40–42]. The grounding of the tanker Exxon Valdez on Prince William Sound's Bligh Reef in Alaska and the consequent spill of about 40,000 m^3 of crude oil was the starting point for a timeline full of changes in the maritime domain.

In this context, the presented BN model is tested against the Exxon Valdez case through the input of evidence (hereafter symbolized as "e") that was identified in the official accident report by the Alaska Oil Spill Commission [43]. For the purposes of this chapter the background and actions of the officer on watch (OOW) at the time of the accident are further examined and scrutinized; there was a 3rd mate in the bridge.

30.5.1.1 *Actual accident of the Exxon Valdez grounding*

The state of "manning levels" at the time of the accident can be set as "low" (evidence1, or e1) due to the fact that the 3rd mate was alone at the bridge in violation of the company policy and of regulations. The presence of another crew member at the bridge would have contributed to the decreasing of the danger of having the ship off course, and of the overall failure of corrective actions.

Judging the nature of the task of an OOW the “type of task” at the time of the accident can be filed as “tedious” (e2).

The “management demands” at the period of the grounding of Exxon Valdez can be evidenced as “demanding” (e3); this can be justified from the geographical location that the ship was sailing in, and from the fact that the 3rd mate should not have been left alone on the bridge at that time.

A strong emphasis is given on the evidence related to the node of “hours awake”. According to the official investigation report the 3rd mate “may have been awake and generally at work for up to 18 to 20 hours preceding the accident” (e4). Interesting enough, if the 3rd mate had had better knowledge of the consequences of fatigue to his performance he would have been aware of the risk staying awake, and he would have asked from the 2nd mate to replace him.

30.5.1.2 *Optimal scenario for the Exxon Valdez grounding*

Trying to depict the importance of the role of fatigue in this accident an optimal alternative scenario was developed and tested. In effect, “manning levels”, “type of task”, and “management demands” are evidenced as “normal” (e1, e2, and e3, respectively). For the “hours awake”, the existing legislation (i.e. STCW) was strictly followed, according to which the value of 8 hours of uninterrupted sleep could be factored in the model (e4), under the prescription that no crew member should stay awake for more than 16 hours within a day.

30.5.1.3 *Comparison between the Exxon Valdez actual and optimal scenarios*

The results of the actual and optimal scenarios for the Exxon Valdez grounding are depicted in Fig. 30.4. Accordingly, in the actual scenario the red zone of fatigue has been already reached with the integration in the model of the evidence related to the “type of task”. However Fig. 30.4 also shows that the posterior probability of the occurrence of fatigue for seafarers can be significantly reduced if the right measures and practices are adopted and promoted. The maritime industry should focus on a healthy and human-friendly environment that will contribute to safe and efficient maritime operations.

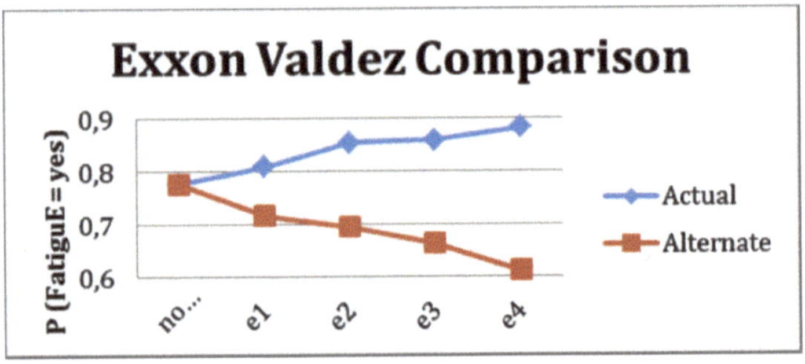

Fig. 30.4 Comparison between the Exxon Valdez actual and optimal scenarios

30.5.2 *Sensitivity analysis*

A multi-faceted sensitivity analysis was performed to define the stability and capability of the presented BN to calculate the resulting changes in a rational manner. To proceed with this analysis and to rule for the reliability and precision of the model, specific nodes were selected and a gradual change on the prior probability distributions from -50% to $+50\%$ with respect to their original values was decided and implemented.

The choice of the aforementioned nodes was based on the international literature and on elements of subjective knowledge (experience) regarding the factors that influence fatigue onboard vessels [44]. Taking into account the particularities of the maritime domain the nodes of noise, light, vibrations, vessel motions, and management demands were selected for the sensitivity analysis of the BN. In this direction, the rationale and results are presented for the changes performed on the node of noise; similar analysis has been done and for the rest of the selected nodes.

The existence of noise has an impact on almost all onboard human-centered activities. The machinery equipment is usually fitted and/or based on the hull of the ship within an overall structure of limited space (i.e. the vessel). As a result, the noise coming mainly from the operation of the main engines and generators of the ship may affect the working environment in numerous of its compartments, such as the bridge, engine-room, or the accommodation. Exposure to intense and continuous noise can cause both short-term and long-term hearing problems, distraction of attention, impairments in situational awareness, and various types of fatigue.

The results of the sensitivity analysis on noise are presented in Table 30.1.

[1]% change of the posterior probability
[2]This state describes the worst possible scenario for Noise.

Table 30.1 Sensitivity analysis of the BN with regards to noise

State	Prior Probability for the node of noise	Posterior Probability for	Change for probability of fatigue (%)[1]
-50%	0.075	0.775	0.25
0	0.150	0.777	0.00
+50%	0.225	0.780	0.38
W.S.[2]	1.000	0.808	3.90

According to the following Fig. 30.5 and to the implemented sensitivity analysis, the changes in the level of noise onboard ships seem to have a contribution, yet a small one, to the occurrence of fatigue on crew members. Even though intensive noises can disrupt sleeping patterns and create a fatigue inducing work environment it is clear from the results that noise (through the introduction of evidences in the BN) is not the steam engine that can solely drive the emergence of fatigue onboard ships.

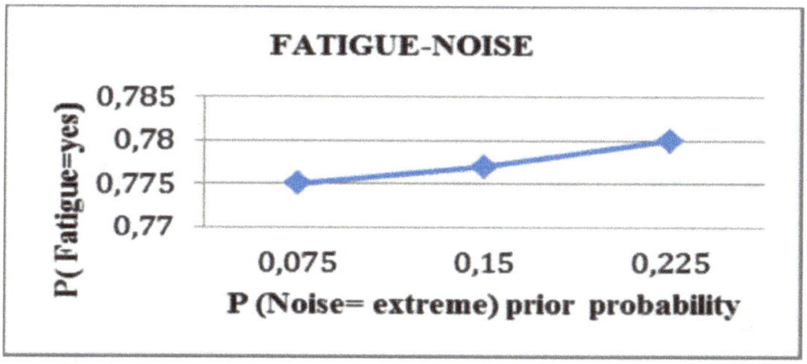

Fig. 30.5 Distribution of fatigue in relation to noise

Respectively, the results from the implemented sensitivity analysis on the parameters of light, vessel motions, vibrations, and management demands are presented hereafter in brief. In effect, the existence of light seems to affect fatigue significantly as it is increased by 7.5% when "poor" lighting is imported as evidence in the BN. The sensitivity analysis yields vessel motions as the most significant parameter, in terms of contributing to fatigue. In particular, the changes (evidencing) of vessel motions can solely raise the probability of having fatigue on the crew members by 8.1%; this high percentage might be explained by the impact that the movements of ships have on both working and sleeping environments. Similarly to noise, changes in the onboard vibration levels seem to have a limited impact on the response of fatigue. However, the presented BN appears to be sensitive in terms of fatigue, to changes regarding the node of "management demands". This is a rather expected due to the human-centered theme of the analysis, the complexity, highly-demanding tasks, and the dynamic nature of the maritime domain.

30.6 Conclusions

Significant accidents, severe consequences, high complexity and current trends in safety-critical industries shift gradually the focus from technological and other aspects on the role of human element; in effect, the emphasis on HRM in high-risk domains is given to human factors, and the way it influences the resilience capability of the organization. According to moderate estimations, the human element is responsible for almost 80% of the causes of all accidents in safety-critical domains. CRM is one of the best practices applied in the effort to strengthening human efficiency; the integration of non-technical skills in CRM provide for a holistic solution to the problem of safety.

The presented BN aims at the estimation of the (posterior) probability of the occurrence of fatigue through the use of evidence(s). The implementation of sensitivity analysis testified for the valid structure, the reliability and the precision of the model.

Through the application of the BN model it is shown that human physiology, including mainly sleep conditions and circadian rhythms, is the corner-stone for fatigue-free procedures; problems with these two parameters can lead to increased levels of fatigue. In the demanding environment of safety-critical domains with operations of 24 hours per day for seven days per week the effort of achieving non-disturbance of sleep and normal circadian rhythms is of crucial importance. It is a fact that even in an ideal working environment, the reduced sleeping quality or the existence of drowsiness will eventually lead to fatigue and therefore low level performance from operators; this can have a serious impact on safety and therefore it should attract the attention of all involved stakeholders.

Bibliography

[1] Kermally, (2006). *Ritilin Kings*, Booksurge Publishing.
[2] Crech M., Horberry T., and Koester T., (2008). *Human Factors in the Maritime Domain.*
[3] Reason J., (1990). *Human Error.* Cambridge: Cambridge University Press.
[4] Flin R., O'Connor P., and Crichton M. (2008). *Safety at the sharp end. A guide to Non-Technical Skills.* Ashgate, USA.
[5] Hetherington C., Flin R., and Mearns K., (2006). Safety in shipping: The human element. *J Safety Res*, 37 (4): 401–411.

[6] Cole M. and Davies R., (2001). Fatigue, health and injury offshore: A survey. In *M. Hanson (Ed.), Contemporary Ergonomics*. London: Taylor & Francis.
[7] Ventikos N.P., Lykos G.V., and Padouva I., (2012). People may not change but perhaps their behavior can? *Behavior Based Safety process in short sea shipping safety*. SSS 2012, Lisbon, Portugal.
[8] Vagias N., (2010). *A Bayesian Network Application for the Prediction of Human Fatigue in the Maritime Industry*. National Technical University of Athens, Athens, Greece.
[9] Coren S., (1996). *Sleep Thieves*. New York: Free Press
[10] Maas J., (1998). *Power Sleep*. New York: Villard.
[11] NRC. Three Mile Island Special Injury Group, (1980). *Human Factors Evaluation of Control Room Design and Operator Performance at Three Mile Island-2*, Volume 1. Washington, DC: US Department of Commerce.
[12] Reason J., (1987). The Chernobyl errors. *B Brit Psychol Soc*, **40**, 201–206.
[13] Cullen D., (1990). *The Public Inquiry into the Piper Alpha Disaster*, (Cm 1310) London: HMSO.
[14] Klein G., (1998). Sources of Power. How People Make Decisions. Cambridge: MIT Press.
[15] Akerstedt T. (2000). Consensus statement: Fatigue and accidents in transport operations. *J Sleep Res*, **9**, 395.
[16] Everson C.A., Bergmann B.M., and Rechtschaffen A., (1989). Sleep deprivation in rat: III. Total sleep deprivation, *Sleep*, **12**, 13–21.
[17] Shen J., Barbera J., and Shapiro C.M., (2005). Distinguishing sleepiness and fatigue: focus on definition and measurement. *Sleep Med Rev*, **10**, (2006), 63–76.
[18] Roehrs t., Burduvali E., Bonahoom A., Drake C., and Roth T., (2003). Ethanol and sleep loss: A 'dose' comparison of impairing effects. *Sleep*, **26**, 981–985.
[19] USCG, (2006). *Crew Endurance Management*. Commandant Instruction 3500.2.
[20] Åkerstedt T., (2006). Sleepiness and Circadian Rhythm Sleep Disorders. *Sleep Med Clin*, **1**, 17–30.
[21] NASA Ames Fatigue Countermeasures Program. Website accessed 12-5-10. `http://humanfactors.arc.nasa.gov/zteam/fcp/FCP.pubs.html`.
[22] Sherry P., (2000). *Fatigue Countermeasures in the Railroad Industry: Past and Current Developments*. Association of American Railroads 2000.
[23] Sherry P. and Philbrick K.E., (2004). *The Effects of Individualized Actigraph Feedback on Fatigue Management in Railroad Engineers*. NCIT Final Report.
[24] Caldwell J.A., Caldwell J.L., and Schmidt R.M., (2008). Alertness management strategies for operational contexts. *Sleep Med Rev* (2008) 12, 257–273.
[25] Psarros G., Skjong R., and Eide M.S., (2009). Under-reporting of maritime accidents. *Accident Anal. Prev., 42*, (2010) 619–625.
[26] Smith A.P., (2007). Seafarer fatigue where next? *International Transport Workers' Federation*.
[27] Psaraftis H.N., Caridis P., Desypris N., Panagakos G., and Ventikos N., (1998). The Human Element as a Factor in Marine Accidents. *Proceedings of the 1st Greek Maritime Transport Conference*, Piraeus, Hellas.
[28] Fahlgren G., (2011). *Human Factors*. AuthorHouse.
[29] Maersk, (2005). *The behavioral safety process can help you. Loss Prevention. Beacon*. Website accessed 12-02-09. `http://www.skuld.com/upload/News\%20and\%20Publications/Publications/Beacon/Beacon\%202005\%20186/The\%20behavioural\%20safety.pdf`.
[30] Flin R., Martin L., Goeters K., Hoermann J., Amalberti R., Valot C., and Nijhuis H., (2003). Development of NOTECHS (Non-Technical Skills) system for assessing pilots' CRM skills. *Human Factors and Aerospace Safety*, **3**, 95–117.
[31] Jensen F.V., (1996). *An introduction to Bayesian networks*. UCL Press, London, UK.

[32] Jensen F.V., (2001). *Bayesian networks and decision graphs.* Springer-Verlag, New York (USA).

[33] Pearl J., (1988). *Probabilistic Reasoning in Intelligent Systems: Networks of Plausible Inference.* Morgan Kaufmann Publishers, Inc., San Mateo, CA, 1988.

[34] Ji Q., Lan P., and Looney C., (2005). *A Probabilistic Framework for Modeling and Real Time Monitoring Human Fatigue.* University of Nevada, Reno.

[35] Ji Q., Lan P., and Looney C., *Information Fusion with Bayesian Networks for Monitoring Human Fatigue.* University of Nevada, Reno.

[36] IMO, (2001). Guidance in Fatigue Mitigation and Management.

[37] Calhoun S.R., (2006). *Human Factors in Ship Design: Preventing and Reducing Shipboard Operator Fatigue.* University of Michigan Department of Naval Architecture and Marine Engineering U.S. Coast Guard Research Project.

[38] Ventikos N.P., Lyridis D.V., Lykos G.V., and Logothetis T., (2010). Identifying and assessing Non-Technical Skills on Greek Maritime Officers: The story under the spotlight. HPAS 2010, 16-18 June, Glasgow, Scotland, UK.

[39] Smith A.P., Allen P., and Wadsworth E.J., (2006). Seafarers' fatigue: the Cardiff Research Programme. Report to MCA and HSE.

[40] Moore A.W., (2001). Bayes Nets for representing and reasoning about Uncertainty. School of Computer Science, Carnegie Mellon University.

[41] Moore-Ede M., Heitmann H., Guttkuhn R., Trutschel U., Aguirre A., and Croke D., (2004). Circadian Alertness Simulator for Fatigue Risk Assessment in Transportation: Application to Reduce Frequency and Severity of Truck Accidents. *Aviat Space Envir MD* Vol. **75**, No. 3, Sec II, March 2004.

[42] Moore W.H., (1992). The Grounding of Exxon Valdez: An examination of the Human and Organizational Factors. Society of Naval Architects and Marine Engineers, Northern California Sec, Student Paper Presentation, 1–15.

[43] Exxon Valdez oil spill trustee council., (1990) Alaska Oil Spill Commission. Website accessed 23-04-10 `http://www.entandallergy.com/nad/`.

[44] Breieriova L. and Choudhari M., (1996). *An Introduction to Sensitivity Analysis.* MIT System Dynamics in Education Project.

Index

C. Kahraman (ed.), *Computational Intelligence Systems in Industrial Engineering*,
Atlantis Computational Intelligence Systems 6, DOI: 10.2991/978-94-91216-77-0,

FSC
www.fsc.org
MIX
Papier | Fördert
gute Waldnutzung
FSC® C083411

Kolibri 360 GmbH
Ferdinand-Jühlke-Straße 7
99095 Erfurt, Deutschland
produktsicherheit@kolibri360.de